PRINCIPLES OF PATHOLOGY
AND BACTERIOLOGY

THIRD EDITION

PRINCIPLES
OF PATHOLOGY
AND BACTERIOLOGY

THIRD EDITION

BY

R. A. WILLIS
D.Sc., M.D., F.R.C.P., F.R.C.S., F.R.A.C.P., F.R.C. Path., Hon.LL.D.(Glasgow)

EMERITUS PROFESSOR OF PATHOLOGY IN THE UNIVERSITY OF LEEDS;
CONSULTANT PATHOLOGIST TO IMPERIAL CANCER RESEARCH FUND, LONDON

AND

A. T. WILLIS
D.Sc., M.D., M.R.A.C.P., Ph.D., M.R.C.Path., M.C.P.A.

DIRECTOR, PUBLIC HEALTH LABORATORY, LUTON

NEW YORK
APPLETON-CENTURY-CROFTS
Division of Meredith Corporation

LONDON
BUTTERWORTHS

First Edition .	.	.	.	.	1950
Second Edition	.	.	.	.	1961
Third Edition .	.	.	.	.	1972

©
Butterworth & Co. (Publishers) Ltd.
1972

Suggested UDC No. 616

ISBN 0 407 62751 0

Printed and bound in Great Britain by R. J. Acford Ltd., Industrial Estate, Chichester.

To the Memory of Matthew J. Stewart

PREFACE TO THE THIRD EDITION

IN THE ten years since the second edition appeared, while the main principles of pathology have not altered, there have been additions of detail in our knowledge on many subjects, especially in microbiology and virology, in which also many changes of nomenclature have been introduced. These additions and changes are reflected in considerable modification of the chapters on the viral and microbic diseases and on the allergic and auto-immune diseases, parts of which have been rewritten. In other chapters also, alterations and additions of details have been made, and more up-to-date references to supplementary reading have been included. Some of the less satisfactory old figures have been deleted and some new ones added. The joint authorship is now justified by the extent of the additional microbiological matter. We are convinced that the usual divorce of teaching of pathology and microbiology is artificial and unfortunate for the medical student; we have done our best to integrate them.

We thank Mr. John Harrison of Luton and Dunstable Hospital for production of some of the new figures.

HESWALL AND LUTON   R. A. WILLIS
May, 1971   A. T. WILLIS

PREFACE TO THE FIRST EDITION

The main object of the book.—Thirty years ago, as an undergraduate student, I felt the need of a comprehensive outline of pathology expounded from its general principles, and without that artificial and repetitive subdivision of the subject into " general " and " special " parts which is still customary. My subsequent experience as a teacher of pathology convinces me that I was not peculiar in that respect ; and this book is an attempt to meet such a need. It aims to embrace the whole of pathology, at least as far as a senior medical student can reasonably be expected to assimilate it ; but descriptions of lesions in the various organs are included in the accounts of *the general processes and principles* which they exemplify. I firmly believe that this is the proper approach to the subject and the one which students can best appreciate.

Order of subjects.—The order in which the main processes of pathology are presented is the one which appears to me to be the most logical and to afford the most natural transition from normal anatomy and physiology. Uncomplicated repair is described first, because it is the most fundamental and physiological of all the processes studied by pathologists. Then follow inflammations evoked by microbic and other agents, and surveys of the ways in which inflammatory diseases vary according to the tissues affected and the particular bacteria responsible. The properties of bacteria, viruses and other parasites, and the phenomena of immunity, are discussed in detail sufficient for an understanding of pathological processes, but without any intention of providing a substitute for text-books of bacteriology and parasitology. The effects of extraneous foreign bodies and poisons and the effects of nutritional and metabolic disturbances then receive general consideration, but again without presuming to supplant text-books of toxicology, nutrition or pathological chemistry. The local effects of impaired circulation are next discussed, followed by an outline of disorders of the blood itself. The general accounts of tumours of each main class are amplified by descriptions of the main regional examples within the class. No excuse is offered for the length of the chapter on epithelial tumours ; this is no more than is appropriate to the importance of these tumours in human pathology. Consideration of endocrine disturbances is deliberately deferred until after the description of neoplasms, because so many endocrine diseases result from neoplasms of the ductless glands. So also, consideration of obstruction, dilatation and other mechanical effects in hollow organs is placed near the end of the book, because most of these are but the consequences of inflammatory and reparative diseases, parasites, foreign bodies or tumours, which have already been discussed in earlier chapters. A chapter is then allotted to a residuum of sundry diseases, the very nature of which is uncertain and which therefore cannot be confidently placed in earlier sections of the book. Although this method of disposal of these diseases is not as neat as that of the " special " pathologist, who can pop them into regional pigeon-holes, the present arrangement at least has the merit of focusing attention on the fact that their causes and nature are unknown. The final chapter is a general outline of the importance of antenatal factors in pathology and of the embryological principles necessary

ix

to an understanding of malformations. But, because teratology is not of great importance to students, no detailed description of particular malformations is attempted.

Special pathology.—While the main object of the book is to present general principles, I have tried to incorporate all essential details regarding regional peculiarities of particular diseases, and to make the information easily accessible by means of plentiful cross references and a full index. To each chapter is appended a short list of selected papers and text-books which are recommended to students wishing to amplify their knowledge of particular subjects. In these ways, I hope that the needs of the student in " special " pathology are adequately met.

Illustrations.—Most of the 300 photographs and diagrams are intended to illustrate general principles rather than diseases of particular organs. I have deliberately introduced many low-power hand-lens magnifications, for they form a valuable and too-little-used bridge between gross and microscopic pathology. Relatively few photographs of gross specimens are included, because the student can become familiar with naked-eye appearances only by examining surgical, necropsy and museum specimens.

The three appendices.—These contain notes on the meaning of names, the history of pathology, and the art of observation. To encourage a man to use his eyes and mind is to do him a greater service than to cram his head with facts.

Acknowledgements.—While the book is based largely on notes prepared for student lectures and elaborated during the last 20 years, its final preparation was carried out in the last 3 years while I held the post of Pathologist to the Royal Cancer Hospital, London. Dr. J. W. Whittick, my assistant there, has my special gratitude ; his careful and critical reading of the whole of the manuscript led to many improvements in it. Some sections were read also, and thereby improved, by Miss J. C. Tolhurst, Mr. K. B. Burnside, Professor Sunderland and Dr. W. King, all of Melbourne, Australia. Figures 85, 105, 186, 191–194, 219, 221, 225, 228, 233, 246, 256, 262 and 272 are from specimens in the museum of the Royal Cancer Hospital, and I am grateful to Miss J. Hunt and her staff for the photography of these. Many of the photomicrographs also are from sections prepared by Mr. C. G. Chadwin and photographed by Mr. L. A. Cowles, of the Pathology Department of the Royal Cancer Hospital, to both of whom I am much indebted. Plate X is the work of Miss Patricia Leicester, who helped me also in the preparation of Plate VIII. For access to the specimens depicted in Figs. 27–30, 48, 50, 51, 57, 63, 64, 106, 121, 130, 183, 266, 276 and 277, I am indebted to the Council of the Royal College of Surgeons, London. Figures 53, 66, 73, 74 and 79 are from specimens kindly lent by Professor H. E. Shortt of the London School of Hygiene and Tropical Medicine ; Figs. 164–166, from sections prepared by Sir G. Lenthal Cheattle ; Figs. 99 and 100, from specimens lent by Professor J. Gough of Cardiff ; Figs. 8 and 9, from sections lent by Dr. H. A. Sissons ; and Figs. 88, 89 and 96–98, from sections lent by Dr. J. W. Whittick. I am indebted to Professor E. R. Clark of Philadelphia for permission to use Fig. 5, and to Dr. Beatrice Pullinger and Sir Howard Florey for Fig. 6. Miss M. Walters greatly lightened my task in the preparation of the manuscript, and my wife helped in proof-reading and in making the index.

No writer could be more fortunate in his relations with his publishers than I am.

RUPERT A. WILLIS

LONDON, *June*, 1950

TABLE OF CONTENTS

xiii

INDEX

CHAPTER 1

WHAT IS PATHOLOGY ?

DEFINITION AND SUBDIVISIONS OF PATHOLOGY

Pathology is the science which studies the causes of, and the structural and functional changes accompanying, disease. It thus includes—(*a*) *aetiology*, the study of the causes of disease, (*b*) *pathological anatomy*, the study of the structural changes in disease, including *pathological histology* or *histopathology*, and (*c*) *pathological physiology*, the study of the alterations of function and metabolism in disease, including *pathological chemistry* or *chemical pathology*.

RELATIONSHIPS OF PATHOLOGY TO OTHER SUBJECTS

(1) To anatomy and physiology
Anatomy and physiology study normal structure and functions, i.e. the structure and functions of the genetically sound body maintained in a healthy or harmless environment. " Health " or " normality ", however, is difficult to define ; even the most healthy of us have some minor inherited imperfections or predispositions which we would be better without, and we are all constantly exposed to harmful environmental circumstances which leave their marks on us— physical injuries, chemical poisons, microbic infections, nutritional defects or excesses. However, anatomists and physiologists have succeeded in defining the main structural and functional norms or standards, anything outside which we recognize as clear evidence of disease or unhealth and therefore in the province of pathology. *Pathology is extended anatomy and physiology ; it is the study of all that the body and its tissues can be and do under all possible abnormal conditions.*

(2) To special sciences studying causes of disease
The general study of the causes of disease, *aetiology*, is a branch of pathology. But some of the groups of causes are so complex that they have become the subjects of special subsidiary sciences. Thus, *bacteriology* studies the properties of the many bacteria or microscopic vegetable parasites which can cause disease ; *parasitology* studies the numerous animal parasites of man and animals ; and *toxicology* studies the properties of chemical poisons and their effects. The pathologist must have a general knowledge of the main principles of these special sciences ; but he need not, indeed cannot, be versed in all their details. *Genetics*, the study of heredity, is another special science the principles of which are important for the pathologist ; for some diseases are transmitted genetically, and others, though acquired as the result of environmental factors, are more apt to develop in some families than in others because of an inherited predisposition, called a " diathesis ".

1

(3) To the clinical arts

Medicine, surgery, obstetrics, and their many branch specialities (gynaecology, dermatology, ophthalmology, radiology, etc.) are not sciences but practical arts concerned with the diagnosis, prognosis and treatment of sick patients. The intelligent practice of these arts of course necessitates a sound knowledge of the relevant parts of several sciences—physics, chemistry, anatomy, physiology and pathology. The best practitioners (and by " best " is meant " best for their patients ", not " best for their pockets ") are they whose practice is based most firmly on the principles and progress of these sciences, especially pathology. Medical practice not so based—and there is much of it—is little better than quackery or witchcraft.

All three of the main functions of the clinician, diagnosis, prognosis and treatment, depend on a knowledge of pathology. *Diagnosis* is the identification of the disease from the symptoms and signs it has produced in the individual patient and from the results of special laboratory tests. Symptoms and signs—the former noticed by the patient himself, the latter detected by the examining physician—are the manifestations of the altered structure and function which constitute the pathology of the case. Symptoms and signs can be understood and properly interpreted only in the light of pathology. Throughout his career, both as student and doctor, the wise practitioner will assiduously cultivate the habit of relating symptoms and signs to pathological changes. The greater his knowledge of these changes, the better diagnostician he will be. For this reason, he will never lose touch with the laboratory and necropsy room, but will ensure that he actually sees the particular lesions removed surgically from his patients or revealed by post-mortem examination. Thus he will maintain his knowledge of real, as opposed to imaginary, pathology ; and this cannot fail to improve his interpretation of symptoms and therefore his diagnosis in future patients. Throughout this book, the writers deliberately mention those main symptoms and signs which accompany, and are explained by, the lesions under discussion. *Prognosis*, the art of predicting the likely outcome of disease in the particular patient, clearly depends on a sound knowledge of pathology ; for it requires on the one hand an understanding of the natural history or progress of that disease in general, and on the other assessment of its severity and the extent of its lesions in the particular case—both of which are in the province of pathology. *Treatment* also often depends on pathology ; for example, the operability or otherwise of a particular case can be decided only from a knowledge of the nature and distribution of the pathological changes present.

COMPARATIVE AND EXPERIMENTAL PATHOLOGY

(1) Comparative pathology

Comparative pathology is the study of the same or similar diseases in different species of animals. Animals suffer from injuries, microbic and other parasitic diseases, tumours and nutritional disturbances identical with or closely similar to those of man ; and study of the progress and pathology of a disease in susceptible animals has often shed light on the human disease. In another way, too, animal pathology is of special importance to human pathology ; some human microbic infections are acquired by transfer from infected animals, e.g. bovine tuberculosis, plague, anthrax and hydrophobia. Mammalian diseases are of course the most closely similar to those of man ; but the scientific student of pathology finds as great an interest also in the diseases of all classes of the animal kingdom. Birds, fishes, amphibians and reptiles suffer from their own peculiar varieties of tuberculosis and other bacterial infections ; tumours occur in all classes of vertebrates and even in some invertebrates, e.g. the honey-bee and the oyster ; much has been learnt from the study of regeneration and repair in amphibia and invertebrates ; and the fundamental phenomena of inflammation can be usefully seen in even so lowly a creature as the earthworm.

(2) Experimental pathology

Experimental pathology greatly amplifies our knowledge of naturally developing diseases in man and animals, and this in the following different ways. (i) When a disease has been of unknown causation, its cause has often been discovered by ascertaining how to reproduce it experimentally in animals. This applies not only to most of the microbic infections, but also to vitamin deficiency diseases, endocrine diseases, many kinds of toxic and nutritional disturbances, and to some kinds of tumours. (ii) Animal experiments have often elucidated the routes of entry of particular infections in the human body, e.g. of the pneumococcus into the lungs, or of the virus of infantile paralysis into the central nervous system. (iii) When a patient is suffering from a disease of obscure nature, appropriate animal tests may enable its cause to be identified ; e.g. a fluid or discharge from a lesion suspected of being tuberculous may, on injection into a guinea-pig, give the animal obvious and easily identifiable tuberculosis, thus establishing the diagnosis. (iv) The progress of a disease and the development of its lesions can often be followed more completely in affected animals killed at different stages than in human cases. For example, the development of tuberculosis in guinea-pigs inoculated with tubercle bacilli, or of skin cancers in mice painted with carcinogenic substances, or of cirrhosis of the liver in animals given carbon tetrachloride, can be examined at all stages in large series of animals killed at selected intervals. (v) The effectiveness of a method of treatment or prevention can often be tested on suitable experimental animals ; e.g. the protective value of immunization against bacterial infections, or the value of particular drugs in the treatment of established infections. Experiments on animals also enable us to assess the importance of such general factors as impaired nutrition, vitamin deficiencies, or blood loss, on the susceptibility to, or the progress of, particular infections or other diseases. (vi) Breeding experiments have clarified the part played by genetic factors in certain inherited diseases or in diseases with an inherited predisposition.

For the foregoing reasons, the modern pathologist needs to know not only the structural and functional changes produced by disease in human tissues, but something also of the corresponding diseases in animals, and of the scope of experimental work in elucidating the causes of disease and the progress of pathological lesions.

P.P.B.—2

DIFFICULTIES ENCOUNTERED BY BEGINNERS IN PATHOLOGY

Many a student, coming fresh from his study of normal anatomy and physiology, which seem so precise and clear, finds pathology a perplexing subject. There are three main reasons for this, namely (*a*) the diverse appearance of diseased tissues, (*b*) the diverse causation of disease, and (*c*) uncertainties as to the nature and causes of some diseases. It will be helpful to consider these points more fully.

(1) The diverse appearance of diseased tissues

While anatomy and physiology deal with nearly invariable or standard structures and processes, the abnormal structures and processes studied by pathology are very diverse in both nature and degree. The organs and tissues of two normal human beings closely resemble each other, but diseased organs and tissues are variable in their naked-eye and microscopic appearance. No two examples of a disease are ever exactly alike. To take concrete examples, if you have learnt the gross and microscopic structure of the lungs of one healthy man, then, except for minor and negligible differences, you know the structure of the lungs of all other healthy men. But this does not apply to diseased lungs ; pneumonia in the lung of one man looks very different from that in another, according to its extent, stage, severity and the particular germ which has caused it ; so also tuberculosis of the lung is infinitely variable in its appearance, extent and complications ; fatal cancer of the lung in one man may appear as a tiny focus found only with difficulty, and in another as a huge mass almost completely replacing the organ and perhaps many other intrathoracic structures as well. Not only does each kind of disease appear in many different guises and stages ; but two different diseases may closely resemble each other in gross appearance and even in microscopical structure. For example, from the naked-eye appearance of a mass of abnormal tissue in a lung it may be impossible to say with certainty whether it is due to simple pneumonia, tuberculosis or cancer. The beginner in pathology naturally finds this diversity of appearances very confusing; but let him take heart ; all pathologists once felt just as confused as he, and even the most experienced can still make mistakes when confronted by unusual examples of quite common diseases. Constant experience in seeing and handling specimens is the only way to become proficient in pathological anatomy ; therefore the student should take every possible opportunity of seeing operation, post-mortem and museum specimens. He cannot see too many of even the most common and obvious lesions.

(2) The diverse causation of disease

Pathology includes a very difficult aspect from which anatomy and physiology are exempt—that of causation. The pathologist's task is not ended when he has described the structural and functional changes in tissues ; he must then identify, if he can, the abnormal factors which evoked those changes. This task is sometimes easy, sometimes difficult and sometimes impossible. To revert to the lung as an example ; the pathologist sees in a diseased area the gross and microscopic appearances characteristic of tuberculosis ; it is usually a simple matter to prove this by demonstrating the presence of tubercle bacilli. On the other hand, seeing changes characteristic of pneumonia, he knows that some microbic infection has been at work ; but it may be difficult or impossible to name the causative microbe, which might be any one of half a dozen different species and which may not be easily identifiable in the pneumonic tissue. A great deal of a pathologist's time is devoted to detective work, i.e. to the search for parasitic, chemical or other agents responsible for lesions the causation of which is not at once obvious. Now the causes of disease are many and diverse ; the pathologist must become familiar with them all and with the kind of changes they evoke in living tissues. This complex subject of causation or aetiology is a great source of confusion to the beginner, who cannot at once appreciate in which diseases our knowledge of causation is clear and definite and in which it is still obscure. This brings us to the third point, namely—

(3) Uncertainties as to the nature and causes of some diseases

Our knowledge of the causes of some main classes of disease is very full, though still not complete ; this applies, for example, to the great majority of parasitic diseases, whether due to viruses, bacteria, fungi, protozoa or metazoa. On the other hand, our knowledge of the causes of tumours, though advancing, is still fragmentary. There are many other common diseases the causation of which is often unknown or obscure, e.g. nephritis, " cirrhosis " of the liver, gastric and duodenal ulcers, many blood diseases and many endocrine diseases. Because these diseases are of unknown or doubtful causation, many opposing hypotheses have been advanced regarding them ; and it is very difficult for the beginner in pathology to distinguish between established facts and hypothetical speculations. One of the greatest services the teacher can do is to make this distinction plain in every subject discussed. He should often tell his students, " These are our speculations, but we really do not know ". In this book we try to distinguish clearly between established knowledge and mere speculation on controversial topics.

THE CAUSES OF DISEASE

Man is a very complex organism set in a very complex physical, chemical and biological environment. His mechanical ingenuity and his physical restlessness, in mechanized industry, in traffic and in war, make him much more liable than other species to physical accidents. He is exposed to great changes of temperature and humidity ; he is bathed in sunlight or immersed in fog. The air he breathes (especially in workshops, mines and populous cities) contains many kinds of gaseous impurities and dusts. The food he eats is extremely complex and diverse ; he consumes also many artificial chemical substances such as drugs, flavouring agents, dyes and antiseptics. He encounters and contends with many other species of living creatures, some of which have become adapted to a parasitic residence in his body. His crowding in great cities makes some of these parasites a special menace, spreading quickly as they do from victim to victim in epidemics.

In man's complex environment we find the causes of most of his ills. But man himself is a variable creature. Men differ intrinsically from one another, not only in the shapes of their noses and the colour of their hair and the quality of their brains, but also in their susceptibility to disease. The intrinsic differences between men are partly inborn or genetic, and partly acquired because of differences of nutrition and other environmental factors. We must therefore group the causes or antecedents of disease into two main classes as follows :

I : ABNORMAL ENVIRONMENTAL FACTORS

(1) *Physical injuries or trauma:*   Mechanical, thermal, chemical, radiational, electrical.

(2) *Parasites:*   Bacteria, fungi, viruses, protozoa, metazoa.

(3) *Harmful inanimate substances:*   (*a*) Inert, insoluble material or "foreign bodies".
    (*b*) Soluble poisonous or toxic substances.

(4) *Nutritional abnormalities:*   (*a*) General ; deficiency or excess of particular food substances, vitamins, minerals or water.
    (*b*) Local ; from impaired blood supply.

(5) *Carcinogenic or tumour-evoking agents :*   Chemical, physical, parasitic.

II : ABNORMAL CONSTITUTIONAL FACTORS

(1) *Genetic or inherited factors:* determining or predisposing to the development of particular diseases.

(2) *Non-genetic factors* predisposing to particular diseases, e.g. previous defective nutrition or previous disease.

A few explanatory comments will clarify this tabulation.

(1) Abnormal environmental factors

(*a*) Physical injury or trauma

This is the most obvious and direct kind of injury inflicted by environmental factors. Such factors may be purely mechanical, producing incised or lacerated wounds, contusions (bruises), fractures of bones, dislocations of joints, or tearing of muscles, ligaments or tendons. Thermal injuries comprise burns produced by subjecting tissues to excessive temperatures, and chilblains and frost-bite produced by subnormal temperatures. Caustic chemical substances, such as strong acids or alkalis, destroy tissues as effectively as heat and produce chemical burns. Various kinds of radiations damage tissues, as in sunburn, X-ray burns and radium burns. Electrical discharges may produce local burns or may disorganize internal parts and cause instantaneous death.

(*b*) Parasites

Plant or animal parasites capable of residing and multiplying in the tissues of human and other hosts, form the largest class of extrinsic causes of disease. We speak of them as *pathogens*, i.e. disease-producers ; and in ascending order of size they include viruses, bacteria, fungi, protozoa and metazoa. *Viruses* are the smallest pathogenic organisms; many of them are ultramicroscopic, while others are just visible microscopically. About 40 distinct human diseases are known to be due to viruses; these include such important ones as influenza, smallpox, measles, and infantile paralysis. *Bacteria* are microscopically easily visible unicellular vegetable organisms, about 60 distinct species of which produce human diseases ; these include boils and other purulent (i.e. pus-producing) infections, tuberculosis, typhoid fever, diphtheria, tetanus and plague. *Fungi*, closely allied to bacteria, but of filamentous or multicellular colonial types, cause about a score of distinct human diseases, e.g. thrush, ringworm and various other skin infections. *Protozoa* which are pathogenic for man include the amoeba of amoebic dysentery, the four types of plasmodia or sporozoa which cause malaria, and the flagellate trypanosome of sleeping sickness. *Metazoa* or multicellular animals parasitic for man may reside on the surface of the body, e.g. mites, ticks or fleas ; or in the alimentary canal, e.g. tapeworms or hookworms ; or they may infest the tissues or blood stream, e.g. hydatid cysts, trichina, or filaria.

(c) *Harmful inanimate substances*

Many kinds of harmful foreign substances may enter the body through the respiratory or alimentary passages or through wounds. Some of these are insoluble or chemically inert particulate masses which are harmful only mechanically, and are called " foreign bodies ", e.g. a peanut or grass-seed inhaled into the lungs, a swallowed bone or denture obstructing the oesophagus or intestine, or a needle or fragment of glass which has penetrated and become embedded in the tissues. A rather special class of extraneous foreign substances which are insoluble or only slightly soluble comprises those which are inhaled as fine dusts. These include carbon, silica and several kinds of silicates, and the diseases of the lungs due to them are collectively called " pneumoconioses ". Ingested or inhaled noxious substances which are more or less soluble are of enormous variety; their sources, chemical properties and effects are the subject matter of the science of toxicology. Some of them have long been recognized as occupational or accidental poisons ; others are drugs which, though used in treatment, sometimes produce serious toxic effects ; others are present in small quantities in our food or the atmosphere and their toxic properties are less clearly recognized.

(d) *Nutritional abnormalities*

Nutritional disturbances may be general, affecting the body as a whole because of alterations of diet ; or they may be local, affecting only part of the body, usually because of locally impaired blood supply. Examples of general nutritional disturbances are starvation and its effects, vitamin deficiency diseases such as rickets and scurvy, and mineral deficiency diseases such as iron-deficiency anaemias and iodine-deficiency goitres. Local nutritional disturbances are mainly due to interferences with the blood supply of parts. Clearly, if the arterial blood supply of a vascular part is cut off completely, the part will die ; if the supply is much reduced, the tissues of the part will show varying degrees of atrophy or degeneration. Arterial obstruction is the essential cause of many common pathological lesions, e.g. gangrene or death *en masse* of the distal parts of limbs, atrophy or softening of the brain, atrophy or degeneration of the myocardium. Bulky tumours which have outgrown the available blood supply, or have obstructed their own vessels, often show extensive degenerative changes or death of tissue.

(e) *Carcinogenic agents*

These comprise chemical, physical or parasitic agents which can evoke neoplasms or true tumours. Chemical agents of this kind include certain polycyclic aromatic hydrocarbons, certain azo-dyes and certain aromatic amines.

Carcinogenic physical agents include ultra-violet rays, X-rays and gamma-rays. A few special parasites are capable of evoking neoplasia in the tissues infested by them ; they probably do so by producing chemical carcinogens in the tissues. Some peculiar tumours in birds and animals are produced by filterable virus-like agents which multiply in the tumours as they grow.

(2) Abnormal constitutional factors

(a) Genetic or inherited causes of disease

The causes of most diseases are wholly or mainly environmental ; in only a small number are genetic peculiarities the sole or main cause. A bullet, a burn or a bacterium will play havoc with you no matter what your ancestry ; and there are no genes which will protect you from starvation or malaria or tape-worms. However, there are a few diseases which are due entirely to inherited chromosomal anomalies, and which develop irrespective of the environment. Examples of such purely genetic disorders include—familial jaundice accompanied by abnormal fragility of the red blood corpuscles ; the bleeding disease, haemophilia, which is transmitted by females but affects males only ; colour blindness, cleft iris, retinal tumours in infants, and some other anomalies of the eyes ; webbed digits, supernumerary digits and some other malformations of the limbs ; hereditary ataxia, and several other rare diseases of the central nervous system.

(b) Non-genetic factors predisposing to disease

It is but common-sense and everyday knowledge that people's resistance or susceptibility to particular acquired diseases depends on their previous nutrition and state of health. General undernourishment and vitamin deficiencies predispose to tuberculosis and other bacterial infections. One bacterial disease often predisposes the patient to another, e.g. influenza or whooping cough predisposes to pneumonia. On the other hand previous attacks of many infective diseases, e.g. smallpox, measles, typhoid fever and diphtheria, leave behind a more or less permanent immunity which protects the person from subsequent attacks.

This is the place to mention the importance of *mental factors in the causation of disease*. In the first place, through their ignorance, carelessness and lack of hygiene, mentally inferior persons—whether inferior by inheritance or training, nature or nurture—are more liable than others to various injuries, infections and nutritional disturbances. In the second place, in otherwise healthy people, severe mental strain or emotional upsets may produce important bodily effects culminating in disease. Thus, hurry, worry and emotional stress are well recognized factors in the causation of gastric and duodenal ulcers, arterial disease and hypertension, and exophthalmic goitre.

THE MAIN KINDS OF PATHOLOGICAL CHANGES IN TISSUES

Living tissues, when injured by the extrinsic causes just outlined, may display one or more of the following distinct kinds of pathological changes :

(1)	Repair	(4)	Neoplasia
(2)	Hyperplasia	(5)	Retrogressive changes
(3)	Inflammation	(6)	Developmental malformations

(1) Repair

Repair is proliferation of tissues to fill a breach caused by injury or disease. The simplest examples are seen in the healing of wounds and the union of fractures. Such gaps in the tissues become filled in by new reparative tissue consisting of young multiplying blood vessels and fibroblasts (or, in bone, osteoblasts) produced by multiplication of these elements from the surrounding tissues. Later this young reparative tissue changes into a scar in soft tissues, or effects bony union of a fractured bone.

(2) Hyperplasia

Hyperplasia is proliferation of cells of a particular kind either (a) as a compensatory response to loss of tissue of the same kind, or (b) as a result of disturbed hormonal control of the tissue. Examples will make these meanings clear. If liver tissue is lost, either by excision or by damage by poisons, new liver cells are freely produced by mitotic proliferation from surviving liver tissue ; removal of part of an endocrine gland is usually followed by proliferation of the cells of the remainder so that the normal amount of the tissue is almost or quite restored ; loss of blood is followed by hyperplasia of the red bone marrow and increased haemopoiesis. These are examples of compensatory hyperplasia or regeneration. Such compensatory regeneration is allied to repair, from which, however, it differs in the specificity of the regenerated cells and in the fact that the regeneration is not restricted to the site of tissue loss. The function of repair is to fill a breach ; the function of regenerative hyperplasia is to compensate for the loss of cells of a specific kind. The second kind of hyperplasia, that due to hormonal disturbances, is seen in the endocrine glands or in organs which are under endocrine control, such as the breast and uterus. The cyclical changes in cellular proliferation and function in the cells of the breast or of the endometrium depend on cyclical variations in the balance of various hormones. Clearly then, an improper balance of the hormones concerned may bring about abnormal proliferations of the special cells of the breast or endometrium even when there is no proper physiological call on them.

(3) Inflammation

Inflammation is the immediate vascular and exudative reaction of tissues to injuries. When tissues are injured by physical or chemical agents or by microbic parasites, they become inflamed, red and swollen. The redness and swelling are due to dilatation of the small blood vessels and to exudation of fluid and leucocytes from the blood into the tissue spaces. These changes subside when the products of the damaged tissues have been absorbed and the invading microbes have been destroyed. Inflammation and repair, though often seen together, should be distinguished from one another ; the one is a prompt vascular and exudative response, involving no cellular multiplication, and its function is removal of the damaging agent and of the products of tissue damage ; the other is a slower proliferative response, the function of which is to fill the breach created by the injury in the tissues.

(4) Neoplasia

Neoplasia is an abnormal mode of growth of tissues which exceeds and is uncoordinated with that of the normal tissues, and persists in the same excessive manner after cessation of the stimuli which caused it. The abnormal mass of tissue produced is called a neoplasm or true tumour. Tumours form a very important class of pathological lesions ; dangerous types of them are popularly called " cancers ". The distinctive feature of tumours is that they continue to grow progressively, irrespective of the structural and functional requirements of the body. In this they differ fundamentally from masses of reparative or hyperplastic tissue, in which of course the cellular multiplication is limited in amount and duration in accordance with structural and functional needs. Neoplasia involves a permanent irreversible change in the cells, manifesting itself in excessive multiplication, and transmitted indefinitely to the descendants of the affected cells. The extrinsic causes of this persistent excessive habit of growth are called carcinogenic agents or carcinogens. Those so far discovered include various special chemical substances (e.g. certain hydrocarbons, azo-dyes and organic amines) and radiations of short wave-lengths (ultra-violet light, X-rays and gamma-rays). The external causes of many kinds of tumours are still unknown, but more of them are gradually being discovered. Of the all-important cellular change which they evoke and which is the real secret of the neoplastic habit of growth, we as yet know next to nothing.

(5) Retrogressive changes

This is a very heterogeneous group of changes in cells and tissues, resulting from many different causes—the action of poisonous substances, disturbances of nutrition or oxygen supply, metabolic disturbances, etc. The most extreme examples are those of necrosis or death of tissues *en masse*, whether from deprivation of blood supply, e.g. gangrene of the limbs or softening of the brain from occlusion of the arteries, or from the action of poisons, e.g. necrosis of the liver from carbon tetrachloride. Short of total necrosis, however, tissues which are starved or poisoned may survive but may show a great variety of degenerative structural changes—e.g. intra-cellular or extra-cellular accumulations of various lipoid substances, abnormal proteins, glycogen, pigments or calcium salts. Or they may show wasting or atrophy of the more specialized parenchymatous cells —secreting epithelia, muscle fibres or nerve cells.

(6) Developmental malformations

These form a special class of structural abnormalities, due to disturbances of development of the young organism either from damaging environmental agencies or from genetic defects. Damage inflicted on a young embryo, whether by chemical poisons, microbic toxins or nutritional defects, may kill it outright ; but, if it survives, it may have suffered such injury to growing parts that these now develop imperfectly. There may be arrest of development of particular parts which may remain stunted and deformed, or which may fail to coalesce as they should ; or there may be abnormal persistence of embryonic structures which normally disappear, e.g. the thyroglossal duct or the vitelline duct ; or there may be extensive deficiencies of large parts of the body ; or, if the disturbing agencies have acted very early in embryonic life, there may be duplication, bifurcation or other anomalies in the formation of the primary axis, leading to double monsters. Malformations of these various kinds are due much more often to harmful external conditions than to genetic anomalies ; but a few special kinds, e.g. syndactyly (webbed fingers) and some other malformations of the digits, are often genetic in origin.

CELLULAR PATHOLOGY

Just as *cells* are the normal structural and functional units studied by the anatomist and physiologist, so they are the abnormal units studied by the pathologist. Inflammation is evoked by injury to cells, and its characteristic changes are in cells, namely, vascular endothelial cells and leucocytes; repair and hyperplasia are co-ordinated, and neoplasia unco-ordinated, proliferations of cells; degenerative changes and necrosis are due to nutritional and toxic injuries to cells. The concept of *cellular pathology* was first set forth by Virchow just over a century ago (1858); it has been fully elaborated, in the light of modern research, by Sir Roy Cameron. Along with histochemical and microchemical studies, the electron-microscope is playing an increasing part in exploring both normal and abnormal cellular structure and function, and is revealing distinctive pathological changes in cell membranes, mitochondria and other organelles of cells, beyond the range of the ordinary microscope. New sciences of normal electron-microscopy and electron-cytopathology are springing up.

SUPPLEMENTARY READING

Cameron, G. R. (1952). *Pathology of the Cell.* Edinburgh and London; Oliver & Boyd. (An encyclopaedic work.)
 – (1956). *New Pathways in Cellular Pathology.* London; Edw. Arnold. (A brief outline which all students should read.)
Symposium in honour of the centenary of Virchow's " Cellular pathology " (1858–1958). *J. Clin. Path.,* **11**, 463.

CHAPTER 2

REPAIR AND REGENERATION FOLLOWING PHYSICAL INJURIES

INTRODUCTION

IN THIS chapter we shall consider the simplest of all pathological processes, the repair and regeneration of tissues following physical injuries uncomplicated by bacterial infection. Indeed the capacity of tissues to repair themselves after being wounded, bruised or fractured may be regarded as physiological rather than pathological. For wounds are common incidents in the lives of all classes of animals and plants, and the capacity to repair them is clearly a necessary self-preservative function without which none but the most sheltered individuals would ever attain adulthood. If no wound ever stopped bleeding or ever healed, man and most other vertebrate species would become extinct in a single generation.

Different classes and species of living things differ very greatly in their capacity to repair injuries and regenerate lost parts. In many plants and many of the lower invertebrates, a small fragment can regenerate a whole organism ; e.g. cuttings or slips from small terminal branches or twigs are the horticulturalist's usual means of propagating many kinds of trees and shrubs ; *Hydra* and many other coelenterates may be cut up into fragments and each fragment will grow into a complete organism ; earthworms may be cut in half and each half will regenerate a new head or tail as required. Structurally more complex creatures do not show this power of total or nearly total regeneration ; nevertheless some species have retained the ability to grow complete new parts ; thus crabs can regenerate complete claws when these have been lost, snails can regenerate eyes, and, even among vertebrates, newts and allied amphibia can regenerate lost limbs or tails, and some lizards can grow new tails.

Mammals do not regenerate complex parts such as eyes or limbs ; but they possess great powers of regeneration of many individual tissues, especially vascular, connective, bony, haemopoietic and most epithelial tissues. As pointed out in Chapter 1, it is convenient to distinguish two types of regeneration following injuries in mammals, namely, (*a*) *repair*, which is the proliferation of tissues at the site of injury to fill the gap, and (*b*) *compensatory regeneration*, which is the proliferation of tissues with specialized functions, e.g. liver epithelium or haemo-poietic cells, to compensate for those destroyed. The local repair of physical injuries is effected by the proliferation of vascular and connective (or, in bone, bony) tissue to fill the breach, and of the overlying epithelium if a surface is broken. We will deal first with these phenomena of local repair, and will then consider the regenerative capacity of other kinds of tissues.

REPAIR OF A CLEAN INCISED WOUND, OR HEALING BY FIRST INTENTION

A cut through the skin or the skin and subcutaneous tissues is the simplest and commonest of injuries, familiar in every cut finger and every surgical incision. The man in the street knows as well as the surgeon that if such a cut is clean and if its two edges are brought and held together, it usually heals quickly with little or no inflammation, and leaves little or no scar. This is often spoken of as primary or direct union or " healing by first intention ". Let us examine this familiar process a little more closely.

The injury inflicted by a clean aseptic (i.e. bacteria-free) incision is slight, consisting only in the destruction of the tissue elements in the actual plane of section. In the epidermis the cut destroys a few epithelial cells ; in the dermis it destroys a few connective tissue cells and fibres, and it severs a number of small blood vessels and lymphatics from which blood and lymph escape onto the cut surfaces and clot there ; and in the subcutaneous tissues, it destroys also a few adipose tissue cells, freeing their fat contents. Of course, if the wound is deeper, it may sever other kinds of tissues also—muscle, tendon, bone or large blood vessels. Repair of these particular tissues will be dealt with presently ; here let us consider what happens in those commonly severed tissues, skin and sub-cutaneous tissue.

If the cut surfaces are brought and held accurately together (by adhesive tape applied to the surface, or by sutures if the cut is deep and tends to gape), almost immediately they become cemented to each other by a thin intervening layer of fibrin coagulum from the extravasated blood or lymph. The less of this the better, since large gaps occupied by clot take longer to heal and also invite bacterial infection. During the next few days, fibroblasts and vascular endothelial cells on each side of the cut multiply by mitosis and grow across the fibrin film, re-establishing continuity of the tissues. The fibrin is absorbed and replaced by new collagen fibres laid down by the young ingrowing fibroblasts ; and endothelial vascular sprouts growing across from each side unite with one another and form new capillary vessels. At the same time the small amount of organic debris (fat, protein, etc.) from the cells destroyed by the cut is absorbed, partly by undergoing chemical disintegration and solution in the tissue fluids and partly by being ingested by wandering phagocytic cells. Within a few days, the clearing-up process and re-establishment of tissue continuity across the plane of cleavage may be so complete that no trace of the line of union may be visible. If apposition of the cut surfaces has been imperfect, however, gaps between them are filled with rather more blood or fibrin clot ; consequently more fibroblasts and vascular endothelial cells multiply to replace this and re-establish tissue continuity ; with the result that finally a visible white layer of fibrous scar tissue comes to mark the plane of the incision. Continuity of the surface epithelium is restored by proliferation of the epithelial cells on either side of the line of cleavage.

REPAIR OF AN OPEN WOUND OR ULCER BY GRANULATION, OR HEALING BY SECOND INTENTION

An ulcer is any breach in an epithelial surface exposing the underlying tissues. The simplest kind of ulcer is an open traumatic wound. In the repair of ulcers the breach in the tissues is filled up from the base by the formation of a pink vascular soft tissue with a granular surface, called *granulation tissue*. While this young reparative tissue derives its name from its appearance in healing wounds, an identical tissue is concerned in many other reparative processes— in the union of torn deep structures unconnected with surface wounds, in the organization of masses of blood clot and fibrin, and in the healing of areas of destruction in tissues resulting from bacterial and other inflammations. The structure of granulation tissue, shown in Plate I and Figs. 1–4, is therefore of great

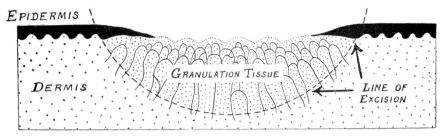

FIG. 1.—Diagram of a granulating excised wound of the skin.

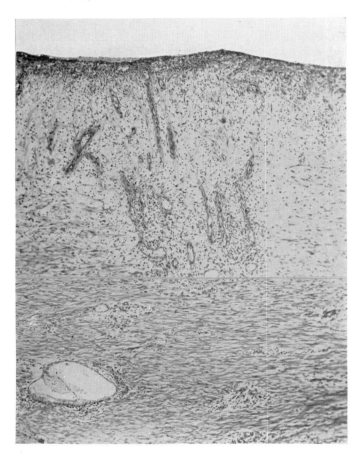

FIG. 2.—Photograph of granulation tissue.   Compare with Plate I.
($\times$ 60.)

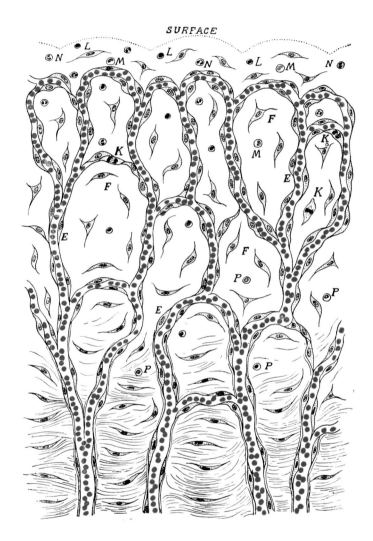

PLATE I.—Diagram of structure of granulation tissue : *E*, capillary endothelial cells ; *F*, fibroblasts ; *K*, cells in mitosis ; *L*, lypmho-cytes ; *M*, macrophages ; *N*, granulocytes ; *P*, plasma cells. Note horizontal orientation of fibroblasts and fibres in the deeper part. Compare with Fig. 2.

importance.   It consists of new-formed blood vessels and proliferating fibroblasts growing up from those of the subjacent tissues.   The new-formed blood vessels arise as sprouts of proliferating endothelial cells ; at first these sprouts are solid tendrils of cells, but later they acquire a lumen and blood flows through them.   The general direction of growth of new vessels is vertically upwards from the base of the ulcer, but the main vertical vessels also send off side-sprouts which unite with those of their neighbours to produce a complex capillary plexus often in the form of a series of arches.   The mul-

tiplying fibroblasts between the vessels are plump fusiform or stellate cells which at first are loosely and irregularly arranged and unaccompanied by collagen fibres, resembling fibroblasts in tissue cultures *in vitro*.   Later, however, as the granulation tissue gets older, the fibro-blasts lay down collagen fibres which, along with the cells, tend to be orientated

FIG. 3.—Vessels of granulation tissue injected (Billroth). Compare with Fig. 5.

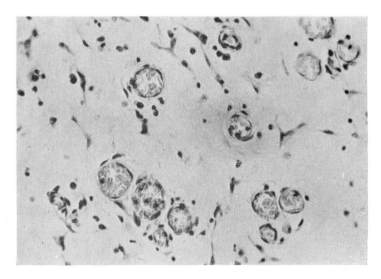

FIG. 4.—Capillaries of granulation tissue seen in cross-section.   (× 250.)

parallel to the surface, i.e. at right angles to the direction of the main up-growing vessels.   In a granulating wound these different stages of healing can often be seen together, the more superficial and younger parts of the granulation tissue being still non-fibrous and unorganized, the deeper and older parts showing different degrees of fibrous organization.   A zone of the pre-existing tissues around and subjacent to the healing ulcer also shows proliferating vessels and fibroblasts.   The orientation of the cells and fibres in granulation tissue has been shown to depend on the directions of mechanical stress in the tissue (see over).

While the granulation tissue is growing up from the base of the wound, the epithelium at its margins is growing in over the granulating surface as a thin film which gradually thickens.   This eventually clothes the entire surface as a uniform layer deficient in papillae, hair follicles, sweat glands and sensory nerve endings.   When epithelialization is complete, the granulation tissue ceases to grow ; thereafter it becomes less and less vascular and more and more fibrous, until eventually it is converted into an almost avascular, densely fibrous scar.

Since open granulating wounds are seldom quite free of bacteria, granulation tissue often shows signs of mild inflammation, by containing a variable number of leucocytes and discharging a small amount of pus from its surface.   But the less the bacterial contamination, the less these signs of inflammation ; and bacteria-free granulation tissue consists almost entirely of proliferating blood vessels and fibroblasts.

THE STUDY OF LIVING GRANULATION TISSUE

The knowledge gained from examination of stained sections of granulation tissue has been greatly amplified by microscopical study of reparative tissue in the living animal, as described by the Clarks and their collaborators.   Into

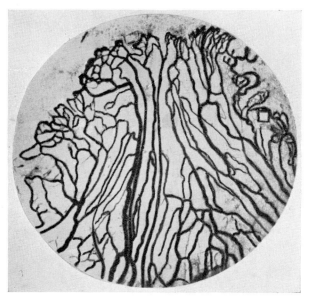

FIG.   5.—New-formed blood vessels growing into a transparent chamber in a rabbit's ear (from Clark *et al.*). Compare with Fig. 3. (× 14.)

punch holes in the ears of rabbits these workers inserted small flat transparent chambers consisting of thin sheets of mica and celluloid separated by a space 40μ to 75μ deep.   The granulation tissue from the wound margins grows into this space from all sides and finally fills it.   The thin layer of living tissue can be closely studied microscopically through the transparent walls of the chamber ; all stages of repair, from the early sprouting of young blood vessels to the final formation of fibrous scar tissue can be observed ; and the fate of individual vessels and cells can be traced (Fig. 5).

This technique has confirmed much that was learned earlier from stained sections. The new vessels grow first as solid sprouts of endothelial cells, which later become hollow and carry a blood flow. These growing vessels advance into the chamber in the form of a capillary plexus often with a series of arcades. Multiplying fibroblasts grow into the chamber along with the vessels ; at first they lie in all directions and produce no collagen fibres, but later they lay down collagen fibres and these become orientated in definite directions determined by mechanical factors. The cells and fibres arrange themselves concentrically around a solid obstruction such as the small blocks which keep the two sides of the chamber apart ; or they come to lie in the direction of lines of traction. The importance of mechanical stresses in determining the orientation of fibro-blasts is seen also in tissue cultures. By cultivating fibroblasts in a thin film in a quadrangular frame, and by exerting traction on this in various directions, it is found that in regions free from directional stress the cells and fibres lie at random in all directions, but that in regions under tension they become arranged in the direction of the lines of tension and they also multiply more rapidly in these regions.

The transparent chamber technique has greatly extended our knowledge of the later differentiation of blood vessels in reparative tissue. All the young vessels which grow into the chamber at first have the structure of capillaries or simple endothelial tubes. Later, however, some of them, carrying blood into the new tissue, acquire the structure of arterioles, developing straight thick walls and losing most of their lateral connections. When a new-formed vessel becomes an artery, the first indication of this is that some of the extravascular fusiform cells, formerly identical with fibroblasts, arrange themselves circumferentially around the vessel. These cells soon exhibit contractility, and, according to the Clarks, acquire a vaso-motor innervation from new-formed nerve fibres which have grown into the tissue. Thus the indifferent perivascular fibroblasts or mesenchymal cells have differentiated into smooth muscle fibres, and the vessel has become an innervated arteriole. Other vessels in the capillary plexus, carrying blood away from the tissue in the chamber, grow wider and acquire the structure of small veins. All of these vessels are highly plastic, readily undergoing modi-fications of structure in response to thermal or mechanical stimuli or to alterations of blood flow. Thus, small veins may revert to a plexus of capillaries ; so may arteries, their encircling cells losing their contractility and resuming the character of indifferent perivascular cells.

The transparent chamber technique has shown also that reparative tissue contains new-formed lymphatics, separate from the blood capillaries and sprouting from the lymphatics of the surrounding tissues. These are difficult to identify in ordinary sections of granulation tissue, but they can be demonstrated by injecting opaque or coloured material into the lymphatics of the part (Fig. 6).

SCAR TISSUE

A *scar* or *cicatrix* is a mass of dense fibrous tissue produced by the healing of a wound or of an area of destruction in soft tissues from any other cause. Like normal fibrous tissues such as ligaments and fasciae, scar tissue consists of

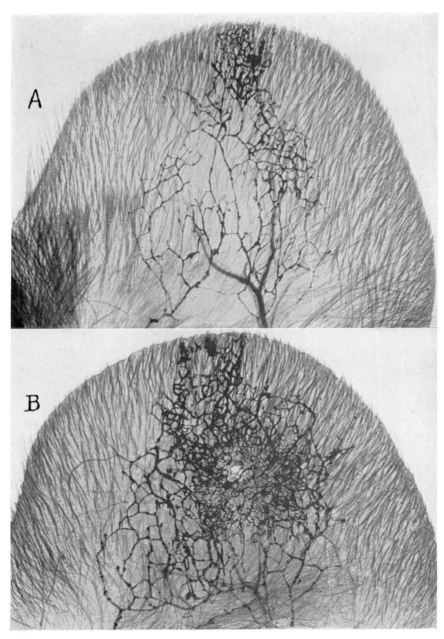

FIG. 6.—Proliferation of lymphatics in injured tissues. A, cleared normal mouse ear
with lymphatics injected. B, mouse ear 21 days after making a turpentine abscess,
which perforated the ear ; a dense new network of lymphatic capillaries surrounds
the hole. (From Pullinger and Florey.) (× 8.)

bundles of collagen fibres with interspersed connective tissue cells or fibrocytes. The older the scar, the denser the fibrous tissue ; and in many old scars the fibres undergo hyaline change, i.e. they become welded together into a homogeneous structureless mass of hard collagenous material. As we have seen, the orientation of the fibres follows the lines of tension in the repairing tissues ; so that, in the skin for example, the main bands of fibrous scar tissue run parallel to the surface and tend to radiate from main points of attachment such as bony prominences. Since scar tissue once formed is almost non-pliable and inextensible, and indeed contracts as it ages, it is very important that injured parts should be kept in proper positions during their repair. For example, following a burn of the front of the neck, the neck should be kept extended during healing, so that scarring shall not result in a permanent flexion deformity. Or, following a fracture or dislocation of the elbow joint with tearing of surrounding soft tissues, the elbow should not be kept extended but flexed to about 90° and semi-pronated ; so that, should there remain some permanent limitation of joint movements because of scarring in the torn tissues, the joint will be in its most useful position. Much of the work of the orthopaedic surgeon is concerned with the posturing of injured parts in their most advantageous positions during repair.

Mention must be made here of *keloid*, an overgrown scar forming a hard irregular projecting mass. This may follow a wound, burn, a vaccination mark, or even such trifling injuries as insect bites or pin pricks. A keloid, unlike an ordinary scar, does not diminish and contract with time, but on the contrary it tends to increase in bulk. If it is excised, further excessive keloid tissue usually appears ; and the patient may develop multiple keloids. These characters show that keloid formation depends on some idiosyncrasy peculiar to the individuals affected, but what this is we do not know. An overgrown scar is not always a true keloid ; it may result from smouldering inflammation in the scar due to mild bacterial infection or to residues of poorly absorbable suture material.

Occasionally, for reasons not clearly understood, a scar becomes partly converted into bone, an instance of osseous *metaplasia* of connective tissue (see p. 36). The metaplastic bone may develop bone marrow with haemopoietic and adipose tissue cells.

THE ORGANIZATION OF A BLOOD-CLOT

Blood is often extravasated into injured tissues ; its absorption and the relation of this to repair require consideration here. A diffuse extravasation of blood into tissues is called an *ecchymosis* ; it may result from diffusion of fluid blood from a bleeding spot into surrounding tissues, or from the local rupture of many small vessels by a blow, as in a contusion or bruise. The extravasated blood in an ecchymosis eventually undergoes complete absorption and leaves no trace behind. The red corpuscles disintegrate, freeing their haemoglobin, which undergoes rapid chemical alteration, into brown iron-containing pigment or *haemosiderin* which gives the Prussian-blue test and other tests for iron, and yellow iron-free pigment or *haematoidin* which is identical with biliverdin and bilirubin. These pigments in varying stages of disintegration give a bruise its splendid colour changes. Some of the pigment is removed by being taken into solution in the tissue fluids, but most of it is taken up by wandering mononuclear phagocytes (macrophages), many of which loaded with brown and yellow pigment granules are to be seen in sections of bruised tissues.

While diffusely extravasated blood is quickly and completely absorbed, a large circumscribed collection of blood-clot or *haematoma* in the tissues cannot be disposed of so simply. An aseptic haematoma is an inert mass of dead organic matter ; and the space occupied by it is a breach in the tissues. Into this breach young vascular reparative tissue grows from its periphery, penetrating into the blood clot and replacing it *pari passu* with its removal. The invasion and replacement of the clot by ingrowing reparative tissue is called *organization* of the clot. The organizing tissue has the structure of granulation tissue ; it consists of advancing arcades of capillary blood vessels accompanied by proliferating fibroblasts, and it differs from the ordinary granulation tissue of an open wound only in that it contains many wandering phagocytes laden with the products of disintegration of the extravasated blood—pigments, cholesterol and other lipoid substances—which are being removed. The organizing tissue later becomes less vascular and more fibrous, and finally forms a scar in which areas of unabsorbed pigment often remain. Sometimes, if the haematoma has been a large one, its replacement by organizing tissue may be incomplete and a cavity containing brownish fluid may remain in the centre of the fibrous scar. Sometimes, too, parts of a haematoma may become calcified or ossified.

THE REPAIR OF TORN TENDONS AND LIGAMENTS

If a tendon is cut or torn across, provided the gap between the cut ends is not too great, it will be filled by new properly orientated tendinous tissue, completely reconstituting the injured tendon. This is only a special instance of repair with orientation of fibrous tissue in the direction of tension, as already described. The regeneration of tendons has been fully studied experimentally. If a tendon in an animal is cut, the space between the cut ends is filled at once by blood clot and tissue fluid. Into this, as into the transparent chamber in a rabbit's ear, young capillary vessels and plentiful fibroblasts grow from the surrounding soft tissues. The fibroblasts and the fibres they produce are at first chaotic, but they soon orientate themselves in parallel bundles in the direction of the tendon. That this orientation of the new tissue is due to the pull exerted by the muscle on the proximal cut end of the tendon, is shown by the fact that if the muscle also is cut so as to abolish the pulling effect. no properly orientated tendinous tissue develops in the gap. Further, if a thread is inserted through the regenerating tissue in the gap so as to exert a slight lateral pull on it, then its fibres orientate themselves transversely to form a strip of tendinous tissue, but at right angles to the proper direction. In the repair of torn ligaments following sprains or dislocations of joints, the natural lines of traction similarly determine the orientation of the new fibrous tissue in directions proper for the ligaments concerned ; so that, if the parts are kept in good position, the normal structure of the ligaments will be perfectly regenerated. This complete restitution of severed tendons and ligaments recapitulates what happened during their normal development. During embryonic foetal and infant life the orientation and strength of the developing tendons and ligaments were determined by the mechanical stresses to which they were subjected. When the tendon or ligament is injured in the adult, the young plastic tissue which repairs it is subjected to the very same stresses and it differentiates accordingly.

THE REPAIR OF A FRACTURED BONE

This may be described in three stages, which merge into each other : (1) granulation and periosteal proliferation, (2) the formation of young bone or callus, (3) consolidation and modelling of the new-formed bone.

(1) The stage of granulation and periosteal proliferation

Following a *simple fracture* of bone, i.e. one with which there is no breach of the overlying skin and therefore no bacterial infection, the gap between the broken ends of the bone becomes occupied at once by a mass of coagulated blood and tissue fluid. Into this there gradually grows from the periosteal tissues and from the broken ends a young reparative tissue resembling granulation tissue and consisting of capillary blood vessels and a swarm of multiplying periosteal and osteal fibroblasts. It replaces and organizes the coagulum, just as a haematoma is organized. It is plainly visible in fractures one or two weeks old, in which it appears as a soft pink gelatinous layer everywhere clothing the fractured bone surfaces and inner surface of the periosteum.

At the same time that the fracture gap is being filled in by young reparative tissue from the injured tissues around, the periosteum of the bone surfaces for some distance away from the site of fracture also shows reactive changes. It increases in thickness, becomes softer, its cells multiply rapidly, and new-formed blood vessels develop in it. These changes are greatest close to the fracture, where the proliferating periosteum merges with the young organizing tissue sprouting into the fracture gap. The changes diminish away from the fracture site, but they are often recognizable 2 or 3 inches from it. They are especially prominent and extensive if the fracture has been oblique or comminuted (i.e. splintered into multiple fragments), or if excessive movement takes place during repair. The endosteal or medullary tissues of the bone also show proliferation for some distance on either side of the fracture.

The new tissue which grows into the fracture gap and the proliferating periosteum of the neighbouring parts of the bone together form a mass of young repair tissue enveloping and provisionally uniting the broken bone ends (Fig. 7a). Although at first the structure of this mass resembles that of granulating or organizing tissue derived from non-osseous soft tissues, its later differentiation into young bone or callus reveals its special properties. This is not surprising ; for most of its cells have proliferated from the periosteum or from the bone itself, i.e. from bone-forming cells.

(2) The formation of young bone or callus

Callus is the whole mass of proliferating and differentiating reparative tissue enveloping a fracture. In the enveloping mass osteoid or young bony tissue soon appears. Bundles of collagen fibres are laid down, and, along with these, trabeculae of a hyaline ground substance often called osseo-mucin. These form an irregular meshwork with richly vascularized connective tissue between them.

At first they are uncalcified, but calcium salts are rapidly deposited in them, converting them into rigid bone trabeculae, the cells enclosed in them and clustered along their margins now being recognizable as differentiated osteoblasts. The new bone so formed, like newly-formed membrane bone in the embryo, at first

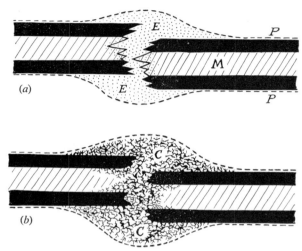

FIG. 7.—Diagram of repair of fractured bone : (*a*), early stage, with fracture gap occupied by blood clot and exudate *E* ; (*b*), later stage, with much new-formed woven bone in the callus *C*.   *P*, periosteum ;   *M*, medulla.

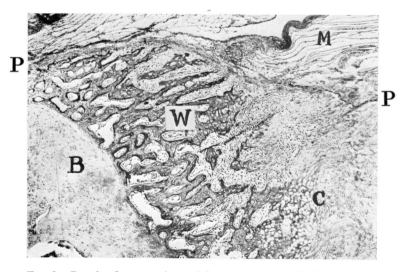

FIG. 8.—Repair of an experimental fracture in a rat : B, bone cortex ; P, periosteal layer ; W, new-formed subperiosteal woven bone ; C, an area of cartilage in the callus ; M, surrounding muscles. (Preparation by H. A. Sissons.) (× 60.)

has no lamellar or Haversian arrangement ; it is laid down in the young reparative tissue as a delicate meshwork appropriately called woven bone. It develops in all parts of the mass of callus—in the organizing fracture gap, in the medullary cavity of the broken ends, and especially in the proliferating periosteal tissue enveloping the ends (Figs. 7b and 8). Areas of cartilage also differentiate in parts of the proliferating callus, especially if this is subjected to movement ; and the cartilage then undergoes ossification by ingrowth of osteoid tissue, just as occurs in endochondral ossification during development (Fig. 9).

In a single mass of callus, all of the stages of repair can often be found—proliferation of young undifferentiated reparative tissue, the direct differentiation

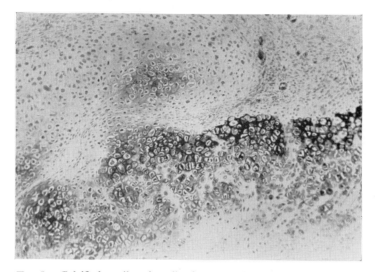

FIG. 9.—Calcified cartilage in callus in a rat. (H. A. Sissons.) (× 60.)

of this into osteoid or woven bone, the differentiation of cartilage, endochondral ossification of this new cartilage, and also the modelling of the new bony tissue by osteoclastic absorption and lamellar deposition, now to be described.

(3) Consolidation and modelling of the new-formed bone

Woven bone in a reparative mass of callus is laid down rapidly and without respect to mechanical requirements. Much of this *provisional callus* must be removed, and much of it must be strengthened, before repair is complete and the bone is able to bear its normal burdens. The modelling of the provisional callus is brought about by the deposition of lamellae of new bone by osteoblasts in places requiring strengthening, and the absorption of already deposited bone by the activity of osteoclasts in places where bone is not needed. The factors determining this deposition and absorption are the mechanical stresses to which the callus is subjected, namely weight-bearing and the thrusts and pulls of the muscles of the part. It was these factors which moulded the bone to its adult form during its normal development; the same factors now mould it to a similar form during its repair. Just as the strength and orientation of healing tendons and other fibrous tissues are determined by the tensions to which the healing tissue is subjected, so the quantity and orientation of bone in callus are determined by the strength and directions of the thrusts to which it is exposed while it is being consolidated. Where the callus experiences a strong thrust, there the osteoblasts are stimulated to lay down lamellae of new bone in appropriate directions : where the callus experiences no thrusting force, there osteoblastic activity is not maintained, but on the contrary osteoclasts abound and remove superfluous trabeculae.*

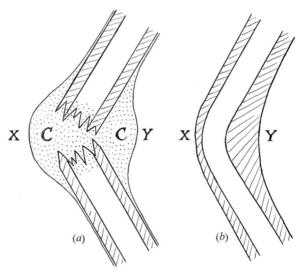

FIG. 10.—Diagram of repair of angulated fracture. (*a*) Early stage, with fracture gap occupied by young callus *C*. (*b*) Final modelling, with a strong bony buttress in the concavity *Y*, and thinned cortex on convexity *X*.

A good illustration of the importance of mechanical stresses in the modelling of callus is seen in the healing of a fracture of a weight-bearing bone with an angular deformity. In the early stages of repair callus develops as usual all around the broken ends (Fig. 10a). During the later modelling of this, on the convex side (X) of the angular deformity, the mechanically useless callus and also the projecting broken edges of the original bone shaft are absorbed. But on the

---

* It will be as well to state in parenthesis here that *all* bone deposition, normal or abnormal, is the result of osteoblastic activity, and that *all* bone absorption, normal or abnormal, is the result of osteoclastic activity. There is no structural evidence whatever in favour of the view that bone can be absorbed by *halisteresis*, i.e. solution and removal of calcium salts without the agency of osteoclasts. Both pathological and normal histology show also that osteoblasts and osteoclasts are *not* immutable self-perpetuating species of cells, but that they are only different functional variants of cells of the same kind and that they can change into one another in response to changes in their environment. Where these cells, in response to surrounding conditions (especially mechanical stresses), are functionally concerned in producing alkaline phosphatase and in laying down new bone, they are osteoblasts ; where, in response to surrounding conditions, they are functionally concerned in bone absorption, they are osteoclasts ; and where they are functionally neutral, engaged neither in the formation of new bone nor the removal of old, they are the quiescent osteocytes of the stable bone in which they lie. The osteoblast of to-day may be an osteoclast to-morrow and vice versa, and this reversal of function probably amounts to no more than a slight alteration in the nature or quantities of enzymes formed by the cells.

concave side (Y) of the angulated fracture, the callus remains and consolidates into a strong bony buttress. The formation of this buttress and the orientation of the new bone in it are clearly determined by the weight-bearing and muscular thrusts exerted on the two parts of the bone shaft and transmitted through the callus in the angle. The final result is a healed shaft with a smoothed-off thin cortex on the convexity of the deformity and a wide dense cortex on its concave side (Fig. 10b).

REPAIR OF CARTILAGE

It is probable that already differentiated cartilage cells do not multiply. Following a fracture of or incision into cartilage, young reparative fibroblastic and vascular tissue proliferates from the perichondrium to fill the defect. In young animals whose skeletal tissues are still growing, this perichondrial granulation tissue can produce much new cartilage and true cartilaginous union can take place. But in adults chondrification of the reparative tissue is slight ; it usually becomes fibrous, sometimes bony. The fact that injuries to adult cartilage so seldom undergo cartilaginous repair is all the more remarkable, since new cartilage is often produced in the callus of fractured bones.

INJURY, REPAIR AND REMODELLING OF BLOOD VESSELS

(1) The sealing of severed vessels

A wound of any vascular tissue of course always severs many small blood vessels, and sometimes some large ones also ; more or less bleeding results. If, as in the superficial cuts in the skin with which we are all familiar, only small vessels are divided, the bleeding usually ceases in a few minutes, especially if the edges of the wound are held firmly together. The bleeding stops because fibrin clot from the shed blood forms and seals up all the open cut ends of the vessels. If a larger vessel has been severed, the bleeding from it is more profuse and prolonged—from a vein a steady flow at low pressure, from an artery pulsatile squirts at high pressure. If the vessel is only of 1 or 2 millimetres calibre, the bleeding from it may still cease spontaneously ; for the vein collapses and fibrin clot seals it over, or the muscular wall of the cut artery contracts strongly and obliterates its lumen or so reduces it that it can be effectively occluded by clot. If, however, the vessel is a larger one, contraction and clotting may not suffice to seal it, and it must be artificially clamped or ligated to arrest the haemorrhage. Blood-clotting is a rapidly effective emergency means of stopping up the mouths of severed vessels, but it is only a temporary one ; permanent healing of the vessels is brought about by the replacement or organization of the clot by proliferating fibroblasts and new capillary vessels. Let us consider first blood-clotting and then the organization of clots in damaged vessels.

(2) The clotting of blood

The coagulation of shed blood is so essential and familiar that it is dealt with as a normal process by physiologists. Only a bare outline of the physiology of coagulation need be given here. When blood is shed, its soluble plasma-protein *fibrinogen* is irreversibly converted into insoluble *fibrin*. This change is brought about by an enzyme *thrombin*, which is not present in circulating blood but develops quickly in it after it is shed.   Thrombin forms from an inactive precursor *prothrombin*, which is manufactured in the liver (vitamin-K is essential to its formation) and is present in circulating blood plasma.   The conversion of prothrombin into active thrombin in shed blood takes place rapidly in the presence of calcium ions and *thromboplastins* (or *thrombokinase*) which are liberated from the platelets of shed blood and also from damaged blood cells and tissues. The formation of thromboplastin depends on the presence of several other substances in the blood, the best known of which is alpha-prothromboplastin, deficiency of which is the cause of haemophilia (see p. 406).

Recently coagulated blood in a test tube is a soft friable jelly ; it consists of a fine meshwork of fibrin threads entangling the corpuscles and serum.   If it is allowed to stand, the fibrin shrinks to form, along with the corpuscles, a tough mass, while the blood serum is expressed as a clear yellow fluid.   Similar changes take place in rapidly formed clots in the body, whether these are inside vessels or are extra-vascular haematomas ; these contract, express their serum which is quickly absorbed, and form tough masses or layers of fibrin with entangled corpuscles.   The contraction and toughness of fibrin clot clearly makes it well adapted for sealing up small severed vessels and cementing cut surfaces together.

(3) Occlusion of vessels by thrombosis following injury : collateral circulation

The formation of fibrin clot inside blood vessels is called *thrombosis*, and an intra-vascular clot is a *thrombus*.   Thrombosis may take place from many kinds of diseased states of the walls of vessels ; here we are concerned only with that resulting from mechanical injuries—the cutting, crushing or ligation of vessels. The tiniest thrombi of all are the fine fibrin threads which plug up the lumina of severed capillaries.   These occupy the vessels for only microscopic distances from the cut surfaces, because the abundant anastomosis of the capillaries offers many alternative routes for the blood stream (Fig. 11).   As already described under healing by " first intention ", the subsequent junctional growth of new capillaries across the plane of the cut soon restores the vascular continuity of the capillary plexus.

But, when larger vessels—arteries or veins— have been severed or ligated, columns of stagnant blood within the vessels for considerable distances proximal and distal to the occluded spot may coagulate.   The extent of the thrombi depends on the positions of the main branch or tributary vessels connected with the occluded one and on the positions of the main anastomoses in the neighbourhood.   Thus if, in the arterial tree shown in Fig. 12, the branch artery C is ligated or crushed or severed at L, the now stationary columns of blood in it and its branches will undergo thrombosis proximally to the point P determined by the position of the commencement of the branch artery D, and distally to the points XX determined by the positions of the anastomoses between the branches of C and those of the neighbouring arteries B and D.   Later, as described below, the thrombosed blood in C undergoes organization, sometimes converting the vessel into a mere fibrous cord, sometimes resulting in its partial or complete recanalization (i.e. re-establishment of a lumen and blood-flow through it).

The vascular connections shown in Fig. 12 are relatively simple ; sometimes the occlusion of a large artery or vein involves the establishment of much more extensive deviations of the blood stream through collateral (i.e. alternative anastomotic) vessels, in some of which indeed the direction of flow is often reversed. The development of such a collateral circulation may involve great modifications in the size and structure of the vessels ; previously small unimportant branch arteries may have to assume the functions of main ones, and the

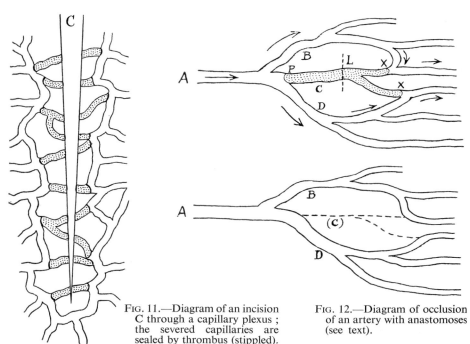

Fig. 11.—Diagram of an incision C through a capillary plexus ; the severed capillaries are sealed by thrombus (stippled).

Fig. 12.—Diagram of occlusion of an artery with anastomoses (see text).

greater flow and greater pressure of blood within them is accompanied by corresponding increase in the muscular and elastic tissue of their walls. Even the wall of a vein, anastomosed to an artery either experimentally or as the result of a penetrating wound involving both (arterio-venous aneurysm), may show increase of its muscular and elastic tissues so that its structure approaches that of an artery. This plasticity of the structure of blood vessels in response to changes of pressure and volume of flow of the blood within them is not surprising in view of what we have already learnt from the changes observed in transparent chambers inserted in rabbits' ears. These showed how plastic are the new-formed small vessels of repair tissue and how responsive to changes in local circulatory requirements. The changes seen in the development of a collateral circulation show that fully formed normal vessels also are capable of great structural modification.

(4) The organization of intra-vascular thrombi

Thrombi within vessels are organized, i.e. replaced by young fibro-vascular reparative tissue, in a manner similar to extravasated haematomas. After speedy absorption of the fluid serum and shrinkage of the fibrin clot containing the corpuscles, capillary sprouts accompanied by proliferating fibroblasts grow into the clot from the vessel walls. The capillary sprouts come from the vasa vasorum and grow through the intima into the clot. As the new tissue grows in, the fibrin and disintegrated blood corpuscles are absorbed ; phagocytes carrying away haemosiderin and haematoidin are to be found in the organizing tissue and in the vessel walls. Finally the whole of the clot is replaced by a core of new tissue which then loses its capillary vessels and shrinks, along with the former walls of the vessel, to form a fibrous cord.

Sometimes, however, the vessel becomes recanalized, i.e. it re-acquires a lumen carrying a blood-flow. This is most likely to happen if the thrombosed segment of the vessel was not completely occluded by the thrombus, parts of its lumen remaining patent as a slit between the thrombus and the intima and carrying a thin stream of blood between side-branches of the vessel. The endothelial cells of the intima lining the patent slit multiply and grow over the surface of the thrombus, lining any crevices of it and participating in the formation of capillary sprouts growing into it. These meet and unite with new-formed capillaries growing into the thrombus from other parts of the vessel walls ; and the capillary meshwork throughout the thrombus may then be greatly modified by the stream of blood passing into it from this source. A patent, though often irregular or tortuous track may eventually develop along the length of the thrombosed segment, which may thus be partly restored as a blood-carrying channel.

THE REGENERATION OF EPITHELIA

As pointed out in Chapter 1, we should distinguish between (a) repair of surface epithelia, covering over a breach in the tissues, and (b) compensatory regeneration of glandular epithelia, resulting from loss of secretory tissue of a particular kind.

(1) Repair of surface epithelia

Many surface epithelia normally shed cells from the surface and replace them by mitotic multiplication of the remaining cells. Thus, the squamous cells of the skin, mouth, oesophagus and vagina are constantly being shed, and the basal cells of the stratified epithelium are constantly multiplying to make good the loss. The continuous growth of hair and nails affords familiar evidence of the normal multiplication of the epithelium of the skin ; and the moulting and replacement of the feathers in birds is an interesting example of seasonal periodicity in epithelial regeneration. The respiratory and alimentary mucous membranes also are constantly losing some of their cells and replacing them by mitotic proliferation of neighbouring cells ; mitotic figures are especially plentiful in the intestinal epithelium. The endometrial epithelium is shed and regenerated with each menstrual cycle.

In view of the normal regenerative replacement of most surface epithelia, it is not surprising that wounds and other breaches in epithelial surfaces are readily re-clothed by new epithelium growing in from that around. We have already seen how new epidermis grows in over a granulating wound of the skin ; and the same applies to wounds of any of the mucous membranes. Wherever, by accidental injury or surgical operation or inflammatory disease, raw surfaces are left in any hollow viscus, the neighbouring epithelium—whether it be bronchial, gastric, intestinal, biliary or urinary—will grow over and epithelialize the granulating areas. As many kinds of anastomotic surgical operations show, even structurally different epithelia, growing over a granulating surface, can meet and form a sound junction. Thus if the surgeon relieves a patient with pyloric obstruction by making a short-circuit opening between the stomach and a loop of small intestine (gastro-enterostomy), the raw margins of the artificial opening or stoma become completely epithelialized by the gastric and intestinal mucosa growing together. Or if the surgeon has to relieve an obstruction of the rectum by bringing a more proximal part of the colon to the surface of the abdominal wall and making an artificial opening in it (colostomy), then the epithelium of the colonic mucosa and the epidermis grow over the raw edges of the opening and unite with one another.

The layer of new epithelium which grows over a denuded surface is sometimes of a simple kind, devoid of specialized glands and appendages. Thus the epidermis clothing a scar usually forms a more or less even thin layer, devoid of papillae, hair-follicles and sebaceous and sweat glands. On the other hand, it has been shown experimentally that, following removal of strips of gastric mucosa, the regenerated mucosa may be as complex as the normal, with gastric glands as well as superficial investing epithelium. So also, following thorough curettage of the uterus, in which the surgeon may scrape away all the endometrial tissue except the deepest tips of some of the glands lying in the myometrium, a complete normal endometrium regenerates from these.

Regenerating epithelium spreading over a raw surface often shows many cells in mitosis.

(2) Compensatory regeneration of specialized glandular epithelia

Many kinds of glands which normally show very little loss and replacement of their cells, and rare mitoses, nevertheless possess great powers of regenerative proliferation following loss of the tissue by injury or disease. Thus, following experimental removal of two-thirds of a dog's liver, the remaining third will grow so that in a few weeks the animal has regained a normal amount of liver tissue. The regenerated tissue may be unanatomical in both its gross shape and its lobular pattern, but it is functionally healthy tissue with appropriate relationships to its new-formed bile-ducts and blood-vessels. There appears to be no limit to the regenerative power of liver tissue ; repeated excisions of large parts of the organ in one animal can be followed by repeated regeneration. Many mitotic figures are seen in liver-cells in the regenerating tissue. Similar compensatory hyperplasia following partial removal occurs in the salivary glands, pancreas, thyroid, adrenal and other ductless glands. If one kidney is removed, the remaining kidney increases in size ; this is due, not to an increase in the number of nephrons, but to hyperplasia of the epithelium of existing convoluted tubules. Three-quarters or more of an animal's kidney tissue can be removed and the animal can survive with compensatory enlargement of its remaining fraction of a kidney.

In all of the foregoing cases, compensatory regeneration is effected by mitotic multiplication of the remaining cells of tissues which normally multiply only rarely in adults. The capacity for multiplication, normally nearly dormant in the cells, is easily evoked by the increased functional demand made on the remaining cells when part of the tissue has been removed. In response to the stimulus of extra work, the cells multiply; and they continue to do so until enough of them have been produced to satisfy the demand.

It is worth noting here that epithelial tissues which are capable of compensatory regeneration following physical injury or removal often show similar regeneration following partial destruction by poisons, bacterial infections or other diseases. Examples will appear in later chapters.

HAEMORRHAGE AND THE REGENERATION OF BLOOD

Loss of blood by haemorrhage from any cause is comparable with a wound in solid tissues ; the lost fluid and the lost cells are restored, the latter by regenerative proliferation of the haemopoietic tissues. These tissues, like the surface epithelia, normally show constant regenerative proliferation to replace the corpuscles which are constantly being lost from the blood. The restoration of blood cells following a severe haemorrhage involves only a sudden acceleration of the normal processes of haemopoiesis.

A rapid severe haemorrhage at once reduces the volume of circulating blood, with consequent fall of both venous and arterial blood-pressures, diminished cardiac output, and increase in the heart rate. Following cessation of the haemorrhage, the plasma volume is quickly restored by the flow of fluid into the blood from the tissue spaces. This dilutes the blood, so that the concentrations of the plasma proteins and the red-cell-count and haemoglobin percentage fall. Restoration of the plasma proteins usually takes several days, for these have to be manufactured by the liver. The restoration of haemoglobin and red corpuscles is slower still and may take several weeks. During the regeneration of the corpuscles, an increased number of immature cells (reticulocytes) appear in the blood and some nucleated red cells (normoblasts) may also appear ; the average diameter of the red cells and the amount of haemoglobin per cell (colour index) are diminished. There is usually a slight leucocytosis.

Repeated small haemorrhages may produce little or no significant change of plasma volume or plasma protein level, for restoration of fluid and proteins may keep pace with their loss. But a severe anaemia (fall of red-cell-count and haemoglobin) may be produced in this way, because, as restoration of the red corpuscles is slower, their continued loss has a cumulative effect. The blood may show great and progressive fall in the red-cell-count and haemoglobin percentage (e.g. to less than 50 per cent), and many immature and abnormal red corpuscles may appear, including reticulocytes, nucleated cells, poikilocytes, and cells showing polychromatic staining or basophilic stippling. In such a case, the red bone-marrow shows evidence of excessive erythropoiesis, and hyperplastic dark red marrow may extend into parts of the skeleton where normally only yellow marrow is present, e.g. the shafts of the femur and humerus. (The main causes of chronic post-haemorrhagic anaemia are mentioned on p. 400.)

INJURY AND REPAIR OF NERVOUS TISSUE

(1) The powers of proliferation of the various cells of the nervous system

Adult (i.e. differentiated) nerve cells cannot multiply ; hence, when nerve cells, central or peripheral, are destroyed by injury or disease, they are permanently lost and cannot be replaced.   The repair of neurones is restricted to regeneration of the axons of peripheral nerves when these have been severed, provided that the nerve cells from which these axons arise are still intact.   In the central nervous system, no regeneration of axons occurs ; fibre tracts which have been destroyed in the brain or spinal cord are not regenerated, even though their cells of origin are uninjured.   Neuroglial cells in the central nervous system and neurilemmal (Schwann) cells in peripheral nerves multiply freely at injured spots and play important parts in repair.

(2) Degeneration and regeneration following section of a peripheral nerve

Let us consider the changes in a motor nerve fibre in a transected peripheral nerve, as shown in Plate II. Degenerative changes occur in the neurone both proximal and distal to the section.

(a) Degenerative changes proximal to the section

The nerve cell of the cut fibre shows temporary changes ; its Nissl granules break up into a fine dust and may disappear (chromatolysis) ; the Golgi apparatus fragments and diminishes ; the cell swells and becomes more rounded ; and its neurofibrils may disappear. These changes are most marked about 2 weeks after section of the fibre ; later they gradually recover and the cell structure returns to normal. For a short distance proximal to the point of section, the neuraxon and its medullary sheath degenerate. If the injury has been close to the cell, or if the axon has been torn and not cleanly cut, the cell may completely degenerate and disappear.

(b) Degenerative changes distal to the section

Within a few days the whole of the distal parts of the cut neuraxon disintegrate and the medullary sheath breaks up into oily droplets. This *Wallerian degeneration** is complete in about 3 weeks. During this time and for some weeks later, macrophages (mononuclear phagocytes) invade the neurilemmal tube and take up the fatty debris, thereby assuming the form of " foam cells ", which remove the debris.

(c) Proliferation of Schwann cells

Soon after the section, the Schwann cells around the two cut ends of the nerve proliferate actively into the gap between the cut ends. Those of the entire degenerating distal segment also multiply and fill up the neurilemmal tube at the same time as the degenerated myelin is being removed. The neurilemmal tube is thus converted into a Schwann-cell column.

(d) Regeneration of nerve fibres

Soon after recovery of the proximal segment, the central cut end of each

---

* A. V. Waller of Faversham, England, was the first to show, in 1850, that the distal part of a cut nerve degenerates while its proximal stump remains relatively intact. Waller correctly maintained therefore that the nerve fibres are merely prolongations of their cells and depend on them for their survival.

axon sprouts out and gives rise to many small fibrils. These penetrate into the junctional mass of proliferated Schwann cells and so into the Schwann-cell columns occupying the peripheral neurilemmal tubes. Within a few weeks of the injury, the tubes just distal to the cut are found to contain varying numbers— some 1 or 2, some as many as 25—of young nerve fibrils growing distally. The rate of growth of new fibres down the neurilemmal tubes varies with the distance down the nerve ; in proximal parts of main nerves it is about 3 milli- metres a day ; but more distally the rate diminishes, until in the furthest parts of the limbs it is only about 0·5 millimetres a day. Fibrils from one axon may enter several different neurilemmal tubes, and the multiple fibrils in one tube may come from several different axons. In a mixed nerve, therefore, a tube may have both motor and sensory fibrils growing down it. A motor fibre growing down a motor tube will eventually reach the end-plate in a muscle fibre, which it will successfully re-innervate. Similarly, a sensory fibre growing down a sensory tube will eventually reach a sensory end-organ. It is this attaining of an appropriate destination which determines the survival of the down-growing fibre ; the first fibre in each tube to reach an appropriate destination survives and becomes functional ; all the others disappear. The one surviving function- ally appropriate fibre enlarges, acquires a medullary sheath and refills the neuri- lemmal tube.

After a motor nerve has been interrupted by crushing close to its muscle, re-innervation of the muscle can take place within a few weeks and almost com- pletely. But when a muscle has been denervated for long periods (because the nerve has been injured more severely or far away from the muscle), its eventual re-innervation is likely to be less complete. However, denervated muscle fibres, though suffering progressive atrophy, persist as striated fibres for 2 years or longer, during which they are capable of re-innervation and recovery, should regenerating motor nerve fibres reach them.

Within this period, the prospects of successful re-innervation of muscles and of good functional recovery depend mainly on whether or not it is possible to obtain accurate re-apposition of the cut ends of the nerve. If so, then many regenerating nerve fibres will grow down their appropriate paths to their proper destinations. But if a segment of nerve has been destroyed by the injury, or if the cut ends have been separated during early regeneration, then many of the fibres will grow haphazardly down inappropriate paths, the competition between appropriate and inappropriate fibres for the occupation of available neurilemmal tubes will be great, and the ultimate recovery of muscles (especially far distant ones) will be correspondingly small.

After crushing injuries which do not sever the continuity of a nerve, regenera- tion of the nerve and recovery of function are speedier and more complete than after complete transection. This is because the general pattern of fasciculi in the nerve is still retained in continuity and many of the neurilemmal tubes are still intact and ready to receive their own proper regenerating axons.

(e) Regeneration of autonomic nerves

Autonomic nerves regenerate like somatic ones, the new axons growing from the cut ends of fibres which are in continuity with nerve cells in the central nervous system or in the sympathetic ganglia or the more peripheral visceral ganglia.

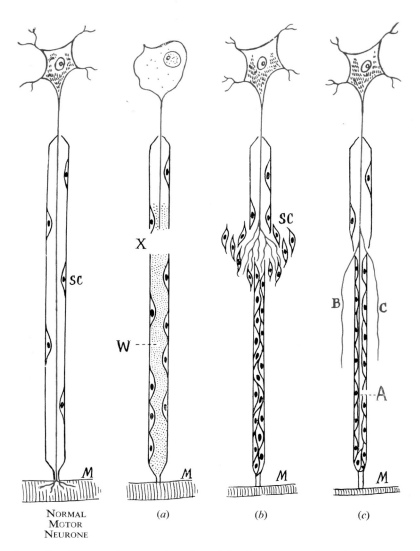

NORMAL MOTOR NEURONE	(a)	(b)	(c)

PLATE II.—Diagram of regeneration of a peripheral motor nerve fibre, the nerve cell and axon shown red. SC, Schwann cells. M, muscle fibre ; myelin sheath left blank, and nodes of Ranvier omitted. (a) : Fibre recently cut at X; Wallerian degeneration of myelin W (stippled), accompanied by some proliferation of Schwann cells; swelling and degenerative changes in nerve cell. (b) : Recovery of nerve cell ; multiple regeneration sprouts from proximal cut end of axon ; proliferation of Schwann cells at gap, and in shrunken peripheral neurilemmal tube ; progressive atrophy of muscle fibre. (c) : Regeneration of axon A nearly complete ; presently it will reach and re-innervate muscle fibre ; unsuccessful collateral branches B and C will then atrophy and disappear. (Note : the successful fibre A does not necessarily come from the nerve cell which originally supplied this neurilemmal tube ; it may have come from another, and perhaps functionally inappropriate, cell.)

(To face page 32)

If the appropriate ganglia are left intact, autonomic denervation of blood vessels, glands and other tissues may be followed by complete re-innervation. It may be recalled here that, in the new vascular tissue growing into a transparent chamber in the rabbit's ear, some vessels differentiate into small arteries and acquire a vaso-motor supply from ingrowing nerve fibres.

(f) The formation of amputation " neuromas "

Sometimes, following section of a nerve without restoration of continuity between the cut ends, and especially in the stump of a limb following amputation, a bulbous mass develops on the central cut end of the nerve (Fig. 13). The name " neuroma", usually applied to this, is inappropriate, for the swelling is not a true tumour. It is produced as the result of the regenerative but disorderly outgrowth

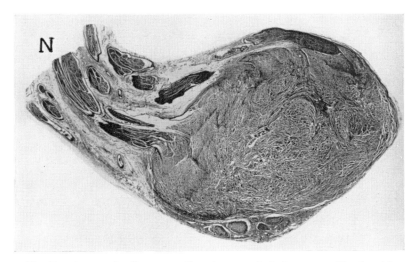

Fig. 13.—Amputation " neuroma " on the cut end of a large nerve N. (× 4.)

of nerve fibres which are undirected because there is no peripheral segment of nerve with neurilemmal tubes to receive the new fibres. The result is a tangled mass made up of outgrowing nerve fibres, proliferated Schwann cells from the cut end, and fibroblasts forming much fibrous tissue. Such " neuromas " may go on growing slowly for a long time ; and, if they are in exposed places liable to pressure, they may be very painful and may need to be excised. Sometimes a fusiform swelling, essentially the same as an amputation " neuroma " develops on an intact nerve as the result of local injury or irritation.

(3) Injury and repair in the central nervous system

In the central nervous system any destruction of nerve cells or tracts of nerve fibres is permanent and irrecoverable. Nerve-fibre tracts of the brain or spinal cord which are cut off from their cells of origin, or whose cells of origin are destroyed by injury or disease, undergo Wallerian degeneration and absorption by phagocytic foam cells,* and no new nerve fibres are regenerated.   It is probable that the inability of central nervous tracts to regenerate is related to the absence of Schwann cells.   The permanently degenerated tracts can be well displayed in sections by the Weigert-Pal method, which stains myelin sheaths bluish-black but leaves degenerated tracts unstained (Fig. 14).

While injuries to nerve cells and fibres in central nervous tissue are irreparable, the neuroglial cells around an injured area proliferate and produce an excess of neuroglial fibres, so forming a neuroglial scar or area of *gliosis* which knits the injured tissues together.   The neuroglial cells chiefly concerned are the astrocytes ; these enlarge, multiply, and develop coarser fibres, which become orientated in

(a) 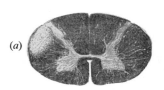 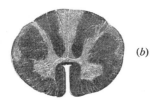 (b)

FIG. 14.—Degeneration of tracts in spinal cord following injury (Weigert-Pal stain). (a) Descending degeneration of one crossed pyramidal tract following a unilateral injury of the brain. (b) Ascending degeneration of columns of Goll and spino-cerebellar tracts on both sides following an injury of the cord at a lower level.

appropriate directions.   Of course, an injury of the brain or cord usually involves some blood vessels and connective tissues, so that areas of vascular and fibrous repair are often present along with glial scars.

INJURY AND REPAIR OF MUSCULAR TISSUE

It is generally believed that adult (i.e. fully differentiated) muscle fibres, like nerve cells, do not multiply ; and it is certainly true that, when muscle fibres, striated or unstriated, are extensively destroyed by injury or disease, no significant regeneration of new fibres takes place.   Such injuries heal by granulation and the formation of fibrous scar tissue.

However, some regeneration of incompletely damaged muscle fibres does occur.   Thus, following transection of skeletal muscle fibres, from the cut ends of the fibres bulbous processes of the sarcoplasm sprout out, accompanied by many nuclei.   These nuclei appear to be produced either by a local congregation of the pre-existing nuclei of the fibre or by amitotic division, since mitotic figures

* In lesions of the central nervous system, pigment-laden or fat-laden phagocytes are often called " compound granular corpuscles " ; they are wandering histiocytes or mesenchymal phagocytes, which are found in the resting state as elongated or branched cells along the course of blood vessels and connective tissues of the nervous system.   These have often been called " microglial " cells, a confusing and unnecessary name, since the cells are just the same as the histiocytes of other tissues.   And, just like the same cells in other tissues, when active in the phagocytosis of lipoid substances or pigments, they enlarge, assume rounded forms, and migrate actively.

are not found in the regenerating sprouts (Fig. 15). At first these sprouts consist of undifferentiated cytoplasm, but later they develop cross-striations in series with those of the undamaged part of the fibre. They do not separate from the old fibre to form new ones.

Le Gros Clark has shown experimentally that, if preformed sarcolemmal sheaths are available, the regeneration sprouts from damaged skeletal muscle fibres will grow freely into these, and that under special circumstances extensive regeneration of an injured voluntary muscle is possible in this way. Thus, if a large part of a rabbit's thigh muscle is destroyed by crushing or by excising and

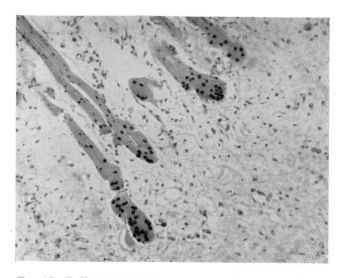

Fig. 15.—Bulbous regeneration sprouts at the ends of skeletal muscle fibres following a wound of muscle. (× 100.)

replanting a portion of it, the damaged but still surviving muscle fibres around the destroyed portion send long regeneration sprouts into it ; these grow inside the original sarcolemmal tubes, *pari passu* with the phagocytic removal of the products of the destroyed sarcoplasm, and they eventually differentiate into striated fibres replacing those destroyed. Complete structural and functional restitution of a large area of muscle is possible in this way. If a block of muscle is excised and then replanted at right angles to its former direction, the regeneration sprouts also turn at right angles to their proper direction to grow along the misplaced sarcolemmal tubes, so that the regenerated fibres come to lie transversely to the axis of the muscle.

Following their destruction by injury or disease, adult cardiac muscle fibres show no signs of regeneration ; myocardial injuries heal by fibrosis only. In infants, however, after myocardial damage by diphtherial or other bacterial toxins, mitotic figures have been seen in the muscle fibres, suggesting that young cardiac muscle may possess limited powers of regeneration.

It is generally believed that adult non-striated muscle fibres are incapable of regeneration, and it is true that destructive injuries of the uterus, bladder and other muscular viscera usually heal by fibrous scars.   However, there are certain facts which should make us hesitate to conclude that smooth muscle has no regenerative powers, namely (a) as we have already noted, according to the Clarks' experiments with transparent chambers in the rabbit's ear, the differentiation of new arterioles in granulation tissue involves the new formation of smooth muscle fibres from the perivascular cells ; (b) after the healing of surgical wounds in hollow viscera (e.g. excisions or anastomosis operations on the gastro-intestinal tract), there is often very little sign of fibrous union where the muscle coats have been joined, the muscle often appearing smoothly continuous ; and (c) smooth muscle fibres certainly multiply in the common muscular tumours or leiomyomas of the uterus and other organs.

METAPLASIA IN REPARATIVE TISSUE

Metaplasia is the transformation of proliferating cells of one kind into cells of another kind under pathological conditions, e.g. of fibroblasts into cartilage or bone, or of glandular into squamous stratified epithelium.   Three instances have already been mentioned in this chapter, namely the ossification of scars and haematomas and the formation of cartilage in callus.   Such transformations show that connective tissues possess wider powers of differentiation than they ever display under normal conditions ; under abnormal conditions the proliferating cells may undergo divergent differentiation into kinds different from their parent cells.

Metaplasia is always a change into cells of a related kind ; it never produces cells of a totally different kind.   Thus, proliferating fibroblasts and vascular endothelial cells may produce, not only fibrous tissue and new vessels, but also other differentiated mesenchymal tissues, namely cartilage, bone, mucoid connective tissue, adipose tissue and even haemopoietic cells, but they never produce epithelium or nervous tissue.   Conversely, epithelium of one kind may change into epithelium of another kind, e.g. the epithelia of the respiratory, alimentary or urinary passages may become stratified and squamous ; but epithelium never becomes converted into connective tissues.

A striking instance of metaplasia in reparative tissue is the formation of bone from connective tissue in the wall of the bladder, ureter or renal pelvis during the healing of wounds in these organs.   Thus, if a gap in the bladder wall is closed surgically by inserting a transplant of fascia lata, the bladder epithelium grows over and epithelializes the inner surface of the graft and the granulation tissue which springs up around it, and these connective tissues often become converted into a plaque of bone.   Similarly, if a piece of bladder or renal pelvis is transplanted experimentally into subcutaneous tissue or into a muscle, the epithelium proliferates and produces a cyst, and around that part of the cyst where the new epithelium grew to clothe the granulating connective tissue, a shell of bone forms in this tissue.   If a rat or rabbit has the pedicle of one kidney ligated, the whole of the renal parenchyma dies from lack of blood supply ; but parts of the damaged pelvic wall survive, its epithelium regenerates, and the proliferating connective tissue around it undergoes ossification.   It is clear from these results that when the transitional stratified epithelium of the urinary tract is undergoing regenerative proliferation, it exerts a special influence on the neighbouring multiplying connective tissue, inducing it to ossify.

Bone formed by metaplasia, whether in reparative or other pathological tissues, may develop bone-marrow with fat cells, haemopoietic cells and megakaryocytes.   It seems clear that these various marrow cells are not immigrants but that they too arise by metaplastic change in young proliferating mesenchymal (endothelial and fibroblastic) cells in the new-formed bone.   Thus, proliferating mesenchymal cells in the adult may possess all or most of the potencies of embryonic mesenchyme and can, under appropriate conditions, differentiate into any or all of the adult tissues which are derived from mesenchyme.

It remains to add here that metaplasia, in connective tissues or in epithelium, is much less common in simple uncomplicated repair than in lesions where there is chronic irritation and inflammation or than in tumours.

REPAIR AND REGENERATION OF TISSUES FOLLOWING INJURIES OTHER THAN TRAUMATIC

The reparative and regenerative behaviour of tissues following simple physical injuries exemplifies their behaviour following ALL kinds of destructive injury. This fundamental principle will appear again and again in our study of inflammatory, toxic and nutritional diseases.   Some simple illustrations will show its applicability.

(*a*) The simplest way to destroy an area of skin is to cut it out with a knife ; in this chapter we have studied the reparative behaviour of the tissues following such an injury.   But the same area of skin might have been destroyed by a burn, or by a caustic chemical substance, or by a bacterial infection such as a boil or a carbuncle ; the reparative changes after these more complex injuries are essentially the same as after simple trauma, but they are complicated and modified by inflammatory and other changes in the damaged tissues.

(*b*) The simplest way to destroy an area of liver is to excise it.   But a similar quantity of liver tissue might have been destroyed by a microbic infection (e.g. yellow fever), or by any one of a hundred selective liver poisons (e.g. carbon tetrachloride, chloroform, trinitrotoluol, etc.) or by particular dietary deficiencies. Regeneration of new liver tissue takes place after injury by these complex means just as it does after simple excision.

(*c*) The simplest injury of bone calling for repair is fracture ; but a bone may be injured also by many kinds of bacterial infections, or by vitamin deficiency, or by parathormone excess.   During recovery from any of these abnormalities, the bone and its periosteum show reparative growth and remodelling comparable with those seen in fracture-callus.

(*d*) The simplest way to destroy an area of brain is to remove it with a knife or a bullet ; the same area might have been destroyed by an abscess, or by a tumour, or by degeneration consequent on senile changes in the arteries supplying the area.   The tract degenerations wrought by these lesions will be just the same as those following simple traumatic destruction, and local phagocytic activity and neuroglial proliferation also may be common to them all.

We have purposely studied first the regenerative changes in various tissues in response to simple physical injury, because here we see these changes uncomplicated by inflammation, the presence of bacterial or other parasites, or persistent toxic or nutritional damage. In succeeding chapters we will study these more complicated kinds of injuries ; but we will often find that recovery from them involves proliferation, organization and remodelling of tissues identical with or closely similar to those described in this chapter.

TRAUMATIC TRANSPLANTATION OF TISSUES

Occasionally, as the result of injury, a piece of tissue is dislodged from its normal position and survives and grows in an abnormal position. The best known instances are these :—

(i) *Epidermal implantation cysts.*—A penetrating wound sometimes drives a piece of skin into the subcutaneous or deeper tissues, even into a bone, and a cyst develops from the dislodged epidermis.

(ii) *Transplanted endometrium.*—Following an operation on the uterus, dislodged fragments of endometrium sometimes survive and grow in the abdominal scar. The transplanted tissue forms multiple small cysts into which periodic menstrual bleeding occurs, so that the swelling enlarges and becomes painful with each menstrual cycle.

(iii) *Traumatic transplantation of splenic tissue.*—Rarely, following traumatic rupture of the spleen, fragments of splenic tissue spilt into the peritoneal cavity engraft themselves on the peritoneal surfaces and form multiple small spleens.

The foregoing are instances of naturally occurring traumatic *autografting*, i.e. transfer of tissue in the same individual. Skin-grafting also is an example of deliberate autografting. It remains to add here that, experimentally, many other kinds of epithelium and other tissues can be autografted. *Homografting*, i.e. the transplantation of tissue from one creature to another of the same species, is much less successful, unless the donor and recipient both belong to the same closely inbred stock of animals. *Heterografting*, i.e. transplantation from one species to another, except under certain peculiar and highly artificial conditions, is never successful.

SUPPLEMENTARY READING

Bowden, R. E. M., and Gutmann, E. (1944). "Denervation and re-innervation of human voluntary muscle", *Brain*, **67**, 273.

Burrows, H. (1926). "Implantation dermoid of the terminal phalanx of the thumb", *Brit. J. Surg.*, **13**, 761.

Clark, E. R., *et al.* (1931). "General observations on the ingrowth of new blood vessels into standardized chambers in the rabbit's ear, and the subsequent changes in the newly grown vessels over a period of months", *Anat. Rec.*, **50**, 129.

— *et al.* (1934). "Microscopic observations in the living rabbit of the new growth of nerves and the establishment of nerve-controlled contractions of newly formed arterioles", *Amer. J. Anat.*, **55**, 47.

— and Clark, E. L. (1940). "Microscopic observations on the extra-endothelial cells of living mammalian blood vessels". *Amer. J. Anat.*, **66**, 1.

Clark, W. E. Le Gros. (1946). " An experimental study of the regeneration of mammalian striped muscle ", *J. Anat.*, **80**, 24.

Dickson, W. E. C. (1947). " Accidental electrocution : with direct shock to the brain itself ", *J. Path. Bact.*, **59**, 359. (A remarkable example of the internal disruption of tissues produced by electrocution.)

Florey, H. W. (1954). " Healing ". Lecture 27 in *Lectures on General Pathology*. London; Lloyd-Duke.

– and Harding, H. E. (1935). " The healing of artificial defects of the duodenal mucosa ", *J. Path. Bact.*, **40**, 211.

Gillman, T. (1968). " Healing of cutaneous wounds ". *Glaxo volume*, **31**, 5. (Good photographs and diagrams).

Hamrick, R. A., and Bush, J. C. (1942). " Autoplastic transplantation of splenic tissue in man, following traumatic rupture of the spleen ", *Ann. Surg.*, **115**, 84.

Monti, A. (1900). *The Fundamental Data of Modern Pathology. The Doctrine of Cellular Proliferation: Its Influence on the Pathology of Man.* New Sydenham Society, London. (A most interesting early review of growth, repair and inflammation.)

Pullinger, B. D., and Florey, H. W. (1937). " Proliferation of lymphatics in inflammation ", *J. Path. Bact.*, **45**, 157.

Shaw, A. F. B., and Shafti, A. (1937). " Traumatic autoplastic transplantation of splenic tissue in man ", *J. Path. Bact.*, **45**, 215.

Sunderland, S. (1945). " The intraneural topography of the radial, median and ulnar nerves ", *Brain*, **68**, 243. (Recovery following section of a nerve depends greatly on the intraneural pattern at the site of injury and on whether or not accurate re-apposition of the cut ends is possible.)

– (1947). " Rate of regeneration in human peripheral nerves ", *Arch. Neurol. Psych.*, **58**, 251.

– (1947). " Observations on the treatment of traumatic injuries of peripheral nerves ". *Brit. J. Surg.*, **35**, 36.

Willis, R. A. (1962). *The Borderland of Embryology and Pathology*. London; Butterworths. (Chapters 13 and 14 give detailed outlines of repair and of metaplasia.)

CHAPTER 3

INFLAMMATION

DEFINITION

THE ORDINARY non-technical meaning of " inflame " is to " set ablaze " or to " catch fire ". The application of the term in pathology is metaphorical, but very appropriate ; for everyone is familiar with the " fiery " signs exhibited by inflamed tissues—the redness, heat and pain of a boil, an infected wound or a sore throat. An intelligent layman, asked to define " inflammation ", might say that it is " the redness, heat, pain and swelling of irritated or damaged tissues ", and this would be a very good definition. The more technical definition of the pathologist is similar to the layman's, but is based on a closer study of the histological changes which account for the redness, heat and swelling. *Inflammation is the immediate vascular and exudative reaction of living tissues to injury.*

The inflammatory reaction is a beneficial one, in that its effect is the removal of the products of tissue injury, and in the case of bacterial inflammations the destruction also of the bacteria and neutralization of their products. The body has two main means of opposing invading microbic parasites, namely (i) phagocytic cells of several kinds, and (ii) antibodies in solution in the blood plasma and tissue fluids.* Inflammation results in the migration of phagocytes and the escape of plasma from the circulating blood into the spaces of the injured tissues. This is called *exudation*, and the exuded fluid and migrated leucocytes together constitute an *inflammatory exudate*.

Inflammation has often been defined simply as " the reaction of tissues to injury ". But this definition is too wide, since it embraces all the phenomena of repair and regeneration. Inflammation and repair are often mingled, but they should be thought of as distinct processes—inflammation being the immediate response to an injury or irritant, having as its function destruction or removal of the injurious agent, and repair being the subsequent proliferation of tissues to replace those destroyed by the injury. Inflammation consists of a series of changes in blood vessels with exudation of some of their contents, and it does not include any cellular proliferation ; while repair is essentially proliferative. This distinction between inflammation and repair is not merely theoretical ; both of these processes are often seen in pure form. Thus, as described in the previous chapter, bacteria-free wounds, fractures or other physical injuries evoke only mild transient inflammatory reactions, after which healing continues uncomplicated by inflammation. On the other hand, there are some inflammations which, because they destroy no tissue, end in simple re-absorption of the exuded fluid and leucocytes and are unaccompanied by any reparative proliferation ; a mosquito bite is a trivial, and uncomplicated lobar pneumonia a severe. inflammation of this kind.

* Anti-teleologists, see Appendix A, p. 677.

FACTORS MODIFYING INFLAMMATION

All inflammatory reactions show fundamentally similar vascular and exudative changes ; but the details and results vary according to (i) the character and intensity of the injurious agent, (ii) the nature of the tissues involved, and (iii) the duration of the inflammation and the degree to which it is mingled with repair. Let us consider these factors.

(1) Character and intensity of the injurious agent

There is an important difference between physical and parasitic agents as causes of inflammation. When tissues are injured by purely physical agents (by mechanical means, destructive chemical substances, heat, electrical discharges or X-rays), a certain injury is inflicted and then ceases. The resulting inflammatory reaction is excited by the products of destruction of the tissues themselves, and when these have been removed the inflammation subsides and repair proceeds. But in that great group of tissue injuries due to living parasites, especially bacteria, the injurious agent is self-reproductive and the injury it inflicts is progressive and is maintained as long as the parasite continues to thrive. The inflammatory response therefore persists until either the parasites are destroyed (or rendered impotent) or the host succumbs. Bacterial and other microbic parasites injure tissues by producing poisonous metabolites or toxins ; while it is possible that some of these toxins may have a direct effect in exciting inflammation, this is evoked chiefly by products of the damaged tissues themselves, just as is the inflammation following physical injuries (see below). The difference between physical and microbic injuries is that the former cause an evanescent inflammation due to the products of sudden tissue damage which is not repeated, while the latter cause a sustained inflammation due to continuously produced tissue damage.

Further, parasitic inflammations differ greatly according to the nature of the parasite. Some bacteria, especially the pathogenic cocci, usually excite a very rapid intense reaction, an *acute inflammation*. Other organisms, such as those of tuberculosis, leprosy and syphilis, usually produce a slow smouldering reaction, a *chronic inflammation*. The kinds of leucocytes which accumulate in the tissue differ with different microbes ; thus neutrophil polymorphonuclear leucocytes predominate in the acute inflammations due to cocci, and in some chronic infections, e.g. actinomycosis ; while mononuclear macrophages, lymphocytes and plasma cells predominate in most chronic inflammations.

(2) The nature of the tissues involved

(i) In *solid tissues*, the inflammatory exudate accumulates in the tissue spaces, producing a general swelling or oedema of the tissues.

(ii) When *natural cavities* are involved, much of the exudate collects in the cavities. Thus in peritonitis, pleurisy or pericarditis, effusions into these serous cavities often develop. In meningitis the inflammatory exudate is mixed with the cerebrospinal fluid. In inflamed lung, the exudate collects in and plugs up the alveoli, so that the pneumonic lung becomes solid and airless.

(iii) *Mucous membranes* when inflamed often secrete excess of mucus which mingles with the inflammatory discharges, to which desquamated epithelial cells are also added. Such inflammations with mucoid discharges are called catarrhs, e.g. nasal, bronchial, gastric, intestinal catarrh.

(iv) In *blood-containing cavities*, the heart or vessels, inflammation is often accompanied by deposition of clot or thrombus over areas of damaged intima. In the heart such thrombi form " vegetations " on the inflamed valves or other parts of the endocardium (see Figs. 31, 32 and Plate V). In inflamed blood vessels, occlusion by spreading thrombosis is often seen.

(v) When *bone* is severely inflamed, the inflamed part dies. Whereas a dead part of a soft tissue is quickly separated from the living, the separation of dead from living bone is a slow process, because of the physical nature of the tissue. Hence untreated osteomyelitis runs a prolonged course (see Plate IV and Figs. 27-30).

(3) The duration of inflammation and its mingling with repair

The terms " acute " and " chronic " refer only to the duration of inflammation ; they mean " rapid " and " slow ". They do not refer to the severity or seriousness of the inflammation : a mosquito bite or a slight superficial scald excites an acute inflammation, though a mild and trifling one ; while tuberculosis, a chronic inflammation, is often a very serious and destructive one. Acute inflammation is the vascular and exudative reaction to speedily acting injurious agents ; the tissue damage may be slight or severe, but it is inflicted rapidly. Chronic inflammation is the vascular and exudative reaction to slowly acting injurious agents ; again, the tissue damage may be slight or severe, but it is inflicted slowly. Acute inflammations not only arise rapidly, but if recovery ensues, they often subside equally rapidly. Conversely, chronic inflammations not only develop slowly, but recovery from them is also slow.

In rapidly arising and rapidly subsiding acute inflammations, repair is fairly distinct from and follows the inflammation, but in slowly developing or protracted, i.e. subacute or chronic, inflammations, supervening reparative changes are necessarily present along with those of the persistent inflammation. This gives chronic inflammations their main characters.

ACUTE INFLAMMATION

(1) The microscopical study of acute inflammation

By the direct microscopical study of inflamed living tissues, Waller (1846), Lister (1855) and Cohnheim (1867) observed all the vascular and exudative phenomena of acute inflammation. For this purpose they used the web of the frog's foot, the frog's mesentery, the tadpole's tail, and the bat's wing, and they evoked inflammatory changes by mechanical injury, heat or chemical irritants.*

---

* In a letter in 1855 Lister described how, following the application of hot water to a frog's foot there was great dilatation of the arteries " and at first an enormously increased flow of blood : the capillary network becoming far more red than natural, and each capillary coming to admit 3 or 4 blood corpuscles abreast instead of only one." Later, " the capillaries became distended and stuffed with the red corpuscles, and the blood was first retarded, then stagnant. Thus with the simplest of stimulants, *heat*, I traced the process of inflammation from the beginning in I believe a more satisfactory way than it has ever been traced before. I often uttered involuntary exclamations of delight during the time ". Lister did not observe the migration of leucocytes from the vessels ; this had been observed earlier by Waller and was more fully described later by Cohnheim.

The changes which they observed have been repeatedly confirmed, using bacterial infections as well as physical and chemical injuries ; and the stationary pictures seen in sections of fixed tissues correspond with the various stages of inflammation as seen in living tissues.

Two main stages are seen in acutely inflamed tissues, (a) a stage of vascular dilatation accompanied by acceleration of the blood flow, and (b) a stage of retardation of the flow accompanied by migration of leucocytes and exudation of fluid from the vessels.

(a) The stage of vaso-dilatation with accelerated flow

Almost at once following the application of the irritant, there occurs dilatation of all the small blood vessels of the part, including arterioles, capillaries and venules, and many capillaries which were previously collapsed and invisible open up and carry a stream of blood (Fig. 16). Except for very mild inflammation

FIG. 16.—Inflammatory hyperaemia. Diagram to compare vessels in normal peritoneum (a) with those in inflamed peritoneum (b), as seen in intact mesentery.

due to a single application of a rapidly dissipated irritant, inflammation rarely subsides at this stage. Usually there has been actual structural damage or destruction of some or many of the cells of the tissue, and the inflammation proceeds to stage (b).

(b) The stage of vaso-dilatation with retardation of flow, emigration of leucocytes and exudation of fluid

The dilatation of the vessels persists but the blood stream slows, and in some of the capillaries and venules it may later cease entirely, the blood becoming stagnant. As the stream slows, the normal zonation of the blood into an axial core of corpuscles and a peripheral zone of clear plasma disappears ; the axial corpuscular stream expands until the cells touch the vessel walls, to which the corpuscles, especially the white corpuscles, tend to cling. The vascular endothelial cells themselves become visibly swollen. The polymorphonuclear leucocytes hugging the endothelium exhibit amoeboid movement and migrate through the walls of the capillaries and venules between the endothelial cells to reach the extra-vascular tissue spaces. A few red corpuscles also may escape passively in the wake of the emigrating leucocytes ; this is often spoken of as " diapedesis", though this term (meaning " a footing through ") might more appropriately be applied to the active migration of the leucocytes. In the tissue spaces the

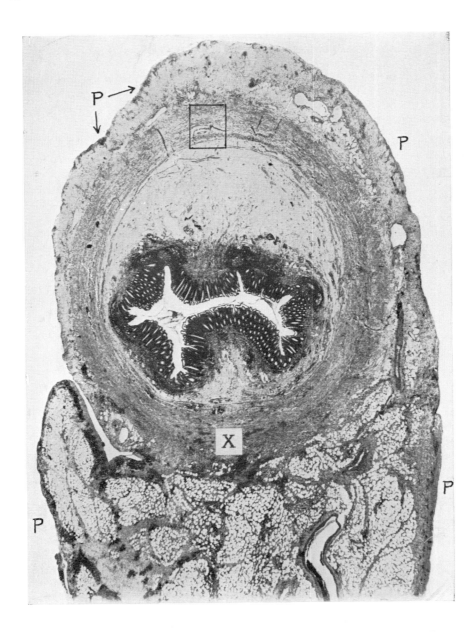

FIG. 17.—Cross section of an acutely inflamed appendix and meso-appendix. Note oedematous swelling of the submucosa, separation of the layers of the muscle coat, deposits of exudate on the peritoneal surfaces PP, and dark patches of haemorrhagic exudation around X. The area in the rectangle is shown in Fig. 18. (× 10.)

leucocytes move about and phagocytose injured cells and certain kinds of bacteria if these are present. During this stage of inflammation, modified blood plasma or " inflammatory lymph " also exudes from the vessels into the tissue interstices, distending these interstices, separating the cells and fibres, and so producing a general swelling or oedema of the tissues. This fluid contains the plasma proteins, including fibrinogen, in concentrations nearly as great as in blood plasma itself. The fibrinogen which it contains coagulates to form fibrin, which appears as a fine microscopic meshwork throughout the interstices of the inflamed tissues or as a visible opaque layer of coagulum on free surfaces such as the serous membranes. These changes are depicted in Figs. 17, 18 and 19.

(c) *The changes summarized*

Briefly then, the changes of acute inflammation are :

Stage a: Vaso-dilatation (hyperaemia) with acceleration of flow.

Stage b: Vaso-dilatation (hyperaemia) with retardation of flow, accompanied by
    (i) loss of zonation of the blood stream,
    (ii) emigration of leucocytes from the vessels,
    (iii) exudation of fluid plasma, from which fibrin is deposited.

The essential results are the escape of leucocytes and plasma into the tissue spaces ; the other changes may be regarded as subserving the liberation of these elements.

(2) Analysis of the changes in acute inflammation

(a) *What actually excites the inflammatory reaction?*

It was formerly assumed that this reaction is directly excited by the irritant chemical substances or the soluble toxins of the bacteria causing the inflammation. But of recent years there has been modification of this view. Clearly, the immediate cause of inflammations evoked by simple mechanical injury or by heat cannot be any extrinsic chemical or bacterial irritants but must be products of the damaged tissues themselves. Hence, it has been asked, may not *all* inflammation be due to the same products, whether the damage be inflicted by mechanical, thermal, chemical or microbic agents ? We know that when cells are damaged or destroyed, some of the products of disintegration of their protoplasm, especially their proteins, are soluble " toxic " substances. One of these, histamine, when injected into the skin, produces a triple response consisting of (a) local dilatation of capillaries and venules by direct action, (b) widespread dilatation of surrounding arterioles from a local axon reflex, and (c) local increased permeability of the capillary walls, with increased exudation of fluid and oedema or wheal formation. The similarity of these changes to those of inflammation is obvious. A triple response exactly similar to that produced by histamine follows many mild injuries —pricks, scratches, slight scalds, electric stimuli, insect bites, or injections of tissue extracts. The simplest explanation of this uniformity of action of many different injuries is that they all liberate the same chemical substance which causes the triple response, and that this is histamine or some closely similar " H "-substance. While there is no doubt that histamine or " H "-substances are partly responsible for the early vascular changes and fluid exudation of inflammation, we know also that " leukotaxine " and other peptides liberated in injured

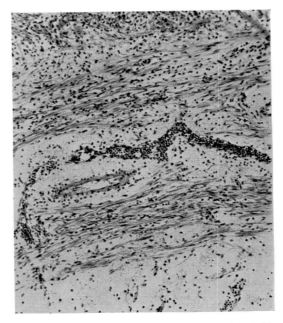

FIG. 18.—Enlarged view of the rectangular field marked in Fig. 17, showing separation of the muscle fibres by oedema, a blood vessel packed with polymorphonuclear leucocytes, and scattered leucocytes throughout the tissues. (× 100.)

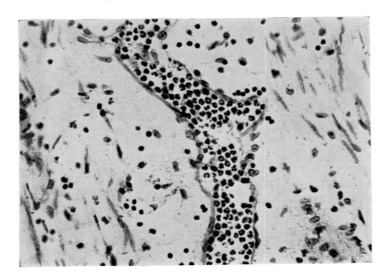

FIG. 19.—Enlarged view of the vessel in Fig. 18, showing the polymorphonuclear leucocytes crowding its lumen and migrating through its walls. (× 350.)

tissues are equally important, as shown by the work of Menkin, Spector, and others. Precise identification of the substances concerned is still in progress; but, whatever their nature, the general concept is the same, namely, that products of the damaged tissues themselves are the immediate excitants of the inflammatory response. It has been supposed that some bacterial toxins may have a similar direct action on the vessels; but it is probable that bacteria, like non-living injurious agents, evoke inflammation, mainly or wholly, by damaging or destroying cells, whence inflammation-evoking substances are produced.

(b) *How is the vaso-dilatation with acceleration in the first stage brought about?*
This question has been answered in the previous paragraph. If we assume that histamine or some substance with a similar action on vessels is liberated in the injured tissues, this causes dilatation of capillaries and veins by a direct paralysing action on their walls. At the same time, probably by a local axon reflex, small arteries in and near the injured area dilate, thereby admitting more blood into the capillaries and accelerating its flow.

(c) *How is the retardation of flow in the second stage brought about?*
It seems contradictory that the stream should slow while the vessels remain dilated. The explanation is undoubtedly that there are local changes in the viscosity of the blood in the dilated vessels or (what amounts to the same) in the adhesiveness of the endothelium of the vessel walls. We have already noted that, in direct microscopical study of inflamed tissues, the corpuscles, especially the leucocytes, cling to the vessel walls, as if the surfaces of these or of the corpuscles themselves were sticky. Since we know that the vessel walls *are* greatly altered as regards their permeability to the proteins and other constituents of the plasma, it is quite likely that they are altered also as regards the surface tension or adhesiveness of their swollen endothelial cells. Probably then, the retardation of flow is due to changes in the colloidal surface properties of the capillary walls, brought about by the direct action on them of those same substances which are the immediate excitants of the inflammatory reaction—histamine or other " H "-substances or peptides. As the flow slows and the capillaries become stuffed with corpuscles, and as fluid exudes from the vessels, the viscosity of the blood within them no doubt increases, still further retarding the flow.

(d) *The formation and functions of the exuded fluid*
The exudation of plasma-like fluid from the vessels into the tissue spaces is undoubtedly due to increased permeability of the vessel walls, brought about by the direct action of histamine or " leukotaxine " or other peptides. This is clear, not only from the demonstrable presence of such substances in injured tissues and inflammatory exudates, but also from observations on the accumulation in inflamed tissues of dyes and other substances injected into the circulation. Injected substances, e.g. trypan blue, to which capillary walls are normally impermeable, are found to traverse inflamed capillaries and to accumulate in inflamed tissues. Two distinct functions of the exuded fluid have been suggested, (a) that it brings abundant soluble antibodies (which, as we shall see later, are modified globulin molecules) into the inflamed tissues to neutralize microbes and their toxins, and (b) that the fibrin precipitated in the tissues acts as a barrier which tends to localize the infection. While most pathologists are agreed on the defensive

value of (*a*), there are wide differences of opinion as to the importance of the
" fibrin barrier ". Menkin supposed it to be the main factor opposing the spread of
bacteria and their toxins from the initial focus of infection; but others think that
antibodies and leucocytes are more important than fibrin in localizing infections,
and the nature of the bacterial toxins plays a great part in determining whether
infections will or will not spread from the initial focus.

(*e*) *The migration and functions of the leucocytes*

The classical work of Metchnikoff on *phagocytosis* (from 1884 to 1900)
established its fundamental importance in inflammation and in the defence of
the body against microbic parasites and particulate foreign matter generally.
He showed how amoeboid cells in the connective tissues and blood of invertebrates as well as vertebrates engulf and digest bacteria and other solid particles,
and he developed the doctrine of inflammation as a means of bringing by the
blood stream and liberating into an injured area a great army of phagocytes
capable of removing or destroying bac-
teria and other foreign particles. The
phagocytosis of bacteria by neutrophil
polymorphonuclear leucocytes can be
seen in almost any smear of pus (Fig. 20).
There is no doubt of the correctness
of Metchnikoff's view that the inflammatory migration of the leucocytes from
the vessels is due to *chemotaxis*, i.e. an
attraction of the leucocytes by chemical
substances in the injured tissues. The
older view was that these substances were
the toxins of invading bacteria ; but it is
now clear that products of injury of the
tissues themselves also have chemotactic
effects, since aseptic (bacteria-free) dead

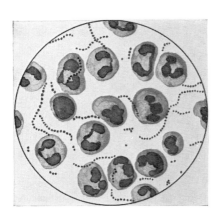

FIG. 20.—Smear of pus from streptococcal peritonitis. Most of the cocci
are free, but a few have been phagocytosed. (× 1000.)

tissue killed mechanically or by heat or
by deprivation of blood supply attracts
leucocytes, and since tissue extracts
are chemotactic. Menkin's " leukotaxine "
and other peptides in injured tissues and inflammatory exudates are strongly
chemotactic for polymorphonuclear leucocytes. Histamine does *not* cause any
significant emigration of leucocytes. While damaged tissue products are certainly
responsible in part for the inflammatory migration of leucocytes, it is probable that
some bacterial products also are potent. Leucocytes appear in great numbers in
certain infections, and different bacteria evoke leucocyte responses different in both
degree and kind, some calling forth polymorphonuclear cells only and others
mononuclear leucocytes.

(3) The local signs of acute inflammation

The four " cardinal signs " of acute inflammation, enumerated in the first
century A.D. by the Roman Celsus, namely *tumor*, *rubor*, *dolor* and *calor*,
are easily understood from the microscopic changes in inflamed tissues.

(i) *Swelling* is due to both the vaso-dilation and the exudation, especially the latter. The tissues contain more blood, more leucocytes, and above all more interstitial fluid than normally. Microscopically, the cells, fibres and other elements of the tissues are often widely separated by inflammatory fluid, coagulated fibrin and collections of leucocytes. The swelling is mainly inflammatory oedema (i.e. fluid accumulation) ; and the fluid can often be visibly squeezed away by local pressure, so that the oedema " pits on pressure ".

(ii) *Redness* is due to the vaso-dilatation or hyperaemia. In the early stage, when the dilated vessels are carrying a rapid stream, the redness is bright scarlet, fades easily on pressure and returns at once when the pressure is removed. In the later stages, when the stream is retarded, the redness is darker and purplish due to deoxygenation of the slowly moving blood, and the colour fades less easily on pressure and returns more slowly. In the skin over a boil or carbuncle, both stages are often visible ; there is a bluish central area where the changes are oldest and the blood sluggish or stagnant, and a bright red peripheral zone where more recent inflammation encroaches on healthy tissue and the circulation is rapid.

(iii) *Pain* is attributable largely to pressure exerted on nerve-endings by the exudate. Hence the pain often " throbs " with the pulse, it is more severe with inflammation in confined rigid tissues than in lax ones, and it is aggravated by pressure (i.e. there is *tenderness*). It is probable that the pain is also partly due to direct irritation of nerve-endings or nerves in the inflamed tissues.

(iv) *Heat* is not produced in inflamed parts, the temperature of which is never higher than that of the blood. An inflamed superficial part of the body, e.g. in the skin, is hotter than the surrounding skin only because the hyperaemia brings more blood and therefore more heat to that part. Its temperature is therefore closer to blood-temperature than normal parts of the skin, which under most conditions are much cooler than the interior of the body. An area of inflamed skin where the circulation is much retarded, and which is therefore bluish in colour, may be no warmer, indeed may be cooler, than the surrounding skin.

(v) *Impairment of function* is a fifth important result of inflammation and must be added to the four classical signs. Inflamed parts are automatically kept at rest or their functions in other ways diminished or suspended. Thus, an inflamed joint or muscle is kept stationary, in pleurisy or pneumonia the respiratory movements on the affected side are restricted, in peritonitis peristalsis is diminished or ceases, an inflamed eye resents the light and is kept closed. Allied to this depression of activity in inflamed tissues is the reflex protective guarding of them by neighbouring muscles. Thus, in peritonitis the overlying abdominal muscles, though not themselves inflamed, are hypertonic or rigid, preventing external pressure from being exerted on the inflamed structures ; in arthritis or osteomyelitis, the surrounding muscles are similarly reflexly on guard, minimizing movement and splinting the part. This conception of the automatic resting and protection of inflamed tissues is of wide application and great importance in both the interpretation of symptoms and the proper planning of treatment.

(4) The termination of acute inflammation

According to the nature and severity of the injurious agent, and the health or otherwise of the tissues, acute inflammation ends in one or another of the following ways :

(i) *Resolution.*—This is simply subsidence of the inflammation and absorption of the exuded elements. Many mild inflammations, especially those due to non-living physical and chemical agents, end in this way. So also do some severe bacterial infections ; e.g. uncomplicated pneumonia often ends by liquefaction and absorption of the inflammatory fluid and cells which have been poured out into the alveoli, and the lung returns to normal.

(ii) *Suppuration or formation of pus.*—Many acute bacterial (especially coccal) inflammations are suppurative, i.e. they produce visible collections of fluid pus, which is a mixture of inflammatory exudate, dead and living polymorphonuclear leucocytes and dead and living bacteria. The leucocytes are called " pus-cells " ; phagocytosed bacteria are often found within them (Fig. 20). On a free surface suppurative inflammation produces a purulent discharge ; in a firm solid tissue it results in a confined local accumulation of pus called an *abscess* ; in a tissue of loose texture like the subcutaneous tissues or areolar tissue planes, and especially when due to certain kinds of bacteria, e.g. streptococci, suppurative inflammation may spread widely, producing diffuse sheets of pus, a condition called *cellulitis*. Once a collection of pus has formed, it is rarely absorbed, because it contains living bacteria and much organic debris on which they may continue to multiply as in a culture medium. Hence the inflammation is maintained and the pus goes on accumulating and the abscess enlarging, until it " points " and discharges on an adjacent surface or is opened surgically. The wall of an abscess cavity consists of inflamed granulation tissue, which, after the abscess discharges, fills up the collapsed cavity and organizes into fibrous scar tissue, just as in a healing wound. A collection of pus which has become confined within a pre-existing cavity—a viscus or a serous cavity—is called an *empyema*, e.g. of the pleura, appendix, gall bladder, maxillary antrum, or middle ear.

(iii) *Necrosis.*—Necrosis (death of tissue en masse) results from some severe microbic inflammations. We have already noted that acutely inflamed bone usually dies ; and examples of inflammatory necrosis in soft tissues are the core of a carbuncle or severe boil, the gas gangrene of muscles invaded by *Clostridium Welchii* (anaerobic myonecrosis) and the gangrene of the lung sometimes seen in pneumonia in debilitated people. (*Gangrene* means massive necrosis or death of tissues where putrefactive bacteria are present).

(iv) *Repair or organization.*—Repair follows any inflammation in which there has been destruction or loss of continuity of the tissues, e.g. in an abscess cavity as mentioned above. When inflammation results in a great deposition of in-flammatory fibrin, as on the serous surfaces, organization of this often takes place by the growth of new vessels and multiplying fibroblasts into it. Repair and organization following inflammatory destruction of tissue are closely similar to repair and organization following simple physical injuries, as described in Chapter 2. The principal difference is that, whereas following simple trauma inflammatory phenomena are slight and repair is seen in an almost pure form, in tissues the seat of sustained inflammation from microbic infection repair begins while the inflammation is still in progress, so that the reparative tissue is itself inflamed, as in an abscess wall. The mingling of repair with inflammation is least in rapidly subsiding acute inflammations, greatest in protracted chronic inflammations.

CHRONIC INFLAMMATION

(1) Causes of chronic inflammation

Chronic inflammation may be defined as the slow smouldering reaction excited in tissues by mild but persistent injurious agents. However, no sharp distinction can be made between " acute " and " chronic " inflammations ; it is merely a question of the rate of progress and duration of the injury. Bacteria which usually excite acute inflammation may, because of loss of their virulence or because of the development of a grade of immunity in their host, cause a chronic inflammation, as in chronic gonorrhoea or chronic pyogenic osteomyelitis. Conversely, bacteria which usually produce chronic inflammation may, from increase of virulence or diminished resistance of the host, cause a rapidly progressive acute inflammation, as in acute or " galloping " tuberculosis of the lungs.

Chronic inflammation may be caused by repeated slight trauma or chemica irritation. Chronic ulcers of the stomach and duodenum exemplify non-specific chronic inflammation maintained by chemical factors (acid secretion) in an injured tissue. However, the most important chronic irritants are microbic, e.g. the bacilli of tuberculosis and leprosy, the spirochaete of syphilis and the fungus of actinomycosis.

(2) Tissue changes of chronic inflammation

In a chronically inflamed tissue we may conveniently distinguish the following features :

(i) *Purely inflammatory changes.*—The vascular and exudative changes are similar to those of acute inflammation, but less in degree and often obscured by reparative changes. There is exudation of fluid and emigration of leucocytes from the vessels ; but, in most chronic inflammations most of the leucocytes are of the mononuclear series—lymphocytes, monocytes and plasma cells—and these cells multiply locally in the tissues as well as coming from the vessels. Sometimes eosinophil leucocytes collect in great numbers. However, in chronic pyogenic inflammations, e.g. actinomycosis, chronic gonorrhoea or chronic osteomyelitis, the polymorphonuclear leucocytes predominate just as in acute inflammations.

(ii) *Reparative proliferation.*—There is slow but persistent proliferation of fibroblasts and new blood vessels with the formation of a chronically inflamed granulation tissue. The fate of this new tissue varies. In some chronic inflamma-tions it undergoes progressive fibrosis, but in others it is poorly vascularized and suffers degenerative changes. In syphilitic lesions poor vascularity results mainly from thickening of the walls of the new vessels which are formed, many of which suffer occlusion ; in tuberculosis, for some obscure reason, very few new vessels develop in the proliferating tissue, which therefore is ill-nourished from its inception.

(iii) *Degenerative changes.*—These are frequent in many kinds of chronic inflammatory lesions, especially those of tuberculosis and syphilis, in which the poorly vascular unhealthy granulation tissue often undergoes a slow progressive necrosis almost as fast as it develops. The lesions of tuberculosis and syphilis thus often show a central area of yellow cheesy debris, and a peripheral zone of still surviving granulation tissue accompanied by marginal fibrosis.

(iv) *Macrophages and foreign-body giant cells.*—These are commonly present in chronic inflammatory lesions, especially associated with areas of degeneration. They phagocytose both the bacteria or other microbic parasites responsible for the disease and the fatty substances resulting from degeneration of the tissues.

The large masses of proliferating, fibrosing and degenerating tissue which often develop in chronic inflammations are sometimes spoken of as " granulomas ", e.g. " tuberculoma ", " syphiloma ", " leproma ". This is an unfortunate terminology, tending to confuse these lesions with true tumours.

THE LEUCOCYTES AND PHAGOCYTES FOUND IN INFLAMED TISSUES

It will be convenient to enumerate here the types, names and sources of the various kinds of leucocytes and phagocytes which accumulate in areas of inflammation (Fig. 21) :

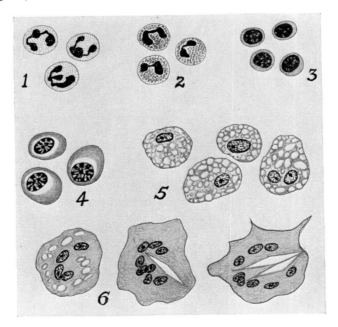

FIG. 21.—Leucocytes and phagocytes found in inflamed tissues. 1, neutrophil granulocytes ; 2, eosinophil granulocytes ; 3, lymphocytes ; 4, plasma cells ; 5, macrophages filled with engulfed fatty material—" foam cells " ; 6, foreign-body giant-cells with included fatty droplets and crystals. (× 1000.)

(a) *Neutrophil granulocytes or polymorphonuclear leucocytes* (" *neutrophils* " *or* " *polymorphs* ")

These are the most abundant of the leucocytes derived from the bone marrow. They are active phagocytes, especially for cocci and some other kinds of bacteria, and were called by Metchnikoff *microphages*. In areas of acute inflammation they migrate from the blood through the vessel walls in great numbers, and in suppurative inflammations they are " pus cells ". Inflammations characterized by the migration of many neutrophil granulocytes from the blood are often accompanied by leucocytosis due to increased production of these cells from the bone marrow.

(b) *Eosinophil granulocytes* (" *eosinophils* ")

These cells collect in large numbers in some chronically inflamed tissues, e.g. around some chronic gastric ulcers and in some mildly inflamed appendices; and occasionally the " pus cells " from an acute abscess are found to include many eosinophils. They also congregate in great numbers in the bronchial walls in asthma, and in other irritated tissues in patients who show sensitization (or allergy) to foreign proteins (see p. 287) ; and these patients may also show eosinophil leucocytosis (eosinophilia). In many metazoal parasitic diseases like-wise, eosinophils collect around the parasites in the tissues, and eosinophilia may also be present (see Chapter 19). Eosinophils are rarely or never phagocytic; their main function appears to be to congregate in areas of histamine formation and neutralize its effects.

(c) *Mononuclear leucocytes*

These comprise *lymphocytes*, small and large, and *monocytes*. Many patholo-gists distinguish sharply between these two types of cells ; but transitions between them are seen both in normal blood and in areas of chronic inflammation, in which they often congregate in great numbers, and it is possible that they are variants of cells of the same kind. The lymphocytic forms are only sluggishly motile and non-phagocytic, and one of their main functions is the formation of antibodies (see p. 282). Monocytes are actively amoeboid and phagocytic.

(d) *Plasma cells*

These structurally distinctive cells are easily recognized by their rounded eccentric nuclei with coarsely granular chromatin pattern and the characteristic staining properties of their hyaline cytoplasm, the perinuclear zone of which is often paler than the peripheral. Plasma cells occur normally in the lymphoid tissues, in which transitional forms between them and lymphocytes occur. They also collect, usually along with lymphocytes, in many chronically inflamed tissues, e.g. in syphilitic lesions, in mildly inflamed granulation tissue, and in the dermis in chronic inflammations of the skin. They are not phagocytic, but are concerned in the production of antibodies (see p. 282).

(e) *Macrophages*

This name embraces all actively motile mononuclear phagocytes. They engulf many kinds of foreign particles, including various bacteria, especially bacilli, carbon or silica particles, dyes, free fatty and lipoid material, and disintegrating blood pigment (haemosiderin and haematoidin). We have already mentioned them as scavengers of degenerating myelin in nervous tissue and of pigments and lipoids in haematomas ; and in later chapters we will meet them again and again in various inflammatory and degenerative diseases. Macrophages arise in several different ways, (1) by mobilization of the *histiocytes* (fixed macrophages) which are present in all connective tissues, (2) in lymphoid tissue, spleen, liver and bone marrow, by proliferation and mobilization of the fixed reticulo-endothelial phagocytes; (3) possibly by the conversion of lymphocytes into monocytes in areas of phagocytic activity, and (4) by migration of monocytes from the blood into the tissues. Because of the diversity of appearances assumed by macro-phages and histiocytes, they have been called by many different names. These include—macrophages, histiocytes, clasmatocytes, wandering cells, adven-titial cells, polyblasts, microglial cells (in the central nervous system), " dust cells " and " heart failure cells " when scavenging dust particles or extravasated blood pigment in the air-sacs or bronchi, " foam cells " when laden with engulfed lipoid material, " melanophores " or " chromatophores " when laden with engulfed melanin particles, " lepra cells " in leprosy, " epithelioid " or " endo-thelioid " cells in tuberculosis, and " Aschoff cells " in rheumatic fever.

(f) Multinucleated phagocytes or " foreign-body " giant-cells

Where free lipoid droplets or crystals or particulate foreign bodies which are not readily absorbable are present in chronic inflammatory and degenerative lesions, macrophages frequently fuse together to form multinucleated syncytia surrounding the foreign substances. These syncytia, which are spoken of as " foreign-body giant-cells ", sometimes have scores or even hundreds of nuclei, which are often situated peripherally. Some workers believe that foreign-body giant-cells may develop, not only by fusion, but also by amitotic nuclear division of macrophages, but from the appearances observed it is clear that fusion is their main, if not their only, mode of formation.

THE PART PLAYED BY LYMPHATICS IN INFLAMMATION

The volume of lymph flowing from an acutely inflamed part by its main efferent lymphatics is greatly increased. Hence, although the lymphatics in the inflamed tissues themselves are not easily seen, it is probable that they, like the capillary blood-vessels, are dilated and that they carry an accelerated flow. In bacterial inflammations, the lymph flowing from the inflamed area contains bacterial toxins and sometimes bacteria themselves, and these cause inflammation—lymphadenitis—in the regional lymph glands to which they are carried ; e.g. lymphadenitis of the cervical glands from tonsillitis, of the tracheo-bronchial glands from pneumonia, and of the axillary glands from a whitlow (a pyogenic infection of a finger). If considerable numbers of virulent bacteria escape from the primary focus by the lymphatics, they may cause abscesses in the inflamed lymph glands ; or they may cause acute inflammation—lymphangitis—along the course of the main efferent lymphatics themselves, so that in a case of whitlow, for example, tender red streaks may be seen extending up the skin of the fore-arm.

We have already noted (p. 17) that in the reparative tissue resulting from injury or destructive inflammation plentiful new-formed lymphatics sprout from those around. Such lymphatic proliferation probably participates in the develop-ment of most chronic inflammatory lesions.

THE CONSTITUTIONAL EFFECTS OF MICROBIC INFLAMMATIONS

So far we have considered only the changes occurring locally in inflamed tissues. Many bacterial and other microbic inflammations, however, also produce remote effects in other tissues.   These effects are :

    (1) Toxaemia ;
    (2) Fever ;
    (3) Toxic changes in parenchymatous cells ;
    (4) Leucocytosis ;
    (5) The production of antibodies.

(1) Toxaemia

Toxaemia means the presence in the blood of soluble toxins or poisonous products of pathogenic bacteria or viruses. As we shall see in the tenth and subsequent chapters, it is principally by their toxins that microbic infections injure the body and produce disease. The constitutional symptoms of most infective diseases are largely due to the absorption of toxins from the inflamed tissues into the blood. These symptoms include fever and its well-known accompaniments, acceleration of the pulse rate, headache, pains in the limbs, loss of appetite (anorexia), mental lethargy, and sometimes delirium or convulsions, especially in children.

(2) Fever or pyrexia

This is the simplest means of measuring the severity of toxaemia. It results partly from increased metabolism and heat production in the tissues, and partly from diminution in the loss of heat by sweating and radiation from the skin ; and the diminished cooling from the skin depends in turn on an alteration of the heat-regulating centre in the brain-stem. In pyrexia, this thermostatic centre is set for a higher temperature than normal. Why this should be so, or how bacterial toxaemia causes this change, we do not know. Clinically, the sudden onset of fever is accompanied by a feeling of chill, and by shivering, which is sometimes violent and is then called a *rigor*. The sudden decline of a high fever, as at the end of a malarial attack or at the crisis of pneumonia, is accompanied by a feeling of heat and by profuse sweating.

(3) Toxic changes in parenchymatous cells

Severe or prolonged toxaemia brings about visible changes in the cytoplasm of the parenchymatous cells of the liver, kidneys, myocardium and other organs. These changes are designated *cloudy swelling* and *fatty " degeneration "*. In cloudy swelling the cells are swollen and their cytoplasmic structure is obscured by the presence of many small granules, most of which are of protein nature but some of the coarser of which are fatty or lipoidal. The appearances are an exaggeration of the normal fine granularity of the cytoplasm, and are due to damage and swelling of the mitochondria, with corresponding interference with their oxidative and other enzymes. Fatty " degeneration " denotes a further degree of cellular injury wrought by toxaemia ; in addition to cloudy swelling the cytoplasm shows many small or large droplets of free fats or lipoids. The term

" degeneration " is not strictly correct ; the fat-laden cells are not disintegrating but are so injured that they are unable to metabolize properly the fats brought to them by the blood stream (see p. 345).   This toxic injury of parenchymatous cells is an important factor in producing symptoms and in determining the outcome of an illness ; e.g. a frequent cause of death in pneumonia, typhoid fever or other severe infection is circulatory failure due to direct toxaemic damage to the myocardium.

(4) Leucocytosis

Many acute bacterial inflammations. especially those due to cocci, are accompanied by neutrophil polymorphonuclear leucocytosis.   Thus in a case of pneumonia, appendicitis, cellulitis or abscess, the leucocyte count may rise from the normal 5,000 or 10,000 per c.mm. to 20,000 or 30,000 or more.   Clearly this must be due to the effect on the leucopoietic bone marrow of soluble substances absorbed from the inflamed area.   Most workers have supposed these substances to be the bacterial toxins themselves, but it is possible that certain products of the injured tissues may be responsible.   " Leucotaxine " and other chemotactic peptides which cause the migration of leucocytes in inflamed tissues, are apparently *not* the leucocytosis-producing factors.

(5) The production of antibodies

This important result of many infective inflammations will be discussed fully in Chapter 21.

SUPPLEMENTARY READING

Archer, R. K. (1963). *The Eosinophil Leucocytes.* Oxford; Blackwell.
Cameron, G. R. (1932). " Inflammation in earthworms ",   *J. Path. Bact.,* **35,** 933.
 — (1934). " Inflammation in the caterpillars of *Lepidoptera* ", *J. Path. Bact.,* **38,** 441.
Cappell, D. F. (1930). " The cellular reactions following mild irritation of the peritoneum in normal and vitally stained animals, with special reference to the origin and nature of the mononuclear cells ",   *J. Path. Bact.,* **33,** 429.
Cheatle, G. L. (1936). " Inflammation : Hunter's views and modern conceptions ", *Brit. Med. J.,* **1,** 1148.
Clark, E. R., and Clark, L. C. (1935). " Observations on changes in blood vascular endothelium in the living animal ", *Amer. J. Anat.,* **57,** 385. (Clear evidence by the transparent chamber technique that the endothelial cells of inflamed blood vessels become sticky.)
Curran, R. C. (1967). " Recent developments in the field of inflammation and repair ", in *Modern Trends in Pathology* (Ed., T. Crawford). London; Butterworths.
Florey, H. W. (Ed.) (1954). "Inflammation ", in *Lectures on General Pathology.* London; Lloyd-Duke.
Fried, B. M. (1938). " Metchnikoff's contribution to pathology ",   *Arch. Path.,* **26,** 700. (An interesting historical outline.)
Menkin, V. (1956). *Biochemical Mechanisms in Inflammation,* Springfield, Illinois; Charles C. Thomas.
Spector, W. G. (1959). " Endogenous mechanisms in the acute inflammatory reaction ", in *Modern Trends in Pathology* (Ed., D. H. Collins). London; Butterworths.
 — and Willoughby, D. A. (1968). *The Pharmacology of Inflammation.* London; English University Press (Med).

CHAPTER 4

INFLAMMATION AND ITS SEQUELAE IN PARTICULAR SITES : SEROUS MEMBRANES

CHAPTER 3 contained a preliminary note pointing out that the nature of the affected tissues is an important factor influencing the appearances and results of inflammation. In the next six chapters we will examine this factor more closely and will consider some typical inflammatory diseases in particular tissues —serous membranes, lung, heart and blood vessels, bone, meninges and epithelial surfaces. We begin with acute inflammations of serous membranes, because these show very clearly the vascular and exudative changes of inflammation and the subsequent organization of fibrinous exudates.

CAUSES OF ACUTE INFLAMMATIONS OF SEROUS MEMBRANES

With few exceptions, pleurisy, pericarditis and peritonitis are due to microbic (mainly bacterial) parasites, and in most cases these come from diseased neighbouring viscera.

Pleurisy (to be consistent, pathologists should call it " pleuritis ") is usually secondary to bacterial infections of the underlying lung, e.g. pneumonia, lung abscess or tuberculosis. Occasionally it results from extension through the diaphragm of subdiaphragmatic infections, such as hepatic, subphrenic or perirenal abscess, or from penetrating wounds of the thorax. The commonest bacteria causing pleurisy are pneumococcus, streptococcus, staphylococcus and tubercle bacillus.

Pericarditis is most commonly due to rheumatic fever, the cause of which is still under discussion (see Chapter 22). Other causes of pericarditis include pneumococcal, streptococcal, staphylococcal and tuberculous infections, most often spreading from the neighbouring pleura or lymph glands.

Peritonitis most often results from acute bacterial inflammations of the abdominal viscera, e.g. appendicitis, cholecystitis, salpingitis ; or from perforation of diseased viscera so that their infected contents are spilt into the peritoneal cavity, e.g. perforation of ulcers of stomach or duodenum, or of tuberculous or typhoid ulcers of the intestine. The most frequent bacteria in peritonitis are those which are present in the alimentary contents, namely the pyogenic cocci, *E. coli*, *Fusiformis* species and *Cl. Welchii*. The pneumoccocus and tubercle bacillus are less frequent.

Tuberculosis of the serous membranes, though often sudden in onset, has special characters and runs a chronic course ; it is considered in Chapter 12. We are concerned here with acute inflammations, and will take first as an example a simple non-suppurative serositis, such as a fibrinous pleurisy accompanying pneumonia or a fibrinous rheumatic pericarditis.

THE CHANGES IN ACUTE NON-SUPPURATIVE SEROSITIS

These may be described conveniently in three stages, which, however, merge into one another.

(1) Early stage : hyperaemia and oedema of the serous membrane

This stage corresponds to that seen during direct examination of experimentally induced inflammation in the living frog's mesentery or web, as described in the previous chapter. The serous membrane shows great hyperaemia, migration of polymorphonuclear leucocytes, and exudation of fluid which at first causes interstitial swelling or oedema of the membrane and then commences to discharge onto its surface.

(2) Stage of exudation into the serous cavity

Migrated leucocytes and exuded inflammatory fluid continue to pour out of the inflamed membrane into the serous cavity. The fibrinogen in this exudate is deposited as a white layer of fibrin on both the visceral and parietal serous surfaces, while the serous fluid of the exudate collects between these two layers. Most of the leucocytes are enmeshed in the fibrin. The thickness of the fibrin deposit varies greatly—from an almost microscopic layer sufficient only to cause a blurring or fine sandpapered appearance with loss of the normal sheen of the serous surface, to a layer of white or cream coagulum 2 or 3 centimetres thick. For the first few days after its deposition, before its organization by ingrowing blood vessels and fibroblasts has advanced far, the fibrin layer can easily be peeled or brushed off the serous surface. Later, it becomes more and more intimately united to the tissues beneath by organizing tissue, and when forcibly detached during life, it leaves a bleeding surface. Fibrin deposits in pleurisy or peritonitis are usually smooth-surfaced, but in the pericardium the cardiac movements churn the fibrin up and give it an irregular shaggy surface (Fig. 22). The movement of fibrin-covered visceral and parietal layers of pleura or pericardium against one another can often be heard through the stethoscope or felt

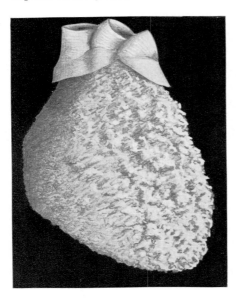

FIG. 22.—Sketch of a shaggy heart in rheumatic pericarditis

by the fingers as a *friction rub*, which in the pleura is synchronous with the respiratory movements, in the pericardium synchronous with the heart beat. The amount of serous fluid which collects in the cavity varies greatly ; there may be little or none, the visceral and parietal fibrin layers being contiguous with one another, a condition very conducive to a friction rub ; or there may be pints of serous fluid separating the two layers, so that no friction rub is possible.

(3) Stage of organization and adhesion formation

The condition of the serosal endothelium on a surface covered by fibrin deposit is probably an important factor determining the fate of the fibrin. Where the endothelium has been destroyed, the conditions are comparable with those of

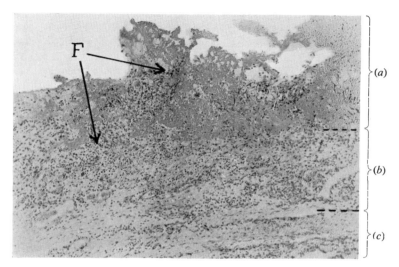

FIG. 23.—Section of rheumatic pericarditis. (*a*), shaggy layer of fibrin ; (*b*), layer of vascular granulation tissue, growing from epicardium (*c*), and invading fibrin layer, e.g. at F. (× 60.)

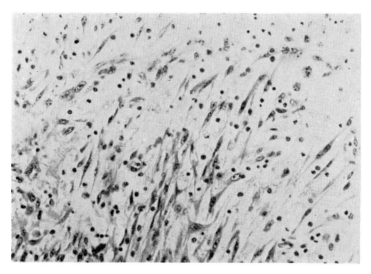

FIG. 24.—Enlarged view of fibroblasts in granulation tissue in Fig. 23. (× 200.)

an open wound on an epithelial surface, and new-formed granulation tissue consisting of blood vessels and multiplying fibroblasts grows into the fibrin from the subjacent tissues (Figs. 23, 24). Where the endothelium is intact, how-ever, it is probable that much or all of the overlying fibrin undergoes solution and re-absorption without organization. It is surprising how, following severe inflammation of serous membranes with much exudation, the inflammation often resolves, leaving only slight patchy adhesions. However, where the organization of fibrin deposits progresses, the serous membrane eventually shows fibrous thickening or the formation of fibrous adhesions between the visceral and parietal layers. Such adhesions acquire a smooth covering of serosal endothelium derived by proliferation of the intact endothelium of the neighbouring surfaces. If there has been widespread damage of the serous surfaces, the final result is sometimes complete adhesive obliteration of the serous cavity by fibrous fusion of the visceral and parietal layers.

VARIATIONS IN ACUTE INFLAMMATION OF SEROUS MEMBRANES

We have considered an acute non-purulent fibrinous serositis as an average example of inflammations in serous membranes. However, the details of such inflammations vary according to the severity of the infection and the nature of the responsible agent. There may be much or little fibrin, much or little serous fluid, and some bacterial infections produce purulent exudates (pus).

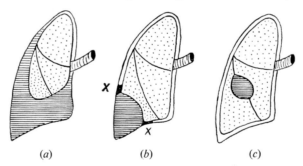

FIG. 25.—Types of empyema. (a) a free or unconfined empyema with general collapse of lung ; (b) an encysted basal empyema confined by adhesions *XX* ; (c) an encysted interlobar empyema.

For descriptive purposes we therefore distinguish the following varieties of inflammations of serous membranes : (i) *Serous inflammations*, with abundant outpouring of nearly clear fluid, poor in fibrin and leucocytes, and leaving few or no adhesions after its absorption ; (ii) *fibrinous inflammations* with abundant fibrin and not much accumulation of fluid ; (iii) *purulent inflammations* with great emigration of leucocytes accompanying severe bacterial infections, leading to pus formation ; the pus may be widespread in the serous cavity as in general peritonitis, or it may become localized and walled-off by granulation tissue and developing adhesions, forming an abscess in one part only of the serous cavity (Fig. 25). It has become customary to speak of a pleural abscess or purulent pleurisy as an *empyema,* but somewhat inconsistently this name is not applied to purulent pericarditis or peritonitis.

Inflammations of intermediate type of course occur—e.g. sero-fibrinous, sero-purulent or fibrino-purulent. These names do not denote any fundamental differences ; they only describe the nature of the exudate found in the particular case, and this depends on the nature and severity of the infection. The exudate may change in character during the progress of the disease ; e.g. the pleurisy accompanying pneumococcal pneumonia is always at first fibrinous or sero-fibrinous, but in some cases it later becomes purulent.

CHAPTER 5

INFLAMMATION AND ITS SEQUELAE IN PARTICULAR SITES: PNEUMONIA

ANY INFLAMMATION of lung tissue is, to the pathologist, a *pneumonia* or *pneumonitis;* but clinically these names are applied chiefly to coccal and other acute or subacute bacterial infections of the lungs. In reading a description of the changes produced by these infections, it must be remembered that this applies to cases untreated by penicillin, sulphonamides or other anti-bacterial remedies. These substances greatly alter the course of many lung infections ; and nowadays the student seldom sees a lung showing the classical changes of pneumococcal lobar pneumonia allowed to run its natural course. We will consider this kind of pneumonia first, and then lobular or bronchopneumonia—the former a specific pneumococcal disease, the latter embracing a variety of infections.

LOBAR PNEUMONIA

(1) Definition

Lobar pneumonia is an acute pneumococcal inflammation affecting almost uniformly a large part, often a lobe, of a lung. It is best to restrict the term to the specific pneumococcal disease and not to apply it (as some writers have done) to other kinds of extensive pneumonia of lobar distribution, many of which are really instances of confluent bronchopneumonia.

(2) Causation

(i) *The essential cause* is the pneumococcus, of which there are several types (p. 120). This organism is frequently present in the upper respiratory passages of healthy people ; and from here, under suitable conditions, it invades the lungs via the trachea and bronchi, and multiplies in the alveoli and in the interstitial tissues of the septa.

(ii) *Predisposing causes* include debilitated general health, as from cancer, chronic nephritis or alcoholism, and also temporary diminution of resistance from exposure to cold, fatigue or starvation.

(3) Pathological changes in the lung tissue

(*a*) *Stages*

The changes in the lung tissue are best described in four stages, which, however, merge into one another (Plate III) :

(i) *Stage of hyperaemia.*—This lasts only a few hours. The usual changes of acute inflammation occur in the interalveolar septa, but the alveoli still contain air.

(ii) *Stage of red consolidation,* or " red hepatization ", so called because the lung is solid, heavy and airless and resembles liver in its consistence. The lung is still hyperaemic and red ; but it is now airless, because fibrinous inflammatory

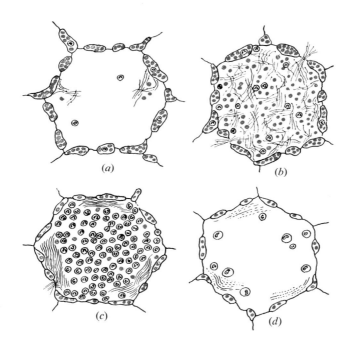

(a) (b)

(c) (d)

PLATE III.—Diagram of stages of pneumonia ; a single alveolus is represented : (*a*), stage of hyperaemia ; little fibrin and few cells have escaped into the alveolar cavity ; (*b*), stage of red consolidation ; alveolus filled with fibrin meshwork and escaped red cells ; (*c*), stage of grey consolidation ; fibrin consolidated on the septal walls, and many polymorphonuclear leucocytes accumulated in the alveolus ; (*d*), stage of resolution ; air-sac re-aerated, macrophages removing cellular debris, fibrin disappearing, hyperaemia subsided.

(*To face page 62*)

exudate, with polymorphonuclear leucocytes and escaped red corpuscles, has filled up all the alveoli and small bronchi. Each alveolus is occupied by a solid plug of coagulated fibrin with enmeshed leucocytes and red cells, and the fibrinous plugs in neighbouring alveoli are often seen to be connected with one another through the septal pores. The solid lung remains red for several days.

(iii) *Stage of grey consolidation.*—The fibrin contracts against the alveolar walls, many more leucocytes collect in the alveoli, many of the escaped red cells disintegrate, and the hyperaemia of the vessels in the septa is less pronounced ; so that the colour of the solid lung changes from red to grey.

(iv) *Stage of resolution.*—An uncomplicated case of lobar pneumonia ends in resolution, which usually commences within 2 weeks of the onset. The fibrin in the alveoli liquefies, probably through the action of enzymes liberated from the leucocytes, and the liquefied material is partly absorbed and partly coughed up. That much of it is absorbed into the circulation is shown by a pronounced rise in the urinary excretion of nitrogenous waste products at this time. In addition to polymorphonuclear leucocytes, many macrophages also participate during resolution and play a prominent part in the phagocytosis of pneumococci and exudate. The hilar and tracheo-bronchial lymph glands are enlarged and contain many active macrophages, showing that absorption takes place partly by the lymphatics. The lung gradually becomes aerated again, and its structure usually returns completely to normal.

(b) *Distribution of the pneumonic changes*

The inflammation affects a large mass of lung uniformly and has a well defined margin ; other parts of the lungs are unaffected. The disease involves an

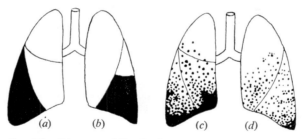

FIG. 26.—Diagram of distribution of pneumonia : (*a*) anatomically lobar pneumonia ; (*b*) non-anatomically lobar pneumonia ; (*c*) severe bronchopneumonia, confluent at base ; (*d*) less severe bronchopneumonia.

anatomical lobe or two lobes, or sometimes an extensive but not anatomically lobar portion of a lung (Fig. 26 *a* and *b*).

There is evidence that in many cases the infection of the lung tissue starts at a particular spot and quickly spreads from this to the surrounding lung tissue, partly by the flow and respiratory propulsion of infected exudate within the alveoli and bronchi, partly through the interstitial tissues. The infection probably spreads easily from alveolus to alveolus through the septal pores.

Because of the lobar or lobe-like extent of the pneumonia, there is always concomitant inflammation of the overlying pleura ; so that the disease is a pleuro-pneumonia, and the patient usually has severe pleural pain at the onset, often with a friction rub.

(4) The local symptoms and signs of lobar pneumonia

Most of the local symptoms and signs of the disease are readily understood in terms of its pathology. The pleural pain and friction rub have just been mentioned ; the pain often being much aggravated by the frequent irritative cough which is present. In the early stages the cough expels only scanty tenacious sputum which is often blood-streaked or " rusty ", because it comes from a hyperaemic lung in which the alveolar exudate contains escaped red blood corpuscles. Later, in the stage of grey consolidation and especially during resolution, more plentiful yellowish sputum containing many leucocytes is coughed up. Pneumococci, both free or phagocytosed by leucocytes, are to be seen in great numbers in stained smears of the sputum. Consolidation of the lung produces dullness to percussion over the affected area, accompanied by characteristic changes in the transmitted breath sounds heard through the stethoscope. Because a large area of lung is functionless, and because the patient breathes shallowly to restrict the movement of the affected side and minimize pleural pain, and also because he is feverish and toxaemic, the respiratory rate is greatly accelerated —perhaps 30 or 40 per minute instead of the normal 16 or 18.

(5) Systemic changes accompanying lobar pneumonia

(i) *Toxaemia and fever* are severe. At the onset there is often a rigor (severe shiver) due to a sudden rise of temperature to 102° or 104° F. or more. The temperature remains high, with daily oscillations, for 7 to 10 days or longer ; then often falls rapidly to normal, i.e. the fever ends by " crisis ". This usually takes place about the same time that resolution begins, or a day or two earlier. Sometimes the fever and toxaemia end, not by " crisis ", but by " lysis ", i.e. gradually over a period of several days. Other effects of the toxaemia during the height of the fever include acceleration of the heart rate, delirium, fall of blood pressure, and perhaps slight albuminuria. In fatal cases toxic degenerative changes are visible in the parenchymatous cells of the myocardium, liver, kidneys, adrenals and other organs.

(ii) *Bacteriaemia** is demonstrable, i.e. pneumococci are present in the blood stream and can be found by blood culture, in the early stages of most cases of lobar pneumonia. This does not seriously affect prognosis, but it shows us how remote pneumococcal infections may arise.

(iii) *Polymorphonuclear leucocytosis* occurs, the neutrophil leucocyte count in the circulating blood often rising to 30,000 per c.mm. or more.

(6) Complications of lobar pneumonia

These can be grouped conveniently as follows :
         Local: (i) in lung, (ii) in pleura ;
         General: (iii) toxaemic, (iv) metastatic.

---

* *Bacteriaemia* means the presence of bacteria of any kind in the blood. *Septicaemia* is sometimes used as a synonym for bacteriaemia, but is more correctly reserved for severe or persistent blood infections by pyogenic cocci.

(*a*) *Local*

(i) *Pulmonary complications.*—These are easily remembered as being the possible terminations of acute inflammation other than resolution. Occasionally, for some unknown reason, the exudate fails to absorb, the fibrin plugs in the alveoli become organized by the ingrowth of vessels and fibroblasts from the septa, and the lung remains airless and slowly becomes fibrous and contracted, a condition called *chronic fibrous pneumonia* or " carnification " of the lung. Or, in debilitated patients, such as alcoholics or diabetics, an area of lung instead of resolving may break down to form an *abscess* or may suffer necrosis or *gangrene*.

(ii) *Pleural complications.*—Pneumococcal pleurisy is usually fibrinous or sero-fibrinous, ending harmlessly by resolution or the formation of a few unimportant flimsy pleural adhesions. But sometimes it becomes purulent, forming an empyema (see Chapter 4), a complication which arises more frequently during convalescence than during the height of the disease.

(*b*) *General*

(iii) *Toxaemic complications.*—The presence of toxic changes in the parenchymal cells of various organs has already been mentioned. The chief cause of adult deaths from lobar pneumonia is *circulatory failure* due mainly to toxic damage to the myocardium. Other occasional toxaemic complications include *toxic hepatitis* and *toxic nephritis*.

(iv) *Metastatic infections.*—Since pneumococci often enter the blood stream from the diseased lung, the occasional development of pneumococcal inflammations in other tissues is readily understandable. The chief of these are endocarditis, pericarditis (which, however, may also be due to direct spread from the affected pleura), arthritis, meningitis, otitis media and peritonitis. Such complications appear usually during the convalescent stage. They are frequently suppurative and serious.

LOBULAR PNEUMONIA

(1) Definition and causation

Lobular pneumonia is any inflammation of the lungs of patchy irregular distribution. It is of two kinds according to the source of the bacteria causing it, namely (1) *bronchopneumonia* due to spread of infection from the bronchi and bronchioles, and (2) *haematogenous lobular pneumonia* due to blood-borne bacteria from some remote source, as in typhoid fever or a streptococcal septicaemia. The former is much the commoner, and its causes require further explanation.

Bronchopneumonia is a common and serious complication of :

(*a*) *the infectious fevers of childhood*, e.g. measles, whooping cough, scarlet fever and diphtheria ;

(*b*) *bronchitic diseases in adults*, e.g. influenza, and bronchitis in the elderly ;

(*c*) *obstructive bronchial diseases*, e.g. tumours, foreign bodies and bronchiectasis ;

(*d*) *aspiration of foreign matter* into the lungs, especially vomitus and mucus while under a general anaesthetic—*post-operative pneumonia*.

(*e*) *hypostasis*, i.e. congestion and oedema of the lungs, especially of the bases, due to impaired circulation, as in old or debilitated bed-ridden people.

The bacteria causing bronchopneumonia are therefore very varied ; they include pneumococci, streptococci, staphylococci, Friedländer's bacillus, *H. influenzae* and the whooping-cough bacillus ; the infection is often mixed and pneumococci frequently participate.

In the following paragraphs we will consider the changes in the commoner disease *bronchopneumonia ;* those of lobular pneumonia due to blood-borne infections are generally similar.

(2) Pathological changes in the lung tissue

Each involved patch of lung passes through stages generally similar to those of lobar pneumonia ; but the changes develop irregularly, at different times for different patches, are of variable duration, often slow in resolving, and often suppurating. The affected lung usually shows patches in all stages, some in the early hyperaemic stage, some solid and red, some solid and grey, some resolving, and some becoming purulent to form scattered small abscesses. Some severe bronchopneumonias, e.g. those complicating epidemic influenza, are haemorrhagic.

(3) Distribution of bronchopneumonic changes

In mild cases, the consolidated areas are small, few, scattered and discrete. In severe cases, they are many and densely sown, and may extend and coalesce to form large but irregular areas of consolidation—confluent bronchopneumonia. Usually the basal lobes are more severely affected than the upper lobes, and the disease is often bilateral. As older patches resolve, new patches develop, so that the distribution of lesions may change during the course of the illness, the duration of which is indefinite. Pleurisy is not a constant accompaniment.

(4) The local symptoms and signs of bronchopneumonia

The ways in which these differ from those of lobar pneumonia are easily understood from the pathological differences between the two diseases. Pleurisy, almost always present in the early stages of lobar pneumonia, is often absent in early bronchopneumonia. Cough in bronchopneumonia varies greatly according to the nature of the co-existing bronchial infection, such as whooping cough or influenza. The sputum also varies in amount and character. Because of the small size and scattered distribution of the patches, the signs of consolidation to percussion and auscultation may be slight and difficult to detect. The respiratory rate varies with the extent and severity of the disease.

(5) Systemic changes accompanying bronchopneumonia

(i) *Toxaemia and fever* vary with the extent of the disease and the nature of the infection. Their onset is often gradual, their duration indefinite—weeks or even months—and their offset gradual. In severe or prolonged cases toxic changes are present in many organs.

(ii) *Bacteriaemia* is not usually present, except in the late stages of fatal cases.

(But haematogenous lobular pneumonia is, of course, a complication of a bacteri-aemic disease, e.g. typhoid fever or a coccal septicaemia.)

(iii) *Leucocytosis* often occurs but is variable in degree.

(6) Complications of bronchopneumonia

(i) *Pulmonary complications.*—Cases of long duration may drag on for weeks or months as *chronic bronchopneumonia*, resulting in patchy fibrosis of the lungs and bronchi, and later bronchial dilatation (bronchiectasis). *Abscesses* of the lung often develop in bronchopneumonia resulting from septic states of the upper respiratory passages, e.g. cancer of pharynx or larynx, or from inhaled foreign bodies, or from bronchial obstruction. *Gangrene* of parts of the lung may also occur in such cases.

(ii) *Pleural complications.*—Empyema may develop, though less commonly than after lobar pneumonia. It is usually due to a mixed infection, e.g. pneumococcus and Pfeiffer's bacillus, or pneumococcus and streptococcus.

(iii) *Toxaemic complications.*—Circulatory failure due to myocardial damage is a main risk in severe or prolonged cases.

(iv) *Metastatic infections.*—These are unusual because most cases are not bacteriaemic ; but in chronic bronchopneumonias with bronchiestasis or chronic lung abscess or empyema, metastatic *cerebral abscess* sometimes occurs.

SUPPLEMENTARY READING

Hadfield, G., and Garrod, L. P. (1947). *Recent Advances in Pathology*. London; Churchill. Chapter X. (Contains an outline of experimental results in the production of pneumonia.)

Hoyle, L., and Orr, J. W. (1945) " The histogenesis of experimental pneumonia in mice ", *J. Path. Bact.*, **57.** 441.

Macgregor, Agnes R. (1960). *Pathology of Infancy and Childhood*. Edinburgh and London; E. & S. Livingstone. (Chapter XVIII contains a brief account of pneumonia in children; some good photographs.)

CHAPTER 6

INFLAMMATION AND ITS SEQUELAE IN PARTICULAR SITES: OSTEOMYELITIS

ANY INFLAMMATORY disease affecting bone and marrow is entitled to be called *osteomyelitis*, but this name is usually restricted to inflammations due to pyogenic bacteria. In most cases these bacteria are blood-borne from some distant infective focus, but similar inflammations follow direct local infection of bones from compound fractures or penetrating wounds. We will describe first the commonest variety of the disease, namely acute osteomyelitis due to blood-borne coccal infection in a child or adolescent, and will reserve for later consideration its less common varieties. This description, like that of pneumonia in Chapter 5, depicts the changes in a typical case of the disease allowed to run its natural course, unmodified by penicillin or other anti-bacterial drugs. Nowadays we rarely see specimens of untreated osteomyelitis as extensive and advanced as many of those preserved in old museums before the advent of efficient surgery and chemotherapy. But description of the untreated case is essential to a clear understanding of the natural history of the disease.

ACUTE BLOOD-BORNE OSTEOMYELITIS IN CHILDHOOD

(1) Causation

(i) *The essential cause* is infection by pyogenic bacteria. The chief of these is *Staphylococcus pyogenes*; less common offenders are *Streptococcus* and *Pneumococcus*. These reach the bone from some distant focus by way of the blood stream ; sometimes the original focus is evident—a boil, impetigo of the skin, or tonsillitis ; sometimes the source of infection is obscure. The early stages of the disease are usually septicaemic, and the responsible bacterium is easily identifiable by blood-culture.

(ii) *Predisposing causes* are (*a*) *age*, and (*b*) *injury*.

(*a*) *Age*. The special proneness of children and adolescents to blood-borne infection of bone is almost certainly due to a special susceptibility of the young growing bone of the metaphyses, in which osteomyelitis of childhood almost invariably starts.

(*b*) *Injury*. There is often a history of a slight recent injury of the affected bone or limb—a bruise, sprain or jar. This probably causes a small haemorrhage in the young metaphyseal tissue, so providing a nidus for the lodgement and growth of the blood-borne bacteria.

(2) The pathological changes in the bone

These may be described in 5 stages as shown in Plate IV ; these stages of course overlap.

(1) *The early inflammatory focus in the metaphysis.*—The usual changes of acute inflammation take place in the young vascular growing tissue of the shaft

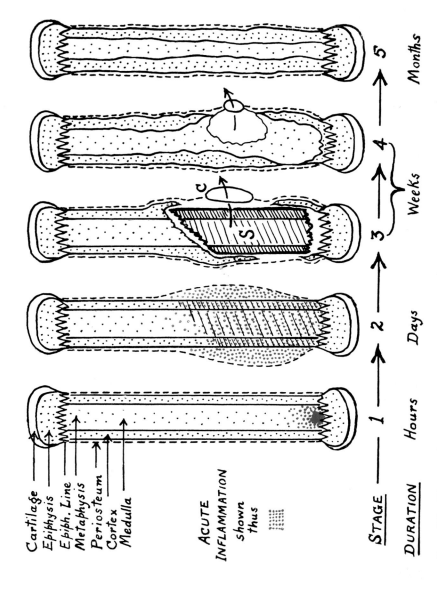

Cartilage
Epiphysis
Epiph. Line
Metaphysis
Periosteum
Cortex
Medulla

ACUTE
INFLAMMATION
shown
thus

STAGE — 1 → 2 → 3 → 4 → 5

DURATION   Hours   Days   Weeks   Months

PLATE IV.—Diagram of stages of osteomyelitis as described in text. *S*, sequestrum ; *C*, cloaca. Stages 3 and 4 really coexist.

(*To face page 68*)

close to the epiphyseal line. The long bones of the limbs are the ones most commonly affected ; osteomyelitis of the bones of the trunk, hands or feet is less common.

(2) *Extension of the inflammation, suppuration, and necrosis of bone.*—The infection and acute inflammation spread rapidly from the initial metaphyseal focus—up the medullary cavity of part or the whole of the shaft, and via the Haversian canals through the cortical bone to the periosteum. Usually within a few days the inflammation has reached its limits of spread, and pus is appearing in both the medullary cavity and subperiosteal tissues. The inflamed periosteum is separated from the bone by the purulent exudate, and the bone itself dies because of deprivation of its periosteal blood supply and inflammatory thrombosis of its medullary and Haversian vessels. At operation necrosed bone may be recognized by its smooth white stripped surface and by its failure to bleed. At the end of this stage, then, we have a length of necrosed bone shaft (still in continuity with the living remainder of the shaft) lying in an abscess cavity enclosed at the sides by detached inflamed periosteum and limited at one end by the intact epiphyseal cartilage and at the other by inflamed medullary tissue at the limit of spread along the shaft.

(3) *Evacuation of pus, demarcation and sequestrum formation.*—The abscess is evacuated either by surgical incision or sometimes by spontaneous perforation through the periosteum. In either case one or several *cloacae* are left in this membrane, through which pus continues to discharge externally for weeks or months during separation of the necrosed bone. The inflammation subsides to a chronic type during this period. The dead area of bone soon becomes marked off from the living by a deepening groove of erosion. This is produced by the osteoclastic and enzymic activity of the young vascular granulation tissue which develops from the interstices of the adjacent living bone and from the periosteum where this is still in contact with the bone surface. This erosion at the line of demarcation continues until the dead bone is completely separated from the living as a *sequestrum*. The margins of a sequestrum and such parts of its surface which were in direct contact with periosteal granulation tissue have a pitted worm-eaten appearance ; but where its surface was denuded of periosteum and bathed in pus, it remains permanently non-eroded and smooth (Figs. 27 and 28). The separation of a sequestrum is a slow process, often lasting many weeks or months ; and while it is taking place, reparative proliferation of the periosteal sheath or *involucrum* is also in progress. Hence, stages (3) and (4), depicted separately in Plate IV, really coexist.

(4) *The formation of a bony involucrum.*—The inner surface of the inflamed periosteal sheath produces a great quantity of vascular granulation tissue, in which much new porous or woven bone develops. This involucral ossification, which is very similar to the formation of callus around a fracture, is already well advanced before the sequestrum is separated ; but, after separation of the sequestrum and its spontaneous discharge or surgical removal, repair proceeds much more rapidly. The bony involucrum is continuous with the periosteum clothing adjacent parts of the intact bone shaft. These also are roughened by periosteal new formation of bone, just as is the shaft for some distance away from the site of a fracture (see Fig. 7). The new bone formed by an

FIG. 27.—Osteomyelitis of whole tibia showing sequestrum of shaft partly encased by involucrum. From a drawing in John Hunter's collection preserved at the Royal College of Surgeons, London. The catalogue description reads : " Two views of a tibia which had become wholly dead from the epiphysis at the knee to the joint at the ankle; to supply the place of which nature was casing it round with living bone, beginning first at the two extremes and shooting towards the middle, but not yet joined ".

FIG. 28.—Another example from John Hunter's collection of drawings. " Two views of a tibia the lower end of which is become wholly dead but not regularly all round but slanting very obliquely upwards on the posterior surface. The separation is almost completed although not visible on an external view. The ossific inflammation with the ossifying of the granulations has produced a considerable deal of new bone on the surface of the living especially near the line of separation ".

involucrum is more abundant, and the roughening and thickening of the neighbouring bone shaft by periosteal proliferation is greater and more extensive, than in a simple fracture callus. The bony involucrum encloses a cavity which was formerly occupied by the sequestrum, but which is now gradually filled up by involucral granulation tissue. During its obliteration, this cavity continues to discharge a diminishing amount of pus through the cloacae, which remain as rounded apertures in the bony involucrum (Fig. 29).

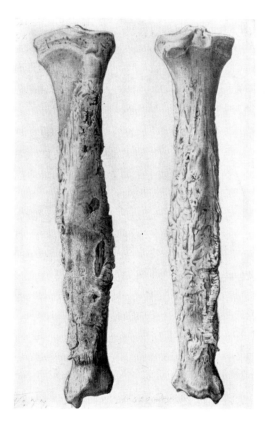

FIG. 29.—Osteomyelitis of tibia showing involucrum, cloacae and sequestrum. From a drawing in John Hunter's collection, preserved at the Royal College of Surgeons, London. The catalogue description speaks of the involucrum as " ossified granulations where they had shot over the dead bone underneath, forming a partial bony case, thro' the deficiencies of which is seen the original bone now become dead and exfoliating ".

(5) *Completion of healing and remodelling of the new bone.*—The sequestral cavity having been filled in by organizing tissue and the cloacae having closed, the bony involucrum now constitutes a new piece of shaft continuous with the intact part. In response to the mechanical stresses and strains to which this new shaft is subjected, much of the excessive porous new-formed bone is removed by osteoclasts and other parts are strengthened and consolidated ; so that finally a remodelled new shaft is completed, sometimes remarkably like the normal shaft in general outline and architecture, though often somewhat thickened and irregular. Here also, the remodelling is comparable to that of a fracture callus.

(3) Local complications of osteomyelitis

(i) *Acute suppurative arthritis.*—This occasionally develops by spread of the infection from the metaphysis into the joint. Usually, however, the epiphyseal cartilage and the strong attachments of the periosteum to its margins offer an effective barrier preventing infection of the joint. Arthritis is more likely to occur where the epiphyseal line and therefore the metaphysis are partly intracapsular in position, as in the hip joint, than where these are wholly extracapsular.

(ii) *Destruction of the epiphysis.*—If the infection does transgress the epiphyseal line, the epiphysis suffers acute inflammation and necrosis, so that the subsequent growth of that end of the bone is arrested.

(4) Systemic effects and complications of osteomyelitis

(i) *Toxaemia.*—This is severe in the early stages of the infection when the inflammation is extending and the pus is still confined. There are high fever, rapid pulse, sometimes delirium, and high polymorphonuclear leucocytosis.

(ii) *Septicaemia.*—Septicaemia (bacteriaemia) is frequently present during the early stages of the illness and may persist and prove fatal.

(iii) *Pyaemia and metastatic abscesses.*—Pyaemia is due to the liberation of infected thrombi from the large venous sinuses in the inflamed marrow ; it produces metastatic abscesses in the lungs, endocarditis, pericarditis, and abscesses in other organs. This condition is usually fatal, but sometimes chronic pyaemia with recurring metastatic osteomyelitis of other bones develops, lasting months or years and occasionally ending in recovery.

(iv) *Amyloid or lardaceous disease.*—This complication (p. 352) may occur in cases with chronic suppurating sinuses of long duration ; but this is now rare.

CHRONIC OSTEOMYELITIS

The foregoing account describes an " ideal " case of acute osteomyelitis, with the formation of a single large sequestrum, after removal of which we have supposed that permanent healing has ensued. The " ideal " case, however, is rare ; and there are the following possible chronic variations of the disease :

(i) *Persistently recrudescing osteomyelitis.*—During stages (4) and (5) depicted above, there is ample opportunity for pockets of infection to persist in the deeper parts of the large healing cavity within the involucrum, or for small detached sequestra to be enclosed and to act as infected foreign bodies. For these reasons smouldering inflammation often persists for many years, during which sinuses communicating with one or more unhealed cloacae may continue to discharge, or they may heal up and then re-open to the accompaniment of repeated febrile attacks caused by local acute " flare-ups " and extensions of the osteomyelitis. In such cases, small sequestra may continue to be formed and discharged at intervals. As the result of its reparative proliferation, the epidermis around the chronic sinus openings sometimes grows deeply into the sinus tracks, even into the interior of the involucrum ; these deep ingrowths of epithelium still further hinder repair within the cavity, and occasionally they eventually suffer cancerous change. A patient with chronic recrudescent osteomyelitis may die from septi-caemia or metastatic infections resulting from an acute recrudescent attack, or from amyloid disease consequent on long-standing suppperation.

(ii) *Insidious chronic osteomyelitis.*—Instead of being acute in onset, osteo-myelitis is sometimes chronic from its beginning. Little or no pus may be formed and there may be no sequestra. The main changes are great thickening and sclerosis of the bone in response to the low-grade smouldering inflammation within it, the thickening being due to successive layers of new periosteal bone, often plainly visible in radiographs. The shaft thus develops an enlarging fusiform swelling, which may be mistaken clinically for a tumour. The swelling is painful, and the patient usually has slight fever and leucocytosis ; but the distinction from a tumour can sometimes be made only after careful and repeated X-ray examination or by surgical operation and pathological examination (bacterio-logical culture and histological study) of a piece of excised tissue.

(iii) *Brodie's abscess.*—This is a special localized form of chronic osteomyelitis, the result of a lowly virulent staphylococcal infection, usually in a young adult. Its site is most often the upper end of the tibia, less often the ends of other long bones. The abscess cavity is usually small, centrally situated in the bone, and surrounded by densely sclerosed bony tissue. There is usually no sequestrum, and in the most chronic cases the pus may be scanty and sterile. The overlying bone surface may be thickened and irregular from periosteal proliferation.

OSTEOMYELITIS IN PARTICULAR SITES

(i) *Osteomyelitis of the skull* may result from either blood-borne infection or from local injury or infective foci, e.g. an inflamed frontal sinus or mastoid. The pericranium is easily stripped up by the accumulating pus and extensive necrosis of the outer table of the vault with sequestrum formation ensues. If the inner table remains intact the patient may recover, but if it is involved fatal meningitis usually supervenes.

(ii) *Osteomyelitis of the mandible* is usually due to local infection spreading from an inflamed tooth socket. The entire mandible may suffer necrosis and may be discharged or removed as a sequestrum ; and a peculiarity here is that there is little or no development of a bony involucrum, so that reformation of a useful mandible does not take place.

(iii) *Osteomyelitis of the vertebrae* is usually due to blood-borne infection, but is commoner in adults than children. It is a rare but very dangerous disease because its systemic effects are severe, because the deep situation and shape of the bones makes their surgical drainage difficult, and because spread of the infection often leads to meningitis, cellulitis of the neck, mediastinitis, empyema, or retroperitoneal abscesses.

(iv) *Osteomyelitis of the pelvic bones* is also very serious because of the severity of its systemic effects and because it spreads to the pelvic tissues and viscera.

OSTEOMYELITIS FROM LOCAL INFECTION

The pathology of osteomyelitis due to the local introduction of pyogenic bacteria is generally similar to that of the blood-borne disease, but with variations due to the kind of bacteria, the way in which they are introduced into the bone, and the particular bone affected. Reference has already been made to infection of the skull and mandible from local causes.   If a compound fracture or a gun-shot or other penetrating wound carries virulent cocci into the medullary cavity of a bone, the resulting osteomyelitis may be very similar in severity and results to that produced by blood-borne bacteria.

But more superficial wounds, reaching only the periosteum or cortex of the bone, may lead to only local superficial inflammation (*periostitis*), with perhaps the formation of a small *sub-periosteal abscess* and of a sequestrum consisting of only a flake of cortical bone.   The separation of small cortical sequestra and the subsequent repair of a bone following local injury were closely studied experiment-ally by John Hunter about 1780.   By means of a cautery Hunter produced necrosis of small areas of the cortex of the metacarpal bones of the ass and examined the bones at varying intervals after the injury: his beautiful series of preparations, still preserved in the museum of the Royal College of Surgeons in London, show all stages in the demarcation and separation of the sequestrum and the development around it of a rampart of new-formed periosteal bone (Fig. 30).

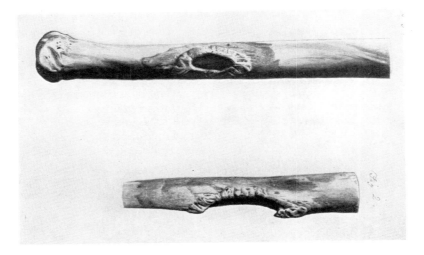

FIG. 30.—John Hunter's experiments on exfoliation of dead bone ; from his collec-tion of drawings in the Royal College of Surgeons, London. " Two bones of the leg of an ass which had been cauterized at different times before the death of the ass. One where the bone had not made its escape, the other where it has ".   Note the rampart of involucral new bone formed around the exfoliated sequestrum.

Osteomyelitis of an amputation stump.—This condition is rarely seen nowadays, but was a very frequent complication of surgical amputation in the pre-aseptic days of surgery.   The open medullary canal became infected during the operation, and the resulting osteomyelitis spread for a short or great distance up the shaft of the bone stump, producing a characteristic " ring " sequestrum. This had a clean cut distal end, with the marks of the amputation saw visible on it, and a short or long eroded tapering proximal end where the sequestrum had been separated and cast off from the living bone stump.

Contiguity periostitis and osteitis.—The bones contiguous with a chronic inflammatory lesion often show proliferative reaction of their periostea with consequent thickening and roughening of the bone surfaces by deposits of new subperiosteal bone. A good example is seen in the tibia and fibula underlying a chronic varicose ulcer of the leg. The inflamed bone surface may actually form the floor of the ulcer, and may then present a well defined plateau-like eminence corresponding in shape to the ulcer. Other parts of the bone shafts, even many inches away from the ulcer, may also show irregular bony outgrowths due to irritative proliferation of the periosteum.

SUPPLEMENTARY READING

Blacklock, J. W. S., and Rankin, W. (1935). " An unusual case of bone regeneration after complete diaphysectomy on two occasions ", *Brit. J. Surg.*, **22**, 825.

Collins, D. H. (1966). *Pathology of Bone*, Chapter 11. London; Butterworths. (A posthumous work, prepared by O. G. Dodge).

Greig, D. M. (1931). *Clinical Observations on the Surgical Pathology of Bone*. Edinburgh; Oliver & Boyd. (A beautiful series of specimens is depicted.)

Knaggs, R. L. (1926). *Inflammatory and Toxic Diseases of Bone*. Bristol; Wright. (Contains a well-illustrated account of osteomyelitis.)

CHAPTER 7

INFLAMMATION AND ITS SEQUELAE IN PARTICULAR SITES: HEART AND BLOOD VESSELS

AN IMPORTANT feature of inflammation of the walls of blood-containing cavities is the deposition of thrombus from the contained blood onto the damaged endothelial surfaces. In vessels inflammatory thrombi often occlude the lumen and spread for variable distances along it ; in the heart they form projecting masses called " vegetations " attached to the inflamed endocardial surfaces. During their formation and before there has been time for their organization, parts of such clots, especially if resulting from progressive bacterial inflammation, may become detached and carried away as emboli in the blood stream, to be arrested in and to occlude vessels in some remote part. Thrombi dislodged from tributaries of the portal vein are arrested in the liver ; those dislodged from systemic veins or from the right side of the heart are arrested in the pulmonary arteries ; those dislodged from the left heart chambers or from pulmonary veins are carried into the arterial circulation and produce embolism in the brain, spleen, kidneys or other parts. If the patient escapes the embolic and other dangers of the diseases causing inflammatory thrombosis, the thrombi undergo organization by young fibroblastic and vascular tissue growing in from the walls of the heart or vessel. Let us consider examples of *endocarditis*, *phlebitis* and *arteritis*—inflammations of endocardium, veins and arteries respectively.

ENDOCARDITIS

(1) Classification and causes of endocarditis

The types and causes of inflammatory disease of the endocardium may be classified as follows :

Type of Endocarditis	Causes
(i) Bacterial or infective endocarditis (a) Acute bacterial endocarditis (malignant or ulcerative endocarditis) (b) Subacute bacterial endocarditis	Gross bacterial infections. Haemolytic streptococci, pneumococcus, staphylococcus, gonococcus, etc. Usually *Streptococcus viridans*; occasionally other bacteria, e.g. *H. influenzae*.
(ii) Acute simple (" non-bacterial ") endocarditis	Usually rheumatic fever, occasionally other infective diseases, e.g. scarlet fever. Generalized lupus erythematosus (see p. 310)
(iii) Chronic sclerosing endocarditis	Rheumatic fever ; syphilis.

(i) *Bacterial endocarditis.*—By this term we mean an endocarditis in which the causative bacteria are abundant and easily seen in smears or sections of the large vegetations which develop on the inflamed endocardium. If the bacteria are of virulent and destructive kinds, e.g. the pyogenic cocci, the disease is rapidly fatal in a few days or weeks, and we speak of *acute* or *malignant bacterial endo-carditis*. But if the bacteria are less virulent, e.g. non-haemolytic varieties of streptococci, the disease may last months or years, and we call it *subacute bacterial endocarditis*. The source of the bacteria causing bacterial endocarditis is some-times clear, sometimes obscure. The acute form is often a complication of some obvious inflammatory disease elsewhere in the body,—a boil, osteomyelitis, tonsillitis, puerperal infection, pneumonia or gonorrhoea. In the subacute form, the primary focus of infection is seldom obtrusive, but there is good evidence that it is often dental or tonsillar (see p. 119). The bacteria causing endocarditis reach the valves by the blood stream ; there has been some discussion as to whether they settle directly on the valve surfaces or enter the valves via the coronary circulation ; the former seems the more likely.

(ii) *Simple* (non-bacterial) *endocarditis.*—In some cases of vegetative endocardi-tis, especially that due to rheumatic fever, bacteria are either not present at all or are very scanty and demonstrable only with difficulty in the vegetations. More-over, the vegetations are small, and the disease is not ulcerative. Hence we speak of *simple endocarditis*, or of " toxic endocarditis " in the belief that it is excited by bacterial toxins only. (The cause of rheumatic fever is discussed in Chapter 22.)

(iii) *Chronic sclerosing endocarditis.*—This is an aftermath of simple acute, especially recurrent, endocarditis of the aortic or mitral cusps, or of syphilitic inflammation of the aortic valve. The diseased valves show slow fibrosis, con-traction and often calcification, leading to progressive valvular incompetence or stenosis. (The degenerative disease atheroma may lead to similar sclerosis of the valves.)

(2) Acute bacterial endocarditis

(i) *The endocardial lesions.*—The affected valve cusps or other parts of the endocardium show the usual changes of acute inflammation, namely hyperaemia, emigration of polymorphonuclear leucocytes, and exudation of inflammatory fluid with swelling of the tissues. Large irregular friable vegetations develop on the damaged endocardial surfaces ; these consist of mingled thrombus (both blood corpuscles and fibrin) and colonies of bacteria (Fig. 31 and Plate V, Fig. 1). The bacteria multiply readily in the blood-clot, as in a culture medium, and they may form one-half or more of the bulk of the vegetations. Little or no organiza-tion of the vegetations takes place, because the lesions are rapidly progressive and destructive. The affected valves or adjacent heart walls suffer gross destruction (ulceration) ; and this may extend deeply into the myocardium and may even penetrate into other chambers of the heart. Extension of the ulceration to the mural endocardium and to the chordae, with rupture of the latter, often occurs. Clinically, the bulky vegetations and rapid destruction of valve cusps may cause loud, rapidly changing bruits on auscultation, and may lead to quickly progressive valvular incompetence, dilatation of the heart chambers and heart failure.

(ii) *Situation of the lesions.*—The disease almost always starts on valve cusps, more often the mitral or aortic than the tricuspid or pulmonary valves ; but the preference for the left cardiac chambers is less pronounced than in subacute bacterial or in simple rheumatic endocarditis. Bacterial endocarditis often spreads from the valves to neighbouring parts of the heart walls, but it rarely starts here.

(iii) *Remote results.*—Since the vegetations are large and friable, large emboli are often detached from them to produce gross infarction of other organs. This occurs in the spleen, kidneys, brain, etc., when the disease affects the left cardiac chambers ; in the lungs, when it affects the right cardiac chambers. Very large pieces may be detached, sufficient to cause fatal cerebral or pulmonary embolism or to occlude the main brachial, renal or mesenteric arteries. Since the emboli carry many virulent bacteria, the infarcted areas are infected, and, if the patient survives long enough, may break down into abscesses, or metastatic meningitis.

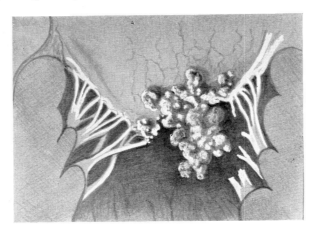

FIG. 31.—Sketch of acute bacterial (streptococcal) endocarditis of mitral valve. Note bulky vegetations, destruction of parts of valve and chordae, and hyperaemia of remainder of valve. (Compare with Fig. 32.)

osteomyelitis or other inflammations may be set up. Because of the grossly septicaemic and pyaemic state of the blood stream, blood-cultures give constantly positive results, the spleen is enlarged and loaded with bacteria, and all the organs show severe toxic changes. The disease is often rapidly fatal; penicillin and other anti-bacterial drugs are of value in some cases, but are ineffective in others, probably because the bacteria are sheltered in large masses of clot.

(3) Subacute bacterial endocarditis

(i) *The endocardial lesions.*—The vegetations are not so bulky and friable as in the acute type of bacterial endocarditis ; and, since the disease is of long duration, partial but incomplete organization of the bases of the vegetations occurs, so that these are more firmly attached to the endocardium. A vegetation thus often consists of a basal mass of inflamed organizing granulation tissue, clothed by an unorganized layer of fibrin clot containing bacterial colonies. During septicaemic stages of the disease, streptococci (of non-haemolytic viridans type) are plentiful in the vegetations, and destruction of tissue progresses ; during more quiescent phases, there are fewer bacteria in the vegetations and partial healing occurs. The affected valves suffer progressive destruction, and this often spreads to the heart walls and chordae, which may become extensively studded with masses of vegetations. Clinically, there are often slowly changing bruits ; and progressive valvular deficiency embarrasses the heart and causes hypertrophy and dilatation of chambers and eventual cardiac failure.

(ii) *Situation of the lesions.*—The disease affects the mitral and aortic valves much more often than those of the right side of the heart. It is clear that previous rheumatic endocarditis predisposes the valves to subacute bacterial infection ; this disease often extends from the mitral cusps onto the posterior atrial wall over an area where rheumatic endocarditis is commonly seen. There is also evidence that developmental anomalies predispose to subacute bacterial infection, which is seen with disproportionate frequency in the aortic valve when this has only two cusps instead of three, and which is apt to attack a patent ductus arteriosus.

(iii) *Remote results.*—Although large emboli are sometimes detached, causing gross embolic infarction of the brain, spleen, kidneys or other parts, in general the emboli are smaller and more numerous than in acute bacterial endocarditis. We often find rather characteristic appearances in the spleen and kidney due to repeated showers of small emboli. The spleen is enlarged, firm, shows multiple small infarcts of different ages, and often perisplenitis and adhesions. The kidneys are enlarged, may show scattered small infarcts, and in addition many small red spots (" flea-bitten kidney ") due to tiny bacterial emboli having been arrested in the glomerular capillaries and having caused haemorrhage into Bowman's capsules and the periglomerular tissues. The condition is spoken of as *focal embolic nephritis ;* and it explains why the patient's urine usually contains a few red blood corpuscles throughout the disease. Small haemorrhagic spots, due to tiny infective emboli, often appear also in the skin, especially of the fingers and toes, where also small tender nodules, called Osler's nodes, may appear. Because *Streptococcus viridans* is of relatively low virulence, the embolic lesions of endocarditis due to it do not suppurate ; the disease is not pyaemic in the strict sense. It is, however, intermittently bacteriaemic ; blood cultures are positive at some times but negative at others. Since the disease is a febrile infection of long duration, the patient becomes weak and anaemic ; and severe toxic changes are found in the organs at necropsy. Before the advent of penicillin, subacute bacterial endocarditis, like its acute counterpart, was always fatal. By antibiotic treatment, however, many cases can be cured, though in some of these the patient eventually dies of heart failure due to the extensive destruction of the valves wrought by the infection before it was overcome.

(4) Acute simple (non-bacterial) endocarditis

Rheumatic fever is by far the commonest cause of vegetative endocarditis in which bacteria are not obviously present. Similar " toxic " endocarditis is occasionally seen accompanying other febrile diseases.

(i) *The endocardial lesions.*—Rheumatic endocarditis is only one manifestation of widespread rheumatic inflammation of the heart. The rather peculiar characters of this inflammation are described more fully in Chapter 22. Here it will suffice to say that in the valves and other parts of the endocardium there is a subacute inflammation with oedema, increased vascularity and infiltration of the tissues by lymphocytes and mononuclear leucocytes as well as polymorphs. There are also small foci of necrosis in the valve tissues, surrounded by accumulations of inflammatory cells and proliferating fibroblasts and histiocytes ; these are called Aschoff bodies. Where the inflamed valve cusps touch during their closure, the friction detaches the injured endothelium, and vegetations develop on the denuded surface. These vegetations appear as a row of small pale firm warty (" verrucose ") nodules attached to the opposed surfaces of the inflamed valves along their lines of closure (Fig. 32). They consist of condensed masses of platelets and fibrin deposited from the blood on the damaged endothelial surface. Soon, however, new-formed vessels and fibroblasts grow into the thrombus from the inflamed

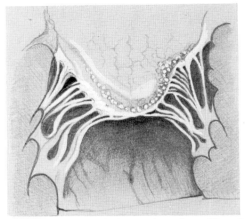

FIG. 32.—Sketch of rheumatic endocarditis of mitral valve. (Compare with Fig. 31.)

cusp, so that each vegetation now consists of a little mound of granulation tissue surmounted by a cap of fibrin and platelets (Plate V, Fig. 2). When the active inflammation subsides, the vegetations become wholly organized into fibrous tissue, which, along with the fibrosis of the cusps as a whole, leads to their thickening, contraction and deformation. In the acute stage, the roughening of the valve margins by vegetations causes faintly audible bruits. Later, fibrous contraction causes valvular stenosis or incompetence, with louder bruits, progressive hypertrophy and dilatation of the appropriate heart chambers and eventually heart failure.

 (ii) *Situation of the lesions.*—Careful microscopical examination has shown that rheumatic endocarditis is a widespread process affecting the whole endocardium. Usually, however, vegetations visible to the naked eye develop only on the valves, where the surfaces are abraded by friction. Hence, also, the valves of the left side of the heart show vegetations more often than those of the right side, because of the higher pressures moving the left-sided valves. The mitral valve is affected rather more frequently than the aortic ; but in many cases both are affected. The tricuspid valve shows vegetations much less frequently,

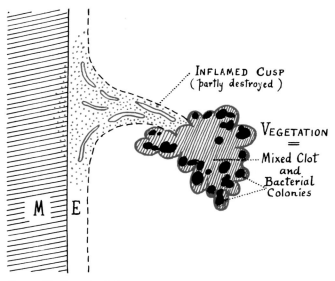

PLATE V, FIG. 1.—Diagram of structure of bacterial endocarditis.
E, endocardium, endothelium of which is shown by the interrupted
line ; M, myocardium. Red stippling denotes area of acute
inflammatory infiltration, traversed by new-formed blood vessels.
(Compare with Plate V, Fig. 2.)

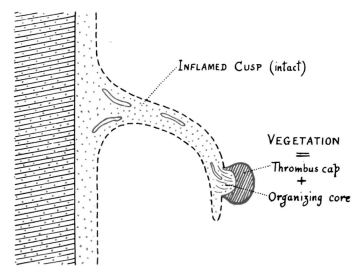

PLATE V, FIG. 2.—Diagram of structure of rheumatic endocarditis.
The sparse red stippling is to suggest widespread mild inflammation.
(Compare with upper diagram.)

(To face page 80)

and the pulmonary cusps rarely. The most severely affected part of the mural endocardium is an oval area on the posterior wall of the left atrium above the posterior cusp of the mitral valve. In severe cases this area may show a finely granular or frosted appearance due to a thin mat of minute vegetations, and when the inflammation subsides fibrosis often leaves a visibly puckered white area of thin scar tissue here. This is of special interest because, if bacterial (streptococcal) endocarditis later supervenes in a patient who has had rheumatic endocarditis, this area is often colonized by the streptococci and covered by vegetations. Mural bacterial endocarditis in this situation is strong evidence of previous rheumatic infection.

(iii) *Remote results.*—Since the vegetations are small and not friable, no gross embolic phenomena occur, though probably some minute emboli may be detached into the circulation. There is no bacteriaemia, and blood cultures are negative.

(5) Chronic sclerosing endocarditis

(i) *The endocardial lesions.*—As a result of repeated attacks of rheumatic endocarditis, or of atheromatous degenerative changes similar to the common atheroma of the large arteries, or of syphilitic inflammation, the valve cusps and orifices undergo varying degrees of fibrous thickening, contraction, calcification and distortion, causing valvular incompetence with or without stenosis. Clinically, bruits are heard on auscultation, the mechanical embarrassment imposed on the myocardium leads first to hypertrophy and later to dilatation of the chambers (see Chapter 37), and to eventual cardiac failure.

(ii) *Situation of the lesions.*—Chronic sclerosing endocarditis is almost restricted to the mitral and aortic valves and is rarely seen in the tricuspid or pulmonary. Rheumatic sclerosis affects both the mitral and aortic valves, the former more than the latter ; atheromatous sclerosis also affects both valves, the aortic more than the mitral ; syphilitic sclerosis is almost restricted to the aortic orifice, where it is usually associated with syphilitic aortitis or aneurysm of the aortic arch. On examination of an old sclerosed mitral valve, it is often difficult to be sure whether rheumatic fever or atheroma was mainly to blame; calcification may occur in either.

(iii) *Remote results.*—These are mainly the effects of circulatory embarrassment and eventual failure. Although there are no vegetations to give rise to emboli, embolism does sometimes occur from a failing heart with chronic valvular disease ; the emboli come from stagnation thrombi which have formed in the dilated chambers, usually the auricles.

PHLEBITIS

(1) Causes of thrombo-phlebitis

All of the small vessels in an inflamed tissue are involved in the inflammation ; and, as described in Chapter 3, in some of the capillaries and venules the retarded blood stream may cease entirely. The stationary blood in these vessels usually clots—an instance of inflammatory thrombosis on a microscopic scale. When we speak of *thrombo-phlebitis*, however, we mean inflammation in the wall of a larger vein, large enough to be a potential source of emboli detached from the clots within it. Such thrombo-phlebitis of a large vein develops either by extension of inflammation directly through its wall from surrounding inflamed tissues, or by spreading thrombosis commencing in its tributaries and extending into it. Because of the thinness of their walls, veins are much more liable to inflammatory thrombosis than arteries.

The principal bacteria responsible for spreading phlebitis are the ordinary pyogenic cocci, and any focus of pyogenic inflammation may initiate it. The following are some important examples of infective thrombo-phlebitis of large veins :

(i) *Thrombo-phlebitis of the venous sinuses of the dura mater.*—The commonest example is inflammation of the transverse sinus resulting from the spread of pyogenic infection from the middle ear. From acute or chronic otitis media the infection often extends to the mastoid antrum and to the surrounding bone, whence it may spread inwards to affect the dura mater and the sigmoid part of the transverse sinus. The phlebitis may extend from here into the internal jugular vein, even for several inches down the neck ; and at necropsy the sinus and vein may be found distended by mingled blood-clot and pus. Similar thrombo-phlebitis sometimes involves the petrosal and cavernous sinuses, the source of infection being osteomyelitis of the temporal bone from middle-ear disease, an infected sphenoidal air sinus, or phlebitis of the angular and ophthalmic veins spreading in from a boil or other infection of the face. Thrombo-phlebitis of the superior longitudinal sinus is a rare complication of infection of the frontal air sinus.

(ii) *Puerperal and post-abortal thrombo-phlebitis.*—If, during childbirth or abortion, streptococci or other bacteria infect the placental site in the uterus, spread of the infection to the uterine wall and parametrium may ensue. An important route of spread of such puerperal infections is by thrombo-phlebitis starting in the uterine veins and extending to those of the utero-vaginal and pampiniform plexuses and to the hypogastric and iliac veins and even to the inferior vena cava.

(iii) *Thrombo-phlebitis of the portal veins or pylephlebitis.*—This fatal condition arises most usually as a complication of appendicitis, but occasionally it results from inflammation of other organs in the area of portal venous drainage, e.g. diverticulitis or inflamed piles. Starting in the appendicular or other tributaries of the portal vein, the thrombosis extends upwards into the larger veins, in some cases eventually into the portal vein itself, and its branches in the liver. The inflamed veins are distended by masses of infected thrombus and pus, and the liver contains multiple abscesses. Sometimes while the phlebitis is still restricted to the peripheral tributaries of the portal vein, emboli of infected clot are carried in the portal blood stream to the liver, producing multiple abscesses, a condition called *portal pyaemia*.

(2) The results of thrombo-phlebitis

The results of infective phlebitis vary according to the nature and severity of the bacterial infection, the rapidity of spread and the extent of the phlebitis, and the efficacy of chemotherapy or other treatment.   In many cases the infection is virulent and rapidly spreading, the vein walls are acutely inflamed, the clots, which are loaded with bacteria, soften and become purulent, and fatal pyaemia quickly ensues.   In other cases, with less virulent infections, extension of the disease is slower, the clots are firmer, and the patient survives for several weeks before fatal pyaemia supervenes.   Sometimes the infection is milder and is overcome by the patient with or without the aid of anti-bacterial remedies ; organization of the thrombi then follows.

(i) *Pyaemia.*—This old name literally means " pus in the blood ", a description which is not far from the truth.   The bacteria causing a phlebitis multiply in the masses of unorganized clot just as in a culture medium ; the infected clot grows by accretion and spreads along the vein lumen, exciting acute inflammation in the vein walls in contact with it ; polymorphonuclear leucocytes from the inflamed vein walls migrate into the infected clot in great numbers, the clot softens, breaks down and may be converted into frank pus ; fragments of softened purulent clot are broken off and carried away into the blood stream, and are arrested in the small vessels of remote parts, where they set up metastatic abscesses.   Thus pyaemia is best defined as the presence of fragments of infected clot (rather than " pus ") in the blood stream.   Pyaemic abscesses from thrombo-phlebitis in the portal venous system occur in the liver ; those from thrombo-phlebitis of systemic veins occur chiefly in the lungs, though some tiny emboli may traverse the pulmonary circulation, to pass on into the systemic arteries and set up metastatic abscesses in the kidney, myocardium, brain or other parts.

(ii) *Organization of intravenous thrombi.*—This, following a phlebitis which subsides, is similar to that following thrombosis from injury or other causes (see p. 27).   From the vein walls, the thrombus is invaded by new-formed granulation tissue, which eventually converts the vessel into a fibrous cord, or which may lead to its partial recanalization.

ARTERITIS

(1) Definition

The term " arteritis " is applicable to any true inflammation involving the walls of arteries.   It should not be applied to non-inflammatory hypertrophic and degenerative changes in arteries ; these common and important lesions are described in Chapter 26.

(2) The main general results of arteritis

Inflammations of arteries have results of two main kinds, namely (1) thrombosis and occlusion, and (2) weakening and aneurysm formation.

(i) *Thrombosis and occlusion.*—As might be expected, small arteries traversing inflamed tissue, or suffering inflammatory changes of their walls from other causes, often show complete thrombosis of the contained blood.   The thrombus in the occluded vessels later undergoes organization by in-growing blood-vessels and fibroblasts, sometimes with partial recanalization of the vessels.   The tissues supplied by the occluded arteries suffer diminution or loss of their blood supply (ischaemia) with resulting degenerative changes or death (see p. 372), unless an adequate collateral supply is established.

(ii) *Weakening and aneurysm formation.*—If the wall of an artery of large or medium size is weakened by inflammation, but its lumen remains patent, the thrust of the arterial blood pressure will produce a local bulging or a more general dilatation of the weakened wall. This constitutes an aneurysm, which may continue to enlarge until it finally bursts, producing a serious or fatal haemorrhage. This often happens in large unoccluded arteries traversing inflamed tissues, e.g. in an infected wound, in a tuberculous cavity in the lung, or in the floor of a gastric or duodenal ulcer ; it happens also with syphilitic inflammation of the aorta or other large arteries.

(3) The kinds of arteritis

Apart from the general similarity in their results just described—thrombosis and occlusion on the one hand, weakening and aneurysm formation on the other—the various kinds of arteritis have little in common. Let us list the named varieties and briefly consider each of them in turn. They are :

(i) Contiguity arteritis ;
(ii) Syphilitic arteritis ;
(iii) Rheumatic arteritis ;
(iv) Polyarteritis nodosa ;
(v) Thrombo-angiitis obliterans ;
(vi) " Temporal " arteritis.

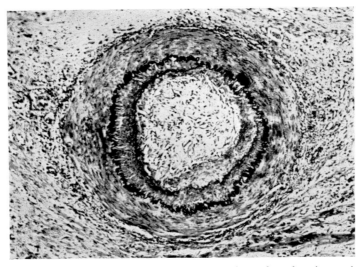

FIG. 33.—Endarteritis obliterans in an artery beneath a chronic gastric ulcer. (Elastic tissue stain.) (× 85.)

(i) *Contiguity arteritis.*—This is a convenient name for the inflammation of arteries traversing or lying alongside inflamed tissues. Either of the results referred to above can ensue. Many small arteries in the walls of inflamed wounds, abscesses or ulcers, or in chronic inflammatory masses, suffer occlusion, either by thrombosis or by endarteritis obliterans—a proliferative thickening of the intima which may itself go on to total occlusion of the vessel or in which occlusion by thrombosis may supervene (Fig. 33) ; thrombosis is followed by organization. But if a larger unoccluded artery is exposed in an inflamed wound, ulcer or cavity, inflammatory weakening of its wall may lead to localized dilatation and rupture (see Fig. 37). This is the usual source of serious haemorrhages from gastric and duodenal ulcers, inflamed accidental or surgical wounds, ulcerating tumours of the skin or mucous membranes, and tuberculous cavities in the lungs or kidneys.

(ii) *Syphilitic arteritis.*—This is described in Chapter 16. It may lead to occlusive endarteritis and thrombosis in small arteries, but in large arteries it is an important cause of weakening of the media and the development of aneurysms.

(iii) *Rheumatic arteritis.*—For details the reader is referred to p. 294.

(iv) *Polyarteritis nodosa.*—This is a rare febrile disease of unknown cause, probably allergic in nature, usually in young adults, with diffuse or nodular acute inflammation of many arteries, chiefly medium-sized and small arteries especially in the viscera. The vessel walls show areas of necrosis, fibrinous exudate, accumulation of neutrophil and eosinophil polymorphonuclear leucocytes, plasma cells, lymphocytes and macrophages. Some arteries show multiple small aneurysms which rupture and produce scattered patches of haemorrhage in the tissues; other vessels suffer occlusive thrombosis, with multiple small infarcts in the organs, and later organization of the thrombi. Veins adjacent to inflamed arteries show phlebitis. According to the main distribution of the lesions, the symptoms are very varied; they may be predominantly alimentary, cardiac, cerebral, muscular or renal. The disease is often fatal, from emaciation, anaemia, haemorrhages, and infarction of the heart, brain or other organs, but spontaneous recovery sometimes occurs.

(v) *Thrombo-angiitis obliterans (Buerger's disease).*—This is a non-febrile chronic disease of unknown cause, almost restricted to young and middle-aged men, and affecting chiefly the arteries of the limbs. The arteries show patchy inflammation, endarteritis obliterans, and much thrombosis, in which later organization and partial recanalization occur. The accompanying veins also show phlebitis and become matted to the arteries by fibrous tissue. The resulting ischaemia of the limbs causes intermittent claudication (lameness due to pain brought on by muscular activity), loss of arterial pulse, trophic changes and gangrene of the distal parts. Buerger's disease may not be a single entity.

(vi) *"Temporal" arteritis.*—This is a rare chronic disease of unknown cause, almost restricted to old people (over 60). The main symptoms are muscular and joint pains, stiffness of neck and jaw, painful mastication, persistent severe headache, and tender thickened temporal arteries—from which the disease derives its name. The lesions are not restricted, however, to the temporal arteries, but are found also in other arteries of the head and neck, in the cerebral and retinal arteries, and sometimes in those of the limbs and viscera. The changes include patchy infiltration of the media and adventitia by leucocytes, chiefly lymphocytes and macrophages, areas of necrosis in the media, collections of foreign-body giant cells (it has been called " giant-cell arteritis " and has even been mistaken for tuberculosis), with endarteritis obliterans and thrombosis in smaller arteries. Veins are unaffected.

SUPPLEMENTARY READING

Clawson, B. J., Bell, E. T., and Hartzell, T. B. (1926). " Valvular disease of the heart, with special reference to the pathogenesis of old valvular defects ", *Amer. J. Path.*, **2**, 193. (A beautifully illustrated account.)

Dible, J. H. (1958). " Organisation and canalisation in arterial thrombosis ", *J. Path. Bact.*, **75**, 1.

Gilmour, J. R. (1941). " Giant-cell chronic arteritis ", *J. Path. Bact.*, **53**, 263.

Hadfield, G., and Garrod, L. P. (1947). *Recent Advances in Pathology.* London ; Churchill. Chapter VIII. (A valuable account of rheumatic and bacterial endocarditis.)

Harrison, C. V. (1948). " Giant-cell or temporal arteritis ", *J. Clin. Path.*, **1**, 197.

Heptinstall, R. H., Porter, K.A., and Barkley, H. (1954). " Giant-cell (temporal) arteritis ". *J. Path. Bact.*, **67**, 507.

MacIlwaine, Y. (1947). " The relationship between rheumatic carditis and subacute bacterial endocarditis ", *J. Path. Bact.*, **59**, 557.

Rhoads, C. P. (1927). " Vegetative endocarditis due to meningococcus ", *Amer. J. Path.*, **3**, 623.

Von Glahn, W. C. (1926). ' Auricular endocarditis of rheumatic origin ", *Amer. J. Path.*, **2**, 1.

CHAPTER 8

INFLAMMATION AND ITS SEQUELAE IN PARTICULAR SITES : THE MENINGES

THIS CHAPTER is concerned with *suppurative meningitis;* tuberculosis, syphilis, torulosis and virus infections of the meninges will be dealt with later. In sup-purative meningitis, the usual acute inflammatory changes take place in the pia mater and arachnoid, and the exuded inflammatory fluid and emigrated leucocytes are poured out into the cerebrospinal fluid, the pressure of which is raised, and which becomes turbid from admixed leucocytes, bacteria and flakes of fibrin. Much of the fibrin of the exudate is deposited on the meningeal surfaces especially about the base of the brain ; and the deposits around the foramina of Luschka and Magendie impede the outflow of cerebrospinal fluid from the ventricles and cause them to become distended (internal hydrocephalus). As a result of this damming up of infected fluid in the ventricles, the ependyma and choroid plexuses become inflamed, and the exudate from this acute ependy-mitis still further increases the intra-ventricular pressure. Since the perivascular Virchow-Robin spaces of the brain and cord are prolongations of the subarach-noid spaces, they also share in the inflammation and are occupied by inflammatory exudate, so that in sections of the cerebral tissues the vessels are seen to be surrounded by cuffs of leucocytes.

THE CAUSES OF SUPPURATIVE MENINGITIS

The bacteria which cause acute meningitis are the following :

(i) *Meningococcus.*—This is the cause of " epidemic " meningitis or " cerebro-spinal fever ", but it also causes isolated cases of meningitis especially in infants and children. It resides in the nasopharynx of carriers, whence it is transferred by inhalation to susceptible persons, in whom it reaches the meninges by way of the blood stream.

(ii) *Pyogenic streptococci and staphylococci.*—These reach the meninges chiefly by direct extension from suppurative lesions of the head or face, by far the most important of which is otitis media, less frequent sources being frontal or sphenoidal sinusitis, and infected wounds, boils or cellulitis of the face or scalp. Sometimes blood-borne streptococcal or staphylococcal infection of the meninges occurs from bacterial endocarditis, puerperal infection or some other septicaemic disease ; or from a metastatic abscess in the brain secondary to a pulmonary abscess or empyema (see Chapter 5).

(iii) *Pneumococcus.*—Pneumococcal meningitis arises either as a blood-borne metastatic infection following pneumonia, or as the result of direct extension from pneumococcal otitis media.

(iv) *Other bacteria.*—Bacteria which occasionally produce acute meningitis include *H. influenzae*, gonococcus, typhoid bacillus, and anthrax bacillus.

Once the meninges have become infected, by whatever route, by pathogenic bacteria, these multiply and are quickly carried by the circulating cerebrospinal fluid to infect other parts of the membranes. It is easy to understand, therefore, how in most cases widespread diffuse inflammation of the leptomeninges rapidly ensues, which in the most virulent infections, especially the streptococcal, is fatal in a few days if untreated. Anti-bacterial drugs, however, greatly modify the course of the disease, reducing a virulent acute inflammation to a subacute or chronic one or curing it completely.

PATHOLOGICAL ANATOMY OF SUPPURATIVE MENINGITIS

(1) Naked-eye appearances

The appearances of the brain and meninges vary somewhat according to the acuteness of the disease and the bacterium responsible. In hyperacute or fulminant cases (chiefly meningococcal), in which the patient dies within 24 or 48 hours of the onset, the chief changes are hyperaemia and oedema of the meninges and brain surface, and there may be little or no visible fibrinous exudate. This applies also to anthrax meningitis, which kills very quickly and which is prominently haemorrhagic.

However, in most cases of more than 48 hours duration, a deposit of fibrinous or fibrino-purulent exudate is plainly visible on the cerebral and meningeal surfaces. It forms an opaque grey, yellow or greenish layer, which obscures the vessels, nerves and sulci, and is especially abundant about the brain stem, the basal aspect of the cerebrum, the cerebellum, and the dorsal side of the spinal cord. The deposits around the brain stem may plaster over and occlude the foramina of exit of the fourth ventricle. The cerebrospinal fluid, both in the arachnoid spaces and in the ventricles is turbid or in advanced cases frankly pus-like. The ventricles are distended, and the ependymal surfaces and choroid plexuses are inflamed.

(2) Microscopic appearances

The exudate on the surfaces of the brain, cord and meninges consists of fibrin with entangled polymorphonuclear leucocytes, accompanied in later stages by lymphocytes and macrophages. Bacteria are usually numerous, and are found both within leucocytes and free. The Virchow-Robin spaces around the vessels entering the brain and cord from the meninges are filled and distended by leucocytes and fibrin, and the perivascular " cuffs " of inflammatory cells may extend deeply into the nervous tissues. In cases of long duration, e.g. those kept alive by means of anti-bacterial drugs, pronounced inflammatory changes, arteritis and phlebitis, may occur in the vessels of the meninges and brain. Toxic degenerative changes take place in many of the nerve cells of the cerebral cortex.

(3) Changes in the cerebrospinal fluid

Most of the changes in the fluid are what might have been predicted from those in the meninges. They are as follows :

(i) *Pressure.*—This is increased to 300 millimetres of water or more (normal 80–180 millimetres), and the fluid flows rapidly from a lumbar puncture needle.

(ii) *Appearance.*—The fluid is turbid or obviously purulent.

(iii) *Cytology.*—Many polymorphonuclear leucocytes are present.

(iv) *Bacteriology.*—The responsible bacteria are easily found in direct smears or can be grown on culture media.

(v) *Proteins.*—Serum proteins (of which the normal fluid contains only a trace) are greatly increased ; tests for globulin (e.g. Nonne-Apelt test with saturated solution of ammonium sulphate) are strongly positive.

(vi) *Chlorides.*—Normally about 750 milligrams per 100 cubic centimetres, these are reduced.

(vii) *Sugar.*—Normally about 100 milligrams per 100 cubic centimetres, this is greatly reduced, usually absent, probably because it is used up by the bacteria during their growth.

In suppurative meningitis, the chemical changes (v)–(vii) are of no great interest in diagnosis, because the other changes in the fluid make the diagnosis obvious. In certain other diseases, however, the chemical tests are of diagnostic value, so that it is always as well to include them in considering the cerebrospinal fluid in any disease (compare with pp. 147 and 261).

TERMINATION AND RESULTS OF SUPPURATIVE MENINGITIS

(i) *Death in the acute stage.*—With the exception of meningococcal meningitis, almost all untreated cases of acute bacterial meningitis die within a few days from a combination of raised intracranial pressure, central nervous damage and toxaemia. Even in rapidly fatal cases, some internal distension of the ventricles is often apparent.

(ii) *Chronic internal hydrocephalus.*—A patient may survive the acute stage of meningococcal meningitis (or of other kinds of bacterial meningitis if treated with suitable anti-bacterial substances) ; but he may gradually develop during subsequent weeks or months progressive internal hydrocephalus, due to obstruction of the foramina of Magendie and Luschka by fibrous adhesions resulting from organization of fibrin deposits around the brain stem and in the cisterna magna. If the patient is a child and the sutures still ununited, he may develop a huge head, with a greatly distended brain, and paper-thin calvarial bones widely separated by membranous tissue only (p. 632).

(iii) *Recovery.*—From meningococcal meningitis, especially if treated, and occasionally from other kinds of bacterial meningitis, complete recovery without hydrocephalus may take place. Some recovered patients, however, have permanent residual sequelae due to injury to cranial nerves, e.g. deafness, facial paralysis, or squint.

SUPPLEMENTARY READING

References Chapter 10 (p. 112), and the following :

Biggart, J. H. (1949). *Pathology of the Nervous System. A Student's Introduction.* Edinburgh and London; E. & S. Livingstone. Chapter V.

Cairns, H., and Russell, D. S. (1946). " Cerebral arteritis and phlebitis in pneumococcal meningitis ", *J. Path. Bact.*, **58,** 649.

Cumings, J. N. (1954). " The cerebrospinal fluid in diagnosis ". *Brit. med. J.*, **1,** 449.

CHAPTER 9

INFLAMMATION AND ITS SEQUELAE IN PARTICULAR SITES: MUCOUS MEMBRANES

MOST INFLAMMATIONS of mucous membranes are due to bacteria or other microbes entering them directly from their surfaces. These microbes are of many kinds, e.g. streptococci, pneumococci, the viruses of influenza and the common cold, typhoid and paratyphoid bacilli, dysentery bacilli, *Entamoeba histolytica*, tubercle bacillus and leprosy bacillus—to name only a few. In this chapter we are not concerned with the many specific bacterial infections affecting mucous membranes, but only with those general features of mucosal inflammation which depend on the structure and functions of these membranes. Such features may be considered conveniently under the following headings :

(1) Catarrh, non-ulcerative inflammation with a free mucoid discharge.
(2) Peculiarities of mucosal ulceration, with special reference to ulcers of the stomach and duodenum.
(3) The extent of mucosal inflammations.
(4) Obstruction of ducts and viscera as it affects inflammation in them.
(5) The defensive properties of epithelial surfaces.

CATARRH

A catarrh is an inflammation of a mucous membrane characterized by a free mucus-containing discharge. The mucous membranes lining most of the alimentary canal, most of the respiratory tract, and parts of the uro-genital tracts

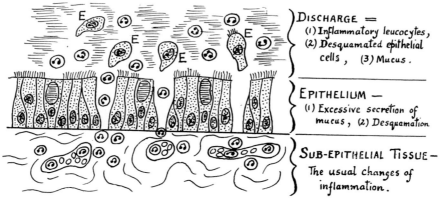

DISCHARGE =
(1) Inflammatory leucocytes,
(2) Desquamated epithelial
   cells , (3) Mucus .

EPITHELIUM —
(1) Excessive secretion of
   mucus, (2) Desquamation

SUB-EPITHELIAL TISSUE—
The usual changes of
inflammation.

FIG. 34.—Diagram of catarrh.

(e.g. urethra, cervix uteri) are clothed by epithelium which is itself mucus-secreting or possesses special mucous glands. When these mucous membranes are inflamed, they secrete excessive mucus from both their mucous glands and their surface goblet cells. Many of the surface epithelial cells are also detached and cast off in the discharge, which thus consists of a mixture of inflammatory exudate and

91

leucocytes from the sub-epithelial tissues, and mucus and desquamated cells from the mucosal epithelium (Fig. 34). In the early stages of a catarrhal inflammation the discharge is predominantly mucoid or muco-serous, later it becomes yellow and muco-purulent because of the increasing amount of admixed inflammatory exudate and leucocytes. The common cold or acute rhinitis has made everyone familiar with these features of catarrhs. Essentially similar changes occur in uncomplicated catarrhs in other situations—bronchitis, laryngitis, otitis media, sinusitis, conjunctivitis, gastritis, enteritis, colitis, cholangitis, cervicitis and urethritis.

ULCERATION OF MUCOUS MEMBRANES, WITH SPECIAL REFERENCE TO " PEPTIC " ULCERS

Some characteristically ulcerative infections of mucous membranes are caused by specific microbic parasites, e.g. the tubercle bacillus, typhoid and paratyphoid bacilli, dysentery bacilli, and *Entamoeba histolytica;* these will be dealt with in later chapters.

But, in addition to these grossly ulcerative specific diseases, mucous membranes, especially of the alimentary tract, often show small evanescent ulcers from trauma or from mild non-specific infections. We are all familiar with such abrasions or small infective ulcers in the mouth or throat, and at necropsy small ulcers or " erosions " are often to be seen in the gastric or intestinal mucosa. Most of these small breaches heal up promptly ; but, under unfavourable conditions, they may fail to do so and may give rise to serious progressive lesions. Such unfavourable conditions are of two kinds, (*a*) virulent microbes, e.g. pyogenic cocci or tubercle bacilli, which happen to be present may gain an entry to the tissues through the little erosion and cause characteristic lesions, or (*b*) the local conditions may be unfavourable to healing, so that the little erosion enlarges and becomes a chronic ulcer. The commonest instance of (*b*) is chronic ulceration of the stomach or duodenum, in which the main factor preventing healing is the acidity of the gastric secretion. Let us consider this very important lesion more fully.

(1) Acute peptic ulcers of stomach and duodenum

Acute ulcers of the stomach are often found at post-mortem examination of persons dying of infective and toxic diseases. They occur anywhere in the organ, are usually multiple, and may be very numerous and tiny, forming widespread haemorrhagic erosions. The larger ones, which clearly develop by extension of minute erosions, measure a few millimetres or even a centimetre or more in diameter ; they are oval or circular, have shelving edges, and usually involve the mucosa and submucosa only. Despite their small size and superficial nature, they may cause severe, even fatal, haemorrhage from eroded arteries ; and sometimes they extend through the muscle coat and perforate into the peritoneal cavity. The exact mode of origin of acute gastric ulcers is uncertain ; they may start as foci of infective thrombosis or embolism in the blood vessels of the mucosa. There is no doubt that most acute ulcers heal rapidly, but a few persist and become chronic ulcers.

(2) Chronic ulcers of stomach and duodenum

(i) *Situation.*—By far the majority of chronic ulcers start in one or the other of two " ulcer-bearing areas ", namely (i) a narrow strip of gastric wall along the lesser curvature from just below the cardia to a point about 3 centimetres above the pylorus, and (ii) the anterior and posterior walls of the first part of the duodenum (Fig. 35). The pylorus itself is the site in only a small proportion of cases—about 5 per cent. The greater curvature, fundus and cardia are rare initial sites of chronic ulcers. Two or more chronic ulcers are not unusual ; these may be gastric only, or duodenal only (" kissing " ulcers of the anterior and posterior walls are common), or coexisting gastric and duodenal ulcers.

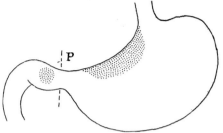

Fig. 35.—The main sites of "peptic" ulcers.

(ii) *Relative frequencies.*—Chronic ulcer is more frequent in the duodenum than in the stomach. Gastric ulcer is almost equally common in men and women, but duodenal ulcer is much more frequent in men than in women. Ulcers are rare in childhood and adolescence ; the incidence is highest in the third and fourth decades.

(iii) *Structure.*—Chronic duodenal ulcers are round or oval. So, at first, are gastric ulcers ; but as these enlarge they often extend onto the anterior and posterior gastric walls to become saddle-shaped or irregular. Duodenal ulcers are seldom more than 1 or 2 centimetres in diameter ; gastric ulcers may attain much larger sizes, up to 10 centimetres or more in diameter. The characteristic structure of chronic gastric ulcers is shown in Figs. 36 and 37. The main naked-eye features are the punched-out crater with thickened over-hanging edges, the smooth or slightly irregular congested floor which is often covered by a thin grey layer of necrotic debris, the complete destruction of the muscle coat which ends abruptly at the sides of the crater, and the thick layer of dense fibrous tissue forming the base of the ulcer and extending for some distance in the surrounding tissues. The fibrous floor of a large ulcer often lies, not in the wall of the stomach itself, but in the substance of neighbouring parts—pancreas, liver or abdominal wall—to which the stomach has become adherent as the ulcer enlarged (Fig. 37). Microscopically, the chronically inflamed tissues in the sides and base of the ulcer show, in addition to extensive fibrosis, inflammatory oedema, collections of leucocytes, especially lymphocytes and plasma cells and often many eosinophil leucocytes, and obliterative changes in blood vessels, both endarteritis obliterans and thrombosis with organization. In cases in which there has been recent haemorrhage, dilated or ruptured arteries are often plainly visible in the ulcer floor (Fig. 37).

(iv) *Healing and relapse.*—Chronic ulcers frequently heal, either spontaneously or following treatment. We know this, not only from the histories of patients and from gastroscopic examinations, but also because in surgically excised stomachs or at necropsy we often find healed gastric scars with puckered areas of fibrosis replacing the muscle coat and covered by flat or irregular mucosa

devoid of normal folds.   In the duodenum such scars are often accompanied
by bulging hemispherical pouches due to mechanical distension of the walls
between the fibrous scarred areas.   That healed ulcers may break down again
is shown both by the histories of patients and by the surgical or necropsy finding
of recent ulceration in old scarred areas.

FIG. 36.—Chronic gastric ulcer.   Note extensive fibrosis in walls and floor.   M, muscle coat;
V, thrombosed vessels.   (× 5.)

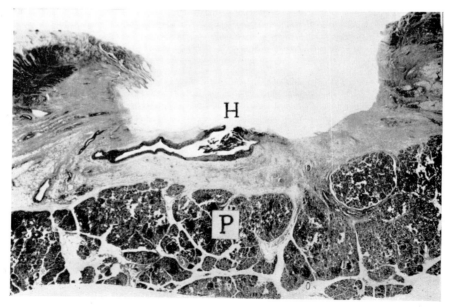

FIG. 37.—Large chronic gastric ulcer with fibrous base adherent to pancreas P.   Fatal
haemorrhage occurred from rupture of large artery at H.   (× 4.)

(v) *Complications.*—(i) *Haemorrhage,* manifested by haematemesis or melaena or both, is the commonest serious complication of chronic ulcer. It comes from the rupture of eroded unobliterated aneurysmal arteries exposed in the floor of the ulcer. These arteries may be small unnamed branch vessels, in which case the bleeding may be of only moderate amount and may cease by closure and organization of the burst vessel ; or they may be large main arteries such as the right or left gastric, gastro-duodenal or splenic artery, in which case the haemorrhage is often fatal. (ii) *Perforation into the peritoneal cavity,* more frequent with duodenal than with gastric ulcer, is a serious complication. Sudden severe symptoms are caused by the spilling of acid gastric contents into the peritoneal cavity, where acute peritonitis is set up, first by the chemically irritant fluid, later by bacterial infection. (iii) *Perforation into neighbouring organs* occasionally occurs into the gall-bladder, bile duct, transverse colon, small intestine, or even into the pericardium from an ulcer high on the lesser curvature which has penetrated through the diaphragm. These rare complications are seen only with large deeply eroding ulcers the floors of which have extended beyond the stomach walls. (iv) *Obstruction to the passage of gastric contents* occurs usually in the first part of the duodenum or at the pylorus, where it is due partly to muscular spasm and partly to fibrotic narrowing and results more often from duodenal than from pyloric gastric ulcer. Obstruction occasionally develops in the stomach itself, due to hour-glass constriction resulting from a large saddle-shaped ulcer of the lesser curvature. (v) *Carcinoma* occasionally arises at the edge of a chronic gastric ulcer (see p. 492).

(3) Chronic peptic ulcers in other situations

Chronic ulcers similar to those of the stomach and duodenum sometimes arise in other situations, *always in acid-bathed mucous membranes.* These specially situated ulcers are as follows :

(i) *Gastro-jejunal or jejunal ulcer.*—Following the operation of gastro-jejunostomy for duodenal ulcer, it is not uncommon for a chronic ulcer to develop at the margin of the orifice between the stomach and jejunum (stomal ulcer), or in the jejunum just opposite the stoma or a short distance down its distal loop. Such anastomotic ulcers are especially apt to develop in patients with hyperchlorhydria ; they rarely occur after gastro-jejunostomy for gastric ulcer or cancer. These facts show plainly the importance of the acidity of the gastric secretion in their causation. A gastro-jejunal ulcer, like a gastric one, may bleed or perforate into the peritoneal cavity, and it is apt also to adhere to neighbouring structures, to form a large inflammatory mass around a relatively small ulcer, and to perforate into the transverse colon forming a gastro-colic or gastro-jejuno-colic fistula.

(ii) *Oesophageal ulcer.*—Chronic peptic ulceration of the oesophagus may occur in association with a gastric or duodenal ulcer or independently. The ulcer is usually single, and it is always situated at the lower end of the oesophagus. It may perforate into the mediastinum, pericardium or peritoneum, or may cause fatal haemorrhage from the aorta ; often it heals, leaving a fibrous stricture.

(iii) *Peptic ulcer in a Meckel's diverticulum.*—This developmental diverticulum, which is a residue of the intestinal end of the vitello-intestinal duct of the embryo, is sometimes lined, not by intestinal mucosa, but by gastric mucosa with a characteristic structure and acid secretion.   Peptic ulceration may occur in this and the ulcer may bleed or may perforate.   One of the commonest causes of otherwise symptomless bleeding from the bowel in a child or adolescent is an ulcerated Meckel's diverticulum.   We may note here that vitello-intestinal remains may persist also at the umbilicus in the form of a closed cyst or an open pocket of mucous membrane.   Occasionally this cyst or pocket is lined by gastric mucosa, from which an excoriating acid secretion is discharged onto the skin.

(4) The causes of chronic peptic ulceration

There is no doubt that chronic ulcers start as acute ulcers which then fail to heal.   Our problem, then, is two-fold—to discover the causes of acute ulcers, and to discover the factors which prevent their healing.   As already stated, the causes of acute ulcers are uncertain, though there is no doubt that some of them at least are related to infective and toxaemic diseases.   But, once established, why do some of these persist and become chronic ulcers ?   Four possible factors call for discussion, namely bacterial infection, mechanical and thermal injury, the acid gastric secretion, and disturbed innervation of the stomach.

(i) *Bacterial infection.*—Some workers have suggested that peptic ulceration is due to a specific streptococcal infection arising from infective foci in the teeth or tonsils, and have claimed in support of this view the improvement that follows removal of such foci.   There is no real evidence to uphold this hypothesis.

(ii) *Mechanical and thermal injury.*—There is no doubt that the stomachs of " civilized " men are subjected to repeated physical and thermal insults.   Many of us gobble our food down in an unmasticated state, along with sharp vegetable husks and seeds and fragments of bone ; we cheerfully swallow our food and drinks at temperatures of 60° or 70° C., which would badly scald our skin ; some of us add chemical insult to injury in the form of pungent spices, strong solutions of alcohol and drugs of untold variety.   It has been suggested that the ulcer-bearing strip along the lesser curvature of the stomach is subjected to such injuries to a greater degree than other parts of the organ.   This may be the case, especially as regards the temperature factor ; but the dietetic-insult hypothesis is less plausible as an explanation of why the first part of the duodenum should be so prone to ulceration.

(iii) *Acid gastric secretion.*—The importance of the acidity of the gastric secretion as a factor in preventing the healing of ulcers is shown by many facts. The chief of these are : (*a*) that peptic ulcers occur only in areas of mucosa bathed by gastric secretion ; (*b*) that hyperchlorhydria is present in the great majority of patients with duodenal ulcers and in a smaller proportion (about one-third) of patients with gastric ulcers ; (*c*) that the symptoms of ulceration are relieved and that healing is encouraged by the administration of alkalis or by other methods which neutralize or minimize gastric acidity ; and (*d*) that, while experimental injuries to the gastric or duodenal mucosa in animals normally heal quickly, they can be converted into chronic ulcers by artificially raising the gastric acidity in various ways.   There is no doubt, then, that hyperchlorhydria is a potent factor in preventing the healing of gastric and duodenal ulcers, once these are established.

(iv) *Disturbed innervation of the stomach.*—Cushing and others have shown that stimulation of the hypothalamus may cause haemorrhagic gastric erosions or ulcers, an effect attributable to over-action of the vagus, the secretory and vasomotor nerve of the stomach. We have good evidence that there exists a special type of person, usually a male, with a predisposition to peptic ulceration —a type with hypersecretion, hyperchlorhydria and hypermotility with rapid emptying of the stomach—characters that again point to over-action of the vagus or " vagotonia ". It is also well known that emotional disturbances greatly alter gastric secretion and motility ; and this makes plausible the suggestion that the great frequency of peptic ulcers in modern civilized communities, especially among doctors and other busy professional people, is due in part to mental stress and hurry. All of the foregoing facts and ideas hang well together and point strongly to the importance of disturbances of nervous (especially vagal) control in the causation of gastric hypersecretion and of chronic peptic ulceration.

Briefly then, we may state our still incomplete knowledge of the causation of peptic ulceration as follows. We do not know for certain the initial causes of the erosions and acute ulcers which may later persist as chronic ulcers ; they may well be due to physical, thermal or chemical injuries or to small infective or vascular lesions. We do know that, given a breach in the gastric or duodenal mucosa, persistent hyperchlorhydria can prevent its healing and can convert it into a chronic ulcer; and there are strong grounds for believing that disturbances of gastric innervation are often primarily to blame for such hyperchlorhydria.

Finally, it may be noted that the adjective " peptic " which is often applied to gastric and duodenal ulcers, though valuable in that it focuses attention on the gastric secretion, is not strictly correct. It is the acidity, not the peptic activity, of this secretion that matters.

THE EXTENT OF MUCOSAL INFLAMMATIONS

Bacterial or virus inflammation of a particular mucous membrane often extends to the mucous membranes of neighbouring parts. Such extension is largely due to progressive surface infection by the microbes which have multiplied in and been discharged from the initially infected area. In the familiar complications of the common cold or coryza, we all know how an infection starting as a tonsillitis or rhinitis may later spread to produce also laryngitis and bronchitis, or how it may spread to the para-nasal sinuses, usually one or both of the maxillary sinuses, or to the Eustachian tube and middle ear. Gonococcal infection in the male may at first be limited to the anterior urethra, but may later extend to the posterior urethra, prostate and even to the ductus deferens and epididymis. In the female, it often extends from the cervix to the endometrium and Fallopian tubes, there producing acute or chronic endometritis and salpingitis.

However, the spread of any given infection over neighbouring mucous membranes is usually limited to a definite territory. Each kind of bacterium or virus has its own favourite habitat beyond which it seldom extends. Thus, while the viruses and bacteria of common colds readily spread to and infect the whole of the upper respiratory passages, they do not spread from here to produce inflammation of the oesophagus or stomach. Gonococci may inhabit and inflame the whole male or female genital tract, but they seldom spread into the bladder. Other bacterial species also show similar preferences for particular habitats in mucous membranes ; e.g. the typhoid bacillus and the tubercle bacillus colonize the ileum much more heavily than either the jejunum or large intestine, while the dysentery bacilli and *Entamoeba histolytica* restrict their attack almost exclusively to the mucosa of the large intestine.

It is important to appreciate that the bacteria or viruses causing inflammation of a mucous membrane are usually not confined to the surface only but that they often penetrate also into crypts and glands connected with it.   For example, the streptococci causing a tonsillitis do not remain only on the tonsillar surface where the redness and exudate are most obvious on inspection ; they also colonize and multiply in the depths of the tonsillar crypts, where indeed they are more dangerous, since it is from here that they may invade the underlying tissues to cause an abscess (called in this region a " quinsy ") or a cellulitis.   So also, gonococci in the male genital tract are not restricted to the urethral surfaces, but often extend also into the urethral glands of Littré, into the bulbo-urethral glands of Cowper, and into the prostate and seminal vesicles.   Gonococcal infection of the female genital tract involves not only the vaginal and cervical surfaces, but also the urethra and urethral glands, the mucous glands of the cervix and sometimes Bartholin's glands.

Glandular and crypt infection is of special importance because the bacteria are relatively sheltered here ; they are not washed away in the catarrhal discharge as readily as those on the surface, and antiseptic irrigations or antibacterial drugs applied to the surface do not reach them.   Hence they are likely to lurk here long after the surface infection has cleared up, causing smouldering inflammation in glands and ducts, and recurrent inflammation of the whole mucous membrane when suitable conditions arise.   Lurking gonococcal infection in the prostate or Cowper's glands, or in Bartholin's or the cervical glands, is the main factor causing chronic and recurrent gonorrhoea.

OBSTRUCTION OF DUCTS AND VISCERA AFFECTING INFLAMMATIONS IN THEM

In ducts or in narrow viscera, such as the appendix or Eustachian tube, otherwise simple catarrhal inflammations are often complicated and rendered much more serious by obstruction of the lumen.   Such obstruction may result from inflammatory swelling of the mucosa itself or from foreign bodies in the lumen.   The result is a damming-up of the infected contents and inflammatory exudate in the organ, to form an abscess in its distended cavity—sometimes called an *empyema*. Some examples will explain the changes.

(a) Eustachian obstruction with otitis media

We are all familiar with the " blocked-up " feeling in the ears which often accompanies a bad cold.   This is due to inflammatory swelling of the Eustachian orifice produced by the pharyngitis.   If the infection spreads into the Eustachian tube and middle ear, the mucosal swelling in the narrow tube prevents free discharge of the catarrhal exudate from the middle ear into the pharynx, the exudate accumulates in the middle ear, and earache results. If the infection is a mild one, the inflammation may subside and the accumulated discharge may escape into the pharynx without further mishap. But if the infection is a more virulent one, especially streptococcal or pneumoccocal, the obstruction caused by the Eustachian mucosal swelling grows worse, muco-pus accumulates under pressure in the middle ear and mastoid antrum, forming an abscess or empyema in this cavity, and the resulting severe earache is relieved only when the pus escapes through a spontaneous rupture or surgical incision of the tympanic membrane.

(b) Obstructive appendicitis

Acute appendicitis is a common and important disease, due usually to streptococci, colon bacilli and anaerobic bacteria from the bowel contents. In the absence of obstruction of the lumen of the appendix, the inflammation may be relatively mild and restricted to the mucous membrane—*catarrhal appendicitis*— but it may extend to the other coats and produce also local fibrinous peritonitis. If, however, as is often the case, the lumen becomes obstructed either by inflammatory swelling of the mucosa or more often by faecal concretions, free discharge into the caecum is prevented, so that *obstructive appendicitis* results. The appendix becomes distended with pus or a foul mixture of faecal matter and pus—*empyema*— very often the organ becomes gangrenous and perforates—*gangrenous appendicitis*. *Perforation* through a gangrenous spot sometimes occurs very rapidly and causes serious widespread peritonitis ; but if it takes place less rapidly, and if the appendix is already surrounded by peritoneal adhesions from previous attacks of inflammation, a local *appendicular abscess* often results. If the bacterial infection of an obstructed appendix is mild, the organ may gradually become distended to a large size by mucopurulent fluid or almost pure mucus. forming a *mucocele*. This is one form of *chronic appendicitis*, which may result also from persistent smouldering infection in a fibrosed or kinked organ which has been the seat of previous acute attacks. Needless to say, such a smouldering infection is apt to flare up at any time into an acute attack, chiefly as the result of obstruction of the lumen. The names which have been italicized in this paragraph do not denote distinct diseases, but only different forms and results of appendicitis, —forms and results which depend partly on the virulence of the bacteria present but also very largely on whether or not the lumen of the organ is obstructed.

(c) Obstructive cholecystitis and cholangitis

What has just been said of the appendix may be applied also to the gall-bladder. This organ is liable to bacterial infection ; resulting inflammation may be catarrhal, suppurative, gangrenous or perforative ; and obstruction of its neck or of the cystic duct by impacted gall stones, or less often by fibrous stenosis, is an important factor which frequently aggravates an otherwise mild infection. As in the appendix also, very mild inflammation coupled with obstruction of the outflow may result in slow distension of the gall-bladder by mucus— a mucocele. Usually, however, if bacterial infection is present, obstruction leads to acute suppurative or gangrenous cholecystitis, requiring urgent operation. Obstruction of the common bile duct by gall stones may similarly lead to ascending bacterial infection of the hepatic ducts—ascending cholangitis—and to multiple abscesses in the liver.

(d) Diverticulitis of the colon

It is quite common to find in the descending and pelvic colon of the middle-aged or old person multiple small pouches of mucosa bulging through the muscle coat into the surrounding fatty tissue—an acquired condition called *diverticulosis*. As time goes on, the diverticula enlarge and hard pellets of faecal matter collect in them. Bacterial infection sometimes supervenes in one or several of these pouches, and if the mouths of the inflamed pouches become obstructed by the faecal pellets, the pouch walls may perforate and acute inflammation may extend into the surrounding fatty tissues or peritoneum, producing a local abscess or a peritonitis. Or, chronic diverticulitis may lead to fibrous thickening of the bowel wall, adhesions, and intestinal obstruction ; and the disease may closely simulate cancer of the colon, both clinically and in the naked-eye appearance of the bowel.

(e) Pyosalpinx and hydrosalpinx

These mean distension of the Fallopian tubes by pus or watery fluid respectively. They result from bacterial infections, usually gonococcal but sometimes streptococcal, which have caused obstruction of the narrow uterine ostium by inflammatory swelling or by fibrosis and of the fimbriated orifice by adhesions, so that inflammatory exudate accumulates in the closed-off tubal lumen. Hydrosalpinx results from mild non-purulent infections ; pyosalpinx from more severe ones. Sometimes an abscess is not confined within the tube alone, but lies in a cavity formed by adhesions between the tube, broad ligament and ovary—a *tubo-ovarian abscess*. The gonococci or non-haemolytic streptococci which are the usual causes of salpingitis tend to die out in an old chronic pyosalpinx, the pus from which is therefore often inspissated and sterile.

(f) Other obstructions

The principle exemplified in the foregoing instances—namely that obstruction in ducts or narrow viscera dams up infective exudates and often aggravates the inflammation—is of wide application in many other parts of the body—in the salivary glands, where duct blockage by a salivary calculus leads to abscess ; in the para-nasal sinuses, free discharge from which is often impeded by inflammatory swelling of the mucosa around their orifices ; in the lungs, where bronchial obstruction often leads to suppurative pneumonia or abscess ; in the breast, where bacterial mastitis often causes abscesses because the multiple small ducts do not give infective discharges a free exit ; in the urinary tract, where obstruction by calculi, inflammatory strictures or prostatic disease often leads to serious ascending infection of the bladder, ureters and kidneys—ascending pyelo-nephritis ; in the uterus, where blockage of the cervical canal by a tumour or by inflammatory fibrosis may cause distension of the organ by pus—pyometra. The present chapter should be read in conjunction with Chapter 37, in which other effects of obstruction in hollow organs are discussed.

THE DEFENSIVE PROPERTIES OF EPITHELIAL SURFACES

The skin, conjunctiva and mucous membranes form a mechanical barrier to the entry of bacteria. Skin secretions are bactericidal to some bacterial species;

and tears, and to a less degree, nasal mucus and saliva, contain a bactericidal enzyme called *lysozyme*. In the nose, mouth and respiratory tract, mechanical removal of organisms is brought about by the flow of lachrymal secretion, mucus and saliva, by the movement of cilia, and by coughing; and normally the contents of the bronchi are sterile. In the alimentary canal, the plentiful secretion of mucus probably plays an important part in preventing microbic infection of its mucosae (*see* Florey). Gastric acidity rapidly kills many bacteria, but not the tubercle bacillus, bacterial spores, or enteric and dysentery bacilli. Following partial gastrectomy, *Staph. aureus*, *E.coli* and *Cl.Welchii* may all be present in the stomach remnant. Peristalsis and mucus secretion in the small intestine help to minimize its bacterial flora. In the vagina, the acid reaction maintained by *Lactobacillus acidophilus* (p. 172) inhibits the growth of many other bacterial species.

Injuries or other pathological changes often interfere with the skin or mucosal barrier and allow bacteria to gain access to the underlying tissues. Burns, cuts and abrasions may become infected; the inhalation of irritant smokes or fumes, or virus infections of the respiratory tract such as influenza or the common cold, may be complicated by bacterial infections; and Vitamin–A deficiency (p. 362) may cause structural changes in mucous membranes and so predispose to bacterial infection.

SUPPLEMENTARY READING

Florey, H. W. (Ed.) (1954). " The secretion of mucus and inflammation of mucous membranes ". In *Lectures on General Pathology*. London; Lloyd-Duke.
Gordon-Taylor, G. (1937). " The problem of the bleeding peptic ulcer ", *Brit. J. Surg.*, **25**, 403. (Some good illustrations of ulcers with burst blood vessels.)
Hurst, A. F., and Stewart, M. J. (1929). *Gastric and Duodenal Ulcer*. London; Oxford University Press. (A full and well-illustrated account.)
Magnus, H. A. (1946). " The pathology of simple gastritis ", *J. Path. Bact.*, **58**, 431.
Stewart, M. J. (1931). " Precancerous lesions of the alimentary tract ", *Lancet*, **2**, 617. (Contains much information about gastric ulcers.)

CHAPTER 10

THE PATHOGENIC BACTERIA : INTRODUCTION

THE PREVIOUS chapters described the general phenomena of inflammation and the ways in which these are modified by the nature of the tissues affected. We have now to consider one by one the particular bacterial and other microbic parasites, and the special characters of the injuries they inflict and the inflammatory reactions they evoke. We will consider only the main features of the structure, cultural requirements, metabolism and pathogenicity for animals of the various micro-organisms, and a simple classification convenient for the clinician and pathologist.

Bacteria may be found almost anywhere. They are present in the air, in all kinds of soil, from desert sands to the mud of the ocean bed, and in all types of water from stagnant marshes to fresh springs. They are present in food and refuse heaps; on the streets, on floors, and ceilings; on plants, on the body surface, and in the visceral cavities of man and animals. Almost no object can be thought of that does not harbour bacteria.

Some bacteria are always present in a particular place, and constitute what is called its *normal bacterial flora*. Thus, soil, water and food all have a natural flora. The human body too has its normal bacterial inhabitants, but, since the structure of the body is so diverse, the normal flora of one part differs from that of another. As would be expected, in some parts of the body there are normally no organisms present and such situations are said to be *sterile*. The organisms which make up the normal body floras are for the most part harmless, but may cause disease under certain circumstances.

The fact that some bacteria are able to produce diseases in man and animals whilst others are not, has led to their subdivision into two great classes, namely a *saprophytic* group and a *pathogenic* group. The great majority of bacteria are free-living saprophytes and do not cause disease. Thus, in the genus *Bacillus*, 25 species are recognized, only one of which (the anthrax bacillus) is a human pathogen; and in the genus *Pseudomonas*, *Ps. aeruginosa* is the only pathogen for man amongst the 149 described species.

Saprophytes go to make up the normal floras of a variety of situations and are often not only harmless but are essential for the existence of other forms of life. Saprophytic organisms which form part of the normal body floras are called *commensals* when they are harmlessly present in their normal environment. Under certain circumstances, however, some commensal bacteria may cause disease, and these are referred to as *potential pathogens*.

The pathogens, a relatively small group of bacteria, are those capable of producing disease. Many of these bacteria produce diseases in both man and animals, whilst others produce disease in one or other only. An organism which is pathogenic for a single species such as man may be called an *obligate parasite* for man, whilst those organisms which normally can exist only in the presence of living tissue are called *strict parasites*. Thus, parasitic bacteria are those which find conditions most favourable to growth in association with living cells, while saprophytes live on dead organic matter, or even on inorganic matter. While all pathogenic organisms are parasites, there are many species of parasitic bacteria which are not necessarily disease-producing, for example, those which make up the normal body floras. Indeed, in some cases the organism and its host live in symbiosis; *Lactobacillus acidophilus* (Döderlein's bacillus), for example, is a normal inhabitant of the adult vagina, where, by fermentation of glycogen, it produces a sufficiently low pH to inhibit the growth of other bacterial species, and thereby reduces the risk of infection.

Pathogenic micro-organisms may be classified broadly into four main groups : (*a*) *Bacteria* or *Schizomycetes* ("fission fungi"), unicellular, or occasionally multicellular, vegetable parasites; (*b*) *Fungi* proper, multicellular mould-like and yeast-like parasites; (*c*) *Viruses* and *Rickettsiae*, extremely minute micro-organisms, smaller than bacteria, and some of them ultra-microscopic and filterable; and (*d*) *Protozoa*, unicellular animal parasites. This and the following six chapters will deal with diseases caused by the first group of microbic parasites, the bacteria; diseases due to parasites of the other three groups will be considered in later chapters.

THE MORPHOLOGY OF BACTERIA
(Fig. 38)
(1) The main shapes of bacteria

Bacteria are divisible into four broad groups according to their shapes, namely (i) *cocci* or spherical bacteria, (ii) *bacilli* or rod-shaped bacteria, the rods being either straight or slightly curved, (iii) *vibrios* or curved rods, and (iv) *spirochaetes* or spiral-shaped bacteria.

Cocci which grow in irregular clusters are called *staphylococci*, those which grow in chains *streptococci*, and those which form pairs *diplococci*. The average diameter of cocci is about 1μ.

Bacilli vary greatly in size and shape. The largest measure up to 8μ in length, e.g. the anthrax bacillus ; the smallest are only about 1·5μ long, e.g. the whooping cough bacillus. Some bacilli are straight and have rectangular ends, e.g. the anthrax bacillus; some may be slightly curved, e.g. the tubercle bacillus; some are ovoid, resembling elongated cocci, and are sometimes called "cocco-bacilli", e.g. the bacillus of whooping cough. Markedly curved rods are called vibrios, e.g. the "comma bacillus" of cholera.

"Spirochaete" is a general name applied to a group of motile spiral bacteria. According to the number and closeness of the coils of the spiral, and certain other minor differences, several different genera have been distinguished, e.g. *Treponema, Leptospira, Borrelia*, but the morphological differences are small.

(2) Bacterial capsules

Certain bacteria, e.g. the pneumococcus and Friedländers bacillus, show an outer capsule distinct from the main body of the cell and differentially stainable (Fig. 58). The degree of capsule development is largely dependent on environmental factors; pathogenic bacteria show their greatest capsule formation when growing in susceptible animal tissues or in culture media specially enriched with animal protein. Some capsules, e.g. of the pneumococcus, are composed of complex polysaccharides; others, e.g. of the anthrax bacillus, are of protein nature. Encapsulation is related both to virulence and to antibody formation.

(3) Bacterial spores

Some bacterial species develop highly-resistant spherical or oval intracellular bodies called " spores ", one in each bacterial cell. These should not be confused with the reproductive spores of the higher fungi; bacterial spores are not reproductive structures. In bacteria of different species, spores may occupy central,

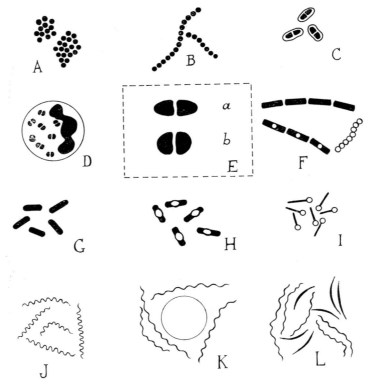

FIG. 38.—The shapes of bacteria. A, staphylococci; B, streptococci; C, pneumococci with capsules; D, gonococci in a pus cell; E, comparison of shapes of pneumococci (a) and gonococci (b); F, anthrax bacilli and " linked " spores; G, *Clostridium Welchii*; H, *Cl. sporogenes*; I, *Cl. tetani*; J, spirochaetes of syphilis; K, spirochaetes of relapsing fever along with a red corpuscle; L, spirochaetes and fusiform bacilli of Vincent's angina. (× 1600, except E which is three times this magnification.)

sub-terminal or terminal positions, e.g. that of the tetanus bacillus is terminal (" drum-stick " bacillus) and that of *Clostridium sporogenes* is central or sub-terminal. Spores withstand chemical and physical injuries better than the vegetative (proliferating, non-sporing) form of the bacterium; they withstand prolonged drought or high temperatures which would soon kill the latter. Under favourable conditions the spore germinates into a vegetative bacillus. Spores are stained with difficulty and therefore appear clear following ordinary stains; but, once stained, they resist decolorization, and this is the basis of special methods of staining for spores.

(4) Bacterial flagella and motility

Many bacteria, chiefly bacilli, are actively motile by means of thread-like processes or flagella, which can be demonstrated by special stains or by electron microscopy. Some bacteria show a single terminal flagellum, e.g. the vibrio of cholera ; others have multiple flagella distributed around the bacterial cell, e.g. the typhoid bacillus and *Proteus* species.

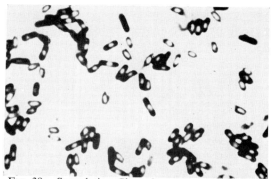

Fig. 39.—Sporulating *Clostridium bifermentans*. The oval spores do not take the Gram stain and appear as clear bodies in the bacilli. Some free spores are seen. (× 1,660).

Among the spirochaetes motility depends mainly on a spiral rotation of the whole organism or on flexion or lashing movements ; but delicate flagella or flagella-like structures are also present in some species, e.g. *Treponema pallidum*.

(5) Reproduction of bacteria

Bacteria multiply by simple fission ; the cell enlarges or elongates and then divides. Division of bacilli or cocco-bacilli is always transverse to the long axis. In some species the daughter cells separate completely ; in others they remain loosely attached in pairs, chains or clusters. The daughter cells of the anthrax bacillus tend to remain attached to one another so that long chains of bacilli result. The " chinese letter " arrangement of diphtheria bacilli is due to attachment of the cells to one another at one corner only. Reproduction occurs so rapidly that as many as 7.5×10^{13} organisms may be developed from a single bacterium in 24 hours.

(6) Bacterial pleomorphism

Most cocci are uniform in size and shape ; but many kinds of bacilli show considerable variation or *pleomorphism*, even under the same conditions. For example *E. coli*, whose average length ranges from 2 to 4μ, often shows filaments up to 10μ long and also short cocco-bacillary forms. So also the tubercle bacillus ranges in length from 1μ to 6μ or more, may be straight or curved, beaded or plain. Under unfavourable conditions, e.g. in old exhausted cultures, many degenerated or " involution " bacterial forms occur which are not met with in healthy cultures, and these unhealthy bacteria also show abnormal staining properties. Pleomorphism may be induced in a number of ways, e.g. by exposure to antibiotics.

STAINING PROPERTIES OF BACTERIA

Synthetic organic dyes are used to stain bacteria and so to render them plainly visible in smears or sections of tissues.   Some bacteria stain uniformly ; some stain unevenly, so that a banded or beaded appearance is produced, e.g. the diphtheria bacillus and the tubercle bacillus ; some show polar staining in which the ends of the bacillus stain deeply but its centre only lightly, e.g. the plague bacillus ; and some, especially the diphtheria bacillus, show metachromatic granules which stain more deeply than, and differently from, the rest of the bacterium.   Apart from these peculiarities of staining, however, there are two differential staining reactions which are of special importance in the identification and classification of bacteria, namely *Gram-staining* and *acid-fastness* as shown by Ziehl-Neelsen staining.

(*i*) Gram's stain

Gram introduced a staining method which has become of the first importance in distinguishing the various groups of bacteria.   The method depends on the fact that when bacteria are stained with certain aniline dyes, such as gentian-violet or methyl-violet, and are then treated with a solution of iodine in potassium iodide (Lugol's solution), in some species the iodine solution mordants (i.e. fixes) the stain, so that decolorization by subsequent treatment with alcohol is prevented.   Other species of bacteria, after similar treatment, are decolorized by the alcohol.   Bacteria which retain the stain are Gram-positive ; those which fail to retain it are Gram-negative.   Most bacteria fall into one or other of these two classes ; though some species are more strongly Gram-positive (i.e. they resist decolorization longer) than others.   As might be expected also, degenerated or involuted bacteria, of species which are normally Gram-positive, often lose their ability to retain the stain and appear Gram-negative.

It is remarkable that so little attention has been paid to the author of a staining method of such fundamental importance.   Gram is not mentioned in Garrison's encyclopaedic " History of Medicine ", nor in S. G. Paine's article on Bacteriology in the last edition of the *Encyclopaedia Britannica*.   Hans Christian Joachim Gram was a Danish physician, born in Copenhagen in 1853.   He devised his staining method while working in Berlin with Friedländer in 1884.

(*ii*) Acid-fastness as shown by Ziehl-Neelsen stain

Some bacilli are difficult to stain and remain uncoloured by the ordinary staining methods, but when stained by strong hot solutions they then strongly resist decolorization by acids.   Such bacilli, which include the tubercle and leprosy bacilli, are said to be " acid-fast ".   These organisms have a high lipoid content, and the difficulty of staining them and their acid-fastness when stained are attributed to a special unsaponifiable waxy substance which has been called " mykol " or " mycolic acid " and which exhibits acid-fastness in the free state.   Bacilli from which waxy substances have been extracted by suitable solvents are no longer acid-fast.

CULTURAL CHARACTERS OF BACTERIA

Details of the cultivation of bacteria are considered in Appendix D ; but certain general aspects of this subject, which are important in classification and in interpreting pathogenic effects, must be mentioned here.

(1) Nutritional requirements of bacteria

The growth of bacteria depends, of course, on an adequate supply of suitable food materials, and these requirements vary greatly from species to species. Many bacteria require ready-made organic substances (amino-acids, carbohydrates, etc.) as the source of their carbon and nitrogen supplies ; and many of them have become dependent also on ready-made enzymes and growth factors, so that they can live only as parasites in suitable hosts, i.e. they are strict parasites.

However, by providing all the nutritive substances needed by particular bacterial species, most of these can be grown on appropriate fluid or solid culture media *in vitro*.

A great deal is known about the living processes of bacteria; the sorts of food material and vitamins they require, the ways in which they respire and multiply, how they obtain their energy and the nature of their waste products. As with all forms of living things, bacteria grow, reproduce and die; they respire, metabolize food substances and " excrete " the products of metabolism.

Bacterial metabolic processes are of great importance in many ways. Sewage disposal, for example, depends on the fact that bacteria convert insoluble carbohydrates, fats and proteins into soluble, odourless compounds. The fertility of the soil depends on similar bacterial activity, since plants cannot utilize these complex food materials. The preparation of butter and cheese is initiated by microorganisms which sour the milk; other bacteria and moulds are responsible for ripening of cheese. Beer and alcohol are manufactured by fermentation processes in which yeasts are allowed to metabolize carbohydrates. The production of antibiotics is an important instance of organismal activity. These are some of the bacterial metabolic activities which are useful to man. Other products of metabolism are harmful to man, and are of special importance in clinical bacteriology. These harmful products are called bacterial toxins, and more attention will be paid to them later.

Our knowledge of the food and oxygen requirements of bacteria makes it possible to grow most organisms artificially in the laboratory. These artificial *in vitro* growths are called *cultures* and the mixtures of food materials on which they are cultivated are *culture media*. Artificial culture of bacteria is a very important and convenient method for their study and forms a large part of clinical bacteriology.

(2) Atmospheric requirements of bacteria

On the basis of their oxygen requirements, bacteria may be divided into four groups: (i) *obligate anaerobes* which will grow only in the absence of free oxygen, e.g. the tetanus and gas-gangrene bacilli, (ii) *facultative anaerobes*, which are indifferent as regards free oxygen and will grow either aerobically or anaerobically, e.g. most of the Gram-positive cocci, the typhoid bacillus, and many others, (iii) *obligatory aerobes*, which will grow only when supplied with free oxygen, e.g. the tubercle bacillus, gonococcus and meningococcus, and (iv) *microaerophiles*, such as the actinomyces, which grow best in the presence of small amounts of oxygen.

A few bacteria, e.g. the gonococcus, require an atmosphere containing a small percentage of carbon dioxide ; while others, e.g. some strains of staphylococcus, produce their toxins best in the presence of carbon dioxide.

(3) Temperature requirements of bacteria

Each pathogenic bacterial species has an " optimum temperature " at which it grows best.   In most cases this is close to the natural body temperature of the usual host, i.e. 37° C. for most bacteria pathogenic for man.   But there is usually a temperature range of several degrees, within which growth proceeds quite well. Some species, e.g. *E. coli*, have a wide temperature range (15°–45° C.) ; others, e.g. the gonococcus and meningococcus, a very restricted range (30°–38° C.). Below the minimum, although growth ceases, the bacteria usually remain viable, and stock cultures are thus often conveniently kept at low temperatures.   Above the maximum temperature at which a bacterium will grow, it usually dies quickly. The spore-bearing species are exceptions to this rule ; spores resist high temperatures, e.g. the spores of the anthrax bacillus may survive boiling for an hour or even longer.

BACTERIAL TOXINS

(1) Definition

Toxins are those bacterial products which are injurious to tissues and by the action of which bacterial infections produce their ill effects.   Toxins and their mode of action are therefore of the greatest interest to the pathologist, for it is they which are responsible for the structural and functional changes which he studies.

(2) Exotoxins and endotoxins

Toxins are classified broadly into 2 main groups, (i) exotoxins and (ii) endotoxins.

(i) *Exotoxins.*—These are poisonous substances which are liberated freely from the living bacteria into the host tissues or culture medium in which the bacteria are growing.   Exotoxins can therefore be separated by growing the bacteria in fluid media and then filtering the culture ; the filtrate contains the toxin. Bacteria which produce exotoxins include the diphtheria bacillus, tetanus bacillus, botulism bacillus, the bacilli of gas gangrene, some dysentery bacilli, and some staphylococci and streptococci.   Exotoxins are characteristically produced by Gram-positive organisms; Gram-negative species are less active in this respect. Exotoxins are antigenic, so that corresponding antibodies against them can be prepared which completely neutralize their activity.   Exotoxins are protein in nature and are generally unstable, being easily destroyed by oxidation, chemical agents and heat, and they can be rendered non-toxic but still capable of stimulating the production of antibodies that neutralize the toxins.   Such detoxification is easily effected by treating a toxin with formaldehyde, and the resulting antigenic, non-toxic product is called a *toxoid*.   This is of great importance in the prevention and treatment of some bacterial intoxications, since patients can be immunized with toxoid (for example, against tetanus and diphtheria), and toxoids are used to prepare antitoxic sera in the horse for prophylactic and therapeutic use in man.

Exotoxins are named according to their pathological effects on the tissues, irrespective of the bacteria producing them.   Thus a *haemolysin* is a toxin which produces lysis of red blood corpuscles, a *leucocidin* is a toxin which kills leucocytes, a *necrotoxin* is one which causes necrosis of tissues, an *erythrogenic toxin* is one which produces an area of redness or erythema when injected into the skin, and a *lethal toxin* is one which kills quickly following injection into the blood stream. Other specific names are given to toxins which are known to act on particular substrates. Thus we recognize *lecithinase, lipase, hyaluronidase, coagulase* and so on. Many toxins are able to produce a number of effects; the α-toxin of *Clostridium Welchii*, for example, is a lecithinase which also exhibits lethal, necrotizing and haemolytic activity, but it is probable that these effects are merely reflections of its lecithinase activity. Special mention should be made of the *neurotoxins* of *Cl. tetani* and *Cl. botulinum* which have been highly purified and obtained in a crystalline state. These are the most potent poisons known. W. E. van Heyningen, in his monograph on bacterial toxins, mentions that 1 mg. of a purified sample of botulinum toxin will kill 1,200 tons of animal matter; 7 ounces would kill the human population of the world.

It seems probable that many of the soluble substances produced by bacteria, such as coagulase, lecithinase, fibrinolysin and hyaluronidase, play an important part in the invasiveness of bacteria, and in determining the special characters of the tissue reactions to different infections.

(ii) *Endotoxins.*—These are toxins which remain attached to the protoplasm of the living bacteria and are not liberated into a culture medium. Hence they cannot be separated by simple filtration, but must be prepared by breaking up the bacterial cells or by extracting them with suitable solvents. In the body of an infected host, endotoxins are liberated presumably only when bacteria are disintegrated. Endotoxins are, in general, less selective in their action than exotoxins, and though they may cause death of animals on intravenous inoculation, no special pathological changes are associated with the intoxication. They are also much less toxic than exotoxins. They are produced chiefly by Gram-negative bacteria. Endotoxins are also more stable than exotoxins, they are not easily toxoided, and though they are antigenic, they are very incompletely neutralized by antitoxic sera. Unlike the exotoxins, they appear to consist of complexes of phospholipid, polysaccharide and protein.

Endotoxins play a leading part in producing the *bacteraemic shock syndrome*, a state of prostration and hypotension accompanying some bacterial infections; hence it is also called *endotoxic shock*. Because it was once thought to be caused only by Gram-negative organisms, another synonym was " Gram-negative shock "; but some cases are certainly due to Gram-positive bacteria.

CLASSIFICATION OF BACTERIA

For discussion of the attempts at precise taxonomic classification of bacteria, advanced works on bacteriology must be consulted. This is not of much significance for the pathologist and clinician, whose needs are adequately met by a simple grouping according to the main shapes and staining properties of bacteria. The following table of bacteria pathogenic for man gives the ordinary names, along with some commonly used synonyms, and the infections produced. A few non-pathogens of importance to the clinical bacteriologist are also included. The student who feels overawed by this long list of special names used by bacteriologists should read the note on " Bacteriological nomenclature " in Appendix A.

GENERA & SPECIES	SYNONYMS	DISEASES
GRAM-POSITIVE COCCI		
Staphylococcus		
Staph. aureus	Staph. pyogenes	Boils, carbuncles, wound infection, food poisoning, enterocolitis
Staph. albus	Staph. epidermidis	
Staph. citreus		
Streptococcus		
Strep. pyogenes	β-haemolytic or Group A streptococci	Tonsillitis, scarlet fever, erysipelas, wound infection, etc.
Strep. viridans	α-haemolytic streptococci; green-forming streptococci	Subacute bacterial endocarditis
Strep. faecalis	Faecal streptococcus; enterococcus	Wound infection, urinary-tract infection
Anaerobic streptococci		Puerperal and wound infections, deep-seated abscesses
Pneumococcus		Lobar pneumonia, upper respiratory infections, meningitis, peritonitis
Diplococcus pneumoniae	Strep. pneumoniae	
GRAM-NEGATIVE COCCI		
Neisseria		
N. gonorrhoeae	Gonococcus	Gonorrhoea, ophthalmia neonatorum
N. meningitidis	Meningococcus	Meningitis (cerebro-spinal fever)
N. pharyngis		
N. catarrhalis		
GRAM-POSITIVE BACILLI		
(i) ACID-FAST BACILLI		
Mycobacterium		
M. tuberculosis	Tubercle bacillus, Koch's bacillus	Tuberculosis
M. bovis		
M. leprae	Leprosy bacillus	Leprosy
M. ulcerans		⎫ Ulcers of the skin
M. balnei		⎭
M. smegmatis	Smegma bacillus	
M. phloei	Timothy-grass bacillus	
Anon. mycobacteria	Atypical mycobacteria	
(ii) AEROBIC SPORE-FORMERS		
Bacillus		
B. anthracis	Anthrax bacillus	Anthrax
B. subtilis	Hay bacillus	
B. cereus		
(iii) ANAEROBIC SPORE-FORMERS		
Clostridium		
Cl. Welchii	Cl. perfringens	Gas gangrene (anaerobic myonecrosis), food poisoning
Cl. oedematiens	C. Novyi	Gas gangrene ⎫
Cl. septicum		Gas gangrene ⎪
Cl. fallax		Gas gangrene ⎬ (Anaerobic myonecrosis)
Cl. Sordellii		Gas gangerne ⎪
Cl. histolyticum		Gas gangrene ⎭
Cl. bifermentans		⎫ Associated with gas gangrene
Cl. sporogenes		⎭
Cl. tetani	Tetanus bacillus	Tetanus
Cl. botulinum		Botulism
(iv) OTHER GRAM-POSITIVE BACILLI		
Corynebacterium		
C. diphtheriae		Diphtheria

GENERA & SPECIES	SYNONYMS	DISEASES
GRAM-POSITIVE		
BACILLI *continued*		
C. Hofmanni	} Diphtheroid bacilli	
C. xerosis		
Listeria		
List. monocytogenes		Meningitis in new-born, etc.
Erysipelothrix		
Erysip. rhusiopathiae		" Erysipeloid "
Lactobacillus		
L. acidophilus	Döderlein's bacillus	
GRAM-NEGATIVE		
BACILLI		
Salmonella		
S. typhi	Typhoid bacillus	Typhoid fever (enteric fever)
S. paratyphi A		
S. paratyphi B	} Paratyphoid bacilli	} Paratyphoid fever (enteric fever)
S. paratyphi C		
S. typhi-murium		Food poisoning
S. enteritidis		Food poisoning
Shigella		
Sh. dysenteriae		Dysentery
Sh. Flexneri	} Dysentery bacilli	Dysentery
Sh. Boydii		Dysentery
Sh. Sonnei		Dysentery
Escherichia		
E. coli	*B. coli*	Wound and urinary-tract infections; gastro-enteritis in infants
Klebsiella		
Kl. pneumoniae	Friedländer's bacillus	Pneumonia, upper respiratory infections
Kl. rhinoscleromatis		Rhinoscleroma
Kl. aerogenes	*Aerobacter aerogenes*	
Proteus		
Pr. mirabilis	*B. proteus*	Wound and urinary-tract infections
Pseudomonas		
Ps. aeruginosa	*Ps. pyocyanea*	Wound and urinary-tract infections
Haemophilus		
H. influenzae	Pfeiffer's bacillus	Meningitis, ? chronic bronchitis
H. Ducreyi	Ducrey's bacillus	Soft chancre
H. aegyptius	Koch-Weeks bacillus	Conjunctivitis
Moraxella		
Mor. lacunata	Morax-Axenfeld bacillus	Conjunctivitis
Bordetella		
Bord. pertussis	*Haemophilus pertussis,* whooping-cough bacillus	Whooping cough
Brucella		
Br. abortus		Undulant fever
Br. suis		Undulant fever
Br. melitensis		Malta fever
Pasteurella		
P. pestis	*Yersinia pestis*	Plague
P. Tularensis		Tularaemia
P. pseudotuberculosis	*Yersinia pseudotuberculosis*	
P. haemolytica		
P. septica		
Vibrio		
V. cholerae	Comma bacillus	Cholera
Fusiformis		
F. fusiformis		Vincent's angina
F. necrophorus		Necrobacillosis
Loefflerella		
Loeff. mallei	Glanders bacillus	Glanders
Loeff. pseudomallei	Whitmore's bacillus	Melioidosis
Bartonella		
Bart. bacilliformis		Oroya fever

GENERA & SPECIES	SYNONYMS	DISEASES
SPIROCHAETES		
Treponema		
Tr. pallidum	*Spirochaeta pallida* (Schaudinn and Hoffman)	Syphilis
Tr. pertenue		Yaws
Borrelia		
Borr. recurrentis	Obermeier's spirochaete	⎫
Borr. Duttoni		⎬ Relapsing fever
Borr. Novyi		⎭
Borr. Vincenti		Vincent's angina
Leptospira		
L. icterohaemorrhagiae		Weil's disease
L. canicola		Canicola fever
Spirillum		
Spir. minus		Rat-bite fever (Sodoku)

It is important to recognize that, while some of the named species of bacteria (e.g. *N. gonorrhoeae*, *V. cholerae*, *Strep. viridans*) are highly specific and show little variation between different isolated strains, other species' names (e.g. *Strep. pyogenes*, pneumococcus, *Cl. Welchii*) embrace a number of variants or different groups and types. Some of these types are so distinct in their cultural, toxigenic, immunological and other properties, that they are given special type or group numbers or letters; e.g. pneumococcus types I, II, etc., *Cl. Welchii* types A, B, C, etc., *Strep. pyogenes* groups A, B, C, etc., and so on. Some organisms, such as *Staph. aureus* and *S. typhi*, may also be differentiated into *bacteriophage types*. Phages are viruses which are able to attack and lyse bacterial cells. This method of typing organisms is of importance in the epidemiological study of infection. (*See* p. 686.)

SUPPLEMENTARY READING

Ash, J. E., and Spitz, S. (1945). *Pathology of Tropical Diseases: An Atlas*. Philadelphia and London; W. B. Saunders.

Boltjes, T. Y. K. (1948). " Function and arrangement of bacterial flagella ", *J. Path. Bact.*, **60**, 275. (Many beautiful illustrations, including electron photomicrographs.)

Burnet, F. M. (1940). *Biological Aspects of Infectious Disease*. Cambridge; Cambridge University Press. (An admirable and unusual presentation of the general principles of infection and immunity.)

Cruickshank, R. (Ed.). (1965). Eleventh revised edition. *Medical Microbiology*. Edinburgh and London; E. & S. Livingstone.

Florey, H. W. (1945). " The use of micro-organisms for therapeutic purposes ", *Brit. Med. J.*, **2**, 635. (An interesting outline of the history of antibiotics.)

Gillies, R. R. and Dodds, T. C. (1965). *Bacteriology Illustrated*. Edinburgh and London; E. & S. Livingstone.

Hare, R. (1956; 2nd edition 1963). *An Outline of Bacteriology and Immunity*. London; Longmans.

Howie, J. W. (1956). " Bacteriology in modern medicine ", *Lancet*, **2**, 951. (All students and teachers should read the pertinent comments on nomenclature, some of which are cited in Appendix A.)

Nichol, H. (1939). *Microbes by the Million*. (A " Pelican Special ") London. (An instructive as well as entertaining account of bacteria in general and of fermentative and putrefactive bacteria in particular; a most unusual book which every student should read.)

van Heyningen, W. E. (1950) *Bacterial Toxins*. Oxford; Blackwell.

Wilson, G. S., and Miles, A. A. (1964). *Topley and Wilson's Principles of Bacteriology and Immunity*. Fifth edition, London; E. Arnold. (A comprehensive work designed mainly for bacteriologists but of great value also to pathologists and students for reference on special points.)

CHAPTER 11

COCCI AND THE LESIONS THEY PRODUCE

THE READER will have noticed that many of the diseases cited in the preceding chapters as typical examples of inflammation in particular sites are caused by cocci. This choice of examples is significant; it is because nearly all species of pathogenic cocci evoke closely similar typical acute inflammatory reactions. However, as already noted nowadays, many bacterial inflammations are greatly modified by treatment with antibiotics, sulphonamides or other chemotherapeutic agents. This particularly applies to the diseases caused by cocci, which, as a result of effective antibacterial agents, are now much less frequent than formerly and are only occasionally seen in their natural form unmodified by treatment. The student must constantly recall that the diseases considered in this chapter, and in later ones, are described as running their natural untreated courses.

STAPHYLOCOCCI

The staphylococci were discovered in pus by Koch in 1878, and were first cultured on solid media by Rosenbach in 1884. They are abundant on the skin, in the mouth, nose and intestinal tract, and are found in varying numbers in dust, water, milk, domestic utensils, clothing, and in the pus of the lesions they produce. They are Gram-positive cocci, about $1 \cdot 0\mu$ in diameter, which tend to grow together in clusters (compare with the streptococci). They do not form spores, are non-motile and non-capsulated, and are facultative anaerobes. They grow freely on all ordinary media, producing domed, opaque and pigmented colonies, by which three species are distinguished: (i) *Staphylococcus aureus* which forms golden-yellow colonies, (ii) *Staph. albus* which forms white colonies, and a rare species (iii) *Staph. citreus* which forms lemon-yellow colonies. Not all strains of staphylococci are pathogenic to man, and it has been found that pathogenic strains are easily identified by their ability to coagulate human plasma and to degrade deoxyribonucleic acid. Most strains of *Staph. aureus*, and a few strains of *Staph. albus*. are coagulase-positive, but, irrespective of their pigmentation, coagulase-negative staphylococci are rarely, if ever, able to cause disease. For convenience, coagulase-positive staphylococci are often referred to as *Staph. pyogenes* to indicate that these strains are pathogenic. *Staph. pyogenes* is present in the anterior nares in about 50 per cent of the population, a fact of importance in the occurrence of staphylococcal infections, particularly in hospitals.

STAPHYLOCOCCUS PYOGENES

(1) Toxins

All strains of *Staph. pyogenes* produce a number of exotoxins. Cultures and culture-filtrates are haemolytic, they will kill leucocytes, they produce tissue necrosis when injected into the skin, and they kill quickly when injected intravenously. Hence, *haemolysin, leucocidin, necrotoxin* and a *lethal toxin* are present. Certain strains of staphylococci, when grown under suitable conditions, produce one or more thermostable enterotoxins which cause acute vomiting and diarrhoea

113

when ingested by man. These strains are the cause of staphylococcal food-poisoning. Other soluble substances produced by *Staph. pyogenes* include a *coagulase*, a *deoxyribonuclease*, a *fibrinolysin*, a *hyaluronidase* and a *lipase*.

(2) Tissue injury and inflammatory reaction

The injury and reaction caused by staphylococci can be well studied by injecting suitable doses of the cocci intravenously into rabbits, killing the animals at times ranging from 6 to 48 hours later, and preparing micro-sections of the lesions in the kidneys, heart and other tissues. Similar early lesions are sometimes observed in man as the result of a staphylococcal septicaemia.

Early lesions, 12 or 18 hours old, show a small colony of cocci surrounded by a zone of tissue necrosis, due no doubt to the direct necrosing effect of the exotoxin ; and in the surrounding tissues an acute inflammatory exudation with abundant polymorphonuclear leucocytes quickly ensues. The coccal colony multiplies, the necrosis extends, and the inflammatory reaction increases. The necrotic tissue is invaded by leucocytes and is softened and digested, so that a fluid *abscess* is formed. Many of the leucocytes are destroyed just as the tissue cells are ; but fresh leucocytes swarm into the area and engulf many of the cocci. When the necrosis ceases to extend peripherally, the surrounding fibroblasts and vessels proliferate to form a wall of granulation tissue lining the abscess cavity. Following liberation of the pus, the few remaining cocci are usually soon destroyed, the granulating wall fills the former abscess cavity, and healing ensues.

Most staphylococcal lesions are of this localized character ; only occasionally do they spread diffusely as a cellulitis. Their tendency to localization may be due in part to the production of coagulase by the cocci, the effect of which may be to aid the development of the inflammatory fibrin barrier around the lesion.

(3) The diseases caused by staphylococci

Staphylococci cause furuncles and carbuncles of the skin, wound suppuration, mastitis and breast abscesses, tonsillitis and quinsy, bronchopneumonia and abscess of the lung, puerperal infections, urinary sepsis, and, after their entry into the blood stream, osteomyelitis, abscesses in the kidneys or other organs, acute endocarditis, and septicaemia or pyaemia.

A *furuncle* or " boil " is a typical example of a localized staphylococcal lesion. It starts in an infected hair follicle or sebaceous gland, and shows the changes already described in the development of a staphylococcal abscess. A necrotic core of tough fibrous dermal tissue is usually discharged along with the pus through the skin opening where the abscess " pointed " and broke. If the infection spreads more widely in the dermis, the lesion is called a *carbuncle* ; it may reach several inches in diameter, multiple discharging openings are formed, and there is often a large irregular " slough " or piece of necrotic tissue which must be separated and removed before healing can occur. The cocci discharged from a furuncle are apt to infect other hair follicles and so cause recurring crops of " boils ".

Staphylococcal bronchopneumonia, osteomyelitis, endocarditis, thrombophlebitis and meningitis, have already been described in earlier chapters.

(4) Isolation and identification

Staph. pyogenes is easily isolated from most pathological material by direct culture on any of the common media. Pigmentation does not occur if cultures are grown anaerobically. Surface cultures have a characteristic rancid smell. Many strains are haemolytic on horse-blood agar, and lipolytic on egg-yolk agar, and show proteolysis on heated-blood agar. The pathogenicity of the organism is demonstrated by the slide or tube coagulase test or by the production of deoxyribonuclease (see Appendix D).

Isolation of the staphylococcus from faeces or the intestinal contents is often made difficult by the presence of other organisms which tend to overgrow and obscure its colonies. This is overcome by inoculating the material into 10 per cent NaCl broth or on 10 per cent NaCl agar. These media almost completely inhibit the growth of other organisms and are thus selective for staphylococci.

The presence of staphylococci in the blood stream is determined by culturing a specimen of blood in glucose broth. These organisms grow well in most broth media in common use.

In suspected cases of deep-seated staphylococcal infection, when material is not readily available for culture, e.g. osteomyelitis, an anti-staphylolysin test on the patient's serum may aid diagnosis. In this test, the presence and amount of antibody to the α-haemolysin of *Staph. aureus* is determined. A rising titre is suggestive of staphylococcal infection.

STREPTOCOCCI

These are Gram-positive cocci, about 1.0μ in diameter, which grow in chains (compare with the staphylococci). Some strains produce capsules, but none of them produce spores and all are non-motile. Except for a group of anaerobic streptococci (*see* below) they are all facultative anaerobes. Streptococci are widely distributed in man and may be isolated from such sites as the skin, mouth, nose and intestines, and in discharges from the lesions caused by them.

The facultative anaerobic streptococci are divided into three types according to the reactions they produce when grown on horse-blood agar: (i) *Alpha-haemolysis* is produced in the presence of free oxygen and is seen as a greenish halo in the medium beneath and around the colony. It is not true haemolysis since the red cells in the discoloured zone are not lysed. (ii) *Beta-haemolysis* may be produced under aerobic or anaerobic conditions. Beneath and around the streptococcal colony there is complete lysis of the red cells in the medium, so that the colony is surrounded by a clear zone sharply demarcated from the rest of the medium. (iii) *Gamma-haemolysis* means the absence of any detectable change in the fresh-blood agar medium. Because the truly haemolytic (β-haemolytic) streptococci are the most important clinically, including *Strep. pyogenes*, they are discussed first.

STREP. PYOGENES AND OTHER HAEMOLYTIC STREPTOCOCCI

There are a number of different beta-haemolytic streptococci which are able to cause disease in man and animals. They are divisible into 15 groups (A-Q, I and J being excluded) according to the presence in them of one or other of a number of polysaccharide antigens (C substances) which were discovered by Lancefield. Lancefield grouping is carried out by means of a precipitin test in which a soluble extract of the organism is allowed to react with antisera to the different polysaccharide C substances. In clinical medicine the most important beta-haemolytic streptococci are those which belong to group A, since it is almost exclusively these organisms which cause infections in man.   For this reason group-A haemolytic streptococci are commonly referred to as *Streptococcus pyogenes*.

(1) Toxins

Like *Staph. pyogenes*, *Strep. pyogenes* produces a number of exotoxins. Many, but not all, strains produce *erythrogenic toxin* which is responsible for the rash and other systemic manifestations of scarlet fever. A *fibrinolysin* and *hyaluronidase* are produced; and two distinct haemolysins are recognized, *streptolysin S* an oxygen-stable haemolysin which is inactivated by heat, and *streptolysin O* a heat-stable haemolysin which is inactivated by oxygen.   Streptolysin O is antigenically related to the oxygen-labile haemolysins of *Cl. Welchii* (theta-toxin) and *Cl. tetani* (tetanolysin).

Antibodies to a number of these toxins may be produced during streptococcal infections.   In the case of the erythrogenic toxin, production of antibodies results in active immunity to subsequent attacks of scarlet fever (see below).   Antibodies are also produced to streptolysin O; and use may be made of the antistreptolysin-O titre in the diagnosis of rheumatic fever.

(2) Tissue injury and inflammatory reaction

Some pyogenic streptococcal infections are localized, like those due to staphylococci ; but many others are more diffuse and accompanied by less tissue necrosis.   Most diffusely spreading inflammations are streptococcal—e.g. most cases of cellulitis, lymphangitis, and rapidly extending thrombo-phlebitis, and all cases of erysipelas.   The diffuse character of the lesions appears to be due to the ease with which streptococci multiply along tissue spaces, lymphatics and blood vessels, and perhaps also to the potency of the fibrinolysin and hyaluronidase produced by these bacteria. The inflammatory reaction excited by streptococci is sometimes a typical pyogenic one, with great outpouring of fibrinous exudate and polymorphonuclear leucocytes and the formation of pus.   In other infections, however, the exudate is not purulent but rather serous, sero-purulent or haemorrhagic.

(3) The diseases caused by haemolytic streptococci

Haemolytic streptococci cause a great variety of suppurative and non-suppurative inflammatory diseases.   Some of these diseases are not always due to streptococci but can be caused also by staphylococci or other bacteria or by mixed infections, e.g. wound infections, abscess, cellulitis, lymphangitis and lymphadenitis, tonsillitis, otitis media and its complications (thrombo-phlebitis, meningitis, cerebral abscess), appendicitis and its complications (abscess, peritonitis, portal phlebitis), bronchopneumonia and empyema, acute ulcerative endocarditis, osteomyelitis, puerperal infections, septicaemia and pyaemia.   The proportion of cases of these diseases which are due to *Strep. pyogenes* varies from disease to disease—e.g. a high proportion of cases of cellulitis, lymphangitis, tonsillitis and puerperal infection, a lower proportion of localized abscesses, bronchopneumonias and cases of osteomyelitis.   Many of these diseases have been described in earlier chapters.   Since the advent of penicillin serious infections due to haemolytic streptococci have become much less frequent.   It remains to give further details of two exclusively streptococcal diseases, namely erysipelas and scarlet fever. (The streptococcal causation of rheumatic fever and of nephritis is discussed in Chapter 22.)

(a) Erysipelas

It was in erysipelas that streptococci were first discovered, by Fehleisen in 1883. For many years the responsible organism was called *Streptococcus erysipelatis*, but it does not differ in any specifiable way from other haemolytic streptococci and it is now regarded as identical with *Streptococcus pyogenes*. The cocci enter the skin through a wound or abrasion, sometimes so small as to escape notice, and invade the lymphatics of the dermis. Erysipelas is most common in the skin of the head and neck, but may occur in any part of the body. The affected area of skin is smooth, tense, red, hot and tender ; and it has a clearly defined raised border. Microscopically, the dermis shows the usual changes of acute inflammation—hyperaemia, oedema and many extravascular polymorphonuclear leucocytes. Many streptococci are present in the lymphatics ; the disease is in the main an extending lymphangitis of the dermal lymphatic plexus. The inflammation continues to spread for several days, then ends quickly by resolution ; suppuration does not occur. We do not know why the infection is non-suppurative, and why it remains confined to the dermis and does not extend to the subcutaneous tissue. In the latter situation, streptococcal infection almost always causes a suppurative cellulitis.

(b) Scarlet fever

Scarlet fever (or scarlatina) has long been recognized as a specific infectious disease characterized by acute streptococcal inflammation of the tonsils and fauces, a distinctive erythematous skin rash, and a considerable risk of acute nephritis during convalescence. A typical scarlatinal rash occasionally accompanies haemolytic streptococcal infections other than tonsillitis, e.g. infected wounds or puerperal infections. But streptococci are not present in scarlatinal skin or in the acutely inflamed kidney of the convalescent; and the relationship between the streptococcal focus of infection and these other manifestations of the disease has only been clarified within recent years.

(i) *Scarlatinal toxin and the Dick test.*—In 1923 Gladys and George Dick produced scarlet fever in volunteers by swabbing their tonsils with a culture of haemolytic streptococci isolated from cases of the disease. Later they showed that culture-filtrates of these streptococci contain a special erythrogenic exotoxin which produces a local erythema when injected intradermally into persons susceptible to scarlet fever, while immune convalescents show no reaction. This *Dick test* is still sometimes used to demonstrate susceptibility or immunity; a positive erythematous reaction to a suitable injection of the toxin shows that the subject has no immunity; the negative test in convalescents from scarlet fever denotes their acquired immunity.

(ii) *The Schultz-Charlton reaction.*—Schultz and Charlton found that if serum from a convalescent scarlet fever patient is injected intradermally into a patient with early scarlet fever, this causes a blanching or extinction of the rash around the site of injection. The convalescent's serum will also neutralize Dick toxin *in vitro* and abolish its erythrogenic effect. It is clear therefore that the scarlet fever rash is caused by circulating erythrogenic toxin, and that the immunity to the action of this toxin which obtains in convalescence is due to a specific antitoxin in the convalescent's serum which neutralizes the toxin.

(iii) *Immunization against scarlet fever.*—A subject who is Dick-positive (and therefore susceptible to scarlet fever) can be made Dick-negative (and therefore resistant to scarlet fever) by repeated small injections of Dick toxin. Immunization of susceptibles in this way has proved of great value in reducing the incidence and severity of scarlet fever in school children and in nurses in fever hospitals. Further, the serum of horses immunized by repeated doses of streptococcal toxin can be used to treat early cases of scarlet fever, in whom it reduces the fever, ameliorates the symptoms and curtails the course of the disease. It is important to note that the immunity which results from the injection of Dick toxin is against the toxin and not against the organism. However, immunization is now superseded by antibiotic therapy.

(iv) *Mode of action of streptococci in scarlet fever.*—From the facts just outlined, it is clear that in an uncomplicated case of scarlet fever, the streptococci themselves are restricted to the primary focus of acute inflammation, usually the tonsils but occasionally some other focus, and that the rash is due solely to the effects of circulating toxin on the skin. The nephritis of convalescence also is a purely toxic lesion, due to reaction in the kidney following the excretion of circulating toxins. The diverse pathogenicity of streptococci appears to depend on two distinct sets of properties, their toxin-producing capacity and their invasiveness. Strains of streptococci which produce scarlet fever are highly toxigenic, but not highly invasive, since in most cases they remain confined to the throat. By contrast, streptococci which produce spreading infections, such as cellulitis, puerperal infections and septicaemia, do not usually produce a scarlatinal rash or nephritis, and are to be regarded as highly invasive but only moderately toxigenic, at least as regards erythrogenic toxin. The invasive powers of streptococci are probably connected with their other toxins, namely haemolysin, leucocidin, fibrinolysin and hyaluronidase. But, scarlatinal streptococci are not devoid of invasive power; they sometimes extend locally to cause otitis media, meningitis or bronchopneumonia, or by the blood stream to cause arthritis, osteomyelitis or septicaemia.

(4) Isolation and identification

Strep. pyogenes will not grow well on simple media but grows readily on fresh-blood agar and heated-blood agar, and in serum broth. In broth media it characteristically produces long chains. On fresh-blood agar colonies are 0·5–1·0 mm. in diameter, and are opaque domes with a dry matt surface. Individual colonies are surrounded by a zone of beta-haemolysis which is sometimes best developed if cultures are incubated anaerobically. This is due to the fact that the haemolysis produced by some strains is due chiefly to the oxygen-labile haemolysin (streptolysin O). On heated-blood agar many strains of *Strep. pyogenes* cause slight bleaching of the medium. The presence of *Strep. pyogenes* in pathological material is demonstrated by direct plating on fresh-blood agar and heated-blood agar, the fresh-blood agar cultures being incubated anaerobically if possible. In cases of septicaemia the organism is easily isolated from the blood stream by culture of blood in glucose broth.

Strep. pyogenes can be divided into a large number of sero-types (1, 2, 3, etc.) according to their different protein antigens, demonstrable by agglutination tests with appropriate antisera; but such typing, though of value in epidemiological studies, is unnecessary in routine clinical bacteriology. Certain specific sero-types cause acute nephritis.

(5) Other beta-haemolytic streptococci

Lancefield groups B–Q are of little importance in human clinical bacteriology. Some of them occur as commensals in man, and occasionally some of them, e.g. those of group D which are normal inhabitants of the bowel, may play a pathogenic part in infections of the intestine, urinary tract, or wounds. Unlike *Strep. pyogenes* these organisms grow readily on simple media and on MacConkey agar; on the latter medium colonies are pink in colour due to fermentation of the lactose. On fresh-blood-agar colonies are large (1–2 mm. in diameter) and are smooth shiny translucent domes, often surrounded by large zones of haemolysis. Non-haemolytic variants are common, however, and these are often referred to as *Strep. faecalis* or enterococci. Group D streptococci also differ from *Strep. pyogenes* in their resistance to heat; they will survive exposure to 60°C. for half an hour.

STREPTOCOCCUS VIRIDANS

(α-HAEMOLYTIC STREPTOCOCCI)

Alpha-haemolytic streptococci (*Strep. viridans*) occur in the normal mouth as well as in mild localized inflammatory lesions of the mouth cavity—tonsillitis, gingivitis (pyorrhoea), and infections of the teeth and teeth sockets ; and they are the commonest cause of subacute bacterial endocarditis. There is little doubt that this serious disease is in most cases a complication of a smouldering oral infection ; for its victims frequently show such infections, and it is well known that, following tooth extractions for root infections or pyorrhoea or accompanying an acute flare-up of a chronic tonsillitis, streptococci identical with those in the mouth lesions can sometimes be isolated from the blood by means of blood cultures. The pathology of subacute bacterial endocarditis has been described in Chapter 7.

Isolation and identification.—*Strep. viridans* will not grow well on simple media such as nutrient agar but grows readily on enriched media such as fresh-blood agar and heated-blood agar, and in serum broth. In fluid cultures the organism characteristically forms chains. On fresh-blood agar colonies are 0·5–1·0 mm. in diameter, and are smooth transparent domes, each surrounded by a zone of alpha-haemolysis. On heated-blood agar colonies are surrounded by zones of bleaching; the medium beneath and surrounding the colonies is a greenish-cream colour in sharp contrast to the surrounding chocolate-brown medium. *Strep. viridans* must be differentiated from the pneumococcus which produces similar changes on fresh and heated-blood agar (see p. 121). From such sites as the mouth and throat *Strep. viridans* is easily identified by direct platıng of a swab on fresh and heated-blood agar. For isolation of the organism from the blood in cases of subacute bacterial endocarditis blood cultures are made in glucose broth.

Other Streptococci

(1) Non-haemolytic streptococci

Gamma or non-haemolytic streptococci are plentiful in the mouth, throat and intestine. Most of them are merely commensals, but some strains are pathogenic. Though their classification is uncertain, many of them are non-haemolytic variants of one or other of the Lancefield groups. *Strep. faecalis* (see above) is an example.

(2) Anaerobic streptococci

Although the anaerobic streptococci have not been satisfactorily classified, many strains are pathogenic for man. They are an important and common cause of puerperal infection, and, in association with aerobic pyogenic cocci, they are the cause of a gangrenous infection of muscle, anaerobic streptococcal myositis. In puerperal infections the organisms gain entry into the uterus from the vagina, where they form part of the normal bacterial flora. Secondary invasion of the blood stream is not uncommon. Anaerobic streptococci are isolated from pathological material by direct plating on fresh and heated-blood agar. The cultures are incubated anaerobically. Colonies are about 1 mm. in diameter, and are smooth transparent domes. They are non-haemolytic on fresh-blood agar. The organisms are isolated from the blood stream by inoculating blood into glucose broth and incubating the culture anaerobically.

THE PNEUMOCOCCUS

The pneumococcus, also called *Streptococcus pneumoniae*, is an ovoid or lanceolate Gram-positive coccus, which in infected tissues or sputum appears characteristically in pairs and has a thick capsule, but which in culture often forms chains and has a less distinct capsule. The long axes of the paired ovoid cocci lie in the same straight line, unlike those of the *Neisseria* (q.v.). Fraenkel and Weischelbaum independently discovered the pneumococcus in 1886, in the sputum of patients with lobar pneumonia, in which it is present in great numbers. Pneumococci are often present in the mouths and throats of healthy persons.

(1) Toxins

The pneumococcus produces an oxygen-labile haemolysin, but it is doubtful if it plays much part in the pathogenic effects of this organism. These appear to be due more to its powers of invasion and to endotoxins liberated from disintegrating cocci. The virulence of pneumococci depends in some way on the presence of their polysaccharide capsules; non-encapsulated cocci are avirulent.

(2) Tissue injury and inflammatory reaction

Little is known of the way in which the pneumococcus injures the tissues ; there is usually no visible destruction of tissue in the inflamed lung, meninges or other parts. It excites a typical acute inflammatory reaction, which in the lung usually resolves completely and seldom leads to suppuration. In other tissues, however, pneumococcal inflammations, e.g. empyema, otitis media, meningitis, peritonitis, are often suppurative.

(3) Diseases caused by the pneumococcus

The principal diseases caused by the pneumococcus are the following :

(i) *Lobar pneumonia*, a specific pneumococcal disease, and its local and metastatic complications—empyema, pericarditis, endocarditis, meningitis, arthritis, etc. (See Chapter 5.)

(ii) *Some cases of bronchopneumonia* and their complications (see Chapter 5).

(iii) *Some catarrhal inflammations* of the upper respiratory tract, para-nasal sinuses or middle ear, and their complications, e.g. thrombo-phlebitis of the dural venous sinuses, meningitis, etc. (see pp. 82, 87).

(iv) *Pneumococcal meningitis* sometimes appears also as a primary infection unassociated with any other focus of pneumococcal disease, especially in children.

(v) *Pneumococcal peritonitis* also appears as a primary disease, especially in female children, in whom the infection probably reaches the peritoneum from the vagina by way of the genital tract.

(4) Isolation and identification

The pneumococcus grows well on media containing blood or serum but does not grow well on simple media. On fresh-blood agar each colony is surrounded by a zone of alpha-haemolysis, and on heated-blood agar colonies are associated with

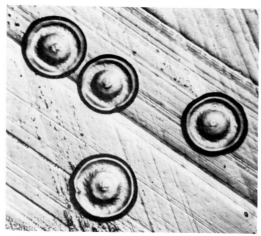

Fig. 40.—" Draughtsmen " colonies of *Streptococcus pneumoniae* growing on horse-blood agar. ($\times$ 24).

zones of bleaching of the medium. In young cultures (16–18 hours) colonies are indistinguishable from those of *Strep. viridans*. With further incubation, however, the colonies enlarge and become flattened, appearing as flat discs on the medium. Often the flattening occurs mainly in the central portion of the colony so that it has an outer raised rim and a central flat depression—this is the so-called " draughtsman " colony (Fig. 40). Disc and draughtsman-like colonies are characteristic of the pneumococcus and serve to differentiate it from *Strep. viridans*. Some strains of pneumococci do not produce these characteristic colonies, and their colonial morphology is then identical with that of *Strep. viridans*. There are, however, a number of other distinguishing features, the most important of which are the *bile solubility test* and the *optochin test* (see Appendix D).

In sputum and pus from lesions produced by the pneumococcus, typical lanceolate diplococci are seen in Gram-stained films of the material. Capsules may be demonstrated by special stains. The organism is isolated by direct plating on fresh and heated-blood agar. In cases of pneumococcal meningitis, many pus cells are present in the cerebrospinal fluid. These are best demonstrated by staining films with Leishman stain, which also shows the morphology of the organism. Films are prepared from a centrifuged deposit of the cerebrospinal fluid.

Many types of pneumococci have been distinguished, mainly on immunological evidence which proves that the composition of the polysaccharide capsule differs from type to type. Types I, II and III are the most distinct and are responsible for the majority of cases of lobar pneumonia. The remaining types are often referred to collectively as " Group IV ".

GRAM-NEGATIVE COCCI

The genus *Neisseria* consists of Gram-negative cocci, about 1·0μ in diameter, which appear characteristically in pairs. They are non-motile, non-sporing and non-capsulated, and are strict aerobes. They are strict parasites of man. The two pathogenic members of the group, *N. gonorrhoeae* and *N. meningitidis*, are delicate organisms which die out readily at temperatures much below 37°C.

THE GONOCOCCUS

The gonococcus (*Neisseria gonorrhoeae*) was discovered in gonorrhoeal pus by Neisser in 1879. The cocci are Gram-negative, oval or kidney-shaped, and occur in pairs with their opposed surfaces flattened or concave, so that their long axes are parallel. The gonococcus is a strict parasite of man; it is found only in the genito-urinary system of patients with gonorrhoea, in the remote complications of that disease, and in conjunctivitis due to accidental infection by gonorrhoeal discharges.

(1) Toxins

There is no evidence that the gonococcus produces any true exotoxins. A thermostable endotoxin, capable of causing tissue necrosis in and death of injected mice or rabbits, can be extracted from dried ground-up gonococci.

(2) Tissue injury and inflammatory reaction

We do not know how the gonococcus injures the tissues and excites inflammation. In uncomplicated gonorrhoea, the infection is a superficial one, spreading along the epithelial surfaces and exciting a purulent discharge without inflicting much obvious damage on the epithelium or underlying tissues. If, however, the infection spreads to the sub-epithelial tissues, it occasions focal necrosis of tissue and abscess formation.

(3) Gonorrhoea

Gonorrhoea is a venereal disease usually transferred by sexual intercourse, and affecting primarily the male or female genital tract.

(i) *Male genital tract.*—In the male genital tract gonococci quickly colonize the mucous membrane of the anterior part of the urethra, evoking a catarrhal inflammation with a profuse purulent discharge. Later, the discharge diminishes, the number of gonococci in it becomes less, and secondary infecting bacteria are often present—pyogenic cocci, coliform bacilli, etc. In some cases the infection extends to the prostate, seminal vesicles, ductus deferens and epididymis. Gonorrhoea is the commonest cause of epididymitis, and of sterility in the male. Persistence of gonococci in the prostate and in Cowper's and Littré's glands results in chronic gonorrhoea or " gleet ". In many cases, especially if promptly and adequately treated, the infection remains superficial ; but in other cases, especially chronic ones, it extends to the peri-urethral tissues producing peri-urethral abscesses or smouldering chronic inflammation with fibrosis and subsequent urethral stricture.

(ii) *Female genital tract.*—In the female genital tract, the cervix and urethra are the usual sites of infection, which may later extend to Bartholin's glands, the endometrium, Fallopian tubes and pelvic peritoneum. The symptoms are often trivial, and the woman is unaware of the infection until the advent of some complication such as salpingitis or Bartholin abscess. Chronic salpingitis, pyosalpinx, hydrosalpinx and chronic pelvic peritonitis are frequent results ; and these are a common cause of sterility.

(iii) *Metastatic gonococcal lesions.*—In either sex, gonococci sometimes enter the blood stream in small numbers, and are carried to distant tissues where they set up local inflammatory lesions. The chief of these are acute or chronic arthritis, usually of the knee-joint or ankle-joint, and usually non-suppurative ; less common are acute ulcerative endocarditis (p. 77) and meningitis (p. 87).

(iv) *Accidental gonococcal infections of children.*—These are of two kinds. (*a*) *Ophthalmia neonatorum*, or ulcerative gonococcal conjunctivitis of the new-born infant, is contracted from the mother's genital tract during parturition. By causing ulceration and subsequent opacity of the cornea, it used to be a frequent cause of blindness ; but this is usually prevented nowadays by applying a solution of one of the sulphonamide drugs to the conjunctival sac of the infant immediately after birth. (*b*) *Vulvo-vaginitis* in little girls is usually contracted accidentally from towels or bedclothes. In a hospital or school it may spread from child to child by way of improperly sterilized towels, and may be difficult to eradicate. Peritonitis and arthritis are occasional complications.

(4) Isolation and identification

The gonococcus is comparatively slowly growing, requiring 48 hours' incubation at 37°C for the development of recognizable colonies. It does not grow on simple media, but grows well on heated-blood agar containing 20 per cent of whole horse blood. The colonies are about 1 mm. in diameter, and are semi-translucent shiny domes. In common with other *Neisseria*, the gonococcus is able to oxidize tetramethyl-*p*-phenylene diamine from a colourless form to a coloured oxidized form. This reaction is used in identifying colonies of gonococci in plate cultures, and is called the oxidase test.

The gonococcus is most commonly encountered in infections of the genital tract. In males smears are made directly from the urethra; in females separate smears are made from the urethra and the cervix. These are stained by Gram's method. In positive cases typical Gram-negative diplococci are present together with many pus cells. Many of the organisms lie within the polymorphonuclear leucocytes, the cytoplasm of which is studded with ingested cocci (Fig. 41). Owing

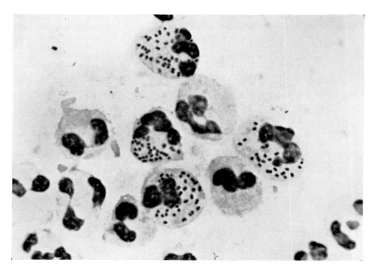

Fig. 41.—Urethral smear from a case of gonorrhoea, showing many gonococci within pus cells. (× 1250.)

to the exteme sensitivity of the gonococcus to cold and to drying, it is best to transfer material for culture directly from patient to the culture medium. The pus is collected on a platinum loop and plated on the gonococcal medium which has been warmed to 37°C. The culture is returned immediately to the incubator. When the patient is not near a bacteriological laboratory, swabs of suspected pus are placed in Stuart's transport medium, a soft agar medium which maintains the viability of gonococci during their carriage. Swabs in transport medium are plated onto gonococcus chocolate agar immediately on arrival at the laboratory. Since the presence of CO_2 greatly favours the growth of some strains of gonococci, it is an advantage if cultures are incubated in a closed jar containing 10 per cent of the gas in air. The gonococcus may be finally identified and distinguished from other *Neisseria* by sugar fermentation tests using glucose, maltose and sucrose; the gonococcus ferments glucose only, with the production of acid, but no gas.

The gonococcal complement fixation test (G.C.F.T.), which demonstrates the presence or absence of antibodies to the gonococcus in the patient's serum, is sometimes used for diagnosis in the chronic stage of the disease.

THE MENINGOCOCCUS

The meningococcus (*Neisseria meningitidis*) was discovered in the cerebrospinal fluid of patients with meningitis by Weichselbaum in 1887. The cocci closely resemble gonococci both in shape and in cultural reactions; they are distinguished from gonococci by slight cultural differences, especially by the fact that they ferment both glucose and maltose, while gonococci ferment glucose only. Meningococci are strict parasites of man only. They are found in the nasopharynx of carriers, who may be either convalescents who have recovered from meningitis or healthy persons. Normally the number of such carriers is small (about 5 per cent), but it is markedly increased during a meningitis epidemic. Epidemics occur chiefly in closely packed communities, e.g. armies or crowded institutions, where droplet infection from coughing or sneezing is readily transferred by inhalation from person to person. Under such conditions a few meningococcal carriers may start an epidemic. Sporadic cases occur, chiefly in young children. Four immunological groups (A-D) of meningococci are distinguished, group A strains being the ones commonly causing epidemics.

(1) Toxins

Potent toxic substances can be obtained from cultures ; they are believed to be, not true exotoxins, but endotoxins which are liberated by the rapid autolysis of the cocci in culture. Meningococci injected into animals do not usually multiply, yet they cause haemorrhages in various tissues and kill the animals ; these effects appear to be wholly due to endotoxins freed from the bodies of the cocci. Injections of dead cocci have the same toxic and lethal effects as injections of living cocci.

(2) Tissue injury and inflammatory reaction

There is little evidence of necrosis in the meninges or other tissues infected by the meningococcus ; but doubtless the tissues are damaged by the potent endotoxin just referred to. The inflammation excited by it is a typical acute one, with hyperaemia, outpouring of fibrinous exudate and migration of polymorphonuclear leucocytes which take up the multiplying cocci in great numbers.

(3) Meningitis and other meningococcal lesions

(i) *Meningococcal meningitis.*—Infection of the meninges is secondary to invasion of the blood stream by the organism. Meningococci are often demonstrable in the blood in early cases of meningitis, especially very acute ones, and they have also been found in the petechial (spotted haemorrhagic) rash which occurs in this disease. Occasionally meningococcal arthritis or other metastatic inflammatory foci occur in cases of meningitis. The pathology of the meningitis itself has already been considered in Chapter 8.

(ii) *Meningococcal septicaemia without meningitis.*—This occasionally occurs. It is accompanied by fever, skin rashes, joint pains and sometimes effusions, and leucocytosis ; and meningococci may be demonstrable by blood culture. Unless treated, the condition may last for months, and the patient may remain remarkably well. Meningitis may or may not supervene. Some of these patients have ulcerative meningococcal endocarditis.

(4) Isolation and identification

The meningococcus is distinguished from the gonococcus by its sugar fermentation reactions. The two organisms closely resemble one another morphologically and culturally, although the meningococcus is rather less fastidious than the gonococcus.

In cases of meningitis a diagnosis can usually be made by the microscopic examination of films prepared from a centrifuged deposit of the cerebrospinal fluid, and stained by the Gram and Leishman methods. Many pus cells are usually present together with the Gram-negative diplococci, many of which lie within the polymorphonuclear leucocytes.

Culture of the centrifuged deposit on fresh-blood agar and heated-blood agar is carried out as soon as possible, bearing in mind the cold-sensitivity of the organism. If only a few organisms are seen in the stained films it is advisable to make an additional culture in heated-blood broth, from which subcultures can be made later. In cases of meningococcal septicaemia without meningitis the organism is recoverable from the blood stream by culture in glucose broth.

NON-PATHOGENIC NEISSERIA

These organisms are not well classified. Two species, *N. catarrhalis* and *N. pharyngis*, are commonly recognized. They are normal inhabitants of the upper respiratory tract and may also occur in the genital tract. They are easily distinguished from the pathogenic *Neisseria* by their sugar fermentation reactions and by the fact that they grow well on simple media.

SUPPLEMENTARY READING

References Chapter 10 (p. 112), and the following :

Bigger, J. W. (1937). " The staphylococci pathogenic for man ", *Brit. Med. J.*, **2**, 837.

Dick, B. M. (1928). " Staphylococcal suppurative nephritis (carbuncle of the kidney) ", *Brit. J. Surg.*, **16**, 106. (Good coloured figures.)

Turner, P., and Dent, R. V. (1949). " Fulminating meningococcal septicaemia ", *Brit. Med. J.*, **1**, 524.

Williams, R. E. O., Blowers, R., Garrod, L. P. and Shooter, R. A. (1966). Second edition. *Hospital Infection: Causes and Prevention.* London; Lloyd-Duke.

CHAPTER 12

ACID-FAST BACILLI AND THE LESIONS THEY PRODUCE

THE ACID-FAST BACILLI OR MYCOBACTERIA

THE GENUS *Mycobacterium* consists of slender bacilli which stain with difficulty, but which, when stained, resist decolorization with acid. The degree of acid-fastness of different species varies. Some of them, e.g. the tubercle bacillus, also resist decolorization with alcohol, i.e. are alcohol-fast. The first acid-fast bacillus discovered was the leprosy bacillus, by Hansen in 1874. The human tubercle bacillus was discovered by Koch in 1882.

While there are only three well-known species that cause human disease, a number of others are occasional pathogens for man and several are animal pathogens. With the mammalian types first, these are:

(i) *M. tuberculosis* (human type) is a strict parasite; it causes by far the majority of cases of human tuberculosis, and some cases in monkeys, dogs and parrots. By subcutaneous or intraperitoneal injection it is highly pathogenic for the guinea-pig, producing disseminated tuberculosis of lymph nodes, spleen, liver and lungs in 6 weeks (one of the tests for suspected material, e.g., sputum, pleural effusion, urine, cerebrospinal fluid). It does not grow on ordinary media, but grows slowly on special media containing enrichments such as serum, egg and glycerine.

(ii) *M. bovis* is a strict parasite, producing tuberculosis in cattle, horses, pigs, goats and cats, and a variable proportion of cases of human tuberculosis. By injection it is also highly pathogenic for the guninea-pig and rabbit, the susceptibility of the latter being an important test for distinguishing bovine from human tubercle bacilli. The bovine bacillus is broadly similar to the human in its cultural requirements, (but see Table below).

(iii) *M. leprae* is the cause of leprosy, in the lesions of which the bacilli are very plentiful. It has not been cultured, and it is non-pathogenic for laboratory animals.

(iv) *M. ulcerans* and *M. marinum* (*balnei*)—the swimming bath mycobacterium —are the cause of chronic ulcers of the limbs in man.

(v) *Anonymous, atypical* or *indeterminate mycobacteria* comprise a variety of strains isolated from human and other sources; some are pathogenic, others are saprophytes or commensals. The human pathogens are found in respiratory diseases, especially pneumoconiosis, and in cervical lymphadenitis.

(vi) *M. microti*, which causes tuberculosis in voles, is only mildly pathogenic for guinea-pigs, rabbits and calves, and is of no importance in human pathology.

(vii) *M. leprae-murium* causes a leprosy-like disease of rats. The bacillus has not been cultivated, and is non-pathogenic for other laboratory animals.

(viii) *M. paratuberculosis* (*Johne's bacillus*) causes a chronic enteritis in cattle and sheep, but is non-pathogenic for laboratory animals.

127

(ix) *M. avium* causes tuberculosis in birds and occasionally in pigs and other domesticated animals, and it is reported to have infected man on very rare occasions. It is non-pathogenic for guinea-pigs.

(x) *Acid-fast bacilli of cold-blooded animals* are present in tubercle-like diseases in fish, amphibia and reptiles ; they are non-pathogenic for mammals and birds. They grow readily on ordinary media at room temperatures.

(xi) *Non-pathogenic acid-fast bacilli* are found in milk, butter, water, grass, faeces, manure, smegma, etc. They grow readily on ordinary media. Smegma bacilli (*M. smegmatis*) may occur in urine and must be distinguished from tubercle bacilli. Acid-fast bacilli in water may multiply in the interior of laboratory taps whence they may contaminate staining solutions and glassware and may be mistaken for tubercle bacilli.

The mycobacteria are strict aerobes and are non-motile, non-capsulated and non-sporing. They are Gram-positive but are often difficult to stain by this method.

MYCOBACTERIUM TUBERCULOSIS AND M. BOVIS

The main distinguishing features between *M. tuberculosis* and *M. bovis*, and the non-pathogenic mycobacteria, are summarized in the accompanying table.

Some characteristics of acid-fast bacilli

Organism	Growth rate	Stimulation by		Pathogenicity		Niacin production	Growth on simple media
		Glyercol	Pyruvate	rabbit	guinea-pig		
M. tuberculosis	+ +	+	−	+	+ + + +	+ + +	−
M. bovis	+	−	+	+ + + +	+ + + +	±	−
Saprophytic mycobacteria	+ + + +	......	......	−	−	......	+

Isolation and identification

Microscopic examination of pathological material may reveal the presence of acid-fast bacilli, and this is often sufficient to establish a diagnosis. Their absence from microscopic preparations, however, does not exclude the possibility of tuberculosis.

In pulmonary tuberculosis a specimen of sputum may be smeared directly on a slide, or the sputum may be concentrated before the smear is made. Concentration may be carried out by treating the specimen with a solution of sodium hydroxide, which liquefies the mucus and destroys cellular material and many bacteria, but does not harm the tubercule bacilli. After treatment the specimen is neutralized with hydrochloric acid and is centrifuged; films are then prepared from the centrifuged deposit. Smears are stained by the Ziehl-Neelsen method and are carefully searched for the presence of acid-fast alcohol-fast bacilli. The organisms are often present only in small numbers, and at least three separate specimens should be examined before a negative finding is recorded. When present they appear as red rods, 3–5μ long, against a blue-stained background material. They are often slightly curved and tend to stain irregularly so that they present a beaded appearance. They commonly occur in pairs, and in small groups in which the organisms lie together in parallel bundles.

In tuberculosis of the urinary tract films are prepared from the centrifuged deposit of a 24-hour or early morning specimen of urine, and are stained and examined as described above (see p. 149).

In tuberculous meningitis films are prepared from a centrifuged deposit of the cerebrospinal fluid, or from the clot should one develop in the specimen of fluid (see p. 147).

Biopsy material such as a lymph gland or piece of synovial membrane is best examined histologically for evidence of tuberculous infection. The organisms may sometimes be seen, however, in smears prepared from ground-up tuberculous tissue.

An alternative to the Ziehl-Neelsen staining technique is the use of fluorescent microscopy in which films are treated with a fluorescent dye. Examination by ultraviolet light shows the organisms as bright fluorescent rods against a darker background.

Culture.—Special solid egg-containing media such as Löwenstein-Jensen medium are used for the isolation of the tubercle bacillus. This medium contains glycerol, which enhances the growth of *M. tuberculosis*, and malachite green which suppresses the growth of unwanted contaminants. Pyruvate is sometimes included in the medium to stimulate the growth of *M. bovis*. The fact that glycerol and pyruvate enhance the growth of *M. tuberculosis* and *M. bovis* respectively may be used as a point of distinction between the two (see Table, p. 128).

Owing to the slow growth rate of the tubercle bacillus the medium is prepared in screw-capped bottles to prevent its dessication during prolonged incubation. Specimens such as faeces and sputum which contain many other organisms are concentrated as described above prior to culture, but this is unnecessary with " sterile " specimens such as cerebrospinal fluid and biopsy material. Typical colonies of *M. tuberculosis* and *M. bovis* are not developed in less than 4–6 weeks at 37°C. Colonies of the former are typically heaped-up, with a granular dry surface and a "bread crumb " or " cauliflower " appearance; they are usually buff-coloured. On the other hand, colonies of *M. bovis* are smooth and flat and of a creamy colour.

Animal inoculation.—The pathological material, which has been concentrated if necessary, is injected intramuscularly into the right hind legs of a pair of guinea-pigs. These animals are highly susceptible to infection by both *M. tuberculosis* and *M. bovis* and show evidence of infection in 6–8 weeks if viable organisms are injected. The first guinea-pig is killed after 6 weeks and is examined for evidence of tuberculosis, particular attention being paid to the inguinal and para-aortic lymph glands, the spleen, liver and lungs. If the first guinea-pig shows no evidence of infection, the second animal is killed and examined 8 weeks after inoculation. If the post-mortem appearances of tuberculosis are present, films are made from some of the lesions to ensure that acid-fast alcohol-fast bacilli are present. This is necessary, because guinea-pigs are susceptible to infection by *Pasteurella pseudo-tuberculosis*, a Gram-negative, non-acid-fast bacillus, which produces lesions very similar to those of tuberculosis.

TUBERCULOSIS

Tuberculosis is a chronic inflammatory disease involving any tissue of the body and due to *M. tuberculosis* or *M. bovis*. As the name implies, the disease typically produces focal inflammatory nodules or " tubercles ".

(1) Entry of tubercle bacilli into the body

On rare occasions the foetus *in utero* receives tubercle bacilli from its mother via a tuberculous uterus and placenta—*congenital tuberculosis*. Rarely also, accidental inoculation of the skin by tubercle bacilli occurs, e.g. in butchers handling tuberculous meat. Save for these two modes of infection, which are so infrequent as to be negligible, tubercle bacilli enter the body in only two ways, by inhalation and by ingestion.

Inhalation of bacilli of human type, in infected droplets coughed up by patients with pulmonary tuberculosis or in dust contaminated by dried sputum, is the commonest mode of infection. The inhaled bacilli may infect the tonsils and be carried thence to the upper cervical lymph glands, but more often they reach the bronchi and lungs to produce lesions there and in the tracheo-bronchial glands.

Ingestion of raw milk containing bovine bacilli from tuberculous cows is a serious source of infection of young children in communities whose dairy herds have much bovine tuberculosis. Where the herds are healthy, however, or where the milk is pasteurized, the incidence of infection by the bovine bacillus is usually very small. Human bacilli in the air may also be ingested as well as inhaled. Ingested bacilli may enter the tonsil and cervical lymph glands ; but more often they pass down to the small intestine (their viability is not affected by the acid gastric juice) to enter the lymphoid patches of the mucous membrane whence they pass to the mesenteric lymph glands.

There are thus *three primary fields* of infection by tubercle bacilli, (i) tonsils and cervical lymph glands, (ii) lungs and mediastinal lymph glands, (iii) intestine and mesenteric lymph glands (Plate VI, Fig. 1). In cases of tuberculosis of the cervical glands, although the tonsils may show little clinical evidence of progressive disease, if they are removed and examined microscopically, small tubercles are often found in them. Similarly, a small clinically quiescent tuberculous focus in a lung may be the source of gross disease in the bronchial and mediastinal lymph glands. In children with tuberculosis of the mesenteric glands and peri-toneum, the original point of entry of the infection in the intestinal mucosa often shows no visible lesions at all.

Tubercle bacilli are non-motile. How, then, do they reach the regional lymph glands from their portal of entry in the tonsillar, respiratory or intestinal mucosa ? They are taken up and carried there by wandering macrophages via the lymphatics These cells act as scavengers of all sorts of foreign particulate matter on the respiratory and alimentary mucous surfaces ; for example they gather up inhaled carbon particles from the lungs and bronchi and carry them to the bronchial and mediastinal lymph glands, which thus become heavily carbonized in all city dwellers. So also, inhaled silica particles, extravasated

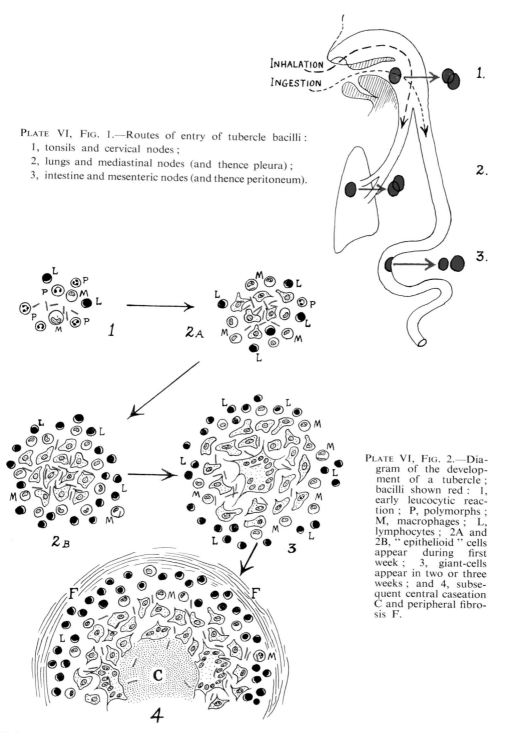

PLATE VI, FIG. 1.—Routes of entry of tubercle bacilli:
1, tonsils and cervical nodes;
2, lungs and mediastinal nodes (and thence pleura);
3, intestine and mesenteric nodes (and thence peritoneum).

PLATE VI, FIG. 2.—Diagram of the development of a tubercle; bacilli shown red: 1, early leucocytic reaction; P, polymorphs; M, macrophages; L, lymphocytes; 2A and 2B, " epithelioid " cells appear during first week; 3, giant-cells appear in two or three weeks; and 4, subsequent central caseation C and peripheral fibrosis F.

(To face page 130)

blood pigment, or experimentally introduced dyes or other materials are transported by pulmonary phagocytes to the glands ; and it is only to be expected that tubercle bacilli should be similarly dealt with.

(2) The fate of the primary lesions

Before the great fall in the incidence of tuberculosis in many Western countries during the last two decades, in most urban communities post-mortem search disclose evidence of old healed tuberculous lesions in one or more of the three primary fields of infection in the majority of adults. Thus it was clear that most people became infected with tubercle bacilli during childhood or youth. But the majority did not develop progressive disease; they overcame the primary infection at an early stage. There is good evidence (to be outlined later) that this successful suppression of the initial infection confers increased resistance against subsequent infection. In a minority of cases, however, bodily resistance is insufficient to check the initial or subsequent infections, the lesions in lungs or lymph glands progress, and frequently bacilli are disseminated to other parts of the body.

(3) Factors influencing susceptibility

From what has just been said it will be clear that individual resistance or susceptibility is a factor of prime importance in determining whether or not a given person will develop progressive clinically important disease. The metaphor is a good one : " The soil is even more important than the seed " ; the " seeds " germinate to produce active disease only when the tissue " soil " is suitable for them. Factors which influence susceptibility to tuberculosis are of the following possible kinds:

(i) *Acquired immunity.*—This results from slight previous infections, and is discussed later. Suffice it to say here that if a young child, or an adult who has never been exposed to infection before, suddenly inhales or ingests a large dose of virulent tubercle bacilli, he usually develops rapidly progressive disease.

(ii) *Debility and malnutrition.*—There is no doubt that general health and nutrition are of the greatest importance in determining individual resistance to tuberculosis. Overcrowding, poor food, vitamin deficiencies and insufficient sunlight diminish bodily resistance to many infections, especially tuberculosis. The prevention of this disease is therefore largely an economic and sociological problem. The main requisite in its treatment is to build up general health and nutrition in every way. Certain chronic illnesses, e.g. diabetes, chronic nephritis and alcoholism, also predispose to active tuberculosis.

(iii) *Respiratory infections.*—Infections such as colds, influenza, bronchopneumonia, whooping cough or measles, sometimes activate previously quiescent tuberculous foci in the lungs.

(iv) *Pulmonary silicosis.*—This strongly predisposes to chronic pulmonary tuberculosis, which therefore is specially prevalent in miners and others whose occupations expose them to silica dust—" miner's phthisis " (p. 316).

(v) *Inherited predisposition.*—It used to be supposed that children of tuberculous parents were more susceptible than others ; but this is probably not so.

It is true that these children acquire active tuberculosis more frequently than others and that therefore the disease " runs in families ". But this is due largely or wholly to the greater chance of cross infection of children in these families or to nutritional or other environmental conditions common to the family ; true genetic factors probably play little or no part.

(4) Tissue injury and reaction in tuberculosis

Just how tubercle bacilli injure the tissues we do not know. Exotoxins have not been demonstrated in cultures. It is significant that dead bacilli, killed by heat, provoke the formation of small nodular lesions closely resembling true tubercles. Presumably the dead bacilli disintegrate or are taken up and destroyed by phagocytic cells, with the liberation of endotoxins which then act characteristically on the tissues. Probably the same occurs in true tubercles due to living bacilli but in these the reaction is sustained and progressive because multiplication of the bacilli continues as fast as or faster than their destruction, and so the supply of material from which endotoxins are liberated is maintained.

(5) The structure of early tubercles

The reaction excited by tubercle bacilli is similar in all tissues ; it appears as a minute inflammatory nodule or granuloma with a characteristic structure. By studying the development of early tubercles experimentally in animals (e.g. in the guinea-pig's peritoneum after intraperitoneal inoculation of suitable doses of bacilli), the stages of this development can be followed and timed accurately. Disseminated lesions in man have a similar structure and pass through similar stages. For descriptive purposes, let us consider their development in 4 stages, remembering that these overlap and pass into one another (Plate VI, Fig. 2 and Figs. 42-45).

(a) Mild transient acute inflammatory reaction

During the first few days, the tissues around the little colony of bacilli show slight hyperaemia, and leucocytes, both polymorphonuclear and mononuclear, collect at the spot. Some of the bacilli are taken up by the leucocytes, and some of the leucocytes perish.

(b) Congregation of mononuclear and " epithelioid" cells and lymphocytes

Mononuclear phagocytes continue to accumulate and phagocytose many bacilli ; polymorphonuclear leucocytes almost disappear. By the end of a week the centre of the little nodule has come to consist of a collection of large pale irregular and branched cells which have been called " epithelioid " cells, a bad name, since the cells do not really resemble epithelium. There has been much dispute as to the precise origin of these cells, whether from immigrated macrophages, tissue histiocytes, vascular endothelia, or fibroblasts. In view of the fact that many of these kinds of mesenchymal cells are certainly intermutable in proliferative lesions, this dispute is pointless. The " epithelioid " cells of tubercles are proliferating mesenchymal cells, probably derived from any of the foregoing kinds of cells, and swollen and abnormal in appearance because they are sick cells which are being acted on by the toxins of the multiplying bacilli within and around

them. About the core of " epithelioid " cells, a zone of lymphocytes and monocytes is present ; these cells have migrated from surrounding blood vessels and
lymphatics and have also multiplied locally.

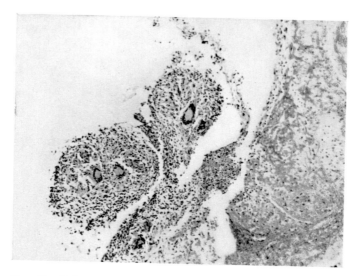

FIG. 42.—Tubercles on the pial surface of the brain in a case of tuberculous meningitis. (× 70.)

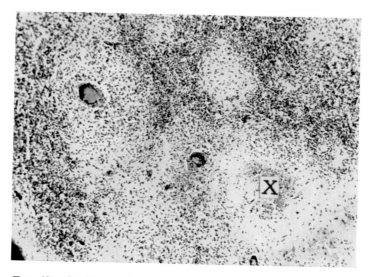

FIG. 43.—Confluent tubercles in a lymph gland ; early caseation
at X. (× 70.)

(*c*) *Giant-cell formation*

In some human, and more animal, lesions the centres of the enlarging tubercles may continue to consist of mononucleated " epithelioid " cells only ; but in most human tubercles more than 2 weeks old multinucleated giant cells appear. These are large irregular cells with short processes and many peripheral nuclei. They are formed by fusion of " epithelioid " cells, and are similar in appearance to

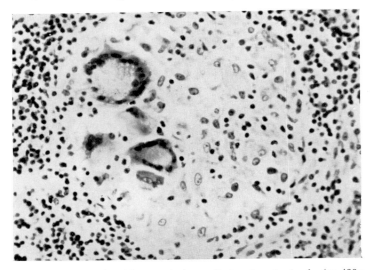

FIG. 44.—A tubercle with several giant-cells in a lymph gland. (× 400.)

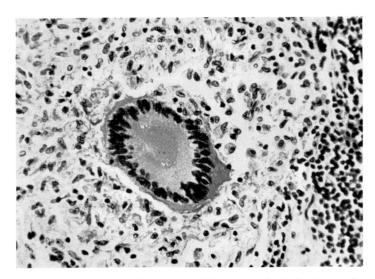

FIG. 45.—A large tuberculous giant-cell, showing well the finely granular cytoplasm. (× 350.)

the foreign-body giant cells which develop around foreign particles or lipoid material in other inflammatory and degenerative lesions. Indeed the giant cells of tubercles *are* foreign-body giant cells which have formed around tubercle bacilli or some of their degenerative products in the tissues ; they often contain demonstrable bacilli or fatty granules or droplets. At this stage then a typical tubercle consists of three zones : (*a*) one or more central giant cells, (*b*) a zone of swollen pale irregular mesenchymal ("epithelioid") cells, and (*c*) a peripheral zone of lymphocytes and monocytes. The whole tubercle, often spoken of as a "giant-cell system", now measures a millimetre or more in diameter and is visible to the naked eye as a tiny greyish semi-translucent nodule in the tissues.

(*d*) Central caseation and peripheral fibrosis

The tubercle, unlike foci of chronic inflammation of most other kinds, is quite devoid of blood vessels ; the "epithelioid" and giant-cell tissue may indeed be regarded as avascular granulation tissue, which is unhealthy because it is avascular and because it is infested with tubercle bacilli. As the tubercle enlarges, the "epithelioid" and giant cells in its centre undergo a slow degeneration and necrosis cell by cell, to form a solid yellow mass of granular fatty debris. Because this resembles cheese, it is spoken of as "caseous" material, and this kind of degeneration is called "caseation". To the naked eye the tubercle, now 3 or 4 weeks old or more, appears as an opaque yellow nodule 2 or 3 millimetres in diameter. In its peripheral zone more lymphocytes and macrophages have accumulated, and, external to these and mingled with them, proliferating fibroblasts of the surrounding tissues are establishing a wall of young fibrous tissue. The structure of the tubercle is now fully developed ; surrounding a central mass of caseous debris there is a zone 0·5 to 1 millimetre thick of unhealthy tuberculous granulation tissue consisting of "epithelioid" cells and giant cells, and external to this a zone of congregated lymphocytes and monocytes and proliferating fibroblasts. Further enlargement of the tubercle does not alter this characteristic structure ; the layer of tuberculous tissue continues to caseate centrally as fast as it extends peripherally. Tubercle bacilli are present in both the unhealthy granulation tissue and in the caseous matter ; but they are seldom numerous, and careful search is often needed to discover them in sections or smears.

(6) Later growth of tubercles

The small caseous tubercle enlarges in two ways, (*a*) by concentric peripheral extension and increasing central caseation, as just described, and (*b*) by the development of new satellite tubercles in the surrounding tissues and their fusion with the parent focus. The formation of these satellite foci is due to the transport of tubercle bacilli from the initial focus by wandering macrophages. These, having carried bacilli a certain distance, have failed to destroy them and have themselves perished and liberated the bacilli in new territory, where they multiply and evoke the development of new foci of disease. As the satellite tubercles enlarge, they coalesce with each other and with the original tubercle to form a *conglomerate tubercle ;* this continues to grow by peripheral extension and by the formation of still further satellite tubercles. However large such a conglomerate lesion becomes—and it may be inches in extent—it consists of caseous debris surrounded by the usual narrow zone of still living tuberculous tissue, external to which is a variable zone of accumulated leucocytes and fibrosis.

In such progressing lesions the central caseous matter may soften and liquefy, forming a " *cold abscess* ". The contents of this is not true pus but merely liquefied fatty debris mingled with a few degenerating cells ; it usually contains very few tubercle bacilli, but sometimes they are numerous. If a caseous focus or cold abscess extends to an epithelial surface (e.g. skin, bronchus, or renal pelvis), it discharges and forms an ulcer or sinus lined by tuberculous tissue.

(7) The healing of tubercles

If the patient's resistance is good, the extension of tuberculous foci, small or large, is often arrested ; the zone of unhealthy granulation tissue diminishes and disappears, and the zone of peripheral fibrosis consolidates and forms a fibrous capsule to the quiescent lesion. Non-caseous tubercles undergo complete fibrosis. When caseous matter is present, it is not absorbed ; it loses water, becomes dry and friable, and often attracts calcium salts and becomes converted into a hard calcareous mass. An old healed lesion thus appears either as a mass of fibrous scar tissue often with gritty areas scattered through it, or as a larger calcified mass encapsulated in fibrous tissue. Guinea-pig inoculation with, and culture of, the material from such lesions show that in some of them the tubercle bacilli have all been exterminated ; the lesions are therefore truly healed. But in other quiescent lesions a few viable bacilli sometimes persist for long periods ; and subsequent malnutrition or other factors adverse to the patient's resistance may result in multiplication of these bacilli and reactivation of the diseased foci.

(8) Acute diffuse tuberculosis

An exception to the rule that tuberculosis is a chronic disease with nodular lesions of characteristic structure is seen in severe rapidly extending lesions in people with very poor resistance. Such lesions occur most often in the lung as " galloping consumption " or " caseous pneumonia ", but similar lesions may occur in other tissues. They show diffuse accumulations of mononuclear and polymorphonuclear leucocytes and epithelioid cells, which undergo prompt caseation, without the development of tubercles, giant cells or fibrosis. The caseous change spreads rapidly through the diseased tissues, so that within a few weeks a whole lobe of a lung may be converted into a structureless homogeneous yellow mass of necrotic material. Acute lesions of this kind usually teem with tubercle bacilli.

(9) The spread of tuberculosis in the body

Tubercle bacilli may spread in the body by any of the available routes, namely— (i) tissue spaces, (ii) lymphatics, (iii) the blood stream, (iv) coelomic and cerebrospinal cavities, and (v) epithelial cavities.

(i) *Spread in tissue spaces.*—We have already noted how, from established tuberculous foci, bacilli are carried by phagocytes into the surrounding tissues, where they cause the development of new satellite foci.

(ii) *Spread by lymphatics.*—As we have seen, tubercle bacilli entering the body through the tonsils, lungs or intestine are often carried via the lymphatics to the regional lymph glands.   Once in the lymphatic system, they frequently spread from gland to gland, probably partly by carriage in phagocytes and partly in the lymph stream itself.   Lymphatics are also important routes of local spread within organs, e.g. lungs, kidneys or intestine.   The lymph glands draining a tuberculous part are frequently affected, and sometimes the intervening lymphatics are plainly visible as solid nodular yellowish strands completely filled and distended by tuberculous tissue—tuberculous lymphangitis.   Good examples of this are often seen in the lymphatics connecting tuberculous ulcers of the small intestine with neighbouring mesenteric lymph glands.   The lymph glands regional to a tuberculous joint, though they are often not palpably enlarged, are nevertheless frequently tuberculous.   This fact is sometimes of diagnostic value ; e.g. if tuberculosis of the knee joint is suspected, this diagnosis may sometimes be con-firmed by excising an inguinal lymph gland and examining it microscopically, culturally and by guinea-pig inoculation.

(iii) *Spread by the blood stream.*—Tubercle bacilli enter the blood stream in several ways, namely by the extension of tuberculous foci into the walls of blood vessels, or into the thoracic duct, or by tuberculous lymphangitis of other lymphatic tributaries of veins.   In micro-sections of tuberculous organs it is not unusual

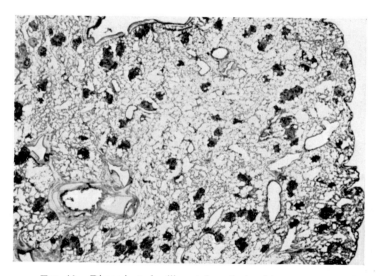

FIG. 46.—Disseminated miliary tuberculosis of lung.   (× 4.)

to see tuberculous granulation tissue extending into the lumen of small veins ; and occasionally grossly visible examples of this are encountered, e.g. a caseous focus in a bronchial or mediastinal lymph gland may extend and rupture into a pulmonary vein or even into the innominate vein or superior vena cava.   The escape of many bacilli at once into the circulation results in the condition known as " *generalized miliary tuberculosis* ", which is usually fatal in a few weeks and in which many small discrete tubercles, all of about the same age and size (usually 0·5 to 2 millimetres) are found widely scattered through the lungs, spleen, liver, kidneys and other organs (Fig. 46).* The disease is accompanied by pyrexia, anaemia, loss of weight and other general symptoms not referable to any particular organ ; clinically it is difficult to diagnose, and may be confused with typhoid fever, bacterial endocarditis or septicaemia. Indeed it *is* a tuberculous septicaemia ; blood cultures for tubercle bacilli are often positive. Miliary tubercles may be present in the choroid, and ophthalmoscopic examination is therefore important in diagnosis. Occasionally, generalized blood-borne tuberculosis runs a less severe course, lasting months or years, with recovery in rare instances. When only a few bacilli escape into the blood stream, only a few metastatic lesions—perhaps only one—may develop in remote parts. These may heal or may appear later as isolated progressive lesions of a joint, bone, kidney or other part. From people suffering from pulmonary tuberculosis or other apparently localized tuberculous disease, blood cultures for tubercle bacilli are occasionally positive, proving the intermittent escape of bacilli into the blood stream from such lesions. A high proportion of subjects dying of pulmonary tuberculosis show gross or microscopic lesions in other organs due to blood-borne bacilli.

(iv) *Spread in coelomic or cerebrospinal cavities.*—Tuberculous pleurisy frequently accompanies pulmonary lesions. Tuberculous pericarditis is less common ; it usually arises by extension from diseased mediastinal lymph glands. Tuberculous peritonitis is usually secondary to tuberculosis of the intestine and mesenteric glands, less often to tuberculous salpingitis. In these coelomic infections, if bacilli escape freely into the cavity, scattered tubercles develop on many parts of the serous membrane. In other cases, adhesions develop and restrict this transcoelomic dissemination. In tuberculous meningitis, bacilli are carried widely in the cerebrospinal fluid and scattered tubercles develop on many parts of the pial, arachnoidal and dural surfaces (see p. 147).

(v) *Spread in epithelial cavities.*—Tubercle bacilli from an established lesion are often carried in the contents of hollow viscera to set up secondary lesions remote from the primary one. Within the lungs themselves, the transfer of bacilli in the sputum is a very important mode of spread of the disease : the infected sputum, aspirated into previously healthy bronchi and lung tissue, sets up new areas of tuberculous bronchopneumonia. Tuberculous ulceration of the larynx, or more rarely of other parts of the upper respiratory passages or mouth cavity. is always due to bacilli implanted from tuberculous sputum. In the very sick patient swallowed sputum commonly causes tuberculous ulceration of the intestine. Tubercle bacilli discharged into the urine from a tuberculous kidney frequently produce lesions in the ureter and bladder. Tuberculous salpingitis often causes tuberculous endometritis.

By one or other of these several routes tubercle bacilli may reach and set up lesions in any part of the body. In considering the regional pathology of tuberculosis, let us remember that the changes are similar in all tissues. In any situation we may see the formation of small tubercles, their coalescence to form large caseous areas surrounded by a narrow zone of pale tuberculous granulation tissue, liquefaction of caseous matter to form cold abscesses, ulceration and discharge of caseous foci on neighbouring surfaces with the formation of

---

* " Miliary " means " like millet seed " ; it refers to the presence of numerous small discrete tubercles, which, however, vary in size from case to case.

tuberculous cavities or sinuses, and, in very chronic or healing cases, calcification of caseous material and fibrosis of the surrounding tissues. The special pathology of tuberculosis is merely this general pathology modified in details by the structure of the particular tissues affected.

(10) Tuberculosis in the respiratory system

(a) Primary infection and reinfection

Almost all respiratory tuberculosis is due to direct inhalation of human tubercle bacilli. On the first occasion on which bacilli are inhaled, usually during childhood, they set up a small primary focus (a Ghon focus) in the lung, whence bacilli are carried to the hilar lymph glands and produce caseous lesions there. In some children this primary infection progresses and quickly leads to fatal generalized miliary tuberculosis, but in most patients it is recovered from and the lesions become healed or quiescent. Because of the increased resistance resulting from this successful dealing with the initial infection, subsequent re-infections pursue a different course ; they usually produce chronic lesions restricted to the lung and unaccompanied by fresh lymph-glandular disease, and, in contrast to the primary Ghon focus which may be situated almost anywhere in the lung, reinfective lesions usually start at or near the apex.

(b) The modes of extension of bacilli in the lung

The first tubercles to develop from either a primary infection or a reinfection are probably interstitial in position, resulting from inhaled bacilli which are taken up from the air spaces by phagocytes and carried a certain distance into the adjacent septal tissues or lymphatics. These interstitial tubercles enlarge and coalesce, and the infection spreads to the surrounding lung tissue in three ways, (a) by interstitial and lymphatic extension and the formation of satellite tubercles, as already described, (b) by local intra-alveolar and intra-bronchial spread, and (c) by aspiration of bacilli into neighbouring or remote bronchi and alveoli, setting up tuberculous bronchopneumonic lesions in fresh areas. Every new area of disease becomes a focus of further spread in these three ways.

c) The appearances of chronic tuberculous lungs

These are very diverse. The lesions may be restricted in extent or widespread throughout both lungs ; they may be rapidly progressive, or chronic, fibrous or calcified ; extension may have been mainly local within the alveoli, or mainly lymphatic and interstitial, or mainly aspirational and bronchopneumonic ; and one lung may show in different parts every variety of lesion. In the usual chronic but progressive form of the disease, the apical or subapical tubercles enlarge and coalesce to form an area of *caseous conglomerate tubercle;* this soon involves a neighbouring bronchus, through which the caseous matter is expelled, leaving a ragged-walled *cavity* lined by caseating tuberculous tissue. Peripheral to this are scattered *satellite tubercles* formed by lymphatic spread, and a variable zone of *fibrosis.* Coalescence of the satellite foci with the main focus proceeds and the cavity enlarges, its discharging contents forming the sputum which the patient expectorates and which contains few or many tubercle bacilli. Small or large blood vessels are exposed in the wall of the enlarging cavity, and dilatation and rupture of their eroded walls causes the more or less profuse haemorrhages (*haemoptyses*) which are so common in this disease.   While the disease is extending, the walls of the enlarging cavity are soft, yellowish and ragged.   In long-standing chronic cases, the walls become smoother and fibrous and often trabeculated by the more resistant bronchi and vessels.   In the wall of a cavity apertures of communication with one or several bronchi will be found.   Old cavities may attain a large size, sometimes excavating the greater part of a lobe.   In addition to cavities produced by tuberculous erosion and excavation of lung tissue, there

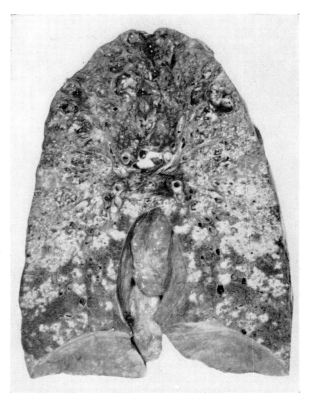

Fig.   47.—Advanced   pulmonary   tuberculosis,   showing multiple cavities in upper parts and confluent tuberculous bronchopneumonia in basal parts of lung.

are sometimes others produced by bronchial obstruction and dilatation—*bronchiectasis*.   Such bronchiectatic cavities are lined by inflamed mucosa, which however may become partly destroyed by tuberculous ulceration.   Tuberculous or bronchiectatic cavities may suffer *secondary infection* by pyogenic cocci, pneumococci or other bacteria—a serious complication which is accompanied by a persistent swinging pyrexia or " hectic fever " and is liable to cause amyloid disease.   In many advanced cases more or less extensive areas of patchy or confluent *tuberculous bronchopneumonia* develop in other parts of the lung or in both lungs, and these caseate, cavitate and extend like the original lesion.

The one case may show all of the varieties of lesions italicized in this paragraph, as well as the *pleural lesions* now to be described. The oldest lesions and largest cavities are usually in the apical parts of the lungs, and the most recent ones in the basal parts (Fig. 47).

(d) Pleural and parietal lesions

Pleural changes almost always accompany pulmonary tuberculosis. They occur in the following forms, or any combination of them. (i) *Simple fibrous thickening of the visceral pleura* may be seen overlying a focus of disease, especially at the apex of the lung. (ii) *Adhesions,* local or widespread, between the lung and parietes may result from attacks of clinically evident pleurisy or may develop silently without obvious pleural symptoms. (iii) *Pleurisy with effusion* is a common complication of, and sometimes the first indication of the existence of, tuberculosis of the lungs. The extension of a small active tuberculous focus to the pleural surface of a non-adherent lung often causes a sudden attack of pleurisy with effusion of clear serous or sero-fibrinous fluid into the pleural cavity. The effusion usually contains only a few cells, mainly lymphocytes, and very few tubercle bacilli, so that these are difficult to find microscopically and even guinea-pig inoculation may give a negative result. In spite of frequently negative findings of this kind, it is certain that most cases of pleurisy with effusion of unproved nature are really tuberculous and result from small pulmonary lesions many of which subsequently heal. (iv) *Tuberculous empyema* is a cold abscess enclosed by pleural adhesions, and due to a pleural focus undergoing progressive caseation. (v) *Pneumothorax* occasionally results from perforation of a tuberculous focus through non-adherent pleura with sudden escape of air into the pleural cavity and collapse of the lung. This may be partial or complete according to the presence or absence of adhesions. Pleural effusion may accompany pneumothorax, giving *hydro-pneumothorax,* or, if secondary infection is present, *pyo-pneumothorax.* (vi) *Broncho-cutaneous fistula* occasionally develops if a pleural collection already communicating with a bronchus also extends externally through the chest wall. (vii) *Tuberculosis of the ribs or sternum* sometimes results from direct extension of underlying pleural and pulmonary lesions.

(e) Acute pneumonic or " galloping " tuberculosis

This exception to the rule of chronicity has already been referred to ; it is an overwhelming infection in a patient of poor resistance. It shows in an extreme degree rapid extension of caseating bronchopneumonic lesions over much or the whole of one or both lungs. Sometimes the disease is mainly *lobar* in distribution, usually in the upper lobe, often with a sudden onset and close clinical simulation of lobar pneumonia in both symptoms and physical signs. Only as the disease progresses and the signs of massive lobar consolidation begin to change to those of softening and cavitation may the tuberculous nature of the disease be suspected. At necropsy the affected lobe or lobes are found to be massively converted into solid, bloodless, yellow caseous material, usually with early softening and cavity formation in the apical part. More frequently, acute phthisis is *bronchopneumonic* in distribution, with yellow patches of caseous pneumonia scattered through both lungs and often confluent and cavitating in one or both apices. Like the lobar form, acute bronchopneumonic phthisis may appear suddenly, by rapid extension of disease from a small focus not previously suspected ; or it may develop terminally in a patient with advanced chronic phthisis. Pulmonary tuberculosis in childhood is usually of the acutely progressive type.

(*f*) *Chronic fibroid phthisis (Phthino, Gk. = decay)*

All chronic tuberculous lesions are accompanied by some fibrosis in the surrounding tissues, and around healing or quiescent foci fibrosis is conspicuous and achieves encapsulation of the lesions. In cases of very slowly progressive phthisis, fibrosis becomes a prominent feature ; the peribronchial, septal and pleural tissues become greatly thickened, bronchi become compressed and distorted, obstructive bronchiectasis appears, areas of lung parenchyma become collapsed and shrivelled ; the diseased lung, which is always densely adherent to the chest wall, contracts and produces displacement of the mediastinum and heart and asymmetrical deformation of the thorax, and the opposite lung becomes enlarged and emphysematous. Extension of the tuberculous lesions, embedded in fibrous tissue, is slow ; and the patient may have fairly good health for many years, interspersed with febrile attacks due to intermittent flare-up of tuberculous lesions or of secondary infection in old cavities or dilated bronchi. Discharges from these maintain a variable amount of sputum, in which a few tubercle bacilli are present often along with many other bacteria. Chronic fibroid phthisis often develops in cases of silicosis (see p. 316).

(*g*) *Tuberculosis of the larynx or other parts of the upper air passages*

Tuberculosis of the larynx is always secondary to tuberculosis of the lungs, the laryngeal mucosa being infected from the bacillus-laden sputum. Tubercles develop in the sub-epithelial tissues, producing pale nodular thickening of the mucous membrane ; these break down to form irregular ulcers with ragged undermined edges, which slowly deepen and extend, sometimes reaching the cartilages and causing widespread necrosis of them. Early lesions, however, may heal as the pulmonary disease and general health improve. Similar lesions occasionally occur in the pharynx, mouth or nasal cavity.

(11) Tuberculosis in the alimentary system

Except for scattered miliary tubercles in cases of generalized dissemination by the blood stream, tuberculous lesions in the stomach, duodenum, pancreas, liver or biliary tract are very rare. Alimentary tuberculosis is almost restricted to the intestine, principally the ileum and ileo-caecal junction. In children it is sometimes a primary infection ; but it is much more frequent as a secondary infection from swallowed sputum in cases of advanced phthisis. The tubercles develop especially in Peyer's patches, where they coalesce, caseate and break down, forming irregular ulcers with undermined edges and yellowish bases. As these enlarge, they become elongated transverse to the long axis of the gut and they may completely encircle it, the extension following the lymphatics (Fig. 48). For this reason also, lines of tuberculous lymphangitis or scattered tubercles are almost always visible in the serous coat of the bowel or adjacent mesentery, and the neighbouring mesenteric lymph glands are enlarged and tuberculous. Only occasionally do ulcers perforate and cause peritonitis. Sometimes they produce fibrous stricture and intestinal obstruction. A special type of primary intestinal lesion, not accompanied by active phthisis, is *chronic ileo-caecal tuberculosis*, in which a large fibrous mass with scattered caseous areas involves the wall of the caecum,

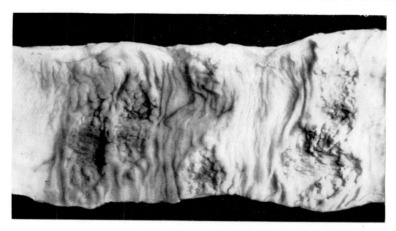

Fig. 48.—Tuberculous ulceration of small intestine, showing characteristic irregular ulcers spreading circumferentially. (Museum of Royal College of Surgeons, London.)

ileo-caecal junction, appendix and adjacent peritoneum and lymph glands. This causes intestinal stenosis and obstruction, and is often diagnosed clinically as carcinoma. Except for this kind of ileo-caecal lesion, tuberculosis of the large intestine is rare.

(12) Tuberculosis in serous membranes

(a) *Tuberculous pleurisy* has already been sufficiently described on p. 141.

(b) *Tuberculous peritonitis*

This affects chiefly children and young people, and is usually due to direct extension from some intra-abdominal tuberculous focus, in the intestine, mesenteric lymph glands or Fallopian tubes. It assumes one or other of three forms, according to the extent and severity of the infection. (i) *An acute form*, with rapidly developing widespread tubercles, is rare. (ii) *A chronic ascitic form*. This differs from the acute form only in its rate of development. Numerous discrete small tubercles develop all over the peritoneal surfaces and are accompanied by an abundant serous or slightly turbid effusion into the peritoneal cavity. Bacilli are often difficult to discover in the fluid—but cultures or guinea-pig inoculations are usually positive. (iii) *A chronic adhesive form*. More or less extensive fibrous adhesions develop around caseating areas of disease, binding loops of intestine and other viscera together and often producing intestinal obstruction or fistulae.

(c) *Tuberculous pericarditis*

This results from extension of tuberculosis from the lung and pleura or from the mediastinal lymph glands. Both the visceral and parietal pericardial surfaces develop a layer of tuberculous tissue, and between the two layers there is either an effusion, turbid or blood-stained, or a layer of solid yellow caseous matter (Fig. 49). Very chronic cases show much fibrosis and dense adhesion of the parietal pericardium to the heart. A considerable proportion of cases of chronic adhesive pericarditis are tuberculous.

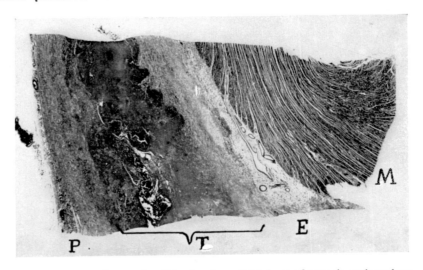

Fig. 49.—Tuberculous pericarditis, showing a thick layer of caseating tuberculous tissue T between parietal pericardium P and epicardium E. M, myocardium. (× 4.)

(13) Tuberculosis in bones and synovial tissues

(a) Incidence and sites

Tuberculosis of bones and joints is commonest in children 2 to 15 years old, in whom it affects especially the vertebrae and the joints or epiphyseal ends of the long bones of the lower limb—the hip, knee and ankle in that order of frequency. Adults are less frequently affected, and the principal sites of the disease in them are shoulder, elbow, wrist and knee. The special susceptibility of children to skeletal tuberculosis and the predilection of the disease for certain sites are attributed to the growth activity of the parts and their liability to slight injuries (compare with osteomyelitis, p. 68). Bone and joint tuberculosis are so often combined that they must be considered together. It is uncertain whether tuberculous arthritis usually results from extension of a focus of disease in the epiphysis or starts in the synovial membrane; undoubtedly both occur. In bones, the initial focus is much more often in the epiphysis than in the diaphysis. The disease is of course always secondary to tuberculous lesions elsewhere, though these are often not clinically apparent. The bacilli reach the bone or synovial tissues via the blood stream. The proportions of human and bovine infections differ greatly in different communities; in those with infected herds of cows 20 to 40 per cent of cases have been due to bovine bacilli, while in those with clean tuberculin-tested herds almost all are due to human bacilli.

(b) Pathology of tuberculosis in bone

While tuberculous tissue similar to that seen in other situations develops in the Haversian and marrow spaces, the bone trabeculae in and around the diseased area suffer progressive absorption.   The bone appears to be slowly

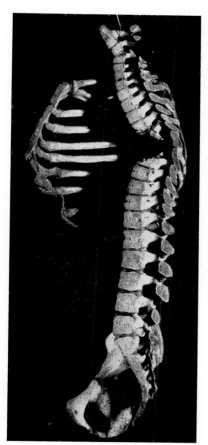

and irregularly eroded by the extending tuberculous tissue, a process called *caries* (decay) (Fig. 50).   As the tuberculous tissue extends it undergoes caseation and often liquefies to form *cold abscesses*.   These increase in size and extend into the surrounding soft tissues where they may form collections of ounces or pints of acellular creamy fluid ; e.g. from a focus of disease in the spine a huge psoas abscess may develop in the psoas sheath down which it may extend to present in the groin.   Discharge of a cold abscess onto the surface leaves a chronic tuberculous sinus. Small sequestra of eroded isolated bone fragments are often formed and may be found in the contents of the cold abscess. Little or no new bone is formed around tuberculous lesions ;   repair, if it occurs, is by fibrous tissue.   When, however, mixed pyogenic infection is admitted to an area of diseased bone through an open sinus, it may excite much new bone formation from the periosteum around the diseased area. Long-standing mixed infection of such lesions often ends in amyloid disease.

(c) Pathology of tuberculosis in joints

Whether it is infected directly through the blood stream or from a neighbouring focus in the bone, the synovial membrane of a tuberculous joint quickly becomes studded with tubercles, which enlarge and coalesce to replace the membrane by extending areas of tuberculous granulation tissue.   This erodes and undermines the articular cartilages, separating loose flakes of them, and produces carious erosion of the bone ends.   The ligaments and periarticular tissues become oedematous, and eventually cold abscesses track through them and form

FIG. 50.—Spinal tuberculosis in a child, seen in median section ; a preparation of John Hunter's preserved in the Royal College of Surgeons, London. The bodies of several vertebrae in the mid-dorsal region are destroyed, with consequent collapse and angulation of this part of the spine.

discharging sinuses through the skin.   In the more chronic, less destructive forms of tuberculous arthritis, multiple ovoid or flat white pellets known as " melonseed " or " rice-grain " bodies may form in the joint cavity.   These develop by accretion of fibrin and albuminous material around nuclei of detached synovial or cartilaginous fragments, and they are moulded by mutual pressure and move-ment. ("Melon-seed bodies" are not peculiar to tuberculosis, but occur also in other forms of chronic arthritis, bursitis or tenosynovitis, e.g. in rheumatoid and osteo-arthritis, or in the traumatic pre-patellar bursitis known as "house-maid's knee".)

(d) Tuberculosis of the vertebrae (Pott's disease)

This merits special notice because the spine is the commonest site of tuber-culosis of bone, and the most serious. The disease starts in a vertebral body in any part of the spine. The bone suffers carious destruction, which often leads to its collapse and resulting dorsal angulation (kyphosis) of the spinal column (Fig. 50). The disease frequently extends under the anterior and lateral ligaments to erode neighbouring vertebrae ; or it extends through the ligaments into the surrounding soft tissues, producing psoas, lumbar, retroperitoneal, mediastinal, retropharyngeal or lateral-cervical cold abscesses. A common and serious complication is involvement of the spinal cord with consequent partial or total *paraplegia* (paralysis) below the diseased level. This results from compression and vascular disturbance of the cord by a tuberculous accumulation beneath, and displacement of, the posterior longitudinal ligament, or less commonly from extension of tuberculous tissue through the ligament into the extradural space. It probably never results from a spinal angular deformity alone.

(e) Tuberculous dactylitis

This rather rare disease affects either phalanges, metacarpals or metatarsals. It usually starts in the centre of the diaphysis, the destruction of which is often accompanied by expansion of and new bone formation by the periosteum, so that the bone assumes a fusiform outline. Cold abscess formation, sinuses and second-ary infection may ensue ; but sometimes the infection is less destructive and the lesion heals.

(f) Tuberculous tenosynovitis

This is most common in the large flexor sheath at the wrist, but it occurs also in other tendon sheaths. The synovial membrane becomes thickened and studded with tubercles, and the sheath is distended by serous or sero-caseous fluid in which "melon-seed" or "rice-grain" bodies may develop. At first the tendons remain unaffected, but later in badly caseating disease there may be destruction of one or more of them.

(g) *Tuberculous bursitis* of the subacromial, ileopsoas or gluteal bursa occasionally develops without any disease of the neighbouring joints.

(14) Tuberculosis in the central nervous system

(a) Tuberculous meningitis

This common and serious disease is always due to blood-borne bacilli from a focus elsewhere in the body. It is most frequent in children, in whom it is usually part of generalized miliary tuberculosis ; it is less common in adults, as a complication of phthisis or some other tuberculous lesion, which may or may not be clinically apparent. At necropsy, a scanty or abundant fibrinous exudate is found in the sub-arachnoid spaces and on the brain surfaces, especially on the basal aspect and around the brain-stem ; and tubercles are usually visible when searched for on the pial surfaces, along the vessels in the Sylvian and other larger fissures, and on the inner surface of the basal dura mater, especially over the clivus and pituitary region.   The foramina of Luschka and Magendie are partially occluded by the tuberculous exudate, the ventricles are distended by opalescent or slightly turbid cerebrospinal fluid (internal hydrocephalus), and tubercles can sometimes be seen on the choroid plexuses or ependymal surfaces.   Microscopi-cally, tubercles of varying ages and of characteristic structure, with or without caseation, are found on the diseased surfaces and in the exudate ; and similar tubercles and exudate are often present in the Virchow-Robin spaces around the vessels entering the brain from its surface (see Fig. 42).   Similar changes are present in the spinal meninges.   *Clinically,* the symptoms are those of a subacute slowly progressive meningitis, which is usually fatal in a few weeks.   *The changes in the cerebrospinal fluid* are easily understood from the pathology of the disease and are briefly as follows : pressure raised ; appearance, clear or opalescent, a fine cobweb-like fibrin coagulum often separating on standing for a few hours ; cytology, an excess of cells, usually about 100 to 300 per cubic millimetre, most of which are lymphocytes, but with polymorphonuclear cells also in the early stages or in acuter cases ; tubercle bacilli usually few and difficult to find, but often discoverable by careful search of the cobweb coagulum, or by culture or guinea-pig inoculation ; raised protein content and positive tests for globulin, and markedly reduced chloride content down to 600 or 500 milligrams per 100 cubic centimetres.

(b) Tuberculous masses (" tuberculomas ") of the brain

Occasionally, blood-borne bacilli produce localized slowly developing caseating tuberculous masses in the brain or cord, just as in other tissues.   These may measure a few millimetres or even 2 or 3 centimetres in diameter, and may be either solitary or multiple.   As such a " tuberculoma " enlarges, it may cause symptoms like those of cerebral tumour ; or, after a long period of slow growth and clinical dormancy, it may erupt into a ventricle or into the subarachnoid space and produce tuberculous meningitis.   Some workers believe that most cases of tuber-culous meningitis arise in this way, from originally solitary small foci in the brain or meninges.

(15) Tuberculosis in the urinary tract

(a) Tuberculosis of the kidney

In cases of *generalized miliary tuberculosis,* the kidneys always share in the dissemination and contain scattered tubercles in both cortex and medulla.   The more common and important lesions of *chronic renal tuberculosis* start from similar tubercles, solitary or few in number, due to bacilli blood-borne from a primary focus in the lungs or elsewhere, though this is often not clinically active.   The initial renal tubercle, usually in the cortex, enlarges and produces satellite tubercles in the surrounding kidney tissues ; these coalesce with the parent focus, to produce a caseating conglomerate mass.   This may attain a diameter of 1 or 2 centimetres, while still being confined to the kidney substance and hence not producing any changes in the urine.   More often, however, it soon involves the pelvis at some point, discharges its caseous contents into the urine, and now appears as a cavity lined by yellow tuberculous tissue and communicating with the pelvis (Fig. 51). From now on, the urine is mixed with variable amounts of caseous debris, leuco-cytes and tubercle bacilli ; the bacilli in the urine infect other parts of the pelvic mucosa, whence additional ulcerative lesions develop and extend into the kidney substance. Haemorrhages from eroded vessels often occur, causing attacks of haematuria, usually only slight or moderate in amount but sometimes profuse. By peripheral lymphatic extension from established lesions, tubercles develop in the neighbouring cortex and beneath the renal capsule ; until eventually the whole kidney may be affected. At the same time, the bacilli in the urine infect also the ureteric and vesical mucosa, leading to ulceration and thickening of the walls of these organs, and often to ureteric obstruction with consequent distension of the renal pelvis by infected urine and caseous material (tuberculous pyonephrosis). Sometimes, if the infection is a mild one, complete ureteric occlusion occurs, the diseased kidney becomes functionless and no longer a source of continued discharge of bacilli to the bladder, and it eventually forms a

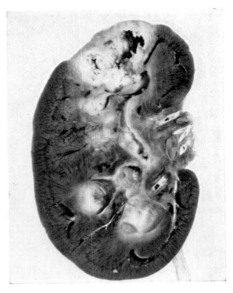

FIG. 51.—Renal tuberculosis. The upper pole of the kidney contains a large ragged caseating cavity which communicates with the pelvis. (Museum of Royal College of Surgeons, London.)

multiloculated fibrous sac filled with inspissated caseous partly calcified masses devoid of bacilli. However, such spontaneous cure is unusual ; in most cases, well established tuberculous lesions in the kidney progress, maintain urinary infection and necessitate surgical removal of the diseased organ.

(b) Tuberculosis of the ureter

We have just noted infection of the ureter from the tuberculous kidney. This is almost always present in cases of actively discharging renal tuberculosis. From the tuberculous pelvis, the disease extends directly into the upper ureteric wall via tissue spaces and lymphatics ; but, much more important, bacilli in the urine infect the mucosa of the lower parts of the ureter also, producing ulceration here and fresh foci of spread into its walls at all levels. It is thus common to find that the entire ureter of a tuberculous kidney is greatly thickened, rigid, infiltrated by tubercles, and its lumen narrowed. For this reason, the surgeon performing nephrectomy usually removes as much of the ureter as possible, even as low as its entry into the bladder.

(c) Tuberculosis of the bladder

Tuberculosis of the bladder, like that of the ureter, is usually secondary to renal tuberculosis, but occasionally the bladder may be infected from the seminal vesicles and prostate. When secondary to renal disease, vesical infection usually appears first around the orifice of the affected ureter and spreads thence over the trigone. The mucous membrane becomes congested and oedematous, and later small shallow ulcers develop ; tubercles are sometimes visible around the ulcers. If the diseased kidney and ureter are excised and the source of constant reinfection thereby removed, the lesions in the bladder often heal. In cases in which nephrectomy is neglected or impracticable, tuberculous infiltration extends through the whole wall of the bladder, which becomes thickened, contracted and inelastic ; and the infection may then spread up the opposite ureter to the other kidney.

(d) The urinary changes with tuberculosis of the urinary tract

These are haematuria, pyuria and the presence of tubercle bacilli. (i) *Haematuria*. Blood corpuscles are often present in small numbers and are detected by microscopic examination ; more or less profuse haematuria may also occur. (ii) *Pyuria*. Leucocytes, both polymorphonuclear cells and lymphocytes, are present, usually in moderate numbers, so that the urine is only slightly turbid ; but in severe cases, or if mixed infection of the bladder has supervened, the urine may be heavily loaded with pus cells. (iii) *Tubercle bacilli*. In films of the deposit stained by the Ziehl-Neelsen method, tubercle bacilli are sometimes easy, sometimes difficult, to find. Failure to discover them by direct examination should be followed by culture and guinea-pig inoculation. It has sometimes been observed that, during the course of a tuberculous lesion elsewhere in the body, tubercle bacilli are found in the urine without any other evidence of renal tuberculosis, and it has been inferred that the kidney can at times excrete tubercle bacilli from the blood without suffering infection. This inference is almost certainly erroneous; thorough microscopic examination of the kidneys from such cases will often disclose small tubercles. Nevertheless, this observation is of interest in showing that tuberculous bacilluria is consistent with the absence of any gross progressive lesion. It is also possible that minute lesions of this kind may sometimes heal and leave no trace behind.

(16) Tuberculosis in the genital systems

(a) Tuberculosis in the male genital tract

Any part may be affected, from the testis to the prostate. There has been some dispute as to how often male genital tuberculosis is due to spread from the urinary tract ; but, while this probably occurs in some cases, there is little doubt that the infection of the genital organs often takes place directly by way of the blood stream. The usual site of such initial infection is the epididymis, which becomes enlarged, firm and caseous, while the testis and cord are at first unaffected. Later, however, the disease extends to these parts both by the lymphatics and within the duct lumina ; the testis becomes studded with coalescing tubercles, and the vas deferens and cord become thickened and nodular. The infection may reach the seminal vesicles and prostate ; and it is possible that some cases of tuberculosis of the bladder and urinary tract arise in this way. Caseating lesions in the epididymis sometimes extend to and discharge through the skin of the scrotum ; while those of the vesicles or prostate occasionally discharge into the rectum.

(b) Tuberculosis of the female genital tract

The usual initial lesion is a tuberculous salpingitis due to blood-borne bacilli ; but sometimes infection of the tubes is secondary to an established tuberculous peritonitis. Tubercles develop in the mucous coat which becomes swollen and proliferates ; caseous matter collects in and distends the tube ; tubercles often appear on the serous coat ; and, as a result of local tuberculous peritonitis, adhesions develop between the tube and ovary and other parts. Tuberculosis of the ovary or of the uterus is usually secondary to tubal disease.

(17) Other sites of tuberculosis

(a) The skin

Tuberculosis of the skin, due in most instances to blood-borne bacilli, is not common. Though generally similar in its pathology in most cases, it assumes a variety of appearances which have been given special names (often very long ones) by the dermatologists.

(i) *Lupus vulgaris* is the least uncommon form of cutaneous tuberculosis. It is a slowly spreading infection of the dermis, in which groups of typical giant-cell tubercles form small nodules visible in the stretched or compressed skin as " apple-jelly " nodules. Lupus is most common in the skin of the face or neck, but may occur in any part of the body. Squamous-cell carcinoma of the skin sometimes supervenes on long-standing lupus.

(ii) *Disseminated miliary tubercles* of the skin are occasionally seen in children with acute generalized miliary tuberculosis.

(iii) *Rare chronic tuberculides* (i.e. tuberculous skin lesions) specially named by the dermatologists include lichen scrofulosorum, acne scrofulosorum, acne agminata, erythema induratum (Bazin's disease), and Boeck's " sarcoid " (see below). Pathologically these are not distinct diseases, but merely different variants of the one disease, cutaneous tuberculosis.

(iv) *Scrofuloderma* is the name applied to the local tuberculous infection of the skin around chronic sinuses connected with underlying lesions, e.g. in lymph glands, bone or epididymis.

(v) *Tuberculous wart* develops at the site of local inoculation of tubercle bacilli in the skin, e.g. in butchers or pathologists.

(b) The breast

Tuberculous mastitis is a rare lesion, which produces a slowly developing firm nodular mass in the breast, usually mistaken clinically for carcinoma. Diagnosis depends on histological examination and the demonstration of tubercle bacilli ; the latter is particularly important, since chronic coccal mastitis often produces focal lesions containing foreign-body giant-cells which may easily be mistaken for tubercles.

(c) The adrenal glands

The commonest cause of chronic adrenal insufficiency or *Addison's disease* is bilateral fibro-caseous tuberculosis of the glands (p. 607). There is often no other focus of active tuberculosis in the body.

(d) The spleen

In generalized miliary tuberculosis the spleen is always studded with tubercles. Not uncommonly, multiple caseous or calcified nodules, 2–5 millimetres in diameter, undoubtedly old healed tubercles, are found scattered through the spleen in persons dying of other diseases (Fig. 52). Large active tuberculous lesions in the spleen are extremely rare.

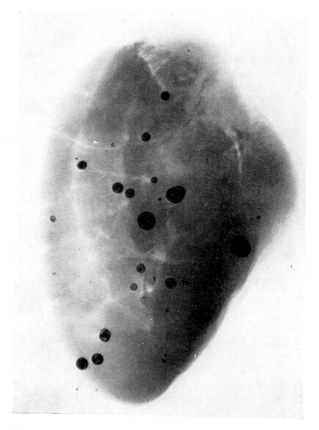

FIG. 52.—X-ray photograph (unretouched positive print) of a spleen containing multiple healed calcified tuberculous nodules.

(18) Boeck's " sarcoidosis "

The bad name " sarcoid " was applied by the dermatologist Boeck to the skin lesions of what was later recognized by Schaumann to be a widespread chronic nodular granulomatous disease affecting also lymph glands, bone marrow, spleen, lungs and other organs. The disease is now usually referred to as " sarcoidosis ", Boeck's disease, or Besnier-Boeck-Schaumann disease. Histologically, the lesions in all tissues consist of nodular collections of epithelioid cells and giant-cells, exactly resembling non-caseating tubercles. The lesions develop slowly and regress by fibrosis and hyalinization, while other nodules develop ; there is often little or no interference with general health ; and the disease runs a benign protracted course which may terminate in complete recovery. Although tubercle bacilli are often not demonstrable in the lesions either in sections or by guinea-pig inoculation, and although the patient often has a negative Mantoux test, there are good grounds for believing that most—possibly all— cases of " sarcoidosis " are really examples of chronic miliary tuberculosis in subjects of high resistance. These grounds are—(a) the similarity of structure and distribution of the lesions of " sarcoidosis " and those of proved chronic miliary tuberculosis, (b) the occasional discovery of bacilli in cases which had previously been called " sarcoidosis ", (c) the well-known rarity of demonstrable bacilli in many very chronic tuberculous lesions, and (d) the fact that a negative Mantoux test is given by some cases of chronic miliary tuberculosis in which the diagnosis is not in doubt. Probably the negative Mantoux test of most " sar- coidosis " patients is actually related to their high resistance to tuberculosis. For further discussion, see Cameron and Dawson (1946) and Scadding (1960 and 1967).

(19) Immunity and allergy in tuberculosis

Our knowledge of the defence mechanism of the body against tubercle bacilli is still very incomplete, and opposing views on this subject make it a particularly confusing one for the student. The following brief outline of it is appropriate here, though it anticipates some of the contents of Chapter 21 which should be read in conjunction with it.

(a) General facts indicative of acquired immunity in tuberculosis

We have already touched on some of these. Tuberculosis acquired during the first two years of life usually runs a rapidly fatal acute course terminating with generalized miliary dissemination ; hence the importance of protecting young children from sources of infection. In older children or adults, who have encountered tubercle bacilli in non-lethal doses previously (and most adults in many communities have done so), there is a much greater resistance to the disease, which therefore is usually chronic. When isolated native communities are sub- jected to infection for the first time, they show extreme susceptibility, and the disease often assumes an epidemic form with a high proportion of acute rapidly fatal cases. Striking instances were recorded of the American Indians, Senegalese, some of the Pacific Islanders, and the people of the Outer Hebrides. From the foregoing facts we may infer that people who have never inhaled or ingested any tubercle bacilli have a poor resistance to the disease; while those who do receive small doses of bacilli, insufficient to cause progressive disease, develop an increased resistance. Exposure to tubercle bacilli is therefore a desirable thing, provided that it does not occur in early childhood and that the dose of bacilli is not excessive. Whatever the immunizing effect may be which follows exposure to infection early in life, it is far from absolute and it often fails to protect against massive reinfection later on.

(b) Koch's phenomena

Koch discovered that the reaction of an already tuberculous guinea-pig to a fresh injection with tubercle bacilli is different from that of a normal guinea-pig to a primary infection.   If a non-tuberculous guinea-pig is inoculated subcutaneously with tubercle bacilli, in 10 to 14 days a tuberculous nodule forms at the site and, in a few weeks, it ulcerates and continues to discharge until the animal dies of generalized tuberculosis later.   But if a tuberculous guinea-pig is given a second subcutaneous injection of bacilli, one of two things may occur according to the extent of the lesions already present and the dose of the second inoculation of bacilli.   If the primary lesions are extensive or the second dose of bacilli is a large one, the animal often dies in 6 to 48 hours.   But if the primary infection has not yet spread widely and the second inoculation is small, the animal survives and in the course of 2 or 3 days the skin at the site of this inoculation becomes haemorrhagic and necrotic, forming a dead slough which is later thrown off leaving a clean ulcer which soon heals.   The reason why the guinea-pig with extensive lesions from the primary infection dies quickly after re-infection is found to be because this causes similar acute reaction around each of the established tuberculous foci, from which a rapid and fatal absorption of tuberculous products takes place.   These results—namely that the tuberculous animal reacts differently from the normal animal to a new tuberculous infection and that such re-infection excites violent changes in and around existing tubercles—constitute " Koch's phenomena ".   Koch's phenomena are obtained also when dead tubercle bacilli are used for the second inoculation.   Dead bacilli injected into a normal guinea-pig produce only a mild temporary local inflammation and no constitutional effects at all.   But a small dose injected into a tuberculous guinea-pig produces local necrosis of the skin and some general toxaemia ; a large dose kills the animal in 6 to 48 hours.   The tuberculous animal is thus hypersensitive to the chemical constituents of tubercle bacilli.   This hypersensitiveness is spoken of as *allergy* and the sensitized animal is said to be allergic to these substances. Clearly, the next step, and one which Koch himself took, was to attempt to isolate the chemical components of tubercle bacilli responsible for these phenomena.

(c) Tuberculin

Koch grew tubercle bacilli in glycerine-broth, then killed the culture by heat and concentrated and filtered it.   The product, known as Koch's " Old Tuberculin ", is a mixture of whatever tuberculous exotoxins there may be, the products of dissolution of the dead bacilli (endotoxins), constituents of the culture medium and about 50 per cent glycerine.   Other tuberculins were also made, e.g. by grinding up the bacilli ; and later, purified forms of " old tuberculin " were prepared by growing the bacilli on special synthetic media and by chemical separation of the specific tuberculo-protein responsible for the skin reaction. The potency of tuberculins is measured by their effects on inoculation into sensitized (tuberculous) guinea-pigs.   A normal guinea-pig is very little affected by intradermal or subcutaneous injections of large amounts of tuberculin ; but injection of quite a small quantity into a tuberculous animal may kill it within a few hours, or, if it survives, a local area of skin necrosis develops.   In the animals which are killed by the injection, there is intense congestion at the site of the injection, and congestion and haemorrhages around all tuberculous foci in the lymph glands, spleen, liver or other organs.   Thus the tuberculin has produced all of the features of Koch's phenomena, namely (*a*) a local reaction at the site of injection, (*b*) a focal reaction around the tubercles in the tissues, and

(c) a constitutional or toxaemic reaction terminating fatally. Similar effects are seen in man. A new-born infant, which has never been infected with tubercle bacilli, suffers no ill effects from large subcutaneous injections, e.g. 1 ml., of old tuberculin; a healthy adult who has encountered and overcome tubercle bacilli can take 0·01 ml. without suffering more than mild malaise and limb pains; but this amount injected into a tuberculous patient causes a severe reaction, with high fever, dyspnoea, vomiting, aggravation of existing tuberculous lesions, and inflammation at the site of the injection. Non-tuberculous animals can be sensitized to tuberculin by injections of dead tubercle bacilli.

(d) Tuberculin tests in diagnosis

Since subcutaneous or intradermal injections of tuberculin excite local and general allergic reactions in the tuberculous but not in the healthy person, these can be used as a test for the diagnosis of the presence of tuberculous infection. The original *von Pirquet scratch test* has now been superseded by the *Heaf test* in which a drop of old tuberculin is applied to a small area of skin and is pricked into it with a multiple-puncture gun, and the *Mantoux intracutaneous test* in which a small intradermal injection of old tuberculin is given. Both tests give similar results; in the sensitized subject an area of swelling and redness develops at the site within a few hours and attains its maximum in 24 to 48 hours. A positive reaction does not necessarily denote active tuberculosis; a high proportion of adults give reactions, because of previous subclinical infection. The tests are of greater diagnostic value in infants and young children, in whom strongly positive reactions often do denote active tuberculous infections. Negative tests in adults are also of value; in most cases they mean an absence of previous tuberculous infection, and therefore a lack of immunity and a special susceptibility to tuberculosis. Hence, it is unwise for a tuberculin-negative nurse to work in a sanatorium, where the risk of infection is great.

The Mantoux test is carried out by injecting dilutions of tuberculin intracutaneously. Owing to the risk of serious reactions developing in patients who have been previously exposed to the tubercle bacillus, a high dilution of tuberculin (0·1 ml. of 1 in 10,000) is first injected, and the result read after 48 and 72 hours. If there is no reaction, a second intracutaneous test is performed using 0·1 ml. of 1 in 1,000, or 0·1 ml. of 1 in 100 tuberculin. A patient is considered to be Mantoux-negative only if he fails to react to the 1 in 100 dilution.

(e) The relation between allergy and immunity

There is little doubt that allergy (hypersensitiveness to tuberculo-protein) and immunity (acquired resistance) to tuberculosis are closely related. We know that animals which are mildly or moderately infected with tuberculosis are both hypersensitive and more resistant to reinfection than normal animals. We know that in human beings the absence of allergy to tuberculo-protein is coupled with a high susceptibility to tuberculosis ; the infants of civilized, and the infants and adults of uncivilized or isolated communities, who give negative tuberculin tests, are liable to contract severe acute forms of tuberculosis. There is good evidence also that, in civilized urban communities, strongly tuberculin-positive persons are less likely to contract active disease than weakly positive or tuberculin-negative persons. There are many individual exceptions to the foregoing statements, exceptions which no doubt depend on the state of existing lesions, the risks of heavy infection or reinfection, and nutritional and other environmental factors. But, in general, allergy and immunity in tuberculosis are linked. The tuberculin reaction is the classical example of " delayed hyper-sensitivity ", which is mediated by lymphocytes or other cells that have acquired specific reactivity to foreign protein, and which is not accompanied by antibodies in the blood serum. (See Chapter 21).

(f) Active immunization

Though the value of vaccination against tuberculosis remains to be finally assessed, it does seem that some degree of protection can be conferred. For this purpose a living attenuated strain of *M. bovis* is employed which is referred to as " B.C.G. " (Bacille Calmette-Guérin) after the two French workers who first advocated its use in 1908. Attenuation of the organism is effected by prolonged growth on a bile-glycerol-potato medium. Immunization is carried out by injecting 0·1 ml. of B.C.G. preparation intracutaneously with a tuberculin syringe. Under no circumstances should Mantoux-positive patients be vaccinated with B.C.G. The vole bacillus, which is non-pathogenic to man, may be used as a vaccine against tuberculosis, as an alternative to B.C.G.

LEPROSY

Leprosy is a chronic inflammatory disease caused by *M. leprae*; it is still prevalent in many tropical countries, but has almost disappeared from most parts of Europe. In spite of its evil reputation, it is not highly contagious and is acquired only after long and close contact with the leprous. The mode of transfer from person to person is not known. *M. leprae* has not been cul-tured, and the disease has not been transmitted to animals.

Leprous lesions consist mainly of collections of large mononuclear cells (macrophages and histiocytes) loaded with great numbers of bacilli and called " lepra cells " (Fig. 53) ; multinucleated giant cells, formed by fusion of mononu-clear cells and similarly loaded with bacilli, also occur but are smaller and less numerous than the giant cells of tuberculosis. Leprosy bacilli are found also within the endothelial cells of capillaries and lymphatics, and in the epithelial cells of the epidermis and skin glands when the disease involves the skin.

Clinically, two forms of the disease, the *maculo-anaesthetic* and the *nodular* are recognized ; pathologically, however, these are not sharply distinct. The maculo-anaesthetic form presents flat patches of discolouration and depigmen-tation of the skin, and patchy anaesthesia due to infiltration of underlying nerves by leprous tissue. Any part of the body and any nerve may be affected, especially those of the limbs and face. Later, the nerves may be wholly destroyed, causing muscular paralysis and wasting, atrophy and trophic ulceration of the skin and great distortion of the limbs. In the nodular form of the disease, raised nodular masses slowly develop in the affected areas of the skin, especially of the face, nose and ears, and in the mucous membrane of the nasal cavity and larynx ; and in advanced cases other internal organs are affected. Slow but extensive ulceration of the nodules may ensue. Sometimes the lesions heal, the bacilli die out, and fibrous scar tissue remains.

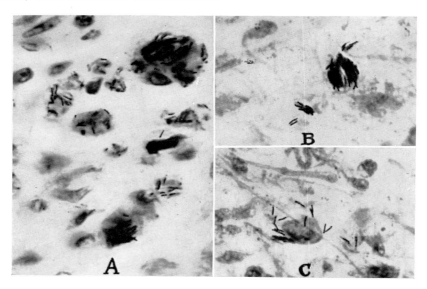

FIG. 53.—Leprosy ; Ziehl-Neelsen stain : A, from a section of leprous skin, showing lepra cells loaded with bacilli ; B and C, from a smear of nasal discharge, showing bacilli in lepra cells and free. (× 1250.)

INFECTIONS BY OTHER MYCOBACTERIA

MacCallum, Tolhurst, Buckle and Sissons in Australia, and Norden and Linell in Sweden described cases of chronic ulceration of the skin in man due to a myco-bacterial infection. In both cases the organism was distinct from the tubercle bacillus and the leprosy bacillus. The Australian and Swedish strains are similar in many ways but are named differently, *M. ulcerans and M. marinum (balnei)* respect-ively. They grow on the same media as the tubercle bacillus, but at an optimum temperature of 30–33°C. They are almost non-pathogenic for the guinea-pig, but are highly pathogenic for mice in which they produce chronic ulceration of the extremities.

The anonymous mycobacteria may be divided into two categories; presumptive pathogens repeatedly isolated from pathological material that has been collected with care to exclude contamination, and " casual " strains isolated only sporadi-cally. Human infections attributable to pathogenic strains include cervical lym-phadenitis, especially in children, less commonly pulmonary infections, and rarely blood-borne focal infections such as osteomyelitis.

SUPPLEMENTARY READING

References Chapter 10 (p. 112), and the following :
Butler, R. W. (1935). " Paraplegia in Pott's disease, with special reference to the pathology and aetiology ", *Brit. J. Surg.*, **22**, 738.

Cameron, C., and Dawson, E. K. (1946). "Sarcoidosis—a manifestation of tuberculosis", *Edinb. Med. J.*, **53**, 465.

Cappell, D. F. (1936). "Some points in the histological diagnosis of tuberculosis", *Edinb. Med. J.*, **43**, 134.

Cave, A. J. E. (1939). "The evidence for the incidence of tuberculosis in ancient Egypt", *Brit. J. Tuberc.*, **33**, 142.

MacCallum, P., Tolhurst, J. C., Buckle, G., and Sissons, H. A. (1948). "A new mycobacterial infection in man", *J. Path. Bact.*, **60**, 93.

Macgregor, A. R., and Green, C. A. (1937). "Tuberculosis of the central nervous system, with special reference to tuberculous meningitis", *J. Path. Bact.*, **45**, 613.

Marks, J. (1964). "The anonymous mycobacteria". In *Recent Advances in Clinical Pathology* (Ed. S. C. Dyke.), Series IV, (Recent advances series), London; Churchill.

Norden, Å., and Linell, F. (1951). "A new type of pathogenic *Mycobacterium*". *Nature*, **168**, 826.

Philip, R. (1932). "Koch's discovery of the tubercle bacillus", *Brit. Med. J.*, **2**, 1.

Scadding, J. G. (1960). "*Mycobacterium tuberculosis* in the aetiology of sarcoidosis". *Brit. Med. J.*, **2**, 1617.

– (1967). *Sarcoidosis*. London; Eyre & Spottiswoode.

Stewart, M. J. (1927). "Healed tuberculoma of the cerebellum", *J. Path. Bact.*, **30**, 577.

Webster, R. (1941 and 1942). "Studies in tuberculosis", *Med. J. Australia*, 1941, Vol. 1, 661; and Vol. 2, 49, 217 and 583; and 1942, Vol. 1, 160.

CHAPTER 13

GRAM-POSITIVE BACILLI AND THE DISEASES THEY PRODUCE

THE PATHOGENIC CLOSTRIDIA

THE ORGANISMS of the genus *Clostridium* are large Gram-positive spore-bearing bacilli, many of which are motile, and some capsulated. A few species are aerotolerant, but most are obligate anaerobes. The anaerobic methods used in their cultivation are described in Appendix D. The spores produced by these organisms may be oval or spherical, and may be placed centrally, subterminally or terminally within the bacillus, usually distorting its shape (compare with the aerobic spore-formers).

The clostridia are widely distributed in nature, being present in soil, water, street dust, etc., and some species are normal inhabitants of the intestinal tract of man. All the species are natural saprophytes, but a few are able to colonize the tissues and produce disease in man and animals. The pathogenic clostridia produce a variety of powerful exotoxins (neurotoxins, haemolysins, lecithinases, lethal toxins, necrotoxins, and so on), and it is these bacterial poisons which are responsible for the severe systemic effects of the clostridial diseases. *Clostridium tetani* is the cause of tetanus in man and animals, *Clostridium botulinum* is the cause of botulism, and several clostridia are able to cause anaerobic myonecrosis (gas gangrene), the most important being *Clostridium Welchii*.

Classification of clostridia

Though there are some 90 recognized species of clostridia, only the ones of clinical importance need be considered here. Identification of the commonly occurring clostridia presents much less difficulty than is generally supposed, and the pathogenic species are easily recognized. Identification depends on a variety of tests, the most important of which are summarized in the accompanying table. The saccharolytic power of an organism is determined by its ability to ferment the sugars glucose, lactose and sucrose with the production of acid and gas, while its proteolytic properties are determined by its ability to digest simple and complex proteins such as gelatin and the proteins of milk. Lipolytic and lecithinase activity are demonstrated by growing the organism on egg-yolk agar. Lipolysis causes an opacity in the medium beneath the growing colonies and the development of an iridescent film, or pearly layer, over the colonies; both the opacity and the pearly layer consist of fatty acids. Lecithinase activity causes a wide zone of opacity in the medium which extends well beyond the edge of the growing colony (the *Nagler effect*) and there is no associated pearly layer. In some cases bacterial and colonial morphology are of importance in identification; *Cl. tetani*, for example, produces a characteristic surface growth and presents a typical microscopic appearance (see p. 164). Animal inoculation experiments are essential finally to prove the toxicity of the pathogenic species.

administration of antibiotics, and (3) possible administration of polyvalent anti-gas-gangrene serum.

(4) The pathology of gas gangrene

Gas gangrene usually develops quickly after the infliction of the wound, from a few hours to 2 or 3 days later. In the worst fulminating cases, it may involve a whole limb within 8 or 12 hours, with severe rapidly fatal toxaemia. In less severe cases, it spreads within an affected muscle or group of muscles, making them oedematous, necrotic and gaseous within a few days. The least serious cases are those in which muscles are not involved, the disease being limited to skin and subcutaneous tissues; here the infection usually remains localized, clears up quickly, and does not produce dangerous toxaemia.

Serious gas gangrene is essentially *an infection of muscle*, commencing at the wounded site and spreading rapidly thence into the healthy muscles. An affected muscle becomes necrotic, discoloured, at first friable, later swollen by oedema and bubbles of gas ; the usual signs of inflammation and suppuration are absent. Microscopically, the fibres become swollen, structureless, and separated from their sheaths by fluid and gas. The infection spreads from end to end of the affected muscle, and may later extend through its fascial sheath to neighbouring muscles. The involved part of the limb is tense and swollen, crepitates on pressure, and is tympanitic to percussion ; the wound exudes watery blood-stained fluid and gas bubbles, and emits a musty or putrid odour. The patient shows signs of severe toxaemia—pyrexia, vomiting, rapid pulse, fall of blood pressure, sometimes with anaemia and slight jaundice due to toxic haemolysis of blood cells. Only in the terminal stages do the bacilli invade the blood stream; and at necropsy on such cases all the organs and tissues are found spongy with gas bubbles and swarming with bacilli, especially the carbohydrate-rich liver, and the blood too is gaseous and full of bacilli.

It remains only to add that occasionally *puerperal or post-abortal infection due to Cl. Welchii* develops at the placental site in the uterus or from cervical lacerations. This produces a serious or fatal spreading infection with gas formation, intravascular haemolysis and severe toxaemia, essentially similar to gas gangrene.

(5) Isolation and identification of organisms

The diagnosis of gas gangrene is usually made on clinical grounds before it is confirmed bacteriologically. The organisms may be seen in Gram-stained films prepared from excised tissue or from the wound exudate. Pathological material is cultured on a variety of media including fresh-blood agar and egg-yolk agar, and, after anaerobic incubation, plates are examined for appropriate surface colonies. *Cl. Welchii* forms discrete circular colonies 2 mm. in diameter and is usually haemolytic on horse-blood agar (Fig. 54). The other species produce colonies with irregular margins, and some of them, especially *Cl. septicum*, tend to spread over the surface of the medium. Facultative anaerobes are common contaminants in gas gangrene infections and should not be confused with anaerobic species in the plate cultures. Robertson's cooked meat broth is a useful fluid medium for the culture of clostridia, since it will support the growth of most species in the ordinary incubator without the use of other anaerobic methods. Identification of the species depends on the various tests outlined in the table above.

Of particular importance are the reactions produced by the different species on egg-yolk agar, *Cl. Welchii, Cl. oedematiens, Cl. bifermentans, Cl. Sordellii* and *Cl. sporogenes* all producing changes in this medium. *Cl. Welchii* produces a

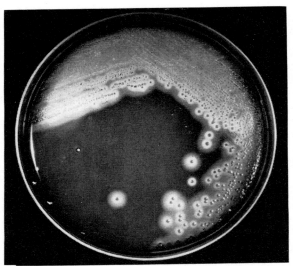

FIG. 54.—*Clostridium Welchii* growing on horse-blood agar. The colonies are surrounded by a central zone of complete haemolysis and an outer zone of partial haemolysis.

typical *Nagler effect*, the growth being surrounded by a wide zone of opacity in the medium, which is due to the lecithinase activity of *Cl. Welchii* α-toxin. Similar lecithinase opacities are produced by *Cl. oedematiens, Cl. bifermentans* and *Cl.*

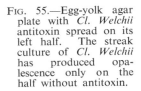

FIG. 55.—Egg-yolk agar plate with *Cl. Welchii* antitoxin spread on its left half. The streak culture of *Cl. Welchii* has produced opalescence only on the half without antitoxin.

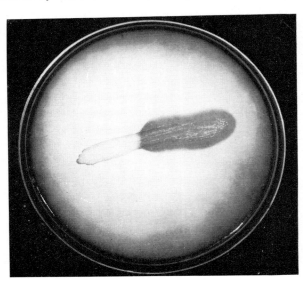

Sordellii. Antitoxins prepared against these different lecithinases inhibit their activity, and use is made of this in the half-antitoxin Nagler plate. In the case of *Cl. Welchii* for example, a little *Cl. Welchii* type A antitoxin is spread over half the surface of an egg-yolk agar plate, and the plate is then streaked with a culture of the organism at right angles to the antitoxin area (Fig. 55). During incubation the organism grows and produces its lecithinase which diffuses out into the medium. On the antitoxin-treated half of the plate, the lecithinase is neutralized so that no opacity is produced. Lipolysis, with its restricted opacity and pearly layer is produced in egg-yolk agar by *Cl. oedematiens* type A and *Cl. sporogenes*.

The action of a clostridium on milk may be determined either by growing the organism in sterile milk as a fluid culture, or by culturing it on an agar plate medium containing 10 per cent of milk. Proteolysis causes zones of clearing in milk agar, while in fluid cultures the milk is usually first clotted and the clot is then rapidly digested. Strongly saccharolytic organisms such as *Cl. Welchii* produce very little change in milk agar, but in fluid milk a characteristic " stormy clot " is produced; the organism rapidly ferments the lactose in the milk with the production of acid, which causes the milk to clot, and gas, which then disrupts the clot.

TETANUS

Tetanus is a specific clostridial infection, usually of a wound, characterized by severe spasmodic contraction of muscles due to the effect of the exotoxin on the nervous system.

(1) The tetanus bacillus

Clostridium tetani is a motile Gram-positive anaerobic bacillus, distinguished from the other *Clostridia* by its large spherical terminal spores which give the bacillus its " drum-stick " appearance (Fig. 38). Tetanus bacilli and spores are found in cultivated soil and in animal (and sometimes human) faeces.

(2) Tetanus toxin

Culture filtrates contain an extremely powerful neurotoxin which is called *tetanospasmin* because of its effect in producing the violent muscular contractions or spasms which are characteristic of tetanus. Injected into guinea-pigs or mice, extremely minute doses of the toxin produce typical tetanic convulsions which kill the animals within 24 hours. The spasms may appear first in groups of muscles near the site of injection (" local tetanus "), just as may happen in infections in man. Tetanus neurotoxin acts on the central nervous system, where it appears to diminish or abolish synaptic inhibition. The route by which the toxin reaches the susceptible cells has been the subject of dispute for many years, but it now seems certain that the toxin passes from the lesion to the central nervous system by way of the regional motor nerve trunks, and then up the spinal cord.

Tetanolysin is an oxygen-labile haemolysin which is closely related to other oxygen-labile haemolysins such as streptolysin O and *Cl. Welchii* θ-toxin.

Tetanus antitoxin is prepared by immunizing horses with increasing doses of tetanus toxoid and crude toxin. Antitoxic serum is of great value as a prophylactic measure, when given soon after the infliction of tetanus-prone wounds. It is of less value as a therapeutic agent once tetanus has developed.

Active immunization against tetanus may be carried out by giving two doses of tetanus toxoid subcutaneously or intramuscularly at an interval of 6 weeks, followed by a booster dose 9–12 months later. Maintenance reinforcing doses of 1 ml. are then given every 5–10 years, or when the person sustains a potentially dangerous wound.

(3) Modes of infection with tetanus bacilli

Tetanus bacilli or spores may gain access to the tissues in the following ways :

(i) *By accidental wounds*.—Dirty wounds, contaminated with soil, manure or other foreign bodies, are the commonest source of tetanus ; but tetanus may result also from small otherwise trifling wounds, e.g. a cut or puncture contaminated with dirt.

(ii) *By surgical wounds*.—Occasionally post-operative tetanus results from infection of surgical wounds. Some defect in theatre sterility usually accounts for this.

(iii) *Tetanus neonatorum*.—This occasionally results from infection of the umbilicus of the new-born infant.

(iv) *Post-abortal tetanus*.—Infection of the uterus with tetanus bacilli has been observed following septic abortion

(4) The pathology of tetanus

Except for the presence of a tetanus-infected wound and the convulsive results of the absorbed exotoxin, tetanus has no pathology. The bacilli remain localized to the initial site of infection and do not invade the tissues ; there are no signs of general toxaemia ; and the disease is produced entirely by the highly specific action of the absorbed toxin on the nervous system. In fatal cases, death is due to exhaustion and to asphyxia resulting from spasm of the respiratory muscles.

(5) Isolation and identification of *Cl. tetani*

The diagnosis of tetanus is made on clinical grounds. Bacteriological confirmation may be obtained by microscopic and cultural examination of material from a wound which is the obvious focus of infection. Gram-stained films may show the presence of typical " drum-stick " bacilli, but the organism may be present in such small numbers as to escape detection. Material from the wound is cultured on a variety of media which, after anaerobic incubation, are examined for the presence of tetanus bacilli. On solid media *Cl. tetani* produces a fine continuous film of growth. The organism does not ferment any sugars and is only slightly proteolytic. Once isolated, confirmation of its identity is obtained by animal inoculation experiments.

BOTULISM

Botulism is a rare but very dangerous form of food poisoning due to the contamination of preserved foods (ham, sausage, canned meats and vegetables) by the Gram-positive, anaerobic, sporing, motile bacillus, *Clostridium botulinum*. In culture media and in contaminated foods, this bacillus produces a powerful neurotoxin. The disease is not due to the formation of toxin by the bacilli in the intestine, but to absorption of toxin already formed in the contaminated food. *Cl. botulinum* is not parasitic at all—neither multiplying in the intestine nor invading the tissues; it is a pure saprophyte, widely distributed in soil and found on fruit, vegetables, leaves, etc. Its spores withstand boiling for several hours, and insufficient heating during canning or bottling of foods is the usual cause of botulism.

Symptoms appear within a few hours of eating the poisoned food. They are mainly those of paralysis of motor nerves, especially oculomotor and pharyngeal paralysis and fatal paralysis of respiration.

The bacteriological diagnosis of botulism is made by demonstrating the presence of botulinum neurotoxin in suspected food and in the contents of the alimentary tract. Culture of suspected material may give positive results; but the isolation of the organism is, alone, of no significance, for we have noted that the disease is a pure intoxication, not an infection. Six types of *Cl. botulinum* are recognized (types A-F) which differ from one another in the antigenicity of their neurotoxins. Types A, B, E and F are the ones responsible for botulism in man. The cultural reactions of *Cl. botulinum* are very similar to those of *Cl. sporogenes*, except that types C, D and E are non-proteolytic. Surface colonies are large and show a marked tendency to spread. On egg-yolk agar the organism is strongly lipolytic.

DIPHTHERIA

Diphtheria is a specific bacterial infection of a mucosal or skin surface, usually of the throat and usually of a young child, characterized by the local formation of an adherent membrane due to superficial necrosis of the affected surface, and by the production of a powerful exotoxin which is absorbed and which seriously injures the myocardium and other tissues.

(1) The bacillus of diphtheria, *Corynebacterium diphtheriae*

The *Corynebacteria* are Gram-positive, non-motile, non-sporing, non-capsulated bacilli with characteristic beaded staining properties and metachromatic granules, and a striking tendency to produce club-shaped and other involution forms in culture. The bacilli are often arranged in parallel rows (palisade formation) and in V and Z forms due to incomplete separation of bacilli from one another following division. They are facultative anaerobes and some are strict parasites of man.

C. diphtheriae is the only species which is pathogenic for man, whilst others such as *C. Hofmanni* and *C. xerosis* occur as commensals in the throat. The species are distinguished from one another morphologically and biochemically. Diphtheria bacilli differ morphologically from other species in that they are long and slender, and they usually show well-marked metachromatic granules which may be demonstrated by Neisser's staining method. In addition, *C. diphtheriae* produces a demonstrable exotoxin, whereas *C. Hofmanni* and *C. xerosis* are non-toxigenic. Corynebacteria other than the diphtheria bacillus are commonly referred to collectively as diphtheroids.

Corynebacterium diphtheriae was first observed in diphtherial membrane by Klebs in 1883 and was isolated and fully studied by Löffler in 1884; hence it was once called the *Klebs-Löffler bacillus* (or " K.L.B."). The bacillus is a strict parasite; it is found only in the membrane of diphtherial lesions, or in the nasal cavities or throats of " carriers ".

(2) Diphtheria toxin

In diphtheria the bacilli remain superficially localized on the tonsils, nasal or laryngeal mucosa or other site of infection, and do not penetrate into the underlying tissues. Apart from the local inflammation and the membrane associated with it, all the symptoms and dangers of diphtheria are due to the powerful exotoxin produced by the bacilli, and absorbed into the blood stream.

The exotoxin is produced in abundance when diphtheria bacilli are grown in suitable fluid media ; and subcutaneous injection of suitable doses of the toxic filtrate into guinea-pigs or rabbits produces local and general results similar to those produced by injection of living cultures of diphtheria bacilli, and generally similar also to those seen in fatal diphtheria in man. Thus, if a guinea-pig is given a sufficient dose of toxin to kill it within a few days, the following changes are found. The site of the injection shows an area of haemorrhagic oedema and inflammation, often with a central area of necrosis. The regional lymph glands are swollen and congested. The pleural and peritoneal cavities usually contain some effused fluid, which may be clear, cloudy or blood-stained. The adrenals are severely congested or haemorrhagic. Microscopically, the parenchyma cells of the adrenals, heart muscle, liver and kidneys show more or less severe cloudy swelling or fatty degenerative changes and even patchy necrosis.

Antitoxin is readily produced by immunizing horses with increasing doses of toxoid and toxin; and the antitoxic serum protects animals against the lethal effects of the toxin and is of great curative value in human diphtheria.

(3) The pathology of diphtheria

(i) *The local lesion.*—The bacilli lodge and multiply on a mucous membrane, most often on the tonsils or throat, less often in the nasal cavity or larynx and trachea, and occasionally on the conjunctiva, vulva, vagina or the surface of a wound. The exotoxin produced by the bacilli kills the superficial cells of the mucous membrane, and the bacilli multiply in the necrotic layer and produce more toxin and extending superficial necrosis. Products of the necrotic cells excite acute inflammatory changes in the underlying tissues, and the fibrin from the inflammatory exudate binds the necrotic layer into a firm pale membrane, the so-called " false membrane " which is characteristic of the disease. Where this is formed on a squamous stratified epithelium, as on the fauces or tonsils, its attachment is firm and it resists peeling; but on surfaces clothed by pseudostratified respiratory epithelium, as in the larynx or trachea, the attachment of the dead membrane is looser and it often comes to lie almost free in the air passages, whence it may be coughed up as a cast or may become impacted in the larynx causing asphyxia. In children with untreated diphtheria, membrane formation often spreads from the tonsils to the larynx and trachea, but this seldom occurs in adults. Swelling of the upper cervical lymph glands results from local absorption of the toxin. Nasal diphtheria is a special type of the disease, often confined to the nose, often unilateral, often running a mild course of several weeks duration, often escaping diagnosis, and therefore a great danger to other children. Beware of the child with a one-sided " cold in the nose ".

(ii) *Early toxaemic results.*—In a case of average severity, the patient has a moderate fever, rapid pulse, general malaise and often mild albuminuria. In severe cases, these symptoms are exaggerated, or there may be extreme prostration with subnormal temperature and respiratory failure. Most fatalities not due to laryngeal or tracheal obstruction by membrane or to supervening streptococcal bronchopneumonia, result from circulatory failure caused by the direct effect of the toxaemia on the cardiac muscle. Toxaemic injury of the adrenals, kidneys and liver also contribute.

(iii) *Late toxaemic results.*—During convalescence, and usually 1 to 3 weeks after the acute stage of the disease has subsided, various muscular paralyses appear in 10 or 20 per cent of cases. The commonest are pharyngeal and palatal paralysis causing impairment of speech, dysphagia and regurgitation of swallowed fluids through the nose, and ciliary or ocular paralyses causing impaired accommodation, squint or ptosis of the upper eyelid. Paralysis of trunk or limb muscles is rare. The paralyses are due to well-marked damage to the corresponding motor nerves, which show patchy Wallerian degeneration of their myelin sheaths. It is believed that the nerves directly absorb and are locally poisoned by the diphtheria toxin, and that this accounts for the anatomical relationship between the usual site of infection, the fauces, and the commonest nerves to be affected, those of the palate and pharynx. The injured nerves usually recover and the paralyses clear up in two or three weeks.

(4) Isolation and identification of *C. diphtheriae*

Direct films made from suspected diphtheritic lesions are usually unsatisfactory. Material (usually throat and nose swabs) is inoculated on fresh-blood agar, heated-blood agar and Löffler's inspissated serum. The fresh and heated-blood agar plates help to identify other organisms present, while Löffler's medium, which is simply heat-coagulated ox serum containing 1 per cent glucose broth, produces a luxuriant growth of diphtheria bacilli in 18–24 hours. Films, stained by Neisser's method, are made from the Löffler's slope culture and are examined for the characteristic bacilli with metachromatic granules.

Media containing potassium tellurite are also commonly used for the isolation of *C. diphtheriae*. In addition to being partially selective for diphtheria bacilli, they serve to differentiate them into three types—*gravis, mitis* and *intermedius*—according to their colonial appearances. This is of importance since *gravis* strains produce more complications and a higher mortality than *mitis* strains; and practically all cases of diphtheria which are fatal or severe in spite of antitoxin treatment are due to bacilli of the *gravis* or *intermedius* types.

The isolation of diphtheria bacilli should be followed by a virulence test in guinea-pigs to determine the toxigenicity or otherwise of the strain. Alternatively, toxigenicity may be determined by use of the Elek plate (see Appendix D).

(5) Immunity to diphtheria

Immunity to diphtheria means immunity to the toxin, and this depends on the body's possessing a certain minimum amount of antitoxin. Immunity may be acquired naturally or artificially, and may be determined by means of the Schick test.

(i) *Natural immunity.*—New-born infants are usually immune, because they have received transplacentally a supply of ready-made antitoxin from their mothers ; but this immunity is only temporary and in most cases disappears within a few months.   Thereafter the child is highly susceptible to diphtheria until he develops his own supply of antitoxin, either as a result of having an attack of diphtheria or by encountering mild sub-clinical infections with diphtheria bacilli which end abortively, or by artificial immunization.   Most adults have thus acquired and thereafter retain sufficient antitoxin immunity to make them insusceptible to diphtheria.

(ii) *Artificial immunization.*—A child who has been exposed to the risk of diphtheria can be protected by a dose of antitoxic serum ; but this protection by ready-made antitoxin, as in the new-born infant, is evanescent.   Stronger and more permanent immunization can be achieved by giving graded injections of toxoid, i.e. diphtheria toxin which has been so modified (by formaldehyde) that its toxicity is reduced without destroying its immunizing properties.   Two or three injections of suitable toxoid preparations evoke sufficient antitoxin formation to make most children immune within a few weeks.   In countries where wholesale immunization of young children has been undertaken, this has resulted in a great reduction in both the number of cases of diphtheria and the number of deaths from this disease.   A variety of vaccines for active immunization against diphtheria is available. One commonly used is " triple antigen ", a mixture of purified diphtheria and tetanus toxoids and whooping-cough vaccine.

(iii) *The Schick test.*—Schick introduced the following simple test for the presence of a protective amount of diphtheria antitoxin in the blood and tissue fluids.   A small amount of toxin is injected intradermally.   If the person has an adequate amount of antitoxin in his blood, the injected toxin is promptly neutralized and produces no effect.   But if little or no antitoxin is present, a positive reaction occurs, consisting of an area of redness and swelling which appears within 24 or 48 hours, attains its maximum in about 4 days, persists for one or two weeks, and then subsides.   This positive reaction denotes susceptibility to diphtheria ;   a negative result denotes immunity.   Application of the test to large numbers of people of all ages has confirmed that most young infants possess good immunity (i.e. give negative tests), that this is rapidly lost after the age of 6 months, that about 80 per cent of children aged 2 years are susceptible (i.e. give positive Schick reactions), and that thereafter the proportion of susceptibles steadily falls until adulthood, when only between 5 and 10 per cent remain Schick-positive.   These findings accord accurately with the age incidence of diphtheria.   People who have recovered from an attack of diphtheria are almost always Schick-negative ;   so are those who, like fever-hospital doctors and nurses, come much into contact with diphtheria patients yet do not contract diphtheria.   Rural communities contain a higher proportion of Schick-positive people than do city populations.

(5) Diphtheria carriers

Diphtheria bacilli are transferred directly from one person to another ;   their sources are either people who have diphtheria, or convalescents who still carry bacilli, or carriers who have never had clinical diphtheria.   During convalescence from diphtheria, most patients lose their infection within 3 or 4 weeks, but a few continue to harbour bacilli for many weeks or months, and so remain a danger to their susceptible fellows.   Most fever hospitals therefore make sure, by repeated cultures from the throat and nose, that the bacilli have disappeared before the convalescent is allowed out of hospital. Healthy carriers can be detected only by culture examination of suspects. We noted earlier that nasal diphtheria is often mild, prolonged and easily overlooked, and that such cases therefore constitute dangerous carriers.

ANTHRAX

Anthrax is a dangerous septicaemic infection of animals, which is occasionally transferred to man either by accidental inoculation of the skin or by inhalation.

(1) The genus *Bacillus,* including *Bacillus anthracis*

The choice of bacterial nomenclature is unfortunate in that we have the general term *bacillus* for any rod-shaped organism, and the generic name *Bacillus* for the Gram-positive spore-bearing aerobic bacilli.

These organisms are large bacilli, 6–8μ in length. Some species are motile, some are capsulated, and all form central or subterminal spores which usually do not alter the shape of the organism (compare with the Gram-positive anaerobic spore-bearing bacilli). They are facultative anaerobes which grow readily on simple media, and commonly produce large, rough, irregular, dry colonies. Most species are free-living forms, and are non-pathogenic, and some, such as *Bacillus subtilis* (the hay bacillus) and *Bacillus cereus,* may be encountered as plate contaminants. *Bacillus polymyxa* is the organism from which the antibiotic polymyxin is obtained. The only medically important member of the group is *Bacillus anthracis,* the causal organism of anthrax in man and animals. Other species are often referred to collectively as *anthracoid bacilli.*

The anthrax bacillus was the first bacterium pathogenic to man to be discovered. In 1849 Pollender, a veterinary surgeon, saw bacilli in the blood of animals dead of anthrax; Rayer and Davaine confirmed this in 1850; and in 1863 Davaine reported a full study of the bacilli in the blood and organs and the experimental transference of the disease to healthy animals by injection of infected blood. The discovery of the anthrax bacillus thus preceded that of other pathogenic bacteria by about a quarter of a century, the next pathogens to be discovered being the spirochaete of relapsing fever by Obermeier in 1873, the leprosy bacillus by Hansen in 1874, *Cl. septicum* by Pasteur in 1877, and staphylococci by Koch in 1878.

B. anthracis is a large square-ended non-motile Gram-positive encapsulated sporing bacillus which tends to grow in chains and grows readily on ordinary media. The spores are central in position ; they do not occur in bacilli growing in an animal's blood or tissues, but develop plentifully when the bacilli are discharged from the body or are grown in artificial culture.

The bacillus causes anthrax in sheep, cattle and other herbivores, as well as in man ; and the organs and blood of an animal dead of anthrax teem with bacilli. From such a carcase, unless it is buried deeply or burnt, anthrax spores heavily contaminate neighbouring pasture and so infect fresh animals, by way of the alimentary canal.

Pathogenic properties of B. anthracis

The pathogenesis of anthrax was for long obscure. Because subcutaneous inoculation of the guinea-pig, mouse or other susceptible animal with anthrax bacilli usually leads to an overwhelming septicaemia within 2 days, it was thought that the pathogencity of the bacillus was related to this unique power of rapid multiplication in the host's body. It is now clear that the virulence of *B. anthracis* is due to two unrelated factors, the capsular polypeptide and an exotoxin which has oedematigenous and lethal properties. This toxin is unusual in that it is composed of at least three virtually non-toxic components which act synergistically.

(2) Anthrax in man

The occasional human cases of anthrax which occur are always from some animal source. Infection by way of the alimentary tract is rare in man ; the disease is usually contracted either by inoculation of the skin or by inhalation.

(i) *Inoculation of the skin.*—This occurs mainly from handling carcases, hides or hair of infected animals, less frequently from the use of infected shaving brushes. The resulting lesion, called a " *malignant pustule* ", occurs usually on the face, neck, hand or forearm. It is an area of acute inflammation with a central dark area of necrosis surrounded by a zone of oedema and skin vesicles ; the regional lymph glands are inflamed and swollen. In some cases, the infection remains localized, the necrotic area is cast off as a slough, the bacilli disappear, and healing ensues. But in other cases, severe toxaemic symptoms develop in a few days, and fatal septicaemia may supervene.

(ii) *Infection by inhalation.*—This is called " wool-sorters' disease ", because it occurs chiefly among workers in wool factories who inhale spores in the dust or filaments of wool from infected animals. It also occurs among brush makers, especially those handling goats' hair. The infection commences in the lower part of the trachea or in a large bronchus and produces an acute inflammatory area comparable with a " malignant pustule " of the skin. From this, severe haemorrhagic and oedematous bronchopneumonia rapidly develops, soon ending fatally with septicaemia. Acute haemorrhagic meningitis sometimes develops, both in pneumonic anthrax and in septicaemia resulting from a primary skin infection.

The serum of artificially immunized animals, e.g. Sclavo's ass serum, is used in the treatment of human anthrax and has greatly reduced the mortality from " malignant pustule ". The anthrax bacillus is also sensitive to penicillin which can therefore be used for treatment.

(3) Isolation and identification of *B. anthracis*

Gram-stained smears of the vesicular fluid from a malignant pustule are prepared. The presence of the characteristic large Gram-positive bacilli confirms the clinical diagnosis, but their absence from smears is of no significance. The organism may be isolated by culturing material on ordinary nutrient agar, but it is better to use fresh-blood agar and heated-blood agar, so that any other organisms present are recognized. *B. anthracis* grows rapidly, colonies being 2–3 mm. in diameter in 24 hours, whitish, opaque and dry, with a characteristic " medusa head " appearance. Gram-stained films show that the colony is composed of long chains of bacilli and chains of free spores (Fig 56). Spores, which are readily formed in artificial culture, do not occur in living tissues.

Though it is necessary to carry out further investigations, such as phage-susceptibility or animal inoculation experiments, to prove the identity of *B. anthracis*, the isolation of an organism with the characteristics described above from a suspected anthrax lesion is almost diagnostic.

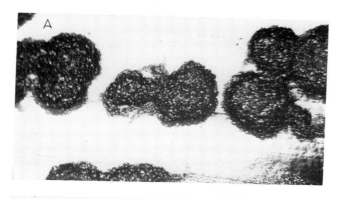

FIG. 56.—*Bacillus anthracis.* A—Colonies on horse-blood agar showing the rough " Medusa-head " appearance (× 10). B—Gram-stained smear showing chains of bacilli, many with developing spores (× 1,660.)

OTHER GRAM-POSITIVE BACILLI

Listeria monocytogenes, the cause of a disease in rabbits in which there is marked monocytosis, is a short motile Gram-positive bacillus. It occasionally causes meningo-encephalitis in man.

Erysipelothrix rhusiopathiae, the cause of " swine erysipelas " and widely distributed also among other animals and fish, is a small non-motile Gram-positive bacillus. It is an occasional cause of infection in man, " erysipeloid " in abattoir workers handling infected meat.

Lactobacillus is a genus embracing a number of closely related " lactic acid bacilli ", so called because they ferment lactose vigorously and are responsible for the souring of milk. Except that they may possibly play a part in producing dental caries, they are not pathogenic. They are mentioned here because they occur normally and predominantly in the alimentary flora of breast-fed infants, and because *Lactobacillus acidophilus* (Döderlein's bacillus) is a normal inhabitant of the adult vagina. In both these situations the lactobacilli maintain an acid reaction which may be of importance in suppressing pathogenic organisms.

SUPPLEMENTARY READING

References Chapter 10 (p. 112), and the following :

Aikat, B. K., and Dible, J. H. (1956 and 1960). " The pathology of *Clostridium welchii* infection ". *J. Path. Bact.*, **71**, 461 and **79**, 227.

Butler, H. M. (1945). " Bacteriological studies of *Cl. Welchii* infections in man, with special reference to use of direct smears for rapid diagnosis." *Surg. Gynec. Obstet.*, **81**, 475.

Govan, A. D. T. (1946). " An account of the pathology of some cases of *Cl. Welchii* infection ". *J. Path. Bact.*, **58**, 423.

McLeod, S. W., Orr, J. W., and Woodcock, H. E. (1939). " The morbid anatomy of *gravis, intermedius* and *mitis* diphtheria ", *J. Path. Bact.*, **48**, 99.

Smith, L. D. S. (1955). *Introduction to the Pathogenic Anaerobes.* Chicago; University of Chicago Press.

Willis, A. T. (1964). *Anaerobic Bacteriology in Clinical Medicine.* Second edition. London; Butterworths.

— (1969). *Clostridia of Wound Infection.* London; Butterworths.

CHAPTER 14

THE GRAM-NEGATIVE INTESTINAL BACILLI

THE GRAM-NEGATIVE intestinal bacilli are short, plump bacilli, non-sporing, and facultative anaerobes. Some species are normal inhabitants of the intestinal tract of man and animals, while the presence of others within the body is always of pathological significance.

Two genera, *Salmonella* and *Shigella*, constitute the *specific enteric pathogens*, and two other genera, *Escherichia* and *Klebsiella*, are intestinal commensals which are only incidentally pathogenic. These four genera are often referred to collectively as the *coli-typhoid-dysentery group*. The fact that *Salmonellae* and *Shigellae* do not ferment lactose, while *Escherichiae* and *Klebsiellae* do, provides a simple means for the separation of the specific pathogens from the others; and the former are often referred to generally as the *non-lactose-fermenters*. The *lactose-fermenters* are also commonly referred to collectively as *coliform bacilli*.

Distinction between lactose and non-lactose-fermenters is effected by culturing the organisms on MacConkey's agar. This medium contains lactose and an indicator (neutral red) so that colonies of lactose-fermenters are coloured pink due to the formation of acid, whilst those of non-lactose-fermenters are colourless. This distinction is of great importance, since the four genera are morphologically almost identical, and their colonial appearances on ordinary media are also very similar. All grow well on simple media. Colonies of *Salmonellae*, *Shigellae* and *Escherichiae* are 2–3 mm. in diameter, and are thick, greyish-white circular domes. *Klebsiella* species produce very large mucoid colonies of irregular shape. *Salmonellae* and *Escherichiae* are motile organisms, whereas *Shigellae* and *Klebsiellae* are non-motile. *Klebsiella* species are capsulated whilst the other genera are non-capsulated. Further differentiation between these genera and their species is made possible by their varying ability to ferment different carbohydrates, and by their agglutination reactions with specific antisera. Differentiation of the *Salmonellae* is based on their somatic, flagellar and virulence antigens (see p. 276).

Because the colon is the normal habitat of *Escherichia* organisms (they are constantly present in large numbers in the faeces of normal individuals), these bacilli are called *typical* or *faecal* coliform organisms. In contrast, the *Klebsiella* bacilli are *atypical* or *non-faecal* since they are present less commonly and in fewer numbers in the faeces, and they also occur naturally outside the animal body. These distinctions, based on the *natural habitats* of the two genera, are of importance in the bacteriological control of water supplies; the presence in drinking water of faecal coliform bacilli is indicative of faecal pollution, whereas non-faecal coliform bacilli may be normally present.

Two other genera of Gram-negative bacilli which may occur in the alimentary flora and which are sometimes of clinical importance are *Proteus* and *Pseudomonas*; and these will be described briefly at the end of this chapter.

TYPHOID AND PARATYPHOID (ENTERIC) FEVERS

Typhoid fever (an old clinical name meaning " typhus-like ", though pathologically the diseases are quite dissimilar) and paratyphoid fever are specific bacillary infections of the intestine accompanied by bacteriaemia and severe toxaemia.

(1) The bacilli of enteric fever

Of the many species of the genus *Salmonella* only 4 cause enteric fever in man, namely, *S. typhi, S. paratyphi* A, *S. paratyphi* B and *S. paratyphi* C. Typhoid and paratyphoid fevers are clinically and pathologically similar, the distinction between them being a bacteriological one. However, paratyphoid fever is clinically milder and of shorter duration than typhoid. In Europe, the incidence of typhoid fever is high in Spain, Portugal and Italy; so that it is not surprising that in recent years in England and Wales about half the cases contracted infection abroad, often as " holiday typhoid ". The only other source of typhoid in Britain today is the immigrant carrier.

(2) The sources of infection

Infection with typhoid or paratyphoid bacilli occurs through the faecal contamination of water or food. These bacilli are essentially parasitic, and cannot exist for long outside the human body ; but they can survive long enough to be transferred by polluted water, milk, ice cream, shell fish, watercress, etc. Before cities had clean water supplies and efficient sewerage systems, typhoid fever was a common and ever-present disease, and severe epidemics were frequent. These were usually due to faecal contamination of water supplies, supplemented sometimes by dispersal of bacilli from the excreta of patients by flies. For the same reasons, before the days of proper sanitation and the introduction of immunization by T.A.B. vaccine, serious outbreaks of typhoid were common in armies.

But nowadays, in most civilized communities, enteric fever is a relatively uncommon disease, and epidemics are infrequent and restricted to a few people in localized areas. The sources of such outbreaks are sometimes from clinical cases of enteric fever, but more often they are from healthy *carriers*. In 2 to 5 per cent of convalescents from enteric fever, the bacilli persist in the body for an indefinite period. This is usually in the gall-bladder and bile ducts, whence bacilli are constantly discharged into the intestine, so that the person is a *faecal carrier*. Less often, he is an even more dangerous *urinary carrier*, due to persistence of foci of bacilli in the renal pelvis. Cholecystectomy sometimes, but not always, cures a faecal carrier.

An outbreak of enteric fever necessitates search for (a) the food conveying the infection, and (b) the carrier or patient infecting the food. It is usually discovered that someone employed in the preparation or distribution of food is a carrier, e.g. in a dairy.

(3) Toxins of the enteric bacilli

It is probable that the enteric bacilli produce no true exotoxins; but culture filtrates contain toxic products from lysed bacilli. Injections of sufficient doses of dead or ground-up typhoid bacilli produce fatal toxaemia in laboratory animals. We therefore suppose that the toxic effects of the enteric infections are due to endotoxins liberated from disintegrated bacilli in the host's tissues.

(4) Pathological anatomy of typhoid fever

We have to consider (*a*) the lesions at the portal of entry of the bacilli, the intestine and mesenteric lymph glands, (*b*) the bacteriaemia which is always present, and (*c*) toxaemic effects of the disease. Paratyphoid infections produce lesions similar to, but in general rather milder than, those of typhoid fever.

(a) The intestinal and mesenteric lesions

The bacilli first invade and multiply in the Peyer's patches of the ileum and the solitary lymphoid nodules of the large intestine. These become swollen and raised above the general level of the mucosal surface (Fig. 57); microscopically, they are congested, oedematous and infiltrated by many macrophages which are derived

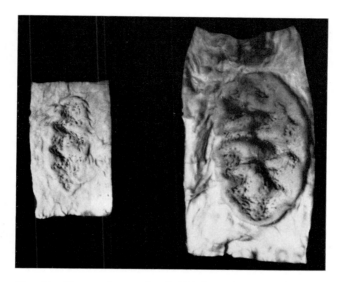

FIG. 57.—Two specimens of typhoid lesions of the small intestine, from John Hunter's collection in the Royal College of Surgeons, London. The lesions show swollen Peyer's patches with commencing ulceration.

both from incoming mononuclear leucocytes and from locally mobilized histiocytes. Few or no polymorphonuclear leucocytes are found. The macrophages engulf bacilli and the products of damaged cells. Similar changes are present in the enlarged mesenteric lymph glands. At the end of a week or 10 days, the enlarged lymphoid patches begin to show central necrosis, with the formation of ulcers and sloughs. Ulceration and the separation of the sloughs proceed throughout the third week of the disease, during which serious haemorrhages may occur or one of the larger ulcers may perforate and cause general peritonitis. The lesions in the Peyer's patches of the lower ileum are always worse and more advanced in their development than those higher up, and haemorrhage and perforation are usually from ileal ulcers not far from the ileo-caecal junction. During the fourth week, the risk of these complications diminishes, the ulcers become clean, and healing ensues. Eventually there is complete restoration of the mucosa, except for loss of parts of the lymphoid patches ; visible scarring is infrequent, and fibrous stricture of the gut very rare.

(b) Typhoid bacteriaemia and its results

From the intestinal lesions the bacilli quickly enter the blood stream. Blood cultures are positive in almost all cases of typhoid during the first week or 10 days, affording a valuable means of diagnosis in the early stages of the fever. The frequency of positive blood cultures diminishes during the second and third weeks, and by the fourth week bacilli are rarely found in the blood. In fatal cases, bacilli are easily demonstrable in the spleen, liver, lungs and other organs. The spleen always shows particularly prominent changes ; it is enlarged, soft, engorged with blood, and microscopically it shows also many macrophages and clumps of typhoid bacilli. Foci of accumulated macrophages, with or without focal areas of necrosis, are found in the liver, bone marrow and other tissues ; these are due to the effects of arrested bacilli, which can often be demonstrated in them. The characteristic skin rash of typhoid fever, consisting of small discrete " rose spots " scattered over the trunk and limbs, is due to similar focal lesions. The granulocytic leucopenia, which occurs in nearly all cases of typhoid fever, is no doubt a direct effect of the widespread lesions in the bone marrow. Occasionally, clinically important typhoid infections of other parts develop during the course of the illness or in the convalescent period. These are typhoid pneumonia, meningitis, osteomyelitis, arthritis, or abscesses in muscles or other tissues. Typhoid osteomyelitis sometimes develops months or even years after the infection. During the course of typhoid fever, bacilli are often excreted in the bile and in the urine ; and we have already noted that in a small proportion of cases colonies of bacilli persist in the biliary or urinary passages, making the patient a dangerous carrier.

(c) Toxaemic effects

In addition to focal lesions due to the actual presence of blood-borne bacilli in various tissues, the typhoid patient suffers toxic damage of myocardial, hepatic, renal and other parenchyma cells, which in fatal cases show severe cloudy swelling and fatty changes. Circulatory failure is therefore a serious risk in badly toxaemic cases, in whom also delirium often bespeaks the poisoning of the cerebral cortex.

(5) Immunity changes in the enteric fevers

(i) *The Widal agglutination test.*—During an attack of typhoid fever, from the second week onwards the patient's blood serum develops increasing quantities of specific agglutinins, i.e. substances which agglutinate or clump typhoid bacilli. The Widal reaction is a simple test for the presence of specific typhoid (or paratyphoid) agglutinins in serum ; it is positive in almost all cases by the end of the second week of the disease, and is thus a valuable diagnostic test. By testing the patient's serum on suspensions of all four kinds of enteric fever bacilli, we can identify the species of the organism infecting him. Carriers are usually Widal-positive. (For further details about agglutinins, see Chapter 21.)

(ii) Prophylactic immunization with T.A.B. vaccine.—Subcutaneous or intracutaneous injections of suitable suspensions of killed typhoid and paratyphoid bacilli have powerful immunizing effects. The Widal tests of inoculated persons become strongly positive; and the wholesale inoculation of the armed forces has greatly reduced the incidence and severity of the enteric fevers among them.

(6) The diagnosis of enteric fever

The clinical diagnosis of enteric fever should always be confirmed bacteriologically, either by culturing the bacillus from the blood, faeces or urine, or by means of Widal tests.

(i) *Blood culture.*—This is of special value in the early stages of the fever, before the clinical signs are definite or the Widal test has become positive, i.e. during the first 7 or 10 days. Blood cultures are made in bile-broth medium, from which subcultures are made on MacConkey's agar as outlined below.

(ii) *Faeces culture.*—The faeces may contain the bacilli at all stages of the disease, but the chances of a positive culture are greatest during the second and third weeks. Separation of the enteric bacilli from the coliform and other intestinal bacteria is effected by plating out a faeces suspension on a suitable selective medium which contains an indicator to enable rapid distinction of colonies of lactose-fermenters from non-lactose-fermenters, and which encourages the growth of the latter (desoxycholate-citrate-agar with lactose and neutral-red). Colonies of non-lactose-fermenters are subcultured and tested for their sugar-fermenting and other biochemical properties. Identification is also facilitated by testing suspensions of the organism with specific agglutinating sera. In addition to direct plating, faeces is inoculated into a selective fluid medium such as selenite or tetrathionate broth. These media inhibit the growth of other Gram-negative bacilli and allow small numbers of *Salmonellae* to grow. After 24 hours' incubation these cultures are plated on the selective solid medium mentioned above.

(iii) *Urine culture.*—Enteric fever bacilli are found in the urine in only a proportion of cases, and they seldom appear there before the end of the second week. When urine cultures give negative results, they should be repeated. The methods are generally similar to those for faeces cultures.

(iv) *Widal tests.*—A positive agglutination test does not necessarily mean that the patient has enteric fever; he may have had this disease previously, or he may have been vaccinated with T.A.B. vaccine. If, however, he has not had vaccine recently, if agglutination occurs in high dilutions of his serum, and especially if his serum shows a *rising titre* of agglutinins on repeated tests, then his present febrile illness is typhoid (or paratyphoid) fever. It must always be remembered that agglutinins are not present in the blood during the first week of the disease.

INFECTIVE GASTRO-ENTERITIS OR FOOD POISONING

Infective gastro-enteritis is due to an infection of the intestine by *Salmonellae* other than those of enteric fever. The most important of these are *S. typhi-murium* and *S. enteritidis*, but many other closely related species, only distinguishable serologically, are also responsible.

Salmonella food poisoning is due to contamination of preserved meats, milk, eggs or other foods, especially those that have been much handled or kept before consumption. Probably the commonest source of the infection is a human carrier, just as in enteric fever, but many outbreaks have been traced to animal sources. The bacilli are common pathogens in many species of animals, including cattle, pigs, rodents and birds. *S. typhi-murium* is the commonest cause of food poisoning in Great Britain.

Infective food poisoning produces vomiting, diarrhoea and colic, accompanied by fever, headache and other signs of toxaemia. In most cases the infection clears up in a few days, with no evidence that the bacteria invaded beyond the intestinal mucosa ; but fatal toxaemia, bacteriaemia or meningitis have occasionally occurred.

Bacteriological diagnosis is made by faeces cultures, rarely by blood culture, or retrospectively by specific agglutination tests with the convalescent's serum. Bacteriological examination of suspected articles of food, if available, should also be carried out. Isolation of the organism is carried out, as described above, for enteric fever organisms.

BACILLARY DYSENTERY

The name *dysentery* is restricted to certain specific infections of the large intestine, characterized pathologically by ulceration of the mucosa and clinically by diarrhoea and the passage of blood and mucus. These infections are of two distinct kinds, bacillary and amoebic. Here we are concerned only with the former ; amoebic dysentery is described on p. 218.

(1) The bacilli of dysentery

Bacillary dysentery is due to organisms of the genus *Shigella*, which are strict pathogens of man and some monkeys. According to their biochemical reactions they are divided into 4 groups: group A, *Sh. dysenteriae*; group B, *Sh. Flexneri*; group C, *Sh. Boydii*; group D, *Sh. Sonnei*. A number of serological types of each species are also recognized. The differences between the various types are not of great importance to the student; all types produce generally similar lesions, though *Sh. dysenteriae* produces a severer form of dysentery than the others, while *Sh. Sonnei* causes a mild form and is the commonest cause of dysentery in Great Britain.

(2) The sources of infection

Like enteric fever, the transfer of dysentery bacilli is usually by the faecal contamination of food or water. Bacillary dysentery is at present commoner in tropical than in temperate climates, but this is largely due to lack of sanitation and to over-population in many tropical countries. Formerly, when European sanitation was poor, dysentery was rampant here also ; and nowadays, wherever large numbers of people are in close contact under poor sanitary conditions, as in some armies, gaols, asylums, or prison camps, dysentery is apt to break out. Flies are an important factor in the spread of large epidemics in the tropics.

In the small outbreaks of dysentery in modern civilized communities, however, carriers and lack of personal hygiene provide the main sources of infection, just as in enteric fever. In Great Britain direct transfer in day nurseries by hand-to-mouth infections is common. In a small proportion of cases, patients recovering from an acute attack of dysentery continue to excrete bacilli in their faeces for months or even years. Some of these carriers, especially carriers of *Sh. Shigae* (*Sh. dysenteriae type* 1), remain poor in health with chronic or recurrent attacks of dysentery, but other carriers remain well; all of them are a potential danger to the community.

During the early stages of acute dysentery, the bacilli are found in very great numbers, sometimes in almost pure culture, in the stools ; later they become progressively fewer.

(3) Toxins of the dysentery bacilli

Bacilli of the *Sh. dysenteriae* group produce in culture a potent neuro-entero-toxin, small intravenous doses of which are fatal for laboratory animals. This toxin, given in graded sublethal doses, evokes the formation of a specific antitoxin which neutralizes the toxin. In spite of this, it is probable that the toxin is not a true exotoxin diffusing from living bacilli, but an endotoxin liberated only on their disintegration. Lysed cultures and dried ground-up bacteria produce similar toxic effects. The other dysentery bacilli are not nearly as toxigenic as *Sh. dysenteriae*. An interesting feature of dysentery toxin is that, even when given intravenously to animals, it often produces diarrhoea and a haemorrhagic condition of the intestinal mucosa, indicating its enterotoxic action.

(4) Pathological anatomy of bacillary dysentery

Dysentery is a widespread acute inflammation of the mucous membrane of the large intestine, characterized by the outpouring of pus, blood and mucus into the bowel, and extensive patchy necrosis and ulceration of the inflamed mucosa. In hyperacute cases, especially seen in children, death from toxaemia may occur within 2 or 3 days, and then the colonic mucosa may show only congestion. In patients who die of toxaemia and dehydration within 2 or 3 weeks, the mucosa shows widespread sloughs and irregular areas of ulceration. Death from exhaustion may occur after weeks or months of subacute or chronic dysentery; and then the large intestine is found almost denuded of its mucosa. Rarely, ulceration extends through the muscle coat and leads to perforation and peritonitis. When severe dysenteric ulcers heal, a good deal of granulation tissue is produced, and there may be much subsequent scarring with strictures of the bowel.

The dysentery bacilli usually remain confined to the inflamed intestine ; they are rarely found in the blood or other organs.

(5) Diagnosis of bacillary dysentery

The clinical diagnosis is confirmed by faeces culture, along the lines described for enteric fever. Non-lactose-fermenters are further identified by fermentation reactions and by tests with specific agglutinating sera. Faeces cultures should be made as early as possible in the disease; later they may prove negative.

Specific agglutinins may appear in the patient's serum after a week or 10 days; but the agglutination results (carried out similarly to the Widal test) are not as clear-cut as in typhoid fever, and they are not used for diagnosis.

COLIFORM BACILLI

We have already noted that the coliform bacilli are made up of two groups of organisms—*Escherichia*, the typical faecal ones, and *Klebsiella*, the atypical non-faecal ones.

Escherichia coli is only occasionally an intestinal pathogen (see below), but is a common cause of urinary-tract infections, wound infections and infections connected with the intestinal tract such as appendicitis.

(1) Urinary infections

Coliform bacilli alone are an important cause of cystitis and pyelitis in female children and pregnant women. Alone, or mixed with other organisms, *E. coli* is almost always present also in chronic urinary-tract infections associated with urethral stricture, prostatic enlargement, vesical calculi or tumours, or paralytic retention of urine.

This is the place to mention the use and abuse of microscopical and cultural examination of urine specimens. Microscopical examination of the centrifuged deposit from *a clean freshly voided mid-stream specimen of urine* is of the utmost value and usually gives much of the information required; the fresh urine contains the bacteria, pus cells, casts and crystalline deposits which were actually present in the bladder. But a stale specimen is often worse than useless; contaminating bacteria of no significance have usually multiplied in it, non-pathological crystalline and other precipitates have formed, and some cellular elements may have disappeared. For these reasons many of the microscopical and cultural examinations of urine specimens as collected in many hospitals are merely a waste of time and culture media. The very frequent report that coliform bacilli have been cultured from urine usually means only that the urine was contaminated by these organisms after it was passed and that they multiplied in it. It is sometimes valuable to perform quantitative bacterial counts on freshly voided specimens of urine. It has been found that bacterial contamination of correctly taken specimens never produces more than 10^4 organisms per ml., while infected urines almost always contain at least 10^5 organisms per ml. Several simple techniques are availabe for such estimations.

(2) Intestinal infections

Coliform bacilli often participate, along with other bacteria, in inflammatory states of the intestinal tract, e.g. appendicitis, diverticulitis or ischio-rectal abscess; infections of the bowel wall resulting from obstruction, strangulation, volvulus or intussusception; and peritonitis or abscesses resulting from any of the foregoing lesions or from perforations. Probably in none of these does *E. coli* play a leading part; it merely follows in the wake of more aggressive pathogens and aggravates their effects, or it participates in mixed infections of devitalized tissue. In such mixed infections, *Proteus* species (commonly *Pr. mirabilis*) and other proteolytic bacteria are often present as well, causing the foul smell that is commonly present in appendical or other abscesses connected with the intestine.

In addition to these types of infection, some strains of *E. coli* cause a true gastro-enteritis in man. Enteric infections by these *enteropathogenic* types is confined almost exclusively to infants under 18 months of age and is most common in bottle-fed babies. Bacteriological diagnosis is made by isolating the organism by direct plating of faeces on fresh-blood agar and MacConkey's agar (note that colonies on MacConkey's agar are pink), and then identifying the strain by agglutination tests with appropriate antisera.

(3) *Klebsiella* infections

Klebsiella bacilli were originally isolated from cases of lobar pneumonia by Friedländer, who believed that the organism was the cause of this disease and called it the " pneumobacillus ". It is now known that Friedländer's bacillus (*Klebsiella pneumoniae*) rarely causes pneumonia; it is present as a commensal

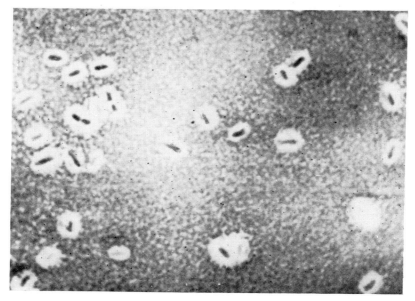

FIG. 58—Capsulated *Klebsiella pneumoniae*. The capsules appear as unstained clear haloes around the stained bacilli. ($\times$ 2,500)

in the upper respiratory tract and the intestines, and may be associated with such infections as empyema, sinusitus and conjunctivitis. It is easily distinguished from the other Gram-negative enteric bacilli by its large mucoid colonies and its possession of a large capsule which is easily demonstrated by special staining methods (Fig. 58).

Klebsiella aerogenes is closely related to Friedländer's bacillus, and is a cause of urinary tract infections. *Kl. rhinoscleromatis*, also akin to Friedländer's bacillus, is found in the lesions of *rhinoscleroma*, a chronic inflammatory disease of the nasal cavity, in Eastern Europe and Asia.

OTHER GRAM-NEGATIVE BACILLI

(1) Proteus

Proteus is the name applied to a group of pleomorphic, motile, Gram-negative, non-sporing and non-capsulated bacilli which are widely distributed in nature, play an important part in the putrefaction of organic matter, and are present in human and animal faeces. Though a number of species are recognized, *Pr. mirabilis* is the one most commonly encountered in clinical material. These bacilli participate in inflammatory lesions connected with the intestinal tract, in infections of the urinary tract, and in infections of wounds.

Pr. mirabilis is a facultative anaerobe and grows well on simple media. On solid media discrete colonies are developed in young cultures, but characteristically the organism swarms over the surface of the medium as a film of growth. Swarming occurs in successive waves so that the film of growth presents an undulating appearance. Swarming does not occur on bile-salt media, and since the organism is a non-lactose-fermenter, its colonies may be confused with those of enteric pathogens when present on MacConkey's agar or desoxycholate-citrate agar. It is easily distinguished from the Gram-negative enteric pathogens, however, by its sugar fermentation reactions and by its ability to decompose urea with the liberation of ammonia; *Salmonellae* and *Shigellae* are urease-negative.

The urease activity of *Pr. mirabilis* is often responsible for napkin-rash in infants. In soiled napkins the organism rapidly breaks down the urine urea, and the liberated ammonia acts as an intense irritant of the skin.

Attention is drawn on p. 268 to the use of *Pr. vulgaris* (strain X19) and *Pr. mirabilis* (strain XK) in diagnostic agglutination tests for typhus fever and scrub typhus respectively.

(2) Pseudomonas

Though the genus is not included in the group of Gram-negative enteric bacilli, *Pseudomonads* are conveniently considered here since they are similar in many ways and occur in the intestinal flora. *Ps. aeruginosa* is the only species of medical importance. It is a motile, Gram-negative bacillus, morphologically similar to *Proteus* species. In artificial culture *Ps. aeruginosa* produces two diffusible pigments, pyocyanin and fluorescin, which impart a greenish-blue colour to the medium. The organism grows readily on simple media. Since it is a non-lactose-fermenter and is sometimes present in normal faeces, it may be confused with the enteric pathogens when growing on MacConkey's agar. It is readily distinguished from them by its sugar fermentation reactions and its pigment production.

Ps. aeruginosa causes suppurative infections of the skin, or of wounds, or otitis media, with the formation of blue pus; on rare occasions it has been present in metastatic abscesses in other parts.

(3) Distinguishing the Gram-negative intestinal bacilli

The following is a simplified scheme for the identification of Gram-negative bacilli which may be present in faeces:

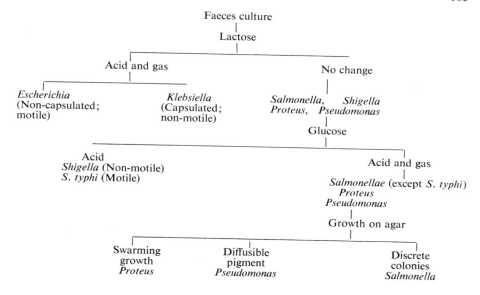

SUPPLEMENTARY READING

References Chapter 10 (p. 112). Ash and Spitz's atlas is particularly useful for the pathology of dysentery and typhoid fever.

CHAPTER 15

DISEASES CAUSED BY OTHER GRAM-NEGATIVE BACILLI

PLAGUE

(1) The plague bacillus and its habitats

The plague bacillus, *Pasteurella pestis*, is a member of the *Pasteurella* group, which comprises a number of species of short oval Gram-negative, non-motile and non-sporing, cocco-bacilli, which characteristically show bipolar staining. Other species in the genus include—*P. septica*, a cause of septic wounds following cat or dog bites, *P. haemolytica* which may be present in human sputum, *P. Tularensis*, the cause of Tularaemia, a plague-like disease in some wild rodents, which is occasionally transmitted to man, and *P. pseudotuberculosis* which causes pseudo-tuberculosis in guinea-pigs (see p. 129). *P. pestis*, which was discovered independently by Kitasato and Yersin in 1894, is a facultative anaerobe, grows well on ordinary media, forming circular white opaque colonies, and exhibits pronounced pleomor-phism in culture. In the tissues it may show a capsule.

P. pestis is a strict parasite of rodents and man. Rats constitute the main reservoir of the disease, and from them or other infected rodents the bacilli are conveyed to man by the bites of rat fleas. Plague has always been prevalent in the tropical and sub-tropical parts of Asia, and the epidemic outbreaks of the " Black Death " in Europe in the fourteenth century were due to the spread to Europe of the black rat, a native of Southern Asia and the chief disseminator of plague. In human epidemics, transfer of bacilli from man to man by inhalation may become much more prevalent than rat-to-man transfer. Plague has now almost disappeared from Europe, partly because of the displacement of the black rat by the brown rat which is less dangerous as a plague reservoir, and partly because of improved public health measures, especially rat extermination and quarantine. Hygienic communities not infested by rats are exempt from plague epidemics.

(2) Pathology of human plague

There are two main forms of plague in man : (*a*) *bubonic plague* and (*b*) *pneumonic plague*. (*a*) The former results from infection by flea bites, whence the bacilli pass rapidly to the regional lymph glands in the groin or axilla and produce an intense acute haemorrhagic lymphadenitis, the swollen glands being called " buboes ". In the majority of cases, fatal septicaemia supervenes ; but in some cases the infection remains localized, the glands soften, suppurate and discharge, and recovery may follow. (*b*) Pneumonic plague results from inhalation of infected droplets from other patients with the pneumonic form of the disease. An acute haemorrhagic bronchopneumonia develops, which rapidly leads to fatal septicaemia.

(3) Diagnosis

Bacilli are usually very numerous in the lesions they produce. Bubonic plague is diagnosed by puncturing the bubo with a hypodermic needle and examining the fluid obtained, by direct smear, culture and animal inoculation. In pneumonic plague, the bacilli are abundant in the sputum. In septicaemic stages, the plague bacillus is readily isolated by blood culture.

CHOLERA

(1) The cholera vibrio and its transfer

Vibrio cholerae, the " comma bacillus ", discovered by Koch in 1883, is the only pathogenic member of the *Vibrio* group, a genus of aerobic, Gram-negative, motile, non-sporing and non-capsulated, curved rods, which grow well on simple media, producing colonies resembling those of *E. coli*. *V. cholerae* is a strict parasite of man and occurs in great numbers, often in pure culture, in the " rice-water " stools of cholera patients. Its growth is favoured by an alkaline environment, and isolation is best effected by culturing the faeces on the alkaline blood-agar medium of Dieudonné. There are several named varieties of *V. cholerae*, the best known being " El Tor ".

Infection is by the ingestion of food or water which has been contaminated from the stools of patients or convalescent carriers. Those in close contact with patients may contract the disease directly, or the bacilli may be carried by flies ; but large epidemics are almost always due to faecal contamination of water supplies.

Cholera is prevalent in Southern Asia, especially India, whence from time to time it has spread to Europe, Africa and America. In India it causes half a million deaths annually.

(2) Pathology of cholera

Cholera is an acute disease characterized by (1) damage to the intestinal mucosa, (2) toxaemia and (3) dehydration. The bacilli remain confined to the intestine and do not invade the blood stream or produce lesions in other organs. The gall-bladder is often infected.

(i) *Damage to the intestinal mucosa.*—This is of a peculiar kind. The mucous membrane of the small intestine, and to a less degree that of the large also, is lax, smooth, opaque and pale, but sometimes shows congestion or scattered haemorrhages. There are no ulcers, and little or no signs of inflammation. Microscopically, the surface of the mucosa is almost denuded of epithelium, detached portions of which, along with mucus, form the white flakes of the " rice-water " stools.

(ii) *Toxaemia.*—The cholera vibrio produces a mucinase, and it contains endotoxins which are liberated from disintegrated bacilli in culture and in the human intestine. These not only damage the intestinal mucosa, but are absorbed into the general circulation. In some cases—of *cholera sicca*—the resulting toxaemia is so severe that it kills the patient before the diarrhoea and dehydration of the disease appear.

(iii) *Dehydration.*—In most cases of cholera, however, dehydration is an even more important factor than toxaemia. As a result of the damage to the intestinal mucosa, there is a rapid escape of fluid from the blood into the bowel with consequent profuse watery diarrhoea. As much as 10 pints of fluid may be lost from the body within 24 hours. The blood becomes concentrated, with rise of its viscosity, specific gravity and red cell count. The tissues become shrunken, the skin wrinkled, the temperature subnormal, the pulse feeble—a state of collapse known as the " algid " (i.e. cold) stage, which often ends fatally in a few hours. In the treatment of cholera the most important measure is therefore to give abundant saline or other fluids intravenously ; properly administered, these greatly reduce the mortality, e.g. from 60 per cent to 10 per cent or less.

UNDULANT FEVER
(Mediterranean or Malta Fever)

(1) The Brucella group of bacilli

The story of the discovery of this group of bacteria is a remarkable one. The group is named after Bruce, an army surgeon in Malta who in 1886 discovered bacteria in stained sections of the spleen of a fatal case of Malta fever, and who a year later successfully cultivated *Micrococcus melitensis* (now *Brucella melitensis*). In 1905 it was discovered that the blood of Maltese goats contained agglutinins for *Brucella;* and it was then found that these animals are a reservoir of infection, excreting the bacteria in their milk without themselves suffering any recognizable illness.

Let us now turn to an apparently unrelated subject, infectious abortion of cows, a disease shown in 1897 by Bang to be due to a small bacillus present in the placenta and accordingly named *Bacillus abortus*. Except for the interference with pregnancy, infected cows do not suffer in health, though they continue to excrete bacilli in their milk.

The first observation connecting Malta fever and bovine abortion was the discovery in 1914 that agglutinins for *Brucella melitensis* were present in the milk and serum of cows in the London area. Closer study then showed that *Brucella melitensis* and Bang's *Bacillus abortus* were in every respect closely similar and were indeed only slightly different types of the one organism. The question therefore arose—Could undulant fever be caused by infections of bovine origin ? Cases were soon found, and have since been recognized in increasing numbers in many countries.

A third type of *Brucella* which occasionally produces undulant fever in man was then discovered, *Brucella suis*, the cause of infectious abortion in pigs.

Briefly then, we now know that undulant fever in man can be caused by any one of three closely related organisms, namely (*a*) *Brucella melitensis*, the caprine type, (*b*) *Br. abortus*, the bovine type, and (*c*) *Br. suis*, the porcine type, (*b*) being the commonest in most communities. The three species are distinguished by biochemical and serological methods.

The *Brucellae* are short Gram-negative cocco-bacilli, non-motile, non-encapsulated and non-sporing. They are aerobic organisms, but growth is favoured by the presence of 5–10 per cent of CO_2. Growth occurs on simple media but is better on fresh and heated-blood agar. Colonies develop slowly, and are small transparent domes, about 1 mm. in diameter after 3 days' incubation at 37°C. They are strict parasites. Human infection occurs by the ingestion of infected milk or by contact with infected animals or meat.

(2) Pathology of undulant fever

As the name denotes, the disease is characterized by a wave-like temperature chart, with alternating febrile periods of 1 to 3 weeks and afebrile periods of a few days. The disease is septicaemic, and during febrile periods blood cultures for *Brucella* are often positive. The bacteria are present also in the urine in many cases. The fever often lasts several months, and is accompanied by anaemia, wasting and neuralgic pains. Recovery usually occurs, but occasional cases are fatal. *Post-mortem*, the bacteria are plentiful in the spleen (which is enlarged), liver, lymph glands, bone marrow and other organs. The fixed phagocytes of the recticulo-endothelial system are laden with engulfed bacilli ; and collections of proliferating macrophages, similar to those in typhoid fever, may be present in many tissues.

(3) Diagnosis of undulant fever

Diagnosis is by *blood culture* and *agglutination tests*. At least 10 ml. of blood should be cultured, and incubation is at 37°C. for 3–5 weeks in an atmosphere containing 10 per cent of CO_2. Blood cultures are positive more often in *melitensis* than in *abortus* infections. Agglutination tests, carried out with a series of dilutions of the patient's serum mixed with suspensions of the three separate types of *Brucella*, are the most reliable means of diagnosis, especially if a rising titre of agglutinins is found on repeated tests. Veterinary workers and others who come into contact with infected animals often have *Brucella* agglutinins in moderate concentration in their sera, doubtless as the result of subclinical infections.

LESIONS DUE TO *HAEMOPHILUS INFLUENZAE*

The genus *Haemophilus* (" blood-lover ") comprises several species of small Gram-negative, non-motile, non-sporing, strictly aerobic cocco-bacilli, some strains of which are capsulated. All species depend for their growth on one or both of two growth factors which are present in blood, namely, X factor or haematin and V factor or co-enzyme I. *H. influenzae* was first discovered by Pfeiffer in the sputum of patients with influenza, and was believed to be the cause of this disease. We now know that true influenza is a virus disease, and that the name " H. influenzae " is therefore a misnomer. Nevertheless this bacillus is frequently present in the air passages in influenza, and it often plays a part along with other bacteria in producing influenzal bronchopneumonia, empyema and other complications. It occurs also in other catarrhal infections of the upper respiratory tract and sometimes in the throats of healthy persons ; and, quite apart from influenza, it is an occasional cause of sinusitis, meningitis (mainly in children), and ulcerative endocarditis. Its causal relationship to bronchitis and bronchiectasis is not clear. Acute epiglottitis in children is almost specifically caused by *H. influenzae*.

On fresh-blood agar, colonies are minute and are easily overlooked. Better growth occurs on heated-blood agar, colonies being colourless transparent domes, about 1 mm. in diameter after 24 hours' incubation. *H. influenzae* grows better in symbiosis with staphylococci, due to the fact that the cocci synthesize co-enzyme I. Thus, in mixed cultures, *H. influenzae* grows better in the immediate vicinity of staphylococcal colonies, an appearance which is described as *satellitism*. Bacterio-logical diagnosis of *H. influenzae* meningitis is made by microscopic examination of stained films (Gram and Leishman), and by culture on fresh and heated-blood agar. Smears from the cerebrospinal fluid deposit show the presence of pus cells and many pleomorphic Gram-negative bacilli, many of which are placed within the pus cells. Variants of *H. influenzae* include *H. haemolyticus*, which produces zones of complete haemolysis on blood-agar, and *H. para-influenzae*, which is independent of the X factor.

H. aegyptius is commonly referred to as the *Koch-Weeks bacillus*. It is almost indistinguishable from *H. influenzae*, and is the cause of acute conjunctivitis.

(*Moraxella lacunata*—the *Morax-Axenfeld bacillus*, may be mentioned here as another small Gram-negative bacillus which causes conjunctivitis, but which is distinct from the *Haemophilus* bacilli.)

SOFT CHANCRE

Another pathogenic species of *Haemophilus* is *H. Ducreyi* (Ducrey's bacillus), the causal organism of " soft chancre ", a venereal infection of the external genitalia or neighbouring parts. A raised ulcerated lesion develops at the site of inoculation, and is often accompanied by suppurative inguinal lymphadenitis. Clinically, it is important to distinguish " soft chancre ", which remains purely local, from the " hard chancre " of syphilis (Chapter 16).

WHOOPING COUGH (or PERTUSSIS)

The cause of whooping cough, a small Gram-negative cocco-bacillus, was dis-covered by Bordet and Gengou in 1906. It resembles *Haemophilus* species in some respects, but does not require the X and V growth factors, and it is therefore now distinguished as *Bordetella pertussis*. Special media, such as the Bordet-Gengou medium, are required for its satisfactory growth. Characteristic small raised colonies, resembling drops of mercury, appear after 3–4 days' incubation. During the first 2 or 3 weeks of the disease, great numbers of bacilli are sprayed out in droplets while the child is coughing, and they can usually be isolated by exposure of a plate of suitable culture medium in front of the child's face or better from a per-nasal swab, but failure to isolate the organism from an obvious case of whoop-ing cough is not unusual. Transfer of the disease from child to child is by inhalation of infected droplets.

The disease, which affects chiefly infants and young children, begins as a febrile upper respiratory catarrh, which for a few days is highly infective ; and then passes into an afebrile stage often lasting several weeks, characterized by paroxysmal coughing, whooping and vomiting. In uncomplicated cases, there are no other signs or symptoms ; but in young children bronchopneumonia, due to *Bord. pertussis* itself or to a mixed infection with pneumococci, is a serious risk and causes many deaths. There are good grounds for believing also that severe whooping cough, even though recovered from, may in some cases initiate progressive bronchiectasis. It may also light up previously quiescent tuberculous foci in the lungs. Because of these risks, prophylactic immunization against the disease by means of suitable vaccines is commonly undertaken, and this may be combined with immunization against diphtheria and tetanus.

LESIONS DUE TO FUSIFORM BACILLI

Organisms of the genus *Fusiformis* are Gram-negative, non-motile and non-sporing anaerobic bacilli, usually long and slender and with tapered ends (fusiform), but short forms also occur. Many strains are found as commensals in the alimentary tracts of man and animals.

F. fusiformis occurs in the mouth in healthy people; but it is found in great numbers along with *Borrelia Vincenti* in Vincent's angina (see p. 209). Bacteriological diagnosis of this infection is made by microscopic examination of Gram-stained or carbol-fuchsin-stained smears, in which plentiful fusiform bacilli and spirochaetes are present.

F. necrophorus, a normal inhabitant of the intestine, may cause or participate in a variety of infections, e.g. post-abortal or puerperal infections of the uterus, pulmonary abscess, cerebral abscess, wound infections, etc. Enriched media and special techniques are required for its culture.

OTHER DISEASES DUE TO GRAM-NEGATIVE BACILLI

These include: (i) *Glanders*, a disease of horses due to *Loefflerella* (formerly *Malleomyces*) *mallei*, which is occasionally transferred to man; (ii) *Melioidosis*, a glanders-like disease of rodents due to Whitmore's bacillus, *Loefflerella pseudomallei*, occasionally contracted by man; and (iii) *Oroya fever*, a Peruvian disease due to a small motile bacillus, *Bartonella bacilliformis*, which invades the red blood corpuscles.

SUPPLEMENTARY READING

References Chapter 10 (p. 112), especially Ash and Spitz's *Pathology of Tropical Diseases: An Atlas* (1945). Philadelphia and London; W. B. Saunders.

CHAPTER 16

DISEASES DUE TO SPIROCHAETES

As ALREADY stated in Chapter 10, the term *spirochaetes* is used here in the general sense of " spiral-shaped bacteria ". The taxonomic distinction between the three genera, *Treponema*, *Borrelia* and *Leptospira*, is based chiefly on the number of convolutions of the organisms.

SYPHILIS

(1) The spirochaete of syphilis

Treponema pallidum was discovered in the fluid from primary syphilitic sores by Schaudinn and Hoffman in 1905. It is a very delicate spiral motile filament 6 to 12μ long with 6 to 12 regular coils. It stains faintly and with difficulty with ordinary stains (hence the name " pallidum "); but it can be demonstrated in dried films by prolonged Giemsa-staining or by Fontana's silver impregnation method, or in tissues by Levaditi's silver impregnation. For the diagnosis of a primary syphilitic sore, however, the best method of seeing the spirochaetes is in the fresh living state in the fluid expressed from the sore, examined immediately by dark-ground microscopy (Fig. 59).

The spirochaete is a strict parasite, and quickly perishes outside the human body. It will not grow on ordinary media, and its cultivation is very difficult and requires special methods. Monkeys and rabbits can be infected by inoculation with the organism.

Spirochaetes are very numerous in the primary syphilitic sore and in the enlarged lymph glands accompanying it ; in the secondary stage, they are widely distributed throughout the body by the blood stream and are plentiful in the lesions in various tissues ; in the tertiary and later lesions they are much less numerous and are difficult to find. In congenital syphilis, they are often abundant, especially in the liver and other viscera.

(2) Modes of infection by syphilis

Infection by *Treponema pallidum* is either (1) acquired or (2) congenital. *Acquired syphilis* results from inoculation of a cutaneous or mucous surface. In the great majority of cases this is a venereal infection due to sexual union with a syphilitic person, and the site of inoculation is somewhere on the genital organs ; but in a small proportion of cases it is an accidental non-venereal infection and the primary lesion occurs in some extragenital situation. Such extragenital infection may occur from kissing, or from handling infected articles ; or it may occur in doctors, students, nurses or dentists from handling syphilitic patients.

Congenital syphilis results from infection of the foetus *in utero* from a syphilitic mother. The special features of this disease will be discussed after acquired syphilis has been described.

190

(3) The four stages of acquired syphilis

The course of syphilis is divided into four stages, namely :

(i) *Primary syphilis:* a period of several weeks, during which the primary sore or " chancre " is developing at the site of inoculation, along with lymphadenitis of the regional lymph glands ;

(ii) *Secondary syphilis:* a period of several weeks or months following subsidence of the primary lesion, when, as the result of dissemination of spirochaetes

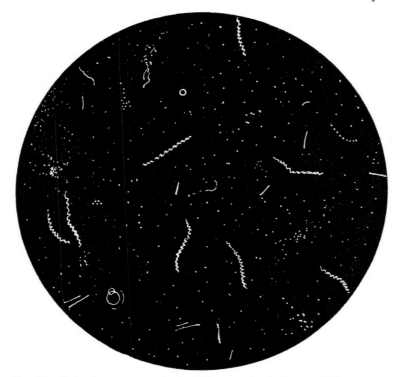

FIG. 59.—Spirochaetes as seen in the living state by dark-ground illumination of fluid from a primary syphilitic sore. (× 1250.)

by the blood stream, the patient feels ill and develops widespread lesions of the skin, mucous membranes and other tissues ;

(iii) *Tertiary syphilis:* an indefinite period after subsidence of the secondary lesions, during which the patient is liable to develop few scattered chronic granulomatous lesions called *gummata ;*

(iv) *Quaternary syphilis* (or *parasyphilis*)*:* comprises late and very distinctive syphilitic affections of the central nervous system, the chief of which are *general paralysis of the insane* and *locomotor ataxia* (or *tabes dorsalis*).

Let us consider the lesions of each of these stages in turn ; remembering, however, that not every case manifests them all. Thus the symptoms of the secondary stage are sometimes so slight that they may pass unnoticed, yet the patient may later develop tertiary or quaternary lesions. The times of appearance of these lesions also vary very greatly ; e.g. gummata may develop within a few months of infection or not for 10 or 15 years, and quaternary disease of the nervous system may appear as early as 2 years from the initial infection or not until 20 or 30 years later.

(4) The lesion of primary syphilis : the hard or Hunterian chancre

(i) *Nomenclature.*—The primary lesion is called a " hard chancre " because of its usual induration, wherein it differs from the " soft chancre " due to Ducrey's bacillus ; but final diagnosis cannot be made solely from the hardness or softness of a chancre, but requires bacteriological confirmation. The primary sore of syphilis is also called a " Hunterian chancre ", because in 1767 John Hunter deliberately inoculated himself with pus from a case of venereal disease and developed a hard chancre.

(ii) *Site.*—As already stated, most primary chancres are venereal infections and are situated on the genitalia. In the male, the usual sites are on the prepuce or glans penis or around or within the urethral orifice, less often the skin of the penis or scrotum. In the female, the usual sites are the labia, vaginal orifice, vaginal vault or cervix ; and, when situated in the vagina or cervix, a primary chancre may easily pass unnoticed. The principal sites of accidentally contracted extragenital chancres are lip, tongue, tonsil, face, fingers and nipple.

(iii) *Naked-eye appearance.*—Usually 2 to 4 weeks after inoculation, the primary lesion first appears as a small raised painless hard papule which slowly enlarges and develops a moist eroded surface or forms an extending ulcer with indurated margins and base. In the absence of antisyphilitic treatment, it persists and perhaps extends for several weeks before finally subsiding and healing, often leaving a barely perceptible scar. Along with the chancre, the nearest lymph glands, usually the inguinal glands, become moderately enlarged and hard, but they do not suppurate.

(iv) *Microscopic structure.*—The lesion is a chronic inflammatory " granuloma ", composed of densely packed lymphocytes, plasma cells and monocytes mingled with proliferating blood vessels, fibroblasts and histiocytes. Apart from the presence of spirochaetes (demonstrated best by dark-ground microscopy of expressed fluid) there is nothing distinctive about the structure of a chancre.

(5) The lesions of secondary syphilis

(i) *General characters.*—In untreated cases, usually about 2 or 3 months after infection and several weeks after subsidence of the primary lesion, the patient begins to feel ill ; he has slight fever, headaches, loss of appetite and anaemia, and inflammatory lesions may appear in various tissues. These lesions are multiple, widespread, symmetrical, pleomorphic in character, usually painless, and usually completely resolving. Microscopically, they are chronic inflammatory foci, generally resembling the primary lesion but smaller, more diffuse and milder ; they contain many spirochaetes and are therefore highly infectious. It is probable that in cases of severe secondary syphilis most tissues contain these scattered focal lesions, but many of these are not within reach of observation. Those which are clinically evident involve the skin, mucous membranes, lymph glands, and occasionally other tissues.

(ii) *Cutaneous lesions.*—These include a variety of generalized skin rashes called *syphilides,* the spots of which may be flat and measles-like (macular) or raised and projecting (papular), or which may become scaly, encrusted or pustular. In most cases they resolve and leave no scars. In moist parts of the skin, especially about the genitals and anus, prominent papillary outgrowths called " condylomata " may develop.

(iii) *Lesions of mucous membranes.*—These are seen best in the mouth and throat. They include non-ulcerated raised white papular lesions called " mucous patches ", and superficial irregular areas of ulceration called " snail-track ulcers ".

(iv) *Lymphadenitis.*—In cases with widespread lesions, there is also generalized mild inflammation of lymph glands which become moderately enlarged, firm and " shotty ", like those of the groin in the primary stage. Besides the cervical, axillary and inguinal glands, the sub-occipital and epitrochlear glands are often palpably enlarged.

Lesions of other tissues which occasionally become clinically evident include iritis, arthritis, epididymitis and periostitis.

After a period ranging from a few weeks to several months, the secondary lesions subside and the patient shows no obvious signs of disease until the appearance of tertiary or quaternary lesions, which may be a few months or many years later.

(6) The lesions of tertiary syphilis

(i) *General characters.*—During the indefinitely prolonged tertiary stage, any organ or tissue may suffer from chronic inflammatory lesions which, though structurally not essentially different from those of secondary syphilis, contrast with them in the following respects—they are irregular and asymmetrical in distribution, often single or few in number, unaccompanied by lymph-glandular enlargement, they do not resolve spontaneously but progress and cause great destruction or fibrosis of the part, and they contain few spirochaetes and so are not highly infectious. Tertiary syphilitic lesions are of two main kinds which, however, are often mingled—(1) a focal " granuloma " with a tendency to central degeneration, called a *gumma,* and (2) diffuse chronic inflammation with fibrosis.

(ii) *The structure of a gumma.*—A gumma is a hard slowly developing " granuloma ", the central part of which shows opaque yellowish degeneration resembling tuberculous caseation, while the peripheral parts consist of pink or grey translucent tissue merging with fibrosis in the surrounding tissues. Gummata range from foci of microscopic dimensions to masses many inches in extent. Microscopically (Fig. 60) the non-degenerated gummatous tissue consists of proliferating fibroblasts, swollen " epithelioid " cells like those of tubercles, a few giant-cells, new-formed small blood vessels, and variable numbers of lymphocytes, monocytes and plasma cells ; it is, in fact, a chronically inflamed granulation tissue. Where giant cells and necrosis are present, it may closely resemble tuberculous tissue ; but, in general, it differs from tuberculous tissue in showing plentiful young blood vessels, fewer giant-cells and less extensive caseous degeneration, and in causing much more fibrosis in the surrounding tissues. The small arteries in and around a gumma usually show pronounced obliterative endarteritis, and also perivascular collections of lymphocytes and plasma cells. When it heals, a gumma leaves in the tissues a densely fibrous scar, enclosing the caseous degenerative products, which may calcify.

FIG. 60.—Section of margin of gumma of liver showing the zone of chronic inflammatory tissue and the central degeneration. (× 80.)

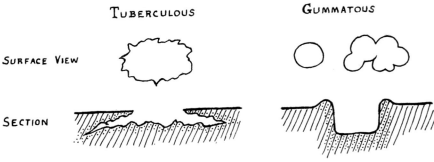

FIG. 61.—Diagram contrasting tuberculous and gummatous ulcers.

On surfaces, e.g. the skin, gummata break down, discharge their necrotic contents, and leave punched-out or serpiginous ulcers, with steep sides, grey necrotic " wash-leather " floors, and indurated surrounding tissues (Fig. 61). When these heal they leave thin depressed wrinkled pale scars with some surrounding brown pigmentation.

(iii) *Diffuse syphilitic inflammation with fibrosis.*—This is quite as frequent as focal gumma formation, or it may occur extensively in parts in which gummata also are present. There is widespread infiltration of the tissues by all of the cellular components of gummata but without focal accumulation or degeneration.

Healing results in much fibrosis and scarring, or in bone thickening and sclerosis.

(a) Tertiary syphilis of the circulatory system

(i) *Syphilitic arteritis.*—The arteries, large or small, in and around any tertiary syphilitic lesion usually show pronounced inflammatory infiltration of their walls and proliferative (often occlusive) endarteritis. But, in addition to this incidental involvement of vessels in lesions of other tissues, syphilis often attacks arteries and periarterial tissues mainly or exclusively. This occurs especially in (*a*) the aorta, and (*b*) the cerebral and spinal arteries.

Syphilitic aortitis usually affects the aortic arch, less often the descending thoracic or abdominal parts of the aorta. The naked-eye appearance is often characteristic ; the interior of the vessel shows raised pearly thickenings of the intima, longitudinal wrinkling and thin depressed areas of scarring ; these lesions are usually distinguishable from the common yellow patches of atheroma which, however, often coexist. Microscopically (Fig. 62) syphilitic aortitis shows patchy infiltration of the media and adventitia by chronic inflammatory cells —lymphocytes, monocytes, plasma cells and fibroblasts. These collect especially around the vasa vasorum in the adventitia and media ; and these vessels increase in number and themselves show obliterative endarteritis. The inflammatory foci lead to patchy loss of the elastic and muscular tissue of the media and to their replacement by fibrous scar tissue. The weakened aorta often suffers dilatation, leading to fusiform or saccular aneurysms. Because of the great importance of syphilitic mesaortitis as a cause of aneurysms in pre-antibiotic times, the arch was the commonest situation of aortic aneurysms, most aneurysms of the arch were syphilitic, and the nearer an aortic aneurysm was to the heart, the more likely it was to be due to syphilis.

Syphilitic arteritis of cerebral or spinal arteries occurs in almost pure form, as well as in more generalized syphilitic inflammation of the meninges. The cerebral arteries most frequently affected are those of the base of the brain, the circle of Willis and its branches ; and the distribution of the changes in them is patchy and irregular. The inflammation affects all coats of the vessels, but proliferative endarteritis is its most important part, because this brings about occlusion of the lumen either by intimal proliferation alone or by supervening thrombosis, with resulting infarction and softening of areas of the brain or spinal cord. Cerebral thrombosis in young adults is sometimes syphilitic; and the destructive effects of syphilitic " transverse myelitis " are largely due to ischaemic infarction following thrombosis of diseased vessels.

(ii) *Syphilitic aortic valvulitis.*—In syphilitic aortitis the sinuses of Valsalva and the aortic ring and cusps are often implicated ; and syphilitic valvulitis is seen only in the aortic cusps. The orifice shares in the general dilatation of the aorta, so that it cannot be efficiently closed by the cusps ; and the cusps themselves show fibrous thickening, rolled edges and increasing immobility, thus aggravating the valvular incompetence. The diastolic regurgitation of blood through the orifice results in a sudden and excessive fall of the diastolic arterial blood pressure and a characteristic " water-hammer pulse ".

(iii) *Syphilitic angina pectoris.*—Involvement of the coronary orifices in syphilitic aortitis often produces " angina pectoris ", i.e. severe attacks of pre-cordial pain brought on by exertion or emotional stress and attributable to myocardial ischaemia. John Hunter suffered from such attacks, and died in one of them in 1793 ; and it is possible that these were due to aortitis resulting from his experimental self-inoculation with syphilis 26 years earlier.

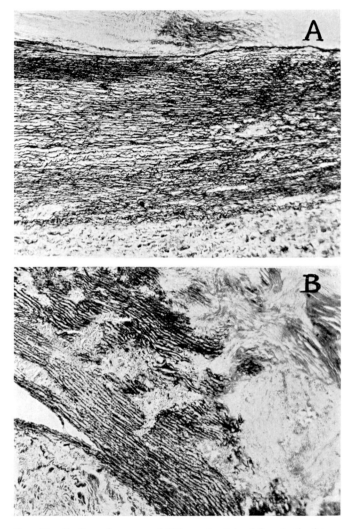

FIG. 62.—Sections from a syphilitic aorta stained for elastic tissue : A, from an almost normal part ; B, from a badly affected part, showing that areas of syphilitic inflammation have caused patchy destruction of the elastic tissue of the media. (× 35.)

(iv) *Gumma of the heart*.—This is very rare. Its commonest situation is the interventricular septum, where it may involve the bundle of His and cause heart block.

(b) *Tertiary syphilis of bones*

Syphilitic lesions of bone, though common in the seventeenth to nineteenth centuries and well represented in old medical museums, are now rarely seen. They occur in three forms : (1) gummata with erosion or caries of the bone, (2) localized sclerosing osteitis without gummata, and (3) diffuse sclerosing osteitis with generalized thickening of the bone. The lesions are often combined, and often multiple in one or several bones.

(i) *Gummata with caries of bone* (Fig. 63).—Gummata may develop in either the periosteum or medulla, more often the former. Periosteal gummata erode the bone to a variable depth, sometimes producing only a superficial serpiginous worm-eaten appearance in the cortex, sometimes destroying the whole thickness of the cortex or the whole thickness of calvarial or other flat bones. Extensive destruction may result in the formation of small isolated sequestra amidst the gummatous granulation tissue. At the margins of eroded areas, there are bony sclerosis and the outgrowth of smooth bosses or irregular osteophytes. Gummatous caries may occur in any bone, especially the skull vault, palate, nasal septum, sternum and tibia. Degenerating gummata in superficially situated bones, like the tibia and skull bones, may break through

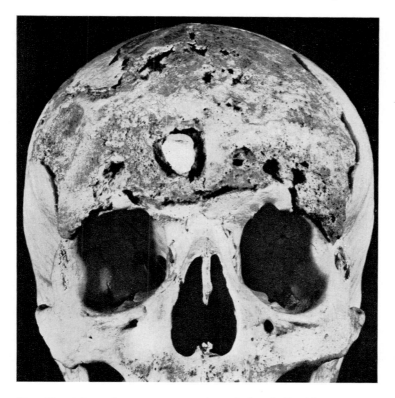

FIG. 63.—Advanced gummatous erosion of the skull with sequestrum formation. (Museum of Royal College of Surgeons, London.)

the overlying skin or mucous membrane and produce characteristic punched-out ulcers with necrotic bone in their bases ; and, through these, secondary pyogenic infection may be superimposed on the syphilitic lesions.   Untreated gummatous disease of the skull (now rarely seen, save in old specimens) may involve the whole vault and destroy large parts of it ; the combination of extensive destruction and surrounding sclerosis and boss formation is especially characteristic here.   Gummatous destruction of the nasal bones and septum produces collapse of the bridge of the nose ; and that of the palate, perforation.   Gummatous ulcers of the shins, a common site, are usually not primarily cutaneous in origin but arise from the tibia.   Syphilitic dactylitis is mainly a destructive central lesion, like tuberculous dactylitis, from which it must be distinguished.

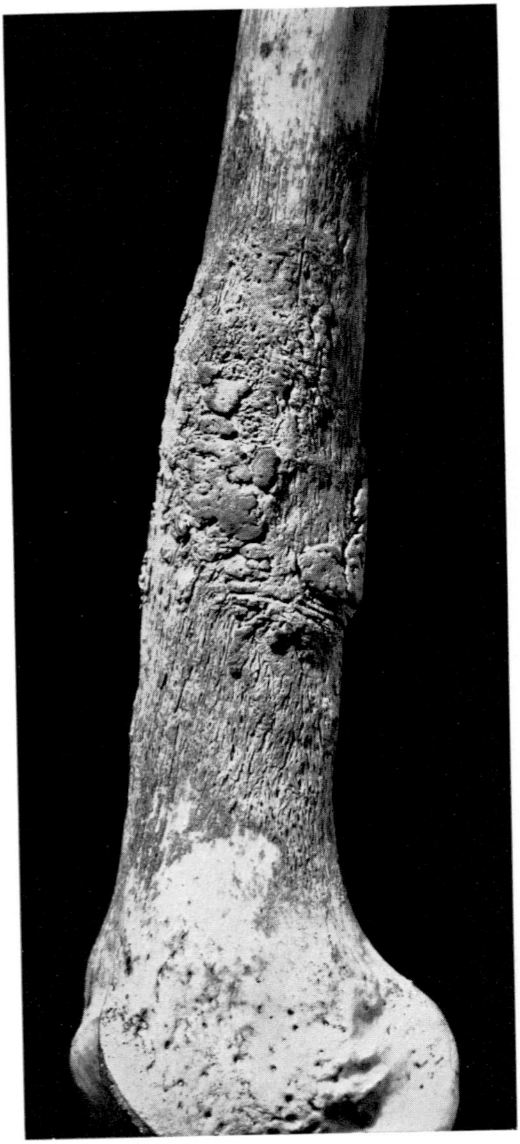

(ii) *Localized sclerosing osteitis.*—Just as in soft tissues, so in bone, syphilitic inflammation may not cause focal gummata with degeneration but may lead to progressive proliferative condensation of the bony tissue and deposition of new dense periosteal bone on the surface.   If this is restricted in extent, the result is a more or less localized boss or fusiform thickening of the bone, often called a " periosteal node " (Fig. 64).

(iii) *Diffuse sclerosing osteitis.*—Bony sclerosis and periosteal new bone formation often takes place throughout much or the whole of a bone, causing dense or spongy, smooth or irregular thickening of the whole bone, and increased density of the medullary spongy bone. The bone becomes thicker, heavier and denser than normal ; radiographs show these features clearly and may also show double contours

FIG. 64.—Syphilitic periostitis with fusiform thickening of lower third of femur ; one of John Hunter's specimens in the Royal College of Surgeons, London.

resulting from the stratification of the periosteal new bone as it is laid down. A long bone affected by diffuse syphilitic osteitis sometimes shows uniform cylindrical thickening of its shaft, sometimes a fusiform thickening greatest in the middle or on one aspect of the shaft.

Figure 65 shows in diagrammatic form the various types of bony changes wrought by tertiary syphilis. It is important to remember that all of these or various combinations of them may be present in one case.

Finally, it is worth while comparing these with the bone lesions of that other important caries-producing granuloma, tuberculosis, as follows : (*a*) Tuberculosis is predominantly bone-destructive, causing caries but little or no bony sclerosis, and healing by fibrous tissue rather than bone ; whereas syphilis is often predominantly bone-formative, causing prominent new bone formation as well as caries, and healing by bony sclerosis. (*b*) Tuberculosis commonly affects the ends of long bones and joints ; whereas syphilis commonly attacks the shafts of long bones, especially the diaphyseal periosteum, and rarely involves joints. Also syphilis frequently attacks the skull, but tuberculosis rarely does so. (*c*) Histologically the two diseases differ markedly ; tuberculosis with its almost avascular granulation tissue composed mainly of " epithelioid " cells and giant-cells and prone to

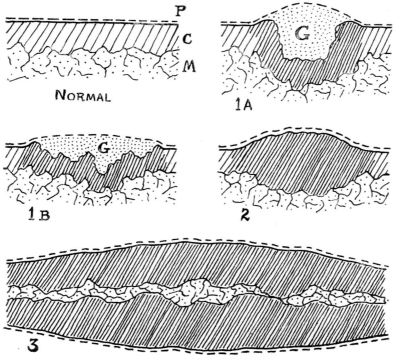

FIG. 65.—Diagram of syphilitic lesions of bone : P, periosteum ; C, cortex ; M, medulla. 1A, localized gumma ; 1B, irregular gummatous erosion ; 2, localized sclerosing periostitis and osteitis ; 3, extensive sclerosing periostitis and osteitis.

progressive caseous degeneration ; syphilis with its vascular granulation tissue composed mainly of proliferating vascular endothelial cells, fibroblasts, osteoblasts, lymphocytes and plasma cells, and prone to progressive osteogenesis.

(c) Tertiary syphilis of the digestive system

Tertiary syphilitic lesions may occur in any part of the digestive organs— rarely the lips, pharynx, oesophagus, stomach, intestines or pancreas, less uncommonly the tongue or liver.

(i) *Tertiary syphilis of the tongue.*—This appears either as (a) a diffuse chronic glossitis or (b) localized gummata, single or multiple. The former is the more common ; it is usually superficial, causing patchy thickening and excessive keratinization of the epithelium (called *leucoplakia**), fissuring of the mucosa, and infiltration of the sub-epithelial tissues by chronic inflammatory cells. It is an important precancerous condition ; between 20 and 40 per cent of cases of lingual cancer show evidence of previous syphilitic glossitis. Sometimes diffuse syphilitic inflammation of the tongue is deep-seated and widespread throughout the muscle, producing general swelling of the organ and limitation of movement. A gumma of the tongue appears as a localized hard swelling which later breaks down to form a characteristic crateriform ulcer with a sloughing base.

(ii) *Tertiary syphilis of the liver.*—This now relatively rare disease also occurs as a diffuse fibrosing inflammation, or as multiple or single gummata, or as a combination of both. Gummata may attain very large sizes in the liver, being sometimes several inches in diameter. Syphilitic fibrosis is usually very irregular in its distribution, and the organ becomes coarsely scarred and lobulated, with relatively normal areas of liver tissue between the bands of fibrous tisse.

(d) Tertiary syphilis of the respiratory system

This too is rare. Gummatous infiltration and ulceration of *the nasal cavity* is usually secondary to lesions in the nasal bones or septum (see p. 198), but it is possible that it may occasionally start in the mucosa. Syphilitic ulceration of *the larynx* leads to great destruction of both the soft tissues and cartilages, and to much fibrosis with deformity and stenosis. Similar lesions may occur in *the trachea or bronchi*, but are very rare. So also are lesions in *the lung*, either gummata or diffuse chronic inflammation with fibrosis.

(e) Tertiary syphilis of the urogenital organs

Gummata and diffuse fibrosing lesions have been seen in all parts of the urogenital organs, but except in the testis they are so rare as to be negligible. Formerly, *the testis* was a fairly common site of gummata, but this is now quite rare in most civilized communities. Gummata start in the testis itself, in which yellow areas of degeneration accompanied by much fibrosis are found, often affecting the whole organ and sometimes spreading to its tunics and to the epididymis. In this respect syphilitic lesions contrast with tuberculous ones, which almost always start in the epididymis.

---

* *Leucoplakia* is not peculiar to syphilis. It is a general term applied to any kind of thickening of stratified epithelia with the development of white opaque patches. Syphilis is only one of the causes—a very important one—of this change in the mouth cavity.

(*f*) *Tertiary syphilis of the nervous system*

Excluding the quaternary lesions to be described presently, syphilis of the nervous system occurs in the following forms, which may be more or less pure or may be combined :

(i) *Syphilitic arteritis.*—(Already considered on p. 195).

(ii) *Syphilitic meningitis.*—This is a diffuse chronic leptomeningitis often with accompanying arteritis of the vessels in the meninges. The membranes and the perivascular Virchow-Robin spaces in the brain itself are diffusely infiltrated by chronic inflammatory cells, and the infiltration may include scattered small gummatous foci. Fibrous adhesions between the brain and dura may develop, and cranial nerves are apt to be strangled in the fibrosing basal membranes. Ocular palsies, with squint and ptosis, are the commonest results, but any cranial nerves may be affected, including the optic nerves with resulting papilloedema or optic atrophy. Other symptoms include headaches, mental confusion, stupor, and localized epileptiform (Jacksonian) convulsions. The cerebrospinal fluid usually shows raised pressure, a moderate number of leucocytes most of which are lymphocytes, positive globulin tests, positive Wassermann reaction, and often a " meningitic " curve with Lange's colloidal gold test (a special test depending on the relative amounts of globulin and albumin present).

(iii) *Localized gummata.*—These may occur in any part of the brain or cord, producing the same symptoms and signs as a tumour.

All of these nervous lesions, once fairly common, are rather rare now, because of the greater efficacy of treatment of syphilis in its earlier stages.

(7) Quaternary syphilis : parasyphilis

The nervous lesions of quaternary syphilis are quite different from the gummatous and diffuse inflammatory changes of tertiary syphilis. They are characterized by degeneration of special groups of neurones, by failure to respond to ordinary antisyphilitic treatment, and by their usually late onset. Whereas tertiary nervous syphilis often develops within a few years of the primary infection, quaternary disease usually does not appear until 10, 15 or 20 years later. It is noteworthy also that patients with quaternary lesions have rarely shown any signs of tertiary lesions. Why quaternary syphilis should have these distinctive characters, we do not know; it has been suggested that a special strain of spirochaete is responsible, but there is no real evidence that this is so.

The quarternary syphilitic diseases are (1) *general paralysis of the insane* or *dementia paralytica*, and (2) *locomotor ataxia* or *tabes dorsalis*. A combination of the two, *taboparalysis*, also occurs.

(*a*) *General paralysis of the insane* (*dementia paralytica*)
Pathology

Changes occur in the brain, leptomeninges, dura mater, and cerebrospinal fluid.

(i) *The brain.*—The brain as a whole is shrunken and underweight, the cerebral gyri are atrophied and the sulci widely open ; the cortex is thin and its pattern blurred ; the ventricles are enlarged and their ependymal surfaces are roughened and granular—"ependymitis granularis". Microscopically, the nerve cells of the cerebral cortex show widespread degeneration, many of them eventually disappearing entirely ; fibre tracts connected with the diseased cells also degenerate ; widespread proliferation of astrocytes and their neuroglial fibres occurs and the atrophied cortex is more cellular than normal because of the increased number of neuroglial cells; the ependymal granulations also consist of foci of proliferating sub-ependymal astrocytes; the Virchow-Robin perivascular spaces of the diseased cortex and other parts contain many lymphocytes and plasma cells. By special stains spirochaetes have been demonstrated throughout the brain.

(ii) *The leptomeninges.*—The pia mater and arachnoid are thickened, oedematous, white and adherent to the diseased cortex beneath. Microscopically, they show fibrosis, excessive new-formed blood vessels, and collections of lymphocytes and plasma cells.

(iii) *The dura mater.*—This is thick, abnormally adherent to the skull, and sometimes presents on parts of its inner surface stratified layers of old or recently extravasated blood, a condition called " pachymeningitis haemorrhagica interna ". (This is not constantly present in, nor peculiar to, syphilitic disease ; but occurs also in other kinds of dural and cerebral disease.)

(iv) *The cerebrospinal fluid.*—The fluid usually appears clear, but it shows an excess of lymphocytes, 50 to 500 or more per cubic millimetre ; tests for globulin are positive ; Lange's colloidal gold test gives a characteristic " paretic " curve ; and the Wassermann reaction is positive.

Clinical features

In its early stages, general paralysis of the insane (often abbreviated to " G.P.I.") usually produces progressive mental changes—confusion, failure of memory, loss of self-restraint, boastfulness, and delusions of grandeur. Speech and writing deteriorate ; epileptiform convulsions occur ; and mentality steadily decays. The reaction of the pupils to light is lost while that to accommodation is preserved (Argyll-Robertson pupils), and the pupils are often unequal. Paresis of muscles appears and becomes more and more crippling, until the patient is bed-ridden, demented and dirty. He dies of exhaustion, bed-sores, urinary infection or bronchopneumonia, usually within 2 or 3 years of the onset in untreated cases. Hyperpyrexial treatment, by malaria or other means, markedly modifies and prolongs the course of the disease.

(b) Locomotor ataxia (tabes dorsalis)
Pathology

The essential change in tabes is slow degeneration of the posterior nerve roots and sensory tracts of the spinal cord and brain stem, and of the optic and occasionally other nerves. Various " trophic " effects of the resulting loss of sensation may also occur ; these will be mentioned along with the clinical features of the disease.

(i) *The posterior nerve roots.*—Between the spinal ganglia and the cord the posterior roots show varying degrees of degeneration of their nerve fibres and myelin sheaths, terminating in some cases in complete atrophy and fibrous replacement. The ganglion cells themselves and their peripheral axons in the mixed nerves are unaffected. The explanation of this remarkable restriction of the degenerative change to the central axons only of the sensory neurones is not known, but spirochaetes have been demonstrated in the degenerating nerve roots and their leptomeningeal sheaths.

(ii) *The spinal cord.*—Tabetic atrophy affects all kinds of sensory fibres entering the cord in the posterior nerve roots, though the different kinds are not always affected equally. Some of the entrant fibres have only short courses within the cord before they arborize around nerve cells of the posterior grey horns from which new axons, not affected by tabes, ascend the cord. Other posterior root fibres, however, namely those subserving touch and proprioceptive sensation, pass up, without relay at the level of entry, in the posterior fasciculi of Goll and Burdach. Tabetic atrophy of these fasciculi is strikingly displayed in cross-sections of the cord stained by the Weigert-Pal or similar methods. When badly affected, the tracts are shrunken, devoid of myelin sheaths, and almost completely replaced by pale glial scar tissue. The disease is named *tabes dorsalis* (*tabes* = wasting) in reference to this atrophy of the dorsal columns of the spinal cord.

(iii) *The optic nerves.*—Similar degenerative changes are responsible for the optic atrophy which occurs in a considerable number of cases of tabes, and which produces pallor of the optic disc viewed through an ophthalmoscope. In some cases, atrophy of the optic nerves is the main lesion and the spinal sensory lesions are relatively slight ; indeed, so slight may they be that quaternary syphilis may present itself as a primary optic atrophy unaccompanied by other evidence of nervous disease.

(iv) *The meninges.*—These show only slight infiltration of the pia-arachnoid sheaths of the posterior nerve roots by lymphocytes and other chronic inflammatory cells. The changes are so mild that they are probably only secondary to, and not the cause of, the degeneration of the nerve roots. The dura mater shows no significant changes.

(v) *The cerebrospinal fluid.*—This appears clear, and shows only a slight excess of lymphocytes, e.g. 10 to 100 per cubic millimetre. It contains a slight or moderate excess of globulin, and may give a characteristic " luetic " curve with Lange's colloidal gold test. It usually, but not always, gives a positive Wassermann reaction.

Clinical features

The symptoms and signs of tabes are easily understood from the pathology of the disease ; they include sensory changes, muscular ataxia and hypotonia, loss of spinal reflexes, ocular changes, and " trophic " lesions.

(i) *Sensory changes.*—In the early stages of degeneration of the posterior nerve roots, irritative sensory symptoms often occur in the form of pains of varying degree in the limbs or trunk, including sharp " lightning pains ", " girdle pains ", and cramp-like pains referred to some viscus (stomach, bladder, larynx) and called " visceral crises ". There may also be excessive sensitiveness (hyperaesthesia) of the skin to touch, and abnormal sensations (paraesthesiae) such as tingling, prickling or crawling sensations. Later, these irritative phenomena are replaced by varying degrees of loss of sensation, both cutaneous and proprioceptive. Anaesthetic areas may be bordered by zones of hyperaesthesia.

(ii) *Muscular ataxia and hypotonia.*—These result from loss of proprioceptive muscle and joint senses conveyed in the posterior columns of the cord. There is early loss of conscious sense of position, as revealed by the finger-nose and knee-heel tests with the eyes closed, and of bodily balance as revealed by a tendency to fall when standing erect with the feet together and the eyes closed (Romberg's sign). Later, the gait and all muscular movements become jerky and uncoordinated ; hence the name " locomotor ataxia ". Since the normal tone of muscles depends on a constant inflow of proprioceptive sensations, the ataxia of tabes is accompanied by muscular hypotonia and weakness, so that the limbs can be placed in abnormal positions.

(iii) *Loss of spinal reflexes.*—Since the spinal reflexes depend on segmental sensory-motor arcs through the spinal cord, severance of the sensory part of these arcs by tabetic degeneration abolishes the reflexes. The knee-jerks, ankle-jerks and plantar reflexes are soon lost ; the abdominal and cremasteric reflexes may persist longer. The micturition reflexes may be disturbed, with resulting retention or incontinence of urine. The sexual reflexes are lost.

(iv) *Ocular changes.*—The pupillary reflexes are also interfered with, and Argyll-Robertson pupil is common. Optic atrophy has already been referred to.

(v) " *Trophic* " *lesions.*—These include Charcot's joints, perforating ulcers, and atrophy of bones. The idea that these depend, even partly, on the loss of special " trophic " nerve impulses subserving the nutrition of tissues, is very doubtful. They are almost certainly due entirely to loss of sensation and muscular control. *Charcot's joint* occurs in a small proportion of tabetics, usually in the lower limb—knee, hip or ankle. It is a painless disorganization of the joint, characterized by effusion into the joint cavity, destruction of cartilages and bone ends, and the formation of extravagant osteophytes and chondrophytes around the margins of the damaged areas. It is really a severe and progressive traumatic osteo-arthritis in an anaesthetic ataxic limb whose owner is unaware of the repeated mechanical injuries he is inflicting on it. The term " Charcot's joint " is often restricted to tabetic arthropathy, but exactly similar joint disease is seen in other anaesthesia-producing diseases, e.g. syringomyelia, myelitis or peripheral neuritis. Charcot's joint is *not* a syphilitic spirochaetal arthritis. *Perforating ulcer* is a deeply penetrating painless chronic ulcer which sometimes develops in the sole of the foot (rarely elsewhere) in a patient with tabes dorsalis (or less commonly with syringomyelia or peripheral neuritis). Like Charcot's joint, it also results from repeated mechanical injuries of which the anaesthetic patient is oblivious ; he remains unaware of his ill-fitting boot or the projecting nail in its sole, and so he goes staggering about on a blister or sore which cannot heal but grows more and more chronic and callous. *Atrophy of bones* seen in the limbs of tabetics probably results from the absence of the normal stimuli of weight-bearing and muscular stresses, because of the ataxia, muscular atonia and lack of proper use of the limbs.

(8) Congenital syphilis

(a) *The mode of infection*

The term " congenital syphilis " is preferable to " hereditary syphilis ", because the disease is *not* inherited in the sense of being due to chromosomal disturbances in the germ cells. Congenital syphilis is always the result of transfer of spirochaetes from an already syphilitic mother to her unborn child ; it is syphilis acquired *in utero*. The mother often shows no signs of active syphilitic lesions and is herself unaware of her infection, which is revealed only by her positive serological reactions. Foetal tissues are clearly an excellent medium for the growth of spirochaetes ; for, if a pregnant mother is syphilitic, whether the infection was contracted during pregnancy or existed previously, the spiro-chaetes congregate and multiply in great numbers in the foetus, in spite of the absence of any clinical disease in the mother. If the woman was recently infected, the syphilitic foetus often dies and is aborted, or it may be still-born at term or may die soon after birth. The organs of such foetuses and infants swarm with spirochaetes and show severe diffuse chronic inflammatory fibrosis and atrophy of parenchyma cells. If the mother has been long affected, her progeny may be born apparently healthy and may survive, but may later develop some or many of the congenital syphilitic lesions now to be described.

(b) Congenital syphilitic lesions in children who survive

Roughly in their order of appearance, these are :

(i) *Early post-natal lesions.*—These include chronic rhinitis or " snuffles ", skin rashes, especially about the buttocks and thighs, fissured sores or " rhagades " about the mouth or anus, enlarged liver and spleen, and general failure to thrive (" marasmus "). The infant may die of malnutrition in a few weeks or months, when its organs will be found to show changes similar to, but less severe than, those in the still-born foetus.

(ii) *Lesions of bones and teeth.*—Syphilitic epiphysitis or osteochondritis produces broad irregular yellowish epiphyseal cartilages due to chronic inflam-matory infiltration of the most recently formed bone at the osteochondral junction. There may be also imperfect ossification of the membrane bones of the vault of the skull, so that these yield on pressure—" cranio-tabes ". Parrot's nodes —smooth bosses on the infant's frontal and parietal bones—which have usually been attributed to syphilis, are almost certainly not syphilitic but due to organizing subperiosteal haemorrhages associated with scurvy (see p. 365). Gummatous disease of the nasal bones with collapse of the bridge, already described under tertiary syphilis, frequently results from congenital syphilis ; it is the end-result of the same nasal inflammation which causes " snuffles " in infancy. Irregular-ities of dentition in both the deciduous and permanent teeth are common ; a particularly characteristic lesion of the permanent teeth is semilunar notching of the biting edges of the upper incisors—Hutchinson's teeth.

(iii) *Lesions of the special senses.*—Deafness may result from inflammation of the cochlea. Interstitial keratitis appears in late childhood or adolescence and may cause serious loss of sight from bilateral corneal opacity.

(iv) *Lesions resembling those of acquired syphilis.*—In addition to the fore-going lesions, which are peculiar to congenital syphilis, victims of this disease may also develop any of the lesions of acquired syphilis (except of course the primary chancre), including those of quaternary type—juvenile general paralysis or locomotor ataxia.

(c) Latent congenital syphilis

Just as in acquired syphilis the infection may for long periods remain quiescent without producing active lesions, so a child of a syphilitic mother may reach adulthood without showing any clinical evidence of the disease. Yet his serological reactions are positive and there are spirochaetes in his tissues, and many years later he may develop an aortic aneurysm or quaternary disease. Thus, the serological tests reveal that, on the one hand, the seemingly healthy mothers of syphilitic children are themselves always syphilitic; and that, on the other, the seemingly healthy children of syphilitic mothers are always syphilitic.

(9) Serological changes in syphilis

Three distinct types of antibody may be present in the serum of patients with syphilis or other treponemal diseases. A *reagin*, which reacts with an antigen composed of alcoholic extract of heart muscle with added lecithin and cholesterol, may be present. It is detected by complement fixation, as in the Wasserman reaction, or by flocculation methods, such as the Kahn test or the slide test of the Venereal Diseases Reference Laboratory. These tests are not specific, and may give positive reactions in some non-treponemal diseases and in some normal persons, e.g. in pregnancy.

A group anti-treponemal antibody, which reacts with a protein antigen extracted from spirochaetes, is detected by the Reiter protein complement fixation test. This is more sensitive and more specific than the reagin tests, giving fewer false positive results.

A third antibody, reacting specifically with *Treponema pallidum* and related organisms, is detected by the treponema immobilization test and by the fluorescent treponemal antibody test. These are almost completely specific, but are less sensitive than reagin tests and may give negative results in primary, early secondary or congenital syphilis.

(10) Diagnosis of syphilis

Summarizing the main points in the diagnosis of syphilis:

In the primary stage, diagnosis is by direct microscopical examination of fresh fluid expressed from the chancre, using dark-ground illumination.

In the secondary stage, the serological tests are almost always strongly positive. Spirochaetes are sometimes demonstrable in the serum from skin eruptions, but this is seldom undertaken.

In the tertiary stage, the reagin and Reiter protein tests are positive in about three-quarters of the cases, and the specific tests for treponemal antibody are usually positive. It is of little use searching for spirochaetes in tertiary lesions.

In the quaternary stage, the reagin tests are positive with both blood and cerebrospinal fluid in nearly all cases of general paralysis and in most cases of locomotor ataxia. The cerebrospinal fluid should always be tested as well as the blood, for it sometimes gives a positive reaction when the blood fails to do so. The tests for specific treponemal antibody performed on blood are almost always positive.

Congenital syphilis with active lesions usually gives strongly positive serological reactions. The mother's blood should also be tested.

With antisyphilitic treatment, the reagin and Reiter tests may become negative, but positive reactions may return when treatment is stopped, thus providing important evidence of the effectiveness of treatment. The tests for specific treponema antibody are little affected by treatment and may remain positive for many years.

YAWS

Yaws (or *framboesia*) is a tropical disease, affecting the natives, especially children, of Central Africa, the East and West Indies, and the Pacific Islands, and caused by *Treponema pertenue*, a spirochaete closely resembling that of syphilis. The disease also resembles syphilis in producing rather similar lesions, in causing positive serological reactions, and in being curable by antisyphilitic drugs. It differs from syphilis in the following ways: it is not a venereal disease, but is transferred by direct contact inoculation of the skin or by biting insects; it is never congenital; and it does not cause quaternary lesions of the nervous system.

The primary lesion is a papule or pustule usually on a limb. The secondary eruption, which appears 2 to 4 weeks later, consists of a widespread crop of projecting encrusted granulomas which have been likened to raspberries, whence the name " framboesia " (*framboise* = a raspberry) ; microscopically, these show hyperplasia of the epidermis and dense infiltration of the dermis by chronic inflammatory cells, especially plasma cells. Tertiary lesions appear months or years after healing of the secondary stage ; they comprise ulcers of skin and mucous membranes and chronic inflammatory lesions of bones.

RELAPSING FEVER

As the name implies, this is an infection characterized by repeated febrile periods, usually 5 to 7 days, separated by afebrile intervals of about the same duration. In different parts of the world, the disease varies in its course, severity, and the number of relapses; and these differences are now known to be due to the fact that several distinct, though closely related, spirochaetes are responsible.

Borrelia recurrentis (or *Borr. Obermeieri*) is notable in the history of bacteriology in being the second bacterial pathogen discovered, the anthrax bacillus being the first (see p. 169). It was first observed by Obermeier in a Berlin epidemic in 1868 (published in 1873); he noticed motile spiral threads in the blood of patients during febrile stages of the disease, their disappearance during afebrile periods, and their reappearance during relapses ; he failed to find them in the blood of normal persons or of patients with other fevers ; and he correctly regarded them as spirochaetes. Obermeier's spirochaete is the usual type found in European relapsing fever ; in other parts of the world relapsing fever is caused also by the closely related *Borr. Duttoni* of Africa, *Borr. Novyi* of America, and others. These spirochaetes are large, easily stained, and plentiful in blood films taken during the febrile period ; diagnosis is thus readily made by blood examination (Fig. 66).

The spirochaetes are transmitted from person to person by external parasites, lice and ticks—*Borr. recurrentis* by the ordinary body louse (*Pediculus corporis*), *Borr. Duttoni* by ticks (hence the alternative name for the disease "African tick fever"). The spirochaetes multiply in the bodies of these ectoparasites, which retain their infectivity for long periods and even pass it on to their progeny.

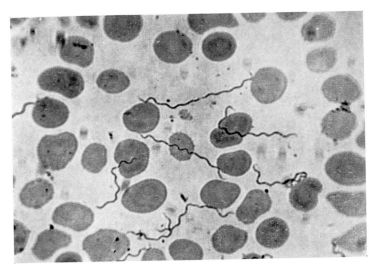

FIG. 66.—*Borr. recurrentis* in a blood film. (× 1250.)

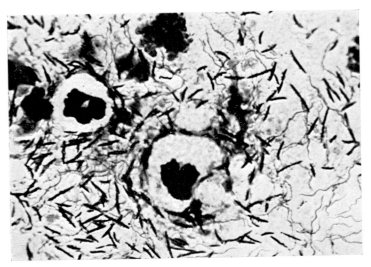

FIG. 67.—Smear from Vincent's angina, showing fusiform bacilli, spirochaetes and pus cells. (× 1250.)

Because its transfer depends on these ectoparasites, relapsing fever is prevalent only where there are low standards of personal hygiene and people are crowded together. It was one of the old " gaol fevers " and " famine fevers ". It has disappeared from Great Britain and most of the countries of Western Europe, but it still occurs in Eastern Europe, and in Asia, Africa and parts of America.

VINCENT'S ANGINA

Vincent's spirochaete (*Borrelia Vincenti*) is found in great numbers along with *Fusiformis fusiformis* (p. 189) in certain ulcerative and membranous inflammations of the throat (*angina* = sore throat), in other spreading necrotic lesions of the mouth such as " noma " or " cancrum oris " in feeble children, and occasionally in similar lesions elsewhere, e.g. gangrenous laryngitis, gangrene of the lung, or gangrenous balanitis. The two organisms are distinct and unrelated, and their constant association is an example of a mixed infection in which both participants are necessary for the production of pathogenic effects (Fig. 67).

LEPTOSPIROSIS ICTEROHAEMORRHAGICA
(*Spirochaetal jaundice, or Weil's disease*)

This is a severe febrile infection which is caused by *Leptospira icterohaemorrhagiae* and characteristically produces jaundice and haemorrhages from mucous membranes (epistaxis, haematemesis, melaena or haematuria) or in the skin. It is especially associated with overcrowding and bad hygienic conditions and was one of the well-known diseases of armies.

Leptospira icterohaemorrhagiae occurs as a relatively common parasite in wild rats and field-mice, without seriously affecting their health; they are excreted in the rodents' urine and so contaminate soil and water. Dogs acquire the disease and aid in its dissemination. Human infection usually occurs through the skin by contact with infected water; hence the frequency of the disease in armies occupying water-logged trenches, in workers in sewers or wet mines, and in fish-cleaners.

In the early stages of Weil's disease the spirochaetes are present in the patient's blood, the infectivity of which can be demonstrated by guinea-pig inoculation. In a week or 10 days, however, he has developed specific agglutinins against the organisms, his blood ceases to be infective, and spirochaetes are excreted in the urine. Hence, early diagnosis is made by intraperitoneal inoculation of guinea-pigs with blood ; later diagnosis, by agglutination tests or guinea-pig inoculation with the deposit from urine. In fatal cases many spirochaetes are present in the kidneys and liver and these organs show severe damage with areas of necrosis.

OTHER SPIROCHAETAL DISEASES

Canicola fever.—Of several other species of *Leptospira* which occasionally infect man, the most important is *Leptospira canicola*, a common pathogen of dogs, which also infects pigs and cattle. This causes a usually mild febrile illness without haemorrhages or jaundice.

Rat-bite fever (Sodoku).—This is an occasional human infection by a small spirochaete-like organism called *Spirillum minus*, following the bite of a rat. It is to be distinguished from another kind of " rat-bite fever " caused by *Streptobacillus moniliformis*, a pleomorphic filamentous Gram-negative bacillus.

SUPPLEMENTARY READING

References Chapter 10 (p. 112), and the following :

Broom, J. C. (1951). " Leptospirosis in England and Wales ", *Brit. Med. J.* **2,** 689.

Greig, D. M. (1931). *Clinical Observations on the Surgical Pathology of Bone.* Edinburgh; Oliver & Boyd. (Contains many beautiful photographs of syphilitic bones.)

King, E. S. J. (1930). " On some aspects of the pathology of hypertrophic Charcot's joints ", *Brit. J. Surg.*, **18,** 113.

Lawson, J. H. and Michna, S. W. (1966). " Canicola fever in man and animals ". *Brit. Med. J.*, **2,** 336.

Perry, W. L. M. (1948). " The cultivation of *Treponema pallidum* in tissue culture ", *J. Path. Bact.*, **60,** 339.

Saphir, A. (1929). " Involvement of medium-sized arteries associated with syphilitic aortitis ", *Amer. J. Path.*, **5,** 397.

Wylie, J. A. H. (1946). " The relative importance of the renal and hepatic lesions in experimental leptospirosis icterohaemorrhagica ", *J. Path. Bact* , **58,** 351.

CHAPTER 17

DISEASES DUE TO FUNGI

INTRODUCTION

AT THE beginning of Chapter 10, we distinguished the bacteria or unicellular " fission fungi " from the fungi proper, which are multicellular mould-like and yeast-like organisms. This distinction, though a practically useful one, is not clear-cut, for many bacteria (e.g. streptococci and anthrax bacilli) character-istically grow in chains or filaments, while some fungi (e.g. actinomyces) are very simple in structure and can be regarded either as colonial " higher " bacteria or as the simplest of the multicellular fungi. The filamentous feltwork of the mould-like fungi is a *mycelium*, and its individual filaments are *hyphae*.

The botanical classification and names of the pathogenic fungi are very confused ; there is no unanimity regarding classification, many of the species have been given several different names, and, worse still, the same name has sometimes been applied to different species by various writers. To add to the student's difficulties, the names applied by dermatologists to some of the fungal infections of the skin have been lamentably lengthy or unfortunate in other respects. The pathologist should not worry himself about the classification of the pathogenic fungi or the implications of the names applied to them by the systematists. Suffice it for him to know what diseases are caused by fungi, the usual names of each of the pathogenic fungi, and enough about their structural characters to recognize them in the lesions they produce.

A disease caused by a fungus is called a " mycosis " (from the Greek, " mykes " = a mushroom, hence a fungus). The most appropriate name for a mycosis is made by adding " -osis " to the root of the genus name of the fungus, e.g. " actino-mycosis " due to *Actinomyces*, " torulosis " due to *Torula*, " aspergillosis " due to *Aspergillus*, etc. But nomenclature does not always follow this practice ; the familiar disease-name and the fungus-name are sometimes quite unrelated and we just have to learn to associate them, e.g. " ring-worm " with *Microsporon* and *Trichophyton*, and " thrush " with *Candida*.

It is not surprising that the skin is, of all tissues, the one most frequently the site of fungus infections, that these are due to a great variety of fungi, and that many of them invade only the epidermis and its appendages and do not penetrate into deeper tissues. These purely superficial fungi can grow in the various layers of the stratified epithelium but have no powers of invading vascular tissues. Some of the skin fungi, however, are more aggressive, invading the dermis also and producing granulomas or even disseminating to other parts of the body. The many mycoses of the skin are collectively called *dermatomycoses*. Other fungus infections are acquired by inhalation or ingestion, and the primary lesions occur in the respiratory or alimentary tracts.

The order of presentation of the fungus diseases, one by one, in this chapter is not botanically systematic. We will begin with actinomycosis because this is the most important of the visceral fungal diseases acquired by ingestion or inhalation, and is due to a structurally simple fungus which is sometimes classed with the "higher" bacteria. After mentioning diseases due to other actinomycetes, we will then deal with other mycoses usually acquired by ingestion or inhalation. Finally we will consider the dermatomycoses, first those which invade sub-epithelial tissues and produce granulomatous lesions, and lastly those which always or usually remain purely superficial. It will be helpful to enumerate in that order the pathogenic fungi and the diseases they produce, as follows :

NAMES OF PATHOGENIC FUNGI	DISEASES
A *The Pathogenic " Ray-fungi "*	
Actinomyces Israeli	Actinomycosis
Nocardia asteroides	Abscesses of lungs, brain, etc.
B. *Other Ingested or Inhaled Fungi*	
Candida (Monilia) albicans	" Thrush ", moniliasis
Cryptococcus neoformans (Torula histol-	
ytica)	Cryptococcosis (torulosis)
Histoplasma capsulatum	Histoplasmosis
Rhinosporidium Seeberi	Rhinosporidiosis
Aspergillus fumigatus	Aspergillosis
C. *Granuloma-producing Dermatomycetes (Skin fungi)*	
Blastomyces dermatitidis	Blastomycotic dermatitis (Gilchrist's disease)
Coccidioides immitis	Coccidioidomycosis
Hormodendrum and *Phialophora*	Chromoblastomycosis
Sporotrichum Schenckii	Sporotrichosis
D. *Superficial Dermatomycetes (Skin fungi)*	
Microsporum (several species)	Scalp " ringworm " in children, etc.
Trichophyton (several species)	Scalp and body " ringworm ", etc.
Epidermophyton (several species)	" Tinea " of various sites
Malassezia furfur	" Pityriasis versicolor''

ACTINOMYCOSIS

(1) The ray fungus

Actinomycosis in man is due to the " ray-fungus ", *Actinomyces Israeli*. A closely similar infection producing " lumpy jaw " in cattle is due to *Actinomyces bovis*. In the lesions the parasite is plentiful and forms large rosette colonies, plainly seen by low-power microscopical examination and sometimes visible to the naked-eye as small yellow granules in the pus. Each colony consists of a central tangled mass of branched filaments which are Gram-positive, bordered by radially arranged club-like outgrowths (Fig. 68). It is these which give the colonies their rosetted appearance and their name of " ray-fungus ". The organism grows slowly on ordinary media under anaerobic or micro-aerophilic conditions. The colonies in culture do not show marginal clubs.

(2) The primary lesions of actinomycosis

The organism occurs normally as a commensal in the mouth, and the primary infection is commonly in the jaw, sometimes in the appendix and caecal region, sometimes in the lungs by inhalation, and very rarely in the skin. In all situations the lesions are similar—slowly developing, ramifying, pyogenic abscesses containing polymorphonuclear pus cells and colonies of the fungus, and surrounded by much chronic inflammatory granulation and fibrous tissue. The suppurating tracks discharge on neighbouring surfaces, forming multiple sinuses or fistulae. The disease runs a very chronic course, often for years; it may remain local, but there is always great risk of its further dissemination.

Primary actinomycotic appendicitis often escapes detection at first. The usual sequence of events is that the appendix is removed for " chronic appendicitis ", temporary recovery ensues, followed by the development of a chronic

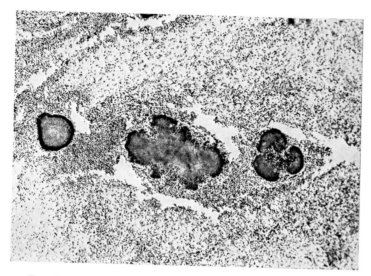

FIG. 68.—Colonies of actinomyces in abscess in liver. ($\times$ 70.)

abscess or discharging sinus, and later by pyaemic abscesses in the liver. The writers have seen several such cases, in one of which the appendix had been saved and subsequent sections of it showed colonies of ray-fungus.

Primary pulmonary actinomycosis is usually mistaken at first for tuberculosis or tumour. The disease often invades the pleura, producing empyema, abscesses in the chest wall or mediastinum, external sinuses, or pericarditis.

(3) Disseminated lesions of actinomycosis

Dissemination of the fungus takes place by the blood stream and constitutes a chronic form of pyaemia. Chronic abscesses, with the same characters as in the primary lesions, may develop in lungs, liver, bones, brain, muscles, kidneys, adrenals or other organs. When the primary infection is in the appendix and caecal region, the metastatic abscesses usually develop first in the liver—a chronic portal pyaemia. In the liver the abscesses often follow the portal tracts and so assume a " fern-frond " distribution. By local spread from affected viscera, large abscesses may develop in any neighbouring serous cavities or large tissue spaces—e.g. loculated intraperitoneal, retroperitoneal, subphrenic or mediastinal abscesses, empyema, pericarditis, or meningitis.

(4) Diagnosis

Pus from the lesion is spread out in a thin layer and searched for tiny yellow granules. These are picked out, crushed, smeared, and stained by Gram's method ; the presence of Gram-positive branching filaments usually settles the diagnosis. For verification by culture, the granules are washed with sterile water, inoculated on suitable media and incubated anaerobically. In pieces of tissue, diagnosis is easily made by finding the characteristic and conspicuous rosette colonies.

NOCARDIOSIS

The generic name *Actinomyces* was used to embrace all species of " ray-fungi "; but systematists now restrict it to anaerobic species, and distinuish aerobic species as a separate genus *Nocardia*. A pathogenic member of this genus is *Nocardia asteroides*, a rare cause of pulmonary and cerebral abscess. " Madura foot " or mycetoma, a chronic inflammatory granuloma occurring in parts of India and Africa, was once thought to be due to a specific Nocardial infection; but it has been found that several different kinds of fungi may produce this disease.

CANDIDIASIS

Candida albicans (also called *Monilia albicans*), a mycelial fungus with yeast-like bodies, is the cause of common " thrush " or aphthous stomatitis in unhealthy children or debilitated adults. The fungus grows on the oral mucous membrane and produces raised white patches composed of desquamated epithelium united by the branching fungal mycelium. The same fungus also causes vaginal thrush, interdigital dermatitis, chronic infections of the nail-beds (onychia), and dermatitis of the thighs and external genitals in diabetics. It is sometimes found also as a secondary invader in the respiratory passages n tuberculosis or other chronic lung diseases; and, very rarely, it may possibly be a primary pathogen causing bronchitis or bronchopneumonia—pulmonary candidiasis. Candidiasis (or less commonly *mucormycosis* due to various moulds) sometimes complicates cortisone, antibiotic or chemotherapy, owing to the suppression of the normal bacterial flora of mucosal surfaces.

CRYPTOCOCCOSIS (TORULOSIS)

Cryptococcus neoformans (*Torula histolytica*) is an encapsulated yeast-like fungus which multiplies almost entirely by budding, and which is the cause of serious infections, especially of the brain and meninges. Although not a common disease, cryptococcosis is probably not as rare as has been supposed; many cases have now been described from all parts of the world.

Infection is by inhalation in most cases ; the primary lesions in the lungs are sometimes large and productive of symptoms, but more often they are small and unsuspected. Metastatic lesions may occur in almost any tissue ; but by far the most important are those in the central nervous system, producing usually symptoms of chronic meningitis, often mistaken for tuberculous meningitis, or symptoms of cerebral tumour due to a large torular collection in the brain.

The lesions consist mainly of masses of the encapsulated yeast-like parasites, the thick mucoid polysaccharide capsules of which give them a slimy gelatinous appearance. Inflammatory reaction on the part of the tissues is usually very

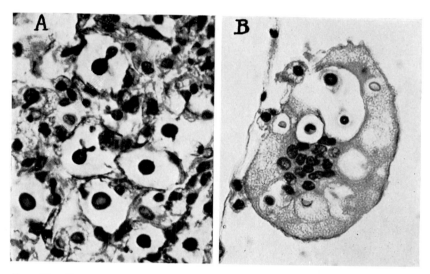

FIG. 69.—Cryptococcosis : A, a collection of torulae in the brain, showing capsules and budding ; B, torulae enclosed in a foreign-body giant-cell in the spinal meninges. (× 470.)

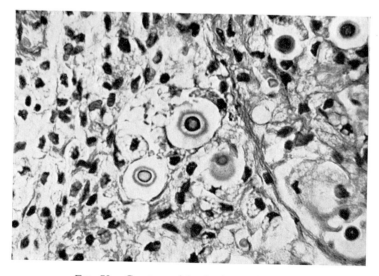

FIG. 70.—Cryptococci in the lung. (× 470.)

slight, consisting only of the aggregation of a few lymphocytes, macrophages or foreign-body giant-cells (Figs. 69, 70). In cases of meningeal infection typical encapsulated yeast bodies can be found by direct examination of fresh cerebrospinal fluid ; sometimes they can also be found in the sputum.

The disease may last months or even years, but is eventually fatal, usually from meningitis. In all cases of atypical chronic meningitis or of supposed " tuberculous " meningitis in which that diagnosis is unproven, the possibility of cryptococcosis should be considered and the cerebrospinal fluid should be examined for yeast bodies. The organisms grow well on ordinary culture media ; but in culture their capsules are much less conspicuous than in the tissues.

HISTOPLASMOSIS

Histoplasmosis is a rare disease due to a small yeast-like fungus, *Histoplasma capsulatum*. It affects many tissues, especially the liver, spleen and lymph glands, in which great numbers of the organisms are taken up by the reticuloendothial cells and macrophages; and it produces a prolonged febrile illness which is usually fatal. The lungs are probably the portal of entry of the fungus into the body, in some cases at least.

RHINOSPORIDIOSIS

This is a polypoid inflammatory disease of the nasal mucosa, or more rarely of the ear, conjunctiva or mouth, occurring chiefly in India. It is due to a peculiar fungus, *Rhinosporidium Seeberi*, which forms cystic bodies up to 2 mm. in diameter, each composed of a cellulose capsule enclosing numerous spores.

ASPERGILLOSIS

Pulmonary aspergillosis is a form of bronchopneumonia in captive birds, especially pigeons and penguins, due to the mould *Aspergillus fumigatus*. It occasionally infects the human lung, producing bronchopneumonic areas and small cavities.

(*Aspergillus* and *Penicillium* are two of the commonest genera of moulds, producing yellow, green or black hairy growths on stale food, culture media, etc. *Pencillium notatum* is the source of the antibacterial substance, penicillin.)

BLASTOMYCOTIC DERMATITIS

Gilchrist's disease, one kind of " blastomycosis ", is due to *Blastomyces dermatitidis*, an encapsulated yeast-like fungus. The infection, which is practically confined to certain parts of America, almost always starts in the skin, though primary inhalation infection of the lungs may occur in some cases. In the skin, chronic pustular and ulcerative lesions develop, with spreading margins containing small abscesses and often accompanied by marked papillary overgrowth of the epidermis. The regional lymph glands are usually not affected. Dissemination by the blood stream may produce widespread abscesses or chronic granulomas in the viscera, especially the lungs.

COCCIDIOIDOMYCOSIS

This disease, another kind of American " blastomycosis ", is caused by *Cocci-dioides immitis*, a yeast-like fungus which also forms mycelium. The primary infection is usually in the lungs, and pyaemic dissemination may occur. The lesions may be suppurative or granulomatous.

CHROMOBLASTOMYCOSIS

Chromoblastomycosis is a chronic slowly progressive granulomatous infection of the skin due to several species of fungi, which in the tissues consist of spherical tough cells distinguished by their brown pigmentation. The lesions are usually on the limbs, and constitute one of the varieties of *dermatitis verrucosa* or " mossy foot ".

SPOROTRICHOSIS

This is a rare cutaneous granuloma, usually of the hand, due to *Sporotrichum Schenckii*, a fungus which produces mycelium in culture but not in the tissues, in which it is very difficult to discover the organism. From the chronic ulcerative primary lesion, there is in most cases an ascending lymphangitis with a series of new lesions.

SUPERFICIAL DERMATOMYCOSES

" Ringworm ", " tinea ", " favus " and " pityriasis " are infections of the epidermis, hairs or nails by non-invasive fungi. There are many species of these parasites, falling into the several genera specified on p. 212. These are among the commonest fungal infections encountered in clinical practice. Laboratory diagnosis is usually easily established by direct microscopic demonstration of fungal elements in skin scrapings, hair or nail. Specific identification is made by culture.

SUPPLEMENTARY READING

Ash, J. E., and Spitz, S. (1945). *Pathology of Tropical Diseases: An Atlas.* Philadelphia and London; W. B. Saunders. (Contains a good section on fungi).

Baker, R. D. (1956). Pulmonary mucormycosis ". *Amer. J. Path.,* **32,** 287.

Browning, C. H., and Mackie, T. J. (1949). *Textbook of Bacteriology* Eleventh edition. London; Oxford University Press. (Chapter 32, a good outline of pathogenic fungi.)

Conant, N. F. *et al.* (1944, second edition 1954). *Manual of Clinical Mycology.* Philadelphia and London; W. B. Saunders. (A concise, well-illustrated handbook.)

Cox, L. B., and Tolhurst, J. C. (1946). *Human Torulosis. A Clinical, Pathological and Microbiological Study.* Melbourne University Press.

Karunaratne, W. A. E., (1936). " The pathology of rhinosporidiosis ", *J. Path. Bact.,* **42,** 193.

Simson, F. W., Harington, C., and Barnetson, J. (1943). " Chromoblastomycosis : a report of 6 cases ". *J. Path. Bact.,* **55,** 191.

Symmers, W. St. C. (1953). " Torulosis ". *Lancet.* **2,** 1068.

CHAPTER 18

DISEASES CAUSED BY PROTOZOA

The following list gives the names of the protozoa pathogenic for man and the diseases they produce, with the more important ones first :

Protozoa	Diseases
Entamoeba histolytica	Amoebic dysentery
Plasmodium vivax	Benign tertian malaria
Plasmodium malariae	Quartan malaria
Plasmodium falciparum	Malignant malaria
Trypanosoma Gambiense and T. Rhodesiense ..	Sleeping sickness
Trypanosoma Cruzi	Chagas' disease
Leishmania Donovani and L. infantum	Kala-azar
Leishmania tropica	Tropical sore or Delhi boil
Balantidium coli	Balantidial dysentery
Toxoplasma gondii	Toxoplasmosis
Pneumocystis carinii	Pneumocystis pneumonia
Naegleria and Hartmannella (amoebae)	Meningo-encephalitis
Giardia (or *Lamblia*) *intestinalis*	Diarrhoea
Trichomonas hominis	Diarrhoea
Trichomonas vaginalis	Vaginitis, urethritis

AMOEBIC DYSENTERY
(*Amoebiasis*)

(1) The amoeba

Entamoeba histolytica is world-wide in its distribution but is much more common in tropical and subtropical than in temperate regions. Like bacillary dysentery and typhoid fever, amoebic dysentery results from ingestion of water or food contaminated by the faeces of patients or carriers, so that it is a disease which is prevalent where sanitary conditions are poor. The amoeba is essentially parasitic; in its amoeboid form it quickly perishes outside the human body, though in its encysted form it can survive in water for as long as a month.

The amoeboid form or trophozoite (" feeding animal ") of *Ent. histolytica* is a typical amoeba 15-60μ in diameter, with broad pseudopodia and a single spherical nucleus which is only faintly visible in unstained specimens. In freshly passed warm faeces, the organisms may be recognized by their active amoeboid motion, and by frequently containing engulfed red blood corpuscles. When cold, they usually become rounded and stationary. In sections of fixed tissues also, amoebae are usually spherical or ovoid, 15-30μ in diameter, each with a single well-stained spherical nucleus 5-8μ in diameter, and sometimes containing recognizable red corpuscles (Fig. 71).

Non-amoeboid encysted forms are also present in the faeces, especially in the chronic stages of amoebic dysentery. These are sharply-defined spherical bodies 10-15μ in diameter, with distinct refractile cyst walls, and with one to four nuclei each. When ingested by a new host, the encysted organisms emerge from their envelopes, each as several small trophozoites ready to invade and feed in the bowel wall.

Entamoeba coli, a non-pathogenic amoeba, sometimes inhabits the intestine, and it and its encysted forms must be distinguished from *Ent. histolytica*. The trophozoites of *Ent. coli* are relatively sluggish, their pseudopodia less prominent, their nuclei are usually distinctly visible, and they do not contain red blood corpuscles. The encysted forms are larger than those of *Ent. histolytica*, and they

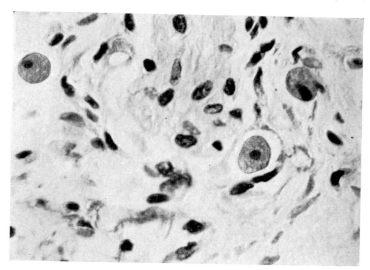

FIG. 71.—Three amoebae in tissue spaces in wall of colon from a case of amoebic dysentery. (× 800.)

often contain 4 to 8 or more nuclei each. Several other non-pathogenic amoebae also occur at times in the human intestine, e.g. the small *Endolimax nana*; these also must be distinguished from *Ent. histolytica*. *Entamoeba gingivalis* occurs in pyorrhoea, gingivitis and dental caries, but is probably a harmless commensal.

(2) The pathology of amoebic dysentery

(a) Intestinal lesions

The amoebae penetrate into the mucosa of the large intestine, partly by their amoeboid motility and probably also by producing a cytolytic toxin or ferment (hence the term *histolytica*). Ulceration of varying severity results—sometimes only superficial, sometimes very extensive and deep. The amoebae are found chiefly in the tissue interstices and small vessels just beyond the necrotic bases and sides of the ulcers. They themselves appear to cause but little inflammatory reaction, but inflammation may be pronounced because of secondary invasion by intestinal bacteria. The ulcers have irregular, ragged, frequently undermined edges, and sinuous tracks extend from them beneath still intact mucosa. In many cases, the caecum and ascending colon are more severely affected than other parts of the bowel ; the appendix may be affected, and the first symptoms may be of appendicitis. In the early stages of acute amoebic dysentery there is frequent diarrhoea with the passage of fluid stools with blood and mucus, in which numerous trophozoites are found. In later chronic stages, the diarrhoea subsides, the stools become less fluid, and they contain few trophozoites but many encysted organisms. In chronic cases, persistent ulceration is maintained ; and " carriers ", who may appear to be in good health, also have mild superficial lesions. Patients with chronic lesions often suffer from acute exacerbations, and these may appear after long intervals (sometimes many years) of quiescence or apparent cure of the infection. Long-standing chronic amoebic dysentery may produce great fibrous thickening of the colon. Tumour-like amoebic granulomas are sometimes produced in the bowel.

(b) Amoebic abscesses of the liver

In the wall of the intestine the amoebae often penetrate into the radicles of the portal vein, whence they are sometimes carried to the liver. Here they produce foci of histolysis and " abscesses ". These form large ragged cavities, often single and never numerous, containing many ounces or pints of thick slimy chocolate-coloured fluid with necrotic debris, altered blood and only a few leucocytes. Amoebae are found mainly in the walls of the abscess, seldom in its contents. The abscess sometimes extends through the diaphragm into the lung where it may rupture and discharge its contents into a bronchus ; or it may rupture into the peritoneal cavity, pleural cavity, pericardium, or adherent intestine or stomach. The incidence of hepatic abscess in amoebic dysentery varies greatly, being highest in the tropics. In Egypt, tropical Asia and the Pacific Islands, liver lesions have been found in 30 to 60 per cent of fatal cases of the disease. But in temperate climates and in communities where more effective treatment of amoebic dysentery is available, the frequency of abscesses of the liver is much lower, perhaps 5 to 10 per cent. This complication may occur in patients with only mild chronic or seemingly quiescent intestinal infection.

(c) Other amoebic lesions

In cases of dysentery, amoebic abscesses, generally similar to those in the liver, sometimes occur in the lung, or rarely in the brain, kidneys or elsewhere. Amoebic infection of the skin around the anus or external genitalia, or amoebic vaginitis, has occasionally been observed.

MALARIA

This important protozoal mosquito-borne infection has been well known since ancient times. It is chiefly a disease of the tropics, where heat, humidity and stagnant water favour the multiplication of mosquitoes ; but it also occurs in subtropical and temperate regions where mosquitoes of suitable species can breed. At one time malaria was prevalent in the fenlands of East Anglia, but with improved drainage and other factors it has now disappeared.

" Malaria " (Italian for " bad air ") was so named because for centuries the disease was believed to be due to a " miasma " or poisonous emanation which arose from swamps and polluted the air. Another name for the disease, expressing the same idea, was " paludism " (from Latin palus = a marsh). In a sense this idea was quite correct ; in malarial areas the air near swamps is polluted, not indeed by a vague poison arising from the swamp, but by swamp-bred mosquitoes carrying the protozoa of malaria from person to person.

Malarial parasites were first observed in human blood by Laveran in 1880. In 1886 Golgi (well known also for his work in neurohistology) discovered that the regularly recurring paroxysms of fever or " agues " coincided with the successive cycles of multiplication of malarial parasites in the blood ; and in 1889 he distinguished between the parasites of tertian and quartan malaria. Mosquitoes were long suspected of being carriers of malaria ; and between 1894 and 1900 Ross, with the advice of Manson, proved this and traced completely the development of the parasites in the mosquito. It was soon discovered that only certain species of *Anopheles* mosquitoes can transfer the parasites of human malaria ; species of the genus *Culex*, though the usual vectors of bird malaria, do not transfer the human disease.

(1) The malarial parasites and their reproductive cycle

The protozoa of malaria belong to the class *Sporozoa*, order *Haemosporidia*, genus *Plasmodium*. Three well-defined species are recognized, (1) *P. vivax*, causing benign tertian malaria, (2) *P. malariae*, causing quartan malaria, and

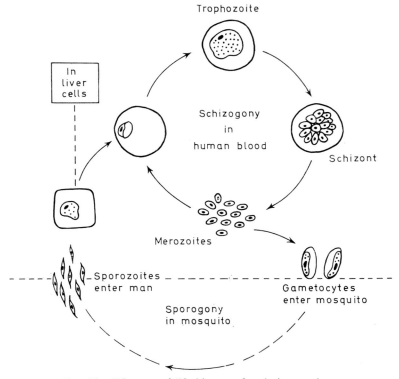

FIG. 72.—Diagram of life history of malaria parasites.

(3) *P. falciparum*, causing malignant malaria. A fourth species or sub-species, *P. ovale*, has been described in a few cases.

The life history of the parasites, shown in Fig. 72, comprises repeated cycles of asexual reproduction or *schizogony* in human blood, and a phase of sexual reproduction or *sporogony* in the body of the mosquito. Let us summarize this life history, starting from the moment when an infected mosquito bites a human being.

(a) Schizogony in human blood

The parasites are injected from the mosquito's salivary glands into the human blood as minute spindle-shaped cells called *sporozoites*. These enter the parenchyma cells of the liver, in which they undergo a *pre-erythrocytic cycle* of multiplication ; after which each parasite enters a red blood corpuscle and becomes an amoeboid *trophozoite*. The trophozoite grows at the expense of the red cell, often assumes a characteristic signet-ring form with a large central vacuole and a dot of nuclear chromatin at one side (Fig. 73*a*), and later acquires brown granules of altered blood pigment. When full-grown the trophozoite is rounded and occupies most of the corpuscle (Fig. 73*b-f*). It now becomes a *schizont* by dividing into a

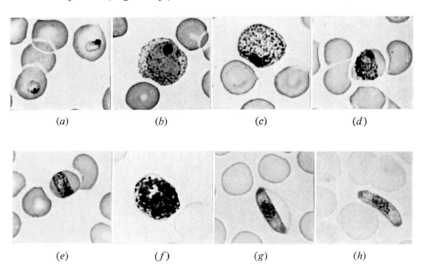

| (a) | (b) | (c) | (d) |

| (e) | (f) | (g) | (h) |

FIG. 73.—Composite photograph of malaria parasites : (*a*), young signet-ring trophozoites of *P. vivax* ; (*b*) and (*c*), large trophozoites of *P. vivax* in enlarged red cells with Schüffner's dots ; (*d*) and (*e*), band-form trophozoites of *P. malariae* ; (*f*), unusually large, heavily pigmented trophozoite of *P. malariae* ; (*g*) and (*h*), gametocytes of *P. falciparum*, one with remains of red cell, the other free. (× 1250.)

number of small ovoid *merozoites*, the number of which varies with the different species of plasmodia. These merozoites enter fresh red cells and become young trophozoites, and so the asexual cycle is repeated. The duration of the cycle varies with the species—48 hours for *P. vivax*, 72 hours for *P. malariae*, 36 to 48 hours for *P. falciparum*. Following a single infection, the individuals of each successive brood of parasites are all at the same stage of development at the same time, and hence the disruption of a large number of red cells takes place

simultaneously and at regularly recurring intervals.   The resulting attacks of fever
or ague thus occur every second day from *P. vivax* infections (tertian malaria), every
third day from *P. malariae* infections (quartan malaria), and rather irregularly
from *P. falciparum* infections.   In the last, schizogony takes place chiefly in the
spleen and liver, and schizonts are seldom seen in the peripheral blood.   (The
terms " tertian " and " quartan " refer to the occurrence of the successive attacks
on the third and fourth days, the day of the previous attack being counted—rather
illogically—as the first).   Multiple infections may produce daily attacks of ague.

(b) Formation of gametocytes in human blood

Some merozoites, instead of going on to repeat the schizogony cycle, form
special male and female cells called *microgametocytes* and *macrogametocytes*
respectively.   The gametocytes of *P. vivax* and *P. malariae* are rounded cells
like large trophozoites, but those of the malignant malarial parasite are very
characteristic elongated crescentic or sausage-shaped bodies (Fig. 73g and *h*),
whence the name *P. falciparum* (="sickle-producing " plasmodium).   The
fully-developed gametocytes are freed into the plasma by destruction of the corpuscles in which they have grown; and unless taken up by a mosquito, they disintegrate in a few days.   But, when a mosquito of a suitable species takes a blood
meal from an infected person, the gametocytes can develop further.

(c) Sporogony in the mosquito

For details of this, special works on parasitology must be consulted.   Suffice
it to say here that in the mosquito the gametocytes mature into male *micro-
gametes* and female *macrogametes*, that these unite to form *zygotes*, that the
zygotes produce multitudes of minute spindle-shaped *sporozoites* which reach
the mosquito's salivary glands where they are ready to infect fresh human beings.
This sexual phase of the life cycle of the parasites in the mosquito is completed
in 7 to 14 days, varying with the atmospheric temperature and the species of
mosquito.

(2) Clinical features of malaria

The incubation period of malaria varies considerably ; the disease may
declare itself soon after infection or not for some weeks or months.   The attacks
of fever or " ague " come on with a severe shiver or rigor, accompanied by headache, vomiting, acceleration of pulse, and rapid rise of temperature to 104° or
105° F.   The temperature remains up for several hours, the patient now feeling
hot ; then it falls rapidly (i.e. by crisis), accompanied by a profuse sweat.   The
patient then feels well until the next attack of fever.

The rigor corresponds to the disruption of red cells and the liberation of
merozoites into the plasma ; during the hot stage the merozoites invade fresh
corpuscles ; during the afebrile interval the new trophozoites are growing in
the corpuscles, and their maturation can be followed in serial blood films until
the fully formed schizonts break down and excite the next attack of fever.

The foregoing description applies principally to tertian and quartan fevers.
The fever of malignant or subtertian malaria is irregular and variable, being
sometimes violent and prolonged, sometimes slight.   Malaria due to *P. falciparum*

is called " malignant " or " pernicious ", because most of the deaths from malaria and the most marked histopathological changes in the disease are caused by this plasmodium.

Relapses after apparent cure or quiescence of the infection for periods of many months or years are common in malaria. During the quiescent period the plasmodia are lying dormant in the liver, spleen and other viscera.

In malarial regions, the natives, as the result of repeated infections, develop partial immunity, showing few or no symptoms, even though plasmodia are present in their blood. Such " carriers " are an important reservoir of the disease.

(3) The pathological anatomy of malaria

(a) Blood changes

In addition to the presence of the *parasites* in the blood, this shows varying degrees of *anaemia* due to the continued destruction of red corpuscles and haemoglobin. Usually there is also a diminution in the white blood corpuscles, *leucopenia*, with a relative increase of large mononuclear cells, which are active in phagocytosing the parasites and liberated blood pigments.

(b) Pigmentation

In long-standing cases, the spleen, liver, bone marrow and other organs show diffuse grey or black pigmentation due to the presence of altered blood pigment (mainly haematin) in wandering macrophages and vascular endothelial cells. Some haemosiderin is also present.

(c) Splenomegaly

In recent acute cases the spleen is moderately enlarged, soft and dark red or brown ; in chronic cases it is greatly enlarged, hard, and greyish-brown or almost black. Microscopically, it shows many infected red cells, segmenting plasmodia, macrophages and endothelial cells loaded with plasmodia and pigments, and often haemorrhagic areas. The enlarged malarial spleen is easily ruptured by slight injuries.

(d) Plasmodial obstruction of small blood vessels

In fatal cases, usually of malignant malaria, the brain, kidneys, gastric and intestinal mucosa, or other tissues may show scattered small haemorrhages due to the obstruction of small blood vessels, especially capillaries, by masses of plasmodia. Severe cerebral symptoms (delirium, convulsions, coma, etc.) may have been produced by such lesions in the brain ; or, in the alimentary canal, they may have caused haemorrhagic diarrhoea.

(e) Blackwater fever

This is the name given to a condition in which a patient with malaria (usually of the malignant type) suddenly commences to pass red or black urine containing haemoglobin and methaemoglobin, due to severe haemolysis of the red cells of the blood. The cause of this haemolysis is not known. Rapid anaemia, jaundice, fever and anuria ensue, and the condition often ends fatally.

(4) Diagnosis of malaria

The only way of making a certain diagnosis of malaria is by finding the plasmodia in the blood. Tertian and quartan infections often give such characteristic and regular attacks of fever that a clinical diagnosis may be confidently made, but this should always be confirmed by blood examination. In malignant malaria and mixed infections, the fever and other symptoms are often irregular and non-characteristic, and correct diagnosis can be made only by blood examination. In any malarial infection severe enough to cause fever, parasites can be found in the blood in almost all cases. Hence we should always doubt the correctness of a diagnosis of malaria in a febrile patient in whose blood no plasmodia can be found. When plentiful parasites are present, thin blood films, stained by Leishman's stain or other usual blood stains, permit easy diagnosis and give the best picture of the organisms. But, when few parasites are present, diagnosis from thick films is quicker and more reliable. (For methods used, books on technique should be consulted.)

The main points in distinguishing between the three kinds of malarial parasites in stained blood films are the following :

	P. vivax (Tertian)	*P. malariae* (Quartan)	*P. falciparum* (Malignant)
Trophozoites.	Small and large rings ; and large irregular mature forms.	Small and large rings ; mature forms smaller than *P. vivax* and often band-shaped.	Very small and larger rings ; large mature trophozoites rare in peripheral blood.
Pigment in trophozoites.	Yellow-brown, in fine grains.	Dark brown, in coarse grains.	Scanty fine granules.
Red cells.	Enlarged, pale, stippled with eosinophilic " Schüffner's dots ".	Normal in size and appearance.	May be shrunken and dark, or pale. May contain irregular basophilic "Maurer's dots."
Schizonts.	Large, mulberry-like, with 12–24 merozoites.	Smaller, rosette-like, with 6–12 merozoites.	8–32 small irregular segments. Rarely seen in peripheral blood.
Gametocytes.	Spherical.	Spherical.	Crescentic, very distinctive, often free in blood plasma.

SLEEPING SICKNESS
(*African trypanosomiasis*)

(1) The trypanosomes

Sleeping sickness is caused by two closely similar species of flagellates, *Trypanosoma Gambiense* and *T. Rhodesiense*. These infect not only man but many other species of domestic and wild mammals, including antelopes, horses, camels, pigs and cattle, which constitute reservoirs of infection. (*T. Brucei* and several other species occur in African animals and are closely related to the human pathogens. It is probable that they became differentiated into distinct species by adaptation to their several hosts.)   The parasites are transferred from man to man or from animals to man by the bites of several species of *Glossina* or " tse-tse " flies, the chief of which is *G. palpalis*.   In human blood and tissues, the trypano- somes appear as elongated actively motile sinuous organisms 8 to 30μ long, each with a longitudinal undulating membrane and a terminal flagellum (Fig. 74).

(2) Sleeping sickness

(a) *The two forms*

Gambian or Mid- and West-African sleeping sickness is prevalent among the natives of the Congo and neighbouring parts of western equatorial Africa. It runs a long course, often of several years, and frequently ends fatally by invasion of the central nervous system.   Rhodesian or East-African sleeping sickness has a more restricted geographical distribution in and around Rhodesia ; and it runs a more severe and rapid course, often ending fatally within a year.

(b) *Pathology*

Following an incubation period of several days after infection by the bite of the tse-tse fly, there is a trypanosomal septicaemia, with the parasites exclusively or predominantly in the blood, accompanied by headache, fever and malaise. Irregular fever continues, the lymph glands and spleen enlarge, and trypanosomes

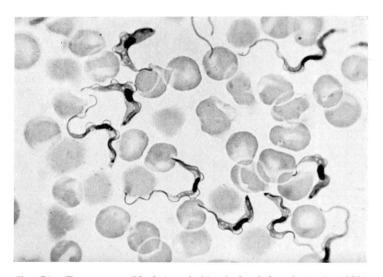

FIG. 74.—*Trypanosoma Rhodesiense* in blood of an infected rat.   (× 1250.)

may be found in the blood during febrile periods and especially in the juice from swollen lymph glands.   Later, signs of special localization of the infection in the central nervous system appear—severe headaches, mental dullness, delusions, increasing sleepiness, and finally coma—and trypanosomes are plentiful in the cerebrospinal fluid.   Microscopically, the brain and spinal cord show chronic inflammatory changes, with collections of lymphocytes and plasma cells in the meninges and perivascular Virchow-Robin spaces, accompanied by neuroglial proliferation and atrophy of nerve cells.

(c) Prognosis and treatment

Early diagnosis is very important ; untreated cases almost always end fatally, but proper chemotherapy in the early stages often effects a cure. The chief trypanocidal drugs are tryparsamide and similar organic arsenical compounds, and Bayer 205 (or germanin) a complex non-arsenical compound.

CHAGAS' DISEASE
(Brazilian trypanosomiasis)

This is a fatal acute or chronic disease, restricted to certain areas of South and Central America, due to infection by a special trypanosome, *T. Cruzi*, which infects many other species of mammals and which is transferred from these to man or from man to man by certain species of bugs. After infection the parasites lose their flagella and undulating membranes, and settle down in various cells of the body, especially the reticulo-endothelial and myocardial cells, where they multiply as small ovoid bodies resembling *Leishmania* (see below). During acute febrile stages of the disease, some of the parasites escape from the tissues into the blood and resume the flagellate form.

LEISHMANIASIS

This is a general name for infections with *Leishmania*, a special genus of flagellate protozoa which in human tissues occur in non-flagellar ovoid or rounded forms. The three pathogenic species, *L. Donovani*, *L. infantum* and *L. tropica*, are closely similar, and indeed may be identical.

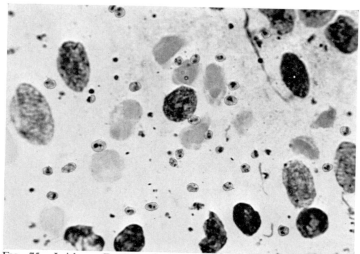

FIG. 75.—Leishman-Donovan bodies in smear from splenic puncture (slightly retouched). (× 1250.)

(1) Kala-azar (Visceral Leishmaniasis)

This is a disease of India and Eastern Asia, characterized by fever, anaemia, emaciation and great enlargement of the spleen. It is caused by *L. Donovani*, which multiplies in great numbers in the macrophages and reticulo-endothelial cells of the spleen, liver, lymphoid tissues and bone marrow, in the form of small ovoid cells 2 by 4μ (Fig. 75). These are called " Leishman-Donovan bodies " after their two independent discoverers (1903). The parasites are transferred by the bites of several species of sand-flies of the genus *Phlebotomus*, in which the organisms occur in flagellate form. Diagnosis is by microscopical examination of fluid or fragments of tissue obtained by puncture from lymph glands, spleen, liver or sternal bone marrow, in which Leishman-Donovan bodies are present in great numbers.

Infantile Kala-azar, due to *L. infantum*, is a closely similar disease of young children in North Africa and the Mediterranean borders.

(2) Tropical sore (Delhi boil or cutaneous Leishmaniasis)

This is a local skin infection due to *L. tropica*, which is transferred by sand-flies. It produces small or large chronic ulcers, which usually heal within a few months. Visceral lesions do not occur. An instructive account of cutaneous Leishmaniasis acquired by direct contagion has been reported (Symmers).

BALANTIDIAL DYSENTERY

Balantidium coli is a large ovoid ciliated protozoon, resembling *Paramoecium*, and measuring up to 100μ long by 70μ broad. From many parts of the world, occasional cases of dysentery due to it have been reported. It penetrates the mucous membrane of the large intestine and causes patchy ulceration closely resembling that of amoebic dysentery. Nests of balantidia are found in the tissues around the ulcers. Diagnosis depends on finding the protozoa in the faeces.

TOXOPLASMOSIS

Toxoplasma gondii is a small oval or crescentic protozoon, which is widespread in man and animals, occasionally causing severe or fatal infections (meningo-encephalitis, pneumonitis, etc.) in human adults, but more often only mild chronic lymphadenitis or symptomless infections only revealed by the presence of anti-bodies. Infection of the foetus *in utero* is a cause of neonatal meningo-encephalitis, hydrocephalus, retinitis, etc.

PNEUMOCYSTIS PNEUMONIA

A rare form of pneumonia, occurring mainly in infants, is due to *Pneumocystis carinii*, a small rounded parasite which develops in clusters inside tiny spherical cysts in the tissues, and which most workers believe to be a protozoon.

PRIMARY AMOEBIC MENINGO-ENCEPHALITIS

Rare cases of acute meningitis or meningo-encephalitis due to amoebae of the genera *Naegleria* and *Hartmannella* have recently been reported (Fowler and Carter; Carter; Symmers).

INTESTINAL AND GENITAL FLAGELLATES

Giardia (or *Lamblia*) *intestinalis*, a flagellate protozoon, is a common inhabitant of the human small intestine. It has been found in great numbers, in either its encysted or flagellate form, in the faeces of some cases of diarrhoea, especially in young children.

Trichomonas hominis is a small pear-shaped flagellate sometimes found in great numbers in the intestinal contents especially in cases of diarrhoea ; but the evidence that it can itself cause intestinal inflammation is inconclusive.

Trichomonas vaginalis, closely similar to *T. hominis*, is frequently found in great numbers in the vaginal discharge in cases of vaginitis. It is also a cause of non-specific urethritis in the male.

SUPPLEMENTARY READING

Alexander, C. M., and Callister, J. W. (1955). " Toxoplasmosis of newborn ". *Arch. Path.*, **60**, 563.

Ash, J. E., and Spitz, S. (1945). *Pathology of Tropical Diseases: An Atlas*. Philadephia and London; W. B. Saunders.

Carter, R. F. (1968). " Primary amoebic meningo-encephalitis ". *J. Path. Bact.*, **96**, 1.

Craig, C. F., and Faust, E. C. (1943). *Clinical Parasitology*. London; Kimpton. (Contains a detailed account of all the protozoal diseases.)

Fowler, M. C. and Carter, R. F. (1965). *See* Carter.

Hamperl, H. (1957). " Variants of pneumocystis pneumonia." *J. Path. Bact.*, **74**, 353.

Hoare, C. A. (1949). *Handbook of Medical Protozoology*. London; Baillière. (An excellent book.)

Lapage, G. (1957). *Animals Parasitic in Man*. (A " Pelican " book.) London. (Contains excellent outlines of the protozoal diseases.)

Stansfeld, A. G. (1961). " The histological diagnosis of toxoplasmic lymphadenitis ". *J. Clin. Path.*, **14**, 565.

Symmers, W. St. C. (1960). " Leishmaniasis acquired by contagion. A case of marital infection in Britain ". *Lancet*, **1**, 127.

— (1969). " Primary amoebic meningo-encephalitis in Britain ". *Brit. Med. J.*, **2**, 449.

Woodruff, A. W. (1954). " Tropical diseases in Britain." *Brit. Med. J.*, **1**, 1030.

CHAPTER 19

PARASITIC METAZOA

EVERY SPECIES of mammal, man included, is liable to infestation by a great variety of metazoal parasites. Most of these are *endoparasites*, residing inside the body of the host, usually in some particular organ or tissue for which the parasite is specially adapted. Others are *ectoparasites*, living only on or in the skin, and many of them coming only into casual contact with man.

The endoparasites of man almost all belong to one or another of three classes of *Helminthes* or " worms ", namely,

(1) the *Nematodes*, " round-worms " or " thread-worms " (Gk., *nema* = a thread)

(2) the *Cestodes* or " tapeworms " (Gk., *kestos* = a strap or girdle)

(3) the *Trematodes* or " flukes " (Gk., *trematodes* = a body with a vent or aperture).

The name " worm " has no exact meaning in zoology ; it is merely a convenient name for lowly invertebrates with an elongated shape, and it embraces an assemblage of diverse classes.

The ectoparasites of man are almost all arthropods of the classes *Arachnida* and *Insecta*. These will be mentioned briefly at the end of this chapter.

GENERAL REMARKS ON THE ENDOPARASITES

(1) Their life cycles

The life histories of the parasitic worms vary greatly. In some species, e.g. *Oxyuris* or the common thread-worm of the intestine, the parasite lives its whole life in the one host, man. Other species have complex life cycles involving two or more hosts, the *final* or *definitive host* being the species which harbours the adult or sexually mature stage of the parasites, and *intermediate hosts* being those in which the parasites pass their larval or immature stages. For most of the species of parasites infesting man he is the definitive host ; for only a few, e.g. the parasite of hydatid disease, is he the intermediate host. For some of the parasitic metazoa man is the only definitive (or intermediate) host ; for others he is only one of several possible hosts, and for some of these he is only an occasional host to a parasite which much more commonly infests other animals. The domestic animals—sheep, cattle, pigs, dogs and cats—often serve as reservoir hosts to parasites which also infest man. The intermediate hosts of the various parasites for which man is the definitive host are of many kinds, including the domestic animals, other mammals, fish, insects, crustaceans and molluscs. In their life cycles some of the parasites have the most incredible and fairy-tale-like adventures, examples of which will appear later.

(2) The main routes and sites of infestation

Metazoal parasites enter the body either by being ingested or by penetrating the skin, the former being much the commoner mode of entry. Because contaminated food is the commonest vehicle for endoparasites, these are most prevalent in primitive peoples who eat much of their animal food raw or whose domestic habits are dirty. From its portal of entry the parasite reaches its final habitat in the body in various ways. In some instances its journey is short and simple, e.g. many ingested parasites have their destination in the intestine. In other instances the parasites make long and circuitous journeys in the body, e.g. hookworm larvae penetrate the skin, pass by the venous blood stream to the lungs, migrate into the alveoli, ascend the air passages to the pharynx, are then swallowed and so reach their destination in the small intestine. This example shows also that not all alimentary parasites enter the body by ingestion. Conversely, we may mention the Guinea or dragon-worm, which, though it lives in the subcutaneous tissues, does not enter the body through the skin but through the alimentary canal.

Most parasites in their definitive hosts show a remarkable organ-specificity, the parasite residing in only one particular organ or tissue for which it is specially adapted. About three-quarters of the species of parasites infesting man live in some part of his intestines ; the remainder live in other tissues, e.g. the bile passages, lungs, lymphatics, or blood. The habitats of larval parasites in intermediate hosts are often less specific ; e.g. in man, hydatid cysts can develop in the liver, lungs, bones, brain or many other situations.

(3) The pathological effects of parasites

These vary very greatly with the species and situations of the parasites ; some e.g. the tapeworms, produce scarcely any ill effects, while others cause serious inflammatory or obstructive lesions in the infested organs or more general toxic effects or anaemia. These pathological effects may be grouped broadly thus :

(i) *Local mechanical effects*, e.g. obstruction of the lymphatics by filariae, obstruction of bile ducts by hydatid cysts or by liver flukes, or obstruction of bronchi by lung flukes ;

(ii) *Irritative effects*, e.g. the perianal irritation due to thread worms, the cystitis or proctitis due to schistosomes, or the irritation of the skin by dragon-worms ;

(iii) *Anaemia*, e.g. from hookworms ;

(iv) *Toxic or allergic effects*, due to absorption of metabolites of living parasites or protein products of dead ones ; these are manifested by hypersensitivity of the host to foreign proteins, asthma, urticaria, other skin lesions, and eosinophilia (see p. 287).

(v) *Nutritional effects* are probably always secondary to the foregoing. A prevalent idea that intestinal worms cause malnutrition directly by robbing the host of food is an error ; the amount of sustenance required by the largest worms is negligible.

(4) Classification of the parasitic worms

Several hundreds of species of endoparasites in man have been identified. Most of these are of little or no interest to the student of pathology, who is not expected to be an expert parasitologist, or to know details of the structure and classification of the parasites, or to identify particular species by their ova in faeces or elsewhere. It is sufficient if he understands the main facts about the life histories and pathological effects of the more important species (as outlined below), and knows where to turn for further information should he need it (e.g. to Craig and Faust's excellent book).

In this Chapter the main parasites of man will be described in the following order ;

I. NEMATODES, NEMATHELMINTHES, *or Round-worms*
  (a) *The Adults Residing in the Intestine*
      1. *Enterobius (Oxyuris) vermicularis* (the common thread-worm)
      2. *Trichuris trichiura (Trichocephalus dispar)* (the common whip-worm)
      3. *Ankylostoma duodenale* ⎤ (the hookworms)
      4. *Necator Americanus* ⎦
      5. *Ascaris lumbricoides*
      6. *Trichinella (Trichina) spiralis*
      7. *Strongyloides stercoralis*
  (b) *The Adults Residing in the Tissues*
      1. *Filaria* or *Wuchereria Bancrofti*
      2. *Filaria loa* or *Loa loa*
      3. *Dracunculus Medinensis* (dragon-worm)
  (c) *Granulomas due to larvae in the tissues*
      1. *Toxocara canis*

II. CESTODES, *segmented* PLATYHELMINTHES, *or Tapeworms*
  (a) *Man the Definitive Host*
      1. *Taenia saginata* (the beef tapeworm)
      2. *Taenia solium* (the pork tapeworm)
      3. *Diphyllobothrium latum* (the fish tapeworm)
      4. Other tapeworms
  (b) *Man the Intermediate Host*
      1. *Echinococcus granulosus (Taenia echinococcus)* (the hydatid tapeworm)

III. TREMATODES, *unsegmented* PLATYHELMINTHES, *or Flukes*
  (a) *The Blood Flukes*
      1. *Schistosoma haematobium*
      2. *Schistosoma Mansoni*
      3. *Schistosoma Japonicum*
  (b) *Other Flukes* (intestinal, liver and lung flukes).

(The general name *Platyhelminthes* merely means " flat-worms " and the two classes of parasites, cestodes and trematodes, embraced by it really have little in common except their flatness and their parasitic mode of life.)

THE NEMATODES, NEMATHELMINTHES OR ROUND-WORMS

These are all elongated, thread-like or cylindrical worms with blunt heads and tapering tails ; the body is all one piece, without the annular segments character-istic of the earthworm and other annelids. The nematodes vary greatly in size ; e.g. the adult *Trichina spiralis* is a barely visible thread about 3 millimetres long, the female dragon-worm (*Dracunculus*) is a narrow thread 3 or 4 feet long, while *Ascaris lumbricoides* is a stouter worm of about the dimensions of an earthworm

and measuring up to 10 inches long (Fig. 76). All nematodes parasitic in man
have separate sexes, and the female is usually larger (in some species much larger)
than the male. For some nematodes man is the only host ; others require
intermediate hosts ; and the life cycles of the various species differ greatly in
their complexity.

(1) *Enterobius* (or *Oxyuris*) *vermicularis* (Fig. 76A)

This is the common thread-worm or pin-worm which may infest the human
large intestine and lower ileum in great numbers, especially in children. The
adult worms attach themselves to the mucosa, especially of the caecum ; but
the mature females, active white threads 8–12 millimetres long, are often passed
in the faeces or migrate through the anus at night to deposit their eggs on the
surrounding skin. Because of the irritation here, the child scratches the area

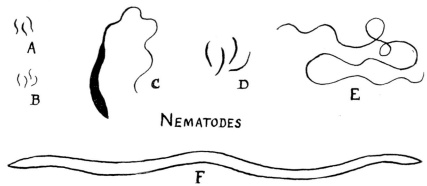

FIG. 76.—Diagram of natural sizes of main nematodes (as lettered in text).

and contaminates his finger-nails with the eggs, which he then carries to his
mouth, thus constantly reinfecting himself, or sometimes infecting others. The
life cycle is thus a simple one, and man is the only known host.

(2) *Trichuris trichiura* (or *Trichocephalus dispar*) (Fig. 76C)

This species, called the whip-worm because of its shape, is another common
intestinal parasite restricted to man. The adult worm is 3 to 5 centimetres long,
with its anterior two-thirds thin and thread-like and its posterior third broader.
Like the thread-worm, its habitat is the caecum and neighbouring parts of the
intestine. The eggs are passed in the faeces and develop in moist soil or water,
whence they may be ingested by fresh human hosts.

(3) *Ankylostoma duodenale* and *Necator Americanus* (Fig. 76D)

These two species of hookworms are closely similar in their structure, life
cycles and pathology, and can be considered together. They cause " hookworm
disease " (ankylostomiasis or uncinariasis) in many tropical and subtropical
parts of the world.

(a) Life cycle

The adult worms, the average length of which is about 1 centimetre, live in the
human small intestine, especially the jejunum, to the mucous membrane of which

they attach themselves often in great numbers. The eggs are passed in the faeces and contaminate the soil, in which, if the temperature and moisture are suitable, they hatch out in a few days to produce actively motile microscopic worm-like larvae which grow to a length of about 0·5 millimetre. When human beings come into contact with infested soil, mud or water, the larvae penetrate the skin, usually of the feet, beneath which they may wander about producing irritation of the skin called " ground itch ", " coolie itch " or hookworm dermatitis. The larval worms burrow into small veins in the dermis or subcutaneous tissues, whence they are carried in the venous blood stream to be arrested in the capillaries of the lungs. Here they migrate into the air-sacs and ascend in the bronchi and trachea to the pharynx, from which they are swallowed. Reaching the intestine, they attach themselves to the mucosa and grow into adult worms. Note that, although the young parasites have a long and adventurous journey, the hook-worms resemble the thread-worms and whip-worms in that man is their only host and there are no intermediate hosts.

The larval development in moist warm soil explains the prevalence of ankylo-stomiasis in semitropical regions and in certain mines and tunnels.

(b) Pathology

(1) *Hookworm dermatitis* occurs at the point of entry of the larvae in the skin. (2) *Pulmonary lesions* are due to migration of the larvae from the capillaries. Scattered small haemorrhages occur in the lungs, and, if the infestation is a heavy one, these are occasionally sufficient to produce a clinically significant broncho-pneumonia. (3) *Anaemia* is the most serious result of hookworm infestation. It results mainly from continuous loss of blood from the intestinal mucosa, the blood being ingested in large amounts by the worms themselves and also being lost by oozing from the punctures inflicted by them. The anaemia is progressive, and may become severe or fatal, especially with heavy hookworm infestations in poorly nourished subjects. The anaemia is usually accompanied by marked *leucocytosis*, often with moderate or great *eosinophilia*.

(c) Diagnosis

Diagnosis is made by finding the ova in the faeces.

(4) Ascaris lumbricoides (Fig. 76F)

Ascaris lumbricoides literally means " the jumper that is like an earthworm ", referring to the worm's vigorous movements. Although called *lumbricoides* because it is similar in size and general shape to an earthworm, it is not really like *Lumbricus;* it is not pink but dull white, it is not ringed, and its skin is tougher than an earthworm's. Ascariasis is found all over the world, especially in warm moist climates and where personal hygiene and the disposal of faeces are poor.

(a) Life cycle

Like the hookworms, young Ascarids have an adventurous career without requiring any intermediate host. The adult worms live freely in the contents of the small intestine, from which they sometimes pass to the stomach or large intestine and are vomited or ejected in the faeces. The eggs are present in great numbers in the faeces, whence they contaminate soil. Here, under warm moist conditions, they develop into embryos, and when ingested by human beings these hatch out in the small intestine, penetrate the mucosa to reach lymphatics or venules, and are carried to the lungs either via the portal blood stream through the liver or via the mesenteric lymph glands and thoracic duct. In the lungs they sojourn for several days, increase in size to 1 or 2 millimetres in length, then migrate into the air-sacs and up the bronchi to the glottis, and so are swallowed. Reaching the intestine for the second time, they now develop into adult worms. Occasionally larval worms from the lungs enter the systemic circulation and are carried to kidneys, brain or other organs.

The pig is the definitive host of a closely similar round-worm, *Ascaris suum;* but it is doubtful if reciprocal cross-infection between man and pig ever occurs. Man is probably infected only with eggs from human faeces.

(b) Pathology

(1) *Lesions produced by migrating larvae* are principally in the lungs, where scattered haemorrhages and bronchopneumonic areas occur, often accompanied by fever, and sometimes by eosinophilia, urticaria, or haemoptysis with larvae in the sputum. The occasional escape of larvae into the general circulation may cause mild or severe embolic lesions, with appropriate symptoms, in the brain, spinal cord, eye or kidneys. (2) *The adult worms* may cause no ill effects, or they may cause colic, diarrhoea, intestinal obstruction due to a matted mass of worms, or appendicitis. Occasionally, they migrate into and obstruct the bile duct or pancreatic duct, causing cholangitis, hepatic abscess or haemorrhagic pancreatitis. If vomited from the stomach they may migrate into the nasal cavity or sinuses, into the Eustachian tube, causing otitis media, or into the larynx causing obstructive asphyxia.

(c) Diagnosis

Diagnosis is usually made by finding the eggs in the faeces, occasionally by finding mature worms in faeces or vomitus, and rarely by finding larval worms in sputum, urine or faeces.

(5) Trichinella (or Trichina) spiralis (Fig. 76B)

This small nematode causes *trichinosis* or *trichiniasis* in many parts of the world, especially where pork is eaten raw or insufficiently cooked.

(a) Life cycle

The adult worms, which are minute white threads measuring up to 4 millimetres long, live in the intestines of man, pigs, rats, cats, dogs and some wild animals. The same animals act both as definitive hosts for the adult worms in their intestines and as intermediate hosts for the encysted larvae which develop in their voluntary muscles. To complete the life cycle, flesh containing these larvae must be eaten by a second host. Animals infect themselves by eating the flesh of their own or other species ; man is infected almost entirely from pork. When infested raw or poorly cooked pork is eaten, the encysted larvae are released and quickly grow into mature worms. The pregnant females burrow into the intestinal mucosa where they deposit viviparously large numbers of motile larval worms in the lymphatics and venules. From here the larvae pass in the circulation, through the lungs, to all parts of the body. They perish in all host tissues except striated muscle, into the fibres of which they penetrate and become encysted. The muscles most heavily infested are usually the diaphragm, inter-

FIG. 77.—*Trichinella spiralis* embryo encysted in muscle. Calcification is commencing at each end of its capsule. (× 60.)

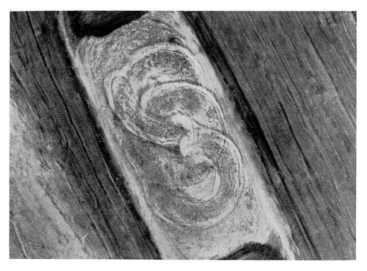

FIG. 78.—Enlarged view of the young worm in FIG. 77. (× 300.)

costal, abdominal, lingual, laryngeal and ocular muscles ; but any voluntary muscle may be affected. The myocardium does not shelter living parasites.

In the muscle fibres the young worms grow and assume a characteristic spiral shape within ovoid capsules (Figs. 77, 78). They remain alive for many years, even 30 years in man, but most of them eventually die and become calcified.

(b) Pathology

(1) *Gastro-enteritis*, with vomiting, diarrhoea and colic, may occur during the period of invasion of the intestinal mucosa by the worms. (2) *The period of invasion of muscles* is characterized by signs of myositis—muscular pains, tenderness, swelling and hardness—often accompanied by oedema of the eyelids, face or other parts, and by fever, cough, anaemia and marked eosinophilia. In tissues other than muscles, although the larvae do not survive, they may cause more or less severe chronic inflammatory changes and foreign-body granulomatous nodules. Thus the heart may show serious myocarditis with infiltration by eosinophil leucocytes, and cardiac failure may ensue. Lesions in the leptomeninges, brain or cord may produce various nervous symptoms.

The clinical severity of the disease depends on the severity of the infestation, which may be mild and cause only slight symptoms, or may be severe and fatal within two or three weeks. In epidemics, the mortality is sometimes high.

(c) Diagnosis

Laboratory confirmation of the clinical diagnosis may be sought in the following ways : (1) *Eosinophilia* is frequent and of much diagnostic value ; eosinophils may constitute from 15 to 75 per cent of the blood leucocytes. (2) *Adult worms* are sometimes to be found in the faeces, but this is unreliable. (3) *Biopsy* of one of the accessible painful muscles affords the best means of diagnosis. At necropsy the affected muscles may contain many fine white spots visible to the naked eye ; the diaphragm is the best muscle to examine for parasites. (4) *Bachman's intradermal test* for sensitivity to an extract of *Trichinella* larvae is useful in diagnosis, especially in mild cases.

(6) *Strongyloides stercoralis*

This small nematode, most prevalent in warm countries, has a complicated and variable life-cycle, resembling that of the hookworms. The minute adult worms invade the intestinal mucosa and deposit their eggs ; these hatch in the tissues, the larvae escape into the intestine and are passed with the faeces to contaminate the soil. Here, under favourable conditions, they grow into adults, which may continue to breed in the soil. But under unfavourable conditions, the young parasites metamorphose into delicate larvae ready to reinfect man. These penetrate the skin, enter the blood stream and are carried to the lungs, whence they ascend the air passages and are swallowed down to the intestine, to repeat the cycle. The main pathological effects are skin lesions at the point of entry of the larvae, bronchitis and bronchopneumonia due to the migration of the larvae in the lungs, diarrhoea due to the lesions of the intestinal mucosa which may be extensive and severe, and leucocytosis with eosinophilia.

(7) *Filaria (or Wuchereria) Bancrofti* (Fig. 76E)

The general name *Filariasis* includes a number of different parasitic diseases caused by related species of filaria worms, the life histories of some of which are still unknown.   The general characters of filariae are these.   The adult worms differ in length in different species, but all are filiform and live in the connective tissues, lymphatics or body cavities of a vertebrate host.   The females produce eggs which are already well developed when laid and which quickly hatch out as delicate motile eel-like embryos called *microfilariae* (Fig. 79).   While circulating in the peripheral blood or tissue fluids, these microfilariae are ingested by blood-sucking insects or other arthropods, in which they develop into more mature larvae, ready to infect the next suitable vertebrate host to be bitten by the insect.

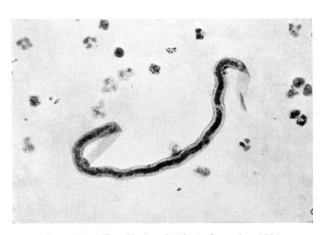

FIG. 79.—Microfilaria of *Filaria loa*.   (× 375.)

(a) Filaria Bancrofti

This nematode (also called *F. nocturna* and *Wuchereria Bancrofti*) is an important parasite of man in most tropical countries.   The adult worms, which measure up to 10 centimetres long, inhabit the lymphatics and lymph glands in various parts of the body, especially the inguinal and lower abdominal glands.   The worms are usually found coiled up tightly in the glands or in distended lymphatics. The females are viviparous, giving birth to huge numbers of microfilariae of average length about 0·25 millimetre.   These pass into the lymph stream and thence into the blood, where they congregate especially in the pulmonary and other visceral capillaries.   They usually appear in the peripheral blood only while the host is asleep, hence the name *nocturna* applied to the parasite.   The microfilariae develop no further unless taken up by certain species of mosquitoes, especially the genus *Culex*, in which they undergo metamorphosis into larvae. These migrate to the insect's proboscis, from which they infect fresh human hosts whom the mosquito bites.   The larvae migrate to lymph glands or lymphatics, grow to maturity there and produce new broods of microfilariae in the new host.

(b) Pathology

The most important results of filarial disease are chronic *lymphadenitis*, *lymphangitis* and *lymphatic obstruction*.   Living, and especially dead, adult worms cause chronic inflammatory changes, with foreign-body giant-cells, collections of leucocytes including eosinophils, and fibrosis. Obstruction of main lymphatics ensues, with consequent lymphatic oedema, and great thickening and prolifera- tion of the tissues affected by the lymph stasis, called *elephantiasis*. If the thoracic duct or its main tributaries are obstructed, the elephantiasis affects especially the external genitalia, and there may also occur *chylous ascites, chylo-thorax* or *chyluria* from rupture of obstructed dilated lymph vessels in the abdominal cavity, pleural cavity or bladder. If the inguinal glands are severely affected, elephantiasis of the lower limbs develops. Elephantiasis of the upper limbs or breasts is relatively rare. The presence of the microfilariae in the blood appears to have no special ill effects.

(c) Diagnosis

Apart from the clinical results just outlined, diagnosis is established by finding microfilariae in blood samples taken between 10 p.m. and 2 a.m.

(8) Filaria loa or Loa loa

This parasite is restricted to the western parts of tropical Africa. The adult worms measure up to 7 centimetres long and live in the subcutaneous tissues, where they wander about producing transient swellings. They are particularly troublesome when migrating beneath the eyelids or conjunctiva. The females are viviparous and the microfilariae (Fig. 79) are found in the blood during the daytime (*Microfilariae diurna*). The intermediate hosts are certain species of biting flies. Eosinophilia is frequent in infested patients.

(9) Dracunculus Medinensis

This nematode, known also as Medina-worm, Guinea-worm, or dragon-worm, was well known to the ancient Egyptians, Greeks and Romans. It is found chiefly in the Nile Valley, tropical Africa and in the Southern Asiatic countries. The adult worms are like long threads measuring up to 120 centimetres in length, which reside in the retroperitoneal and other deep connective tissues and migrate to the subcutaneous tissues. The pregnant female is almost all uterus distended with larvae ; when her head comes into contact with the skin, an ulcerated area develops through which, whenever the skin is immersed in water, the larvae are discharged in great numbers. Further development of the larvae occurs in the small fresh-water crustacean *Cyclops*. When a definitive host (dog, horse, ox, wolf, monkey and many other mammals, as well as man) drinks water con- taining an infected *Cyclops*, the larvae escape, penetrate the wall of the intestines and migrate to the deep connective tissues where they come to maturity in 8 to 12 months.

(10) Toxocara canis

The adult of this nematode lives in the dog's intestine. If ova are ingested by man, they develop into small larvae which penetrate the intestine, enter blood vessels and are carried to various organs where they die and evoke granulomas. These may occur in the retina, with loss of vision; or the patient may develop asthma or eosinophilia.

THE CESTODES, SEGMENTED PLATYHELMINTHES, OR TAPEWORMS

Adult tapeworms live in the intestines of man or other vertebrates, while their larval forms develop in other host species, either vertebrates or invertebrates. The white tape-like bodies of the adult worms (Fig. 80) consist of a series of segments, each of which is called a *proglottis* and contains both male and female sex organs. The head is called a *scolex*, and has suckers and sometimes hooklets whereby the worm attaches itself in the intestine. The lengths of different species range from a few millimetres to 10 metres or more and the number of proglottides from 3 to over 4,000. The anterior proglottides are small and immature, the more posterior ones grow progressively broader and more mature, until the terminal ones consist almost wholly of a uterus distended with mature ova. These ripe segments are either shed whole and passed in the faeces or they discharge their ova while still remaining attached.

All the important cestodes of man require an intermediate host for the further development of the eggs and larvae. This development produces in the intermediate host a cyst-like parasite called a *cysticercus* or bladder-worm which contains a scolex or multiple scolices, but no proglottides. When the infested tissues of the intermediate host are eaten by a definitive host, the young scolices attach themselves in the intestine, develop proglottides and so form adult worms.

Man is the definitive host of all his cestode parasites but one. This exception, *Echinococcus granulosus*, is a very imoprtant one; its cysticercus stage is human hydatid disease.

THE INTESTINAL TAPEWORMS OF MAN

(1) *Taenia saginata*

T. saginata or the beef tapeworm is a common human parasite where raw beef is eaten, e.g. in Eastern Europe, Abyssinia and parts of Asia. The adult is a large tapeworm, sometimes 10 metres long or more, with 2,000 or more proglottides. The head has 4 suckers but no hooks. The mature terminal proglottides, which measure up to 20 millimetres long by 7 millimetres broad, become detached and are actively motile, sometimes migrating through the anus as well as being passed in the faeces.

(a) *Life cycle*

When ova from human faeces are eaten by cattle, especially calves, the larvae hatch out, penetrate the intestinal mucosa, reach lymphatics or small veins and are carried thence in the circulation to be arrested in muscles. Here they develop into small ovoid cysts each measuring up to 1 centimetre in main diameter and containing a scolex—*Cysticercus bovis*. The infestation is often called " beef measles ". The cysts are found especially in the muscles of mastication, tongue, diaphragm, rump, and in the heart. When live cysts are eaten by man, the scolices attach themselves in the upper part of the small intestine and develop into adult tapeworms.

(b) *Pathology*

Some patients have abdominal pains, diarrhoea or mild anaema ; but others harbour tapeworms for years without suffering any ill effects.

(2) *Taenia solium*

T. solium or the pork tapeworm occurs in men who eat raw or poorly-cooked pork. The adult generally resembles the beef tapeworm but is rather smaller and the head has hooks as well as suckers.

(a) *Life cycle*

Man is the chief definitive host and the pig is the usual intermediate host. When the pig ingests eggs from human faeces, the larvae hatch out in the intestine and pass by the blood stream to the muscles where they form cysts, each about 1 centimetre in diameter, and containing a scolex, *Cysticercus cellulosae*. The infested flesh is " measly pork ". When this is eaten by man, the liberated scolices attach themselves by their hooklets and suckers to the intestinal mucosa and grow into tapeworms.

(b) *Human cysticercosis*

On rare occasions, man also serves as the intermediate host, either by infection with ova which he himself has passed or from food or water contaminated by another host. Many cysticerci, similar to those in the pig, develop in the human tissues, especially in subcutaneous tissues, brain, eye, orbit, heart, muscles, and peritoneum.

(c) *Pathology*

As with *T. saginata*, the adult worm may cause abdominal pain and digestive disturbances, or may occasion no obvious ill effects. In human cysticercosis, the cysts form small nodules surrounded by fibrous capsules ; the larvae may eventually degenerate and become calcified. The symptoms depend on the number of cysts and the organs infested ; epileptiform fits or symptoms of brain tumour may accompany cerebral cysticercosis.

(d) *Diagnosis*

Diagnosis of the adult tapeworm depends on finding ripe proglottides in the faeces ; these can be distinguished from those of *T. saginata*. Diagnosis of cysticercosis is made by searching for calcified cysts in X-ray photographs of skin, brain or other parts, or by microscopical examination of an excised nodule.

(3) *Diphyllobothrium latum*

This worm, also known as the broad tapeworm or the fish tapeworm, meaures up to 10 metres long and may have 4,000 proglottides (Fig. 80). The head is very small and has two lateral suctorial grooves or bothria but no hooklets. The worm is widespread, especially in the Baltic countries, central Europe, and other regions where fresh-water fish are an important part of the diet.

(a) *Life cycle*

Man, dog and cat are the main definitive hosts, and there are two intermediate hosts, a crustacean and a fish. Ova discharged in the faeces of the definitive host hatch out in water and liberate minute spherical ciliated larvae. These are ingested by small crustaceans, e.g. *Cyclops*, in which they develop further into elongated larvae. If the crustaceans are now ingested by a suitable fresh-water fish (e.g. pike, perch, salmon or trout), the larvae penetrate the intestinal wall and migrate to the muscles, where they develop still further and become encysted. When the raw or insufficiently cooked flesh of the fish is eaten by man or other definitive host, the larva attaches itself in the intestine and grows to an adult worm.

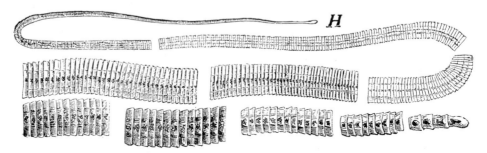

FIG. 80.—A typical tapeworm, *Diphyllobothrium latum*, natural size. *H*, the head.

(b) Pathology

The worm often causes no ill effects or only mild ones. Sometimes, however, especially in Finns, its presence is accompanied by a severe pernicious-like anaemia, which usually improves after expulsion of the worm.

(c) Diagnosis

Diagnosis is by finding the ova or proglottides in the faeces.

(4) Other intestinal tapeworms of man

Of many other species of cestodes which occasionally infest man, the only ones which are frequent enough to be mentioned are :

(i) *Dipylidium caninum*, a common dog and cat tapeworm, sometimes transferred to children who play with these animals, the intermediate hosts being fleas;

(ii) *Hymenolepis nana*, a small tapeworm, prevalent in Southern Europe, U.S.A., etc., especially in children, and unique in that it requires no intermediate host ;

(iii) *Hymenolepis diminuta*, the rat and mouse tapeworm, occasionally found in man, fleas and other arthropods being intermediate hosts.

HYDATID DISEASE
(The cysticercus stage of Echinococcus granulosus)

(1) The life cycle of Echinococcus granulosus

Echinococcus granulosus or *T. echinococcus* (Fig. 81) is the smallest of all the tapeworms, measuring only 3 to 6 millimetres long, and consisting of a scolex and 3 proglottides, the scolex bearing 4 suckers and many hooklets. Dogs (also wolves and jackals) are the definitive hosts, in the intestines of which the worms are often present in great numbers. The commonest intermediate hosts are sheep, cattle, pigs and men ; but horses, goats, camels, monkeys and many other mammals are sometimes infested. Human infestation is less common than in the other principal intermediate hosts, especially sheep, which constitute

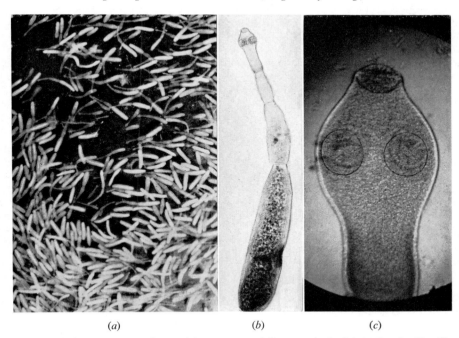

(a) (b) (c)

FIG. 81.—*Echinococcus granulosus* : (*a*), numerous adult worms in dog's intestine, (× 3) ; (*b*), enlarged view of a single worm, (× 15) ; (*c*), head showing four suckers and anterior ring of hooklets, (× 60).

FIG. 82.—An intact hydatid cyst, natural size.

the main reservoir of the disease in regions where it is prevalent, e.g. in Australia, New Zealand, South Africa, Argentine, Iceland, North Wales, etc.

Taenia ova from dog faeces are ingested by man, either in contaminated water or food or as a result of direct contact with the animals. The embryos hatch out in the intestine, penetrate its wall and enter small veins, whence they pass by the blood stream to the liver or elsewhere. Most commonly they are arrested in the liver, in which about three-quarters of all hydatid cysts in man are situated ; but the embryos may traverse the liver and produce cysts in the lungs, brain, bones or other organs. Within the cysts (as described below), daughter cysts and brood-capsules full of scolices develop. When dogs eat carcases or offal from sheep, pigs or cattle infested by cysts, ingested scolices grow into adult worms in the dogs' intestines.

(2) The structure of living hydatid cysts

The most frequent form of a hydatid is a spherical unilocular cyst, usually containing daughter cysts and brood-capsules of scolices ; but under special conditions it sometimes assumes other forms.

The typical spherical hydatid cyst (Fig. 82) grows slowly but may eventually reach a diameter of 6 inches or more. It is enveloped in a thin fibrous capsule derived from the host's tissues. Its own wall consists of an outer chitinous milky-white laminated *ectocyst* (Fig. 83) lined by a delicate inner cellular *endocyst* or germinal layer, and it contains a clear watery fluid with small amounts of salts,

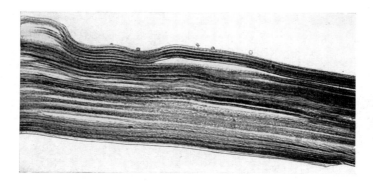

FIG. 83.—Section of laminated hydatid membrane. ($\times$ 300.)

amino-acids and protein in solution. From the germinal endocyst of most hydatids, multiple *daughter cysts* and *brood-capsules* are budded off to become free in the fluid. The walls of the daughter cysts show laminated ectocyst and germinal endocyst, like their parent, and they also may produce brood-capsules or grand-daughter cysts in their interiors. Brood-capsules are tiny thin-walled vesicles, about the size of pins' heads, within which varying numbers of *scolices*, usually 8 to 15, are produced (Fig. 84). The capsules are at first attached to their parent endocyst, but later they become free in the cyst fluid in which they form " hydatid sand ". Most well-developed hydatids produce brood-capsules with scolices, but some fail to do so and remain sterile. The appearance of a hydatid

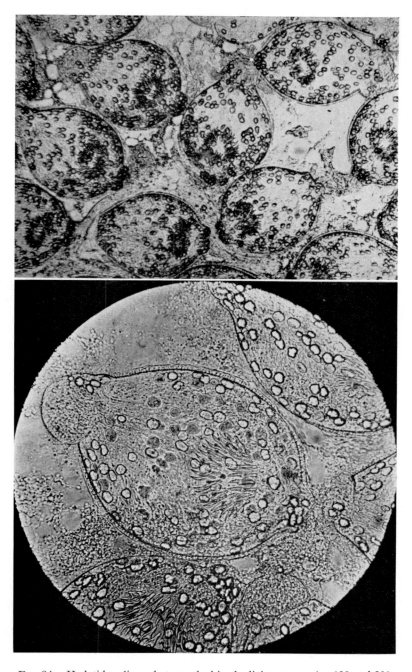

FIG. 84.—Hydatid scolices photographed in the living state. (× 130 and 280.)

cyst filled with its daughter cysts is very characteristic, the latter often resembling small and large grapes and their walls being well described as " grape-skin membrane ".

Unusual forms of hydatid cyst are as follows :

(i) *Irregular cysts* may be formed in confined rigid spaces, especially in bones, in which a cyst may grow into an elongated cylindrical form within the medullary cavity.

(ii) *Exogenous daughter cysts* sometimes develop by external budding from the parent cyst.

(iii) *Alveolar cysts* are a rare type, in which prolific exogenous budding produces an ill-defined honeycomb-like mass which infiltrates the tissues like a malignant tumour.

(3) Secondary changes in hydatid cysts

(i) *Spontaneous cure by degeneration* sometimes occurs, especially in the liver. The hydatid dies, the fluid is gradually absorbed, the cavity shrinks, and its contents becomes putty-like debris in which, however, folded masses of discoloured " grape-skin" membrane may long be recognizable (Fig. 85). Calcification, visible radiographically, often occurs in the walls or contents of old degenerated cysts. The death of the hydatid is sometimes attributable to the diffusion of bile into it from a contiguous bile duct, for degenerated cysts are often bile-stained ; but in other cases there is no bile-staining and the cause of the parasite's death is not known.

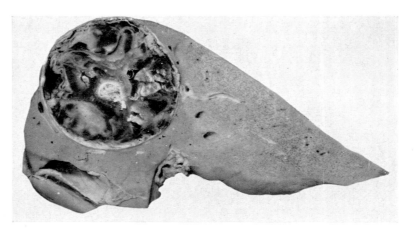

FIG. 85.—Degenerated hydatid cyst of liver ; two-thirds natural size.

(ii) *Rupture of cysts into neighbouring cavities* is an important complication. Cysts in the liver may rupture into the peritoneal cavity, in which multiple abdominal hydatids later develop from the spilt daughter cysts, brood-capsules and scolices. (It has been proved that isolated scolices or brood-capsules can grow into fully-formed cysts.) Hepatic cysts may also rupture into the bile duct or may extend through the diaphragm into the pleural cavity or lung. Pulmonary cysts may rupture into the pleura, or into a bronchus so that the contents are coughed up. Rupture of cysts, especially into the serous cavities, may cause severe allergic reactions, with urticarial rash, fever, dyspnoea or fatal circulatory collapse.

(iii) *Pyogenic infection* sometimes occurs, especially in the lung or liver, so that the cyst becomes an abscess and soon perforates into a contiguous cavity.

(4) The pathological effects of hydatid cysts

The reaction excited in the host tissues by the presence of an uncomplicated hydatid is usually slight, consisting in little more than the development of a thin fibrous capsule around the parasite, just as occurs around inert foreign bodies. But when the cyst is injured, ruptured or infected, the surrounding capsular tissues display pronounced chronic inflammatory changes with collections of macrophages, foreign-body giant-cells, proliferating fibroblasts, and sometimes eosinophil leucocytes.

Because of the slow growth of, and absence of inflammation around, uncomplicated living hydatids, these are clinically symptomless until by their increasing bulk they press on some important structure or form visible or palpable masses, or until one of the complications already mentioned takes place. The local symptoms produced of course depend on the situation of the cyst ; e.g. a palpable mass, obstructive jaundice or ascites from cysts in the liver ; bronchiectasis, obstructive pneumonia, haemoptysis or expulsion of the parasite by coughing, from cysts in the lung ; signs of cerebral tumour from a cyst in the brain ; bony swelling, pathological fracture or compression paraplegia from hydatids in bone ; and so on.

(5) Diagnosis of hydatid disease

Apart from clinical and radiographic evidence, the following laboratory investigations are important in diagnosing hydatid infestation:

(a) The identification of hydatid elements

Membrane, scolices or characteristic hooklets may be identified in suspect material coughed up, or discharged in the urine, or found at operation. Deliberate puncture and aspiration of suspected hydatid cysts for diagnostic purposes should be avoided, because it may precipitate serious allergic reactions from leakage of the cyst contents.

(b) Casoni's skin test

This test for sensitivity to hydatid protein is made by intradermal injection of a small quantity of sterile hydatid fluid ; a positive reaction in the form of a promptly developing wheal appears in sensitized persons. The test is positive in about 80 per cent of cases of hydatid disease.

(c) Complement-fixation test (see p. 280)

This also gives a high proportion of positive results in patients with living cysts.

(d) Eosinophilia

Eosinophilia is present in a small proportion of cases. If present in a person suspected of having hydatid disease, it helps to confirm this diagnosis ; but its absence is of no diagnostic value.

THE TREMATODES, UNSEGMENTED PLATYHELMINTHES, OR FLUKES

The trematodes are parasitic flat-worms of relatively simple unsegmented structure; they are called " flukes " because of their general resemblance in shape to fish of that name, e.g. the common flounder. Most species have complex life cycles, in which the intermediate hosts are usually invertebrates. In man the *schistosomes* or blood flukes are much more important parasites than intestinal, liver or lung flukes. The schistosomes look more like round-worms than flat-worms, because the body is narrow and folded together ventrally (hence, " schistosome "=split body). The sexes are separate, and the male, about 1 centimetre long, carries the female, about twice as long, in a ventral groove in his body, the gynaecophoric canal (Fig. 86). The three species of schistosomes are closely similar in structure and in their life cycles.

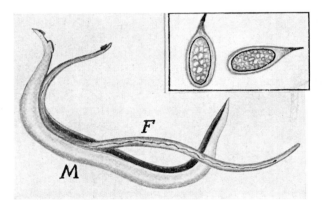

FIG. 86.—*Schistosoma haematobium*, the male *M* carrying the female *F* in the gynaecophoric groove. (× 10.) The inset shows the eggs. (× 180.)

(1) Schistosoma haematobium (Bilharzia haematobia)

This is the parasite of vesical schistosomiasis or Bilharziasis, an ancient disease of the Nile Valley which is still very prevalent in Egypt and the Sudan and other parts of Africa. The worms and eggs were discovered by Bilharz in 1851 in Cairo.

(a) Life cycle

Man is the main definitive host. The adult worms live principally in the vesical and other pelvic veins. The females deposit large numbers of their characteristic eggs (with terminal spines) in the small venules. The spiked ova penetrate these, set up severe inflammatory changes in the bladder wall, and many of them are discharged in the urine. In contaminated water the eggs hatch out, each producing a ciliated pyriform larva or *miracidium*. On contact with water-snails of appropriate species, the miracidia enter these, develop and multiply to produce great numbers of motile fork-tailed larvae called *cercariae* which escape from the snail. When man bathes in the infected water, the cercariae penetrate the skin and burrow into small veins, whence they are carried in the blood stream to the lungs. They squeeze their way through the pulmonary capillaries, pass into the systemic arterial blood, and are carried into many tissues. In most of these they perish (setting up mild inflammatory foci), but those entering the intestinal arteries make their way through the mesenteric capillaries and enter the portal circulation. In the portal branch-veins in the liver, they develop into adolescent worms, which then migrate against the portal blood stream to reach the inferior mesenteric and pelvic veins, where they become sexually mature adults.

(b) Pathology

(i) *A skin rash*, with minute petechiae, develops where the cercariae penetrate the skin, but it is usually mild and clears up in a few days.

(ii) *Migration of the larvae* in the body produces focal inflammatory lesions in lungs, lymph nodes, heart, spleen, liver and other tissues. There may be fever, headache, generalized pains, and palpable enlargement of the liver and spleen, often accompanied by allergic effects, namely urticaria, asthma and marked eosinophilia.

(iii) *Cystitis* is the main lesion of the disease. It results from the irritative effects of the eggs, which produce severe chronic inflammatory changes, often complicated by secondary bacterial infection. The urine contains blood, pus, mucus and schistosome eggs. The disease may extend to the ureters, prostate, urethra, vagina or rectum. In not a few long-standing cases carcinoma of the bladder supervenes.

(iv) *Lesions due to ova in remote organs.* From the pelvic veins eggs are frequently carried in the systemic venous blood to the lungs and in the portal blood to the liver. In the liver they may produce serious chronic inflammatory cirrhosis, accompanied by splenic enlargement and ascites.

(2) Schistosoma Mansoni

This parasite is closely similar to *S. haematobium* but infests and lays its eggs chiefly in the venules in the wall of the large intestine, producing severe chronic colitis with bloody diarrhoea. The ova, which differ from those of *S. haematobium* in having lateral spines, are discharged in great numbers in the faeces. They are also carried in the portal blood to the liver, often producing cirrhosis and splenic enlargement. The bladder and other pelvic organs may also be affected.

(3) Schistosoma Japonicum

This closely similar parasite is the cause of intestinal schistosomiasis in Japan, China and some of the Pacific islands. The parasite lays its eggs chiefly in the tributaries of the superior mesenteric veins, producing severe chronic enteritis of the small intestine. Hepatic cirrhosis from transported eggs is frequent. The bladder and other pelvic organs are not affected.

(4) Liver flukes parasitic in man

(a) *Fasciola hepatica*, the common sheep liver fluke, has occasionally been found in man, infesting the biliary passages.

(*b*) *Opisthorchis felineus*, the cat liver fluke, is a frequent parasite of the bile passages of man also in Prussia, Siberia and other parts of Asia. Infestation results from eating raw or inadequately cooked fish of certain kinds, which are the intermediate hosts.

(*c*) *Clonorchis sinensis*, the Chinese liver fluke, is a common inhabitant of the bile passages of men, dogs and cats in China, Korea and Japan. Infestation is from raw fish, the intermediate hosts.

(5) Intestinal flukes parasitic in man

In certain parts of Asia and Africa the native peoples show infestation by various species of intestinal flukes, due to the consumption of raw foods of various kinds.

(6) *Paragonimus Westermani* (the oriental lung fluke)

Infestation by this fluke is prevalent in Eastern Asia, Indonesia and the Pacific Islands. The adult parasites live principally in the lungs, where they produce nodular and cystic masses, bronchiectasis and bronchopneumonia, but they may also infest other tissues. The eggs are coughed up and discharged in the sputum, or swallowed and passed in the faeces. In water the miracidia infest certain snails, the cercariae from which invade the viscera and muscles of crabs or crayfish. When a definitive host (man or many other mammals) eats these, the larval parasites migrate through the intestinal wall, abdominal cavity and diaphragm to reach the lungs or other parts.

THE ECTOPARASITES OF MAN

(1) Mites and ticks

Mites and ticks constitute the order *Acarina* of the class *Arachnida*.

Sarcoptes (or *Acarus*) *scabiei* or the " itch-mite " is the cause of human mange or scabies. The male and female pair on the skin ; the female then burrows into the skin, laying her eggs as she goes. The eggs soon hatch and the young mites, which mature in about a month, migrate to the surface and pair. Any part of the skin may be affected, but the commonest sites of scabies are the interdigital creases, backs of the hands, axilla, groin, beneath the breasts or on the back.

Demodex folliculorum is a small mite which resides in hair follicles and sebaceous glands, usually without any ill effects but sometimes causing mild irritation.

Harvest mites (*Trombidoidea*) are the cause of " harvest itch ". The young mites abound on grass and bushes in summer and autumn, whence they crawl onto the skin of man and other mammals, take a meal of blood and then drop off, leaving bites which develop into irritable lumps and blisters.

Ticks are large blood-sucking mites of many species. They are of importance to man for three reasons : (1) Their bites may admit pyogenic or other bacteria ; (2) The bites of some species, e.g. *Dermacentor*, produce toxic effects—tick paralysis—which may be fatal ; (3) Several species of ticks act as hosts and transmitters of microbic infections of man, e.g. relapsing fever (p. 207), tularaemia (p. 184), and some Rickettsial infections (p. 267).

(2) Lice

Lice are small parasitic wingless insects. Man is the host of three species : (1) *Pediculus capitis*, the head louse, (2) *Pediculus corporis*, the body louse, and (3) *Phthirus* (or *Pediculus*) *pubis*, the " crab " or pubic louse. The parasites deposit their minute white eggs (or " nits ") on hairs or clothing, where they hatch in a few days.

Lice are important to man for two reasons, namely : (*a*) they produce an irritable dermatitis, or pediculosis, and (*b*) they transmit certain infective diseases, notably typhus and trench fever, and relapsing fever.

(3) Blood-sucking insects

There are many species of insects which bite man and feed on his blood when opportunity offers, but are not wholly dependant on him ; like ticks, they are predatory rather than parasitic. Many of them are important transmitters of bacterial, protozoal or virus diseases.

(*a*) *Fleas*

The human flea, *Pulex irritans*, causes irritable bites, and transmits plague from man to man. Rat fleas are important in spreading plague among rats (p. 184).

(*b*) *Bugs*

The common bed-bug, *Cimex lectularius*, crawls out of its hiding places at night to obtain meals of human blood, and its bites may be troublesome ; but, as far as is known, it is not a disease-transmitter. Other species of bugs, however, transmit *Trypanosoma Cruzi* of Chagas' disease (p. 227).

(*c*) *Mosquitoes*

Various species are the transmitters of malaria (p. 220), filariasis (p. 238), yellow fever (p. 264), and dengue (p. 264).

(*d*) *Other insect transmitters of disease*

Tse-tse flies of the genus *Glossina* transmit trypanosomiasis (p. 226). Sand-flies of the genus *Phlebotomus* transmit Leishmaniasis (p. 228) and the virus disease Pappataci or sand-fly fever (p. 264).

(4) Larval flies parasitic in man

Myiasis is the invasion of living tissues by fly larvae. Some of the species of flies—the warble and bot flies—have larvae which are regularly parasitic in the skin of horse, ox, sheep or man. Sometimes non-parasitic flies lay their eggs on or near wounds or ulcers, which then become infested by the larvae. This is usually quite harmless ; indeed some surgeons have supposed that maggots aid in cleaning dirty wounds. Fly eggs on food occasionally hatch in the intestine, so that living larvae are passed in the faeces—intestinal myiasis.

SUPPLEMENTARY READING

Ash, J. E., and Spitz, S. (1945). *Pathology of Tropical Diseases: An Atlas.* Philadelphia and London; W. B. Saunders. (Contains good accounts of the parasitic metazoa with excellent diagrams of their life-histories.)

Craig, C. F., and Faust, E. C. (1943). *Clinical Parasitology.* London; Kimpton. (Comprehensive accounts of the metazoal diseases.)

Davey, T. H. (1961). Seventh edition. *A Guide to Human Parasitology*. London; H. K. Lewis.

Lapage, G. (1957). *Animals Parasitic in Man*. A "Pelican" book (An excellent outline.)

Woodruff, A. W. and Thacker, C. K. (1964). "Infection with animal helminths". *Brit. Med. J.*, **1**, 1001. (A useful note on *Toxocara.*)

CHAPTER 20

DISEASES DUE TO VIRUSES AND RELATED MICRO-ORGANISMS

GENERAL CHARACTERS OF VIRUSES

(1) Definition

Viruses are minute micro-organisms, which are smaller than ordinary bacteria and which can multiply only within the living cells of suitable hosts. They are resistant to antibiotics and some other bactericidal agents.

Two adjectives which have often been applied to viruses, namely " filterable " and " ultramicroscopic ", have not been admitted to our definition, because for the larger viruses they are not true. Both filterability and visibility depend on size, and the particles of the largest viruses are held back by bacterial filters and are visible by ordinary microscopy.

(2) The sizes of virus particles

The sizes of minute particles, too small to be seen microscopically, can be estimated by means of ultra-filtration through graduated collodion membranes, by centrifugation at high speeds, by ultra-violet photography, and by phase-contrast and electron microscopy. By these methods, the sizes of even the smallest virus particles have been accurately measured. Their diameters range from 300mμ (i.e. 0·3μ) for the largest, the poxviruses, to 28mμ (i.e. 0·028μ) for the smallest, e.g. the virus of poliomyelitis (Fig. 87). Smallpox virus particles have a mean diameter of 150mμ (i.e. 0·15μ) and are just visible by ordinary microscopy.

Virus particles are also called virions or " elementary bodies " and some of the visible ones are specially named after their discoverers, e.g. the *Buist-Paschen bodies* of smallpox and vaccinia. Most virus

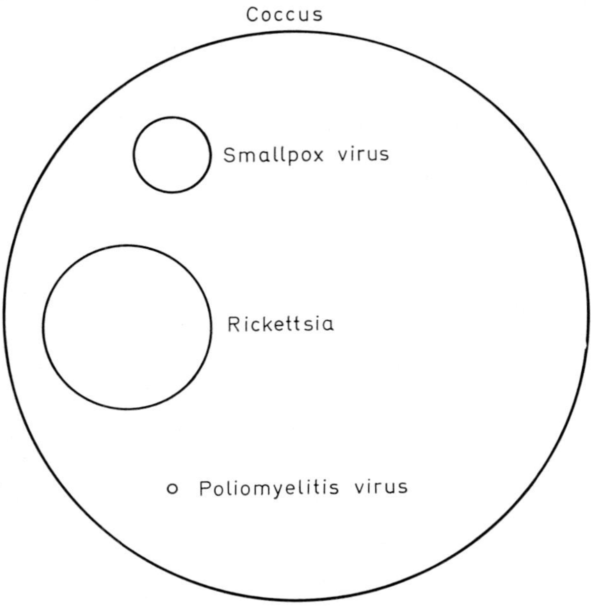

FIG. 87.—Diagram of relative sizes of viruses.

particles, as seen with the electron microscope are spheroidal or have distinctive crystal-like forms, but some are elongated or filamentous.

Rickettsiae and some kindred micro-organisms, which were formerly looked upon as the largest viruses, are now regarded as intermediate between bacteria and true viruses and are grouped separately (see below).

(3) The intracellular habitat of viruses

Although viruses can survive in the free state and can thus be transferred from host to host, they can multiply only within susceptible host cells. They do not grow on ordinary media, probably because they are too small to contain all the enzymes and other metabolic machinery necessary for independent existence. To grow and multiply they must utilize the enzymes and other mechanisms of the host cells. Virus multiplication is a complex process in which viral nuclear material that has entered the host cell diverts the normal activities of the cell to the synthesis of specific proteins and nucleic acids of the infecting virus. The various components of the virus particle are formed separately and are later assembled towards the end of the reproductive cycle. Large numbers of progeny are produced in this way. The new virus particles may escape from the host cell by a slow continuous " leaking " process, or the host cell may burst and release large numbers of virions.

Many viruses are very selective as to the species of the host and the particular kinds of cells in which they will grow. Experimental work on their multiplication and pathogenic effects can therefore be carried out only in a susceptible animal, or in a suitable tissue culture, or in the chorio-allantoic membrane of the developing hen's egg. This last method has proved of great value in studying many kinds of viruses, including those of smallpox, chickenpox, and influenza.

In many virus diseases the cells infected by the virus show characteristic rounded or irregular, hyaline or granular bodies within their cytoplasm or nuclei. These are called " inclusion bodies " ; and in some diseases they have been given special names, e.g. " Negri bodies " in the nerve cells in rabies and " Guarnieri bodies " in the epidermal cells in smallpox and cowpox. There is now little doubt that inclusion bodies are actual colonies of the virus, for in some diseases they have been shown to contain masses of elementary bodies and to be infective for new hosts.

(4) Physical and chemical properties of viruses

Microchemical methods have shown that viruses, like bacteria, contain proteins, carbohydrates, fatty substances and inorganic salts. Like vegetative bacteria too, they are usually destroyed by moist heat at 55° to 60° C. Viruses vary in their sensitivity to chemical disinfectants. The most effective viricides are oxidizing agents such as potassium permanganate and hypochlorites; phenolic disinfectants are active against only a few viruses. Many viruses withstand prolonged drying, and also low temperatures of the order of −70° C. They also survive well in 50 per cent glycerol—which affords a convenient method of storing infected material, for contaminating vegetative bacteria are killed by this treatment. True viruses are insensitive to antibacterial drugs, such as sulphonamides, penicillin and streptomycin.

(5) Immunity phenomena with virus infections

Virus diseases, like the ordinary bacterial diseases, excite immunity reactions in the host, in whose blood specific agglutinins, precipitins, or neutralizing antibodies often appear. These antibodies act principally by " neutralizing " the virions and preventing their entry into the host's cells. Because normal cell membranes are impervious to circulating antibodies, these can have no effect on virus that has already entered host cells. Two other factors may also operate in protecting cells from invasion—a mechanism of *cellular immunity*, which confers resistance on host cells even in the absence of circulating antibody (as occurs in cases of agammaglobulinaemia), and the phenomenon of *interference*, whereby cells that have been invaded by one type of virus are resistant to invasion by other viruses, even immunologically distinct ones. With many of the virus diseases, e.g. measles, yellow fever, poliomyelitis and smallpox, one attack confers long-standing strong immunity. In other diseases, however, e.g. influenza and the common cold, the immunity is only transient.

CLASSIFICATION OF VIRUSES AND VIRUS DISEASES

All kinds of living creatures have their virus diseases. This applies even to bacteria' the *bacteriophages* of which are viruses ranging in size from 10 to 70mμ, which when they infect bacterial colonies cause their lysis. Many bacterial species have their phages, each more or less specific for the particular bacterial host. Use is made of these bacterial viruses for the subdivision of some species of bacteria into phage-types, which is of particular importance in the epidemiological study of staphylococci, *Salmonellae* and *Pseudomonas*, and in the recognition of *Bacillus anthracis* (see Appendix D).

There are many virus diseases of plants, some of them of great economic importance, e.g. mosaic disease of tobacco and "leaf-roll" of potatoes. The many virus diseases of animals are not restricted to vertebrates; bees and silk-worms suffer serious infections of economic importance. As examples of virus diseases of domestic birds and mammals may be mentioned fowlpox, fowlpest, foot-and-mouth disease of cattle, pigs and sheep, and distemper of dogs and cats.

Man suffers from over 40 distinct virus diseases, some rare but others among the most important infections. The classification and naming of the viruses concerned are still unsettled, Linnaean binomial nomenclature has not been adopted, and the present grouping—according to size and morphology of the virions, their chemical and physical properties, susceptible hosts and tissues, and other factors—must still be regarded as tentative. "A formal classification is at present impracticable" (Wilson and Miles, 1966). Regrettably also, some of the names in current use are unscientific nicknames, rather childish inventions which lack root meanings and etymological congruity. But, as the student will encounter them, we give here brief notes about them.

Poxviruses, causing smallpox, cowpox, vaccinia and molluscum contagiosum in man, and some other cutaneous pox diseases in animals, are a group of large antigenically related viruses with a predilection for epidermal cells.

Herpesviruses cause herpes simplex, herpes zoster and chickenpox, the last two being caused by the same virus. They selectively attack epithelial cells, and produce intranuclear inclusions. The virus of cytomegalic disease has also been included in this group.

Myxoviruses, so named because some of them prefer a mucinous environment, include those of influenza and some other respiratory infections, mumps and measles, and some workers include the viruses of rubella and rabies.

Enteroviruses, so called because of their intestinal habitat, include the three types of *poliovirus*, groups A and B (and many sero-types) of *Coxsackie viruses* (named after a small town in

U.S.A.), and many sero-types of *echovirus* (some workers prefer to write "ECHO virus"), an abbreviated nickname standing for "enteric cytopathogenic human orphan". *Rhino-viruses* (nasal viruses) are the most important causes of the common cold. The enteroviruses and rhinoviruses have been classed together as "picornaviruses", another nickname derived from a little-known root "pico" (small) and RNA, these viruses all being small ones containing ribonucleic acid.

Reoviruses, "reo" standing for "respiratory enteric orphan", have been so named because they cause both mild respiratory infections and diarrhoea in children.

Adenoviruses (of many sero-types), so called because they were first found in adenoids, cause respiratory infections, conjunctivitis, and possibly lymphadenitis.

Arboviruses, an abbreviation denoting "arthropod-borne", are so named because they are transmitted by mosquitoes or ticks. They cause yellow fever, dengue fever and several kinds of encephalitis.

Papillomaviruses, also called *papovaviruses* (a particularly obscure nickname, derived from the first two letters of each of three words, "papilloma", "polyoma" and "vacuolating"), cause the common wart in man and a number of other papillary lesions in animals.

Until virological classification and nomenclature become more mature and generally acceptable, it is best to group virus diseases according to their pathology and the main systems or organs affected. We will discuss them under the following heads, Rickettsial and allied diseases being considered separately:

Virus diseases

(a) *Generalized infections with skin eruptions*
    (1) Smallpox, cowpox and vaccinia
    (2) Chickenpox, including herpes zoster
    (3) Measles
    (4) German measles
    (5) Other infections (e.g. ECHO)

(b) *Infections of the respiratory tract*
    (1) Influenza
    (2) The common cold

c) *Infections of the nervous system*
    (1) Poliomyelitis
    (2) Rabies
    (3) Other infections (ECHO, Coxsackie, encephalitis lethargica)

(d) *Arthropod-borne infections*
    (1) Yellow fever
    (2) Dengue and kindred fevers
    (3) Encephalitis—various regional forms

(e) *Local infections of the skin*
    (1) Herpes simplex
    (2) The common wart
    (3) Molluscum cantagiosum

(f) *Other virus diseases*
    (1) Mumps
    (2) Infective hepatitis
    (3) Glandular fever
    (4) Cytomegalic inclusion disease
    (5) Sundry enterovirus infections

Rickettsial and allied diseases

    (1) Classical epidemic typhus
    (2) Endemic typhus
    (3) Scrub typhus
    (4) Other Rickettsial diseases

    (5) Allied diseases
    (a) Psittacosis
    (b) Lymphogranuloma venereum
    (c) Trachoma

GENERALIZED INFECTIONS WITH SKIN ERUPTIONS

(1) Smallpox, cowpox and vaccinia

It is appropriate that our consideration of virus diseases should begin with smallpox, because our modern knowledge of immunity began with Jenner's discovery in 1796 that inoculation with cowpox protects against smallpox. He believed that cowpox was really smallpox, modified and rendered less virulent by passage through the cow. However, the two viruses, though closely related, are distinct; and so also is the virus of vaccinia, which was formerly thought to be identical with that of cowpox—of which it is probably a variant.

(a) *Clinical characters*

(i) *Variola* or *smallpox* is a disease of great antiquity and of widespread distribution. It is estimated to have killed 60 million people during the 18th century. In its severer form it has almost disappeared from England and most other Western European countries ; but it is still prevalent in many other parts of the world, especially in densely populated Asiatic communities—e.g. in India it causes about 50,000 deaths annually. Where severe smallpox is rare, a mild type of the disease, known as " alastrim " or *variola minor* is often found ; and it is probable that the replacement of the severe classical disease by its mild variant is due largely to the widespread practice of protective vaccination.

Classical smallpox is characterized by severe toxaemia with high fever, a distinctive exanthem (skin rash) which is at first papular but later becomes vesicular and pustular and leaves depressed pitted scars (the skin lesions are called " pocks "), and a similar eruption of mucous membranes ; it has a mortality rate of 10 to 30 per cent, from the severe toxaemia or from bronchopneumonia. Alastrim has much milder toxaemic effects, the rash is slight and atypical, and the nature of the infection may easily escape recognition. The mode of infection by smallpox virus is almost certainly by the inhalation of droplets coughed out by patients ; these infect the nasopharyngeal and lower respiratory mucosa, whence the virus is distributed throughout the body by the blood stream. Although the most obvious lesions are in the skin, it is probable that similar lesions occur in many other tissues.

(ii) *Vaccinia.*—The lesion of vaccinia in man is familiar to everyone who has been vaccinated. After an incubation period of a few days, the inoculated skin becomes red and swollen, vesicles containing clear fluid develop in it, later these become pustular, then dry up and form crusts, the inflammation subsides, and a pitted scar is left. Toxaemic symptoms are usually slight, and in most cases there is little doubt that the virus remains localized at the site of inoculation. Rarely, vaccination has been complicated by dangerous *post-vaccinal encephalitis*, with widespread inflammatory foci in the brain and cord ; but whether this is due to the vaccinia virus itself, or to some other factor stimulated to activity by the vaccination, is uncertain. More commonly, in patients with chronic skin diseases, especially children with eczema, *eczema vaccinatum* may develop from contamination of the skin lesions with the vaccinia virus.

(b) *The pathology of the skin lesions of smallpox and vaccinia*

The main changes are in the epidermis, many of the spinous cells of which undergo swelling and necrosis. Spaces form between the cells of the several layers of the epidermis, and by the accumulation of fluid in these they are converted into vesicles. Acute inflammatory changes in the dermis are accompanied by aggregations of leucocytes and purulent change in the fluid of the vesicles which thus become pustules. In mild lesions the continuity of the epidermis is not destroyed, so that on subsidence of the inflammation, no scar is left ; but in severe lesions the epidermis is destroyed through its whole depth, the inflamed dermis is exposed, and healing leaves a pitted scar.

The diseased epidermal cells show rounded Guarnieri cytoplasmic inclusion-bodies, and these and the fluid from the vesicles contain many Buist-Paschen granules, or elementary virus bodies.

(c) Laboratory diagnosis of smallpox

This is always a matter of great urgency. Specimens should be collected and examined by a trained worker who has been properly vaccinated during the preceding 12 months. Direct microscopy of suitably stained material from papules or vesicles usually shows many elementary bodies. The presence of the virus may also be demonstrated rapidly by complement fixation and by agar-gel diffusion precipitin tests, in which vesicular fluid as antigen reacts with high-titre anti-vaccinial serum; this serum is appropriately used because the vaccinia and variola viruses have common antigens. The variola virus may be isolated by inoculation of egg chorio-allantoic membranes with suspensions of vesicular fluid, ground-up skin crusts, or blood serum.

(2) Chickenpox or varicella

This well-known, mild, infectious disease, chiefly of children, is characterized by a superficial vesicular skin eruption which becomes encrusted and which leaves no scars. Both of its names are unfortunate—" chickenpox " because it has nothing to do with chickens (it is *not* the same as fowlpox), and " varicella " (diminutive of " variola "), because it is quite unrelated to smallpox. It is due to a virus, the elementary bodies of which can be demonstrated in the vesicle fluid, and which probably infects new patients by being inhaled.

Herpes zoster or *shingles* (*zoster*, Gk. = girdle; " shingles " from L. *cingulum* = a belt) is due to the same virus as varicella. The vesicular skin eruption is restricted to the distribution of a sensory nerve, usually an intercostal nerve but sometimes the ophthalmic division of the trigeminal, with serious involvement of the cornea, and is often accompanied by severe neuralgic pain, due to inflammation of the spinal or Gasserian ganglion. Herpes zoster occurs mainly in adults, and its localized character is attributed to their having had varicella previously and therefore being partly immunized.

(3) Measles or morbilli

Measles is another well-known, highly infectious disease with a characteristic blotchy skin rash, a spotted rash on the inside of the cheeks (Koplik's spots), and general catarrhal inflammation of the nasal and respiratory mucosa. It occurs chiefly in children, in whom (especially infants) the catarrhal inflammation of the respiratory passages is often complicated by dangerous bacterial broncho-pneumonia. Characteristic giant cells occur in infected tonsils, lymph glands and lungs. By intratracheal inoculation with filtered nasopharyngeal washings from measles patients, the disease was reproduced in monkeys, and the virus was later isolated in tissue culture. Its transfer is by inhalation of infected droplets. One attack usually confers life-long immunity; and immunization can be achieved by injection of a single dose of live attenuated vaccine.

(4) German measles or rubella

This mild, measles-like disease is also due to a virus. It is of special interest because infection of a pregnant woman during the early months of pregnancy is likely to damage the young embryo and lead to congenital opacities (cataracts) of the lenses and to other developmental anomalies of the eyes, brain or heart.

The virus has been shown to be present in these lesions. The recent development of an effective live attenuated rubella vaccine is of interest in that it might be used in advance to immunize prospective mothers and so avoid teratogenic effects.

(5) Other infections

Other virus infections occasionally cause generalized skin rashes; for example ECHO viruses may cause an erythematous rash resembling that of rubella.

VIRUS INFECTIONS OF THE RESPIRATORY TRACT

(1) Influenza

The term "influenza" is often misused. It should be restricted to infection by the specific virus which is now well established as the cause of epidemic influenza. This virus is always obtainable in typical influenza epidemics but is usually *not* present in sporadically occurring colds and other feverish attacks which are often called "influenza".

(a) The virus of influenza

In view of our present knowledge, the long controversy as to whether influenza was caused by *Haemophilus influenzae* (p. 187) has lost all but historical interest. In 1933, Smith, Andrewes and Laidlaw showed that influenza could be induced in ferrets by intranasal instillation of filtrates of the nasal washings from patients with epidemic influenza, that the disease could be transmitted from ferret to ferret, and that ferrets convalescent from the disease resisted reinfection and had a high concentration of protective antibodies in their serum.

The influenza virus does not readily infect mice, but on passage from mouse to mouse its virulence increases, and it then produces fatal pneumonia in this animal. The virus can also be trained to grow on the chick chorio-allantoic membrane. The virus particles have a diameter of about 100mμ. They enter the body by inhalation of droplets coughed out by patients. Three types of virus are recognized—A, B and C. Major epidemics or pandemics are usually due to type A virus; type B strains cause sporadic endemic infections and minor epidemics, type C strains cause mild or subclinical infections and never epidemics.

(b) Clinical characters

Epidemic influenza is an acute fever of sudden onset, often with a shiver, headache, limb pains, and a variable degree of sore throat, catarrhal rhinitis or bronchitis. Some patients have severe gastro-intestinal or nervous symptoms. The disease usually ends in recovery in a few days, but often leaves the patient weak and exhausted for a time. In some epidemics serious bronchopneumonia has been a frequent complication, due to supervening secondary bacterial infection by pneumococci, streptococci or *H. influenzae*. According to the frequency of this complication—probably related to the variable virulence of the virus— influenza epidemics have differed greatly in their mortality.

(c) Pathology

Uncomplicated influenza is a simple catarrhal inflammation of the upper respiratory passages accompanied by a moderate or severe degree of toxaemia.

Fatal cases show supervening bacterial bronchopneumonia ; this is often accompanied by marked oedema and diffuse haemorrhages in the lungs, or, in cases dying later, by patchy suppuration. Other organs show severe toxic changes.

(d) Immunity changes

The presence of antibodies in the serum of convalescents was mentioned above. These protect the convalescent from re-infection for a few months ; but thereafter their concentration steadily falls, and he again becomes susceptible to infection by the nasopharynx even though demonstrable amounts of circulating antibody are still present. No fully effective method of artificial immunization has yet been devised; and any protection afforded by vaccines is short-lived.

(2) Coryza or the common cold

(" Coryza "=Gk. for " running at the nose ".)

Because " colds " are generally similar to influenza, and because they can be transmitted to human volunteers or apes by intranasal instillation of filtered nasal washings from patients, they were long suspected as being virus infections; and this has been fully established. Rhinoviruses are the most frequent cause; but para-influenza, ECHO, Coxsackie, adenoviruses and reoviruses may all cause clinically indistinguishable " colds ", or sometimes more severe infections of the respiratory tract, including pneumonia.

VIRUS INFECTIONS OF THE NERVOUS SYSTEM

(1) Poliomyelitis or infantile paralysis

Poliomyelitis is an epidemic or sporadic disease which chiefly affects children, especially those about 2 or 3 years of age. It attacks only a small proportion of the population, but there is good reason to believe that, during an epidemic, mild or subclinical infection of a large proportion of susceptibles occurs, making them immune to subsequent infection. This is why the disease is rare in adults. The mortality ranges between 10 and 40 per cent in different epidemics, and residual paralysis persists in 10 to 40 per cent of survivors.

The distribution of cases in an epidemic is often irregular, with no clear evidence of transfer from case to case. But this is explained by the facts, that the virus is often spread by healthy carriers, and that only a small proportion of susceptibles develop clinically apparent disease.

(a) The virus and its entry

In most cases, the virus probably enters the body from the pharynx or intestine. It is an enterovirus and is present in the nasopharyngeal secretions and faeces of patients and of healthy carriers. It reaches the central nervous system by way of the nerves from the pharynx or bowel or by the blood stream.

The virus particles are amongst the smallest known, measuring about 28mμ. In central nervous tissue preserved in 50 per cent glycerine at low temperatures, they remain alive for years. The disease can be transmitted to monkeys by intracerebral, intravenous, or intraperitoneal inoculation with infected material, and infectivity tests are performed in this way. The animals develop the typical disease with lesions similar to human ones. Three immunologically distinct types of virus are known—1, 2 and 3; the first is the common epidemic variety.

(b) Clinical characters

The initial symptoms are non-distinctive and are those of a febrile illness, with headache, pains in the back and limbs, and vomiting. In some cases meningitic symptoms—neck pain and stiffness, and rigidity of limb muscles—are pronounced. The cerebrospinal fluid usually shows well-marked changes, namely : raised pressure ; increased cells, 50–500 per cubic millimetre, mainly lymphocytes but with some polymorphonuclear cells in early cases ; excess of protein, with positive globulin tests, but normal chloride content (unlike tuberculous meningitis).

In some cases the disease subsides at this stage without any paralysis developing ; but in others, within 2 or 3 days, flaccid paralysis of various groups of muscles appears. This is irregular in distribution, most frequent in the limb muscles, but in severe cases affecting the trunk also, including the respiratory muscles. Occasionally, muscles innervated by cranial nerves (pharynx, palate, tongue, ocular or masticatory muscles) are mainly affected ; the disease is then appropriately called *polio-encephalitis*. Occasionally, paralysis starts in the lower limbs and rapidly ascends to the trunk, upper limbs and cranial area—a condition called *ascending myelitis*, one kind of Landry's paralysis.

In patients who survive the acute phase—and most do so—some recovery of the paralysed muscles takes place, but varying degrees of residual paralysis may remain, the paralysed muscles undergo atrophy, and deformities of the limbs ensue. Prolonged re-educational and postural treatment is necessary to obtain the best functional result in the affected parts.

(c) Pathology

Mild inflammatory changes may occur throughout the central nervous system, but the most severe lesions are in the anterior grey horns of the spinal cord

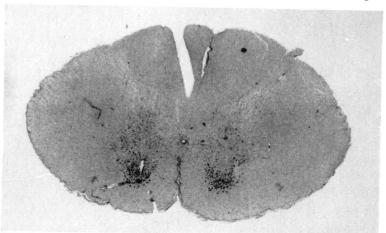

FIG. 88.—Cross-section of spinal cord from a case of poliomyelitis, showing blurring and congestion of the grey matter especially of the anterior horns. (× 10.)

(Figs. 88 and 89), or, in polio-encephalitis, the corresponding motor nuclei of the cranial nerves. The changes are (a) hyperaemia and patchy haemorrhages, (*b*) collections of polymorphonuclear leucocytes and lymphocytes in the Virchow-Robin spaces around the inflamed vessels, (*c*) degenerative changes in the motor nerve cells, ranging from mild cloudy swelling to complete disintegration, and (*d*) inflammatory changes in the neighbouring leptomeninges, accounting for the cerebrospinal fluid changes already mentioned.

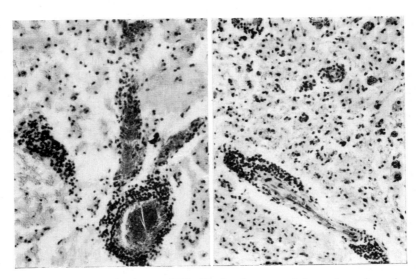

FIG. 89.—From anterior horns of Fig. 88, showing congested vessels, perivascular collections of lymphocytes, and disorganization of tissue. (× 120.)

(*d*) Immunity

One attack of poliomyelitis produces strong permanent immunity ; second attacks are very rare. The serum of convalescents contains specific antibodies capable of neutralizing the virus *in vitro,* and antibodies are often present also in the serum of persons who have never suffered from a clinically recognizable attack of the disease. In urban communities about 75 per cent of children 15 years old show antibodies, indicating that mild subclinical infection is frequent. Vaccination with virus preparations has been shown to confer a high degree of protection against paralytic disease. The Salk vaccine contains the three types of poliovirus, which have been inactivated by formaldehyde; and immunization is by intramuscular injections. The Sabin vaccine is a live attenuated vaccine, which is administered orally. Both vaccines give similar immune responses.

(2) Rabies or hydrophobia

(*a*) Clinical features

Rabies in primarily a disease of animals, especially dogs, cats and other carnivores, but is transmissible to man by the bites or scratches of rabid animals, the virus being present in their saliva. From the site of inoculation the virus spreads along the nerves to the central nervous system, the process occupying an incubation period of 4 to 8 weeks or longer. On reaching the central nervous system, the infection causes violent painful muscular spasms somewhat resembling those of tetanus or strychnine poisoning, and these are fatal within a few days. The name " hydrophobia " (= a dread of water) refers to the victim's inability to drink, because the spasms affect particularly the muscles of deglutition.

Occasionally the rabies virus causes in man an ascending myelitis with paralysis. In one remarkable outbreak of this kind in Trinidad the disease was transmitted by the bites of blood-sucking vampire bats. Paralytic rabies, spread by vampire bats, is prevalent among animals in Brazil.

(b) Pathology

The virus measures about 125mμ. It is present throughout the central nervous system, but is especially concentrated in the brain stem. Its presence excites little or no inflammatory change, but it produces characteristic cytoplasmic inclusions called Negri bodies in the nerve cells, especially of the hippocampus and cerebellum. These are so distinctive that they afford a ready means of diagnosis in a suspect animal. It is uncertain whether Negri bodies are actual colonies of the virus or products of degeneration in infected nerve cells.

(c) Immunity

Rabies is of great interest in the history of immunology, because the successful use of modified living virus as a vaccine for its prevention was Pasteur's crowning achievement. Realizing that the long incubation period of the disease offered an opportunity for immunizing a patient bitten by a rabid animal, Pasteur devised a vaccine consisting of material from the dried spinal cords of infected rabbits, in which the virulence of the virus was gradually attenuated. By repeatedly inoculating dogs with this material, starting with the most attenuated and working up to the more virulent doses, he made the animals refractory to the disease. And finally, in 1885, by the same means, he successfully immunized a boy who had been badly bitten by a rabid dog. In some institutions, Pasteur's vaccine is still used in its original form; in others it has been superseded by other living attenuated or chemically inactivated virus preparations.

(3) Other virus infections of the nervous system

(a) Virus meningitis, also called aseptic meningitis, embraces a variety of acute but usually benign infections, sometimes with signs of encephalitis also, due to a variety of viruses, including mumps virus, poliovirus, ECHO and Coxsackie viruses, and the virus of " lymphocytic choriomeningitis".

(b) Encephalitis lethargica, prevalent after the first world war but rarely if ever occurring since, was probably due to a virus, but this was never proved.

(c) Encephalitis due to arthropod-borne viruses (see next section).

ARTHROPOD-BORNE VIRUS DISEASES

(1) Yellow fever

Yellow fever, formerly a rampant and often fatal disease in Central America, the West Indies and West Africa, has been rendered much less serious since the discoveries of Reed, Carroll and Agramonte early in this century. Largely by experiments on human volunteers, these workers showed that the blood of patients was infective during the first 3 days of the fever ; that the infecting agent was filterable; that mosquitoes of the species *Aëdes aegypti* fed on patients' infective blood during the first 3 days of fever were, after an interval of about 12 days, transmitters of the disease to normal persons bitten by them; and that, by mosquito extermination, yellow fever could be practically eliminated. As a result of this brilliant and courageous work, Panama, once deservedly known as "the white man's grave ", is now free of the disease, and so are many other areas where it used to be dreaded. Many of the earlier investigators of yellow fever, including Noguchi, died of the disease. Later workers have been well protected by immunization. Ultra-filtration shows the yellow fever virus particles to have an average diameter of between 20 and 40mμ.

(a) Clinical features

The disease starts as a febrile toxaemic illness, in which jaundice appears in a few days. The urine is diminished and contains albumin and often blood. In severe cases there are petechial haemorrhages in skin and mucous membranes, and vomiting of black or blood-stained vomitus occurs. The mortality is about 25 or 30 per cent.

(b) Pathology

Besides the jaundice and the haemorrhages in skin and mucous and serous membranes, the principal organs affected are the liver and kidneys. The liver shows severe fatty change and cloudy swelling of its cells and coagulative necrosis of those of the middle zone of the lobules. The cells may show characteristic intranuclear inclusion bodies. The liver tissue is highly infective ; minute amounts of it will infect monkeys. The kidneys show pronounced degenerative changes in the tubular epithelium. The handling of all infective material from a case of yellow fever is very dangerous, and infection can occur through the unbroken skin.

(c) Immunity changes

The serum of convalescent patients or monkeys contains abundant antibodies to the yellow fever virus. These confer prolonged protection against further infection ; and second attacks of yellow fever are almost unknown. Man (or monkey) can be effectively immunized by means of inoculation of a single dose of living attenuated vaccine.

(2) Dengue and related fevers

Dengue or " breakbone fever " is an epidemic febrile disease, prevalent in warm climates, not usually serious, and characterized by severe limb pains, headache, and a skin rash. It is due to a virus which can be demonstrated in the blood of patients, and which is transferred by mosquitoes, especially *Aëdes aegypti*. In various parts of the world, there occur other dengue-like fevers, due to distinct viruses and transferred by insects or other arthropods, e.g. sandfly fever, Rift Valley fever and West Nile fever.

(3) Encephalitis—various regional forms

A number of arthropod-borne viruses cause human encephalitis. Some of these, for example St. Louis encephalitis, Japanese encephalitis and Murray Valley encephalitis, are carried by mosquitoes; others, including several varieties of Central European and Russian encephalitis, are tick-borne.

LOCAL VIRUS INFECTIONS OF THE SKIN

(1) Herpes simplex

(*Herpes*, Gk.=a creeping eruption.)

Patients with pneumonia or other febrile or debilitating diseases often develop small vesicles in or around the mouth and nostrils. These are due to a virus (*Herpesvirus hominis*), the virions of which measure about 125mμ and are visible in the fluid from the vesicles. Intranuclear inclusions called " Lipschutz bodies "

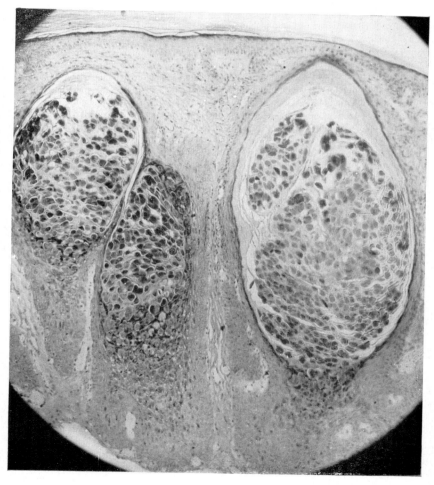

FIG. 90.—Molluscum contagiosum. Section of skin, showing masses of altered epidermal cells containing oval inclusion bodies. (× 120.)

are prominent in infected cells. Much the commonest infection is gingivo-stomatitis, but the virus also produces similar vesicular lesions on the cornea or genitalia, and occasionally a meningo-encephalitis.

(2) Herpes zoster or " shingles " (see p 258).

(3) Warts

The common wart can be transmitted in the human subject by inoculating the skin with wart filtrates, and inclusion bodies occur in the epidermal cells. The way in which warts occur in a crop on a child's hands and then clear away, never to return, also shows their infective nature and the acquisition of immunity to the virus.

(4) Molluscum contagiosum

This is a contagious skin disease characterized by soft wart-like lesions with a distinctive microscopical appearance. The infected cells of the epidermis show large ovoid eosinophil inclusions, called " molluscum bodies ", each about 20μ by 30μ, and found on microdissection to consist of masses of elementary bodies 250mμ in size (Fig. 90). The lesions have been reproduced by inoculating the skin with filtrates.

OTHER VIRUS INFECTIONS OF PARTICULAR PARTS

(1) Mumps

Mumps is a well-known infectious disease, chiefly of children, characterized by fever, bilateral non-suppurative inflammation of the parotid glands, and some-times of the testicles, pancreas, thyroid (see p. 604), or other parts. The disease can be transmitted to monkeys by introducing saliva from a human patient into their parotid ducts; and it can be transferred from monkey to monkey by filtrates of ground-up parotid glands. The virus can be grown in tissue cultures, and attenuated vaccines have been prepared from these.

(2) Infective or epidemic hepatitis

" Catarrhal jaundice " was the name formerly applied to a relatively mild febrile disease accompanied by jaundice, which occurs in epidemics or sporadically, chiefly in children and young people. Recovery occurs in most cases ; and the disease was called " catarrhal " because the jaundice was believed to be due to a simple catarrhal inflammation of the duodenum which spread to the bile duct and caused its temporary obstruction by mucus. This is now known to be erroneous ; study of the liver in the occasional fatal cases and of aspiration biopsy material from living patients shows an acute hepatitis which is often much more severe than the mildness of the symptoms would suggest. The liver cells in the central zone of each lobule show swelling and necrosis, the bile canaliculi are distended by plugs of inspissated bile, and the portal tracts are infiltrated with lymphocytes and monocytes. Prompt recovery, the usual termination of the disease, is accompanied by mitotic proliferation of the healthy liver cells to replace those lost, with complete restoration of normal lobular structure. When recovery is slower, however, there is chronic inflammatory fibrosis of the portal tracts, leading to some degree of fibrosis of the liver. In severe cases the initial necrosis may be very extensive and the patient may die with his liver showing the typical picture of "acute yellow atrophy". Infective hepatitis is caused by a virus which, however, has not yet been isolated.

"Serum jaundice" and "syringe jaundice"

A disease clinically and pathologically similar to epidemic infective jaundice has occurred in many persons 60 to 120 days after injections of human serum for various reasons, e.g. yellow fever vaccine containing plasma, measles or mumps convalescent serum, transfused blood and pooled plasma or serum. It has also followed injections of drugs and has been traced here to infection from imperfectly sterilized syringes or needles. It is possible that the agent causing "serum" and "syringe" hepatitis is distinct from that of epidemic infective jaundice, but further investigation is needed.

(3) Glandular fever or infective mononucleosis

This disease is characterized by fever, generalized enlargement of lymph glands and spleen, pronounced lymphocytosis, and usually complete recovery. It is probably due to a virus, but this is not certain. A peculiar and unexplained feature of the disease is that the patient's serum acquires heterophile antibodies that agglutinate sheep red blood corpuscles, a reaction of diagnostic value (Paul-Bunnell test).

(4) Cytomegalic inclusion disease

This disease is seen mainly in children, especially infants dying soon after birth with anaemia and hepato-splenomegaly, or in older children or occasionally adults with other chronic diseases. Intranuclear and cytoplasmic inclusion bodies occur in greatly enlarged epithelial cells in many organs, especially the salivary glands, for which the cytomegalovirus has a special affinity.

(5) Sundry enterovirus infections

Besides the infections of the respiratory and nervous systems already mentioned, other illnesses that are produced by Coxsackie viruses include: epidemic myalgia or Bornholm disease; myocarditis and pericarditis, mainly in young children; and herpangina, an acute but benign febrile infection of children, with vesicular lesions of the tonsils and pharynx. ECHO viruses are plentiful, and possibly causative, in some cases of diarrhoea in infants and children.

THE RICKETTSIAL AND ALLIED DISEASES

Rickettsia is the generic name applied to a group of related, tiny bacillus-like organisms usually less than 1μ long, which may be looked upon either as minute bacteria or as large viruses. The type species is *Rickettsia Prowazeki* of classical typhus fever, so named in honour of Ricketts and Prowazek, both of whom died of typhus while investigating it. The Rickettsiae live in the alimentary tracts of blood-sucking arthropods (lice, fleas, mites, ticks and bugs), from which the pathogenic species infect man or other mammals. Unlike true viruses, they are susceptible to antibiotic agents; but, like viruses, they cannot be cultivated on ordinary media but multiply only in suitable host cells, tissue cultures or the chick chorio-allantois.

(1) Classical epidemic typhus fever

Classical typhus fever of Europe and Asia is due to *Rickettsia Prowazeki*, which is transmitted by the common louse (*Pediculus*). When a louse feeds on the blood of a typhus patient, the Rickettsiae multiply in its intestine, are discharged in its excreta, and so infect a fresh human host to whom it may be transferred.

(a) The clinical characters of typhus

These are a high fever and other signs of severe toxaemia, followed about the fourth day by a characteristic spotted skin rash. In severe and fatal cases the rash may become haemorrhagic, and delirium or coma supervene.

(b) The lesions

Those found in the human body are not very conspicuous. They consist of endothelial swelling and thromboses in small blood vessels in the skin, brain and other tissues, accompanied by perivascular collections of monocytes and lymphocytes. Rickettsiae are present in the swollen vascular endothelium and in the blood.

(c) Diagnosis

Of great value in diagnosis is the *Weil-Felix agglutination test*. It was found that certain strains of *Proteus vulgaris* (designated X19) were agglutinated by the sera of typhus fever patients, and that this did not happen in any other fevers, except sometimes in *Brucella* infections. The test, which is carried out exactly like the Widal test, is a purely empirical one, for *Pr. vulgaris* has nothing whatever to do with typhus fever.

(2) Murine endemic typhus

This disease is much less serious than classical typhus and due to a distinct organism, *R. Mooseri*, which infects rats and is transmitted to man by the bites of rat fleas. It is endemic in many localities, where it has been given local names, such as Mexican, Toulon, Moscow, Manchurian typhus, etc. The Weil-Felix test is positive.

(3) Scrub typhus

Scrub typhus, also known as *tsutsugamushi* or *Japanese river fever*, was of special importance to our troops in Malaya, the East Indies and New Guinea during the second world war. It is due to a distinct Rickettsia, designed *R. orientalis*, which infects field mice and rats and is transmitted to man by the bite of a jungle mite (*Trombicula*). The disease is characterized by fever, an inflammatory lesion at the site of the bite, enlargement of regional lymph glands, a rash, and a high mortality. The Weil-Felix test with *Pr. vulgaris* X19 is negative but it is positive with strain XK of *Pr. mirabilis*.

(4) Other Rickettsial diseases

These include Rocky Mountain spotted fever due to *R. Rickettsi*, trench fever due to *R. pediculi*, Q fever due to *R. Burneti* (first discovered in Queensland, Australia), and some others. Q fever infection is endemic in cattle and sheep in many parts of the world, and man becomes infected by inhaling contaminated dust, often with resulting pneumonitis, or by drinking contaminated milk. Endocarditis is an occasional complication.

(5) Allied diseases

The micro-organisms of psittacosis, lymphogranuloma venereum and tranchoma differ from the true viruses in a number of ways; they are conveniently dealt with here along with the Rickettsiae, which they closely resemble.

(*a*) *Psittacosis* (from L. *psittacus* = a parrot) is a prevalent infection, often without serious illness, in parrots, parakeets, budgerigars and canaries. These excrete the organism in their droppings, which then dry and become scattered as dust, and man is infected by inhalation. The disease occurs in bird-fanciers, dealers and laboratory workers as a subacute or chronic febrile illness, often accompanied by patchy pneumonia. Most patients recover, but some die. Diagnosis is by intraperitoneal inoculation of mice with the sputum, and subsequent demonstration of the easily visible organisms (300 mμ) in the spleens and livers.

(*b*) *Lymphogranuloma venereum* is a venereal disease, in which an insignificant sore on the genitals is followed in males by swelling of the inguinal lymph glands, which suppurate and form chronic discharging sinuses. In females the infection often spreads also to the peri-anal and peri-rectal tissues, where it may cause stricture. The organisms measure about 150mμ, and can be identified in both human and animal lesions. Diagnosis is aided by the Frei test; an intradermal injection of a sterilized preparation from a lymphogranuloma lesion produces, in an infected patient, an inflammatory focus which persists for a week or longer.

(*c*) *Trachoma* is a chronic granular conjunctivitis due to a micro-organism measuring 200mμ and demonstrable in the discharge. The conjunctival epithelium shows characteristic inclusions.

A NOTE ON MYCOPLASMAL PNEUMONIA

Mycoplasma is a genus of very small pleomorphic bacteria-like organisms of minute coccal and filamentous forms, measuring up to 300 mμ in diameter. They are of widespread occurrence, and have been postulated as possible causes of various human diseases of obscure aetiology. But the only substantiated instance is that of *Mycoplasma pneumoniae* (Eaton agent), which causes " primary atypical pneumonia ". In culture this organism grows slowly to produce tiny colonies on special agar media. It is insensitive to sulphonamides and penicillin, but sensitive to tetracycline, chloramphenicol and neomycin.

SUPPLEMENTARY READING

Ash, J. E., and Spitz, S. (1945). *Pathology of Tropical Diseases: An Atlas.* Philadelphia and London; W. B. Saunders. (Many illustrations of the virus and Rickettsial diseases.)

Brit. Med. Bull. (1959). " Current virus research ". **15.** (Contains 15 excellent review articles.)

Burnet, F. M. (1945). *Virus as Organism.* Harvard; Harvard University Press. London; Oxford University Press. (An informative and unusual series of lectures which all students should read.)

— (1953). " Viruses and Man ". (A " Pelican " book). London. (An excellent account which all students should read.)

Israel, M. (1966). " Viruses ". *Brit. Med. J.,* **2,** 687. (Interesting notes on structure and general properties).

Rhodes, A. J. (1960). " Recent advances in virus infections: the new era in virology ". *Brit. Med. J.,* **1,** 1071.

Smith, K. M. (1942). *Beyond the Microscope.* (A " Pelican " book.) London. (An interesting semi-popular account of viruses, including those of plants.)

Tyrrell, D. A. J. (1965). *Common Colds and Related Diseases.* London; Edw. Arnold.

CHAPTER 21

IMMUNITY

IN PREVIOUS chapters we have already glanced at particular facets of the large subject of immunity—e.g. antitoxin formation in diphtheria, tetanus and scarlet fever ; the agglutination of typhoid and allied bacilli, of the Brucella group of bacilli, and of the spirochaetes of Weil's disease ; the relationship of allergy and immunity in tuberculosis ; the Wassermann test ; and the artificial production of immunity against typhoid fever, smallpox, yellow fever, rabies and other infections by vaccination. In this chapter we have to gather together these and other immunity phenomena and consider the general principles which embrace them.

DEFINITION OF IMMUNITY

At the outset we must distinguish between two quite different kinds of immunity, namely *acquired immunity* and *innate or species immunity* ; immunology, the subject of this chapter, is concerned only with the former.

(1) Acquired immunity

Naturally acquired immunity is the state of increased resistance to infection by a particular pathogenic micro-organism, which the animal body develops as a result of an established infection, and which enables it to resist re-infection. The host animal is of a species which is naturally susceptible to invasion by the particular microbe. When infection occurs, however, the body reacts in such a way that the invasion is checked and overcome, and the enhanced powers of resisting the invader persist after recovery and prevent or minimize the chance of re-infection. This acquired immunity or enhanced resistance is specific for the particular pathogen that evoked it ; and it depends on chemical changes in the host-animal's tissues —changes which result in the presence of specific *antibodies* in solution in the blood plasma and other body fluids. Acquired immunity is often not absolute, but varies according to the degree of infection evoking it and the kind of pathogen concerned. All degrees of immunity are possible, between total susceptibility and total protection.

Immunity may be acquired *naturally*, i.e. as the result of a spontaneous infection, or *artificially*, i.e. as the result of deliberate injection of the particular toxin, bacterium or virus (usually killed or in a modified form) or of already prepared antibodies. Artificial immunity produced by introducing the infective agent itself (generally called " vaccination ") or modified toxin is an *active immunity*, comparable with that of the natural disease, since the treated subject elaborates his own antibodies. Immunity produced by introducing antibodies already prepared by another person or animal, e.g. anti-diphtherial horse-serum, is *passive immunity*, and is more evanescent than active immunity.

271

(2) Innate or species immunity

Innate (species or genetic) immunity means simply the *insusceptibility* of a particular species of animal (or plant) to infection by a particular microbe. For that species, the microbe is non-pathogenic. Thus, the microbes of typhoid fever, syphilis, malaria, measles and influenza are non-pathogenic for most of the lower animals ; while those of canine distemper, avian tuberculosis and rat leprosy are non-pathogenic for man. Innate immunity is not due to specific antibodies, and all we can say of it is that the particular species is by nature an unsuitable host for the particular microbe. No doubt, the factors which determine innate susceptibility or insusceptibility of a host to a particular microbe are fundamentally chemical, just as are those which determine the fertility of a soil for a particular kind of seed ; but we are still ignorant of just what those chemical factors are.

No more will be said here of innate immunity, with which immunology is not concerned. In the heading to this chapter and in what follows, the single word *immunity* is used to mean *acquired immunity* only.

ANTIGENS AND ANTIBODIES

(1) Definitions

An antigen is any substance which, when introduced parenterally into the tissues of an animal, stimulates it to produce specific antibodies. To ensure antibody formation, most antigens must be introduced *parenterally*—i.e. by subcutaneous, intravenous or intraperitoneal injection—because, since antigens are usually proteins, they are digested in the alimentary tract and do not enter the body tissues and fluids in an unaltered state. However, some antigens are effective when inhaled, ingested or absorbed through the skin.

An antibody is a substance which makes its appearance in the blood or other body fluids of an animal, in response to the stimulus provided by the parenteral introduction of an antigen into the tissues, and which reacts specifically with its antigen. Antibodies are named according to the effects observed when they are mixed with the corresponding antigens. Thus an antibody which causes precipitation of the antigen is a *precipitin*, one which agglutinates its antigen is an *agglutinin*, one which causes lysis of bacterial or other cells is a *lysin* (e.g. *bacteriolysin* if it destroys bacteria, *haemolysin* if it destroys red cells), one which facilitates phagocytosis of bacteria is an *opsonin* (Gk.=an appetizer), one which neutralizes a toxin is an *antitoxin*, and one which neutralizes a snake-venom is an *antivenin*.

(2) Nature of antigens

Bacteria and their toxins are the principal naturally acting antigens; and the whole mechanism of immunity has undoubtedly been evolved primarily as a reaction to invasion by pathogenic micro-organisms. However, other foreign cells or soluble proteins also act as antigens when introduced artificially into the body. Even the cells and proteins of another individual of the same species (except in closely in-bred, and therefore genetically homogeneous, animals) are to some extent " foreign " and can act as antigens. This is of importance in blood-transfusion and in tissue-grafting ; e.g. in blood-transfusion the donor's and recipient's red corpuscles must be compatible, and the practical testing for compatibility has led to the identification of a number of different " blood-groups ". A remarkable instance of incompatibility of the red cells of a child for its own mother (dependent on a special antigen called " Rh-factor ") is described on p. 401.

Bacteria or other foreign cells are chemically complex structures, many of the constituents of which are non-antigenic. The surface of a bacterium or any other organized cell is a mosaic of complex organic molecules, some protein, some polysaccharide and some lipoid in nature. Only those molecules which excite the formation of antibodies are antigens, and most of these are either proteins or complexes of proteins with carbohydrates. Some complex poly-saccharides which are a main constituent of many bacterial capsules are also antigenic. Some polysaccharide, lipoid and other substances, which alone are not antigenic, can influence the properties of true antigens when attached to them, and antibodies evoked by such aggregates will cause precipitation of the pure non-antigenic substance in a test-tube. Such substances are called partial antigens or *haptens*.

(3) Nature of antibodies

By fractional precipitation or electrophoresis of the proteins of the sera of immunized animals, it has been found that antibodies, of all kinds, are concentrated in the gamma-globulin fraction. *Indeed, antibodies ARE serum gamma-globulin molecules, which during their formation have been modified in such a way that they possess specific chemical groupings or side-chains which enable them to combine with the corresponding antigens.* Antibodies are not new substances elaborated by the tissues, but modified normal components of the body.

The modifications of the surface structure of the new globulin molecules, which makes them antibodies, may aptly be likened to the cutting of a key to fit a particular lock (Fig. 91). The new globulin molecules acquire a side-chain structure which enables them to combine or interlock with the foreign protein molecules, so that these become blanketed with the animal's own proteins and so rendered innocuous in various ways. If the foreign molecules are in solution, this blanketing results in their precipitation, and the modified globulins are called " precipitins ". If the dissolved protein is poisonous (e.g. a bacterial toxin or a snake venom), its precipitation by the blanketing globulin molecules makes these appear to be a specific antidote to the poison (" antitoxin " or " antivenin ") ; but this is an incidental rather than a designed result, since harmless proteins are treated in just the same way. If the foreign protein molecules are not in solution but in bacterial or other kinds of organized cells, the modified globulin molecules blanket the surfaces of the cells, making them sticky and causing them to clump together, i.e. the globulin molecules act as " agglutinins ". The modification effected in the bacterial surface is often such as to render the bacteria more easily engulfed and digested by phagocytes, i.e. the antibodies act as " opsonins ". Following blanketing of their surface by globulin antibodies, some bacteria disintegrate, so that we speak of the antibodies as " bacteriolysins ".

By chemical, electrophoretic and other methods, several different classes of immunoglobulins are distinguished. These are collectively denoted by the abbre-viation Ig, and the several classes are labelled IgG, IgA, IgM, etc. Antibodies of the IgG class, which are the most abundant, are those most concerned in reacting with soluble antigens such as toxins; while the IgM antibodies are the main ones in bringing about the lysis of bacteria in the presence of complement (see p. 279).

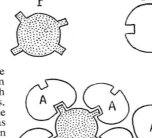

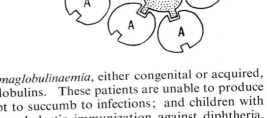

FIG. 91.—The "lock-key" simile applied to antibodies. P, foreign protein molecule (antigen) with characteristic chemical side-chains. A, antibody, i.e. globulin molecule specifically modified to fit side-chains of P. The blanketing of antigen molecules by antibody molecules is shown.

There occur rare cases of *agammaglobulinaemia*, either congenital or acquired, in which the patient lacks gamma-globulins. These patients are unable to produce any antibodies and are therefore apt to succumb to infections; and children with this condition fail to respond to prophylactic immunization against diphtheria, whooping cough, etc.

ANTIGEN-ANTIBODY REACTIONS

Let us now consider each of the following antigen-antibody reactions in turn :

Precipitin reactions
Toxin-antitoxin reactions
Agglutination
Opsonic effects
Bacteriolysis
Lysis of other foreign cells
Complement fixation tests
Anti-viral reactions

(1) Precipitin reactions

The simplest test-tube reaction which can be used in the study of antibody formation is that of precipitation. In the other reactions—agglutination, lysis, etc.—we are dealing with whole cells which may contain many different antigenic components ; but with precipitin reactions we can use a chemically pure single antigen in molecular or colloidal solution. If such a solution, e.g. of purified egg albumin, is injected into an animal, the animal's serum acquires the property of precipitating egg albumin when mixed with a solution of it in a test tube. The maximum precipitation (or flocculation, as it is often called) occurs when the antigen and antiserum are mixed in a certain optimum proportion and concentration and when electrolytes are present in a certain amount. The precipitate consists of antibody and antigen combined, the former contributing by far the greater part of it. Precipitation reactions can also be elegantly demonstrated by various gel-diffusion techniques, e.g. in Ouchterlony agar plates, in which antigen and suspected serum diffuse towards each other, and if the serum contains specific antibody zones of precipitation occur when they meet in the gel. (See also the Elek plate, Appendix D).

Like all antibodies, the precipitin acquired as the result of the artificial introduction of a foreign protein is *specific*, i.e. it will combine with and cause precipitation of that particular protein only. For example, the serum of a rabbit which has received injections of fowl egg albumin will cause precipitation of a solution of fowl egg albumin, but not of duck egg albumin; the serum of an animal treated with the haemoglobin of one species will not react with the haemoglobins of other species; and so on. Practical uses which have been made of this specificity of precipitin tests include the identification of blood stains as human in medico-legal cases, the detection of illegal adulterants in food, e.g. dog flesh in sausages, and the subdivision of *Strep. pyogenes* into the various Lancefield groups (p. 115).

Before leaving the subject of precipitin reactions, we may pause to speculate on what must happen to the microbic proteins which pass into solution in the body fluids of an animal during the natural course of an infective disease. We are justified in supposing that what happens in a test-tube happens also in the living body. The disintegration of dead bacteria *in vivo* must liberate many soluble antigens, and these must excite the production of specific antibodies which bring about precipitation of any further bacterial molecules of the same kind which may be liberated into the animal's tissues. It is probable that any precipitates so formed are, like bacteria themselves, taken up and digested by the fixed phagocytes of the reticulo-endothelial system or by wandering phagocytes in the tissues.

(2) Toxin-antitoxin reactions

Exotoxins constitute a special group of bacterial antigens which are freely liberated in soluble form from the bacteria producing them, both *in vivo* and in culture (p. 108), and which, during the course of the natural disease or as a result of artificial immunization, excite the formation of specifically neutralizing antitoxins. The toxin-antitoxin reaction is only a special instance of precipitin reactions in general; when toxin and antitoxin are mixed in a test-tube, specific precipitation occurs. However, the precipitating antibodies which render exotoxins harmless deserve their special name of "antitoxins", because the dangers of diseases like diphtheria and tetanus depend wholly on the bacterial exotoxins, and therefore recovery from, or immunization against, these diseases depends solely on the successful development or artificial introduction of sufficient exotoxin-neutralizing antibody.

Toxoid antigens.—If a solution of exotoxin, e.g. diphtheria toxin, is suitably treated with formaldehyde or other chemical agents, or by heat, it loses its toxicity but retains its antigenic properties, i.e. it can be used effectively for immunization. Such a detoxicated toxin is called a *toxoid* or *anatoxin*. Formalinized toxoids are extensively used in the immunization of horses for the production of antitoxic sera for treating diphtheria, tetanus and gas gangrene, and in the prophylactic immunization of man against diphtheria (p. 168) and tetanus.

Diagnostic tests.—Based on the specific neutralization of toxin by antitoxin are the *Dick test* for susceptibility to scarlet fever (p. 117) and the *Schick test* for susceptibility to diphtheria (p. 168).

(3) Agglutination

Any foreign cells—bacteria, fungi, foreign red corpuscles or tissue cells—injected into an animal's tissues, will stimulate the formation of specific agglutinins, and cells of the same kind will be clumped and precipitated when mixed in a test-tube with the antiserum so produced. It is, of course, with the agglutination of pathogenic bacteria that we are primarily concerned. This agglutination reaction differs from a precipitin reaction only in the larger size and greater chemical complexity of the particles of antigen. We are observing the behaviour not of protein molecules in solution, but of sensitized cells whose surfaces become blanketed by globulin antibodies. The surface of a bacterial cell usually contains multiple antigenic substances, and the injection of bacteria into an animal induces the formation of a corresponding number of distinct antibodies. However, we need not go deeply into the complex subject of the antigenic structure of bacteria : it must suffice here to note that multiple antibodies are often concerned in the blanketing of the bacterial surface and that the main visible result of this is agglutination. Agglutination of bacteria *in vitro* can be watched in process under the microscope ; but more often such tests are carried out in test-tubes, and the flocculation of the suspended bacteria by the antiserum is observed with the naked eye. Dead bacteria are agglutinated in the same way as living ones. Agglutination of living organisms does not kill them. Agglutination differs from simple precipitin reactions in that, whereas in the latter the bulk of the flocculus consists of globulin antibody, the bulk of the agglutinated flocculus consists of antigen, i.e. the bacterial cells.

(a) Flagellar, somatic and spore agglutinins

The multiplicity of antibodies evoked by bacteria, which has just been referred to, is well exemplified in the case of motile bacteria, e.g. typhoid and related bacilli. These have two distinct sets of antigenic constituents, one set in the flagella and the other in the body or somatic part of the bacteria. For reasons too technical for explanation here, the flagellar antigen is called *H-antigen* and the somatic is *O-antigen* (Fig. 92). When flagellate bacteria are introduced into the body, distinct H and O agglutinins are formed. Differential testing of a serum for H and O agglutinins can be carried out by varying the condition of the bacterial suspension used in the tests, e.g. by inactivating the H-antigen by alcohol or by heat, or by using a non-motile variant of the organism devoid of flagella. The distinction between H and O agglutination is of practical as well as theoretical interest ; the H and O results do not always run parallel, and it is now usual to carry out the Widal test with standard H and O bacterial suspensions. As a rule in enteric infections, both H and O agglutinins are developed, but in some cases only one of these is demonstrable, especially in early

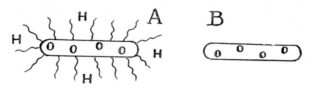

FIG. 92.—Flagellar and somatic antigens : A, a bacillus (e.g. *S. typhi*) with both H and O antigens ; B, a non-motile variant with only somatic (O) antigen.

stages of the disease. In prophylactic immunization also, it has been found that the O-antigens are of greater importance than the H-antigens. In certain virulent freshly isolated strains of typhoid bacilli, the O-antigens include a special component which is lacking in less virulent older cultures : this has been called the *Vi-antigen* (virulence antigen). It is probable that many other bacteria have similar " virulence " antigens ; the capsular polysaccharide-protein complexes of the pneumococcus are similar in nature. Spore antigens are present in the spores of bacteria of the genera *Bacillus* and *Clostridium*, and are distinct from the H and O antigens of the parent organisms.

(b) Group agglutinins

It sometimes happens that two or more related species of bacteria possess certain antigenic constituents in common, e.g. the typhoid, paratyphoid and food poisoning bacilli. Consequently, antisera prepared by using them as antigens will have common agglutinins as well as wholly specific ones ; and in doing agglutination tests with these bacteria we will get some overlapping or " group agglutination ".

(c) Applications of agglutination reactions

Agglutination tests are of practical value in the following three ways :

(i) *In clinical diagnosis.*—The serum from a patient suspected of a particular infection is tested to see if it will agglutinate suspensions of the appropriate bacterium. The classical example is the Widal test in the diagnosis of suspected enteric fever (p. 176); but similar tests are of use in food poisoning (p. 178), undulant fever (p. 187) and Weil's disease (p. 209). In Rickettsial diseases (p. 268) diagnosis by the agglutination of special strains of *Proteus* species depends on the presence of heterologous antibodies, i.e. antibodies which will react with an antigen biologically unrelated to the antigen which actually evoked them.

(ii) *In the identification of species of bacteria.*—An unidentified bacterium which has been isolated by culture is tested with appropriate type antisera. Its specific agglutination by a particular antiserum serves to identify it. Such tests are of value in identifying the various species of the typhoid-colon-dysentery group of bacilli, Brucella bacilli, cholera-like vibrios, etc.,

(iii) *In serological typing of bacteria.*—This is only an extension of (ii).   By the use of animal antisera prepared against particular strains of bacteria as antigens, many serological types can be distinguished within particular species of bacteria, e.g. *Salmonellae*, streptococci, pneumococci, meningococci.   While such typing is sometimes of value in identifying virulent strains and in the epidemiological study of certain infections, students need not venture far into this labyrinthine subject.

(Blood agglutination tests concerned in the naturally occurring blood-groups are not considered here, as they are not comparable with acquired immunity reactions.)

(4) Opsonic effects

(a) *Phagocytosis*

Many instances of the phagocytosis of microbic parasites have been mentioned in earlier chapters—phagocytosis of pyogenic cocci or other bacteria by poly-morphonuclear leucocytes in acute inflammatory lesions, of tubercle bacilli, leprosy bacilli or typhoid bacilli by mononuclear leucocytes and tissue histiocytes. These are but particular instances of a defensive reaction which is universal throughout the animal kingdom from hydra to man.   Every metazoal organism possesses wandering amoeboid cells which engulf foreign particles entering the body, bacteria included.

The mobile phagocytes of the human body are of two kinds, (*a*) the poly-morphonuclear leucocytes, called by Metchnikoff *microphages*, and (*b*) the large mononuclear cells or *macrophages*, which include the monocytes of the blood and similar wandering phagocytes (histiocytes) in the tissues.   In addition to these mobile scavengers, the body has large depots of fixed phagocytes, often spoken of collectively as (*c*) the "*reticulo-endothelial system*", which includes the sinus-endothelial cells of the lymph glands, spleen, liver (Kupffer's cells) and bone marrow.   Foreign particulate matter in the blood stream is removed principally by these fixed phagocytes ; this applies to non-living material such as carbon particles or dyes as well as to living bacteria, protozoa and viruses.

Phagocytosis is thus a non-specific process, which comes into play whenever particulate foreign matter is introduced into the tissues or fluids of the body. The phagocytosis of bacteria is not essentially different from that of inert particles ; and non-pathogenic or avirulent bacteria, injected into the tissues, body cavities or blood, are quickly removed by the mobile or fixed phagocytes.   In the phago-cytosis of pathogenic bacteria, however, specific antibodies, named " opsonins ", are concerned.

(b) *Opsonins*

Virulent bacteria often noticeably resist phagocytosis, both in the body and in test-tube experiments.   Thus, if on a warm microscope stage we mix together a suspension of virulent living pneumococci with living human or rabbit leucocytes, few or none of the cocci will be phagocytosed.   If to such a mixture we add some serum from a healthy person or from a patient with early pneumonia, there will be little or no improvement in phagocytosis of the cocci.   But if we add some serum from a patient with pneumonia who has just passed the crisis, the phagocytes will quickly engulf and digest most of the cocci.   This sudden improve-ment in phagocytosis is clearly due to some specific substances which appear in the patient's blood at or about the time of the crisis. Similar results are obtained experimentally : the sera of animals artificially inoculated with pneumo-cocci or other virulent bacteria contain specific phagocytosis-promoting sub-stances which the sera of untreated animals lack. These substances are called *immune opsonins* or *bacteriotropins*.

The opsonins act, not on the phagocytes, but on the bacteria ; they become attached firmly to the bacteria and cannot be removed by washing. If an opsonin-rich serum from an immunized animal is mixed with bacteria of the same kind as were used as the antigen, and the mixture is then centrifuged, all the opsonin is carried down in the bacterial deposit and the clear serum is now inactive.

Opsonins are almost certainly not a distinct class of antibodies, but are the same kind of modified globulin molecules which under other circumstances may cause agglutination or bacteriolysis. How they promote phagocytosis of the bacteria is not clear ; they may act by neutralizing particular antigenic components of the bacterial capsules which inhibit phagocytosis, or by damaging the capsules and liberating substances which are positively chemotactic for phagocytes, or they may act merely by blanketing the bacterial surface with a layer of innocuous native protein, thus converting the bacteria into so many chemically inert particles which are then engulfed like any other inert particles.

In testing the opsonic activity of sera, free leucocytes have usually been the phagocytes employed. But we know also that the phagocytosis of bacteria from the blood stream by the fixed reticulo-endothelial phagocytes of the body is promoted in the same way. The rate at which the blood is cleared of pathogenic bacteria injected intravenously is much greater in the specifically immunized than in the non-immunized animal ; and in the former, great numbers of phago-cytosed bacteria are to be found in the sinus-endothelial cells of the spleen, lym-phoid tissue and bone marrow, and in the Kupffer cells of the liver. In fact, the fixed reticulo-endothelial phagocytes of the body deal with virulent bacteria in the blood of an immunized animal in much the same way as they deal with non-pathogenic bacteria or inert particulate matter ; and undoubtedly this facilitation of phagocytosis is due to the presence of opsonic antibodies in the blood.

(5) Bacteriolysis

Besides agglutination and facilitated phagocytosis, there is yet a third effect which the serum of an immunized animal may have on the particular species of bacterium, namely direct lysis or destruction. We attribute this to specific *bacteriolysins* or *bactericidins*.

The lysis of bacteria by specifically immune sera differs from the other reactions so far considered (precipitation, detoxication, and agglutination) in that it requires the presence of a normal non-specific unstable component of fresh serum called *complement*. This is not a single substance, but a series of at least five interacting factors. Its existence is clear from the following facts. If an immune serum capable of lysing a particular species of bacterium is kept for several days, or is heated to 55° C., it becomes inactive and will no longer cause bacteriolysis. But, if a little fresh normal (non-lytic) serum is now added to it, this restores its specific lytic power. Hence, the specific antibody was not destroyed by the keeping of the heating; what was destroyed was some other component which is essential to the lytic activity of the antibody, and which is present in all fresh normal sera.   This is complementary to the antibody—hence " complement ".

In bacteriolysis, then, the sequence of events is this : the antibody combines firmly with the bacterial antigen, the complement then unites with the antibody-coated bacterium ; and lysis of the bacterium ensues.   The antibody may be regarded as sensitizing the bacteria to the action of complement, the latter being the essential lytic agent.   Ehrlich, in his original " side-chain " theory of immunity called the antibody " amboceptor ", because he supposed it to act merely as a link between the antigen and complement.   Although this is an over-simplified picture of what happens, it is substantially true.

In addition to specific immune bacteriolysins, normal serum contains a globulin-like substance called *properdin* which lyses many species of bacteria.   It combines with them and sensitizes them to the lytic action of complement.   Some bacteria, notably staphylococci and streptococci, resist such lysis.

(6) Lysis of foreign red corpuscles or other cells

Just as bacteria introduced into the tissues artificially or by disease evoke the formation of specific bacteriolysins, so also the artificial introduction of any other kind of foreign cells leads to the formation of specific *cytolysins*.   In most of the investigations in this field, the foreign cells used as antigens have been red blood corpuscles of other species.   These are especially useful in such studies because they are cells of a uniform size and structure, already freely suspended in fluid, and their lysis—*haemolysis*—is easy to observe because it liberates their coloured haemoglobin which dissolves in and tints the suspension fluid.

If the blood of a sheep is injected into a rabbit, in a week's time the rabbit's serum has acquired the property of lysing sheep's corpuscles, i.e. it contains a specific haemolysin.   This will cause lysis of sheep's cells only if complement also is present.   If the antiserum from the rabbit is heated to 55° C., its comple-ment will be destroyed, and sheep's cells mixed with it will not be lysed.   But they will be " sensitized " by having haemolysin attached to them, and they will dissolve as soon as some fresh complement is added to the mixture.   Hence, a mixture of inactivated (i.e. complement-free) haemolytic serum and red cells of the kind used as antigen is a useful reagent for testing for the presence of complement in any given fluid.   Such a mixture is called a *haemolytic system*, and is much used in complement fixation tests now to be described.

(7) Complement fixation tests

(a) The absorption or fixation of complement

As a result of artificial immunization or a natural infection, the patient's or animal's serum contains specific antibodies.   These may be bacteriolytic, so that a sample of the serum will lyse a suspension of bacteria of the species that caused the infection.   In the process of lysis complement will be used up, and its partial or complete disappearance can be demonstrated by subsequent testing of the solution with an artificial haemolytic system, such as the mixture of inactivated haemolytic rabbit serum and sheep's red cells described in the previous paragraph.   Using a haemolytic system as an indicator, we can thus ascertain whether complement has or has not been " fixed " (used up) as a result of the previous admixture of the patient's serum and bacteria of the species suspected of infecting him. If this serum contained specific bacteriolysins, complement will have been fixed, and the haemolytic system introduced subsequently will therefore show no haemolysis or diminished haemolysis. If the patient's serum contained no specific antibodies, complement will not have been absorbed when his serum was mixed with the suspected bacterial antigen, and the haemolytic system introduced subsequently will undergo normal haemolysis. Sometimes the antibodies in a specifically immune serum may not be demonstrably bacteriolytic, yet, along with antigen, they may absorb complement ; and the presence of such *complement-fixing antibodies*, tested for with a haemolytic system, is also good evidence of the identity of the bacterium causing the infection.

(b) Practical uses of complement fixation tests

Complement fixation tests of this kind, introduced by Bordet and Gengou, are not often used in the diagnosis of bacterial infections, because they are difficult to perform and simpler methods are usually available, such as direct demonstration of the causative bacterium or agglutination tests. However, complement fixation is sometimes of value in the diagnosis of chronic gonorrhoeal infections in which gonococci cannot be found. It is also of value in some metazoal parasitic diseases, e.g. hydatid disease, hydatid fluid with scolices being used as the antigen. The Wassermann test requires separate consideration.

(c) The Wassermann test

In devising his complement fixation test for the diagnosis of syphilis, Wassermann used as antigen a watery extract of syphilitic foetal liver, because this teems with spirochaetes. A mixture of this extract and a syphilitic patient's serum was found to absorb complement strongly. The test proved to be very reliable in the diagnosis of syphilis, and was naturally looked upon as being dependent on the presence in the serum of specific antibodies. But this view was soon upset by the discovery that alcoholic extracts of *normal* liver or heart muscle, or even mixtures of lecithin and cholesterol, could be used as antigens, with even better diagnostic results. Clearly then, the reaction is not dependent on any specific antibodies, but is due to some physical or chemical change in the syphilitic's serum which determines the absorption of complement by the extracts used as " antigens ". The active components of these extracts are lipoid substances.

That the Wassermann reaction is non-specific and may be unrelated to immunity accords with certain other facts about it. Clinically, it is a test of the activity and severity of the infection, not of effective immunization against it; the worse the disease, the stronger the Wassermann reaction. The occasional presence of a positive reaction in a variety of non-spirochaetal diseases (p. 206) is also significant ; evidently these conditions may sometimes bring about serum changes similar to those which much more regularly accompany active syphilis. It is indeed a remarkable coincidence that a test which was originally based on immunity theory, but which may have no relationship to immunity, should nevertheless be so reliable as a means of diagnosis of a particular disease.

(8) Anti-viral reactions

The sera of men or animals immunized naturally or artificially against viruses exhibit properties generally similar to antibacterial sera. With various anti-viral sera, specific agglutination, precipitation and complement fixation reactions have been observed. The inactivation of a virus by an immune serum has often been attributed to " viricidal " antibodies, but apparently complement is not necessary for this effect and its exact nature is still uncertain.

THE PRODUCTION AND FATE OF ANTIBODIES

(1) Site of formation of antibodies

Since the reticulo-endothelial tissues are concerned with the removal of bacteria and precipitated foreign protein from the blood or lymph, we might expect them also to be concerned directly or indirectly with the elaboration of antibodies. This expectation is confirmed by the following facts: in recently immunized animals antibodies are present in the spleen, bone marrow and lymph glands in higher concentrations than in the blood; lymph glands regional to the site of injection of an antigen are specially active in the production of antibody ; and tissue cultures or transplants of spleen, lymphoid tissue, bone marrow or fat from immunized animals may continue to produce antibodies.

As to the cells responsible for antibody formation, it is now well established that in lymph glands, spleen, bone marrow, and local inflammatory lesions, it is the mobile cells of reticulo-endothelial origin—lymphocytes, plasma cells and macrophages—that are mainly concerned. It is possible that the fixed cells of the reticulo-endothelial tissues may also participate, though much less actively. Immunofluorescence studies have proved clearly that soluble antigens are taken up by lymphocytes and plasma cells, and that plasma cells are particularly active in producing antibodies. Animals undergoing active immunization often show many plasma cells in spleen, lymphoid and other tissues, and the amount of antibody produced is roughly proportional to the plasma-cell reaction. Plasma cells are often conspicuous in chronic inflammatory lesions in which immune reactions are in progress, e.g. those of syphilis, rheumatoid arthritis, and rheumatic fever.

Experimental work shows that in rodents the thymus is an important source of immunologically active cells, and that migrant thymic lymphocytes form part of the population of the lymphoid tissues, spleen and bone marrow. The role of the thymus in human immunology is much less clear; this organ is relatively much smaller in man than in rodents, especially in the adult. However, the occurrence of thymic hyperplasia or tumour in some " auto-immune " diseases is noteworthy (see p. 612).

(2) Rate, amount and duration of antibody production

(a) The response to a single dose of antigen

Following injection of an effective dose of antigen, there is a lag period of 4 or 5 days before antibody commences to appear in the blood. The concentration of the antibody then rises rapidly to a maximum, usually 10 to 20 days after the injection. It then falls, at first rapidly and then more slowly, to a more or less sustained level which may be maintained for months or years (Fig. 93A). There are good grounds for believing that this maintained level is due, not merely to persistence of the antibody already formed, but to continued production of fresh antibody by the tissues which form it.

(b) The response to repeated doses of antigen

If, after the concentration of antibody has declined to a nearly steady level, a second injection of antigen is given, the resulting antibody response is quicker, greater and more sustained than after the first injection (Fig. 93B). This enhanced effect is of great interest, because it shows that the antibody-forming cells were

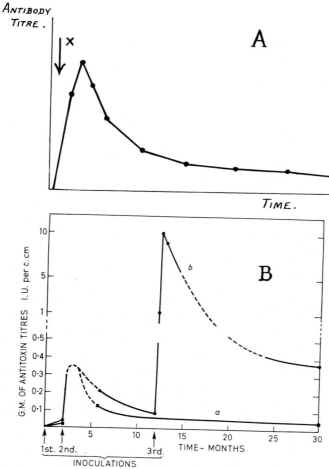

Fig. 93.—Antibody response (A) to a single injection X of antigen, and (B) to repeated injections of tetanus toxoid, (a) to two doses, and (b) to a third dose. (The second figure is by courtesy of D. G. Evans and the Editor of *The Lancet*.)

permanently modified by the first dose of antigen—modified in such a way that they now react more quickly and efficiently to subsequent doses. This fact is of great importance in the artificial immunization of animals and man. For example, in immunizing horses with diphtheria toxoid for the production of antitoxic serum, whereas after the initial injection the antitoxin in the animal's blood rises only slowly and to a relatively low level (less than 1 unit per ml.), on the second dose it rises steeply to 8 or 10 times that concentration, and by repeated doses it can be raised to 1,000 units per ml. or more. So also in immunizing children with diphtheria toxoid, the second and third injections are much more effective than the first in raising the antitoxin level and so establishing immunity. It must be added that the injection of a large dose of antigen into an animal that has already developed some antibody in response to earlier injections often causes a pronounced temporary fall in concentration of the antibody, followed later by a secondary rise. In therapeutic immunization it is important to adjust dosages so as to avoid producing such a " negative phase ".

(c) Antibody responses in natural infections

The examples given above describe the antibody responses to artificially administered antigens. However, the responses in natural infections are closely similar, e.g. the rise and fall of agglutinins in the blood during an attack of typhoid fever are much the same as in an animal given an injection of typhoid bacilli, and a typhoid patient who suffers from a relapse shows an additional step-up of agglutinin level resembling that following a second injection in an animal.

Patients who have developed an active immunity against any particular toxic agent do not possess complete protection against further exposure to it. Second attacks of enteric fever are by no means unknown, and enteric fever and diphtheria may occur in artificially immunized patients. Similarly, an animal which has been immunized against tetanus toxin can be killed if a sufficiently large dose of toxin is administered. An immune subject, however, is less susceptible to infection or intoxication than a non-immune person, so that in any particular population, the mortality and morbidity among the immune are lower than among the non-immune. Bacteriological immunity thus refers to a state of increased resistance to infection or intoxication rather than to complete insusceptibility; and should a bacterial disease occur in an immune individual, the natural course of the disease is likely to be modified so that it is less severe and does not terminate fatally.

Not all bacterial diseases of man produce a sound or lasting immunity. There is apparently little, if any, immunity acquired to gonorrhoea; and in syphilis the infected person is immune to re-infection only while he has the disease (infection-immunity). In neither of these diseases is prophylactic vaccination of any value.

In tetanus, some degree of immunity to tetanus neurotoxin is developed, but it is of low grade and short duration. Active artificial immunization with tetanus toxoid is highly successful, however, and provides a sound basal immunity against future intoxication. Apart from the fact that tetanus toxoid is a good antigen, the value of tetanus immunization lies partly in the incubation period of the disease, which, on the average, is about 10 days. Following infection of an immunized individual, the neurotoxin produced by the organism excites a secondary antibody response in the host. The titre of circulating antibodies commences to rise on about the third day, and by the tenth day (the end of the incubation period) a large amount of circulating antibody is present. In practice this secondary antibody response is ensured by administering a booster dose of tetanus toxoid. In contrast to tetanus, artificial immunization against the toxins of other Clostridia which cause wound infections (the gas-gangrene organisms) does not provide any measure of protection. This is due to the fact that the incubation period of gas gangrene is short (it varies from 8 hours to 5 days, depending on the infecting organism), so that the disease is usually fulminating and may terminate fatally before a secondary antibody response has occurred; immunization in itself is successful, but the nature of the disease does not allow this protective mechanism to become effective.

(3) The distribution of antibodies

Wherever they are formed, antibodies soon enter the blood and are distributed throughout the body. Their concentration remains highest in the serum, but they are present also in lymph, tissue fluid and inflammatory exudates. They are present in the milk of immunized mothers, and are of importance in conferring temporary passive immunity on the suckling infant. Antibodies are also taken up by various tissue cells, which, as a result, are rendered sensitive to the corresponding antigens. We must now consider the phenomena connected with this condition of cellular sensitization.

ANAPHYLAXIS AND ALLERGY

So far, we have considered antigen-antibody reactions as beneficial or protective in nature, the presence of antibody making the individual less susceptible to the harmful effects of subsequent doses of the antigen. We have now to discuss certain experimental and clinical phenomena in which the reverse obtains, the individual showing an abnormal hypersensitiveness to the effects of particular antigens. Although these phenomena are related to one another, it will be convenient to describe them under the following distinct heads :

Anaphylaxis

Allergy to non-bacterial proteins

Allergy in bacterial infections

(1) Anaphylaxis (Gk.= " unguardedness ")

If a normal guinea-pig is given 1 ml. of normal horse serum intravenously, it suffers no ill effects. But if a guinea-pig, which 2 weeks previously has been given a small dose (e.g. 0·1 ml.) of horse serum by any parenteral route, is now given 1 ml. of horse serum intravenously, in a few minutes it becomes restless, dyspnoeic, passes urine and faeces, and often dies. Death is due to acute spasm of non-striated muscle, including the bronchial muscles, with consequent asphyxia. By the same means, acute " anaphylactic shock " can be produced in other animals, but the particular effects differ from species to species. For example, in the dog the cause of death from acute anaphylaxis is not asphyxia but extreme congestion of the liver. Anaphylaxis is specific, i.e. it can be evoked only by the particular antigen which was used to sensitize the animal.

Now the symptoms of histamine poisoning in different species of animals are closely similar to those of acute anaphylaxis. Thus the guinea-pig's bronchial muscle is peculiarly susceptible to the action of histamine, a fatal dose of which kills the animal by asphyxia from bronchiolar spasm ; while in the dog the muscle in the walls of the hepatic veins is very sensitive to histamine, small doses of which produce obstructive congestion of the liver. We believe therefore that acute anaphylaxis is due to the sudden liberation of histamine or histamine-like substances from the tissues as a result of the "shocking" dose of antigen.

Two discoveries give the clue to the nature of anaphylaxis. The first is the discovery that the anaphylactic state can be passively transferred by injecting serum from a sensitized animal into a normal one. Thus, if a guinea-pig is given an injection of foreign protein and after it has developed specific precipitins some of its blood is transferred to an untreated guinea-pig, then, if after 2 or 3 days this second animal is given a sufficient intravenous dose of the antigen, it dies of acute anaphylaxis. Suitable tests prove that the responsible substance which has been transferred is the specific precipitin. The second important discovery is that the tissues of a sensitized animal are themselves specifically sensitized to the antigen. Thus if a strip of involuntary muscular tissue, e.g. from the uterus or intestine, of an actively or passively sensitized guinea-pig is suspended in a bath of Ringer's solution, the addition of a small amount of the specific antigen to the fluid results in strong contraction of the muscle, a result which is not obtained with the muscle of a normal animal or of one sensitized to some other antigen. Thus we have proof that antibody is fixed in the tissues and that it reacts there with added antigen.

From these and other experiments, we conclude that anaphylaxis is the result of an antigen-antibody reaction taking place on or in cells which have taken up (i.e. "fixed") antibody from the blood, that this intracellular reaction injures the cells and liberates histamine or some similar substance, and that the main symptoms of anaphylactic shock are due to the secondary effects of this substance on histamine-sensitive cells throughout the body.

The difference between an immune animal and a sensitized one simply depends on the relative amounts of circulating and fixed antibody. In an animal with abundant antibody in the circulating blood, intravenously injected antigen reacts harmlessly with this in the blood stream and does not reach the tissues ; but if the circulating antibody is insufficient to react with the whole of a large intravenous dose of antigen, the surplus of this reaches the tissues and reacts harmfully with fixed antibody in the tissue cells. Anaphylactic shock is, then, the result of a very unnatural procedure, namely the sudden introduction into the circulation of large amounts of soluble antigen.

Although anaphylaxis is an artificial phenomenon, it has occurred occasionally in man, e.g. following an intravenous injection of antitetanic or other immune serum given to a person who had previously had serum for some other reason. The sensitization in such cases is to the proteins of the horse serum. Subcutaneous or intramuscular injections of serum do not produce severe anaphylactic shock such as follows intravenous administration ; but in a sensitized person or animal they may cause local oedema and inflammation, skin rashes and fever. Such a local tissue reaction to an injection of antigen into sensitized tissues is called the "Arthus phenomenon".

Serum sickness also is essentially an anaphylactic reaction. This is characterized by skin rashes or urticaria, swelling of glands and joints, and fever, coming on 8 to 10 days after a single injection of antitoxic serum in a person with no previous sensitization. The foreign serum protein is eliminated only slowly and some of it is still present in the body after antibodies to it have been formed. The symptoms are produced by reaction between these antibodies and the residual protein antigen.

(2) Allergy to non-bacterial proteins

A naturally developing hypersensitiveness to particular foreign proteins sometimes appears in man or other animals, as in hay fever, asthma and urticaria. This is distinguished from artificially produced anaphylaxis by the name *allergy* (Gk.=" other energy "). Allergy to non-bacterial foreign proteins (" atopy ") is sometimes distinguished from that which develops to bacterial proteins during infective diseases, but the two are essentially the same.

Foreign proteins to which people may become sensitized include those of pollens, dandruff or hair of various domestic animals, feathers, various articles of food (shell fish, crayfish, strawberries, eggs), fungi and their spores, and non-pathogenic bacteria. How these antigens enter the body to sensitize it is not always clear; many of them are inhaled and absorbed from the respiratory tract, others may be absorbed through the skin, and some may escape digestion in the alimentary canal and may be absorbed from here as effective antigens. In the sensitized person antibodies to a particular substance have been elaborated and have become fixed in the tissues, sensitizing them to local applications of the antigen, renewed contact with which results in the liberation of histamine. In many cases, the symptoms are largely limited to the particular regions of application of the antigen, e.g. to the nasal mucosa and conjunctiva in hay fever due to pollens, or to the bronchial system in asthma due to inhaled pollens or animal dusts. But in these patients sensitization is body-wide and is not restricted to the local region of application of the antigen ; for the application of the particular antigen to a scratch in the skin of any part of the body will produce an urticarial wheal indicating the local liberation of histamine. This is made use of in identifying the antigen responsible for asthma, hay fever or other allergic conditions in particular cases : to a series of scratches on the arm a series of protein extracts (from pollens, animal dusts, etc.) is applied, and the skin reactions evoked indicate the substances to which the patient is sensitive.

In many cases of allergic sensitivity, eosinophil leucocytes collect in the reacting tissues and there may also be pronounced eosinophilia. For example, during an acute attack of asthma great numbers of eosinophils are present in the bronchial walls and sputum and the eosinophil count in the blood may rise from the normal 1 or 2 per cent of the total leucocyte count to 10, 20 or even 50 per cent. The main function of eosinophil leucocytes is to counteract the effects of histamine.

In Chapter 19 several references were made to allergic phenomena due to infestation by metazoal parasites, e.g. urticarial rashes accompanying the migrations of larval ascarids or schistosomes in the body, the Casoni test for skin sensitivity in hydatid disease, the Bachman test for sensitivity in trichiniasis, and the eosinophilia accompanying many parasitic infestations. We are now in a better position to understand these effects ; they are allergic reactions to foreign metazoal proteins to which the patient's tissues have become sensitized by the fixation of antibodies which he has developed.

(3) Allergy in bacterial infections

Since hypersensitiveness develops as a result of contact with purely external proteins, we would expect it to develop also as the result of absorption of bacterial proteins during the course of natural infections. The classical and most fully studied example of this is the tuberculin reaction in tuberculosis (see p. 153). This reaction shows that the tissues of the tuberculous subject are sensitized to tuberculo-protein. There is little doubt that similar sensitization to microbic proteins occurs in many other infections, especially subacute and chronic ones. Such sensitization, far from being an undesirable by-product of immunization, probably has protective value, since it determines prompt vigorous inflammatory reaction in tissues exposed afresh to invasion by the particular bacterium or virus. In Chapter 3 (p. 45), we saw that the immediate excitants of the vascular and exudative changes of inflammation are probably products of the injured tissues themselves, including histamine or histamine-like substances. In an animal already immunized against a particular infection, the specific allergic state of the tissues serves to enhance their inflammatory reaction to local re-infection. It is probable, then, that, in all inflammatory reactions caused by bacteria or viruses which the individual has previously encountered and reacted to as antigens, there is an allergic ingredient. But we must be careful not to over-emphasize the importance of this ingredient. Vigorous effective inflammation is excited by microbic infections of kinds to which the individual has not previously been subjected. The capacity of tissues to react to injury by inflammation is a primary and general one, and is not dependent on the previous immunological state of the body. The inflammatory reaction of the immunized animal to a reinfection is not different in kind, but only in intensity and effectiveness, from that of a non-immunized animal.

(4) Cell-mediated allergy—delayed hypersensitivity

We have spoken of the tuberculin reaction as a classical example of " delayed hypersensitivity "—" delayed " in that, unlike the immediate hypersensitivity reactions of asthmatic or hay-fever subjects to the responsible allergens, the reaction does not begin for several hours and does not reach its peak until one or two days later. Also unlike immediate hypersensitivity, delayed hypersensitivity cannot be transferred to a healthy animal by the serum of a sensitized one, and does not depend on the presence of already formed serum antibodies that have become bound in the tissues. It can, however, be transferred by mononuclear cells, mainly lymphocytes, of a sensitized animal. These sensitized cells are presumably carrying bound antibodies or their precursors or may be in the process of producing antibodies. The delay in the appearance of the tissue reaction following introduction of a dose of antigen into the tissues of a sensitized animal is attributable to the time needed for the sensitized lymphocytes and monocytes to collect at the injected site. Cell-mediated delayed reactions not only occur commonly in chronic bacterial diseases, but are mainly responsible also for the rejection of tissue grafts from one animal to another (the homograft reaction), and for the lesions in the organs of animals injected with homologous tissues, as noted in the next section.

The three types of immunological responses—the free production of circulating protective antibodies, immediate hypersensitivity due to the binding of antibodies in the tissues, and delayed hypersensitivity mediated by sensitized lymphocytes and monocytes—should not, however, be thought of as separate processes. They may occur simultaneously or successively in one person, and they are only different aspects or accompaniments of the one basic phenomenon—the response of immunologically competent cells, mainly of the lymphocyte-monocyte series, to the presence of foreign antigens.

AUTO-ALLERGY OR AUTO-IMMUNITY

Antibodies against particular tissues or organs can be obtained experimentally by giving an animal injections of that tissue from another animal of the same species, or in some instances its own tissue; and 2 to 4 weeks later it develops destructive and inflammatory changes in the organ corresponding to the tissue injected. The timing and specificity of this result show clearly that the damage to the organ arises from an allergic reaction in it, probably of the delayed hypersensitivity type mediated mainly by sensitized lymphocytes or other cells.

The discovery that in some human diseases the patients' sera often contain antibodies against particular tissues has led to the view that " auto-immunity " (better " auto-allergy ") may be a factor in the pathogenesis of these diseases. At first sight it seems contrary to nature that an individual's own tissues should ever act as antigens; but there are several ways in which this could happen. Thus, as a result of microbic, chemical or other injury to a tissue, some of its proteins could suffer denaturation or other changes, or could become conjugated with foreign haptens from the bacteria or from extrinsic substances such as drugs; such altered or conjugated proteins might thus be rendered " foreign " and antigenic.

The main diseases in which auto-antibodies are found are described in later chapters, but they may be briefly mentioned here:

Systemic lupus erythematosus Various auto-antibodies are often present, especially antinuclear factor directed against nucleoproteins in many different tissues.

Hashimoto's thyroiditis Antibodies against both thyroglobulin and thyroid cells are found in most cases. Similar antibodies are found in cases of primary myxoedema and in some cases of thyrotoxicosis.

Acquired haemolytic anaemia Auto-haemolysins are regularly present.

Pernicious anaemia Many cases have antibodies against the parietal cells of the gastric glands or gastric intrinsic factor or both.

Rheumatoid arthritis Many patients have an abnormal gammaglobulin, called " rheumatoid factor " and estimated by the Rose-Waaler test; and some workers believe this to be an auto-antibody evoked by altered tissue proteins or inflammatory products. In *Sjøgren's syndrome* many patients have rheumatoid and antinuclear factors.

An interesting feature of these " auto-immune " diseases is their overlap; a considerable number of patients have auto-antibodies of two or more distinct kinds, in some cases with clinical signs of both diseases; various combinations of systemic lupus erythematosus, thyroid diseases and pernicious or haemolytic anaemia occur. Such combinations of auto-allergies suggest that the patients have some underlying immunological anomaly, a suggestion supported by inherited or familial incidence in some cases.

Do the auto-antibodies in the aforenamed diseases play a part in their causation or progress, or are they only a result of tissue damage? It is known that, following myocardial infarction, auto-antibodies against heart tissues may develop; or following chemical injury of the liver, auto-antibodies against liver. These antibodies clearly play no part in causation of the lesions, and it is doubtful if they have any subsequent pathogenic effects. Indeed, on the contrary, it has been suggested that auto-antibodies arising secondarily to tissue damage may aid recovery by promoting the disposal of damaged proteins. However, in several of the auto-allergic diseases, although the primary cause of the tissue damage remains unknown, it is probable that the antibodies that are produced play an important part in continuing to damage the particular tissues; e.g. that in Hashimoto's disease thyroid tissue is being continuously injured by the plentiful auto-antibodies that are continuously produced, that the auto-haemolysins of haemolytic anaemia continuously destroy red cells, and that in systemic lupus erythematosus the nuclei of cells in many organs are being constantly injured by anti-nuclear factor. For many other diseases of obscure aetiology, for which auto-allergy has been suggested as playing a causative or perpetuating part—ulcerative colitis, non-tuberculous Addison's disease of the adrenals, cirrhosis of the liver, diabetes, myasthenia gravis, the several " collagen diseases " mentioned on p. 292, eczema and some other skin diseases, and some cases of male sterility—the evidence for this is less substantial and further work is necessary.

Finally, a misuse of the term " auto-immune " calls for comment. Some bacteria and some human tissues have antigens in common and therefore excite the formation of similar or identical antibodies. Thus, haemolytic streptococci of strains that cause rheumatic fever have some antigens in common with both cardiac muscle fibres and heart valves, and so evoke anti-cardiac antibodies indistinguishable from auto-antibodies against heart tissue; and these may well be involved in the pathogenesis of rheumatic myocarditis and endocarditis. But this is *not*, as some writers have suggested, an example of auto-immunity. The responsible antigens are bacterial, not human.

SUPPLEMENTARY READING

Asherson, G. L. (1967). " Autoimmune disease ". *Brit. Med. J.*, **2**, 417 and 479.

Coombs, R. R. A. (1968). " Immunopathology ". *Brit. Med. J.*, **1**, 597.

Glynn, L. E. and Holborow, E. J. (1965). *Autoimmunity and Disease*. Oxford; Blackwell.

Holborow, E. J. (1967). " An A.B.C. of modern immunology." *Lancet*, **1**, 833, 890, 942, 995, 1049 and 1098.

Ward, F. A. (1970). *A Primer of Immunology*. London; Butterworths. (An excellent outline).

Wilson, G. S., and Miles, A. A. (1964), Fifth edition (2 vols.). *Topley and Wilson's Principles of Bacteriology and Immunity*. London; Edw. Arnold.

CHAPTER 22

INFLAMMATORY DISEASES OF " TOXIC ", " ALLERGIC " OR UNKNOWN CAUSATION

INTRODUCTION

IN THE inflammatory diseases discussed in previous chapters, the responsible bacteria or viruses are known and are demonstrably present in the inflamed tissues. We have now to consider some important diseases in which, while there are good grounds for believing that bacterial infection is a factor in their causation, in some cases at least, the inflamed tissues are devoid of bacteria. The tissue injury in these diseases is believed to be due to the effects of bacterial toxins or other antigenic substances acting on tissues already sensitized to them. We will discuss the following " toxic " or " allergic " diseases:

> Rheumatic fever
> Rheumatoid arthritis
> Other " rheumatic " affections
> " Toxic " nephritis (Bright's disease)
> Generalized lupus erythematosus
> Erythema nodosum
> Polyarteritis nodosa

We will also consider briefly two other inflammatory diseases of unknown cause, namely:

> Ulcerative colitis
> Regional ileitis

(1) Concept of " toxic " inflammation

In the belief that the inflamed tissues in several of these diseases are injured by bacterial toxins coming from foci of infection elsewhere in the body, they have often been spoken of as " toxic " inflammations—" toxic " endocarditis and arthritis in rheumatic fever, " toxic " nephritis following scarlet fever or other infections, and so on. But the word " toxic ", though useful to express the absence of bacteria from the inflamed tissues, does not explain the pathogenesis, nor is it necessarily applicable to all cases of the diseases in question. Thus, while post-scarlatinal nephritis is clearly " toxic " in that it is due to injury sustained by the kidney in consequence of streptococcal toxaemia, the nature of the injury still calls for elucidation, and there are many cases of nephritis the causes of which are unknown. It is quite possible that these also are due to toxic injury from unrecognized bacterial infections, but we do not *know* that this is so. However, the fact that some cases of nephritis are certainly " toxic " justifies us in considering the disease as a whole here.

(2) Concept of " allergic " inflammation

For several of the diseases referred to, there is strong evidence that allergic sensitization of the tissues to bacterial or other antigenic substances plays an important part. Thus, this view looks upon post-scarlatinal nephritis as the result of sensitization of the kidneys to excreted scarlatinal toxin or other strepto-coccal products. Rheumatic fever is attributed to body-wide sensitization to certain streptococcal antigens, with consequent allergic inflammatory reaction of various tissues to further doses of these antigens coming from stretococcal foci in the throat or elsewhere. Because rheumatic fever, rheumatoid arthritis, lupus erythematosus, polyarteritis nodosa, and some other rare diseases (scleroderma, dermatomyositis, and Wegener's granulomatosis of the lungs) all show prominent lesions in connective tissue, including foci of " fibrinoid " degeneration or necrosis, they are sometimes spoken of collectively as " the collagen diseases "; but this is a vague term and of dubious value. We will discuss the " auto-allergic features " with the particular diseases.

RHEUMATIC FEVER

(1) Clinical features

(a) Incidence

Rheumatic fever is widespread in temperate climates, but rare in the tropics. It is more frequent in densely populated cities than in country districts, and among the poorer classes than among the well-to-do. Attacks of the disease occur much more frequently during the winter than the summer. It is rare in children under 5 years of age, but then appears with rising frequency until about the tenth year. After this age the incidence of first attacks declines, but recurrent attacks are common throughout childhood and adolescence. The incidence of the disease has declined during recent years, doubtless because of the decline of streptococcal infections.

(b) Symptoms

The disease varies greatly in its mode of onset and main symptoms. Some cases show an acute onset, often preceded or accompanied by acute tonsillitis, and associated with high fever, acutely painful multiple arthritis, and often with some evidence of endocarditis, pericarditis or pleurisy. In other cases, especially in children, the disease is insidious in onset, with vague pains, loss of weight, increasing pallor due to anaemia, recurrent febrile attacks of tonsillitis, and signs of progressive carditis with or without pericarditis. Few, many or all of these manifestations may occur together, with or without arthritis. Some cases show also the development of crops of painless firm subcutaneous rheumatic nodules in various parts of the body, especially over bony prominences ; these remain for variable periods and then usually disappear. Other cases show chorea (" St. Vitus's dance "), characterized by spasmodic grimacing, other involuntary muscular movements, incoordination of voluntary movements, and emotional instability. Arthritis seldom coexists with chorea ; carditis often does so.

(c) Course

Initial acute attacks usually subside in the course of a few weeks, leaving no disability of the joints but almost always causing some degree of permanent endocardial or myocardial damage. If this is not severe and if no further attacks occur, the child may grow up without any significant cardiac disability—a fortunate outcome which happens in about one-third of the cases. But in the remainder, especially those who suffer from recurrent attacks, more or less severe cardiac lesions persist. If these progress, cardiac failure ensues sooner or later. In childhood, death from rheumatic fever usually follows upon a recrudescence of the disease, often with pericarditis. Death from slowly progressive heart failure due to valvular fibrosis is usually delayed until young adult life.

(2) Pathological anatomy

(a) Lesions in the heart

The most important rheumatic lesions are those in the heart. They affect the endocardium, myocardium and pericardium with about equal frequency and in many cases simultaneously. The changes seen at necropsy are usually a mixture of recent acute and older fibrous lesions, for it is unusual for a patient to die during an initial attack.

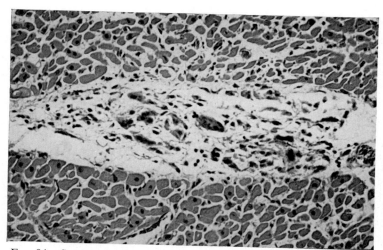

FIG. 94.—Small Aschoff focus in myocardium. (By courtesy of the late Prof. D. H. Collins.) (× 220.)

(i) *Endocarditis.*—Endocarditis is of the simple verrucose type, showing rows of small warty vegetations along the lines of closure of the valve cusps, already described on p. 79 (Fig. 32). In addition to these recent vegetations, the cusps and chordae are often visibly thickened, contracted and rigid, from earlier attacks of the disease. The mitral and aortic valves are affected more frequently and more severely than the tricuspid and pulmonary valves. Inconspicuous mural endocarditis, especially of the posterior wall of the left atrium, usually accompanies the valvular lesions. The left auricular appendage, removed during the operation of mitral valvotomy in cases of mitral stenosis, often shows active endocardial lesions, although most of the patients show no clinical signs of rheumatic fever at the time.

(ii) *Myocarditis.*—The most characteristic lesions are *Aschoff nodules.* These are microscopic or just visible, ill-defined, scattered inflammatory patches, consisting of collections of mononuclear histiocytes or macrophages, large irregular or multinucleated cells derived from these, lymphocytes and plasma cells—all grouped around small foci of degenerated collagen in the perivascular and interstitial connective tissues of the myocardium (Fig. 94). In places these focal lesions coalesce and become diffuse, especially in the more severe acute stages, in which they often show an admixture of polymorphonuclear leucocytes. Similar lesions, in continuity with those in the myocardium, are the basis of rheumatic endocarditis and pericarditis.

(iii) *Pericarditis.*—Rheumatic pericarditis is fibrinous or sero-fibrinous, never purulent (Fig. 22, p. 58). Following subsidence and recovery, patchy or extensive fibrous adhesions between the parietal and visceral layers of the pericardium are left ; sometimes also there are dense adhesions between the parietal pericardium and the sternum, ribs and diaphragm—an adhesive fibrous mediastinitis—with anchoring of the heart to the chest wall and serious mechanical embarrassment of its movements.

(b) Lesions in other tissues

Like the cardiac lesions, those in other tissues show Aschoff foci or similar but diffusely coalescent inflammatory changes. In *joints*, these are found in the synovial membrane and capsular fibrous tissue ; in the cavity turbid fibrinous exudate mingles with the synovial fluid ; but suppuration never occurs, and the arthritis resolves completely and leaves no demonstrable fibrosis. *Subcutaneous nodules* show structureless central necrosis surrounded by inflammatory cells and proliferating fibroblastic connective tissue. *Arteries*, including the aorta and other large arteries and also smaller visceral arteries, often contain Aschoff foci in their media and adventitia, and the aortic arch sometimes shows an endaortitis comparable with rheumatic endocarditis. *The serous membranes*, other than the pericardium, may show fibrinous exudate ; in the pleura this is sometimes sufficient to give clinical signs of pleurisy ; in the peritoneum it is usually slight and clinically insignificant. *The lungs* may contain scattered inflammatory foci in their interstitial tissues, and " rheumatic pneumonia " has been described. In cases of chorea, perivascular inflammatory lesions may be found in *the brain.*

It is clear, then, that in rheumatic fever there is widespread inflammation in many parts of the body, especially in vascular and connective tissues. This inflammation is of a rather distinctive kind ; it tends to occur in foci around patches of altered collagen, it evokes predominantly mononuclear or histiocytic cellular accumulations, and it never leads to suppuration but ends in partial resolution and fibrosis.

(3) Causation of rheumatic fever

(a) Streptococcal infection

It is over 60 years since streptococci were first suspected as the cause of rheumatic fever; but for many years conflicting views were held as to how they exerted their pathogenic effects, and only gradually did it become clear that the organisms are *not* present in rheumatic lesions in the heart, joints and other affected tissues. However, haemolytic streptococci are certainly the cause of the attacks of tonsillitis which usually precede attacks of rheumatic fever, and they frequently persist in the throats of rheumatic fever patients between attacks of the disease. It has also been shown that the blood of patients contains specific antibodies against haemolytic streptococci and their products—anti-haemolysin, agglutinin, precipitin, etc., that following rheumatic fever attacks the concentration of these antibodies rises slowly and steadily and persists much longer than after attacks of non-rheumatic tonsillitis, and that in rheumatic subjects skin-tests often show abnormal sensitivity to streptococcal antigens.

(b) *Allergy*

It is very probable, then, that rheumatic lesions are due to sensitization of the tissues to streptococcal proteins, followed by allergic reaction to fresh doses of these antigens coming from foci of coccal infection in the tonsils. In support of this are the following points: (i) that the slow persistent production of antibodies during the course of a rheumatic attack, as mentioned above, indicates a peculiar immunity response; (ii) that allergic reactions in animals following sensitization to foreign proteins may produce widespread focal lesions in connective tissues resembling those of rheumatic fever; and (iii) that allergic reactions in man, e.g. serum sickness, are sometimes accompanied by arthritis. The possible pathogenic effect of anti-cardiac antibodies evoked by particular streptococcal antigens, which these organisms have in common with heart tissues is noted on p. 290.

RHEUMATOID ARTHRITIS

(1) Clinical characters

Rheumatoid arthritis, also called "chronic infective arthritis", is a disease chiefly of adults between 20 and 40 years old, more often women than men. It usually commences insidiously, but in some cases its onset is acute with fever, leucocytosis and other signs of toxaemia. The small joints of the hands and feet are usually the first affected, but the disease commonly extends later to some or all of the larger joints of the limbs, especially the knees, ankles, elbows and wrists. The affected joints are painful and show pronounced peri-articular swelling, which is rendered more prominent by disuse atrophy of neighbouring muscles. Subcutaneous nodules, resembling those of rheumatic fever, are sometimes present also. The disease runs a chronic course, with exacerbations of the arthritis from time to time, often accompanied by intermittent fever, sweats, leucocytosis, and anaemia. Some patients recover, but many become progressively more and more crippled and eventually bed-ridden. Rheumatoid arthritis in children is usually acute in onset and often accompanied by enlargement of superficial lymph glands and of the spleen—a condition called *Still's disease*.

(2) Pathological anatomy

(a) *Affected joints*

The synovial membrane shows congestion, oedema and villous overgrowth ; microscopically it is infiltrated by chronic inflammatory cells, including macrophages, lymphocytes, plasma cells and proliferating fibroblasts, and shows many new-formed vessels as in granulation tissue. This proliferating chronically inflamed synovial tissue grows over the articular cartilages as a velvety vascular layer or *pannus*, which erodes the cartilage. It also creeps under the edges of the cartilage between it and the bone. The two layers of pannus on the opposing joint surfaces often coalesce, leading to the formation of fibrous adhesions across the joint cavity (fibrous ankylosis); and eventually this may become a bony ankylosis. The fibrous articular capsule and peri-articular tissues share in the inflammation and oedema, especially during exacerbations of the arthritis.

(b) Other tissues

Subcutaneous nodules, when present, have a structure resembling that of rheumatic nodules, and somewhat similar chronic inflammatory lesions are sometimes present in muscles, nerve sheaths, fasciae or lungs. In some cases, necropsy discloses old chronic endocarditis, pericardial adhesions or adhesive mediastinitis.

(3) Causation

The severe active inflammatory character of the joint lesions, the fever, leucocytosis and other signs of toxaemia which often accompany the disease, and the lymph-glandular and splenic enlargement in Still's disease, strongly suggest that it is of infective nature. Some patients have infective foci, usually streptococcal, in tonsils, antra or teeth, and their sera often contain anti-streptococcal agglutinins and precipitins; but such infection is much less regularly found than in rheumatic fever. Bacterial cultures from the joints are usually sterile; but mycoplasmas have been isolated in some cases, a finding of doubtful significance.

There is a possibility—evident from what has been said above—that rheumatic fever and rheumatoid arthritis may be related diseases, the one characterized mainly by progressive structural damage of the heart but only evanescent non-destructive arthritis, the other by progressive destructive arthritis and only occasionally significant cardiac lesions. The difference between the two diseases may depend partly on their different age incidence, or on nutritional or endocrine factors which many clinicians believe to be important in the genesis of rheumatoid arthritis. The frequent presence of the abnormal serum globulin known as " rheumatoid factor " has led to the suggestion that auto-allergy may play a part in the pathogenesis of the disease (see p. 289). But this also is uncertain; and we must conclude that the essential cause or causes of rheumatoid arthritis remain unknown.

(4) Distinction between rheumatoid arthritis and osteoarthritis

The crippling effects of old quiescent rheumatoid arthritis may resemble those of osteoarthritis, but the two diseases have nothing else in common. Osteoarthritis is commoner in men than women, usually occurs after the age of 40, is gradual in onset and progress, is unaccompanied by fever or other signs of infection, its joint lesions are degenerative rather than inflammatory, and it usually affects few joints, or one only, and does not extend to others. The large joints, especially the hip and knee, are those most commonly affected. When the fingers are affected, the disease characteristically produces painless bony knobs on the phalanges at the terminal interphalangeal joints—a common condition called *Heberden's nodes*.

The larger joints show degeneration and splitting of the articular cartilages, consequent exposure of the bone ends which show ivory-like condensation and grooving from mutual friction, a condition called *eburnation* (L., *ebur*=ivory), atrophy and absorption of the underlying bone with consequent shortening, prolific cartilaginous and bony outgrowths called *chondrophytes* and *osteophytes* at the margins of the degenerated articular surfaces, and detachment of pieces of cartilage or hypertrophic synovial fringes to form loose bodies in the joint. The joint may eventually become grossly disorganized and may suffer dislocation ; ankylosis does not occur. In its commonest situation in the hip, the disease is called *morbus coxae senilis.*

Spinal osteoarthritis is common in old people, and is characterized by marginal osteophytes or " lipping " of the edges of the articular surfaces of the vertebral bodies, and sometimes by bony ankylosis by fusion of these lips or by ossification of the surrounding ligaments. It is to be distinguished from the *ankylosing spondylitis* (Marie-Strumpell disease) which occurs in young people, usually men, progresses more quickly and painfully than spinal osteoarthritis in the elderly, is associated with mild fever, is accompanied in a proportion of the cases by rheumatoid arthritis of peripheral joints, and regularly causes bony ankylosis of the spine.

The ordinary osteoarthritis of the elderly is clearly a non-infective degenerative process. It is probably initiated and maintained by mechanical injury of the joints. Severe osteoarthritis sometimes results from improperly treated fracture-dislocations or from less serious but neglected injuries such as fractured patella, detached meniscus, or sprains or minor fractures in the neighbourhood of joints. Once a mechanical disability, even trifling, of a joint is incurred, a vicious circle may readily be established, weight-bearing and movement of the defective joint constantly aggravating and extending the injury. A Charcot's joint in tabes dorsalis or other anaesthesia-producing disease (p. 204) is the extreme instance of chronic osteoarthritis resulting from repeated and unrecognized mechanical insults. In its nature and origin osteoarthritis is therefore sharply distinct from rheumatoid arthritis ; it has been considered here only in order to make this distinction clear.

OTHER " RHEUMATIC " DISEASES

The term " rheumatism " is used so vaguely and in reference to so many painful complaints that it is devoid of any useful meaning. Besides rheumatic fever and rheumatoid arthritis, the name popularly embraces a variety of painful conditions which have been called *fibrositis, myositis, neuritis, neuralgia, sciatica* and *lumbago.* However, the nature and causes of these complaints are diverse and in many cases obscure, so that we cannot describe any unified pathology for them.

Fibrositis and myositis are believed to be the pathological basis of the common " muscular rheumatism " with which most adults in cold or temperate climates are familiar. In a few cases this complaint appears to be related to infective foci, since it has cleared up after removal of infected teeth or tonsils. In other cases its onset appears to be related to sudden unusual movements, and it has been attributed to tearing of fascial or muscular tissue. In most cases, however, no definite causes of " muscular rheumatism " are recognizable, though it certainly varies with the season and the state of the weather. Focal inflammatory lesions in muscles, fasciae and nerve sheaths have been described as being present in cases of "muscular rheumatism"; but opportunities for seeing these are rare.

"*Neuritis*", "*neuralgia*" *and* "*sciatica*" are names applied to similar complaints when the pain has the distribution of a particular nerve. In many cases such pains are evanescent and are no doubt akin to those of "muscular rheumatism". When they are persistent, however, it is important that other possibilities should not be overlooked, e.g. compression of the nerve by a spinal tumour, by a herniated intervertebral disc or by spondylitic osteophytes.

"*Lumbago*" is an even vaguer term, meaning no more than "lumbar pain". When we have excluded such obviously non-rheumatic causes of backache as spinal tumours and tuberculosis, renal or ureteric calculus, inflammations and tumours of the kidney, and diseases of the pelvic viscera, we are left with cases of evanescent or recurrent lumbar pain of uncertain origin, to which the name "lumbago" can more properly be applied. Undoubtedly some of these are cases of "muscular rheumatism", i.e. fibrositis or myositis of the lumbar fasciae or muscles; but others are attributable to spondylitis, sacro-iliac osteoarthritis or a protruding intervertebral disc.

"TOXIC" NEPHRITIS OR BRIGHT'S DISEASE

(1) Terminology

It would be perfectly correct to apply the name "nephritis" to any and all inflammations of the kidney. By custom, however, it has become reserved especially for those inflammations in which no bacteria are present in the injured kidneys. These must be sharply distinguished from microbic infections of the kidney, e.g. abscess, ascending pyelonephritis, tuberculosis, Weil's disease, etc. To emphasize this distinction we use the adjective "toxic". This is not to imply that bacteria play no part in causation; indeed we know that many cases of nephritis *are* related to bacterial infection elsewhere in the body.

At the outset, two announcements, one sad and the other cheering, about the nomenclature of non-infective nephritis must be made. The sad announcement is that in no other subject have clinicians and pathologists invented such a redundant and confusing lot of names. The cheering announcement is that the difficulties of the subject are not nearly as great as the superfluity of names would suggest. There *are* difficulties; but, to begin with, let us take as simple a view of nephritis as possible, even at the risk of over-simplification, and leave the difficulties for later discussion.

The simple view is this. As the result of infections elsewhere in the body, the kidneys may be damaged and they may react to the damage in various ways. (Chemical injury of the kidneys is discussed in Chapter 24.) If the injury is severe and rapidly inflicted, the clinical onset of the nephritis, characterized by oedema, albuminuria and haematuria, is sudden and often obviously related to the infection that caused it; we then speak of *acute nephritis* or *Type I nephritis*. If the injury is less severe and more slowly inflicted, the main effects are again oedema and albuminuria, but the clinical onset is insidious and usually not obviously related to any acute infection; we then speak of *subacute nephritis* or *Type II nephritis*. If the injury inflicted on the kidneys is still milder but more sustained, the patient may at no time have any significant oedema or albuminuria and no symptoms of disease until he begins slowly to develop rising blood-pressure and increasing retention of nitrogenous waste products; he is then said to have *chronic nephritis* or, better, *chronic hypertensive nephritis*. A patient with subacute nephritis may lose his oedema and albuminuria and may then or later develop the signs of chronic hypertensive nephritis. Or a patient with acute nephritis of sudden onset may apparently recover (except that a little albuminuria may persist) and then years afterwards he may develop the signs of chronic hypertensive nephritis. As we shall see later, the hypertension of the terminal stage of nephritis is not peculiar to nephritis but is a supervening secondary result of progressive renal ischaemia which occurs also in many other chronic destructive diseases of the kidneys.

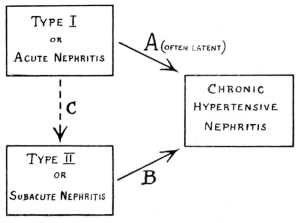

FIG. 95.—Schema of nephritis.

We may thus regard nephritis as a single disease with variants. The variable clinical and pathological pictures presented by it depend on differences in its severity and rate of onset, on the degree to which recovery ensues (probably related to evanescence or persistence of the toxaemic or other causative factors), and on the stage reached by the disease when it comes under observation. The many names applied to Bright's disease refer to variants or stages of the disease and do not imply the existence of several distinct kinds of nephritis. However, because in most cases the cause of acute nephritis is known while that of subacute nephritis is unknown, it is wise to follow Ellis and his co-workers in distinguishing these as Types I and II nephritis. If, pending further knowledge of the causes of subacute nephritis, we adopt this distinction, we have the simple classification of Bright's disease shown in Fig. 95. Any particular case can be placed conveniently in this schema. In addition to those cases which fall into one or other of the three named forms of the disease (those in the rectangles), others are seen in transitional forms denoted by the arrows. Arrows A and B denote common and well-recognized transitions ; arrow C denotes an infrequent transition, the interpretation of which is still in doubt.

Having adopted the above simple schema, the student will be better able to withstand the shock of introduction to the host of names now to be enumerated.

(1) *Acute nephritis* or Ellis's *Type I nephritis* has had the following aliases:

Acute glomerulonephritis, to stress the importance of the glomerular lesions.
Acute haemorrhagic (or *haematuric*) *nephritis*, in reference to the haematuria of the disease.
Post-infectious nephritis, to stress its frequently known bacterial causation.

(2) *Subacute nephritis* or Ellis's *Type II nephritis* has had the following aliases :

Chronic parenchymatous nephritis } because of the prominence of
Chronic tubular nephritis } lipoid changes in the tubular
Degenerative nephritis } epithelium.

Lipoid nephrosis } because of a former belief that the lipoid
Nephrosis } changes were primary, degenerative and un-
Nephrotic nephritis } related to inflammatory Bright's disease.

Chronic hydraemic nephritis } because of the pronounced general-
Chronic hydropigenous nephritis } ized oedema.

(3) *Chronic hypertensive nephritis* has had the following aliases :

Chronic interstitial nephritis } because of the chronic inflammatory
Sclerosing or fibrotic nephritis } fibrosis in the kidneys.

Chronic azotaemic nephritis, because of the retention of nitrogenous waste
products in the blood (azote=nitrogen).

Terminal stage nephritis, because this is the common termination of cases
both of acute and subacute nephritis which fail to resolve.

It must be noted here that, formerly, chronic renal fibrosis secondary to
hypertensive arterial disease was often confused with that due to chronic Bright's
disease. The distinction between these will be discussed presently, after the
characters of nephritis proper have been considered.

(2) Acute nephritis or Type I nephritis

(a) Causation

Acute nephritis is predominantly a disease of childhood and adolescence,
about three-quarters of the cases occurring in the first two decades. In most
cases, the onset is preceded by an acute infection—scarlet fever, tonsillitis, a
cold, otitis media, pneumonia, etc. The interval between the infection and the
onset of nephritis is usually between 1 and 3 weeks. The fact that nephritis
does not appear during the height of the causative infection, but after the toxaemia
has abated, shows that the renal injury is not directly inflicted by circulating toxins,
but is related to an allergic or hypersensitive state induced by the toxaemia. Experimental and histological evidence using fluorescent-antibody technique, show that
antibody-antigen complexes, usually streptococcal, are localized in the glomeruli
of the nephritic kidney.

(b) Clinical picture

The symptoms and signs of acute nephritis, which come on abruptly, are :

(i) *Haematuria*.—Obviously bloody or " smoky " urine is noticed in many
cases, and blood corpuscles are found microscopically in all.

(ii) *Oedema*.—A moderate degree of oedema of the face, back or legs is present
in nearly all cases. (Its mechanism will be discussed in Chapter 26.)

(iii) *Hypertension*.—Raised blood-pressure is present in most cases, and is
often associated with headache and vomiting, and sometimes with convulsions,
temporary loss of sight, or coma, a condition called " encephalopathy ".

(c) Urine changes

(i) *Oliguria*, i.e. diminution in quantity, sometimes amounting to complete
suppression or *anuria* ; the specific gravity is high.

(ii) *Haematuria*, gross or microscopic, as just mentioned, usually accompanied by an excess of leucocytes from the inflamed kidneys.

(iii) *Albuminuria* of greater degree than is accounted for by the amount of blood present.

(iv) *Casts* of the urinary tubules—elongated cylindrical masses composed either of blood corpuscles, desquamated epithelial cells, granular debris, or all of these mixed.

(d) Blood changes

(i) *Red cells and haemoglobin.*—There is usually a mild degree of secondary anaemia, with the haemoglobin percentage down to 75 or less, attributable partly to the preceding infection and partly to loss of blood in the urine. The patient is often pale.

(ii) *Plasma proteins.*—The loss of albumin in the urine results in fall of the plasma proteins to 5 grams per cent or less.

(iii) *Nitrogenous metabolites.*—The blood-urea often rises above 50 milligrams per 100 cubic centimetres, and not infrequently above 100 milligrams.

(e) Course and prognosis

About three-quarters of patients recover completely. A few die in the acute stage, either from rapid circulatory failure with dyspnoea and oedema of the lungs, the result of severe hypertension, or from uraemia due to retention of nitrogenous waste products. A few more show persistent oedema, urine and blood changes and hypertension, and die within a few weeks or months. Between 10 and 25 per cent of cases recover from the acute attack, but show persistent slight albuminuria and a few casts, and develop chronic progressive hypertension and azotaemia many years later ; this is the important group denoted by arrow A in Fig. 95.

(f) Naked-eye appearance of kidneys

This is often not very striking. The organs are slightly swollen and congested, tiny haemorrhagic spots may be present beneath the capsule, the cortical pattern is somewhat blurred, and on close inspection of its cut surface the glomeruli often stand out with undue prominence like tiny pale beads on the congested background.

(g) Microscopic changes (Fig. 96)

(i) *Glomeruli.*—The glomeruli are increased in size, unduly cellular, and relatively bloodless. Their cellularity is due to proliferation of the endothelial cells of the capillaries and to infiltration of the tufts by variable numbers of polymorphonuclear leucocytes. Their bloodlessness is due to capillary occlusion by the proliferating endothelium and to the formation of hyaline fibrin-thrombi in the capillaries. The glomerular cavities contain extravasated blood cells, leucocytes, fibrin and desquamated epithelial cells. In later stages, the epithelium lining the Bowman's capsules proliferates and forms crescentic masses of cells which compress the tufts ; this epithelial proliferation appears always to be evoked by the presence of blood or fibrin in the glomerular space.

(ii) *Tubules.*—The convoluted tubules show varying degrees of cloudy swelling, degeneration and desquamation of their epithelial cells, and the presence of blood corpuscles and casts in their lumina.

(iii) *Interstitial tissues.*—These show congestion, some oedema, and inconspicuous collections of polymorphonuclear leucocytes especially around the Bowman's capsules of severely affected glomeruli.

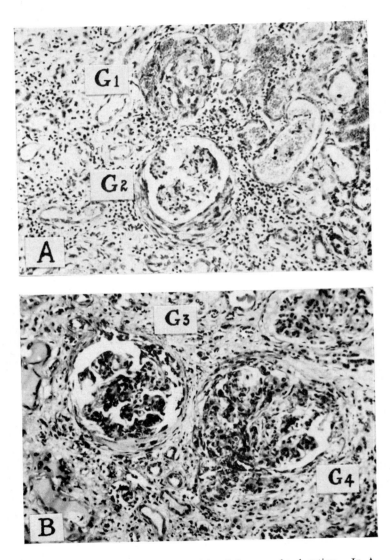

Fig. 96.—A and B. Type I nephritis of three weeks duration. In A, glomerulus G1 shows blood in the capsular space, G2 shows an epithelial crescent, tubules to the right contain blood, and there are many leucocytes in stroma. In B, G3 and G4 show capsular adhesions and many occluded capillary loops. (× 150.)

(3) Subacute nephritis or Type II nephritis

(a) Causation

The causes of subacute or Type II nephritis of insidious onset are unknown. It affects children and adults with about equal frequency. In only a few cases is there a history of previous infection, and in these its relationship to the nephritis is usually doubtful. In some cases immunofluorescence and electron-microscopic studies give evidence of the presence of antibody-antigen deposits, some streptococcal in origin, on the glomerular basement membranes. A few cases of acute nephritis fail to clear up but run a prolonged subacute course with increasing generalized oedema, but the histological picture in these differs from that of Type II nephritis of insidious onset. Hence the reason for the doubt mentioned on p. 299 regarding arrow C in Fig. 95. We must note here that the "nephrotic syndrome" (i.e. generalized oedema, proteinuria and hypoproteinaemia) can be due to other renal diseases, e.g. amyloidosis and systemic lupus erythematosus; but by far its commonest cause is Type II nephritis.

(b) Clinical picture

The onset is insidious with oedema as the only symptom. This may get worse slowly or rapidly, but eventually it becomes severe and widespread, the patient being waterlogged and bed-ridden. There is no obvious haematuria, and no hypertension. (For discussion of the cause of renal oedema, see Chapter 26.)

(c) Urine changes

(i) *Oliguria* occurs : the amount of urine passed is moderately diminished, and its specific gravity is raised or normal.

(ii) *Haematuria* is rarely visible to the naked eye, but microscopically a few red corpuscles are often intermittently present.

(iii) *Albuminuria* is constant and severe, some patients losing as much as 30 grams of albumin daily, and the urine often containing between 1 and 3 grams per cent of albumin and coagulating solid on boiling.

(iv) *Casts* are usually plentiful ; they are either hyaline, epithelial or granular, and they often contain much lipoid material. Free lipoid granules and droplets may also be abundant.

(d) Blood changes

(i) *Red cells and haemoglobin.*—A mild degree of secondary anaemia usually develops, and this along with the oedema gives the patient a characteristic puffy pallor.

(ii) *Plasma proteins.*—These are always reduced, usually to less than 5 grams per cent and sometimes as low as 3 grams per cent. The albumin:globulin ratio is reduced to less than 1:1, sometimes as low as 1:2.

(iii) *Nitrogenous metabolites.*—The blood-urea level is normal, and the urea-concentration and similar tests show that the kidney excretes nitrogenous waste products efficiently.

(iv) *Cholesterol.*—In most cases the blood-cholesterol is markedly raised. The reason for this is not clear.

(e) Course and prognosis

Recovery is rare, especially in adults. About one-third of the cases die in the oedematous stage, usually of bronchopneumonia, pericarditis, or other infections. Most of the remaining two-thirds gradually lose their oedema and albuminuria and develop instead progressive hypertension and azotaemia ; they show the transition denoted by arrow B in Fig. 95. In some of these cases, hypertension and azotaemia develop quickly, within a few months and while some oedema and albuminuria still persist. In others, the symptoms of the subacute phase clear up and those of the hypertensive phase do not appear until many months or years later. Thus, combinations of typical Type II and chronic hypertensive nephritis in all possible degrees and proportions are encountered.

(f) Naked-eye appearance of kidneys

In the oedematous albuminuric stage, the kidneys are usually quite character-istic ; they are "the large pale kidneys" of a "large pale person". Their surfaces are smooth, the capsules strip away easily, the cortex is wide and pale and shows yellow streaks and mottled areas.

(g) Microscopic changes

(i) *Glomeruli.*—A characteristic, though in its early stages not easily detected, lesion accompanies Type II nephritis with gross oedema. The glomerular capillaries remain patent but their walls and the interstitial tissue between them show a progressive hyaline change. At first slight and inconspicuous, this becomes gradually more prominent, until, as the transition to the chronic hyper-tensive stage takes place, the glomeruli are largely occupied by hyaline material and their capillaries occluded. In Type II nephritis, the glomeruli show little evidence of the inflammatory and reparative changes which occur in Type I nephritis of long duration ; there is little or no increase in cellularity of the tufts, no haemorrhage or inflammatory exudate into the capsular spaces, and little or no formation of epithelial crescents.

(ii) *Tubules.*—The epithelium of the convoluted tubules is laden with lipid granules and droplets, especially cholesterol esters, and the tubules contain granular, lipid and epithelial casts. The degree of lipid change varies ; it is greatest in cases with pronounced hypercholesterolaemia ; and, like the latter, the reason for its development is not known.

(iii) *Interstitial tissues.*—These show oedema with resulting separation of the tubules, which largely accounts for the swelling and pallor of the kidney. Collec-tions of lipid-laden foam-cells are sometimes present.

(4) Chronic hypertensive nephritis

(a) Causation

According to their previous history, patients with chronic hypertensive nephritis fall into 3 groups :

(i) Those who have had acute nephritis years previously, with subsequent smouldering latent renal disease—the group denoted by arrow A in Fig. 95.

(ii) Those who have had subacute Type II nephritis previously and have shown the transition denoted by arrow B in Fig. 95.

(iii) Those in whom hypertension and azotaemia have developed insidiously, without any previous history of known acute or subacute nephritis. There are good grounds for believing that many of these cases also are really examples of the transition A, the original acute attack having been mild and unrecognized.

(b) Clinical picture

The symptoms and signs are attributable either to (i) renal failure or (ii) hypertension.

(i) *Renal failure.*—The kidneys' ability to excrete nitrogenous and other waste products and to maintain a proper balance of electrolytes in the blood is progressively impaired. As a result there appears a condition of auto-intoxication called *uraemia*, characterized by headache, dyspnoea, vomiting, muscular twitchings, and terminal coma. This is not due to retention of any single waste product, but results from a combination of *azotaemia* (retention of urea, uric acid, creatinin, etc.) and *acidosis* or diminished alkali reserve (largely due to retention of phosphates and a fall in blood calcium).

(ii) *Hypertension.*—There is a progressive rise of blood-pressure, accompanied by thickened arteries, hypertrophy of the left cardiac chambers, and retinal exudates, haemorrhages or papilloedema. Patients who do not die of uraemia usually die of cardiac failure or of cerebral haemorrhage. They sometimes die of a combination of renal and cardiac failure.

(c) Urine changes

(i) *Polyuria.*—There is abundant urine, but its specific gravity is low (often remaining fixed at about 1·008 or 1·010) because it is poor in solids, the concentrating power of the kidneys being impaired.

(ii) *Haematuria.*—Red corpuscles are present occasionally and then only in small numbers.

(iii) *Albuminuria.*—This is often absent or only a trace of albumin is present intermittently.

(iv) *Casts.*—Casts are few and usually hyaline.

(d) Blood changes

Red cells and haemoglobin, plasma proteins, and cholesterol are usually normal in amount. In later stages of the disease, the blood-urea, creatinin and other nitrogenous metabolites are raised, the blood-urea sometimes rising eventually to 500 or 600 milligrams per 100 cubic centimetres. In the earlier stages, the impaired excretory function of the kidneys is revealed by special urea-clearance and concentration tests.

(e) Course and prognosis

Chronic hypertensive nephritis is a progressive fatal disease, ending in either uraemia, cardiac failure or cerebral haemorrhage. The rate of its progress varies from case to case ; in some it runs a rapid course of a few months, in others it progresses slowly for many years.

(f) Naked-eye appearance of kidneys

In the fully established disease the kidneys are small, tough, red or pale, with a finely or coarsely granular surface, their capsules strip with difficulty and tear the surface, the cortex is narrow and its pattern blurred, the arteries are often visibly thickened, and because of the general shrinkage of the kidney, the peripelvic adipose tissue is excessive. In patients dying during the transitions A or B of Fig. 95, various transitional appearances of the kidneys are seen ; e.g. a patient who has shown transition from oedematous Type II nephritis may still have rather large pale, yellowish mottled kidneys in which, however, fibrosis, granularity and vascular thickening are now also prominent.

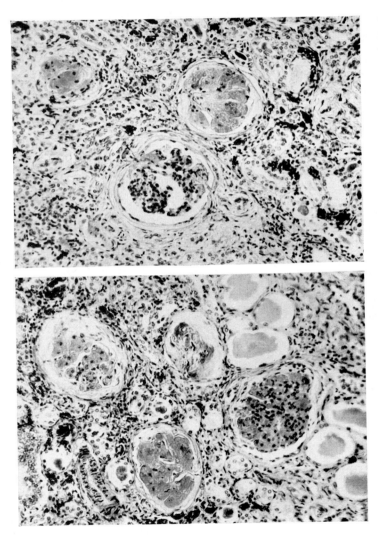

Fig. 97.—Chronic hypertensive nephritis, showing distortion and hyalinization of glomeruli, and capsular and interstitial fibrosis. (× 150.)

(g) Microscopic changes (Fig. 97)

These also differ according to the previous history of the patient (see p. 306) and the rate of progress of the disease ; and experts are often able to distinguish the chronic hypertensive kidney of a previous Type II nephritis from others. We need not make these distinctions here, however, but will be satisfied to enumerate the following main changes of advanced chronic hypertensive nephritis in general :

(i) *Glomeruli.*—These show a wide range of changes, including epithelial crescents, fibrous thickening of Bowman's capsules, adhesions between the capsules and the tufts, fibrosis and capillary obliteration of the glomeruli, and their final conversion into hyaline structureless balls.

(ii) *Tubules.*—These show varying degrees of collapse and atrophy, accompanied in places by cystic dilatation. The epithelium in most tubules is low or flattened ; some fatty change may be present, especially in cases which are transitional from Type II nephritis.

(iii) *Interstitial tissues.*—There is widespread but often patchy fibrosis, diffuse and focal collections of lymphocytes, and thickening of the walls of arteries and arterioles.

(5) The relationship of nephritis and hypertension

We have noted that acute nephritis is accompanied by a rise of arterial blood-pressure which is usually temporary, and that in chronic hypertensive nephritis there is a permanent and progressive rise. We must now inquire more closely into the relationship of renal disease and blood-pressure. This relationship is a complex one which is still not fully understood; but light has been thrown on it by the study of (a) *essential hypertension* in man, and (b) *experimental hypertension* in animals.

(a) Essential hypertension

Essential hypertension (primary hyperpiesia or hypertensive arteriosclerosis) is a condition of maintained high blood-pressure due to primary arterial disease and not secondary to nephritis. The nature of this arterial disease will be discussed in Chapter 26 ; here we are concerned only with its effects on the kidneys. According to these effects, we distinguish two varieties of essential hypertension —(i) *benign hypertension*, of long duration and slow progress, chiefly of middle-aged and old people, usually ending in cardiac failure or cerebral haemorrhage, and accompanied by little or no evidence of significant renal damage (and hence not calling for any further discussion here) ; and (ii) *malignant hypertension*, a rapidly progressive disease, chiefly of young and middle-aged people, characterized by a rapidly mounting blood-pressure and signs of progressive renal damage and azotaemia. The renal changes generally resemble those of chronic hypertensive nephritis, but with severe and characteristic lesions of the arteries, namely a peculiar swelling or "fibrinoid necrosis" of the walls of arterioles, especially the afferent glomerular vessels, and a cellular fibrous endarteritis of the larger interlobular arteries (Fig. 98). Of special interest is the fact that these vascular lesions are not confined to the kidney, but occur also in the arterioles of many other tissues. This fact suggests that malignant hypertension is a primary vascular disease with widespread lesions in the arterial system and serious secondary ischaemic effects in the kidneys.

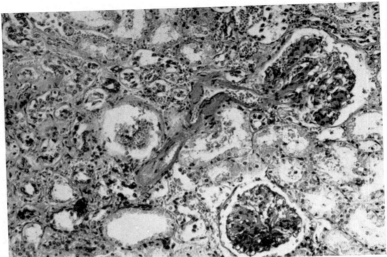

Fig. 98.—" Fibrinoid " change in afferent glomerular arteriole in hypertension. (By courtesy of the late Prof. D. H. Collins.) (× 125.)

(b) Experimental hypertension

Goldblatt and his colleagues discovered that if partially constricting clips were applied to both renal arteries in dogs, permanent hypertension resulted. Some of these hypertensive dogs developed progressive renal failure and showed in organs other than the kidneys arteriolar lesions similar to those seen in malignant hypertension in man. Arteriolar lesions were not present in the kidneys, presumably because the constriction of the renal arteries protected the kidneys from the hypertension. Wilson and Byrom produced permanent hypertension in rats by constricting one renal artery only, and found that arteriolar lesions like those of malignant hypertension developed in the unclamped kidney but not in the clamped one, and also that the unclamped kidney developed glomerular and interstitial changes resembling those of chronic hypertensive nephritis in man. These and other experimental results leave no doubt (i) *that renal ischaemia can cause hypertension*, and (ii) *that hypertension can cause renal changes resembling nephritis*. It is interesting to note here that cases comparable with experimental hypertension by renal arterial clamps have been observed in man : in occasional cases atheromatous stenosis or occlusion of one or both of the main renal arteries has been accompanied by progressive hypertension.

(c) The vicious circle in nephritis

Both in Type I or acute nephritis and in Type II or subacute nephritis in its transitional or hypertensive stage, there are lesions in the kidneys which produce renal ischaemia—in the former endothelial proliferation and fibrin-thrombi occluding the glomerular capillaries, in the latter a much slower occlusion by accumulated hyaline material in and between the capillary walls. It is very probable that the renal ischaemia so produced is the main or only cause of the hypertension accompanying nephritis. If, as in many cases of acute nephritis, the vascular damage in the kidney is evanescent and recoverable, the hypertension subsides. But if sufficient permanent damage has been inflicted, some degree of ischaemia persists, and a vicious circle is established. The hypertension produces added arterial lesions in the kidney, these aggravate the ischaemia and thus heighten the hypertension, and so the renal damage progresses inexorably to renal failure, or cardiac failure, or both. The same end-result is reached from the primarily vascular lesions of malignant hypertension. It may also be reached in a miscellaneous variety of other diseases which damage kidneys, arteries or both, e.g. chronic lead poisoning, chronic bacterial pyelonephritis, pregnancy toxaemia, or prolonged amyloid disease of the kidneys. The anatomical end-result of all of these may be small, tough, granular kidneys, which were formerly indiscriminately grouped together under the name "chronic interstitial nephritis". This name thus embraced a hotch-potch of various renal disorders on which the hypertensive vicious circle had been superimposed. Out of this hotch-potch we have extracted Types I and II nephritis, the genuine non-infective renal inflammations, and essential hypertension, a primarily vascular disease.

(d) The cause of hypertension in renal disease

Why should renal ischaemia produce hypertension ? It has long been known that extracts of normal kidneys contain a powerful vaso-pressor substance, which has been called *renin*. This enzyme, probably produced by the juxtaglomerular cells, though itself non-pressor, acts with a blood globulin (*hypertensinogen*) to produce the vaso-constricting substance *hypertensin* (angiotensin or angiotonin). It is known also that excessive renin is discharged into the blood from ischaemic kidneys; that the blood and tissues contain an enzyme (*hypertensinase*) which destroys hypertensin; and that normal kidney tissue contains also an anti-pressor substance capable of neutralizing the pressor effects of renin. To what extent these important experimental findings are applicable to human renal pathology is still unsettled. Enough has been said here to show both the biochemical complexities of the subject and the promise it holds out of elucidating the mechanism of renal hypertension.

(e) Is essential hypertension possibly of renal origin ?

As already stated, most workers look upon essential hypertension as a primarily vascular disease with secondary effects in the kidney. But its causes are still unknown ; and in view of the biochemical mechanisms referred to in the previous paragraph, we cannot exclude the possibility that some disturbance of these may constitute the primary stimulus responsible for seemingly primary hypertension. Future research must explore this possibility. However, we must not forget other possibly implicated factors, including the innervation of arterioles, adrenaline and other hormones, and sodium metabolism and its adrenocortical control—of special relevance in renal physiology.

GENERALIZED LUPUS ERYTHEMATOSUS

This not very rare disease, which occurs chiefly in middle-aged women, may run an acute or a chronic relapsing course; it often ends fatally, but some cases recover.   Clinically, it shows a variable combination of fever, an erythematous skin rash (especially of the face), arthritis, pleurisy, enlarged lymph glands and spleen, albuminuria and haematuria, signs of endocarditis, and anaemia.   Pathologically, various organs show foci of " fibrinoid " degeneration in connective tissue and small arteries with accompanying inflammatory changes, and often the presence of rounded haematoxylin-staining bodies composed of altered nucleoprotein.   The kidneys frequently show distinctive glomerular changes, namely, foci of necrosis, deposition of hyaline protein material in capillary walls (" wire-loop " capillaries), haematoxyphil bodies, and sometimes arterial lesions resembling those of polyarteritis.   Cardiac involvement takes the form of simple (i.e. non-bacterial) vegetations on valve cusps which show areas of " fibrinoid " degeneration, and this is called Libman-Sacks endocarditis.   Many cases of lupus erythematosus have arthritis, often indistinguishable from rheumatoid arthritis.   An important diagnostic laboratory finding is the presence of " lupus-cells " (LE cells) in the blood and bone marrow;   these are polymorphonuclear leucocytes containing rounded masses of phagocytosed altered desoxyribonucleic acid, essentially similar to the haematoxylin-staining bodies seen in the kidneys and other organs.   The blood serum often contains excessive gamma-globulin;   and special serological studies give evidence of complex auto-immunity changes, with the formation of antibodies against the patient's own nucleoproteins (see p. 289).

ERYTHEMA NODOSUM

This is a mildly febrile disease of unknown cause, characterized by multiple bilateral, tender, red, oedematous swellings of the skin and subcutaneous tissues, usually of the extensor aspects of the legs and sometimes on the arms as well, in patients under 30 years of age, usually females.   The histological changes are those of an extensive subcutaneous and dermal subacute inflammation, which resolves without suppuration or fibrosis.   In many cases, either early pulmonary tuberculosis or chronic streptococcal tonsillitis is present ;   and the most popular view of the nature of erythema nodosum is that it is an allergic reaction to these or other infections.

POLYARTERITIS NODOSA

The pathology of this disease is described on p. 85.   Its causation is uncertain; but many cases have had acute or chronic respiratory infections, usually streptococcal, and have been treated with sulphonamides or other drugs.   Some workers have therefore supposed that the disease is essentially one of allergic hypersensitivity to antigenic combinations of the drugs with tissue proteins; but Rose has summarized the objections to this view and the reasons for suspecting that the hypersensitivity may be to the streptococcal infections themselves.

ULCERATIVE COLITIS

This is a severe chronic inflammation of the mucosa of part or the whole of the large intestine, characterized by irregular ulceration closely resembling that of chronic bacillary dysentery, accompanied by the discharge of blood, mucus and pus, and by progressive emaciation, weakness, secondary anaemia and dehydration.

Fibrous strictures of the intestine sometimes develop. The disease may progress rapidly and cause death within a few weeks or months ; more often it runs a prolonged course with periods of improvement and repeated exacerbations. It may be a microbic infection, but no constantly present specific organism has been identified; and allergic, auto-immune, nervous and other factors have also been suggested as causative.

REGIONAL ILEITIS
(Crohn's disease)

This is a surgically important chronic inflammatory disease of the terminal part of the ileum in young people. It converts the affected segment of gut into a rigid thick-walled tube like a hose-pipe, with a narrowed lumen, but with little or no mucosal ulceration. Microscopically, the thickened bowel wall and adjacent mesentery show oedema, extensive active fibrosis, diffuse and patchy infiltration by leucocytes of various kinds, including lymphocytes, plasma cells, macrophages and polymorphs, and sometimes focal aggregations of " epithelioid " and giant-cells resembling non-caseating tubercles. The neighbouring lymph glands show similar changes with or without tubercle-like foci. The disease is probably an infection, but the causative organism has not been identified.

SUPPLEMENTARY READING

Rheumatic Fever and Rheumatoid Arthritis

Clawson, B. J., Bell, E. T., and Hartzell, T. B. (1926). " Valvular disease of the heart, with special reference to the pathogenesis of old valvular defects ", *Amer. J. Path.*, **2,** 193. (A beautifully illustrated account.)

— and Wetherby, M. (1932). " Subcutaneous nodules in chronic arthritis ", *Amer. J. Path.*, **8,** 283.

Collins, D. H. (1937). " The subcutaneous nodule of rheumatoid arthritis ", *J. Path. Bact.*, **45,** 97.

— (1949). *The Pathology of Articular and Spinal Diseases.* Edinburgh and London; E. & S. Livingstone. (Contains excellent chapters on rheumatic fever, rheumatoid arthritis and osteoarthritis.)

Cruickshank, B. (1958). " Heart lesions in rheumatoid disease." *J. Path. Bact.*, **76,** 223.

Glynn, L. E. and Holborow, E. J. (1965). *Autoimmunity and Disease.* Oxford; Blackwell.

Hart, F. D. (1969). " Rheumatoid arthritis: extra-articular manifestations ". *Brit. Med. J.*, **2,** 131.

Sacks, B. (1926). " The pathology of rheumatic fever. A critical review ", *Amer. Heart J.*, **1,** 750.

Nephritis

Bell, E. T. (1929). " Lipoid nephrosis ", *Amer. J. Path.*, **5,** 587.

Churg, J., and Grishman, E. (1959). " Subacute glomerulo-nephritis ". *Amer. J. Path.*, **35,** 25.

Ellis, A. (1942). " Natural history of Bright's disease ", *Lancet*, Vol. 1, 1, 34 and 72. (A most valuable, simple but detailed account.)

Freedman, P., and Markowitz, A. S. (1959). " Immunological studies in nephritis ". *Lancet*, **2,** 45.

Guthrie, K. J. (1936). " A study of the pathology of nephritis in infancy and childhood ", *J. Path. Bact.*, **42,** 565. (Many good photomicrographs.)

Houssay, B. A., and Braun-Menendez, E. (1942). " The role of renin in experimental hypertension ", *Brit. Med. J.*, **2,** 179.

Payne, W. W., and Illingworth, R. S. (1940). " Acute nephritis in childhood ", *Quart. J. Med.*, **33,** 37.

Lupus Erythematosus and Polyarteritis

Annotation. (1960). "Auto-antibodies and lupus erythematosus". *Brit. Med. J.*, **2,** 1141.

Bardawil, W. A., *et al.* (1958). " Disseminated lupus erythematosus, scleroderma, and dermatomyositis as manifestations of sensitization to DNA-protein ". *Amer. J. Path.*, **34,** 607.

Bywaters, E. G. L. (1956). " What is the evidence of hypersensitivity in the pathogenesis of the so-called collagen diseases? " *Proc. Roy. Soc. Med.*, **49,** 287.

Glynn, L. E. and Holborow, E. J. (1965). *Autoimmunity and Disease.* Oxford; Blackwell.

Hill, L. C. (1957). " Systemic lupus erythematosus ". *Brit. Med. J.*, **2,** 655 and 726. (An excellent and full account.)

Rose, G. A. (1956). " Infections and their treatment in the aetiology of polyarteritis nodosa ". *Proc. Roy. Soc. Med.*, **49,** 289.

Regional Ileitis

Barrington-Ward, L., and Norrish, R. E. (1938). " Crohn's disease, or regional ileitis ", *Brit. J. Surg.*, **25,** 512.

Warren, S., and Sommers, S. C. (1948). " Cicatrizing enteritis (regional ileitis) as a pathological entity ". *Amer. J. Path.*, **24,** 475. (A full account with good illustrations.)

CHAPTER 23

FOREIGN BODIES

By " FOREIGN BODIES " we mean large or small masses of particulate non-living matter introduced into the tissues or cavities of the body. Although microbic and other parasites are foreign particulate things, we do not count them as foreign bodies, because they are living pathogens. Dead parasites, however, e.g. hydatid cysts, filariae, trichinae, schistosome ova or degenerated torulae, fall into our definition of " foreign bodies " and are treated as such by the tissues. Foreign bodies may be either *extrinsic*, i.e. particulate matter introduced into the body from without, or *intrinsic*, i.e. non-living material produced within the body. A slough or a sequestrum is merely dead organic matter and is treated as such by the tissues around ; and a mass of fatty acid or cholesterol crystals produced by degeneration in the tissues themselves is just as much a " foreign body " as if it had been inserted from without.

We will consider foreign bodies in the following order :

A. *Extrinsic foreign bodies*
   - (1) Large foreign bodies in hollow viscera :
     - (*a*) in the alimentary canal,
     - (*b*) in the respiratory tract,
     - (*c*) in the genito-urinary tract ;
   - (2) Inhaled finely particulate matter ;
   - (3) Solid foreign bodies embedded in tissues ;
   - (4) Free gases in tissues.

B. *Intrinsic foreign bodies*
   - (1) Free fats and lipoids in tissues ;
   - (2) Non-lipoidal substances free in tissues ;
   - (3) Calculi.

LARGE FOREIGN BODIES IN HOLLOW VISCERA

(1) Foreign bodies in the alimentary canal

(a) Foreign bodies swallowed by normal people

Almost every mother has had moments of anxiety because a young child has swallowed a bead, button, coin, pin or some small plaything. Only rarely do these cause any trouble ; an article which can be swallowed easily by a child usually traverses the alimentary canal without mishap and is passed in the faeces. Occasionally an adult accidentally swallows a large foreign body, e.g. a denture, which becomes impacted in the pharynx or oesophagus. Components of hastily swallowed food which sometimes cause trouble are bones (especially fish bones), large fruit stones, and large masses of vegetable fibre, e.g. fruit skins. It is astounding what some people will gobble down whole. The writers have seen impacted against a stenosing cancer of the colon a collection of cherry stones along with the intact bones of most of the limbs and trunk of a rabbit ; and large shaggy sea-weed-like masses which were persistently vomited up by an intelligent man proved to be the fibrous remains of almost whole unmasticated oranges !

(b) Foreign bodies swallowed by lunatics

Some insane or mentally subnormal people develop the habit of swallowing all manner of solid objects, some of them of quite astonishing size—coins, spoons, pocket-knives, safety pins, and even butcher's hooks. Since such large objects cannot pass through the pylorus, they accumulate in the stomach and eventually cause obstruction.

(c) Hair-ball (trichobezoar)

Hysterical or other mentally subnormal girls sometimes chew and swallow their own hair. This gradually accumulates in the stomach as a firmly matted mass which eventually fills and distends the organ, forming a complete internal cast of it and causing obstruction. Many animals which lick themselves or their fellows also develop hair-balls, usually spherical, in the stomach or intestines.

(2) Foreign bodies in the respiratory tract

Foreign bodies in the respiratory passages are more common in children than adults, because of their habit of putting things in the mouth. A sudden start or cry may easily lead to inhalation of the object. If this is large, it may become impacted in the larynx and cause speedy death from suffocation, unless tracheotomy is performed quickly. More often, a foreign body arrested in the larynx allows some air to pass but causes hoarseness of the voice and stridor (noisy harsh respiration). Still more frequently, inhaled foreign bodies pass down into the bronchi, more often the right bronchus than the left, and are arrested in the main bronchi themselves or their branches in the lungs. The commonest foreign bodies inhaled are peas, peanuts, beads, bits of wood or bone, and teeth, especially if extracted under general anaesthesia.

When a foreign body is impacted in a bronchus, the neighbouring mucosa becomes inflamed and swollen, so that obstruction of the bronchus is often complete. This causes collapse of the area of lung supplied by the bronchus, because of absorption of the alveolar air. When the foreign bodies are vegetable substances, such as peas or peanuts, these decompose quickly and severe acute bronchitis and bronchopneumonia often appear within a few days. A mineral foreign body may cause less acute effects, and may remain in the bronchus for months or years causing only local chronic inflammation. More often, however, bronchial obstruction causes accumulation and infection of secretions in the bronchi and lung peripheral to the obstructed spot and leads to bronchiectasis and chronic purulent bronchopneumonia or abscess.

" Lipoid pneumonia "

This is the name applied to pulmonary lesions caused by the inhalation of oils. It has followed the administration of cod-liver oil or liquid paraffin to infants or to sick adults with impaired swallowing or cough reflexes, and the use of oily nasal or pharyngeal sprays. Experimental work has shown that olive, sesame and other vegetable oils do not produce serious effects in the lungs, that liquid paraffin and other purified mineral oils cause effects of moderate severity, and that cod-liver and other animal oils produce the worst effects, probably because of their ready hydrolysis with the production of fatty acids. The changes seen in oil-inhalational pneumonia include phagocytosis of oil droplets by macrophages which congregate in the alveoli and bronchioles, inflammatory exudation, the formation of tubercle-like areas with giant-cells, fibrosis, and supervening bacterial bronchopneumonia.

(3) Foreign bodies in the genito-urinary tract

Foreign bodies retained in the bladder include in-dwelling catheters, suprapubic drainage tubes, unabsorbed sutures following operations, or articles introduced by sexual perverts. These not only cause persistent bacterial infection and inflammation, but they also often become encrusted by phosphate and urate deposits which may grow into large calculi. Foreign bodies in the vagina, e.g. pessaries introduced to correct uterine displacements, if retained for long periods, may cause erosion, infection and concretions.

INHALED FINELY PARTICULATE MATTER

Pneumoconiosis is any pathological condition of the lungs caused by the inhalation of dust. The varieties of pneumoconiosis are :
  (1) *Anthracosis*, due to inhalation of carbon ;
  (2) *Silicosis*, due to inhalation of silica ;
  (3) *Sidero-silicosis*, due to inhalation of silica and iron oxide ;
  (4) *Asbestosis*, due to inhalation of asbestos ;
  (5) *Bagassosis*, due to inhalation of " bagasse " or sugar-cane fibre ;
  (6) *Byssinosis*, due to inhalation of cotton dust.
The pneumoconioses due to silica and silicates—(2), (3) and (4)—are by far the most important.

(1) Anthracosis

Anthracosis is found in some degree in all adult city dwellers. The child's lung is pink, the adult's more or less black. The soot particles which blacken our homes also enter and blacken our lungs. Coarse particles are arrested in the larger bronchi, from which they are swept by the action of the cilia and are coughed up. Only fine particles reach the terminal air passages. Scavenging macrophages in the air-sacs and bronchioles take up these particles ; some of the carbon-laden phagocytes are coughed up in the sputum, but others migrate through the alveolar walls into the interstitial, peribronchial and subpleural lymphatics and to the broncho-tracheal and other mediastinal lymph glands. In all of these sites, increasing amounts of free carbon particles are deposited. The most intense degrees of anthracosis are seen among coal-miners, whose lungs, pleura and thoracic and cervical lymph glands are often jet black throughout. In the less extreme degrees of anthracosis, the carbon is seen to be distributed, not uniformly, but in distinct foci ; these correspond in position with the microscopic foci of lymphoid tissue which are scattered throughout the lungs. Carbon alone is innocuous, causing little or no reactive fibrosis in the tissues ; but if the dust also contains some silica, as happens in some coal-mines, then the more serious effects of silicosis are added, constituting *silico-anthracosis*.

(2) Silicosis

Silicosis is a serious occupational disease occurring in miners, stonemasons, metal-grinders and other workers who are exposed for long periods to dust containing fine crystalline particles of silica (quartz or silicon dioxide). These particles are taken up from the air passages by macrophages and deposited in the same situations as carbon particles. Each focus of deposited silica produces active fibrosis in the surrounding tissue and leads to the formation of a densely fibrous *silicotic nodule* which slowly enlarges. When the disease is well established, X-ray examination shows the presence of many discrete small rounded shadows

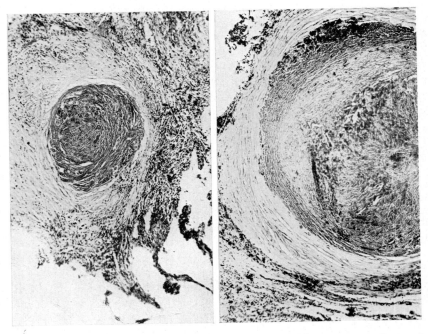

FIG. 99.—Early silicotic nodules in lung, consisting of mingled silica, carbon and fibrous tissue. (× 60.)

throughout the lungs; and at necropsy the organs feel as if studded with shot. On section the nodules are grey or show a pale centre and a grey or black periphery; microscopically they consist of hyaline fibrous tissue enclosing silica and carbon particles in varying relative proportions (Fig. 99). As the disease progresses the nodules enlarge and coalesce, so that large areas of the lungs become solid, tough and airless. In advanced cases the walls of the pulmonary arteries often suffer fibrosis and disorganization, with consequent stenosis or thrombosis of these vessels, obstructed pulmonary circulation and right cardiac failure (p. 630). Adhesions to the chest wall, areas of emphysema, bronchiectasis, and especially supervening chronic tuberculosis often complicate the lung picture. Many cases of silicosis show coexisting tuberculosis, and there are good grounds for believing both that silicosis predisposes to tuberculosis and that pre-existing tuberculosis accelerates the development of silicotic lesions.

While these changes are going on in the lung, similar lesions are developing in the mediastinal lymph glands, often with extension to the cervical and abdominal glands also. The glands may show discrete silicotic nodules, or may be diffusely fibrous and pigmented, or may show more or less distinct areas of silicosis and fibro-caseous or calcified tuberculosis. Transported silica occasionally causes the development of typical silicotic nodules in the spleen.

The effect of silica in producing fibrosis in the tissues is not due simply to its mechanical irritative properties but to its chemical action. Silica is slowly dissolved in the tissue fluids to form silicic acid, and this is a strong irritant which causes necrotic and fibrosing lesions. It has been shown experimentally that if a known amount of finely powered silica is introduced into the tissues, the amount

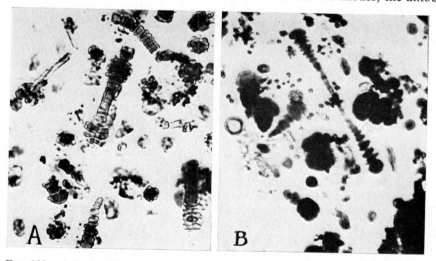

Fig. 100.—Asbestos bodies in lung : A, unstained ; B, stained by Perl's method for iron. (× 800.)

recoverable gradually diminishes, because of the slow solution of the substance. It is found also that very fine particles dissolve much more quickly than coarse ones ; and, correspondingly, fine silica particles produce more severe silicosis than coarse ones and produce it more quickly. Intravenously injected colloidal silica is rapidly fatal. It has also been suggested that in silicosis *sericite*, a hydrated silicate of aluminium and potassium, may play a part in causing fibrosis, but this is doubtful.

It remains to add here that many of the common pleural scars seen at the apices of adults' lungs are silicotic and not tuberculous as was once believed. They are devoid of caseation and calcification, and they contain silica dust.

(3) Sidero-silicosis

Haematite (red iron-ore) miners develop a special form of silicosis in which the lungs contain along with silica considerable amounts of haematite dust (ferric oxide), which gives the organs a brick-red colour. The fibrosis is more diffuse than in simple silicosis, and there is little doubt that the haematite modifies the effect of the silica in some way.

(4) Asbestosis

Asbestos is a fibrous crystalline mineral of variable composition but consisting mainly of magnesium silicate. Workers engaged in crushing or weaving asbestos before the introduction of preventive measures suffered from a characteristic form of pneumoconiosis, with diffuse fibrosis, more marked in the lower than the upper parts of the lungs, diffuse pleural thickening and adhesions, frequent bronchiectasis, and more rapid development than in silicosis. The asbestos is found in the lung in two forms, as free small particles and as peculiar " asbestos bodies ". These are yellow or brown elongated structures, measuring up to 100μ in length, and consisting of a narrow segmented shaft with expanded bulbous ends (Fig. 100). An asbestos fibre lies in the centre of each body, the remainder of which forms an iron-containing sheath around the fibre. How this sheath develops is not clear ; it is believed to be the product of chemical interaction between the asbestos and the tissue fluids around it, its iron content probably being derived from haemoglobin rather than from the asbestos. The presence of asbestos bodies in the sputum is important in diagnosis.

(5) Bagassosis

"Bagasse" is the name applied to crushed sugar-cane fibre after the sugar is extracted. It is used for making boards ; and besides cellulose it contains about 1 per cent of protein and 5 per cent of silica. The inhalation of bagasse dust by workers handling this material in the dry state has caused cases of rapidly developing pneumonia-like disease with mottled areas of opacity in X-ray pictures of the lungs. These usually clear away when the patient ceases handling bagasse ; but in a few cases the shadows have persisted. The nature of the lesions is uncertain ; it has been suggested that they may be a peculiar variety of silicosis ; or that they may be due to allergic sensitization to the inhaled foreign protein.

(6) Byssinosis

Byssinosis or cotton-dust disease is a form of chronic bronchitis with great dyspnoea, and cough with mucoid or muco-purulent sputum, resulting from the inhalation of cotton dust by workers engaged in the carding of raw cotton. The dust contains cotton hairs, fragments of contaminating moulds, other vegetable fibres, and fine sand particles ; which of these is mainly to blame for the disease is uncertain.

SOLID FOREIGN BODIES EMBEDDED IN TISSUES

(1) Large foreign bodies

These include bullets and other missiles, needles, sutures, glass fragments, splinters, or pieces of clothing carried in by penetrating wounds. The tissue reaction to such bodies depends on their composition and whether or not they are accompanied by pathogenic bacteria. A sterile, chemically inert foreign body, such as a bullet, needle, steel wire suture, or glass splinter, may remain embedded in the tissues for years and excite no more than the formation of a thin fibrous capsule ; but organic objects such as wood splinters, pieces of clothing or cat-gut sutures, even if sterile, excite more vigorous phagocytic " foreign-body " reactions, because they contain foreign protein or lipoid substances.

A good example of such " foreign-body " reactions is often to be seen around absorbing surgical cat-gut sutures. Clustered around the suture there are many macrophages and large irregular multinucleated giant-cells formed by their fusion —*foreign-body giant-cells* (Fig. 101). Accompanying these phagocytic cells there may also be variable numbers of lymphocytes, plasma cells and proliferating fibro-

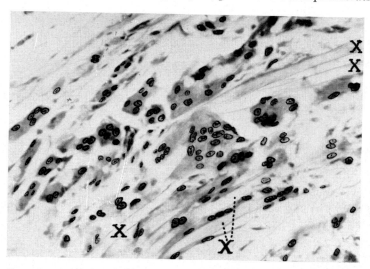

FIG. 101.—Foreign-body giant-cells of various sizes clustered around unabsorbed fibres of cat-gut suture, XX. (× 220.)

blasts. Undoubtedly the solution of the cat-gut is effected by proteolytic enzymes of the phagocytic and other cells. Fragments of the disintegrating suture may be seen completely enclosed within foreign-body giant-cells. When the foreign substance is completely absorbed, the phagocytic cells disperse and nothing but a small focus of fibrous tissue remains. The whole focus of reactive tissue around the foreign body is often called a " foreign-body granuloma " or " foreign-body tubercle ". It is typical of the structure of reactive foci around many other kinds of foreign materials.

(2) Finely particulate matter

The tissue reaction to fine solid particles again depends on their composition.

(i) *Chemically inert substances*, like carbon particles, many metallic dusts or the pigmented earths used in tattooing (e.g. ochre, sienna, vermilion), produce minimal reaction—a slight encapsulating fibrosis.

(ii) *Particles of silica and some silicates* excite much more active fibroblastic proliferation, with or without the formation of foreign-body giant-cells. Such a "siliceous granuloma" may develop in the skin following an injury in which fine sand particles are driven into the dermis, or in operation scars or on the peritoneum or pleura following the use of talc as a dry lubricant for surgical gloves. Talc is a hydrous magnesium silicate with serious irritant effects on tissues. Many examples of peritoneal adhesions or nodular granulomas due to surgically intro- duced talc (sometimes mistaken for tubercles) have now been reported. Its surgical use should be discontinued.

(iii) *Organic particles*, such as fragments of wood or other plant tissues, or the lycopodium spores which were once used on dressings, usually excite marked foreign-body reactions with many giant-cells. Interesting examples of this are sometimes seen following the spilling of alimentary contents into the peritoneal cavity, e.g. from a perforated gastric or duodenal ulcer or an inflamed perforated diverticulum of the colon ; months or years later, " foreign-body tubercles " with macrophages and giant-cells clustered around particles of plant tissue, may be found in the peritoneum. " Grease-gun finger " occurs accidentally in garage workers; lubricating grease under high pressure is forced from the gun through the pores of the skin into the sweat glands, which it disrupts, setting up severe foreign-body reaction in the dermis.

FREE GASES IN TISSUES

Apart from gases generated by bacterial fermentation, as in gas gangrene (p. 159) free gases occur in the tissues in the following distinct conditions :
(1) Escape of air from the lung into surrounding tissues ;
(2) Air embolism ;
(3) Caisson disease ;
(4) Intestinal pneumatosis.

(1) Escape of air from the lungs

Air may escape from the lungs into surrounding tissues in the following ways :

(i) *Pneumothorax* means the escape of air from the lung into the pleural cavity. It has already been referred to as a complication of pulmonary tuber- culosis ; but it occurs also from trauma such as penetrating wounds or crushing injuries with laceration of lung tissue by fractured ribs, or more often from simple rupture of emphysematous bullae of the lung following ordinary exertion. If there is no infection of the pleural cavity, the effects of the air in it are purely mechanical ; the lung collapses, the heart is displaced towards the opposite side, and there is shortness of breath. Usually the air is absorbed quickly and the lung re-expands. If infection is present, the pneumothorax is accompanied by a serous or purulent effusion—*hydro-pneumothorax* or *pyo-pneumothorax*.

(ii) *Surgical emphysema* means the percolation of air bubbles from the lung into the subcutaneous tissues, as a result of a crushing or lacerating injury involving both the chest wall and lung. Pneumothorax is often present as well. The inflated subcutaneous tissue gives a tympanitic (drum-like) note on percussion, and characteristic crepitus (crackling) on palpation.

(iii) *Interstitial pulmonary emphysema* is the escape of air from the alveoli into the interstitial tissues of the lung, whence it sometimes percolates to the hilum, mediastinum and even to the subcutaneous and fascial tissues of the neck. This condition occurs chiefly in children with severe paroxysmal coughing or bronchial obstruction, e.g. with whooping cough, laryngeal diphtheria, or a foreign body in a bronchus.

Sterile air is quickly absorbed from subcutaneous and other tissues and causes negligible reaction in them.

(2) Air embolism

The introduction of small quantities of air into the circulation produces no serious effects, for the gases are soon dissolved and expired. But, if large veins in the neck or thorax are opened during operations or by stab wounds, sufficient air may enter them to cause fatal air embolism. This is because of the negative pressure of the blood in these vessels and the suction effect of the inspiratory movements on their contents. The entrance of the air is recognized by a hissing sound in the wound. Air embolism occasionally occurs also as a result of the entry of air into large veins remote from the thorax, e.g. in the uterine veins following child-birth.

The fatal effects of air embolism are due to the gas being churned up into froth in the right cardiac chambers, with consequent arrest of the circulation. These chambers and the pulmonary arteries are found distended with froth, and, because the entry of venous blood into the heart is prevented, the lungs appear pale and bloodless.

(3) Caisson disease or diver's paralysis

A caisson is a large water-tight case filled with compressed air in which men can work on bridge foundations and other underwater constructions. To keep out the water, air pressure in a caisson must be increased by 1 atmosphere, or 15 lb. per square inch, for every $33\frac{1}{2}$ feet depth. Thus at a depth of 100 feet, a worker in a caisson, or a man in a diving-suit, is subjected to a pressure of 4 atmospheres. Under these conditions the blood and tissues hold in solution much more of the atmospheric gases than under normal pressures. If then the worker is returned too quickly to normal pressures, the excess of gases in solution, especially nitrogen and carbon dioxide, suddenly appear as free bubbles throughout the body; his blood and tissues literally effervesce. The clinical effects include pains in muscles, paralysis of limbs, deafness, vertigo (dizziness), and sometimes convulsions, loss of consciousness, or fatal respiratory and circulatory failure. These effects are largely due to widespread air embolism in the blood vessels, especially of the central nervous system, but disruptive effects of the suddenly liberated gas bubbles in the tissues themselves also play a part. In patients who survive, small or large areas of aseptic necrosis (infarction) may occur in bones, due to the blood supply of these areas having been cut off by nitrogen emboli.

Caisson-workers and divers are therefore decompressed gradually, by means of proper air-locks, so that the excess of gases in solution in the body are discharged slowly through the lungs and no effervescence occurs in the tissues. The safe rate of decompression is about 20 minutes for each atmosphere of pressure.

(4) Intestinal pneumatosis

This is a rare condition characterized by the development of multiple gas-containing cysts in the subserous tissues of the intestine, in particular the ileum, especially along its mesenteric border. The cysts occur in clusters, at first sessile, later pedunculated. The gas is under pressure and its main component is nitrogen. The cyst walls are thin and usually show no reactive changes, but sometimes they are thickened and contain lymphocytes, macrophages and foreign-body giant-cells. The cysts cause no symptoms and are discovered only incidentally at operations for other conditions. Their mode of origin is obscure.

INTRINSIC FREE FATS AND LIPOIDS IN TISSUES

Free fatty and lipoidal substances appear in the tissues in a great many different circumstances, which may be grouped conveniently as follows :

(1) Fat necrosis ;
(2) Fat embolism ;
(3) The phagocytosis of free sebum ;
(4) Chronic inflammations with degeneration ;
(5) Other local degenerative lesions ;
(6) Metabolic disorders.

(1) Fat necrosis

This means the destruction of adipose tissue with the liberation of free neutral fat. It occurs from direct physical injuries such as crushing or incision, as in fat necrosis of the breast ; or from chemical injury, as in fat necrosis of peritoneal and retroperitoneal tissues resulting from spilt pancreatic juice or bile ; or from bacterial infections in adipose tissue, as in some cases of chronic infective mastitis, subcutaneous cellulitis, or peritonitis. The free fat in the tissues is soon surrounded by macrophages and foreign-body giant-cells formed from them, and the fat is split into glycerol and fatty acids, no doubt by the agency of lipases from these phagocytes. The glycerol passes quickly into solution and is absorbed ; the fatty acids, either as such or combined with bases to form soaps, are taken up by the phagocytes. Macrophages loaded with these substances, along with other lipoidal products of the tissues, have a foamy vacuolated structure and are called " foam-cells ". The accompanying giant-cells may show similar foamy cytoplasm or may contain refractile crystals of fatty acids or soaps (see Fig. 21, p. 52).

(2) Fat embolism

Droplets of free neutral fat sometimes enter the blood-stream from injuries in which adipose tissue is in close contact with large veins, especially from extensive fractures involving yellow marrow. On rare occasions, this has also occurred following injury or toxic damage of a fatty liver ; e.g. MacMahon and Weiss saw an alcoholic in whom carbon tetrachloride poisoning had freed so much fat from his fatty liver into the hepatic veins that the blood in the right heart chambers and pulmonary artery contained 60 per cent fat. The principal danger of fat embolism is extensive plugging up of the pulmonary arterioles and capillaries by the oil droplets, causing circulatory embarrassment or supervening broncho-pneumonia. But some of the oil often traverses the pulmonary circulation and may be found also in the small vessels of the brain, kidneys and other viscera ; and in some cases death results from cerebral embolism.

(3) Phagocytosis of free sebum

This occurs if a sebaceous cyst is ruptured and its contents spilt into the surrounding subcutaneous tissues. It occurs also in the walls of intact sebaceous cysts or skin-lined cystic teratomas (" dermoid cysts ") of the ovary or other parts, if the epithelium lining these cavities is partly desquamated and the connective tissues exposed ; the denuded lining becomes replaced by a layer of granulation tissue rich in foam-cells and foreign-body giant-cells. Granulation tissue with many phagocytes engulfing sebum is also seen lining the chronically inflamed, partly skin-lined, hair-bearing sinuses (*pilonidal sinuses*) which sometimes develop at the site of the post-anal dimple (Fig. 102).

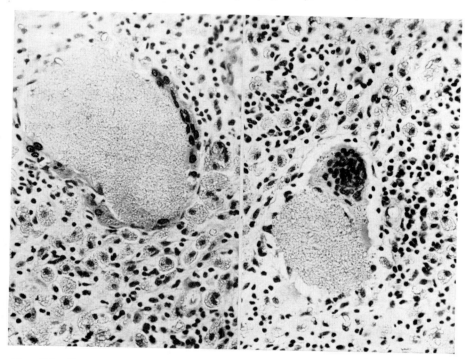

FIG. 102.—Foam-cells and foreign-body giant-cells engaged in the phagocytosis of free lipoids in the wall of a pilonidal sinus. (× 350.)

(4) Chronic inflammations with degeneration

In tuberculous, syphilitic and many other chronic inflammatory lesions, macrophages and foreign-body giant-cells concerned with the phagocytosis of lipoid degenerative products are frequently present. In these lesions phagocytosis of the causative bacteria is also in progress, and it may be impossible to say of any particular giant-cell whether it is engaged in taking up bacteria or degenerative lipoid material. That the phagocytosis of such material is common, however, is shown by its demonstrable presence in droplet form in foam-cells and giant-cells and sometimes by the presence of lipoid crystals in giant-cells.

Lymph glands draining areas of chronic inflammation often contain large numbers of lipoid-laden macrophages which have migrated there.

(5) Other local degenerative lesions

Lipoid-laden foam-cells abound in atheromatous patches in arteries, the degenerating peripheral segments of severed nerves, softened areas of brain due to vascular occlusion, infarcts in other organs, and degenerated areas in tumours. All of these are described in other chapters. Degenerating central or peripheral nervous tissue attracts very numerous macrophages, because of the great amount of free myelin to be removed. Foam-cells are often plentiful also in and around absorbing haemorrhages, where they are engaged especially in the removal of the lecithin of the degenerating red corpuscles.

(6) Metabolic disorders

Metabolic diseases associated with the deposition of lipoid substances in various tissues are described in Chapter 25. The accumulated lipoids are taken up by the fixed reticulo-endothelial phagocytes in the spleen, lymph glands, liver or bone marrow, or by wandering macrophages and foreign-body giant-cells formed from them in other tissues.

INTRINSIC NON-LIPOIDAL SUBSTANCES FREE IN TISSUES

Other products of the body itself which may be extravasated or may accumulate in tissues, and may attract macrophages and foreign-body giant-cells are (Figs. 103, 104) :

(1) *Hairs*, e.g. in the walls of dermoid or teratomatous cysts or pilonidal sinuses ;

(2) *Masses of keratin*, e.g. in squamous-cell carcinomas or degenerated dermoid cysts ;

(3) *Thyroid colloid*, e.g. in Riedel's thyroiditis ;

(4) *Extravasated bile*, e.g. in the liver in obstructive jaundice ;

(5) *Retained mucus*, e.g. in some mucus-secreting tumours ;

(6) *Amyloid*, e.g. in generalized amyloidosis or in local amyloid "tumours " ;

(7) *Sodium biurate crystals* in the tophi of gout.

CALCULI

A calculus (=a small stone, from Lat. *calx*) is a solid concretion formed by precipitation from the liquid contents of a hollow viscus. The composition of calculi naturally varies according to their site of formation and the conditions determining their deposition. They are usually mixed in composition, and do not often consist of a single substance in pure or nearly pure form. Factors which contribute to their formation are—(*a*) excessive concentration of the substances concerned ; (*b*) inflammatory or other lesions of the viscus, causing chemical changes in its fluid contents, often with desquamation or the precipitation of particles of organic matter which provide nuclei around which calculi are deposited ; (*c*) obstructive lesions causing stagnation of the contents of the viscus ; and (d) metabolic disorders accompanied by the accumulation of excessive amounts of particular substances. In view of this complexity of factors concerned in the deposition of calculi, it is not surprising that in many individual cases we cannot specify precisely just what factors have been responsible.

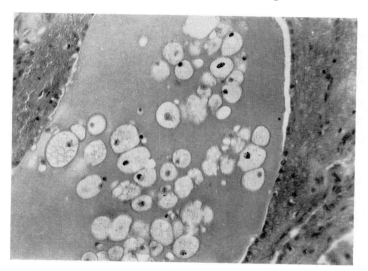

Fig. 103.—Foam-cells in a colloid cyst of the thyroid. (× 300.)

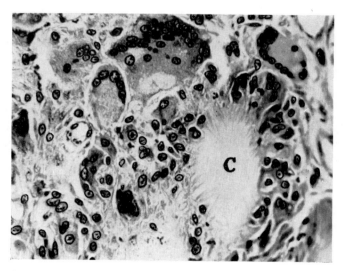

Fig. 104.—Foreign-body giant-cells clustered around deposits of sodium biurate crystals (C) in a gouty tophus. In preparing the section, the crystals have been dissolved out, leaving an empty space with only a vague suggestion of crystalline structure: see also Fig. 112. (× 300.)

(1) Salivary calculi

Salivary calculi (sialoliths) occur chiefly in the ducts of the submandibular gland ; they are much less frequent in the parotid ducts, probably because the parotid secretion is more serous and poorer in salts. The calculi consist of calcium phosphate and carbonate, usually with a small amount of organic matter. Probably they are formed as the result of bacterial infection, by the deposition of calcium salts around nuclei of altered mucus, detached epithelium or bacteria. A salivary calculus is usually single, small and rounded, or elongated and about the size of a date stone. It causes obstruction of the duct, with periodic pain and swelling of the gland, often accompanied by chronic bacterial infection.

(2) Biliary calculi or gall-stones (cholelithiasis)

(*a*) *Types and causes of gall-stones*

Gall-stones are of three kinds—(i) pure pigment stones, (ii) pure cholesterol stones, and (iii) the common mixed stones.

(i) *Pure pigment stones* are the least common ; they are small multiple dark stones, sometimes irregular and friable, sometimes smooth and hard ; they are most frequent in patients who are excreting excessive amounts of bile pigments, e.g. in haemolytic jaundice or pernicious anaemia.

(ii) *Pure cholesterol stones* are pale yellow or colourless, single or multiple, rounded, or ovoid, with a smooth or mulberry surface, and often a beautiful radial crystalline structure on section. Solitary cholesterol stones are most frequent in the gall-bladder in adipose middle-aged women and unaccompanied by any evidence of previous cholecystitis. The cholesterol normally present in bile is held in solution by the bile salts ; and it is easy to imagine how a slight excess of cholesterol or deficiency of bile salts could lead to precipitation of the former during concentration of the bile in the gall-bladder. Multiple cholesterol stones are most frequent in the " strawberry gall-bladder " (cholesterosis), in which the mucosa is studded with bright yellow granules, which are sub-epithelial collections of foam-cells or macrophages loaded with cholesterol esters. These granules often project and become polypoid, and are then detached into the gall-bladder where they may form the nuclei of cholesterol stones. It is probable that this focal cholesterosis is due to mild cholecystitis, with consequent disturbance in the absorptive functions of the gall-bladder as regards cholesterol.

(iii) *Mixed stones* are the commonest. They are composed of bile pigment, cholesterol and calcium salts in varying proportions. They often contain a nucleus of desquamated epithelium, bacteria or other organic matter, and they are usually laminated in structure. In most cases they are multiple or faceted, and may be few or many, large or small. When few or single, they often fill and form a complete cast of the gall-bladder, either as a single large ovoid stone or as a series of 3 or 4 barrel-shaped ones fitting neatly together. Mixed stones are often associated with chronic bacterial cholecystitis, and in many cases it is clear that inflammation and obstruction are the factors responsible for their formation.

(*b*) *Sites of gall-stones*

Gall-stones are far commoner in the gall-bladder than in the bile ducts. This is attributable to the two facts that the bile is concentrated in the gall-bladder and that the gall-bladder is a common site of bacterial inflammation. Stones in the bile ducts have usually arrived there from the gall-bladder ; but they sometimes form primarily in the ducts, usually as the result of infection or obstruction of the common duct. Stones in the ducts are more serious than stones in the gall-bladder ; in the latter site they often lie harmlessly for long periods, but in the former they almost always cause obstruction.

(c) Effects of gall-stones

Impaction of a stone in the neck or cystic duct of a non-infected gall-bladder shuts the organ off from the bile ducts, the bile within it is absorbed, and it becomes distended by colourless mucus secreted by its own epithelium (*mucocele*). If infection is present, obstruction of the neck or cystic duct by a stone will cause

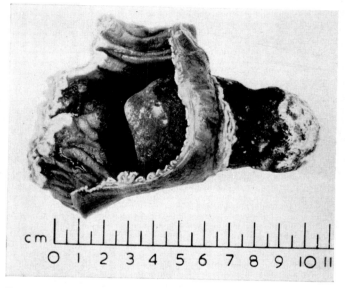

FIG. 105.—A gall-stone 3 inches long which became impacted in, and perforated, the small intestine.

acute cholecystitis (see p. 99). As a result of *chronic cholecystitis*, the gall-bladder may become adherent to the duodenum, into which a *fistula* from the gall-bladder may be formed, and gall-stones may pass through this into the intestine. A large gall-stone reaching the bowel in this way may become impacted in the small intestine and cause intestinal obstruction (Fig. 105). Gall-stones in the common bile duct cause *biliary colic* due to spasmodic muscular contraction of the duct, and *obstructive jaundice*, with distension of the gall-bladder and of all the bile ducts above the site of obstruction ; and, if infection is present, *ascending suppurative cholangitis* and liver abscesses may supervene. A gall-stone impacted in the ampulla of Vater may obstruct the pancreatic duct and cause *haemorrhagic pancreatitis* (p. 623). A further complication of gall-stones is *carcinoma of the gall-bladder*, in cases of which gall-stones are almost always present (p. 503).

(3) Pancreatic calculi

Calculi in the pancreatic ducts are rare. They are single or multiple, vary in size from fine gravel to masses 2 or 3 centimetres in diameter, and occupy any part of the organ, most often the main ducts in the head. They are composed mainly of calcium phosphate and carbonate, and are probably the result of catarrhal inflammation of the ducts. Obstructive dilatation of the ducts and chronic fibrosing pancreatitis are frequently present ; and occasionally an abscess develops around a calculus. Gall-stones often coexist.

(4) Faecoliths

Faecoliths (enteroliths, stercoliths or bowel-stones) are usually friable, putty-like concretions composed of a mixture of faecal matter and calcium and magnesium phosphate. Small faecoliths are frequently present in the appendix and in diverticula of the colon, where they may cause obstruction and inflammation (p. 99). Large faecoliths in the main bowel lumen are found chiefly in constipated old or sick people, in whom intestinal obstruction may result from impaction of almost dry, hard faecal concretions in the rectum. Enteroliths in the small intestine are rare ; they are sometimes found in the loop of bowel just proximal to a chronic stricture.

(5) Urinary calculi

(a) Mode of formation of calculi

Urine is a highly complex and variable solution of many inorganic and organic substances, the former including chlorides, phosphates and sulphates of sodium, potassium, calcium, magnesium and ammonium, the latter including urea, uric acid, urates, creatinine, hippuric acid, various amino-acids, oxalic acid, and traces of a great many other substances. Many of these substances are of low solubility and are easily precipitated from normal urine when this is concentrated or when it is allowed to cool. The nature of the precipitates depends on the reaction of the urine. Phosphates, namely, crystalline ammonium magnesium (or triple) phosphate and calcium hydrogen (or stellar) phosphate, and amorphous calcium and magnesium phosphates, are frequently deposited in alkaline urines. Calcium oxalate, uric acid and urates are frequently deposited in acid urines. We might expect then, that important factors in the deposition of urinary calculi in the body would be the concentrations of these precipitable constituents (largely dependent on the diet and the amount of water ingested), the reaction of the urine, and the presence of inflammatory or other diseases of the urinary tract which may alter the reaction of the urine or add protein or other abnormal organic constituents to it. Experience confirms these expectations.

(b) Types and sites of calculi

Urinary calculi can be classified conveniently thus :

(1) *Primary calculi*, deposited from an acid urine in an otherwise healthy urinary tract, as the result of slow precipitation of normal solutes or as the result of metabolic disorders, namely (i) calcium oxalate stones, (ii) uric acid and urate stones, (iii) cystine stones.

(2) *Secondary calculi,* deposited from alkaline urine in an inflamed urinary tract, on a preformed nucleus of inflammatory exudate, bacteria, a foreign body or a primary calculus, constituting (iv) phosphate stones.

(i) *Calcium oxalate* forms rounded or irregular, mulberry-like or spinous, hard crystalline calculi, usually deposited in the renal pelvis but often entering and becoming impacted in the ureter.

(ii) *Uric acid and urate stones* form in either the renal pelvis or bladder. They are hard but smooth, yellow or brown, rounded, ovoid or irregular, sometimes forming a large cast of the renal pelvis.

(iii) *Cystine stones* are rare. They consist of the sulphur-containing amino-acid cystine which appears as flat hexagonal crystals in the urine of people suffering from the familial metabolic disorder *cystinuria.* In the small proportion of these people who develop cystine renal calculi, these often appear in childhood, are multiple, and recur repeatedly after removal.

(iv) *Phosphate stones* are white or grey, usually friable, and composed mainly of triple phosphate, sometimes with other phosphates or calcium carbonate admixed. They are most frequent in the bladder in cases of cystitis, prostatic enlargement, urethral stricture, or foreign body. They also occur as secondary deposits around urate or oxalate calculi in the renal pelvis, where they form the majority of large " stag-horn " casts.

(*c*) *The effects of urinary calculi*

(i) *In the kidney.*—The ill effects of renal calculi are obstructive and infective. Obstruction of the outflow of urine leads to distension of the pelvis and calices (hydronephrosis) with atrophy and fibrosis of the distended renal tissue around. Supervening infection by pyogenic cocci or coliform bacilli leads to pyonephrosis and pyelonephritis, and to further phosphatic deposits on the calculi.

(ii) *In the ureter.*—A stone in the ureter is almost always a renal stone, usually of calcium oxalate, arrested in its descent. During its passage down the ureter it causes severe renal (more correctly, ureteric) colic due to spasmodic muscular contraction of the ureter, often with haematuria. If arrested, it obstructs the ureter partially or completely and leads to hydronephrosis and hydroureter above the obstruction.

(iii) *Stones in the bladder.*—These may be primary ; but they are more often secondary to cystitis or obstructed vesical outflow, and their effects are submerged in those of the primary disease.

(iv) *Stones in the urethra.*—Such stones are rare ; they have usually migrated there from the bladder, but becoming impacted in the urethra they grow there by phosphatic accretion. They cause urinary obstruction, with distension of the bladder and bilateral hydronephrosis, and often ascending infection with cystitis and pyelonephritis.

(6) Prostatic calculi

These are rare. They are usually small and multiple, but may be large and solitary. They consist of mixtures of calcium phosphate, carbonate and oxalate, usually deposited around nuclei of organic debris or bacteria ; and they are probably the result of a low-grade chronic infective prostatitis.

(7) Phleboliths

Phleboliths (vein-stones) are the result of calcification of sterile thrombi, resulting from stagnation of blood in dilated tortuous veins. They occur in old thrombosed varicose veins of the legs, in varices of the spermatic cord and scrotum (varicocele), and in the parametrial and other pelvic veins. The small round shadows of phleboliths are often seen in X-ray pictures of the pelves of elderly people. Dissection reveals the phleboliths as hard spherical bodies like white shot lying immovably wedged within occluded veins or varicose side-pockets of still patent vessels.

SUPPLEMENTARY READING

Bruce, G. G. (1936). " A case of hair-ball of the stomach and duodenum in a child of 3½ years ", *Brit. J. Surg.*, **23**, 855.

Collins, D. H. (1948). " Bullet embolism : a case of pulmonary embolism following the entry of a bullet into the right ventricle of the heart ", *J. Path. Bact.*, **60**, 205.

Davson, J., and Susman, W. (1937). " Apical scars and their relationship to siliceous dust accumulation in non-silicotic lungs ", *J. Path. Bact.*, **45**, 597.

Gough, J. (1959). " Occupational pulmonary diseases." In *Modern Trends in Pathology* (Ed., D. H. Collins), Chap. 14. London; Butterworths.

Illingworth, C. F. W. (1929). " Cholesterosis of the gall-bladder ", *Brit. J. Surg.*, **17**, 203. (Good figures.)

Mackey, W. A., and Gibson, J. B. (1948). " Siliceous granuloma due to talc : a cause of post-operative peritoneal adhesions ", *Brit. Med. J.*, **1**, 1077. (Papers by Smith and Walker on the same subject appear in the same journal.)

— (1937). "Cholesterosis of the gall-bladder", *Brit. J. Surg.*, **24**, 570. (A beautifully illustrated account.)

MacMahon, H. E., and Weiss, S. (1929). " Carbon tetrachloride poisoning with macroscopic fat in the pulmonary artery ", *Amer. J. Path.*, **5**, 623.

Paterson, J. L. H. (1938). " An experimental study of pneumonia following the aspiration of oily substances : lipoid cell pneumonia ", *J. Path. Bact.*, **46**, 151.

Roberts, G. B. S. (1947). " Granuloma of the Fallopian tube due to surgical glove talc : siliceous granuloma", *Brit. J. Surg.*, **34**, 417.

Russell, D. S. (1941). " Fat embolism and the brain ", *Proc. Royal Soc. Med.*, **34**, 645.

Stewart, M. J., Tattersall, N., and Haddow, A. C. (1932). " On the occurrence of clumps of asbestosis bodies in the sputum of asbestos workers ", *J. Path. Bact.*, **35**, 737.

Symmers, W. St.C. (1955). " Simulation of cancer by oil granulomas of therapeutic origin ". *Brit. Med. J.*, **2**, 1536.

Teoh, T. B. (1963). " A study of gall-stones and included worms, in recurrent pyogenic cholangitis", *J. Path. Bact.*, **86**, 123.

Wakeley, C. P. G., and Willway, F. W. (1935). " Intestinal obstruction by gall-stones ", *Brit. J. Surg.*, **23**, 377.

Ward-McQuaid, N. (1950). "Intestinal obstruction due to food". *Brit. Med. J.*, **1**, 1106.

Watson, A. J. (1937). " Fat embolism : report of a case, with review of the literature ", *Brit. J. Surg.*, **24**, 676. (Excellent figures.)

CHAPTER 24

EXTRANEOUS POISONS

EXTRANEOUS POISONS are chemical substances which, applied to or introduced into the body, have harmful structural or functional effects. Almost all soluble elements or compounds, which are not natural constituents of our diet, are harmful to us if introduced in adequate amounts. The chemical environment of modern civilized man has become exceedingly complex ; he ingests, inhales or comes into contact with a bewildering variety of industrial and technological reagents and by-products, dyes, antiseptics, synthetic flavours, colours and cosmetics, and he consumes large quantities of drugs many of which in adequate doses are poisonous. The study of all kinds of poisons and their effects is the province of *toxicology*, while *pharmacology* also is concerned with the poisonous as well as the beneficial effects of drugs. The pathologist cannot and need not be an expert in these two subjects ; but he should have a general knowledge of their principles and a particular knowledge of the properties and effects of the more common poisons which bring about, slowly or quickly, serious structural changes in the body.

A discussion of poisons may be undertaken in either of two ways : the individual poisons may be classified and dealt with one by one ; or the principal pathological effects of poisons may each be considered, including a statement of the substances known to cause them. We will adopt the second method, because it focuses attention on the pathological changes ; but we will also summarize the main effects of particular inorganic poisons. This chapter will not discuss carcinogenic chemical agents ; these are reserved for Chapter 29.

We will consider in order :

> Locally corrosive and irritant poisons ;
> Poisons which injure the liver ;
> Poisons which injure the kidneys ;
> Poisons which injure haemopoietic tissues ;
> Poisons which affect circulating red corpuscles ;
> Certain metallic and non-metallic poisons.

LOCALLY CORROSIVE AND IRRITANT POISONS

A substance which directly corrodes or destroys tissues to which it is applied is a *corrosive* or *caustic poison*. If you spill on yourself some strong nitric or sulphuric acid or a strong solution of sodium or potassium hydroxide, it will make holes not only in your clothing but in your skin. The destructive injuries inflicted by such substances are spoken of as " chemical burns ". The chief corrosive poisons are strong acids and alkalis, strong solutions of phenol (carbolic acid), cresols (e.g. lysol), oxalic acid, chromic acid and chromates, zinc chloride, copper sulphate, silver nitrate and antimony trichloride. What corrosive poisons

331

do to the skin they will do also to the mucous membranes of the mouth, oesophagus and stomach if they are swallowed ; and in fatal cases, these mucous membranes are found discoloured, necrotic and sloughing. If the patient survives, healing often produces fibrous strictures in the oesophagus.

An *irritant poison* is one which, though it does not destroy tissues outright when applied to them, injures them severely enough to excite acute inflammation. Corrosive poisons in dilute form are irritants ; and other substances which, when ingested in sufficient amounts, produce acute inflammation of the gastric and intestinal mucous membranes include arsenious oxide (white arsenic), antimony tartrate (tartar emetic), mercuric chloride (corrosive sublimate) and other soluble mercury salts, soluble lead, copper and zinc salts, phosphorus, excessive doses of vegetable purgatives such as castor oil, croton oil, colocynth, aloes and podophyllin, and the medicinal dried beetle cantharides.

Irritant gases constitute a special group of poisons, because of their effects on the respiratory system when inhaled. They include chlorine, bromine, fluorine, ammonia, sulphur dioxide, hydrogen sulphide, nitric fumes (nitric and higher oxides of nitrogen), nickel carbonyl ($Ni(CO)_4$), phosgene or carbonyl chloride ($COCl_2$), chloropicrin or trichloronitromethane (CCl_3NO_2), mustard gas and lewisite. The last four have been used in chemical warfare ; mustard gas and lewisite, besides being dangerous by inhalation, cause also severe inflammation and blistering of the skin. The effects of irritant gases on the respiratory tract vary with the nature and concentration of the gas ; they include mild or severe rhinitis, tracheobronchitis, pulmonary oedema and haemorrhages, and supervening bacterial bronchopneumonia.

It remains only to add here that when a patient survives the immediate local corrosive or irritant effects of a dangerous dose of a poison, he is still not out of the wood. He may yet succumb to the toxic effects of the poison following its absorption, effects due to injury to the liver, kidneys, blood or nervous system.

POISONS WHICH INJURE THE LIVER

(1) Hepato-toxic substances

In view of its exposure to all substances absorbed into the portal blood from the alimentary canal, and its detoxicating and other metabolic functions, it is not surprising that the liver should bear the brunt of the toxic effects of many extraneous poisons. It has been shown experimentally that the introduction of small quantities of alcohol, chloroform, carbon tetrachloride, or many other organic substances, into the portal blood-stream will produce rapid massive necrosis of the liver.

The number of known hepato-toxic chemical substances now runs into scores, and doubtless there are a great many more as yet unrecognized. They have been classified in various ways—according to their chemical nature, according to their usual mode of introduction (whether occupational, medicinal, dietetic, or accidental), or according to whether they usually produce massive necrosis of liver tissue or only zonal necrosis restricted to particular zones of each lobule. All of these distinctions are somewhat arbitrary ; but the following grouping of the main substances known to be responsible for liver injury in man is convenient :

> *Mode of administration mainly occupational*
>> Carbon tetrachloride
>> Trinitrotoluol (T.N.T.)
>> Tetrachlorethane ("dope")
>> Chlorinated naphthylene
>> Phosphorus
>
> *Mode of administration mainly medicinal*
>> Chloroform (used as an anaesthetic)
>> Tannic acid (used for treating burns)
>> Cincophen (used for treating gout)
>> Arsphenamine and other organic arsenical drugs
>> Organic gold compounds (used for treating tuberculosis, arthritis, etc.)
>> Sulphonamides (used as anti-bacterial remedies)
>> Other drugs (see Cook and Sherlock)
>
> *Mode of administration mainly dietetic*
>> Alcoholic drinks (identity of hepato-toxic ingredients still unknown)
>> Selenium (in grain grown on seleniferous soils)
>> Copper (?)

(2) Other causes of liver damage

Before we proceed to consider the effects of liver poisons of extraneous origin, it is important to state that these are not the only causes of liver damage. As we have already seen, in some microbic diseases (e.g. yellow fever, infective hepatitis and Weil's disease), the liver is a main site of infection, while in others (e.g. typhoid fever, diphtheria, pneumonia and severe pyogenic infections) there may be serious toxic changes in this organ. Eclampsia—a severe toxaemia of pregnancy, of unknown cause—is accompanied by extensive liver damage. Dietetic protein deficiency as a cause of hepatic necrosis and cirrhosis is discussed in Chapter 25. Thus, a great many different factors are capable of injuring liver tissue. In relatively few cases of liver necrosis or cirrhosis in man are we able to specify the causative factors with certainty ; in the majority the nature and source of the hepato-toxic factors are unknown. It is probable, however, that in many cases they will prove to be extraneous poisons as yet unrecognized in the occupational environments, diets or drugs of the victims, the effects of which may sometimes be conditioned or aggravated by dietary deficiencies.

This is the place to mention the obscure causation of "alcoholic" cirrhosis. While there is no doubt that intemperate beer or spirit drinkers often develop fatty or cirrhotic livers, the factors responsible are obscure. Alcohol itself is probably not the cause ; but alcoholic drinks contain many other organic compounds besides alcohol, and some of these may be to blame. Nor is it impossible that these drinks may contain inorganic liver poisons ; Mallory suspected phosphorus, free traces of which he found in acid fluids in contact with iron or tin vessels. On the other hand, it may be that "alcoholic" cirrhosis results mainly from dietary deficiency (see p. 350).

(3) The degrees of toxic damage to the liver

We need not go into details of the effects of particular hepato-toxins. In general, apart from mild transient toxic damage with subsequent complete restitution, we recognize three descending grades of severity of toxic injury of the liver :

(a) If a large dose of a liver poison is administered at once or over a short period, it may produce rapid extensive necrosis of liver tissue. This is usually fatal within a few days, when the liver is found to be shrunken, wrinkled, soft, and yellow or mottled red and yellow—a condition called " *acute yellow atrophy* ", or better *acute necrosis*.

(b) If the necrosis is less extensive, the patient may survive for a few weeks and then die, when his liver is said to show *subacute necrosis*. The organ is

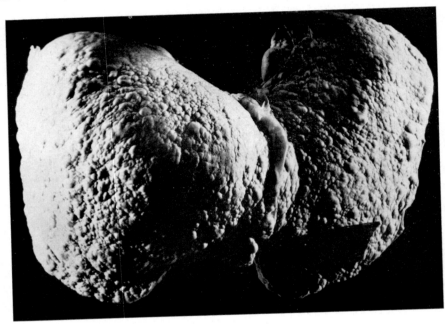

FIG. 106.—Laennec's cirrhosis—" hob-nail liver ".

coarsely irregular ; it shows alternating shrunken areas of cellular fibrous tissue replacing necrotic parts, and projecting nodular masses of regenerating liver tissue growing from surviving parts.

(c) If small doses of an hepato-toxic agent are administered over a long period, the repeated slight injuries lead to death of a few cells at a time, widespread fibrosis, and regeneration of liver cells in small foci throughout the organ. The liver is usually smaller than normal, its surface is more or less evenly studded with fine or coarse nodules (" hob-nail " liver), and on section it shows widespread fibrosis often visibly spreading out from the portal tracts and dividing the cut surface up into an irregular mosaic of large or small areas representing groups of liver lobules. This is the typical *cirrhosis* of the liver, variously called multilobular cirrhosis, portal cirrhosis, atrophic cirrhosis or Laennec's cirrhosis (Fig. 106).

(4) The meaning of " cirrhosis "

The term " cirrhosis " (from Greek *kirros*=tawny) does not denote a well-defined disease, but is applied to any condition of widespread, nearly uniform fibrosis of the liver.   This can result not only from hepato-toxic agents, but also from dietary deficiencies (p. 350), specific microbic infections (p. 333), chronic obstruction or infection of the bile ducts (biliary cirrhosis, p. 623), and the metabolic disease haemochromatosis (p. 356). All of these lead to extensive, more or less uniform hepatic fibrosis, or " cirrhosis ". This name should not be applied to localized fibrotic lesions of the liver, such as the scars of healed gummata or abscesses, or the fibrous areas left by subacute necrosis. (The effects of cirrhosis in producing obstruction of the portal circulation are described on p. 384.)

POISONS WHICH INJURE THE KIDNEYS

It is not surprising that the kidneys should often sustain injury during urinary excretion of poisonous substances from the blood.   The chief of such extraneous substances are the soluble salts of mercury, uranium, chromium, cadmium, bismuth and copper, and, amongst organic poisons, carbon tetrachloride, iodoform, and dioxan (diethylene dioxide). Mercuric chloride, potassium bichromate, and uranium salts have frequently been used to produce renal disease experimentally. The renal changes seen in experimental animals, and also in cases of human poisoning by these substances, are chiefly in the convoluted tubules, the epithelial cells of which show cloudy and fatty degeneration, necrosis, desquamation, and sometimes calcification, followed by regenerative proliferation ; glomerular changes are inconspicuous, but thickening of the basement membrane of the glomerular capillaries has been observed.

Since albuminuria and generalized oedema have been produced in rabbits by uranium poisoning, it is possible that the renal damage caused by this poison is similar to that occurring in Type II nephritis in man. Hence, we must not overlook the possibility that human Type II nephritis, the causes of which are unknown (see p. 303), may in some cases be due to unidentified extraneous poisons. However, there is no evidence to show that this is so.

POISONS WHICH INJURE THE HAEMOPOIETIC TISSUES

Many substances have selective injurious effects on red bone marrow, interfering with erythropioesis, leucopoiesis, thrombocytopoiesis, or all three, and thereby producing in varying degrees and combinations anaemia, leucopenia and thrombocytopenia. According to the nature and dosage of a haemopoietic poison, its effects on the bone marrow and circulating blood may be mild, moderate or severe, slowly or rapidly progressive, recoverable or irrecoverable. If the injury is severe and there is little or no evidence of regeneration of the injured haemopoietic tissues, we speak of *aplasia* of the bone marrow. Toxic aplasia sometimes affects especially the erythroblastic elements, producing *aplastic anaemia* ; sometimes it affects especially the leucoblastic elements, producing *agranulocytosis* ; in many cases it affects all of the haemopoietic elements simultaneously or successively, producing severe *thrombocytopenia* as well as anaemia and leucopenia. Some poisons arrest the proliferation of the primitive haemopoietic cells ; others interfere with the maturation of particular elements at later stages.   The proper interpretation of the effects of haemopoietic poisons necessitates careful examination of the changes in the bone marrow as well as those in the circulating blood.

The known haemopoietic poisons include benzol, trinitrotoluol, nitrobenzene, aniline, mustard gas and the nitrogen-mustards ; salts of radium, thorium and other radioactive metals ; organic compounds of arsenic and gold ; the sulphonamide drugs, dinitrophenol (used for " slimming "), thiourea and thiouracil (used in treating toxic goitre), and sedative drugs of the amidopyrine (Pyramidon) group.   The compounds towards the beginning of this list are especially apt to produce aplastic anaemia ; those towards the end, agranulocytosis.   The damaging effects of radium and thorium on the bone marrow are due to their radioactivity and to the fact that these heavy metals accumulate and are stored in the bones.

It must be added here that aplastic anaemia and agranulocytosis are not always caused by known poisons.   There are many cases of these diseases which are " idiopathic ", i.e. of unknown cause ; some of these are doubtless due to unidentified poisons, while others are due to nutritional, vitamin or hormonal disturbances of haemopoiesis (see Chapter 27).

POISONS WHICH AFFECT CIRCULATING RED CORPUSCLES

In the following diseases, red corpuscles already in circulation are affected by extrinsic poisons :
   (1) Acute haemolytic anaemia ;
   (2) The anaemia of lead poisoning ;
   (3) Carbon monoxide poisoning ;
   (4) Toxic enterogenous cyanosis ;
   (5) Toxic porphyrinuria.

(1) Acute haemolytic anaemia

Certain poisons in adequate doses produce direct destruction of circulating red cells with liberation of haemoglobin into the blood plasma.   These include potassium chlorate, arseniuretted hydrogen (arsine, $As H_3$), pyrogallic acid, nitrobenzene, many snake venoms, the plant substances saponin and ricin, and the drug phenylhydrazine used in the treatment of polycythaemia.   Following poisonous doses of these substances, the haemoglobin freed from the destroyed red cells may be partly excreted by the kidneys and partly taken up by the reticulo-endothelial tissues and converted into bilirubin.   If the blood destruction is rapid, the patient passes much unaltered or altered haemoglobin in the urine (haemoglobinuria and methaemoglobinuria) ; but if the haemolysis is less rapid, most or all of the pigment may be converted into bilirubin, with resulting bilirubinaemia and often visible jaundice.

(2) Anaemia of lead poisoning

Anaemia is one of the earliest indications of chronic lead poisoning.   The red-cell count may be reduced to 3,000,000 per cubic millimetre or lower ; there is a still greater reduction of haemoglobin, so that the colour index is less than 1.

A very characteristic feature of the blood film is the great number of red cells showing punctate basophilia (stippled cells). These are really young red cells or reticulocytes in which the basophilic reticulum has been altered by the lead. The anaemia of lead poisoning is due to the fact that the affected red cells are short-lived and readily destroyed. The high reticulocyte and stipple-cell count is due to active compensatory regeneration of erythroblasts in the bone marrow.

(3) Carbon monoxide poisoning

Carbon monoxide (CO) may be inhaled in dangerous amounts from coal gas (a common means of suicide), from badly ventilated coke or coal fires, or from exhaust gases of petrol engines. Since its affinity for haemoglobin is much stronger than that of oxygen, when inhaled it displaces oxygen from the oxyhaemo-globin of the red cells and forms the very stable cherry-red pigment *carboxy-haemoglobin*. When one-third of a man's haemoglobin is converted into carboxy-haemoglobin, symptoms of poisoning appear; when one-half of it is converted, he is in grave danger. Carbon monoxide reduces the supply of oxygen to the tissues by usurping the place of oxygen in the red cells; it is also a tissue poison, producing muscular paralysis, loss of consciousness, and myocardial degeneration, and, in persons who survive, degenerative changes in the brain with Parkinsonism and mental deterioration.

(4) Toxic enterogenous cyanosis

This rare condition is characterized by chronic cyanosis (a blue or muddy complexion), unaccompanied by polycythaemia or dyspnoea, and due to the presence of methaemoglobin and sulphaemoglobin in the red cells of the blood. It is usually the result of the habitual use of certain drugs, especially acetanilide, phenacetin, or sulphonamides. The patient is usually constipated, and the forma-tion of sulphaemoglobin is attributed to the absorption of H_2S from the stagnant intestinal contents. Enterogenous cyanosis also occurs without a history of drug-taking, apparently as a result of an excessive production of nitrites by bacteria in the intestine.

(5) Toxic porphyrinuria (haematoporphyrinuria or porphyria)

Red urine due to the presence of haematoporphyrin occasionally results from taking certain drugs, e.g. sulphonal, Trional or sulphonamides. There also occur cases of porphyrinuria of unknown cause.

CERTAIN METALLIC AND NON-METALLIC POISONS

It will be convenient to summarize briefly the chief effects of poisoning, especially chronic poisoning, by the soluble salts of certain of the metals and one or two other elements.

(1) Lead

Chronic lead poisoning, also called *plumbism* or *saturnism**, occurs as an occupational disease in smelters, pottery glazers, plumbers, painters, etc., and also

---

* Ancient astrologers credited the planet Saturn with producing a heavy gloomy (lead-like) temperament in those born under its influence; hence also the adjective *saturnine*.

accidentally in persons drinking water or other fluids contaminated from lead pipes, paints or solder.   The following are its chief effects :

(i) *Lead anaemia* (see above).

(ii) *Lead colic* is the commonest symptom.   The patient suffers from repeated attacks of severe intestinal colic with constipation, probably the result of a direct effect of lead on the intestinal musculature.

(iii) *Lead paralysis* usually occurs first in the extensor muscles of the wrists, producing wrist-drop, but it extends later to other muscles.   Though often attributed to " toxic neuritis ", it appears to be a direct effect of the lead on the muscles and to occur chiefly in muscles liable to over-fatigue.   Later, however, the lead also affects the motor fibres to the paralysed muscles and causes varying degrees of degeneration in them.

(iv) *Lead encephalopathy,* a rare result, is characterized by convulsions, delirium or coma.

(v) *A blue line on the gums* is often present, due to a deposit of lead sulphide.

(vi) *Vascular and renal changes,* such as hypertensive arteriosclerosis and renal fibrosis are present in cases of lead poisoning of long duration, perhaps because of a direct effect of the lead on the muscular coats of arteries, analogous to that on the intestinal muscle.

Lead is stored in the body mainly as tertiary phosphate in the bones, where it produces increased density in radiographs, and from which it is eliminated only very slowly.

(2) Mercury

(a) Acute mercurial poisoning

Apart from the direct corrosive effect of mercuric chloride on the upper parts of the alimentary tract, the two chief dangers of acute mercurial poisoning are : (1) suppression of urine and uraemia, due to the effect on the kidneys (see p. 335) ; and (2) ulcerative colitis, accompanied by extensive necrosis and sloughing, due to the excretion of mercury by the large intestine.

(b) Chronic mercurial poisoning

Besides general debility and anaemia, the main effects are : (i) stomatitis, gingivitis, profuse salivation, and loosening of the teeth ; (ii) chronic ulcerative colitis ; (iii) dermatitis, especially from the fulminate of mercury handled in explosives factories ; (iv) muscular tremors ; (v) mental changes, e.g. a peculiar irritability known as " erethism " ; and (vi), in cases of poisoning by methyl-mercury compounds (used as agricultural fungicides), serious degenerative changes in the central nervous system and peripheral nerves, causing ataxia, dysarthria (impaired speech), and constriction of the fields of vision.

(3) Arsenic

(i) *Acute arsenical poisoning* produces acute gastro-enteritis (p. 332) and fatal circulatory and respiratory failure.

(ii) *Chronic arsenical poisoning* produces any or all of the following : (1) dermatitis, pigmentation and excessive keratinization of the skin, due to the fact that arsenic accumulates in and is retained by the epidermis and its appendages ; (2) chronic inflammation of mucous membranes—stomatitis, pharyngitis, laryngitis, conjunctivitis ; (3) peripheral neuritis of the lower or upper limbs or both, characterized by paralysis and atrophy of muscles, and by numbness, hyperaesthesia, and anaesthesia.

(4) Manganese

Chronic manganese poisoning produces degenerative changes in the nerve cells of the basal ganglia, with resulting signs of Parkinsonism, namely tremors and other involuntary movements, muscular rigidity, a mask-like face, and an unsteady gait.

(5) Silver

Chronic silver poisoning (*argyria*) results in widespread grey or black discoloration of the skin and organs due to the deposition of granules of silver in the reticulo-endothelial and connective tissues. Cases have been recorded in which the body contained as much as 100 grams of silver. This appears to be quite inert, causing no toxic effects in the pigmented organs and exciting no foreign-body reaction.

(6) Bismuth

When bismuth preparations were used for treating syphilis, toxic results were not unusual ; these have occurred also from absorption of bismuth from B.I.P.P. (bismuth-iodoform-paraffin paste) applied to infected wounds. The chief effects are nephritis, enteritis and stomatitis, often accompanied by bluish discoloration of the gums.

(7) Beryllium

Workers inhaling beryllium fumes or dusts may suffer from an acute pneumonitis or from chronic focal granulomatous lesions and fibrosis of the lungs.

(8) Phosphorus

Acute phosphorus poisoning produces widespread necrosis of the liver, accompanied by severe fatty changes in any surviving liver tissue and in the renal epithelium, myocardium, striated muscles, and other tissues. Chronic poisoning may produce similar though less severe fatty changes, anaemia, fragility of bones, loosening of the teeth, and bacterial infection and necrosis of the jaws.

(9) Fluorine

Since fluorine enters into the composition of dental enamel, it is not surprising that a high fluoride intake produces abnormalities of the enamel—mottling and discoloration. In addition, however, chronic fluoride poisoning, which occurs in certain parts of India, China and elsewhere as a result of fluoride-rich well-water, produces remarkable changes throughout the skeleton, the bones becoming heavy, thickened and covered with spinous osteophytes. Fluorine accumulates in the bones as calcium fluoride (CaF_2).

It would be out of place here to describe the toxic effects of a great many other substances which are used as drugs or which cause occasional accidental poisoning. These include the large and important group of plant alkaloids, e.g. strychnine, morphine, atropine, hyoscine, cocaine, nicotine, ergotine, curarine, and many others ; the glucosides, e.g. digitoxin, strophanthin, salicin, etc. ; and other poisons of plant or animal origin, e.g. muscarine, cantharadin, snake and spider venoms, etc. For details of these, works on pharmacology and toxicology must be consulted. They are mentioned here only as an indication of the great range of extraneous poisons to which man is exposed.

SUPPLEMENTARY READING

Britton, C. J. C. (1969). Tenth edition. Whitby and Britton's *Disorders of the Blood*. London; Churchill. (Chapters on poisons affecting blood and haemopoietic tissues.)

Cameron, G. R., and Karunaratne, W. A. E. (1936). " Carbon tetrachloride cirrhosis in relation to liver regeneration ", *J. Path. Bact.*, **42**, 1.

– – and Thomas, J. C. (1937). " Massive necrosis (' toxic infarction ') of the liver following intra-portal administration of poisons ', *J. Path. Bact.*, **44**, 297.

– , Milton, R. F., and Allen, J. W. (1943). " Toxicity of tannic acid ", *Lancet*, **2**, 179.

Cook, G. C. and Sherlock, S. (1965). " Jaundice and its relation to therapeutic agents". *Lancet*, **1**, 175.

Dutra, F. R. (1948). " The pneumonitis and granulomatosis peculiar to beryllium workers ", *Amer. J. Path.*, **24**, 1137.

Gettler, A. O., Rhoads, C. P., and Weiss, S. (1927). " A contribution to the pathology of generalized argyria with a discussion of the fate of silver in the human body ", *Amer. J. Path.*, **3**, 631.

Harmon, E. L. (1928). " Human mercuric chloride poisoning by intravenous injection ", *Amer. J. Path.*, **4**, 321.

Lyth, O. (1946). " Endemic fluorosis in Kweichow, China ", *Lancet*, **1**, 233.

Mallory, F. B. (1933). " Phosphorus and alcoholic cirrhosis". *Amer. J. Path.*, **9**, 557.

Moon, V. H. (1934). " Experimental cirrhosis in relation to human cirrhosis ", *Arch. Path.*, **18**, 381.

Stoner, H. B. and Magee, P. N. (1957). " Experimental studies on toxic liver injury". *Brit. Med. Bull.*, **13**, 102.

Turner, G. G. (1939). " Non-malignant stenosis of the oesophagus ", *Brit. J. Surg.*, **26**, 555. (Includes accounts of cases due to swallowing caustics.)

Wheeler, W. I. de C. (1930). " Pigmentation from bismuth absorption after the use of B.I.P.P.", *Brit. J. Surg.*, **18**, 329.

CHAPTER 25

NUTRITIONAL AND METABOLIC DISTURBANCES

THIS CHAPTER is to survey briefly the structural and functional changes which may result from deficiency or excess of the various constituents of our diet or from general or local disturbances of their metabolism in the tissues. As pathologists we are concerned primarily with the visible abnormalities of structure; but these should be correlated with the underlying biochemical changes accompanying disturbed nutrition, a subject which is dealt with admirably in Thorpe's book on Biochemistry. Here we will consider :

Starvation ;
Disturbances of carbohydrate metabolism ;
Disturbances of fat and lipoid metabolism ;
Disturbances of protein and purine metabolism ;
Disturbances of iron metabolism ;
Jaundice (icterus) ;
Disturbances of calcium metabolism ;
Vitamin deficiency diseases.

STARVATION

Starvation may result, not only from simple lack of food, but also from diseases which interfere with the intake or digestion of food, such as cancer of the oesophagus or stomach, pyloric stenosis, or any disease that causes persistent vomiting or diarrhoea. Total deprivation of food and water is usually fatal in 3-7 days ; but, with water, starved persons, especially professional fasters, may survive for several weeks. Partial starvation, including that caused by diseases which interfere with nutrition, may of course be very prolonged. The effects of starvation include : (1) emaciation, (2) anaemia, (3) oedema, (4) ketosis, and (5) other biochemical changes.

(1) Emaciation

The starved person wastes ; the wasting occurs first in stored substances, especially glycogen and fat, and later in the organs and tissues themselves. Liver glycogen is greatly diminished ; adipose tissue throughout the body becomes scanty and almost disappears ; the viscera and muscles all lose weight ; the blood volume is diminished ; and the bones, though not reduced in size, lose some of their calcium. The organs least affected are the central nervous system and heart ; the former remains almost normal to the end, and when the voluntary muscles have wasted to half their normal bulk the cardiac muscle may have lost only 2 or 3 per cent of its weight.

(2) Anaemia

Simple starvation causes only a moderate degree of anaemia ; this is of the hypochromic type, i.e. the haemoglobin is reduced to a greater degree than the number of red cells. There is also a fall in the number of circulating white cells

341

—leucopenia. In cases of slow starvation from alimentary and other diseases, much more severe anaemia is often present ; but factors other than starvation, such as haemorrhage or toxaemia, often contribute to this.

(3) Oedema

Prolonged under-nutrition with deficiency of protein may produce generalized oedema, called *famine oedema*. This is attributable mainly, possibly wholly, to hypoproteinaemia (see p. 349), though it is possible that increased permeability of the capillary endothelium may also result from the nutritional deficiency.

(4) Ketosis

In the metabolism of fats, the final oxidation of fatty acids requires the simultaneous oxidation of carbohydrate. In starvation, when the glycogen stores have been depleted and the body is utilizing only its fat for its energy requirements, incomplete oxidation of fatty acids results in the formation of aceto-acetic acid, β-hydroxybutyric acid and acetone (called collectively *acetone bodies*). The presence of these substances in the blood constitutes *ketosis*, and their excretion in the urine *ketonuria*. Since two of the acetone bodies are acids, their presence in the blood depletes the alkali reserve and constitutes *acidosis*. This acidosis is often an important factor in death from starvation.

(5) Other biochemical changes

Since the *respiratory quotient* (i.e. the ratio $\frac{\text{vol. of } CO_2 \text{ produced}}{\text{vol. of } O_2 \text{ used}}$) is unity for carbohydrate and 0·7 for fat, once the starving person has exhausted his carbohydrate stores and begun to use his fat, his R.Q. falls. When, after prolonged starvation, all the fat reserves of the body have been exhausted, tissue proteins remain as the sole source of energy. Their utilization is accompanied by a pronounced rise in the amount of nitrogen excreted in the urine, most of it as urea ; and when this occurs the patient is in a dangerous state. The excretion of ammonia nitrogen also rises, a reflex result of the acidity of ketosis. The starving body conserves chlorides, and the urinary excretion of chlorides is markedly reduced. The excretion of calcium also is reduced, but the body slowly loses calcium at the expense of the bones, the plasma calcium remaining at a normal level.

DISTURBANCES OF CARBOHYDRATE METABOLISM

To be mentioned briefly here are : (1) abnormalities in the distribution of glycogen, (2) glycosuria, (3) hypoglycaemia, and (4) gargoylism.

(1) Abnormalities in the distribution of glycogen

Glycogen or " animal starch " is the reserve carbohydrate of the body ; it is stored chiefly in the liver and muscles, but is present also in many epithelial cells, cartilage, and especially in embryonic tissues. Its synthesis (glycogenesis) and its hydrolysis to sugar (glycogenolysis) occur very rapidly in the body, the latter under the influence of the enzyme glycogenase (or phosphorylase). Post-mortem glycogenolysis takes place rapidly, so that two or three hours after death, usually little or no glycogen is left in the liver, muscles or other tissues. For its histological demonstration, small pieces of fresh tissue are fixed in alcohol, and sections are stained by Best's carmine method.

The glycogen stores of the body are depleted by violent muscular activity, starvation or diabetes. They are enormously increased in the rare congenital metabolic disorder, *glycogen disease* or *von Gierke's disease*, in which from early life the liver, heart, kidneys and other organs are greatly enlarged and their cells loaded with immobilized glycogen. The cells of some kinds of tumours contain large amounts of glycogen.

(2) Glycosuria

Glycosuria, the presence of more than a trace of glucose in the urine, will result from any condition which raises the blood-sugar level above the renal threshold for sugar or which lowers the renal threshold below the blood-sugar level. The normal renal threshold for glucose is about 180 milligrams per 100 cubic centimetres of blood ; concentrations of blood-sugar higher than this constitute *hyperglycaemia*. The varieties of glycosuria are :

(a) Alimentary glycosuria

Ingestion of a large amount of sugar may cause temporary hyperglycaemia and glycosuria, the rate of absorption of glucose into the portal blood exceeding the rate at which the liver can polymerize it into glycogen.

(b) Adrenaline glycosuria

Adrenaline stimulates glycogenolysis in the liver and so may cause hyperglycaemia. This is the cause of the glycosuria produced experimentally by stimulating the sympathetic nerves supplying the adrenals or by Claude Bernard's puncture of the floor of the fourth ventricle of the brain, or occurring as the result of an injection of adrenaline, anger, fear or anxiety (e.g. in students sitting for examinations), the administration of anaesthetics, or some pathological lesions in the hypothalamic region or brain stem.

(c) Diabetes mellitus

This, the most important disturbance of carbohydrate metabolism, will be described in Chapter 36. It results from deficiency of pancreatic insulin, and is accompanied by hyperglycaemia, glycosuria, depletion of glycogen reserves, and a tendency to ketosis.

(d) Renal glycosuria

In all of the preceding conditions, glycosuria is due to hyperglycaemia. Renal glycosuria is due to a lowered renal threshold for sugar, without hyperglycaemia. There are some people, otherwise quite healthy, who constantly pass small amounts of glucose because of a low renal threshold, e.g. 150 milligrams per 100 cubic centimetres or lower. This condition is also called " benign glycosuria " or " diabetes innocens ". A severe form of renal glycosuria can be produced experimentally by means of the plant glucoside *phlorrhizin*, which poisons the renal epithelium and prevents the reabsorption of glucose from the tubules.

(3) Hypoglycaemia

Hypoglycaemia, an abnormally low concentration of glucose in the blood (e.g. less than 70 milligrams per 100 cubic centimetres), occurs from an injection of 10 or 15 units of insulin in a healthy person or a larger dose in a diabetic, or from the excessive secretion of insulin by an islet-cell tumour of the pancreas (p. 504). Slight hypoglycaemia causes tachycardia and " palpitation ", air hunger, tremor, emotional disturbance and sweating ; severe hypoglycaemia causes muscular incoordination, coma or convulsions.

(4) Gargoylism (Hurler's disease, lipochondrodystrophy)

This is a rare inborn metabolic disorder, characterized by skeletal deformities with dwarfism, a distinctive facies, mental retardation, and enlargement of liver and spleen. Many organs show parenchyma cells and accumulated histiocytes swollen by some stored material, the nature of which is uncertain but is thought to be a mucopolysaccharide possibly linked with lipids.

DISTURBANCES OF FAT AND LIPID METABOLISM

The term " fat " strictly applies only to the fatty-acid esters of glycerol, but it is often used also to embrace all greasy substances extractable from tissues by ether, benzene or other fat-solvents, i.e. all fat-like or " lipoid " substances. These are :

(i) *True or neutral fats*, esters of fatty acids with glycerol, the common fats of adipose tissue being triglycerides of palmitic, stearic and oleic acids ;

(ii) *Fatty acids*, derived from fats by hydrolysis ;

(iii) *Waxes*, esters of fatty acids with higher alcohols other than glycerol, the chief of which in the human body are cholesterol esters, e.g. cholesterol palmitate, stearate and oleate ;

(iv) *Sterols*, alcohols of high molecular weight, the chief of which is cholesterol ;

(v) *Phosphatides*, complex lipids containing phosphorus and nitrogen, including lecithins, kephalins and sphingomyelins ;

(vi) *Cerebrosides*, complex lipids containing nitrogen but no phosphorus.

Although some of these lipoid substances are especially plentiful in certain situations (e.g. cholesterol esters in the blood plasma, adrenal cortex and sebaceous glands, and phosphatides and cerebrosides in the myelin sheaths of nerve fibres), they are present in cells of almost all kinds. Under various pathological conditions, they accumulate and become structurally visible in tissues where normally they are inconspicuous. To identify them, the pathologist uses various special stains, e.g. Sudan III, Nile-blue and osmic acid ; and he examines them by polarized light ; for, unlike neutral fats, cholesterol esters are anisotropic or doubly refractive. Abnormal accumulations of fats or lipids occur in the following conditions :

(1) Adiposity, or excessive adipose tissue ;

(2) Accumulation of neutral fat in other tissues ;

(3) Lipid storage diseases (lipidoses) ;

(4) Local lipoid accumulations.

(1) Adiposity : excess of adipose tissue

When a person becomes " stout ", the extra fat is deposited chiefly in the regions where adipose tissue is normally present, namely the subcutaneous, retroperitoneal, omental, mediastinal, and peri-visceral tissues. Often also, in a very fat person, the epicardial adipose tissue extends to a varying depth into the substance of the myocardium between its muscle fibres ; and the peri-pancreatic adipose tissue extends into the pancreas and separates its lobules. Some otherwise healthy people inevitably grow fat on an ordinary diet ; their adiposity is physiological. A pathological degree of adiposity results from idleness, over-eating of fats and carbohydrates, over-drinking of alcohol, and certain endocrine diseases, e.g. hypothyroidism and some pituitary disorders (see Chapter 36).

(2) Accumulation of neutral fat in other tissues

In many pathological states, e.g. acute infective diseases, severe anaemias, chronic malnutrition, alcoholism, phosphorus poisoning, etc., droplets of free fat appear in the cells of certain tissues where normally little or no fat is visible. This fatty change is sometimes clearly visible to the naked eye. The liver is found to be uniformly pale or mottled, yellow and greasy ; the myocardium may be uniformly pale and flabby, or may show prominent yellowish speckling like a thrush's breast ; the renal cortex is often pale and mottled, contrasting with the darker red colour of the less affected medulla. Microscopically, in the liver

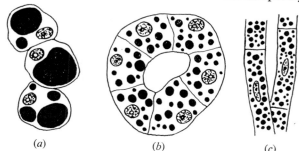

(a)   (b)   (c)

Fig. 107.—Diagram of fatty changes in (a), liver cells ; (b), kidney tubule ; (c), myocardial fibres; as seen in fresh tissue stained with osmic acid.

large fat droplets fill and distend the liver cells, especially those of the peripheral zone of the lobule, so that they resemble adipose tissue cells. In the renal epithelial cells and in the cardiac muscle fibres, the fat is usually in the form of many small droplets (Fig. 107).

It used to be thought that this fatty change in parenchymatous cells was due to degenerative changes in the cytoplasm with the liberation or unmasking of combined fat already present ; hence it was called *fatty degeneration* or *phanerosis* (Gk.=a making visible). This view is no longer tenable ; it has been shown clearly that the fatty liver, kidney and myocardium have acquired a real excess of neutral fat brought to the cells by the blood-stream ; the change is one of *fatty infiltration* or accumulation in the cells. The cells have been so injured by malnutrition, toxaemia or deficient oxygenation that they are unable to metabolize the fats brought to them.

Experimental work has shown that the nitrogenous organic base *choline*, a component of lecithin, is important in the development of fatty infiltration in the liver. Deficiency of lecithin, choline and the related amino-acid methionine in the diet of an animal leads to fatty infiltration of the liver, especially if the diet is also rich in fat ; while administration of these substances tends to prevent fatty infiltration. Such substances are called " lipotropes " or " lipotropic factors ". In the absence of lipotropes, the liver synthesizes extra fat from carbohydrate. (See also p. 350.)

(3) Lipid storage diseases : lipidoses

Excessive accumulations of lipids occur in the following:

(*a*) Gaucher's disease ;

(*b*) Niemann-Pick disease ;

(*c*) Tay-Sachs disease;

(*d*) Hand-Schüller-Christian disease ;

(*e*) Xanthomatosis or xanthelasma of the skin ;

(*f*) Experimental hypercholesterolaemia.

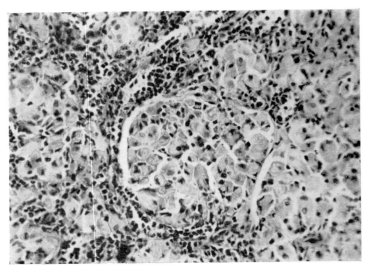

FIG. 108.—Section of spleen in Gaucher's disease. (× 300.)

(a) Gaucher's disease

This is a rare familial disease, commencing in childhood, and characterized by slow progressive accumulation of cerebrosides in the fixed reticulo-endothelial cells of the spleen, liver, lymph glands and bone-marrow. These cells are greatly enlarged, rounded and show pale homogeneous cytoplasm (Fig. 108). The spleen becomes enormously enlarged, the liver and lymph glands less so. The skin becomes pigmented. The cause of the disease is an inborn error of metabolism. It ends fatally, usually during adolescence.

(b) Niemann-Pick disease

Niemann-Pick disease is a somewhat similar but more rapidly progressive and quickly fatal disease of infants, characterized by an accumulation of the phosphatide sphingomyelin in the reticulo-endothelial and other tissues.

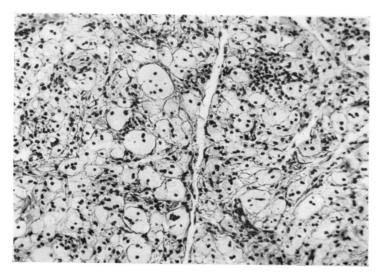

FIG. 109.—Compactly aggregated foam cells from a skin lesion in xanthomatosis. (× 150.)

FIG. 110.—Spaces in fibrous tissue from which cholesterol crystals have been dissolved out in preparation of section ; from a skin lesion of xanthomatosis. (× 70.)

(c) Tay-Sachs disease (Amaurotic family idiocy)

This is a rare familial disease of infants or children, characterized by accumulation of cerebrosides in nerve cells of the central nervous system and retina, with resulting idiocy and blindness.

(d) Hand-Schüller-Christian disease

Commencing in childhood but sometimes prolonged into adult life, this is characterized by localized accumulations of cholesterol and its esters in the histiocytic cells of the bone-marrow, especially of the skull. Clinically, the main features are diabetes insipidus due to involvement of the posterior lobe of the pituitary (see p. 607), protrusion of one or both eyes (exophthalmos) due to deposits in the orbit, and multiple localized areas of rarefaction seen in X-ray pictures of the skull. Other bones also may be affected but there is usually no generalized storage of the lipid in spleen, liver or lymph glands. Microscopically, the bone lesions consist of foam-cells and foreign-body giant-cells laden with lipid, along with variable numbers of eosinophil leucocytes and plasma cells. In some cases the eosinophil cells are very numerous; and the rare " *eosinophilic granuloma of bone* " is believed by many to be only a variant of Hand-Schüller-Christian disease. Another rare, probably related, disorder is called " *Letterer-Siwe* " *disease*. It is doubtful if these diseases are metabolic storage disorders; they are probably granulomatous reactions to some unidentified agents.

(e) Xanthomatosis or xanthelasma of the skin

These are the names applied to rare cases in which multiple yellow nodular masses of lipid-laden foam-cells accumulate in the skin, and sometimes in other tissues also. The lipid consists of cholesterol and its esters, which are engulfed by macrophages and foreign-body giant-cells, or which lie as pure cholesterol crystals in elongate spaces in fibrotic tissue (Figs. 109, 110). Xanthomatosis is seen in cases of diabetes mellitus or chronic jaundice with hypercholesterolaemia ; but it occurs also, again accompanied by hypercholesterolaemia, as an hereditary metabolic disorder. The name " xanthoma " is unfortunate, for the lesions are not tumours.

(f) Experimental hypercholesterolaemia

Hypercholesterolaemia can be induced by feeding rabbits or other animals with excessive amounts of cholesterol. The excess is taken up as esters, chiefly by the fixed phagocytes of the reticulo-endothelial tissues, especially in the liver ; but it is also taken up to some extent by macrophages in other tissues, e.g. in the intima of arteries, where yellow atheromatous patches are formed. Xanthomatosis of the skin has not been induced.

(4) Local lipid accumulations

Little need be said of these here ; examples are described in other chapters, e.g. in subacute nephritis (p. 304), in cholesterosis of the gall-bladder (p. 326), in chronic inflammatory and degenerative lesions (pp. 322–324), and in some tumours (p. 507). In the contents of many sebaceous cysts, branchial cysts, skin-lined teratomatous cysts, and in many old haematomas, free cholesterol crystals separate out in the form of large flat rhombic plates with or without one re-entrant angle (Fig. 111). These shimmer like spangles in the fluid containing them.

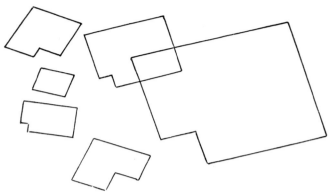

Fig. 111.—Free cholesterol crystals. (× 50.)

DISTURBANCES OF PROTEIN AND PURINE METABOLISM

The following are of interest to the pathologist :
(1) Hypoproteinaemia ;
(2) Amino-acid deficiencies and liver disease ;
(3) Abnormal metabolism of certain amino-acids ;
(4) Gout ;
(5) Amyloid degeneration.

(1) Hypoproteinaemia

Hypoproteinaemia may result from deficient formation of plasma proteins, as in starvation or liver disease, or from loss of protein from the blood, as in nephritis or exudation from extensive burns. For the maintenance of a normal level of plasma protein, adequate protein must be ingested, and not all proteins are equally valuable in this respect. There is evidence that the sulphur-containing amino-acids, especially methionine, are essential for the synthesis of serum albumin by the liver. Hypoproteinaemia due to starvation, direct or indirect, is mainly due to fall in the amount of albumin in the plasma. So also, loss of proteins from the blood by nephritis or exudation reduces the albumin more than the globulin, because the former has the smaller molecule.

The most striking objective result of hypoproteinaemia is oedema, which often appears when the total plasma protein level falls much below 5 grams per 100 cubic centimetres or the albumin level below 3 grams per 100 cubic centimetres. This result depends on the fall of colloidal osmotic pressure of the plasma, to which albumin with its smaller molecules contributes most (see p. 388). For hypoproteinaemia to produce oedema, the body must be supplied with adequate amounts of salt and water ; dehydration from any cause, e.g. by vomiting or diarrhoea, may prevent the development of famine oedema, which may appear as soon as fluids are restored to the body.

(2) Amino-acid deficiencies and liver disease

Prolonged experimental deprivation of choline and the related amino-acid methionine produces, not only fatty infiltration of the liver, but eventually widespread fibrosis resembling human portal cirrhosis. It has therefore been suggested that the great prevalence of cirrhosis in many native communities in Africa and elsewhere is due to chronic malnutrition, especially a deficiency of first-class animal protein, and that this may well apply also to alcoholic cirrhosis.

Experimental cystine deficiency produces a quite different result, namely extensive massive necrosis of the liver resembling acute yellow atrophy in man. Hepatic necrosis does not occur on a cystine-deficient diet if this contains abundant methionine, because the latter is easily converted into cystine. The experimentalists have suggested that hepatic necrosis in man may be due to deficiency of these sulphur-containing amino-acids in the diet; and that, even when necrosis is caused by chemical poisons, this is conditioned by protein deficiency.

While there is no doubt of the influence of these dietetic factors in the causation of liver disease, the importance of toxic factors is equally certain. Let us recall that, whatever the diet and nutritional state of an animal, the introduction of very small quantities of hepato-toxic substances into the portal blood will promptly produce extensive necrosis of liver tissue (p. 332). It is quite probable that in many cases of liver necrosis or cirrhosis in man both toxic and nutritional factors are at work.

(3) Abnormal metabolism of certain amino-acids

Cystinuria is an inborn error of metabolism in which cystine is constantly present in the urine, from which it crystallises out in the form of colourless hexagonal plates, and sometimes forms renal calculi (p. 329). *Cystinosis or Lignac-Fanconi syndrome* is a more serious disorder of metabolism of cystine and other amino-acids, in which these are excreted in the urine and extensive deposits of cystine crystals occur in the reticulo-endothelial tissues of the spleen, liver, bone-marrow, lymph glands and other tissues.

Alkaptonuria is a rare hereditary disorder, in which the body cannot complete the katabolism of tyrosine and phenylalanine, which therefore appear in the urine as homogentisic acid, a related compound which blackens on exposure to air. Ochronosis—a blackening of the cartilages and ligaments, and sometimes of the conjunctivae—may also occur.

Phenylketonuria is an inborn enzymic defect which prevents the normal conversion of phenylalanine into tyrosine, with the result that the blood and tissues have abnormally high levels of phenylalanine and phenylpyruvic acid is excreted in the urine. Atrophy of the cerebral cortex occurs, and most untreated patients become imbecile in childhood.

Oxaluria is a rare inborn error, probably in the metabolism of glycine, characterized by high urinary oxalate excretion and the formation of oxalate calculi early in life, and in some cases by *oxalosis*, the deposition of calcium oxalate crystals in the tissues.

Several other distinctive but rare *amino-acidurias* are known.

Abnormalities of melanin formation are as follows:

(1) *Excessive normal pigmentation of the skin*, e.g. freckles (lentigo) and sun-burn. As in the dark-skinned races, the melanin is present as granules in the cells of the deeper layers of the epidermis.

(2) *Chronic adrenal insufficiency* or Addison's disease (see p. 607) is accom-panied by excessive pigmentation of exposed parts of the skin and of the oral mucosa. Since adrenaline, like melanin, is closely related to tyrosine, there is probably a chemical connection between the pigmentation of Addison's disease and the cessation of the production of adrenaline.

(3) *Melanotic moles and tumours* (see p. 578).

(4) *Melanosis of the colon.* Sometimes in a middle-aged or old person, the mucosa of the large intestine is found to be brown or black. This is due to the presence of many melanin-carrying macrophages in the sub-epithelial tissues. It is probable that this melanin is a product of putrefactive changes in the faeces of constipated subjects, especially those who habitually take purgatives. The condition is harmless.

(5) *Albinism.* This is an inborn anomaly in which there is complete absence of melanin from the skin, eyes and other tissues. The hair is white, the eyes are pink, and the skin fails to pigment even after prolonged exposure to the sun. The albino lacks the enzyme tyrosinase necessary for the formation of melanin from tyrosine or other precursors. Albino individuals or albino breeds of almost all species of mammals and of many birds are known. Human albinism is rare ; it is inherited as a Mendelian recessive character.

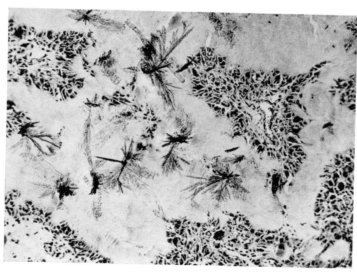

FIG. 112.—Section of gouty tophus similar to that of Fig. 104, except that the sodium biurate has been incompletely dissolved out and has re-precipitated in an atypical crystalline form in the mounted section. (× 100.)

(4) Gout

Gout occurs chiefly in middle-aged men, and appears clinically as an acute or chronic arthritis, most often of the metatarso-phalangeal joint of the great toe, but sometimes also of the fingers, knees or other joints.   It is characterized by the deposition of masses of sodium biurate crystals in the synovial membrane, articular cartilages, joint capsule and periarticular tissues.   Similar deposits occur also in the subcutaneous tissues and cartilages of the ears or eyelids, forming chalky masses or " tophi ".   The overlying skin may ulcerate, exposing the chalky contents of the tophus.   Microscopically, great numbers of macrophages and foreign-body giant-cells collect around the masses of crystals (Fig. 112).

The cause of gout is unknown.   It is clearly a disturbance of purine meta-bolism, for there is an excess of uric acid in the blood (e.g. 4 milligrams per 100 cubic centimetres instead of the normal 1·5 to 2·5 milligrams per 100 cubic centi-metres), and the amount of uric acid in the urine is abnormally low just before an acute arthritic attack but rises sharply during the attack.   Gout is not merely the result of the excessive uric acid in the blood, for no gouty signs appear in uraemic patients in spite of a very high level of blood uric acid.

(5) Amyloid degeneration

(a) Nature of amyloid

Amyloid is a translucent, homogeneous protein material which develops in the connective tissues of affected organs.   Like the protein chondromucoid of the ground-substance of connective tissues, amyloid contains the polysaccharide prosthetic group chondroitin-sulphuric-acid.   It is probable that in amyloid disease the connective tissue proteins, perhaps collagen and elastin as well as chondromucoid, undergo direct conversion into amyloid, of which, however, there is also a great accumulation causing enlargement of the affected organs. Amyloid has characteristic, though somewhat variable, staining properties. In the fresh state it stains a dark brown with weak aqueous solutions of iodine ; Lugol's solution is used for its demonstration in the post-mortem room.   With methyl-violet and similar polychrome aniline dyes it stains pink, and it is selectively stained also by Congo red.   Amyloid disease occurs in two forms, a generalized and a local.

(b) Generalized amyloidosis

This usually results from a long-standing bacterial infection, most often chronic suppuration or tuberculosis, less frequently tertiary syphilis or chronic infective endocarditis.   On rare occasions, it has arisen without any obvious antecedent cause.   A peculiar form of amyloidosis is present in occasional cases of plasma-cell myelomatosis (p. 561).   The most conspicuous deposits of amyloid develop in the liver, spleen and kidneys, but the intestines, lymph glands, adrenals and other organs may also be affected. *The liver* becomes smoothly enlarged, pale, of firm soap-like consistency, and its cut surface has a translucent waxy appearance.   The earliest deposits are laid down around the sinusoids and the

blood-vessels in the portal tracts ; but eventually large confluent areas of homogeneous amyloid are formed, in which most of the liver cells have disappeared
by compression atrophy (Figs. 113, 114).   In *the spleen*, amyloid is usually

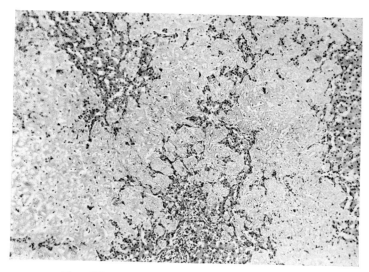

FIG. 113.—Severe amyloidosis of liver.   (× 70.)

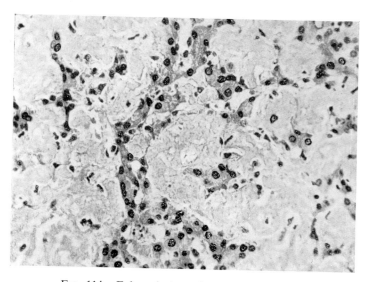

FIG. 114.—Enlarged view of Fig. 113.   (× 350.)

deposited first in the Malpighian nodules, so that the cut surface shows many
discrete translucent foci like sago grains, whence the name " sago spleen "

(Fig. 115). In other cases the deposition is more diffuse. In *the kidneys*, deposits occur in the glomeruli (Fig. 116) and also around the tubules and in the walls of the larger vessels. Polyuria and albuminuria frequently result. In *the intestines*,

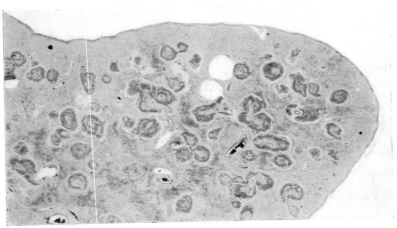

FIG. 115.—Hand-lens view of a section of " sago spleen " stained with Congo red. (× 8.)

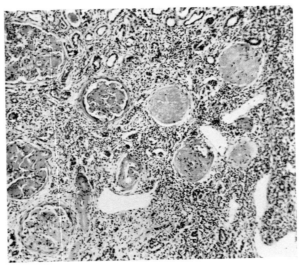

FIG. 116.—Amyloid change in the glomeruli. (× 70.)

the vascular cores of the villi are most affected, with consequent defective absorption and serious persistent diarrhoea. Generalized amyloidosis is a dangerous condition ; recovery from it is rare.

(c) Local amyloid degeneration

This is an uncommon lesion, most frequent in the larynx, trachea or tongue, producing a local smooth swelling which is usually mistaken clinically for a true tumour, and has often been called " amyloid tumour ". The causes of local amyloid degeneration are unknown ; the patient has not suffered from chronic suppurative or other infections, and is usually in good general health. The amyloid replaces, and appears to be directly derived from, the connective tissues of the part ; its staining properties are often atypical.

DISTURBANCES OF IRON METABOLISM

These include : (1) iron-deficiency anaemia, (2) haemosiderosis due to excessive blood destruction, (3) local haemosiderosis, and (4) haemochromatosis.

(1) Iron-deficiency anaemia

Most of the iron in the body is combined in haemoglobin and is concerned with the carriage of oxygen. The iron normally liberated by breakdown of the haemoglobin from disintegrated red cells is not excreted but is retained in organic combination as *haemosiderin* in the spleen, liver and bone-marrow, and is re-utilized in the synthesis of fresh haemoglobin. The daily requirement of iron is therefore small, probably only a few milligrams. A deficient intake of iron depletes the body of its haemosiderin reserve and leads to a simple hypochromic anaemia ; this is curable by the administration of iron salts (see p. 397). Traces of copper are also essential for the utilization of iron in the synthesis of haemo-globin.

(2) Haemosiderosis due to excessive blood destruction

In anaemias accompanied by excessive destruction of red cells, namely per-nicious anaemia, acholuric jaundice, malaria, and anaemias due to haemolytic poisons (p. 336), excessive granular deposits of haemosiderin are found in the spleen, liver, kidneys and other organs—not only in the reticulo-endothelial cells, but also in the liver cells, renal epithelium, and other parenchymatous cells. The haemosiderin gives the staining reactions of iron, e.g. with potassium ferrocyanide in acid solution it forms ferric ferrocyanide (Prussian blue). This reaction (Perls' test) is very useful in pathology ; a piece of liver or other tissue showing haemosiderosis, dropped into this solution, quickly turns blue ; and the same solution applied to micro-sections stains the haemosiderin granules dark blue.

(3) Local haemosiderosis

We have already noted the disintegration of haemoglobin and the phagocy-tosis of haemosiderin and haematoidin in and around haemorrhages and thrombi (p. 19). Similar changes occur around haemorrhagic infarcts (p. 373), areas of haemorrhagic degeneration in tumours, and in tissues the seat of chronic venous congestion.

Congestive pigmentation is frequent in the skin of the legs in cases of varicose veins, especially around old varicose ulcers, and in the lungs, liver and spleen in cases of chronic cardiac failure.   In the congested tissues, a few red cells are constantly escaping from the capillaries, which are not only over-distended but have their endothelial walls injured by the deficient oxygenation.   In *the lungs*, the red cells escape mainly into the alveoli, where the haemosiderin and other products of their disintegration are taken up by scavenging macrophages called " heart failure cells " (Fig. 117).   Many of these are coughed up in the sputum : others remain in the alveoli or migrate back into the interstitial tissues of the lung, and they may congregate in such numbers that they form nodules large enough to throw shadows in X-ray pictures and to be felt and seen at necropsy.   Chronically congested lungs often become brown from the pigment which they contain,

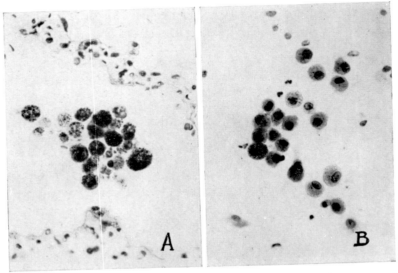

FIG. 117.—" Heart failure cells " : A, in a lung alveolus ; B, in a smear of sputum. (× 375.)

and also diffusely fibrous—a condition known as " brown induration ".   Such lungs give a strong Prussian blue reaction.   Chronic venous congestion in *the liver* produces " nutmeg liver ", in which the lobules, especially their central zones, show varying degrees of distension of the sinusoids, fatty change and atrophy of the liver cells, and pigmentation by haemosiderin and haematoidin. In long-standing cases, some diffuse fibrosis also develops—a condition of congestive cirrhosis.   Chronic congestion of *the spleen*, either from cardiac failure or portal venous obstruction, produces moderate enlargement of the organ, haemorrhages, pigmentation and some fibrosis.

(4) Haemochromatosis

This rare chronic disease, also called *bronzed diabetes*, occurs almost exclusively in adult males.   It is characterized clinically by hepatic enlargement and cirrhosis, diabetes mellitus, and bronze pigmentation of the skin.   The principal pathological change is an enormous accumulation of haemosiderin in most of the organs, in both parenchyma cells and interstitial tissues. In advanced cases *the liver*, which may contain 30 grams or more of iron, is much enlarged, shows a coarsely nodular cirrhosis, is of a dark rusty-brown colour, and gives an intense Prussian blue reaction. Microscopically the liver cells are packed with haemosiderin granules (Fig. 118). In a considerable number of cases, multicentric carcinoma of the liver supervenes (p. 503). *The pancreas* shows similar pigmentation, less intense than in the liver, due to deposition of haemosiderin in the epithelial cells of the acini. Fibrosis is usually not very noticeable ; and, in spite of the diabetic state of the patient, the islets sometimes appear well preserved though pigmented. The upper abdominal *lymph glands* are always very heavily pigmented ; those of

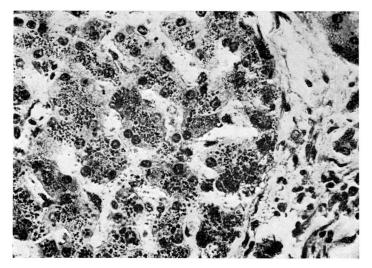

FIG. 118.—Liver cells studded with haemosiderin granules in haemochromatosis. (× 450.)

other groups, less so. Pigmentation, though usually of a less degree than in the organs already mentioned, may be visible in many *other tissues*, especially the cardiac muscle fibres, adrenal glands, parathyroids, intestinal and gastric mucosa, all of which will give the Prussian blue reaction. The bronzing of the skin, however, is mainly due to melanosis of the epidermis. Usually the kidneys and the central nervous system show little or no haemosiderosis.

The cause of haemochromatosis is unknown. It is certainly not due to excessive blood destruction ; and Mallory's hypothesis that it is a manifestation of chronic copper poisoning has not been substantiated. The most acceptable idea is that it is an inborn metabolic disorder which, however, does not usually display itself until middle-age, because it takes many years for the accumulation of the great quantities of iron which are eventually found in the body. There is good evidence that intracellular iron participates, along with various other substances (cytochromes, cytoflavin, glutathione, ascorbic acid, peroxidases, and dehydrogenases), in cell respiration. The metabolic disturbance underlying haemochromatosis may well be due to deficiency of some specific enzymic substance connected with intracellular oxidation.

JAUNDICE
(*Icterus*)

(1) Metabolism of the bile pigments

The normal destruction of senile red corpuscles by the fixed phagocytes of the reticulo-endothelial tissues results in the formation of iron-containing haemosiderin and the iron-free bile pigments *bilirubin* and *biliverdin* which are excreted by the liver. In the intestine the bile pigments are reduced by the action of bacteria to colourless *stercobilinogen* (or *urobilinogen*) ; this is partly converted into the brown pigment *stercobilin*, which is the main pigment of faeces and is identical with the urinary pigment *urobilin*. Part of the urobilinogen escapes excretion in the faeces and is reabsorbed from the large intestine into the blood. Some of this is re-excreted by the liver and some by the kidney.

Largely on the basis of the ven den Bergh test, it has been shown that serum bilirubin fresh from the reticulo-endothelial system differs from the excreted bilirubin in the bile, which is in a conjugated water-soluble form, the former giving an indirect, and the latter a direct, van den Bergh reaction. (For details of the interpretation of this reaction, see Gray.)

(2) Types of jaundice

The amount of bilirubin in the blood is normally very small, less than 0·5 milligrams per 100 cubic centimetres. If it is much increased, it diffuses through the capillaries and stains the skin and other tissues yellow. This staining constitutes *jaundice* (Fr. *jaune*=yellow). Jaundice is of three main types : (*a*) that due to obstruction of the bile ducts, *obstructive jaundice*, (*b*) that due to excessive destruction of red corpuscles, *haemolytic jaundice*, and (*c*) that due to toxic, nutritional or infective injury and necrosis of liver tissue, usually called *toxic jaundice*.

(a) Obstructive jaundice

This is due to mechanical blockage of the bile ducts by stones, tumours of the ducts or pancreas, or other lesions. The retained bile is reabsorbed into the blood, producing deep jaundice accompanied by toxic symptoms and itching of the skin, due to the retention of bile salts as well as pigments (*cholaemia*). The faeces are pale, because of the absence of bile ; the urine is deeply coloured by bilirubin and also contains bile salts ; and the serum gives a direct van den Bergh reaction.

(b) Haemolytic jaundice

Haemolytic jaundice may develop acutely, as a result of haemolytic poisons (p. 336), *Cl. Welchii* infections (p. 161), blackwater fever (p. 224), or an incompatible blood transfusion. Or it develops more slowly from sustained excessive blood destruction in pernicious anaemia or acholuric jaundice. Bilirubin is formed by the reticulo-endothelial tissues more quickly than the liver can excrete it; unexcreted non-conjugated bilirubin therefore accumulates in the blood. The faeces are normal in colour; the urine, though perhaps highly coloured from excess of urobilin, contains neither bilirubin nor bile salts; and the serum gives an indirect van den Bergh reaction.

(c) Toxic jaundice

This results from extensive necrosis of the liver (acute and subacute yellow atrophy) brought about by poisons or dietetic protein deficiencies (p. 334); or from infective inflammations of the liver, e.g. yellow fever, Weil's disease or infective hepatitis (p. 333). The disorganization of the liver tissue results in the escape and reabsorption of bile and also impairment of the liver's power of excretion; the jaundice may be of a mixed character, showing features of both the obstructive and haemolytic types, but is usually mainly obstructive.

DISTURBANCES OF CALCIUM METABOLISM

(1) Normal metabolism of calcium

Calcium is absorbed from the small intestine largely in the form of soluble acid phosphate ($CaHPO_4$). A high acidity of the diet therefore favours its absorption; with neutral or alkaline bowel contents, most of the ingested calcium escapes absorption and is excreted in the faeces as insoluble $Ca_3(PO_4)_2$ and $CaCO_3$. In the blood, calcium is entirely confined to the plasma, and its normal level is remarkably constant, about 10 milligrams per 100 cubic centimetres of plasma or 6 milligrams per 100 cubic centimetres of blood. This is partly in ionisable form and partly in organic combination. The Ca and PO_4 ions in plasma are normally in a state of mutual equilibrium; if one is increased, there is reciprocal diminution of the other, either by excretion in the urine or deposition in bone.

Most of the calcium in the body is deposited in the bones, the inorganic part of which contains 85 per cent of $Ca_3(PO_4)_2$, 10 per cent $CaCO_3$ and nearly 2 per cent $Mg_3(PO_4)_2$. These and the other inorganic salts of bone are probably compounded with one another, as in minerals. The deposition of calcium and other salts in bone is due to the activity of osteoblasts. These bone-formative cells produce an enzyme, alkaline phosphatase, which hydrolyses phosphoric esters brought to the ossifying area in the blood, thereby causing a local increase in the concentration of PO_4 ions and a consequent deposition of $Ca_3(PO_4)_2$. (For the same reason, other metals with insoluble phosphates, e.g. Pb, Sr, Li, Ra and Th, if administered, are deposited in bone and are removed from the bone under the same conditions as Ca.)

Just as the deposition of Ca in bone is due to the enzymic activity of osteoblasts, so its removal is due to the enzymic activity of osteoclasts. Wherever bone is undergoing absorption, whether in the physiological modelling of growing bones, or the remodelling which follows the repair of fractures or the healing of osteomyelitis, or in the pathological rarefactions of hyperparathyroidism or locally destructive tumours or other lesions of bone, the absorption is effected by the agency of osteoclasts. These cells are not of a different species from osteoblasts; osteoblasts and osteoclasts represent the reversed functional phases of cells of the same kind (see p. 24).

Many factors influence the osteoblastic deposition or osteoclastic absorption of calcium salts in bone. These include :

(*a*) The local mechanical stresses and strains to which the bone is subjected (p. 23).

(*b*) The intake of calcium and phosphate, which depends not only on the amounts in the food but also on the reaction of the intestinal contents and on vitamin D.

(*c*) The acid-base equilibrium of the blood, which affects the ratio of calcium ions to phosphate ions in the plasma (p. 359).

(*d*) Vitamin D, which in some way controls the amount of calcium and phosphate which can be utilized (p. 363).

(*e*) Vitamin A also influences the activity of osteoblasts and osteoclasts (p. 362).

(*f*) Parathyroid hormone, which controls the concentration of calcium in the blood (p. 610).

The effects of these several factors have been discovered by the study of abnormal conditions in which one or other of them is excessive or deficient. Normal calcium metabolism and bone structure require that all of these factors should lie within a normal range. The various pathological results of departures from these norms are described on the pages referred to above. Here, we may round the subject off by considering briefly, hypocalcaemia, hypercalcaemia, and local pathological calcification.

(2) Hypocalcaemia

Fall of the concentration of calcium in the plasma occurs from hypoparathyroidism following removal of the parathyroid glands ; vitamin D deficiency (rickets) ; dietary calcium deficiency ; alkalosis, from excessive vomiting, excessive intake of alkalis, or depletion of CO_2 by deliberate forced breathing (hyperpnoea) ; or from severe chronic nephritis. Most of these are discussed elsewhere. The last-named produces hypocalcaemia because of the inability of the kidneys to excrete phosphates ; the plasma phosphate level therefore rises, the intestinal excretion of phosphates is increased, and the excess of phosphates in the bowel combines with most of the ingested calcium, thereby causing virtual calcium starvation of the tissues. If this occurs during childhood, the growth of bones is seriously impaired, causing *renal rickets* or *renal dwarfism*. This condition is accompanied by enlargement of the parathyroid glands, evidently a compensatory hyperplasia evoked by the hypocalcaemia. Even in adults, some cases of chronic nephritis show hypocalcaemia, compensatory parathyroid enlargement, and decalcification of bones. A recently discovered thyroid hormone, *calcitonin*, lowers serum calcium; but we do not yet know if it has any pathological or clinical bearings.

If from any cause the plasma calcium level falls below about 7 milligrams per 100 cubic centimetres, *tetany* (spasmophilia) ensues. This is a condition of muscular hyperexcitability with characteristic spasmodic contractions, sometimes accompanied by other nervous symptoms, and due to the direct effect of the deficiency of calcium ions on muscles and nerves. It is promptly relieved by the administration of calcium.

(3) Hypercalcaemia

An excessive concentration of calcium ions in the blood may result either from hyperparathyroidism with its concomitant mobilization of calcium salts from the bones (p. 610), from extensive destruction of bone by secondary tumours (p. 451), or from excessive intake of vitamin D (p. 364). Sustained hypercalcaemia

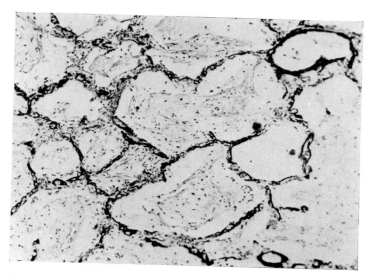

FIG. 119.—Metastatic calcification in the lung, in a case of widespread destruction of the bones by secondary growths from carcinoma of breast. (The air-sacs also show organizing pneumonia.) (× 85.)

from any of these causes often results in *metastatic calcification*, i.e. a deposition of calcium salts in otherwise normal tissues, especially in the septa of the lungs (Fig. 119), the mucosa of the intestine, the walls of arteries, and in the kidneys, where phosphate and carbonate calculi form in the pelves and interstitial calcification takes place in the renal tissue.

(4) Local pathological calcification

This is very common in many degenerative lesions, e.g. old caseous tuberculous foci, atheromatous plaques in arteries, the media of senile arteries (Mönckeberg's degeneration), fibrotic heart valves, old scars, old thrombi (e.g. phleboliths), degenerated hydatid cysts and other parasites, old degenerated goitres, old uterine myomas, a dead foetus extruded into the peritoneal cavity from a tubal pregnancy (lithopaedion or " stone child "). In many of these lesions the calcification takes place in dead material, and is largely attributable to the slow combination of diffusible calcium salts with fatty acids formed by hydrolysis of fats, insoluble calcium soaps being formed which later become changed into phosphate and carbonate. In other lesions, calcium deposition takes place in avascular intercellular material, e.g. in hyaline collagen in a scar or in amyloid material.

Some tumours show characteristic granular calcification, e.g. meningiomas and so-called " psammocarcinomas " (sandy carcinomas). Calcified tissues, e.g. scars or arterial walls, sometimes ossify.

(5) Calcinosis of the subcutaneous tissue

This is a rare condition characterized by the development of multiple scattered calcified masses in the subcutaneous tissues. In some cases these have resulted from phosphate retention due to chronic nephritis, but in the majority the cause is unknown.

VITAMIN DEFICIENCY DISEASES

Vitamins or *accessory food factors* are organic constituents of diet, necessary for growth or the proper functioning of tissues, but present in amounts so small as to distinguish them from ordinary foodstuffs. Their discovery began with the observation that young animals fed on diets of pure carbohydrate, fat, protein, salts and water failed to thrive and finally died. Animal species differ in their vitamin requirements ; and we now know of many vitamins which are important to particular animal species but which are not of proved value to man. Most of the vitamins have now been synthesized. For details of their history, chemistry and functions, works on nutrition and the accounts cited in the reference list below must be consulted. Here, only a bare outline of those important in human pathology can be given. It is convenient to divide them into fat-soluble vitamins, A, D, E and K, and water-soluble vitamins, B group and C.

(1) Vitamin A

Vitamin A or the closely related pro-vitamin *β-carotene* is present in liver, fish-liver oils, milk, butter, eggs, green vegetables and carrots. Deficiency causes:

(i) *Night-blindness* (nyctalopia) i.e. impaired vision on passing from a bright to a dim light, due to delay in the regeneration of the retinal pigment, visual purple (rhodopsin), which contains vitamin A.

(ii) *Xerosis*, or over-keratinization and dryness of epithelia. In the cornea and conjunctiva this constitutes *xerophthalmia* (dry eye) ; in the skin it may produce " *toad skin* " *dermatitis* ; in the mucous membranes of the respiratory, alimentary or genito-urinary tracts, it may cause *squamous metaplasia* and increased liability to bacterial infection.

There is some experimental evidence that vitamin A inhibits the activity of osteoblasts and stimulates that of osteoclasts, deficiency therefore resulting in over-production of new bone.

(2) Vitamin D

Vitamin D, also called *anti-rachitic vitamin* or *calciferol*, was isolated from *ergosterol* irradiated with ultra-violet light ; there are also several other closely related anti-rachitic sterols. These substances are present in fish-liver oils and eggs ; but the best source of vitamin D is sunlight which generates anti-rachitic sterols in the skin. A person's need for dietary vitamin D is inversely proportional to his exposure to sunlight.

Deficiency of vitamin D causes *rickets* (inappropriately also called " rachitis "), of either the infantile or adult type. The essential feature of the disease in children is deficient deposition of calcium phosphate in the bones, especially at the growing metaphyses. These show broad irregular zones of imperfectly calcified or uncalcified cartilage and defective ossification. The consequent irregular growth of soft osteoid tissue produces palpable or visible swellings at the epiphyseal junctions, e.g. the " rickety rosary " of the ribs (Fig. 120). The bending of the

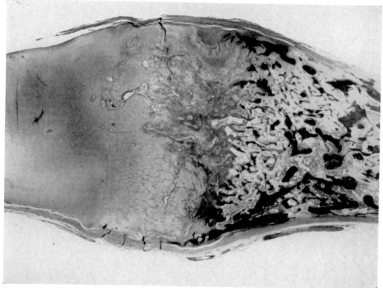

FIG. 120.—Section of costo-chondral junction from an infant with rickets. Note the swelling of the junction, and the excessive width and irregularity of the zone of ossification. (× 6.) (By courtesy of the late Prof. D. H. Collins).

poorly calcified bones produces characteristic deformities—bow-legs, contracted pelvis, and " pigeon-chest ". In active rickets there are subnormal levels of plasma phosphates or calcium or both, and excessive loss of calcium and phosphates in the faeces.

Adult rickets or *osteomalacia* is almost restricted to malnourished pregnant or lactating women. It is due to a combination of vitamin D deficiency, Ca and PO_4 deficiency, and the drain on the patient's calcium stores by the pregnancy or lactation. Occasionally, a similar condition is seen in old neglected women or men who have lived on nothing but biscuits or slops. The bones, especially those of the pelvis, lumbar spine and lower limbs, become decalcified, soft and pliable ; microscopically, they may show little more than a fibrous shell enclosing fatty marrow. Gross deformations and spontaneous fractures are frequent. The parathyroid glands often show compensatory hyperplastic enlargement.

Hypervitaminosis D can be produced experimentally or by therapeutic over-dosage with anti-rachitic vitamin in man. It produces excessive calcium and phosphate levels in the blood, and calcification not only in the bones but in tissues which normally do not calcify, e.g. kidneys, arteries and lungs.

(3) Vitamin E

Vitamin E or *α-tocopherol* is present in many plant-fats, e.g. wheat-germ oil, in egg yolk, and in green vegetables. In rats and other animals, deficiency causes sterility, degeneration of muscles, and some other changes ; but it is still uncertain whether man suffers from vitamin E deficiency.

(4) Vitamin K

Vitamin K, present in green vegetables, has been shown in animals to be essential for the formation of prothrombin. Deficiency of this vitamin underlies certain haemorrhagic conditions due to hypoprothrombinaemia. e.g. in obstructive jaundice and haemorrhagic disease of the new-born (see p. 406). The delayed clotting-time of the blood which often accompanies obstructive jaundice may be due to poor absorption of vitamin K.

(5) The B group of vitamins

The original " vitamin B " has now been shown to include several different substances, namely :

(*a*) Vitamin B_1 (aneurin or thiamin), the anti-beriberi vitamin ;

(*b*) Vitamin B_2 (riboflavin or lactoflavin) which prevents cheilosis, etc.;

(*c*) Vitamin P.P. (nicotinic acid, or niacin) pellagra-preventing vitamin ;

(*d*) Several other factors include vitamin B_{12} and folic acid, important in blood formation (see p. 395).

Factors (*b*), (*c*) and (*d*), which are often found together, are sometimes spoken of collectively as the " vitamin B_2 complex ".

(i) *Vitamin B_1 (aneurin or thiamin).*—Vitamin B_1 deficiency or *beriberi*, common in Asiatic communities who live too exclusively on diets of polished (white) rice, is characterized by polyneuritis of peripheral nerves, muscular weakness, cardiac irregularities, anorexia, and sometimes emaciation (" dry beriberi "), sometimes oedema (" wet beriberi "). Polyneuritic, cardiac or oedematous symptoms may predominate. In Western countries, the polyneuritis which may accompany chronic alcoholism, gastro-intestinal disease, or malnutrition during pregnancy, is also believed to be due to vitamin B_1 deficiency.

(ii) *Vitamin B_2 (riboflavin or lactoflavin).*—Deficiency causes *cheilosis* (sore, cracked lips), and possibly similar lesions in the oral mucosa and cornea.

(iii) *Vitamin P.P. (nicotinic acid or niacin).*—Deficiency causes pellagra, characterized by a peculiar dermatitis, stomatitis and glossitis, diarrhoea and wasting, and mental disturbances. The disease is almost confined to districts where maize is the staple food, e.g. parts of U.S.A., Russia, Egypt and Italy. There is no risk of pellagra on any reasonably mixed diet ; nicotinic acid is present in meats, milk, fish, cereals and potatoes.

(6) Vitamin C

Vitamin C (ascorbic acid or anti-scorbutic vitamin) is plentiful in fresh fruit and green vegetables, and occurs in moderate amounts in potatoes, milk and fresh meats ; but it is easily destroyed by heat. It is stored in the body in the adrenal cortex. Deficiency causes *scurvy*, a disease which used to be very prevalent, especially among sailors on long voyages, but which is now seldom seen in adults. However, infantile scurvy (Barlow's disease) is still seen sometimes in bottle-fed infants whose diet has not been supplemented adequately.

The chief manifestation of both adult and infantile scurvy is haemorrhage—haemorrhages in the skin, gums, mucous membranes, joints, viscera, and especially

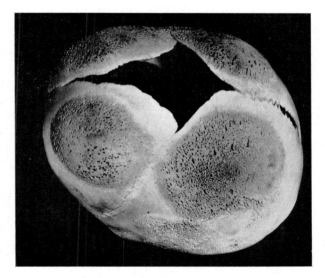

FIG. 121.—Infantile scurvy—Barlow's disease. An infant's calvarium showing multiple raised bosses of new bone where subperiosteal haemorrhages due to scurvy have undergone organization. The specimen is from Sir Thomas Barlow's original collection (1882), and is preserved in the Museum at the Royal College of Surgeons, London.

under the periostium of bones. Subperiosteal haemorrhages are particularly common in infantile scurvy, in both the long bones of the limbs and the flat bones of the skull, which show very tender palpable swellings. The growth of new bone at the metaphyses is also arrested, so that there are abnormally broad zones of epiphyseal cartilage, and characteristic X-ray appearances. In chronic cases, the subperiosteal haemorrhages are slowly organized and deposits of new spongy bone are formed on the original bone. On the outer surface of the vault of the skull these form smooth bosses, which have often been errone-ously attributed to congenital syphilis. Parrot's original specimens showing the supposedly syphilitic " Parrot's nodes " are preserved in the museum of the Royal College of Surgeons, London ; and they show spongy bony bosses similar to those in Barlow's specimens (also preserved there) which are the result of organization and ossification of sub-epicranial haemorrhages (Fig. 121). Other features of untreated cases of scurvy are a pronounced hypochromic anaemia, a risk of bronchopneumonia or other infective complications, and impaired healing of wounds. It is possible that vitamin C deficiency may play a part in delaying healing in some chronic ulcers of the stomach.

There is good evidence that one of the fundamental functions of vitamin C is to facilitate the proper development of intercellular substances in fibrous connective tissue, bone matrix, and the walls of blood vessels. The haemorrhagic tendency and delayed healing in scurvy accord with this idea.

SUPPLEMENTARY READING

Fraser, J. (1935). " Skeletal lipoid granulomatosis (Hand-Schüller-Christian's disease) ", Brit. J. Surg., 22, 800.

Glynn, L. E., and Himsworth, H. P. (1944). " Massive acute necrosis of the liver ", J. Path. Bact., 56, 297.

Gray, C. H. (1957). " Bile pigments ". Brit. Med. Bull., 13, 94.

Harris, L. J. (1947). " All the vitamins ", Brit. Med. J., 2, 681. (A useful brief outline.)

Himsworth, H. P. (1946). " Protein metabolism in relation to disease ", Proc. Royal Soc. Med., 40, 27.

Humphreys, E. M., and Kato, K. (1934). " Glycogen-storage disease. Thesaurismosis glycogenica (von Gierke) ", Amer. J. Path., 10, 589.

Jaffe, H. L., and Lichtenstein, L. (1944). " Eosinophilic granuloma of bone ", Arch. Path. 37, 99.

Price, N. L., and Davie, T. B. (1937). " Renal rickets ", Brit. J. Surg., 24, 548. (A clear account with good figures.)

Russell, D. S., and Barrie, H. J. (1936). " Storage of cystine in the reticulo-endothelial system and its association with chronic nephritis and renal rickets ", Lancet, 2, 899.

Sellers, E. A., Lucas, C. C., and Best, C. H. (1948). " The lipotropic factors in experimental cirrhosis ", Brit. Med. J., 1, 1062.

Sheldon, J. H. (1934). " Haemochromatosis ", Lancet, 2, 1031.

Stewart, M. J., and Hickman, E. M. (1931). " Melanosis coli ", J. Path. Bact., 34, 61.

Symmers, W. St. C. (1956). " Amyloidosis", J. Clin. Path., 9, 187 and 212.

Symposium on Disorders of Carbohydrate Metabolism (Edited by G. K. McGowan and G. Walters). (1969). J. Clin. Path. 22, suppl. 2.

Thorpe, W. V., Bray, H. G. and James, S. P. (1970). Biochemistry for Medical Students. London; Churchill.

Willis, R. A. (1962). The Borderland of Embryology and Pathology. London; Butterworths. (Chap. 10 briefly reviews the inborn errors of metabolism.)

Witts, L. J. (1947). " A review of the dietetic factors in liver disease ", Brit. Med. J., 1, 1 and 45.

CHAPTER 26

CIRCULATORY DISTURBANCES

WITH RARE exceptions, the tissues of the body are pervaded by blood vessels and lymphatics, and tissue nutrition, respiration and other metabolic activities are dependent on a proper local vascular supply. Any serious interference with this is bound to produce structural and functional changes. An important section of pathology, therefore, is that concerned with the causes and effects of local or general impediments to the movement of blood, lymph and tissue fluid. These topics can be discussed conveniently under the following heads :

Obstructive diseases of arteries ;
Local effects of arterial obstruction ;
Aneurysms ;
Venous obstruction ;
Congestive cardiac failure ;
Lymphatic obstruction ;
Oedema ;
Anhydraemia or haemoconcentration ;
" Shock " ;
The crush syndrome.

OBSTRUCTIVE DISEASES OF ARTERIES

An adequate supply of arterial blood is essential to the health of any tissue. Local diminution of a tissue's arterial blood supply below the amount needed to maintain its health is called *ischaemia*. This may be total, causing death of the part ; or partial, causing a variable degree of atrophy and degeneration of parenchyma and replacement fibrosis. The obstructive diseases of arteries which may lead to ischaemia of tissues are :

(1) *arteritis*, embracing all true inflammations of arteries (p. 83) ;
(2) *atheroma* or *atherosclerosis*, a degenerative change in the intima ;
(3) *medial calcification* or *Mönckeberg's sclerosis*, a degenerative change in the media ;
(4) *hypertensive sclerosis*, a generalized arterial disease associated with high blood-pressure ;
(5) *mechanical constriction* of arteries ;
(6) *local arterial spasm* ;
(7) *arterial embolism*.

The term *arteriosclerosis* has no precise pathological meaning ; it has been applied indiscriminately to (2), (3) and (4). The several forms of arteritis are described in Chapter 7.

(1) Atheroma or atherosclerosis

(a) Structure and sites

Atheroma is a very common degenerative change in the intima of arteries. Almost all aortae show it in some degree, and it is common also in medium-sized and small arteries.   In large vessels it appears either as raised, white or yellow plaques still clothed by endothelium, or as irregular " ulcerated " patches where the endothelium has been destroyed and the degenerated atheromatous material is exposed.   This material frequently becomes calcified, and sometimes ossified ; and badly affected arteries thus become deformed, rigid and brittle.   Some thrombus is deposited on ulcerated areas, but occlusive thrombosis of the aorta or its main branches seldom occurs.

Microscopical study of the earliest plaques shows only hyaline fibrous thickening of the subendothelial intimal tissue.   Older yellow plaques show narrow clefts containing cholesterol crystals and other free lipoid substances, or collections of foam-cells loaded with lipoids ; these are usually situated deep in the plaque close to the media (Fig. 122).   Calcification ensues in the

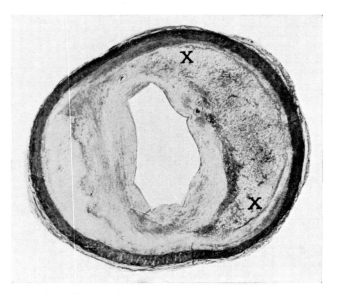

FIG. 122.—Cross-section of main hepatic artery showing great
narrowing of lumen by irregular atheromatous thickening of
intima.   Media darkly stained by elastic tissue stain.   The
mottled crescentic area XX is a mass of foam-cells and
cholesterol crystals.   (× 10.)

degenerated lipoid-laden tissues.   When ossification occurs in a calcified plaque, this becomes vascularized by ingrowth of sprouts from the vasa vasorum of the adjacent media.   Badly atheromatous vessels often show patchy atrophy and degeneration of the media also.

In smaller vessels, such as the coronary or cerebral arteries, the changes are similar but are usually accompanied also by degeneration and splitting-up of the internal elastic lamina.   Atheromatous patches in one of these smaller vessels often come to occupy a large part of the lumen ; so that, when they ulcerate, total occlusion of the vessel by thrombosis is apt to supervene.   This is the usual cause of angina pectoris, myocardial fibrosis or infarction, cerebral softening, and senile or diabetic gangrene of the limbs (see below, p. 372).

Atheroma of the pulmonary arteries is relatively infrequent, and usually of mild degree. It is rarely seen except when there has been prolonged obstruction to the pulmonary circulation and consequent raised pulmonary arterial pressure. Atheroma is common in the aortic and mitral valve cusps, but rare in the pulmonary and tricuspid valves (p. 81).

(b) Causation

The precise cause of atheroma is not known ; but it is certainly related to (i) age, (ii) blood-pressure and (iii) lipoid metabolism. As to age, small atheromatous patches do occur in young people and even in children, but they increase greatly in number and extent during adult life. Severe atheroma is not always accompanied by raised blood-pressure, but in general it is more marked in hypertensive subjects than in those with normal blood-pressure. The importance of the intra-arterial pressure is shown also by the incidence of atheroma in the pulmonary arteries, just referred to. Atheroma develops in rabbits in which hypercholesterolaemia is maintained by a cholesterol-rich diet ; and in man conditions which produce hypercholesterolaemia, e.g. uncontrolled diabetes and subacute nephritis, and probably excessive fats in the diet, are commonly associated also with pronounced atheroma.

It has also been shown that thrombus deposits on the intima of arteries may undergo fibrous organization, may acquire a clothing of endothelium, and may suffer lipoid degenerative changes in their deeper parts, so that they form patches indistinguishable from atheromatous plaques. These changes may occur also in the layers of mural thrombus formed over or around earlier atheromatous areas ; this process of secondary atheromatous change in organizing thrombi undoubtedly plays a part in the progress of arterial atheroma. However, we cannot attribute all atheroma to secondary change in mural thrombi ; many young atheromatous patches are clearly due to alterations in the deeper layers of the intima itself.

(2) Medial calcification or Mönckeberg's disease

This is a common senile degenerative change characterized by conversion of parts of the media into calcified rings or plates. When it is extensive, long reaches of the affected arteries become converted into rigid brittle tubes, called " pipe-stem arteries ". The vessels most commonly affected are the medium-sized arteries of the limbs and the coronary arteries. In itself it is not of great importance, but it is often accompanied by intimal atheroma, with which it participates in producing coronary thrombosis and senile gangrene of the limbs. The cause of ordinary senile Mönckeberg's degeneration is unknown ; but it is of interest that medial calcification occurs in cases of hypercalcaemia, as in hyperparathyroidism or hypervitaminosis D (p. 361).

(3) Hypertensive sclerosis of arteries

(a) Causes of hypertension

Cases of pathological elevation of the arterial blood-pressure fall into one or another of the following groups :

(1) *Primary* (*essential*) *hypertension*, (benign or malignant) of unknown cause ;

(2) *Renal hypertension*, attributable to renal ischaemia (p. 308) ;

(3) *Endocrine hypertension* due to disturbed function of the ductless glands (see pp. 606, 607).

In Chapter 22 we have already discussed (1) and (2), their distinction and possible relationship. Group (1) comprises nearly 90 per cent of all cases of persistently raised blood-pressure, the causes of which are still unknown. It is legitimate, however, to speculate as to what the causes *might* be. Since, in groups (2) and (3), we know that hypertension can be caused by renal ischaemia or endocrine disturbances, we must admit the possibility that " primary " hypertension may depend on renal or endocrine factors. Since we also know that blood-pressure is controlled by afferent nervous impulses ascending from the carotid sinus and aortic arch to the vasomotor centre, and that by section of the sino-aortic nerves persistent hypertension can be produced experimentally, we must not overlook the possibility that nervous disturbances may underlie " primary " hypertension. Future research must discover what parts these several factors may play in this very common disease.

(b) Arterial changes in hypertension

Since it is on the muscular tonus of the small arteries and arterioles that the height of the arterial blood-pressure depends, it is not surprising that the principal changes in hypertension are found in these small vessels. These show hypertrophy of the muscle of the media, reduplication of the internal elastic lamina, and fibrous thickening of the intima. At later stages, degenerative changes frequently supervene ; the hypertrophied media often undergoes fibrous replacement, it and the thickened intima together become hyaline and the lumen becomes greatly narrowed or occluded. In larger arteries also variable degrees of medial hypertrophy, splitting of the elastic lamina, and thickening of the intima may be seen; pronounced atheroma is often added ; and occlusive thrombosis may supervene. In the rapidly progressive arteriolar lesions of malignant hypertension, little or no muscular hypertrophy is seen; the changes are almost entirely degenerative, often amounting to hyaline (or " fibrinoid ") necrosis of the arteriolar walls, accompanied by their partial disruption and by small haemorrhages.

(4) Mechanical constriction of arteries

Serious external compression of an artery occasionally results from the displaced fragments of a fractured bone or from an unreduced dislocation of a joint. A very common and usually harmless kind of mechanical constriction of arteries is that applied deliberately by the surgeon, either by permanent ligature or temporary tourniquet. Yet even this is not wholly free from danger ; a tourniquet left on too long can cause the loss of a limb ; and there are some arteries which cannot be ligated without serious risk to the tissues supplied by them, e.g. the common carotid artery and mesenteric arteries and their main branches.

Mechanical constriction of arteries occurs also when the vascular pedicle of a freely mobile viscus becomes twisted (*torsion* or *volvulus*). Structures liable to torsion are the pelvic colon, loops of the small intestine, the ovary when this is enlarged by cysts or tumours, the great omentum, and rarely the caecum, gall-bladder or testis when these are abnormally mobile. Twisted organs become deeply engorged and haemorrhagic, because while torsion is in progress the veins suffer occlusion sooner than the arteries. The same applies to *strangulation* of parts without torsion, e.g. loops of bowel or other viscera snared in hernial sacs with narrow necks, or by bands of adhesions in the abdominal cavity, or in intussusceptions.

(5) Local arterial spasm

In the following conditions, local impairment of the circulation is brought about by sustained muscular spasm of the arteries :

(a) Arterial spasm induced by cold

We are all familiar with the blanching of the skin on exposure to cold, a reflex effect due to vaso-constriction. Some persons are abnormally susceptible to the vascular effects of cold, and in different ways : in some, the fingers go pale and " dead " ; others develop " chilblains " (erythema pernio), in which reflex vaso-constriction of the subcutaneous arteries is accompanied by vaso-dilatation of the minute superficial arterioles and capillaries. Intense or prolonged chilling of the extremities, as in frost-bite, " trench foot " or " immersion foot ", finally causes death of the part (gangrene), because the circulation is completely arrested as a result of vaso-constriction, structural damage of arteriolar walls, and " silting up " or actual thrombosis of their contents.

(b) Raynaud's disease

This is a rare condition, seen chiefly in women, and characterized by painful attacks of blueness and numbness followed by pallor of the fingers or toes. The attacks are precipitated by moderate cold or emotional disturbances. They may lead eventually to ulceration or gangrene of the skin of the fingers.

(c) Trauma

Injury of the tissues close to an artery sometimes induces severe spasm of the vessel and impairment of the circulation in the limb.

(d) Ergotism

This is chronic poisoning from the consumption of rye infested by the ergot fungus, *Claviceps purpurea*, the alkaloids of which (ergotoxin and ergotamine) have a strong vaso-constrictor action. The principal result of this now rare disease is gangrene of the extremities.

(6) Arterial embolism

An *embolus* (Gk. *embolos*=a plug or stopper) is any abnormal mass of solid, liquid or gaseous matter carried in the blood and large enough to be arrested in some part of the vascular system. *Embolism* is the occlusion of a blood-vessel by an arrested embolus. The only kind of emboli which develop within the vascular system itself are fragments of clot or thrombi, and these are the commonest emboli. Other kinds of emboli, which enter the blood-stream from outside the vessels, include fragments of tumours, clumps of bacteria, protozoal and metazoal parasites, fat, air bubbles, amniotic fluid, and solid foreign bodies. We have already discussed emboli of infected thrombus from osteomyelitis (p. 72), from bacterial endocarditis (p. 76), and from inflamed veins (p. 81), emboli of malarial parasites in capillaries (p. 224), emboli of schistosome eggs in the liver and lungs (p. 249), air embolism (p. 321), and fat embolism (p. 322). Two other important sources of emboli are, (*a*) post-operative thrombosis of veins (p. 378), and, (*b*) intracardiac thrombi, either those which form in dilated heart chambers (p. 81) or those associated with infarction of the ventricular walls (p. 376).

The destination of emboli depends on their source. Emboli from the systemic veins are usually arrested in the pulmonary arteries or capillaries, those from the portal venous system in the portal branch veins in the liver, and those from the pulmonary veins or left heart chambers in the systemic arteries. Rarely, an embolus from a systemic vein by-passes the lungs through a patent foramen ovale or interventricular septum and is arrested in a systemic artery (" paradoxical embolism "). Excepting the portal circulation, then, most embolism is arterial embolism, causing occlusion of small or large arteries in the lungs or in any of the tissues supplied by the systemic arterial blood-stream. The size of the artery occluded by an embolus depends on the size of the embolus. Emboli from thrombosed veins are sometimes large enough to obstruct the main pulmonary arteries, or those from bacterial endocarditis may block the main arteries to the limbs or the main cerebral or mesenteric arteries ; on the other hand, showers of small emboli may occlude many small arterioles or capillaries, as in subacute bacterial endocarditis (p. 79) or the dissemination of malignant tumours (p. 447).

LOCAL EFFECTS OF ARTERIAL OBSTRUCTION

(1) Ischaemic necrosis of tissues

The local effects of arterial occlusion, from any cause, depend on the size of the artery, the functional importance of the tissues supplied by it, whether the occlusion is sudden or gradual, and whether or not an alternative collateral route of supply of arterial blood is available.

The last factor is of special importance ; if an artery is the only one through which a particular area of tissue can receive an adequate blood-supply, then clearly its occlusion will cut off the supply and cause necrosis of the tissue. Arteries of this kind, occlusion of which causes death of the tissue, are called *end-arteries* ; the death of the tissue is called *infarction*, and the dead tissue an *infarct*. End-arteries should not be defined as those which have no anastomosis with neighbouring vessels (like the branches of a tree), for all arteries have at least capillary connections with their neighbours. They are those which anastomose so little that their occlusion endangers the life of the tissues supplied by them. In this sense, the renal, splenic, pulmonary, coronary, cerebral, retinal, mesenteric and hepatic arteries and their main branches are end-arteries. In addition, some of the main arteries of the limbs, e.g. the external iliac, femoral, popliteal and brachial, have anastomotic connections which, relative to their own sizes, are small ; so that their sudden occlusion by injury, embolism or ligature not infrequently causes gangrene of the extremities, especially in old people. Gradual occlusion of these large limb vessels, as by atheroma or compression by tumours, usually causes no ill effects, because there is time for the development of an adequate collateral circulation. The smaller arteries of the limbs, trunk, head and neck and most of the viscera can be ligated with impunity—fortunately for surgery. However, even in tissues which normally have a richly anastomotic blood-supply, necrosis may result from obstruction of a sufficient number of the main arteries supplying them. This is exemplified by senile or diabetic gangrene of the limbs, in which there is widespread and extending thrombosis in many similarly diseased vessels, and by gangrene resulting from cold, Raynaud's disease or ergotism, in which all the arteries of the ischaemic parts share in the constriction. So also occlusion of multiple arteries by thrombosis or embolism can cause necrosis in many other tissues, e.g. pancreas, adrenals, pituitary or bones.

(2) General appearance of infarcts (Figs. 123–126)

A sterile infarct resulting from obstruction of a single end-artery is usually an approximately pyramidal mass of dead tissue, with its apex near the site of obstruction and its base at the periphery of the organ. The colour of the infarct

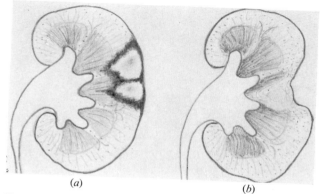

(a)   (b)

FIG. 123.—Sketch of infarcts in the kidney. (a), recent infarcts with haemorrhagic margins; (b), old infarct, showing fibrous contraction and depression of surface.

depends on the tissue affected and the age of the infarct. In the lung young infarcts are dark red and haemorrhagic, because much blood diffuses into the lax dead area from the congested tissues around it. In the kidney, because of the compactness of the tissue, infarcts are usually pale and yellow from the beginning, and congestion and haemorrhage are restricted to a narrow zone at the periphery of the dead area (Fig. 123). Infarcts in the heart or spleen are intermediate in appearance, often showing mingled haemorrhagic and pale yellow areas. With the lapse of time, infarcts in all of these tissues become pale and yellow, as the effused blood is absorbed; they also shrink, because of absorption of fluid and the diffusible products of autolysis and because of later organization, so that they form depressed areas on the surface of the organ. Infarcts of the brain are pale and become soft and fluid, from degeneration and liquefaction of myelin, which is then removed by phagocytes, leaving shrunken depressed areas.

Microscopically the dead tissue of a young solid infarct shows ghostly outlines of the structure of the original tissues ; but later, as autolysis proceeds, it is converted into amorphous granular debris.	In the early stages, the living tissues around the infarct show varying degress of congestion, extravasation of blood and inflammatory reaction.	As this subsides, phagocytes and organizing granulation tissue penetrate into the dead area : small infarcts may eventually be replaced completely by fibrous tissue ; but in large ones a central yellow mass of necrotic material remains, encapsulated by a zone of fibrous tissue.

The foregoing account applies to sterile infarcts.	But if infarction has resulted from an *infected embolus*, e.g. from suppurative thrombo-phlebitis or bacterial endocarditis, the infarct also is infected, it excites acute inflammation in the marginal tissues, and the dead area softens and becomes converted into an *abscess*. This is the way in which, in cases of pyaemia, metastatic abscesses develop in organs with end-arteries.

When ischaemic necrosis takes place in parts which themselves contain or are exposed to putrefactive bacteria, such as the limbs or intestine, we speak of *gangrene*.	The commonest example is the dry gangrene (or " mummification ") of the toes or feet resulting from senile or diabetic atheroma.	The dead tissues become black, dry and hard.	The necrosis slowly extends proximally until it reaches adequately vascularized tissues ; here it stops and a line of demarcation forms between the living and the dead parts.	By the formation of granulation tissue and the activity of phagocytes at this boundary, the dead part is slowly separated and cast off as a slough by the living—unless of course Nature's slow amputation is forestalled by the surgeon.	If the gangrenous tissues contain much fluid, e.g. from oedema, putrefactive changes in them are much more active and offensive and we speak of " moist " gangrene.	Gangrene of the intestine resulting from strangulation or from thrombosis or embolism of the mesenteric arteries is also a " moist " gangrene ; unless prompt surgical excision of the gangrenous part is practicable, it is always fatal.

(3) Effects of ischaemia in particular organs

Our discussion of ischaemia may be concluded by a summary of its causes and effects in particular organs.

(a) *Kidneys*

Infarcts of the kidneys are often found in patients dying of cardiac failure ; they are due to sterile emboli detached from thrombi in the left heart chambers. They occur also in cases of bacterial endocarditis, either acute or chronic, in which they are due to infected emboli.	Only occasionally does renal infarction result from thrombosis in atheromatous renal arteries.	Another rare cause is poly-arteritis nodosa.

A special form of renal infarction is widespread *bilateral cortical necrosis* which sometimes occurs in pregnant or puerperal women, especially in association with retro-placental intra-uterine haemorrhage.	This disease is usually included as one of the " toxaemias " of pregnancy, but the way in which the renal lesion is produced is not clear.	There appears to be complete stasis of the renal cortical circulation but without actual thrombosis.	Similar bilateral cortical necrosis occurs also in non-pregnant subjects, in whom it may be due to arrest of the renal circulation resulting from acute inflammatory blockage of the glomerular capillaries (Dunn and Montgomery).

Elsewhere (p. 307), we have discussed the relationship of general ischaemia of the kidneys to *hypertension*, most strikingly seen in malignant hypertension. In this disease renal ischaemia is due to " fibrinoid " necrosis of the walls and narrowing of the lumina of arterioles throughout the kidney and to intimal thickening in the larger arteries.   Thrombosis and infarction sometimes supervene in places, but are not an important feature.   It should be recalled, however, that hypertension does occasionally result from the renal ischaemia produced by ordinary atheromatous stenosis of the main renal arteries.

(b) Spleen

Splenic infarcts due to embolism are common in cardiac failure and bacterial endocarditis. They occur also from thrombosis of atheromatous splenic arteries, especially in persons with an impeded splenic circulation, from cardiac failure or chronic portal obstruction ; and in greatly enlarged spleens in leukaemia, Hodgkin's disease or lymphosarcomatosis.

(c) Intestines

Occlusion of the main mesenteric arteries is relatively rare and is more often due to embolism (usually from thrombi or vegetations in the left heart) than to local atheroma and thrombosis.

(d) Heart

Occlusion of the coronary arteries is one of the commonest causes of death. It is almost always due to thrombosis supervening in atheromatous vessels, rarely to embolism.   Atheroma and medial calcification of the coronary arteries are very common in middle-aged and old people ; and the myocardium, even in those dying of other diseases, often shows signs of an impaired coronary circulation.   These signs range from patchy areas of fibrosis where, as a result of slow occlusion of small branch arteries, the muscle fibres have atrophied and disappeared, to large areas of total infarction due to sudden occlusion of main arteries by thrombosis. Ischaemic myocardial fibrosis—improperly called " chronic myocarditis "—is an important cause of both chronic and sudden cardiac failure.

As we might expect, the apical parts of the heart suffer much more than the basal parts from the effects of ischaemia ; myocardial atrophy or infarction is most frequent in the walls of the ventricles, especially the apical portion of the left ventricle.

People with serious coronary atheroma often suffer from repeated attacks of severe precordial pain ; some of these attacks are evanescent and due to relative myocardial ischaemia brought on by effort (" angina pectoris ") ; but, in others the pain is more sustained and unrelieved by rest, and is due to occlusive coronary thrombosis. This may kill forthwith, or the patient may survive and an infarct develop in the heart wall. The necrotic area undergoes autolytic softening (myomalacia cordis), and the heart may rupture at the softened area. Or the infarct may undergo slow organization into white fibrous tissue. This non-contractile inelastic area may subsequently yield to the high intraventricular pressure (especially in the left ventricle) and form a cardiac aneurysm which may eventually rupture into the pericardium (Fig. 124).

Myocardial infarction is often accompanied by inflammatory reaction in the overlying epicardium and by a pericardial friction rub ; and later, localized adhesions between the infarcted area and the parietal pericardium may develop. The endocardium lining the inner aspect of the infarct also suffers, and becomes clothed by a layer of thrombus, parts of which may subsequently break off and cause embolic infarcts in other organs, as already described. An old fibrous area of infarction which has dilated to form a cardiac aneurysm also usually contains a mass of laminated thrombus ; and from this too emboli may be freed and carried to other organs.

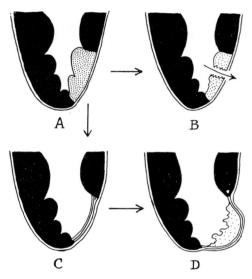

FIG. 124.—Diagram showing effects of cardiac infarction : A, recent infarct (stippled) ; B, rupture of heart through recent infarct ; C, fibrosis of old infarct ; D, cardiac aneurysm containing adherent thrombus ; the aneurysm may rupture.

(e) Brain

The causes and effects of ischaemia of the brain are closely comparable with those in the heart. Atheroma of the cerebral arteries with supervening thrombosis is much more frequent than embolism as a cause of cerebral ischaemia ; and the results range from patchy but widespread atrophy of the brain, accompanied by gliosis, due to slow occlusion of many small vessels, to large areas of infarction and softening due to sudden thrombosis of large arteries. Cerebral softening frequently occurs in multiple areas. The nervous signs and symptoms produced vary greatly according to the extent and situation of the vascular lesions. Extensive thromboses, or those which affect the vital centres, may kill quickly ; but more often the patient survives with varying degrees of hemiplegia, facial or other paralysis, aphasia, mental deterioration, Parkinsonism (mask-like facies and muscular rigidity), ataxia or other signs of cerebellar degeneration. Senile dementia is due to widespread cortical atrophy, usually along with multiple areas of softening ; and in advance cases the brain shows great shrinkage and distortion, wide separation from the skull, dilated ventricles, and a loss of many ounces in weight. Calcification of the degenerated parts sometimes occurs (Figs. 125, 126).

Accompanying the ischaemic destruction of particular parts of the brain, there is degeneration of the corresponding nerve fibres, e.g. the pyramidal tracts, cortical association fibres or cerebellar tracts, as the case may be.

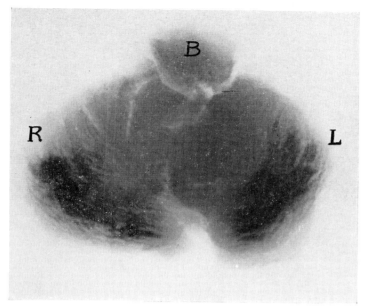

FIG. 125.—X-ray photograph (positive print) of cerebellum RL and
brain-stem B, showing dense shadows due to calcification in both
lobes of cerebellum where extensive thrombosis and infarction had
occurred. The cerebrum also showed multiple areas of softening,
but without calcification. (See also Fig. 126.)

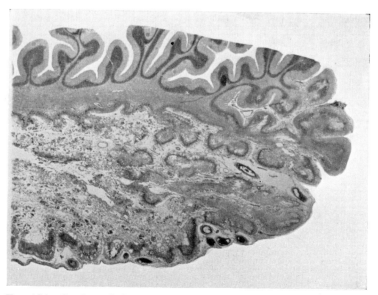

FIG. 126.—Section of the cerebellum of Fig. 125, showing junction of
normal and degenerated parts. ($\times$ 5.)

(f) Lungs

Infarcts in the lungs occur in three distinct classes of cases, namely (i) chronic cardiac disease, (ii) post-operative embolism, and (iii) infective embolism from systemic veins or the right heart chambers.

(i) *Pulmonary infarcts in chronic cardiac disease.*—These may occur with any kind of chronic right cardiac failure, especially that due to stenosis or incompetence of the mitral valve and accompanied by auricular fibrillation. Ante-mortem clots develop in the dilated right atrium, whence they may become detached and carried as emboli into the pulmonary arteries. The infarcted areas of lung become airless, solid and haemorrhagic, the extravasated blood coming mainly from the still patent bronchial vessels. Probably also, in some cases of cardiac failure, pulmonary infarction results from local thrombosis in the pulmonary arteries, consequent on the obstructed flow of blood and the presence of some degree of atheroma in those vessels. Whether due to embolism or thrombosis, the clot in the vessel may extend and cause infarction of a still larger mass of lung tissue. The occurrence of infarction of the lung in cases of cardiac disease is recognized clinically by the sudden onset of pleural pain, dyspnoea, and slight or moderate haemoptysis.

(ii) *Post-operative embolism.*—This serious complication may occur at any time after an operation, but is most frequent in the second week, when the convalescent patient, kept in bed until then, begins to move about. It follows pelvic and abdominal operations much more frequently than others. In only a few cases does the embolus come from thrombosed veins at the site of operation ; much more frequently its source is thrombosis of the veins of the lower limbs, especially those of the calf muscles, less often the main popliteal, femoral or iliac veins. The thrombosed veins show no signs of inflammation, and the causes of the thrombosis are mechanical injury of the veins by the pressure of retractors or packs or bad posture during operation, stasis of the blood in the vessels due to post-operative immobility of the lower part of the body, compression of the leg veins by props or pillows under the knees, old age, and anaemia or general debility. The importance of venous stasis as a cause of post-operative embolism is borne out by the infrequency of this complication in children, in healthy adults who are encouraged to move about and to leave bed soon after operation, and in cases of operations on the upper limbs or thorax, in whom there is seldom any reason for post-operative immobility of the lower limbs. In fatal post-operative embolism, large thrombi may be found completely occluding the main pulmonary arteries ; in less severe cases, occlusion of branch arteries causes single or multiple infarcts in the peripheral parts of the lungs.

(iii) *Infective embolism from systemic veins or right heart chambers.*—This result of infective phlebitis or endocarditis has already been referred to in Chapter 7 (pp. 78, 83). The infected pulmonary infarcts excite acute pneumonitis in the surrounding lung tissue, and frequently soften to form abscesses.

ANEURYSMS

(1) Definition and types of aneurysms

An aneurysm is any abnormal dilatation of an artery. Three structural types of aneurysm are recognized (Fig. 127) ; (*a*) *fusiform aneurysm* consisting of a general dilatation of part of an artery affecting its whole circumference ; (*b*) *saccular aneurysm*, a localized protrusion or diverticulum of an artery not involving its whole circumference ; and (*c*) *dissecting aneurysm* characterized by rupture of the inner layers of the wall of a large artery and splitting of its coats apart by the accumulation of blood extravasated into the wall at the point of rupture. Some aneurysms show combinations of these types of structure, especially (*a*) and (*b*).

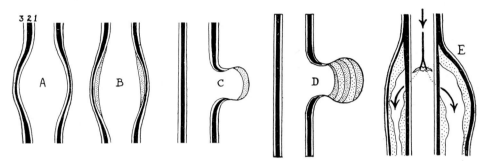

FIG. 127.—Diagram of types of aneurysms. 1, intima ; 2 (thick band), media ; 3, externa Thrombus, stippled. A, fusiform aneurysm without thrombus ; B, fusiform aneurysm with thrombus ; C, saccular aneurysm with little thrombus ; D, saccular aneurysm with much laminated thrombus ; E, dissecting aneurysm splitting media.

(2) Sites

In general, the larger an artery, the greater its liability to aneurysms. The aorta is the commonest site of all, and the frequency of aortic aneurysms is greatest in the aortic arch and diminishes progressively in the descending thoracic and abdominal parts of the vessel. The iliac, femoral, popliteal, innominate, carotid, renal, splenic and other visceral arteries are all much less frequent sites of aneurysms than the aorta. The cerebral arteries, however, especially those of the circle of Willis and its main branches, are a frequent site of aneurysms, some of developmental origin, some due to degenerative arterial disease.

(3) Causes of aneurysms

Apart from inflammatory weakening of the walls of arteries traversing inflamed tissues (p. 84) and wounds of arteries, the causes of aneurysms are (*a*) syphilitic arteritis, (*b*) non-syphilitic degenerative lesions, and (*c*) developmental defects.

(*a*) Syphilis

Syphilis as a cause of aneurysms has already been discussed on p. 195 : it accounts for the majority of fusiform and saccular aneurysms of the aortic arch and for a minority of those of more distal parts of the aorta or of other arteries. Syphilitic aneurysms were much more frequent during the 17th and 18th centuries than they are now ; the writings of Morgagni, John Hunter and others contain very numerous descriptions of them.

(*b*) Non-syphilitic degenerative lesions

These include both atheroma and medial defects. Atheroma itself is not a cause of aneurysm, but it is often accompanied by atrophic and degenerative changes in the media, which weaken the wall and lead to general dilatation or saccular aneurysm. Most cases of dissecting aneurysm are due to faults in the media, which gives way usually as a result of a sudden rise of blood-pressure accompanying physical effort or mental stress ; the tear in the media results also in a tear of the overlying intima, through which blood penetrates into the arterial wall. It is probable also that some dissecting aneurysms begin from atheromatous ulcers. Syphilis is rarely, if ever, the cause of a dissecting aneurysm.

(c) *Developmental defects*

Saccular aneurysms of the circle of Willis or its main branches occur most frequently in young people, even children, in whom there is no other arterial disease. They are attributed to developmental defects in the muscular and elastic tissue of the arteries—defects which are frequently found at the bifurcation or junction of two vessels. The aneurysms occur chiefly at these points, especially on the anterior communicating and middle cerebral arteries, and they are sometimes multiple. Because they arise from developmentally weak spots, they are often spoken of as " congenital aneurysms " ; but it is the mural defects and not the aneurysms themselves which are present at birth, the dilatation developing later.

(4) Structure of aneurysms

(a) *Fusiform aneurysms*

The affected segment of artery, usually part of the aorta, shows general dilatation, and its wall shows the changes of syphilitic arteritis or of non-specific atheroma and medial degeneration. Its lumen may contain no thrombus at all, or its wall may be lined by a stratified thrombus in the form of a tube through which the blood-stream passes (Fig. 127).

(b) *Saccular aneurysms*

Saccular aneurysms differ greatly in size : those on the cerebral arteries may rupture and cause fatal haemorrhage while still only a few millimetres in diameter, while the largest aortic aneurysms may attain diameters of many inches and may contain pints of blood and clot. The walls of the sac rarely contain any recognizable muscular tissue, but consist merely of thickened densely fibrous adventitia. Its contents consist partly of laminated thrombus of various ages adherent to its walls and partly of fluid blood, the relative amounts of the two varying according to the degree of sacculation present and the size of the orifice of communication with the main channel (Fig. 127).

(c) *Dissecting aneurysms*

At the tear in the intima and media, the blood penetrates into the wall of the artery and extends between its layers, splitting them apart for short or long distances. The split lies usually between the layers of the media, but sometimes between the media and a thickened adventitia. It may involve only a part of the circumference ; or it may completely encircle the vessel, so that this consists of a tube within a tube. The split may extend along the walls of branches of the primarily affected artery ; but some lateral branches may persist intact across the dissected channel. The dissection sometimes proceeds for surprisingly long distances, e.g. from the aortic arch (the usual site of origin) right down the aorta to the iliac arteries. The speed of its progress varies : sometimes it takes place rapidly and soon terminates fatally by rupturing at some point or another ; sometimes it extends slowly and symptomlessly, and may reopen into the main lumen at some distant point. When this happens, a blood-stream is established along the dissected channel, which heals and acquires a smooth endothelial lining ; the artery is now a double-barrelled one (Figs. 128, 129). In chronic cases, the proximal part of the dissected channel (in the aortic arch) often becomes greatly distended and saccular, as in Fig. 128.

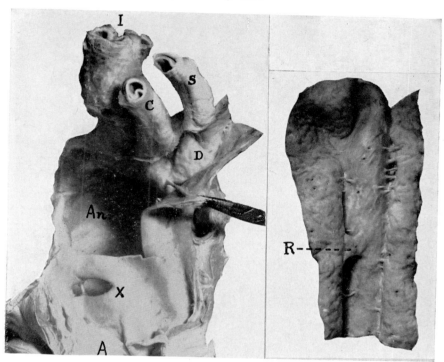

FIG. 128.—Extensively healed dissecting aneurysm of aorta converting this vessel and its main branches into double-barrel channels. A, aortic valves ; X, button-hole opening in posterior wall of ascending arch, where rupture of inner coat first took place and whence dissection started ; An, large healed aneurysmal sac at commencement of dissected channel, exposed by cutting out a window from the intervening wall of the main channel ; D, healed dissected channel of the arch ; I, double-barrel innominate artery ; C, double-barrel left carotid ; S, double-barrel left subclavian. Figure on right is part of descending aorta with healed dissected channel opened out to show bridges traversing it, the largest of which R is renal artery, and others are intercostal and lumbar arteries. The dissected channel re-entered the main one in the left common iliac artery. (See Fig. 129.)

(5) Results of aneurysms

The chief results of aneurysms are as follows.

(a) The formation of a pulsatile swelling

This may be easily visible or palpable when the aneurysm is in the neck, anterior mediastinum, limbs or abdomen ; or it may be seen radiographically when the aneurysm is in the descending thoracic aorta. Most, but not all aneurysms pulsate ; pulsation may be absent when a saccular aneurysm is occupied largely by old adherent clot.

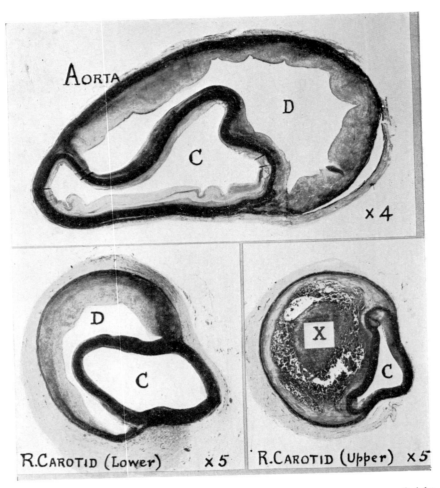

FIG. 129.—Cross-sections of abdominal aorta and of lower and upper parts of right common carotid artery (elastic tissue stain) from same specimen as Fig. 128. C, main channel ; D, healed dissected channel ; X, thrombus at upper limit of dissection where this is still in progress.

(b) Pressure on surrounding structures

Pressure produces many and varied results according to the situation. Some of the chief pressure symptoms of thoracic aortic aneurysms are pain, dysphagia, cough, dyspnoea, laryngeal paralysis, inequality of the pupils, signs of venous congestion in the upper limbs and head, inequality of the pulses and blood pressures in the two upper limbs. It is a good anatomical exercise to work out the causes of these symptoms, or alternatively to work out the particular symptoms likely to be produced by aneurysms in particular situations. Bones which are pressed upon by an aneurysm undergo slow pressure atrophy or erosion (Fig. 130) ; large thoracic aneurysms often erode the vertebrae, ribs or sternum, and may appear externally beneath the skin.

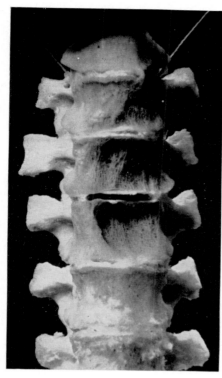

FIG. 130.—Erosion of three thoracic vertebrae by an aortic aneurysm. Note how intervertebral discs resist the pressure better than the bone. Specimen in Royal College of Surgeons, London.

(c) *Rupture*

This is a common and usually fatal termination. Intra-thoracic aneurysms may rupture into the trachea or bronchi, pleural or pericardial cavities, oesophagus, heart, or externally through the skin. Rupture of a cerebral aneurysm usually takes place into the basal cerebrospinal cisternae—*spontaneous subarachnoid haemorrhage*—but sometimes the haemorrhage is into the brain substance, tearing this apart. The first attack may be rapidly fatal ; but in other cases the leakage is slight and is temporarily arrested, to be followed later by several attacks at intervals, before the fatal one.

VENOUS OBSTRUCTION

Obstruction of veins may be caused by thrombosis due to infective phlebitis (p. 81), thrombosis due to non-inflammatory injury and stasis (p. 378), rapid mechanical constriction as in twisted or strangulated parts (p. 370), or slowly developing invasion or compression of the veins by tumours or other neighbouring masses. The effects of venous obstruction on the tissues vary with the degree of the obstruction and the rapidity of its development. In ascending order of severity, we see (1) the development of a collateral venous circulation, with or without cyanosis and oedema of the part, (2) severe congestion, oedema and haemorrhagic infiltration of the tissues, and (3) complete arrest of the circulation and necrosis of the tissues.

(1) Collateral venous circulation

If a large vein suffers gradual occlusion, e.g. by tumour invasion or compression or slowly spreading thrombosis, alternative paths for the venous return may be opened up through collateral vessels. Thus, with slow occlusion of the inferior vena cava, the subcutaneous veins of the trunk become greatly dilated and carry the venous blood of the lower limbs and pelvis to the tributaries of the superior vena cava. Slow occlusion of the superior vena cava or innominate veins leads to the development of prominent subcutaneous collaterals carrying the venous blood of the head and upper limbs caudally to reach the tributaries of the inferior vena cava. Following thrombosis of the axillary, femoral or other large peripheral veins, there occurs a similar development of collateral channels appropriate to the site of the obstruction. In all such cases the return of blood by the collateral route is never as efficient as by the normal route, so that there is some congestion, cyanosis or oedema of the affected tissues, and the peripheral veins are often varicose, i.e. dilated and tortuous.

Slow obstruction of the portal vein, e.g. from cirrhosis or tumours of the liver, leads to the development of large collateral channels of communication between the radicles of the portal and caval systems. These appear in the following situations : (a) around the cardia and lower part of the oesophagus, beneath the mucous membrane of which they project as large thin-walled varices which often rupture and cause serious or fatal haematemesis ; (b) around the anal canal and lower part of the rectum, where the dilated veins may form haemorrhoidal varices or " piles " ; (c) around the liver, where dilated anastomoses develop between the veins of this organ and those of the diaphragm, round ligament and abdominal wall, sometimes being visible as radiating channels around the umbilicus (caput medusae) ; and (d) retroperitoneal anastomoses between the mesenteric veins and tributaries of the inferior vena cava (veins of Retzius). Portal venous obstruction is accompanied also by congestive enlargement of the spleen, congestion of the gastro-intestinal tract and impaired digestion, and ascites.

It is of interest to note here that veins show *structural adaptations to raised intravenous pressure*. Thus, in cirrhosis of the liver, the veins of the portal system show hypertrophy of the circular muscle fibres in their media and thickening of their intima ; while in congestive cardiac failure the caval veins show intimal thickening and hypertrophy of the longitudinal muscle fibres of the adventitia.

(2) Haemorrhagic venous obstruction

Torsion and other forms of *strangulation* of pedunculated organs or tumours has already been mentioned (p. 370). Under these circumstances, the veins suffer rapid constriction while the arterial supply is little or not at all impaired ; the part quickly becomes stuffed with arterial blood, capillaries and small veins rupture, and the tissues become greatly swollen and haemorrhagic. A similar state of the intestine occurs in cases of *mesenteric venous thrombosis*, a condition sometimes attributable to spreading phlebitis from appendicitis or other infective inflammations, but more often seemingly spontaneous and of undetermined cause.

(3) Necrosis from venous obstruction

Clearly, if all the veins draining a part are completely occluded, the circulation through the part will be arrested and the tissues will die. This applies to many twisted and strangulated organs, and to the affected segments of intestine in cases of thrombosis of the mesenteric veins. Extensive cortical necrosis of the kidney has occasionally resulted from arrest of the renal circulation by massive thrombosis of the renal veins. Necrosis of the anterior lobe of the pituitary gland is not unusual in women suffering from severe haemorrhage after child-birth, and is due to thrombosis of the venous and capillary sinuses of the organ. If the patient survives, she develops signs of hypopituitarism (p. 605).

CONGESTIVE CARDIAC FAILURE

Congestive cardiac failure is really a condition of universal venous obstruction, due to the inability of the heart to maintain an adequate flow of blood from the veins through the lungs to the systemic arteries. Whatever the cause—arterial hypertension, chronic valvular incompetence or stenosis, myocardial disease, or obstruction of the pulmonary circulation due to fibrosis of the lungs—the extra load imposed on the heart eventually exceeds the capacity of this organ for compensation by myocardial hypertrophy. The right cardiac chambers then become dilated and unable to receive and transmit the full volume of the returning venous blood, the veins throughout the body become distended, and the tissues show cyanosis, oedema, congestive pigmentation (p. 355), and other results of chronic venous obstruction. The cyanosis is due to the over-depletion of the red corpuscles of oxygen during their retarded passage through the capillaries, so that the blood contains an excessive proportion of reduced haemoglobin. The tissues suffer from oxygen deficiency (anoxia) and excessive accumulation of CO_2 and other metabolites. Stimulation of the respiratory centre results in hyperpnoea, which is often so pronounced and distressing as to constitute dyspnoea or air-hunger, or to necessitate that the patient should be propped up in a sitting position to avoid intolerable respiratory distress (orthopnoea). The cause of cardiac oedema is discussed below.

LYMPHATIC OBSTRUCTION

Obstruction or obliteration of lymphatics frequently occurs from chronic inflammatory diseases or from infiltrating tumours. When the obstruction affects the lymphatics of only a restricted area of tissue, collateral channels suffice to maintain an adequate drainage of lymph from the region. It is only when the obstruction is of wide extent, or involves large lymphatic trunks, that lymph-stasis results, and the affected part becomes oedematous. The oedema resulting from chronic lymphatic obstruction often differs from that of venous obstruction in being more " solid ", i.e. less liable to pit on pressure. Long sustained lymphatic oedema frequently leads also to *elephantiasis*, a condition of great fibrous overgrowth of the skin and subcutaneous tissues. Obstruction of the mesenteric lymphatics, cisterna chyli or thoracic duct may cause *chylous ascites* or *chylothorax*. The principal causes of lymphatic obstruction and oedema are as follows:

(1) Carcinoma

Frequently, the skin overlying a carcinoma of the breast, or occasionally of some other organ, presents the appearance aptly called either " peau d'orange " or " pig-skin " ; the skin is thickened and shows prominent pits which are the orifices of hair follicles exaggerated by the surrounding swelling. This condition is one of lymphatic oedema of the dermis due to cancerous obstruction of subjacent lymphatics. Widespread cancerous invasion of the lymphatics and lymph glands of a part, or extensive surgical removal of lymph glands because of cancer, may lead to extreme oedema of neighbouring parts. This is seen in some cases of carcinoma of the breast, in which, following cancerous infiltration or wide surgical removal of the axillary tissues, the whole upper limb becomes greatly swollen, immobile and painful. Cancerous occlusion of the thoracic duct (p. 444) may cause chylous ascites.

(2) Filariasis

As described earlier (p. 239), this is a common cause of lymphatic oedema and elephantiasis in tropical countries. Any part may be affected, but the usual sites are the lower limbs and external genitals, which may suffer enormous enlargement. Chylous ascites and chylothorax may also occur.

(3) Chronic bacterial inflammations

Occasionally, chronic coccal, tuberculous or other kinds of lymphangitis and lymphadenitis are sufficiently extensive to cause lymphatic oedema of the affected part, usually a limb.

OEDEMA

Oedema is an excess of extravascular fluid in the tissues. In solid tissues it produces swelling which pits on pressure ; in the coelomic cavities it produces large serous effusions—ascites, hydrothorax and hydropericardium ; in the lungs, the oedema fluid effuses into the alveoli, and may drown the patient. For a clear conception of the causes of oedema, we must understand the normal fluid movements between the blood, tissue spaces and lymphatics.

(1) Normal fluid movements in the tissues

The blood capillaries and the lymphatics form two closed systems of vessels ramifying through the tissues. Endothelial cells lining the blood vessels are the only cells bathed by blood, and the only cells bathed by lymph are the endothelial cells lining the lymph vessels. All other tissue cells are bathed by *tissue fluid*, an intermediary fluid conveying nutritive materials and oxygen to, and metabolites and CO_2 away from, the cells (Fig. 131). The blood-pressure at the arterial end of the capillaries is about 35 millimetres Hg, and at the venous end about 10 millimetres ; while the osmotic pressure of the plasma proteins, tending to prevent the filtration of fluid through the capillary walls (which are normally impermeable to proteins), is about 25 millimetres. Hence at the arterial end of the capillaries there is a *net* outward filtration pressure of about 10 millimetres, and fluid is constantly passing out of the blood into the tissue spaces ; while at the venous end there is a *net* return pressure of about the same, and tissue fluid is constantly being reabsorbed back into the blood-stream. A much smaller quantity of tissue fluid percolates through the endothelial walls of the lymphatics and passes away

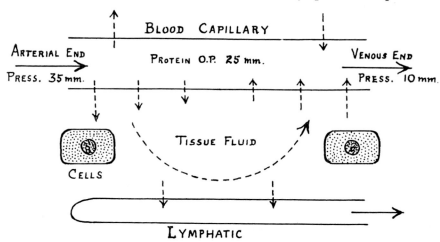

FIG. 131.—Diagram of movement of fluid in tissues.

as lymph. The composition of normal tissue fluid approximates to that of plasma but with a very low protein content. Lymph is similar but its protein content higher, though not as high as that of blood plasma—varying between 2 and 5 per cent.

(2) Factors concerned in the production of oedema

From the foregoing outline of the normal formation and removal of tissue fluid, we can easily specify the possible factors which might bring about an excessive accumulation of this fluid. These factors are : (a) *increase of capillary blood-pressure*, which increases the transudation of fluid from the blood ; (b) *hypo-proteinaemia*, which diminishes the protein osmotic pressure restraining transudation from the blood ; (c) *increased permeability of the capillary walls*, which permits the escape of the plasma proteins as well as electrolytes into the tissues ; and (d) *lymphatic obstruction*, which prevents the removal of tissue fluid via the lymphatics. The following types of oedema observed by the clinician and patholo-gist can all be traced to one or more of these factors.

(a) Oedema due to venous obstruction and cardiac failure

These can be considered together, since they are essentially similar. Factors (a) and (c) above both contribute to the development of the oedema. Raised venous and capillary blood-pressure increases trans-capillary filtration, especially in dependent parts where the excessive venous pressure is greatest. But that the permeability of the capillary walls is also affected is shown by the fact that the oedema fluid is rich in protein. This still further increases transudation from the capillaries into the tissues. The causes of this impairment of the capillary membrane are probably complex, including anoxia, excess of CO_2, alteration of pH, and nutritional disturbance. Perhaps also, the same factors affect the extra-vascular tissue cells and lead to the accumulation of metabolites and increased osmotic retentiveness of the tissues for fluid.

(b) Renal oedema

The oedema of acute nephritis is attributed to toxic injury of the capillary walls throughout the body, aided perhaps by impaired urinary excretion of water and salt. The oedema of subacute (Type II) nephritis is undoubtedly due in a large measure to the pronounced hypoproteinaemia of this disease ; but that this is not the only factor is shown by the appearance and disappearance of the oedema without corresponding changes in the plasma protein level in some cases, and by the fact that, in animals with experimental nephritis produced by uranium or mercury salts, excessive salt in the diet can produce oedema in the absence of significant hypoproteinaemia. These and other facts show that sodium chloride retention is probably an important auxiliary factor in producing the oedema of subacute nephritis. That this disease is not accompanied by significant changes in capillary permeability for proteins is shown by the fact that the protein content of the oedema fluid is very low. Renal oedema, whether due to acute or subacute nephritis, affects the whole body, appearing first in the face as often as in dependent parts ; its distribution is much less determined by gravity than is that of cardiac oedema.

(c) Nutritional or famine oedema

This has already been referred to on pp. 342 and 349. Its main cause is hypoproteinaemia due to dietary deficiency of proteins ; but increased capillary permeability resulting from malnutrition may also play a part.

(d) Allergic and allied kinds of oedema

On p. 287, urticaria due to allergic sensitization to foreign proteins was mentioned. Urticaria or " nettle-rash " (L., urtica=a nettle) is characterized by the sudden appearance of multiple itchy wheals or oedematous swellings in the skin. These resemble the wheals produced by the local injection of histamine, and are doubtless the result of increased capillary permeability for proteins due to the liberation of histamine or histamine-like substances from the interaction of the foreign antigen with fixed antibody in the sensitized tissues. However, the subject is not wholly understood, for similar urticarial wheals, or even universal oedema, are sometimes seen in people with idiosyncrasies to non-protein substances, such as aspirin, salicylates, iodides, bromides and various other drugs. It is therefore possible that the urticaria resulting from eating strawberries, shell-fish, lobster, etc., by certain people is not due to allergy in the strict sense, but to idiosyncrasy to non-protein substances. Factitious urticaria or dermatographia means the production of wheals by stroking the skin of certain susceptible indivi-duals : this is an exaggeration of the normal reaction of the skin to scratching, and is undoubtedly due to the local liberation of histamine or similar " H-sub-stances ". Angioneurotic oedema is a rare form of localized oedema of sudden onset, short duration, and unknown cause, affecting certain susceptible persons and sometimes showing a familial incidence. It usually affects some part of the skin, e.g. the face or a limb ; but sometimes viscera are involved, and asphyxia from oedema of the larynx may occur.

(e) *Inflammatory oedema*

This was discussed in Chapter 3 (p. 45). It is mainly or wholly attributable to increased capillary permeability, and the exuded fluid is rich in plasma proteins.

(f) *Lymphatic oedema*

This was discussed above (p. 385).

ANHYDRAEMIA OR HAEMOCONCENTRATION

(1) Causes of anhydraemia

By *anhydraemia* or *haemoconcentration* is meant diminution in the volume of blood plasma without corresponding diminution of corpuscles, so that the haemoglobin percentage, red cell count and packed cell volume (haematocrit percentage) are raised. This may be brought about in several ways :

(a) *By thermal burns*

In an extensive burnt area a large volume of protein-rich fluid is poured out from the damaged vessels ; and rapidly progressive haemoconcentration occurs due to the loss of plasma, one quarter of the total volume of which may be lost within a few hours. There is also a fall in the concentration of plasma proteins in the blood.

(b) *By chemical burns*

Extensive destruction or blistering of the skin by such substances as mustard gas and lewisite (p. 332) produces anhydraemia identical with that following thermal burns.

(c) *By crushing injuries*

Extensive crushing injuries of tissues, especially muscles, or injuries which by causing prolonged compressional arrest of the circulation in a limb lead to extensive ischaemic necrosis of muscle, are followed by haemoconcentration. As in burns, this is due to a great outpouring of protein-rich fluid into the damaged tissues. The " crush syndrome " will be referred to again presently.

(d) *By subcutaneous injection of hypertonic solutions*

Experiments in animals show that subcutaneous injections of large quantities of strong solutions of salt or glucose produce, by their osmotic effect, great local oedema with consequent rapid anhydraemia. Very little protein is lost from the blood, and the plasma protein level is unchanged ; hence no capillary damage or alteration of capillary permeability has occurred.

(e) *By water deprivation*

Following prolonged thirst, or deprivation of water by severe vomiting or diarrhoea from any cause, the plasma volume is at first maintained by absorption of available tissue fluid into the blood ; but, later, haemoconcentration develops, not only of the corpuscles but also of all plasma constituents, including the proteins.

(2) Effects of anhydraemia

By whatever means it is produced, anhydraemia is accompanied by similar well-defined clinical results. The respiratory rate increases, the pulse rate becomes rapid and the pulse feeble, the blood-pressure falls, the skin becomes pale, cold and clammy, and death may occur from circulatory failure. These results are identical with those which clinicians commonly refer to as " shock ", a term which we must now discuss.

" SHOCK "

In man, any serious collapse resulting from accidental injury, surgical operations, burns, loss of blood, cold or pain may be given this label " shock ". It is important to recognize that " shock " is not a distinct pathological entity, but merely a clinical syndrome—and a rather variable one at that—which may be produced in diverse ways by diverse kinds of injury, and the chief end-result of which is circulatory failure.

Causes of " shock "

Let us enumerate the different factors which may contribute to the production of the " shock " syndrome. In many cases of human injury two or more of these are in operation, simultaneously or successively.

(a) Anhydraemia

As already described this is a major factor in the production of shock from burns, crushing injuries, and severe dehydration.

(b) Haemorrhage

The circulatory collapse resulting from severe haemorrhage is identical with that due to severe plasma loss ; and haemorrhage is undoubtedly a major factor in causing " surgical shock " from injuries and operations. Indeed, in the great majority of cases of human shock (a) or (b) or both together are conspicuously present ; from which we may infer that *the most important single factor in the causation of shock is a sudden reduction of blood volume.*

(c) Traumatic toxaemia

A view formerly widely held, that the essential cause of shock was the production of histamine or allied vaso-paralytic substances in the injured area, has lost favour of recent years. Nevertheless, there are good grounds for believing that toxic substances are produced in, and absorbed from, injured tissues ; the oedema fluid from burns, extracts from injured muscle, and the thoracic duct lymph from animals with crushed lower limbs, are toxic and sometimes lethal to healthy animals. Hence, toxaemia must be admitted as a probable auxiliary factor in many cases of traumatic shock, and further endeavours must be made to identify the responsible substances.

(d) Reflex nervous effects

It is known that severe prolonged pain can bring about a state of circulatory collapse and nervous exhaustion ; and there is no doubt that pain and its reflex effects on the vaso-motor system and on the secretion of adrenaline play an important part in the production of *immediate shock* following such injuries as burns, fractures, or the sudden flooding of the peritoneal cavity from a perforated ulcer of the stomach or duodenum.

(e) Other contributory factors

In particular cases, exposure to cold, bacterial toxaemia from infection of the injured area, fat embolism, excessive fear or other psychical factors, may contribute to the development of the circulatory failure which is the essence of the " shock " syndrome.

THE CRUSH SYNDROME

Following severe injuries which produce extensive crushing or ischaemic necrosis of muscle, the patient may not only develop the usual signs of " shock " but may later suffer from *myohaemoglobinuria* and serious renal damage which may end in fatal uraemia, a condition spoken of as the " crush syndrome ". The sequence of events is as follows : The crushing injury causes prolonged obstruction of the circulation in the limb, so that the muscles suffer partial or complete necrosis with liberation of their myohaemoglobin and other substances ; release from the compression is followed by absorption of these substances into the circulation ; myohaemoglobin is excreted by the kidneys and contributes to the formation of pigmented casts in the tubules, especially the second convoluted and collecting tubules ; at the same time the glomeruli and the epithelium of the tubules suffer injury, and oliguria and azotaemia appear and may terminate fatally in uraemia. The cause of the injury to the nephrons is not clear : the view that it is secondary to tubular obstruction by the pigment casts is almost certainly not correct ; it is much more probably a direct toxic effect of the substances which are being excreted. Similar changes occur in the kidneys in cases of acute intravascular haemolysis from poisons (p. 336), blackwater fever (p. 224), and incompatible blood transfusion ; but in these cases the pigmented casts in the tubules contain haemoglobin and methaemoglobin, not myohaemoglobin.

SUPPLEMENTARY READING

Asher, R. A. J. (1947). " The dangers of going to bed ", *Brit. Med. J.*, **2**, 967.
Bywaters, E. G. L., and Dible, J. H. (1942). " The renal lesion in traumatic anuria ", *J. Path. Bact.*, **54**, 111.
Cameron, G. R. (1946). " Sudden shifts of body fluids ", *Proc. Royal Soc. Med.*, **40**, I.
— (1948). " Pulmonary oedema ", *Brit. Med. J.*, **1**, 965.
— (1954). " Haemorrhage " and " Shock ". In *Lectures on General Pathology* (Ed. H. Florey), Chaps. 31 and 32. London; Lloyd-Luke.
— Burgess, F., and Trenwith, V. (1946). " An experimental study of some effects of acute anhydraemia ", *J. Path. Bact.*, **58**, 213.
Duguid, J. B. (1946). " Thrombosis as a factor in the pathogenesis of coronary atherosclerosis ", *J. Path. Bact.*, **58**, 207.
Florey, H. (1960). " Coronary artery disease ". *Brit. Med. J.*, **2**, 1329. (Reviews the part played by lipoid metabolism.)
Harrison, C. V. (Ed.) (1967), Eighth edition. *Recent Advances in Pathology*. London; Churchill. (Chapter 5 on Diseases of Arteries.)
Hunter, W. C. *et al.*, (1941). " Thrombosis of the deep veins of the leg ", *Arch. Internal Med.*, **68**, 1.
Li, Pei-Lin, (1940). " Adaptations in veins to increased intravenous pressure ", *J. Path. Bact.*, **50**, 121.
Moon, V. H. (1936). " Shock. A definition and differentiation ", *Arch. Path.*, **22**, 325.
— (1942). " The vascular and cellular dynamics of shock", *Amer. J. Med. Sc.*, **203**, 1.

Osborn, G. R. (1963). *The Incubation Period of Coronary Thrombosis.* London; Butterworths. (Contains many beautiful photomicrographs of diseased coronary arteries.)

Rogers, L. (1947). " Ligature of arteries, with particular reference to carotid occlusion and the circle of Willis ", *Brit. J. Surg.*, **35,** 43.

Shennan, T. (1934). " Dissecting aneurysms ", *Med. Res. Council Reports,* No. 193, London.

– H. L. (1937). " Post-partum necrosis of the anterior pituitary", *J. Path. Bact.*, **45,** 189.

Smirk, F. H. (1949). " Pathogenesis of essential hypertension ", *Brit. Med. J.*, **1,** 791.

Steiner, P. E., and Lushbaugh, C. C. (1941). " Maternal pulmonary embolism by amniotic fluid as a cause of obstetric shock and unexpected deaths in obstetrics ". *J. Amer. Med. Ass.*, **117,** 1245.

Thorpe, F. T., and Clegg, J. L. (1936). " Multiple arteriosclerotic aneurysms of the circle of Willis ", *J. Path. Bact.*, **42,** 565.

Young, J., *et al.* (1949). " Discussion on the pathological features of cortical necrosis of the kidney and allied conditions associated with pregnancy ". *Proc. Royal Soc. Med.*, **42,** 375

CHAPTER 27

DISTURBANCES OF HAEMOPOIESIS

HAEMATOLOGY, the study of normal and abnormal haemopoiesis and of the normal and abnormal cytology of the blood, has become a highly specialized branch of pathology and clinical medicine with a complex (and often confusing) terminology of its own.   The aim of this chapter is not to attempt a comprehensive summary of the whole subject, but to give only a simple outline of its main principles.   We will consider these in the following order :

(1) Abnormalities of erythropoiesis.
    (*a*) Deficiency of red corpuscles or haemoglobin : anaemia.
    (*b*) Excess of red corpuscles and haemoglobin : polycythaemia.
(2) Abnormalities of leucopoiesis.
    (*a*) Deficiency of leucocytes : leucopenia.
    (*b*) Temporary excess of leucocytes : leucocytosis.
    (*c*) Leukaemias.
(3) Purpuric and haemorrhagic diseases.

ANAEMIA : GENERAL CONCEPTS

Anaemia, or deficiency of red cells or haemoglobin in the circulating blood, can result from many different causes.   Before discussing these, we must recall briefly the phenomena of normal erythropoiesis and the factors controlling it.

(1) Normal erythropoiesis

The maturation of the precursors of the red cells in bone-marrow is depicted in Plate VIII.   From primitive reticulo-endothelial cells which are believed to be the common stem-cells of all the haemopoietic cells of the marrow, and which are called *haemocytoblasts*, a series of nucleated *erythroblasts* is produced.   The earliest of these are large cells with large immature nuclei and basophilic cytoplasm containing little or no haemoglobin.   The later erythroblasts, produced by mitosis from the early ones, show a progressive diminution in size, their nuclei also become smaller and more condensed, and their cytoplasm becomes increasingly eosinophilic because of its acquisition of haemoglobin.   The most mature erythroblasts, equal in size and haemoglobin content to normal red corpuscles, are called *normoblasts*.   These extrude their nuclei and become non-nucleated *reticulocytes* with a characteristic supravitally stainable reticulum, which become fully mature *erythrocytes* and are discharged into the blood-stream.   The average period of survival of red cells in the circulation is probably about 120 days ; they then suffer fragmentation and phagocytosis by the reticulo-endothelial cells of the spleen, bone-marrow and liver.   The number of erythrocytes in the blood is maintained at a level of between $4\frac{1}{2}$ and 6 million per cubic millimetre, the production of new cells from the marrow keeping pace with the destruction of old ones.

(2) Erythropoietic hyperplasia

In all cases of anaemia, except those due to toxic destruction of the bone-marrow, the marrow shows compensatory hyperplasia of greater or less degree. In severe and sustained anaemia, e.g. pernicious anaemia or that due to repeated haemorrhages, the normal haemopoietic areas of marrow in the bones (namely in the vertebrae, ribs, sternum, diploe, pelvis, and upper ends of femur and humerus) become dark red, and this dark red marrow extends also into bones where normally only yellow marrow is present (Plate VII).

Microscopically, the hyperplastic marrow is found to contain a great excess of erythroblastic cells. Two distinct kinds of erythroblastic hyperplasia occur :

(i) *Normoblastic hyperplasia*, evoked by anaemias due to haemorrhage, haemolysis, or iron deficiency, and characterized by a speeding-up of the production and maturation of erythroblasts and erythrocytes of about normal sizes.

(ii) *Megaloblastic hyperplasia*, evoked by pernicious anaemia and other anaemias due to lack of the haemopoietic principle, and characterized by the production of abnormally large, haemoglobinized erythroblasts, called *megaloblasts*, which mature into abnormally large erythrocytes, called *macrocytes* or *megalocytes*.

In both normoblastic and megaloblastic hyperplasia, haemoglobinization of the immature cells is hastened, so that early erythroblasts with still immature nuclei already contain abundant haemoglobin and have eosinophilic cytoplasm.

(3) Abnormal red cells in the blood-stream

In the blood of the new-born infant a few nucleated normoblasts are normally present ; but in healthy adults the youngest red cells in the blood are reticulocytes (less than 2 per cent), and nucleated cells are never present. In anaemias, however, many kinds of abnormal red cells, including nucleated forms, are discharged into the circulation. These abnormal cells, depicted in Plate VIII, show :

(a) Nucleated forms

Any of the nucleated red cells found in the bone-marrow may appear in the blood. In post-haemorrhagic, haemolytic and iron-deficiency anaemias, *normoblasts* in various stages of maturation are often plentiful. In pernicious and kindred anaemias, *megaloblasts* like those of the hyperplastic marrow, may appear also in the blood.

(b) Abnormalities of size and shape

In anaemia the red cells often show great variation in size, *anisocytosis*. Abnormally large cells are *macrocytes*, and abnormally small ones *microcytes*. Anaemias in which the average diameter of the red cells is greater than normal, e.g. pernicious anaemia, are *macrocytic anaemias* ; those in which the average diameter is less than normal, e.g. most iron-deficiency anaemias, are *microcytic* ; while those in which the red cells are of about normal average size, e.g. many post-haemorrhagic anaemias, are *normocytic*. In haemolytic jaundice the red cells lose their normal flat biconcave shape and become more globular, a condition called *spherocytosis*. In all severe anaemias, some of the red cells are of pyriform or other irregular shapes, *poikilocytosis*.

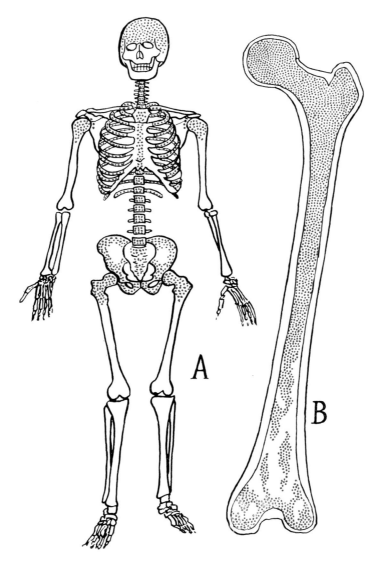

PLATE VII.—Distribution of red bone-marrow : A, in a healthy adult ; B, in femur in a case of moderately severe anaemia.

(To face page 394)

NORMAL
ERYTHROPOIESIS

**MICROCYTIC
ERYTHROPOIESIS**

FROM IRON
DEFICIENCY
etc.

**MACROCYTIC
ERYTHROPOIESIS**

FROM DEFICIENCY
OF HAEMOPOIETIC
PRINCIPLE

EARLY ERYTHROBLAST

LATER ERYTHROBLAST

SMALL
HYPOCHROMIC
NORMOBLAST

NORMOBLAST

MEGALOBLAST —
LARGE HAEMOGLOBIN-
IZED ERYTHROBLAST

RETICULOCYTE

MICROCYTE

NORMOCYTE

MACROCYTE

OTHER ABNORMAL RED CELLS IN ANAEMIAS

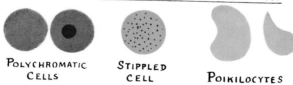

POLYCHROMATIC
CELLS

STIPPLED
CELL

POIKILOCYTES

PLATE VIII.—Normal and abnormal red cells and their development.

(c) Abnormalities of staining

Poorly haemoglobinized but mature cells in anaemic blood films stain faintly, i.e. are *hypochromic*. Immature erythrocytes which are deficient in haemoglobin often take up as much basic stain as eosin and so appear purple or violet, *polychromasia* or *polychromatophilia*. Other immature cells, though eosinophilic, are peppered with basophilic granules, *stippling* or *punctate basophilia*. Supravital staining shows that both polychromatic and stippled cells are really *reticulocytes*, and that in anaemias accompanied by much regeneration of red cells the blood shows pronounced reticulocytosis, even 50 per cent or more of the circulating red corpuscles being reticulocytes.

(4) Substances essential for normal erythropoiesis

The chief substances essential for the proper production of red corpuscles are iron, copper, haemopoietic principle, vitamin C, thyroxine, and adequate proteins.

(a) Iron

Iron in adequate amounts—about 15 milligrams daily for women and 5 to 10 milligrams for men—is essential for the proper haemoglobinization of the red cells in their later stages of maturation, i.e. during the conversion of normoblasts into reticulocytes and red corpuscles. A deficiency of iron causes an anaemia in which the red cells are only slightly reduced in number, but are deficient in haemoglobin and reduced in average size, i.e. a *hypochromic microcytic* anaemia.

(b) Copper

Traces of copper are essential for the utilization of iron in the synthesis of haemoglobin.

(c) Haemopoietic principle (anti-pernicious-anaemia factor)

The discovery of Minot and Murphy in 1926 that pernicious anaemia could be cured by the administration of raw liver showed that liver contained a blood-building substance which was lacking in pernicious anaemia. The subsequent researches of Castle and his co-workers showed that this *haemopoietic principle* (or anti-pernicious-anaemia factor) was produced by the interaction of an extrinsic factor present in the diet, now identified as Vitamin B_{12} or *cyanocobalamin*, and an intrinsic factor (haemopoietin) secreted by the stomach and present in gastric juice. Intrinsic factor is essential for the absorption of Vitamin B_{12} from the intestine and its storage in the liver. While there is evidence against the view that haemopoietic principle *is* simply stored Vitamin B_{12}, there is no doubt that liver extracts owe their anti-pernicious-anaemia activity largely to their Vitamin B_{12} content. Folic acid and other substances are also concerned, along with Vitamin B_{12}, in haemopoiesis. These substances are essential for the proper maturation of early erythroblasts in the bone-marrow. In their absence many large erythroblasts—megaloblasts—are produced in the marrow, and these become abnormally large red corpuscles or macrocytes. The anaemia produced by deficiency of haemopoietic principle (or of Vitamin B_{12} or intrinsic factor) is thus *macrocytic*; and because the big cells each contain more haemoglobin than normal, *hyperchromic*, i.e. the colour index is high.

(*d*) *Vitamin C*

This vitamin has a specific effect on the anaemia of scurvy, causing a prompt reticulocyte response followed by improvement of the anaemia. This anaemia is usually hypochromic and normocytic, but sometimes macrocytic ; and it is believed that vitamin C is a general erythropoietic stimulant, not acting at any particular stage of maturation.

(*e*) *Thyroxine*

Some patients with myxoedema are anaemic and the anaemia responds to treatment with thyroxine. This probably acts, not on any particular stage of maturation, but as a general tissue stimulant.

(*f*) *Proteins*

Clearly, haemoglobin synthesis must depend on an adequate supply of the proper amino-acids ; and there is good evidence, both clinical and experimental, that anaemia can be caused or aggravated by protein deficiency.

(5) Classification of anaemias

According to their main causes anaemias can be classified thus :

(*a*) *Dyshaemopoietic anaemias*, due to impaired production of red corpuscles in the bone-marrow.

    (1) *Due to deficiency of iron :* various hypochromic anaemias.

    (2) *Due to deficiency of haemopoietic principle.*

        (*a*) Pernicious anaemia.

        (*b*) Other macrocytic hyperchromic anaemias.

    (3) *Due to other deficiencies :* of vitamin C, thyroxine, or proteins (referred to above).

    (4) *Due to toxic injury or other disease of bone-marrow :* infections, chemical poisons, radiations, leukaemia, tumours, osteosclerosis, etc.

(*b*) *Post-haemorrhagic anaemia*, due to acute or chronic extravascular loss of blood.

(*c*) *Haemolytic anaemias*, due to excessive intravascular destruction of red cells.

    (1) *Due to haemolytic microbic parasites* e.g. *Cl.Welchii*, malaria.

    (2) *Due to haemolytic poisons :* lead, etc.

    (3) *Due to intrinsic abnormalities of the blood :*

        (*a*) Erythroblastic anaemias of the new-born.

        (*b*) Familial acholuric jaundice.

        (*c*) Other familial haemolytic anaemias.

    (4) *Acquired idiopathic haemolytic anaemias.*

(*d*) *Unclassified anaemias*, e.g. " splenic anaemia ".

While most cases of anaemia can be readily placed in the foregoing scheme according to their *main* cause, it is important to recognize that multiple causes are often operative. Thus gastric carcinoma or other gastro-intestinal diseases may bring about anaemia by causing a deficiency of haemopoietic principle, by interfering with the absorption of iron, by impairing protein digestion and absorption, or by producing metastatic growths in bone-marrow ; and several or all of these factors may operate in a single case. Again, while lead and other chemical poisons cause anaemia chiefly by a direct effect on the red cells, increasing their rate of destruction, these substances probably also impair the production of red cells in the bone-marrow, so that the anaemia is partly dyshaemopoietic.

DYSHAEMOPOIETIC ANAEMIAS DUE MAINLY TO DEFICIENCY OF IRON

Iron deficiency is the main cause of *hypochromic anaemias* in the following conditions :

(1) *Nutritional anaemias of infants*, due to prolonged breast-feeding with poor iron content in the milk, or to improper artificial feeding ;

(2) *Pregnancy anaemia* caused by the foetal consumption of iron when the mother is receiving a barely sufficient amount of iron in her usual diet ;

(3) *Chronic gastro-intestinal diseases*, which interfere with iron absorption, e.g. alimentary carcinoma, obstruction, fistulae, sprue (a tropical diarrhoeal disease of uncertain cause), " coeliac " or Gee's disease in children and idiopathic steatorrhoea in adults (digestive disorders related to each other and to sprue, and characterized by the passage of fatty stools and the development of anaemia), ulcerative colitis, and bacillary or amoebic dysentery ;

(4) *Chlorosis* or " green sickness ", a now rare anaemia which was formerly common among young working women in unhygienic surroundings ;

(5) *Nutritional iron-deficiency anaemia of middle-aged women*, accompanied by achlorhydria or hypochlorhydria, and often by glossitis and dysphagia, the complete syndrome being called the Plummer-Vinson syndrome. Severe cases show also koilonychia, i.e. fragility and spoon-shaped depression of the finger-nails.

Nothing more will be said of these anaemias here. The blood picture in all of them is very similar ; they are all microcytic and hypochromic, with colour index under 1, and any nucleated red cells which appear in the blood are normo-blasts. Administration of iron produces reticulocytosis and improvement of the anaemia.

DYSHAEMOPOIETIC ANAEMIAS DUE TO DEFICIENCY OF HAEMOPOIETIC PRINCIPLE

Haemopoietic principle (anti-pernicious-anaemia factor) requires for its production both dietary extrinsic factor (Vitamin B_{12}) and gastric intrinsic factor (haemo-poietin). Deficiency of either of these factors will cause anaemia of the macrocytic hyperchromic type. Anaemia due to deficiency of extrinsic factor (or of folic acid) is rare, and is practically restricted to Asiatic communities on deficient diets. In Europeans nearly all macrocytic hyperchromic anaemias are due to deficiency of intrinsic factor, and the commonest of such anaemias is classical pernicious anaemia.

(1) Pernicious anaemia (Addison's anaemia ; Biermer's anaemia)

(a) Causation

Pernicious anaemia is a severe macrocytic hyperchromic anaemia which occurs in people of either sex, chiefly between the ages of 40 and 70, and is due to a permanent lack of gastric intrinsic factor. The cause of the loss of intrinsic factor is unknown ; but there is certainly an inherited predisposition to the disease, since it not infrequently shows a familial incidence. We may suppose that predisposed persons inherit a weak gastric mucosa, the secretory activity of which fails at some age over 40. This failure concerns not only the intrinsic factor but also the hydrochloric acid and pepsin ; with rare exceptions, persons with pernicious anaemia show also achlorhydria. The gastric mucosa shows gastritis and atrophy; and there is some evidence that this may be an auto-immune lesion, many patients' sera containing antibodies against gastric parietal cells and intrinsic factor.

(b) Blood picture

A moderately severe, untreated case may show a red-cell count of 2 million per cubic millimetre and a haemoglobin percentage of 50 ; but in more advanced cases the figures may be much lower than this. The colour index is raised, usually to 1·2 or 1·4. The average diameter of the red cells is raised, e.g. to 8 or even 9μ ; but the blood contains many microcytes as well as macrocytes. Besides this anisocytosis, the film shows also poikilocytes, polychromatic cells, stippled cells, normoblasts and megaloblasts. Reticulocytes vary in number with the stage of the disease ; in advanced cases very few are present ; treatment with potent liver or stomach extracts produces a striking reticulocytosis (40 or 50 per cent) as the first stage in the recovery of the blood picture. Leucopenia is usual, the leucocyte count often lying between 2,000 and 4,000 per cubic millimetre; and the fall is due chiefly to a reduction of the granulocytes, so that there is a relative (seldom an absolute) lymphocytosis.

(c) Changes in the tissues

In untreated cases, which are rare nowadays, the following changes are found :

(i) *Results of the severe anaemia.*—All organs except spleen and bone-marrow are pale. The parenchyma cells of liver, kidneys and heart show pronounced fatty infiltration, the heart muscle often exhibiting the thrush-breast appearance (p. 345).

(ii) *Evidence of abnormal erythropoiesis.*—The red marrow is dark and extends into all the bones of the limbs. Microscopically it shows megaloblastic hyperplasia.

(iii) *Evidence of excessive blood destruction.*—The abnormal red cells are taken up and destroyed in great numbers by the reticulo-endothelial phagocytes of the spleen, bone-marrow and liver. The spleen is moderately enlarged. The spleen, liver, kidneys and other organs contain large amounts of haemosiderin, and give a strongly positive Prussian-blue test. The serum bilirubin is increased, and the patient is often slightly jaundiced.

(iv) *Changes in the alimentary tract.*—These comprise atrophic glossitis, with a red smooth tongue ; atrophic gastritis, already mentioned ; and sometimes atrophic enteritis.

(v) *Changes in the central nervous system.*—In a small proportion of cases the spinal cord shows degenerative changes in the tracts of the posterior columns, with or without degeneration of the lateral columns, a condition called *subacute combined degeneration.* During life, this manifests itself by impaired sensation in the limbs, with or without signs of slight spastic ataxic paraplegia and altered reflexes.

(d) Prognosis

Before the discovery of specific liver therapy, the prognosis was very serious, most patients dying within 2 or 3 years. With adequate liver or Vitamin B_{12} treatment, however, most patients can be kept in good health and with a normal blood picture. But their primary deficiency is not cured; they still lack intrinsic factor of their own, and the anaemia recurs if treatment is omitted or is insufficient. They must therefore be kept permanently under observation with blood examination at regular intervals.

(2) Other macrocytic hyperchromic anaemias

(a) Anaemias due to lack of intrinsic factor

Anaemias due to lack of intrinsic factor, and with blood pictures similar to that of pernicious anaemia, sometimes occur in the following conditions :—

(i) *Pregnancy*, in which the deficiency is usually only temporary ;

(ii) *Gastric carcinoma*, with failure of gastric secretion ;

(iii) *Following gastrectomy* ;

(iv) *Diarrhoeal diseases* ; sprue, steatorrhoea, enteric fistula ;

(v) *Diphyllobothrium latus* infestation, which probably causes macrocytic anaemia only in predisposed subjects, since most of the patients show achlorhydria.

(b) Anaemia in liver disease

Liver disease, especially cirrhosis, is occasionally accompanied by a pernicious-like anaemia, which is attributed to failure of the liver to conserve haemopoietic principle, even though this is properly formed.

(c) Achrestic anaemia

This is the name applied to rare cases of pernicious-like anaemia which fail to respond to liver treatment and which are therefore supposed to be due to failure of the bone-marrow to utilize available haemopoietic principle.

(d) Tropical macrocytic anaemia

This is the pernicious-like anaemia of malnourished Asiatics, which is cured by Marmite or other yeast preparations as well as by liver extracts, and which is believed to be largely due to dietetic deficiency of Vitamin B_{12} or folic acid or both.

DYSHAEMOPOIETIC ANAEMIAS DUE TO DISEASES OF THE BONE-MARROW

Anaemia, usually hypochromic and microcytic or normocytic, may result from direct injury of the bone-marrow by the following agents, many of which have been mentioned in earlier chapters :

(1) Bacterial infections

Syphilis, tuberculosis, chronic suppuration, subacute bacterial endocarditis, typhoid fever, etc., probably cause anaemia mainly by toxaemic injury of the erythropoietic tissue ; but nutritional and other factors also contribute.

(2) Chemical poisons

The anaemia resulting from injury to the bone-marrow by chemical agents is dealt with on p. 335.

(3) X-rays and radium

One of the main effects of over-exposure to these agents is anaemia, which in severe cases is aplastic (i.e. non-regenerative) and irrecoverable. All radium and X-ray workers should have regular blood examinations, in order to detect the first signs of any deleterious effects.

(4) Tumours

Any kind of tumour which involves bone-marrow extensively may thereby induce anaemia. Such tumours include leukaemias (pp. 555, 562), myelomatosis (p. 559), and metastatic carcinomas (p. 451). How these growths depress erythropoietic activity is not quite clear; probably it is partly by direct replacement, and partly by nutritional competition. The anaemia evoked by extensive metastatic carcinoma often shows many nucleated red cells and is usually accompanied by pronounced leucocytosis (and hence is called "leuco-erythroblastic"); it is sometimes megalocytic and hyperchromic.

(5) Osteosclerosis

There are two rare diseases, both of unknown cause, in which the marrow cavities of the bones are encroached on and obliterated by overgrowth of the bony tissue, and progressive leuco-erythroblastic anaemia results. One of these occurs in children and is called *osteopetrosis* ("marble-bones" or Albers-Schönberg disease); the other occurs in adults and is called *myelosclerosis* (or myelofibrosis).

It remains to add here that the name *aplastic anaemia* is applied to any dyshaemopoietic anaemia in which the bone-marrow shows little or no regenerative hyperplasia, and few or no new red cells appear in the blood. This arrest of erythropoiesis is usually accompanied by arrest also of leucopoiesis and thrombocytopoiesis, so that there is leucopenia and thrombocytopenia. The condition is usually rapidly progressive and fatal. In some cases the cause of the disease is known, e.g. benzol poisoning, radioactive substances or X-rays, or it occurs as a terminal marrow-exhaustion phase of pernicious anaemia; but more often it is "idiopathic", i.e. its cause is not known.

POST-HAEMORRHAGIC ANAEMIA

This was considered in Chapter 2 (p. 30). It remains only to add here that repeated small haemorrhages may cause a severe degree of anaemia, which is microcytic and hypochromic, and that sometimes the source of the bleeding may not be obvious. Chronic gastric or duodenal ulcer, bleeding haemorrhoids, or ankylostomiasis may produce serious anaemia without the patient being aware of the loss of blood. Other more obvious causes of chronic haemorrhage include menorrhagia (excessive menstrual bleeding), dysentery, haemophilia and the other haemorrhagic diseases described below (p. 405).

HAEMOLYTIC ANAEMIAS

Elsewhere we have dealt with microbic parasites which cause haemolysis, namely *Cl. Welchii* (p. 160), the bacillus of oroya fever (p. 189), and the plasmodia of malaria (p. 224) ; and in Chapter 24 (p. 335) we discussed lead and other haemolytic poisons. There remain for consideration certain haemolytic anaemias which are due to intrinsic abnormalities of the blood.

(1) Haemolytic erythroblastic anaemias of the new-born

The occasional occurrence of severe, often fatal, haemolytic jaundice with anaemia in new-born infants has long been recognized ; but its pathogenesis has been elucidated only recently. The disease has been variously named haemolytic disease of the new-born, *icterus gravis neonatorum*, and *erythroblastosis foetalis*, the last referring to the great number of erythroblasts present in the anaemic blood of the foetus or infant.

(a) Causation

The haemolysis of the foetal blood is brought about by an antibody evoked in the mother by a corpuscular antigen from the foetal red cells. The factors concerned are usually, but not always, the hereditary Rhesus (Rh) factors (so called because they are similar to factors present in the red cells of rhesus monkeys). The usual sequence of events is as follows : a woman who is Rh-negative (i.e. whose red cells are devoid of Rh factor) is pregnant with a foetus which has inherited Rh-positivity from the Rh-positive father ; Rh factor diffuses from the foetal to the maternal blood via the placenta, and acts as an antigen, stimulating the mother's tissues to produce anti-Rh agglutinins ; these diffuse back through the placenta into the foetal circulation and destroy the foetal red cells. From this mode of production, it can be readily understood why the disease is often familial, successive infants of the same mother and father being affected, why the risk to the foetus increases with successive pregnancies, and why transfusion of the mother at any time with Rh-positive blood also increases this risk.

(b) Pathology

The foetus may die *in utero* ; or a full-term infant, in an anaemic oedematous state (hydrops foetalis) may be still-born or may survive only a few hours. The term *icterus gravis* is applied to those cases in which the infant is born alive and soon becomes deeply jaundiced. Some cases recover spontaneously, but a large proportion of untreated cases prove fatal. The pathological changes comprise severe anaemia with many nucleated red cells, jaundice, haemosiderosis, enlargement of the spleen and liver, and foci of extramedullary haemopoiesis in these and other organs. In cases which recover some fibrosis of the liver and spleen may ensue ; and it has been suggested that juvenile cirrhosis may in some cases be a late sequel of congenital icterus. Another complication is bile-staining and resulting degeneration of the basal ganglia and other parts of the brain, a condition called *kernicterus*, which may result in spasticity, athetosis (slow involuntary movements) or mental deficiency.

Anaemia neonatorum (congenital anaemia), appearing soon after birth without jaundice, is in most cases a haemolytic disease essentially similar to icterus gravis but less severe.

(2) Familial acholuric jaundice

This is a rare chronic familial disease characterized by spherocytosis and abnormal fragility of the red cells, excessive blood destruction, splenomegaly, anaemia, and jaundice without bilirubinuria.

(a) Familial incidence and nature of the disease

The disease often affects several members of a family, and may be transmitted through several generations. It behaves as a Mendelian dominant character ; and it affects, and is transmitted by, both males and females. The essential change is the increased fragility and spherocytosis of the red cells, which is an inborn and permanent anomaly—a cellular malformation of genetic origin. The symptoms of the disease—anaemia, jaundice and splenomegaly—are secondary results of excessive destruction of the abnormal cells.

(b) Clinical course

The degree of fragility of the red cells, and therefore of the symptoms, varies. In some cases symptoms appear soon after birth, in others not until childhood or adult life. They may be severe and may progress in a series of " crises " of excessive blood destruction, during which the patient is severely jaundiced and anaemic, febrile, and has a large tender spleen. Or the symptoms may be continuous and mild, consisting of only slight jaundice, anaemia and splenic enlargement. With proper treatment, prognosis is good. Splenectomy is symptomatically curative ; this is because it minimizes destruction of the red cells which, however, still remain abnormally fragile.

(c) Blood changes

The red cells are microcytic and spherocytic ; and their abnormal fragility is demonstrated by their range of haemolysis in a series of salt solutions of different concentrations. The degree of anaemia varies. In mild cases, it is slight, with perhaps $3\frac{1}{2}$ or 4 million red cells per cubic millimetre, Hb. 75 per cent, colour index 0·8 or 1, and film showing only slight anisocytosis along with the characteristic spherocytosis. Reticulocytosis of 10 or 20 per cent or more is almost constantly present. In haemolytic crises, Hb. may drop to 50 per cent or less, many poikilocytes, anisocytes and normoblasts appear in the blood, and reticulocytes may rise to 50 per cent or more.

(d) Pathology

The spleen is enlarged, often markedly so ; microscopically it is engorged with red cells in all stages of destruction. There is haemosiderosis of liver, spleen, kidneys, and other tissues, which give strong Prussian-blue tests. Pigment gall-stones, precipitated from the excess of bilirubin in the bile, are frequently present. There is a high renal threshold for bilirubin ; so that, in spite of the excess of this substance in the blood serum, none appears in the urine. The red bone-marrow is excessive, occupying bones where only yellow marrow is normally present, and microscopically showing great hyperplasia of normoblastic type.

(3) Other familial haemolytic anaemias

Several other forms of anaemia, due to abnormal haemoglobins, include:

(i) *Sickle-cell anaemia*, a severe familial anaemia in negroes, characterized by an abnormal haemoglobin-S and by sickle-shaped red cells.

(ii) *Target-cell anaemia* (Cooley's anaemia or thalassaemia), a severe familial anaemia of children, chiefly of the Mediterranean peoples, characterized by an inherited inability, partial or complete, to form adult haemoglobin-A, and hence by retention of foetal haemoglobin-F.

(4) Acquired idiopathic haemolytic anaemias

There occur rare cases of non-familial acquired haemolytic anaemia of unknown cause. The acute form is sometimes called *Lederer's anaemia*, and the chronic form *acquired haemolytic icterus*. Many of these cases have auto-haemolysins in their blood, and these probably play a major part in producing or maintaining the disease. A similar mechanism underlies *paroxysmal haemoglobinuria* due to exposure to cold, most of the subjects of which have been syphilitic; and the same may apply to *blackwater fever* in malarial cases (see p. 224).

SPLENIC ANAEMIA

Splenic anaemia, Banti's disease and *hepato-lienal fibrosis* are the names applied to a chronic disease characterized by great and progressive enlargement of the spleen, progressive hypochromic anaemia with leucopenia, and cirrhosis of the liver often with haematemesis from oesophageal varices. There has been much dispute as to whether the disease is a primary splenic disorder with supervening hepatic cirrhosis, or only a variety of cirrhosis of the liver with congestive spleno-megaly. Against the second view are the difficulty of explaining a progressive anaemia as the result of cirrhosis, the fact that in some cases great splenomegaly has preceded other symptoms for years, and that in some cases there has been splenomegaly and anaemia of long duration but with little or no cirrhosis. The spleen shows nothing very characteristic—only fibrosis, congestion, and various degenerative changes in the main vessels. Until more is known of the cause of the disease, it will be as well to regard it merely as a syndrome and to leave the anaemia unclassified.

EXCESS OF RED CORPUSCLES AND HAEMOGLOBIN : POLYCYTHAEMIA

Polycythaemia is a general term applied to any condition in which the red-cell count is abnormally high. It is of three distinct kinds :

(i) *Anhydraemic polycythaemia*, a temporary concentration of the corpuscles due to loss of fluid (see p. 389) ;

(ii) *Polycythaemia vera* (primary or splenomegalic polycythaemia, erythraemia, or Vaquez-Osler disease), due to permanent unexplained overgrowth of the erythropoietic tissues, probably neoplastic (see p. 564) ;

(iii) *Secondary polycythaemia* or *erythrocytosis*, due to temporary compensatory hyperplasia of the erythropoietic tissues, consequent on deficient oxygenation of the blood and tissues. This occurs in people living at high altitudes ; and in patients with anoxia from chronic cardiac or pulmonary disease, e.g. congenital abnormalities of the heart, chronic valvular disease, pulmonary fibrosis, stenosis of the pulmonary arteries (Ayerza's disease), etc.

ABNORMALITIES OF LEUCOPOIESIS

(1) Deficiency of leucocytes : leucopenia

The normal range of the leucocyte count is from about 4,000 to 10,000 (possibly 12,000) per cubic millimetre. Counts below 4,000 constitute *leucopenia*. This is almost always due to a fall in the number of granulocytes, especially neutrophil granulocytes, i.e. it is a neutropenia. It occurs from the following causes :—

(i) *Starvation or malnutrition* from any cause (p. 341).

(ii) *Certain infections*, especially typhoid and paratyphoid fevers, undulant fever, acute tuberculosis and malaria. In many other infections, e.g. influenza, mumps, measles, smallpox, leucopenia may occur, but is inconstant and of little diagnostic importance. Overwhelming infections of kinds which usually cause leucocytosis (e.g. pneumonia, meningitis or streptococcal septicaemia) may cause leucopenia.

(iii) *Haemopoietic poisons* (see p. 335).

(iv) *Certain blood disorders* already considered, notably pernicious anaemia and aplastic anaemia.

(2) Temporary excess of leucocytes : leucocytosis

Leucocyte counts much above 10,000 per cubic millimetre are usually abnormal. If temporary and reversible, such excessive counts constitute *leucocytosis*. This is usually neutrophil granulocytosis ; less frequently, it is eosinophil granulocytosis (eosinophilia), lymphocytosis or monocytosis.

(a) Neutrophil granulocytosis

This occurs in many acute infections, especially those due to cocci, e.g. pneumonia, osteomyelitis, furunculosis, appendical and other abscesses, scarlet fever, tonsillitis. These usually produce neutrophil leucocytosis of 20,000 to 40,000 or more per cubic millimetre. A sustained high leucocytosis is of much diagnostic value in cases of suspected deep-seated suppuration, e.g. empyema or perinephric abscess, or in distinguishing inflammatory from other swellings, e.g. subacute osteomyelitis from a tumour of bone (see also p. 73).

Other less important causes of neutrophil leucocytosis, usually of only slight or moderate degree, are various infections other than coccal ; post-haemorrhagic regeneration of blood ; tissue injury, e.g. in fractures, extensive wounds or surgical operations ; malignant disease and other cachectic states ; anaesthetics, and various drugs and chemical poisons.

(b) Eosinophilia

The normal eosinophil count of 150 to 400 per cubic millimetre (1 to 4 per cent of the total count) is exceeded in a variety of conditions, which can be grouped conveniently thus :

(i) *Allergic diseases*, e.g. asthma, hay fever, urticaria, and food sensitivity (see p. 287).

(ii) *Infestations by metazoal parasites* (see Chapter 19). The eosinophil reactions excited by these are probably allergic. *Tropical eosinophilia*, occurring in India and Egypt, is probably due to filarial infestation.

(iii) *Certain skin diseases*, e.g. pemphigus, and various cases of dermatitis and " eczema ", in which allergic sensitization to an infection or foreign protein is probably concerned.

(iv) *Sundry diseases*. Eosinophilia occurs occasionally in tuberculosis, rheumatic fever, convalescence from scarlet fever and other acute fevers, polyarteritis nodosa, in some cases of malignant disease (especially those with metastases), and in Hodgkin's disease, in occasional cases of which the eosinophil count may reach 50 per cent or more of the leucocyte count. *Familial eosinophilia* in otherwise normal people has been recorded.

(c) *Lymphocytosis*

An absolute lymphocytosis is rare. It sometimes occurs in infants in response to infections which in adults cause neutrophil granulocytosis ; and it also occurs in whooping cough, glandular fever, and occasionally in tuberculosis, syphilis, influenza, measles and other infections. In diseases with neutrophil leucopenia, the proportion of lymphocytes (but not their absolute number) is increased ; this is spoken of as a " relative lymphocytosis ".

(d) *Monocytosis*

Monocytes, which normally constitute less than 10 per cent of the leucocytes, are increased in glandular fever, malaria, tuberculosis, typhoid fever, etc. Whether or not these cells are distinct from lymphocytes was referred to on p. 53. If we adopt the view that lymphocytes and monocytes are but variants of cells of the same kind, then monocytosis is only a variety of lymphocytosis ; and it is certainly true that these two types of cells behave similarly in many infections. The leucocytosis in glandular fever is regarded by some haematologists as a monocytosis and by others as a lymphocytosis.

(3) Leukaemias

Leukaemias are considered on pp. 553 and 562. They are the manifestations in the circulating blood of malignant neoplasms of the leucopoietic tissues. Leukaemias are *not* merely extravagant leucocytoses ; they differ intrinsically from leucocytoses in being progressive, irreversible and invariably fatal, in showing many abnormal cells, and in forming metastatic invasive colonies of these cells in various tissues.

PURPURIC AND HAEMORRHAGIC DISEASES

Purpura means the spontaneous appearance of haemorrhages in the skin. It is not a disease, but only a prominent symptom of a great many different diseases which have in common an abnormal tendency to haemorrhage. Such diseases

CHAPTER 28

TUMOURS : DEFINITION AND CLASSIFICATION

DEFINITION

PATHOLOGISTS use the word *tumour* in its restricted sense of a true *neoplasm* and not in its ancient and general sense of any localized swelling, which included also inflammatory and reparative masses, hyperplasias, simple cysts and malformations.

A true tumour or neoplasm is an abnormal mass of tissue, the growth of which exceeds and is uncoordinated with that of the normal tissues, and persists in the same excessive manner after cessation of the stimuli which evoked the change.

(a) " *Excessive uncoordinated growth* "

This is an essential characteristic of neoplasia, distinguishing it from all other kinds of pathologic overgrowth—reparative proliferations, hyperplasias or malformations. In repair and hyperplasia, cellular multiplication is limited in amount and duration in accordance with certain specifiable structural or functional conditions—a breach in the tissues to be filled or a functional deficiency to be made good. A malformation sometimes shows excess of a particular tissue, but this tissue has no power of disproportionate excessive growth ; it grows only with the growth of the part of which it is a blemish. The cells of a neoplasm, however, continue to multiply indefinitely irrespective of any structural or functional requirements, forming an ever increasing superfluous mass of tissue of a particular kind.

(b) " *Growth persisting after cessation of the stimuli which evoked it* "

This striking characteristic of a neoplasm is linked with the previous one. Whereas cellular proliferation in repair and hyperplasia continues only as long as the evoking stimulus persists, a tumour retains its habit of excessive growth independently of its exciting causes. There is ample evidence of this in both human and experimental pathology : e.g. X-ray or tar cancers of the skin may appear many years after their victims have retired from the occupations which were to blame, and experimentally produced tumours may develop long after applications of the chemical or physical carcinogens which caused them have been discontinued. This irreversible neoplastic habit of growth, not dependent on the continued action of the evoking agent, is most strikingly shown by transplantable tumours, which retain their neoplastic properties unimpaired during successive transfers to many other animals. It is clear then that carcinogenic agents act by initially inducing in the cells to which they are applied some sort of irreversible changes of metabolism and proliferation, which are then transmitted indefinitely to the descendants of the initially changed cells—in the primary tumour, in its metastases, in transplants, and also in cultures *in vitro*. The nature of the neoplastic change, the very essence of neoplasia, is still unknown ; we will discuss it further in the next chapter.

THE BASES OF CLASSIFICATION AND NOMENCLATURE

Tumour classification and nomenclature rest on two bases, (1) the *histogenesis*, and (2) the *behaviour* of tumours. The former is the primary and fundamental basis, the latter secondary and arbitrary.

(1) Histogenetic classification

The fundamental basis of the classification of tumours is histogenetic ; i.e. we classify and name them according to the tissues from which they arise and of which they consist. Tumours may arise from any of the tissues of the body, and in most cases the essential parenchyma of a tumour consists of cells of a single type derived from only one tissue. Clearly, each tumour should be named and classified according to the nature of this tissue. The classification of tumours therefore is essentially similar to the histologist's classification of normal tissues. Just as the histologist conveniently groups these in several main classes—epithelia of various types, non-epithelial tissues (connective, vascular, skeletal, muscular and haemopoietic), nervous tissues, etc.—so the pathologist groups tumours into corresponding main classes.

The histogenetic classification of tumours encounters difficulties of two kinds, (*a*) difficulties in the classification and naming of normal tissues, and (*b*) those due to uncertainty regarding the tissues of origin of certain tumours.

(*a*) Difficulties in the naming of normal tissues are exemplified in the use of the familiar name " epithelium ". Everyone agrees that this name is applicable to the epidermis, the linings of visceral mucous membranes and glandular derivatives of these. The lining cells of vascular and coelomic cavities, however, though formerly called " epithelium ", are now usually distinguished as " endothelium ", and the latter sometimes as " mesothelium ". What, then, of the compact layer of coelomic cells on the surface of the ovary ? Is this an " epithelium " or " mesothelium " ? Again, we all agree that the endocrine cells of the thyroid, parathyroid and pituitary are " epithelial " ; but most histologists do not apply this name to the cells of the adrenal cortex or the interstitial cells of the testis. Is ependyma epithelial ; and, if not, are the cells of the choroid plexus epithelial ? Differences of opinion on these points reflect themselves in differences in pathological nomenclature.

(*b*) The second difficulty in histogenetic classification is the purely practical one of not being able to recognize microscopically the nature of some poorly differentiated or anaplastic growths. An example will clarify this point. A malignant tumour springing from thyroid epithelium may show almost perfect thyroid-like structure in parts, and therefore be identifiable microscopically as " adenocarcinoma of the thyroid ". A second thyroid tumour may show glandular architecture, but without colloid formation or any other feature distinctive of thyroid tissue ; it is microscopically identifiable only as an " adenocarcinoma ". A third thyroid tumour may be recognizably epithelial in character, but without any glandular orientation of its cells ; it can be diagnosed no more specifically than " carcinoma ". A fourth tumour may be so diffusely undifferentiated in structure, that the histologist cannot even be certain that it is epithelial in nature and can identify it only as " a cellular anaplastic tumour of undetermined origin ". Thus, while all four tumours are really of the same species, namely

" carcinoma of the thyroid ", the precision of histological diagnosis and naming varies from tumour to tumour.

(2) Behaviouristic classification : innocence and malignancy

Besides the fundamental histogenetic classification, by which tumours are grouped solely according to the kinds of tissues composing them, we also make an additional grouping, less fundamental and more arbitrary but of great practical value, according to their behaviour and clinical results.

The tumours of any given histological group show a wide range of behaviour in their rates of growth, powers and modes of spread, and amount of risk to their victims. Some grow slowly, remain local and circumscribed, do not invade neighbouring tissues, and cause no harm except by virtue of their situation or some accidental complication ; these are typically *benign* or *innocent tumours*. Other tumours of the same cell-type grow rapidly, invade neighbouring tissues, produce remote metastases, and quickly kill their bearers ; these are typically *malignant tumours*. But between these two extremes, many of the histogenetic groups of tumours contain members of intermediate or borderland behaviour, so that sharp separation into innocent and malignant species is not possible. " Innocent " and " malignant " are relative terms used to make a convenient separation of any particular histogenetic series into two sections according to those characters which are of greatest prognostic significance. Fortunately, for practical purposes, most tumours can be placed without much difficulty into either the " innocent " or " malignant " compartments of their histogenetic group ; and often some assessment of their probable degrees of malignancy can be made from their microscopical characters.

The main points of distinction between innocence and malignancy are shown in the following table :

	Innocent tumour	Malignant tumour
(1) Structure	Usually well differentiated and typical of tissue of origin	Often imperfectly differentiated and atypical
(2) Mode of growth	Usually purely expansive, and a capsule formed	Infiltrative as well as expansive, hence not strictly encapsulated
(3) Rate of growth	Usually slow ; and mitotic figures scanty	May be rapid, with many mitotic figures
(4) End of growth	May come to a standstill	Rarely ceases growing
(5) Metastasis	Absent	Frequently present
(6) Clinical results	Dangerous because of : (a) position, or (b) accidental complications, or (c) production of excess or hormone	Dangerous also because of progressive infiltrative growth and metastasis

Each of these points calls for some comment.

(a) Structure

Most innocent tumours attain a high degree of structural differentiation, so that their tissues closely resemble their normal counterparts. A papilloma of the skin consists of overgrown epidermis which is normal in appearance except for exaggerated papillation and cornification ; a papilloma of the bladder is clothed by an even layer of non-cornifying stratified urinary-tract epithelium ; a lipoma consists of structurally typical adipose tissue, a leiomyoma of smooth muscle fibres, and a chondroma of cartilage. Malignant tumours, on the other hand, frequently show *anaplasia*, i.e. imperfect differentiation, of varying degrees. An example of this was given above in the case of thyroid carcinomas ; and the same applies to malignant tumours of all classes. In general the degree of malignancy of tumours is roughly proportional to their degree of anaplasia ; and various attempts at precise " tumour-grading " are based on this principle. But it is important to recognize (*a*) that a single tumour often shows considerable diversity of structure, some parts being well differentiated while other parts are anaplastic, and (*b*) that some malignant tumours attain well-nigh perfect structural and functional differentiation, a subject referred to again later.

(b) Mode of growth

The enlargement of many innocent tumours is purely *expansive*, thrusting adjacent tissues aside and compressing them but not infiltrating them. In this way, adenomas, leiomyomas, lipomas and many other benign growths become sharply circumscribed and encapsulated. But this does not apply to all benign tumours ; some fibromas and meningiomas and most angiomas and osteoclastomas show marginal mingling with the surrounding tissues. The enlargement of malignant tumours, besides being expansive, is always to some degree *infiltrative*, i.e. the tumour cells penetrate into the tissue spaces, lymphatics and blood-vessels of neighbouring tissues. The extent of this infiltration varies greatly ; in some tumours it is restricted to the immediate neighbourhood of the main mass ; in others it is present, sometimes plainly visible to the naked eye and sometimes inconspicuous and detectable only microscopically, for great distances around the main mass. It is this marginal zone of infiltration which is responsible for the local recurrence of many malignant tumours following surgical excision.

(c) Rate of growth

Most innocent tumours grow only slowly and the tumour cells show few mitotic figures. Most anaplastic malignant tumours, on the other hand, grow rapidly and contain many cells in mitosis. However, not all malignant tumours grow rapidly ; many well differentiated adenocarcinomas, squamous-cell carcinomas and scirrhous (i.e. densely fibrotic) carcinomas, grow slowly and show few mitotic figures. Mitosis in tumours is described more fully on p. 436.

(d) End of growth

Most true tumours, both benign and malignant, grow continuously and indefinitely ; but spontaneous cessation of growth and retrogressive changes sometimes occur. This is much more frequent in benign than in malignant tumours ; e.g. adenomas of the thyroid, leiomyomas of the uterus, fibromas, meningiomas, and benign teratomas, not infrequently become quiescent, and undergo fibrosis, calcification or cystic degeneration. Malignant tumours very rarely cease growing spontaneously, but this has occasionally been recorded of papillary ovarian tumours, mammary carcinoma, malignant melanoma, and some others. It is very rare for a malignant tumour to undergo complete and permanent disappearance ; but examples of even temporary and partial retrogression are of great interest.   They remind us that a tumour is not, as we are apt to assume, something foreign to the body, possessing complete autonomy, but a part of the individual's own tissues, subject like them to the metabolic state of the rest of the body and to the largely unknown factors involved in senescence and death.

(e) Metastasis

The development of separate secondary growths in parts of the body more or less remote from the primary growth is always clear proof that a tumour is malignant, no matter how slow its proliferation or how perfect its histological differentiation.   Metastasis depends on the invasion of lymphatics, blood-vessels, or serous or other cavities, and the detachment and transfer of tumour particles. These phenomena will be described in Chapter 30.   Not all kinds of malignant tumours produce metastases : thus rodent carcinomas of the skin are invasive destructive growths of acknowledged malignancy, which however almost never metastasize ; and, while most gliomas of the brain are infiltrative and fatal tumours, and may disseminate in the cerebrospinal spaces, they very rarely produce secondary growths in other parts of the body.   Moreover, in those classes of tumours which frequently do metastasize, e.g., carcinomas of the lung, stomach or breast, there occur members which fail to do so, even after a long period of active growth and the attainment of a great size.

(f) Clinical results

Benign tumours may prove harmful in any of three ways :

(i) *Because of their situation* ; e.g. within the cranial cavity, mediastinum, or orbit ;

(ii) *Because of accidental complications* ; e.g. haemorrhage from, or ulceration and infection of, tumours of mucous membranes ; torsion and strangulation of ovarian or other bulky tumours with narrow pedicles ; or intestinal intussusception resulting from a pedunculated tumour ;

(iii) *Because of the effects of excessive hormone production* ; e.g. acromegaly or giantism resulting from a pituitary adenoma, generalized fibrocystic disease of bone from a parathyroid adenoma, or hypertension from an adrenaline-secreting tumour of the adrenal medulla.

Malignant tumours may produce illness in the same three ways, to which are added also their intrinsically dangerous properties of invasive destructive growth and metastasis.   A malignant tumour seldom grows for long in the body without reaching and seriously injuring some important structure or without suffering some serious complication.   A prevalent idea that malignant tumours bring about a specific " malignant cachexia " by the production of toxic metabolites or by depriving the body of particular nutritive substances, is erroneous. " Cachexia " is simply a state of debility, malnutrition and anaemia ; and malignant tumours often produce these ill effects by causing starvation, haemorrhage, ulceration, bacterial infection, destruction of functionally important tissues, pain, sleeplessness and anxiety.

(g) Conclusion

We have considered at some length the contrasts between " innocence " and " malignancy ", in order to emphasize that these are not two distinct biological properties of tumours. They are merely convenient terms which we use for prognostic purposes. By experience of each histogenetic class of tumours, we have learnt which particular members of that class are likely to remain local and to be curable by local removal, and which ones are likely to spread more or less widely and dangerously. The pathologist greatly helps the clinician in making a prognosis if he can assert unequivocally of a particular tumour that it is " innocent " or " malignant " ; and this he *can* do in the majority of cases. But he also knows that in most histogenetic classes, tumours of inter-mediate behaviour occur, for the adequate prognosis of which the bald pronounce-ment " innocent " or " malignant " does *not* suffice.

IMPROPER BASES OF CLASSIFICATION

It is necessary to consider here, in order to discard, certain other bases which have been suggested for classifying tumours.

(a) Regional classification

To some extent, of course, histogenetic classification is regional ; e.g. gliomas can arise only in the central nervous system, meningiomas only in the meninges, and chordomas only in the spinal axis. But it is the *tissue* of origin, not the *organ* of origin, of a tumour which largely determines its properties.

(b) Embryological classification

Attempts to classify tumours according to the germ-layers from which their component tissues originally arose are valueless. The squamous stratified epithelia of the mouth cavity, oesophagus and uterine cervix are ectodermal, entodermal and mesodermal respectively ; but the squamous-cell carcinomas of these three regions are closely similar in structure and behaviour. Carcinomas of the posterior third of the tongue are *not* of a different species from those of the anterior two-thirds. Indeed the germ-layers are devoid of significance for the student of tumours.

In other minor ways also embryological concepts have been introduced improper-ly into the nomenclature of tumours. Thus it has been a prevalent habit to replace the ending " -oma " by " -blastoma ", the usual intention being to suggest that tumours consist of young proliferating cells of this or that type, e.g. " fibro-blastoma " to embrace all tumours composed of fibrous connective tissue cells, " lymphoblastoma " all tumours of lymphoid tissue, and so on. But in this usage there is a risk of conveying a false idea of embryonic origin. To call a tumour of the adult brain a " spongioblastoma " is to suggest that it arises from embryonic spongioblasts, when in fact it is merely an anaplastic growth derived from adult tissue. So also the word " myoblast " refers specifically to embryonic muscle-formative cells, so that to call a tumour derived from adult muscle a " myoblastoma " is misleading. Objection to the use of the " -blastoma " ending for tumours arising from adult tissues is particularly necessary, because there do occur certain truly embryonic tumours which arise in early life from immature tissues, namely the embryonic tumours of the kidney, adrenal and sympathetic ganglia, retina, and some other tissues. These are properly given the names "nephroblastoma", "neuroblastoma", and "retinoblastoma", because they actually do consist of embryonic renal, neural or retinal cells.

(c) Aetiological classification

Some pathologists have suggested that increasing knowledge of causation will provide a valid basis for classifying tumours. This is a vain hope, because (a) a variety of different carcinogenic agents can evoke identical tumours, and (b) a single carcinogenic agent, when applied to different tissues, can evoke a variety of tumours. Thus, carcinomas of the skin indistinguishable from one another in structure and behaviour may be caused by ultra-violet light, X-rays, arsenic, and a great variety of carcinogenic hydrocarbons. On the other hand, a single carcinogenic hydrocarbon, suitably applied, may produce squamous-cell carcinoma of the skin, adenocarcinoma of the lung, sarcomas of soft tissues, osteosarcoma of bone, and glioma of the brain. It is clear then that *the nature of neoplastic change in any given tissue is largely, if not wholly, independent of the kind of stimulus responsible for it*, and that causation can never afford a basis for the classification of tumours.

THE CLASSIFICATION ADOPTED

The following grouping, based solely on histogenesis and behaviour, is simple and adequate, and will be followed in later chapters.

Group I. Tumours of Epithelial Tissues

(a) *Papilloma*, benign tumour of surface epithelium.

(b) *Adenoma*, benign tumour of glandular epithelium.

(c) *Carcinoma*, a generic name embracing all malignant epithelial tumours.

Group II. Tumours of Non-haemopoietic Mesenchymal Tissues

(a) *Fibroma*, benign tumour of fibrous connective tissue.

(b) *Myxoma*, benign tumour of mucoid connective tissue.

(c) *Lipoma*, benign tumour of adipose tissue.

(d) *Chondroma*, benign tumour of cartilaginous tissue.

(e) *Osteoma*, benign tumour of osseous tissue.

(f) *Osteoclastoma*, benign tumour of osteoclasts.

(g) *Synovioma*, benign tumour of synovial tissue.

(h) *Angioma* (haemangioma, lymphangioma, glomangioma) benign tumour of vascular tissue.

(i) *Meningioma*, benign tumour of meningeal tissue.

(j) *Leiomyoma*, benign tumour of smooth muscle.

(k) *Rhabdomyoma*, benign tumour of striated muscle.

(l) *Sarcoma*, malignant tumour of any of the foregoing kinds of tissue, with an appropriate prefix when the kind is known, e.g. *fibrosarcoma, osteosarcoma, rhabdomyosarcoma*, etc.

GROUP III. TUMOURS OF HAEMOPOIETIC TISSUES

(The peculiar nature of the haemopoietic tissues and their tumours, especially their production of freely circulating cells, justifies their separation from those of the fixed mesenchymal tissues.)

(a) *Tumours of lymphoid tissue :*
  (1) Follicular lymphoma ;
  (2) Lymphosarcoma and lymphatic leukaemia ;
  (3) Hodgkin's disease ;
  (4) Reticulosarcoma.
(b) *Myelomatosis and plasmocytoma.*
(c) *Myelogenous leukaemia and chloroma.*
(d) *Primary polycythaemia.*

GROUP IV. TUMOURS OF NEURAL TISSUES

(a) *Gliomas*, tumours of neuroglial tissue :
  (1) Astrocytic gliomas (astrocytoma and " glioblastoma ") ;
  (2) Oligodendroglioma ;
  (3) Ependymal glioma ;
  (4) Medulloblastoma and neuro-epithelioma ;
  (5) Pinealoma.
(b) *Papillary tumours of choroid plexus.*
(c) *Neurilemmoma*, tumour of Schwann cells.
(d) *Neuroblastoma and ganglioneuroma*, tumours composed respectively of immature or mature nerve cells.
(e) *Chromaffinoma*, tumours of chromaffin tissue.
(f) *Tumours of carotid body and other chemoreceptors—chemodectomas.*
(g) *Retinal and ciliary tumours :*
  (1) Retinoblastoma ;
  (2) Tumours of the ciliary body or iris ;
  (3) Glioma of the retina.

GROUP V. SUNDRY SPECIAL CLASSES OF TUMOURS

(a) *Melanomas*, tumours of melanotic tissues.
(b) *Chordoma*, tumour of notochordal tissue.
(c) *Embryonic tumours of viscera :*
  (1) Nephroblastoma ;
  (2) Hepatoblastoma ;
  (3) Of other parts.
(d) *Teratomas*, tumours composed of multiple tissues.
(e) *Chorion-epithelioma*, tumour of chorionic tissue.

SUPPLEMENTARY READING

Evans, R. Winston (1966). *Histological Appearances of Tumours*. Edinburgh and London; E. & S. Livingstone.
Nicholson, G. W. (1921-1938). " Studies on tumour formation ", *Guy's Hospital Reports*, **71-88.** (Invaluable studies on many aspects of tumours for advanced students, reprinted in book form by Butterworth and Co. (Publishers) Ltd. London, 1950.)
Willis, R. A. (1967). *Pathology of Tumours*. London; Butterworths. (Chapters 28 to 35 of the present work are a summary of this book, to which the advanced student is referred for fuller details and other references on particular subjects.)
  – (1962). *The Pathology of the Tumours of Children*. Edinburgh and London; Oliver & Boyd. (Supplements the previous work.)

CHAPTER 29

THE CAUSATION OF TUMOURS

SINCE 1920 our knowledge of the causation of tumours has advanced rapidly. On the one hand, there has been increasing recognition of occupational and other environmental factors in the causation of human tumours ; and on the other hand, experimentalists have demonstrated the carcinogenic properties of a great number of chemical substances and other agents, many of which are clearly the same as those responsible for occupational tumours in man. In this chapter we will consider first the carcinogenic agents which have been identified, and then in order the parts played by occupation, social class, nation and race, age, and heredity in tumour causation. We will also refer briefly to spontaneous tumours in other animals, and to speculations regarding the nature of neoplastic change.

CARCINOGENIC AGENTS

The word " carcinogenic " is not restricted to carcinoma-producing agents but is used in a wider sense to embrace agents capable of evoking tumours of any kind. There is a great variety of such agents, which can be discussed conveniently under the following heads :

> Carcinogenic hydrocarbons ;
> Other chemical carcinogens ;
> Hormones as carcinogens ;
> Radiations as carcinogens ;
> Parasites as carcinogens ;
> Endogenous carcinogens ;
> Co-carcinogenic and anti-carcinogenic agents ;
> The time factor in carcinogenesis.

(1) Carcinogenic hydrocarbons

Long before pure carcinogenic hydrocarbons were isolated, it had been recognized that people whose occupations exposed them to soot, tar, pitch or mineral oils were specially liable to cancers of the skin. The first clearly recognized occupational tumour was chimney-sweeps' cancer of the scrotum, described by Percivall Pott in 1775 ; and a century later, Volkmann described the first recorded cases of paraffin cancer of the skin. In 1914, by prolonged applications of tar, Yamagiwa and Ichikawa produced carcinomas of the skin of rabbits' ears, the first experimentally produced tumours. In 1922 Leitch proved the carcinogenic properties of mineral oils experimentally.

Since tars and mineral oils are complex mixtures, and since they were found to vary greatly in their carcinogenic potency, attempts were soon made to isolate their specific carcinogenic ingredients. These culminated in 1930 in the synthesis by Kennaway and his co-workers of the first chemically pure carcinogen to be identified, 1 : 2 : 5 : 6-dibenzanthracene. This substance, though not itself a constituent of tar, is closely related to 3 : 4-benzpyrene, a much more potent carcinogen which is the main active agent in tars and mineral oils (Fig. 132). Kennaway and his team also showed that potent artificial tars could be produced by heating skin, muscle, yeast, cholesterol or other organic substances to temperatures over 700° C.

These discoveries gave a tremendous impetus to cancer research. Many new carcinogenic hydrocarbons and other compounds related to benzanthracene have since been synthesized, and tar-cancer research has largely been superseded by more precise researches with these chemically pure agents. Of these, cholanthrene and methylcholanthrene are of special interest because, not only are they very potent agents, but they are chemically related to the bile acids (Fig. 132).

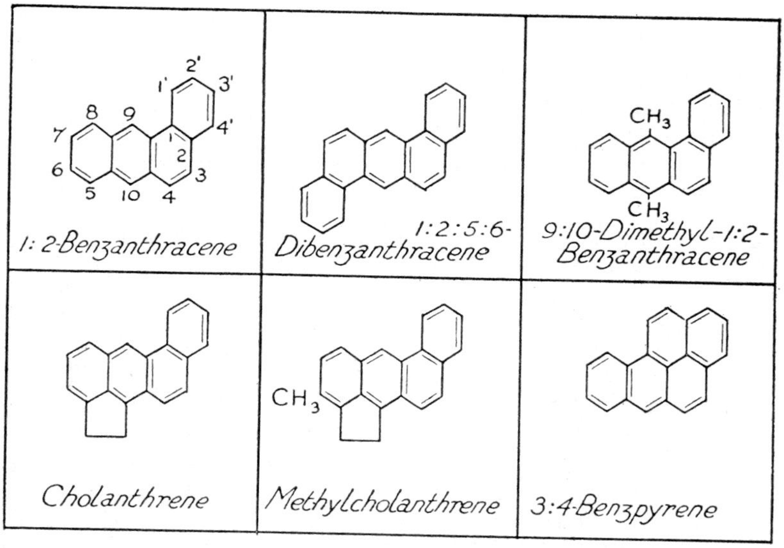

FIG. 132.—Carcinogenic hydrocarbons.

A variety of tumours can be produced by carcinogenic hydrocarbons. The earlier experimental tumours were carcinomas of the skin evoked by painting mice or other animals with solutions of the agents. Subsequently it was found that by suitable subcutaneous, intraperitoneal or intra-osseous injections of the same agents, sarcomas of various types can be produced ; that if subcutaneous injections are made in the mammary regions, adenocarcinomas of the mamma may be evoked ; that carcinomas of the kidney, liver, prostate, lung or other epithelial tissues can be produced in these organs by local implantation of carcinogens, or in subcutaneous transplants of pieces of the organs wrapped around crystals of the agents ; and that gliomas can be evoked by introducing carcinogens intracerebrally. A few instances of alimentary carcinomas probably due to carcinogens given experimentally by mouth have been recorded. It was found also that, in strains of animals of known tumour incidence, the application of carcinogens by any route (subcutaneous, intravenous, by ingestion or by skin painting) may strikingly increase the incidence of a particular kind of tumour to which the strain is spontaneously liable, e.g. lymphosarcoma, leukaemia, mammary carcinoma, or multiple " adenomas " of the lung in mice. These results show clearly that carcinogens may be absorbed, and may then act on a susceptible tissue remote from the site of application. Much important work remains to be done on the absorption, metabolism and excretion of chemical carcinogens.

(2) Other chemical carcinogens

The aromatic hydrocarbons of the benzanthracene series are not the only carcinogenic compounds ; there are several other groups of potent substances (Fig. 133).

(a) Azo-compounds

Several azo-dyes related to scarlet-red, notably o-aminoazotoluene and p-dimethylaminoazobenzene (" butter yellow "), are highly active in producing adenomas and carcinomas of the liver when given orally or by injection to rats

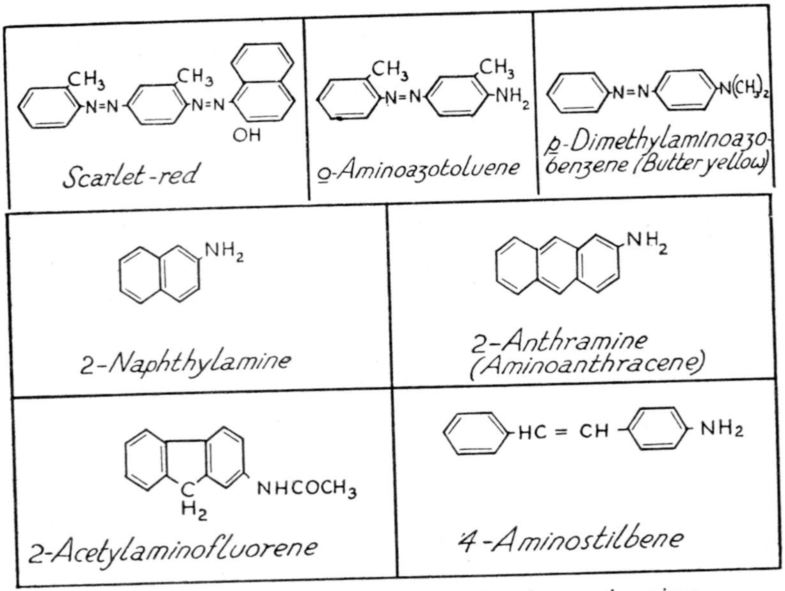

FIG. 133.—Carcinogenic azo-compounds and aromatic amines.

or mice. Some of the animals also develop papillomas and carcinomas of the bladder, doubtless because derivatives of the azo-compounds are excreted in the urine. We do not know of any occupational tumours in man attributable to azo-dyes.

(b) 2-Naphthylamine

Since the end of last century, the occupational incidence of carcinoma of the bladder and renal pelvis in workers in fuchsine and other aniline dye factories has been recognized. These were often called " aniline cancers ", but we now know that the potent agent is not aniline itself but 2-(beta-)naphthylamine ; prolonged administration of this compound to dogs produces bladder tumours.

(c) Other aromatic amines

Oral administration of the insecticide 2-acetylaminofluorene to rats or mice produces epithelial tumours of many different organs, the distribution varying with the species and breed of animal ; these include carcinomas of bladder, renal pelvis, liver, pancreas, lung, mamma, external auditory meatus, intestine and thyroid. Applications of 2-anthramine (aminoanthracene) produce carcinomas of the painted skin, and also remote tumours including carcinomas of the auditory meatus. Recent experiments have shown that 4-aminostilbene and several derivatives are carcinogenic.

(d) Other substances

Other substances which have been found to be carcinogenic include a complex quinoline-styryl compound, urethane, several " nitrogen-mustards " (substances allied to mustard gas), arsenic, and radioactive substances (see below, p. 420).

(3) Hormones as carcinogens

In a number of tissues which are under endocrine control, not only hyperplastic changes but also true tumours have been induced experimentally by appropriate hormonal treatment. This applies to the mamma, uterus, ovary, testis, pituitary gland, and perhaps to the adrenal cortex and thyroid gland.

(a) The mamma *

Before ovarian oestrogenic hormones were isolated, it had been known that removal of the ovaries of female mice greatly reduced the incidence of mammary carcinoma in them during adult life, and also that, while intact or castrated male mice never developed breast cancers, these appeared in a small proportion of castrated animals into whom ovaries had been successfully grafted. In 1932 Lacassagne obtained mammary cancers in both male and female mice by giving large doses of pure oestrogens, a result which was soon confirmed by others. The oestrogens first induce varying degrees of hyperplasia and cystic changes somewhat similar to those of cystic hyperplasia in the human breast ; and then, in a proportion of the mice which varies from breed to breed, carcinoma supervenes. However, oestrogen excess is not the only factor concerned in the experimental production of mammary tumours. These can be produced also by physical injury and obstruction of the mammary ducts, and, as we have already noted, by local injections or remote applications of chemical carcinogens. A further factor of great interest in mice is the " milk factor " or " breast-cancer-producing agent " which is transmitted through the milk of mice of cancerous stock to their progeny (see also p. 421).

(b) The uterus

Animals given large doses of oestrogens show varying degrees of overgrowth and squamous metaplasia of endometrial and cervical glandular tissue, and in a

---

* Since here we are discussing the mammary glands of animals, *mamma* is preferred to *breast*. *Mamma* is the more general term, applicable to all species of mammals, whereas *breast* refers particularly to the human mamma.

few instances genuine carcinomas appear to have supervened. Prolonged treatment may also produce myomas or myoma-like overgrowths of the myometrium.

(c) The testis
Oestrogens applied to male mice produce hyperplasia of the interstitial cells of the testis ; and, in a proportion of the mice, benign or malignant tumours of these cells develop.

(d) The ovary
Granulosa-cell and other ovarian tumours can be evoked experimentally by X-irradiation of the ovaries, by their transplantation to the spleen, and by oestrogen over-dosage.

(e) The pituitary gland
Pituitary tumours have been produced experimentally by oestrogen administration, and by chronic thyroid deficiency brought about by removal of the thyroid, by radio-iodine or by irradiation.

(4) Radiations as carcinogens

(a) Light and ultra-violet rays
The exposure to sunlight of agricultural and other outdoor workers in sunny climates, has long been recognized as a factor in causing carcinoma of the skin in man, especially of fair-skinned persons. Experimentalists have fully confirmed this. Mice and rats, especially albino animals, regularly exposed to excessive sunlight or ultra-violet rays, develop carcinomas of the ears, nose and other relatively hairless parts.

(b) X-rays
Röntgen's discovery of X-rays in 1895 was soon followed by examples of chronic X-ray dermatitis in exposed persons ; and cases of supervening carcinoma of the skin were reported from 1902 onwards. Later, adequate protective measures were adopted, with a corresponding decline in the incidence of accidental X-ray burns and cancers. The carcinogenic action of X-rays was confirmed experimentally on animals. Not only carcinomas, but also sarcomas of the skin or subcutaneous tissues, have followed exposure to X-rays, both in man and experimental animals; so also have carcinomas of the thyroid and sarcomas of bone.

(c) Radioactive substances
In man, occupational sarcomas of bones have resulted from absorption of radium and mesothorium salts used in luminous paints. These heavy metals are selectively taken up by the bones, where they set up a chronic radiation osteitis, which in some cases ends in the development of sarcoma. Instances of carcinoma of the skin following local applications of radium have also occurred. Experimentally, sarcomas of both soft tissues and bones have been produced by locally introduced or absorbed radioactive substances ; and it has been shown that the gamma-rays are mainly or wholly responsible for the carcinogenic effects. Sarcomas have been produced experimentally by injections of Thorotrast (colloidal thorium dioxide) used in radiodiagnosis, and this substance has also caused sarcomas and carcinomas of the liver in man.

(5) Parasites as carcinogens

In considering parasites as possible causes of tumours, we must distinguish between two quite different possibilities, namely : (*a*) a parasite might, like the external carcinogenic physical or chemical agents we have already considered, evoke neoplastic change in a tissue and the tumour then continue to grow without the parasite being present ; or (*b*) an intracellular parasitic virus might not only initiate neoplasia but might continue to reside and multiply in the tumour cells as the essential indwelling cause of their sustained neoplastic growth. Both of these theoretical possibilities must be considered in connection with certain mammalian and avian tumours.

(a) Parasites which only initiate tumour formation

In the rat's liver, *Cysticercus fasciolaris*, the cystic phase of the cat tapeworm, *Taenia crassicollis*, often evokes sarcomas of the surrounding connective tissues. In man, carcinoma of the bladder sometimes supervenes in cases of schistosomiasis ; and it is believed that the tumours are induced by the presence of the parasites. In cases of syphilitic inflammation of the tongue, carcinoma often supervenes ; so that *Treponema pallidum* may be looked upon as a carcinogen for this organ, though its action is probably only an indirect one.

(b) Indwelling filterable agents

These occur in the following kinds of tumours :

(i) *The filterable tumours of birds.*—The Rous sarcoma of fowls and a few other similar avian tumours contain filterable particulate tumour-evoking agents which multiply in the tumours. Cell-free filtrates of tumour tissue promptly induce identical tumours when injected into normal birds. The tumours are not infectious; normal birds kept with tumour-bearing ones or fed with tumour tissue do not develop tumours.

(ii) *Adenocarcinoma of frog's kidney.*—There is some evidence that renal adenocarcinomas in leopard frogs may contain a specific filterable tumour-evoking agent.

(iii) *The Shope tumour of rabbits.*—Certain papillary overgrowths of the skin of cotton-tail rabbits are transmissible to other rabbits by cell-free filtrates. Some of these " papillomas " become carcinomatous ; but filtrates of the carcinomatous tissue are inactive.

(iv) *The " milk factor " in mammary carcinomas in mice.*—Inbred strains of mice differ greatly in the incidence of carcinoma of the mamma. Early workers attributed these differences to purely genetic variations in the susceptibility of the mamma to carcinogenic stimuli ; but later work showed them to depend largely on a non-genetic filterable agent present in the milk and tissues of mice of high cancer strains. Newly born mice of high cancer strains, foster-suckled by females of low cancer strains, show a low incidence of mammary cancer when they become adults ; while infant mice of low cancer strains, foster-suckled by females of high cancer strains, later show a high incidence of mammary cancer.

(v) *Mouse leukaemias.*—Some of these contain a filterable leukaemia-evoking agent.

(vi) *" Polyoma virus ".*—This is a recently discovered agent capable of evoking a variety of tumours in mice and other rodents.

Are filterable agents concerned in the causation of mammalian tumours in general ? Enthusiastic virologists say, Yes ; but it is doubtful. Except for the experimental examples listed above, no evidence of the presence of a filterable agent has been forthcoming for any spontaneously occurring mammalian tumours, in spite of intensive search for such; and on the other hand, we have abundant evidence, already outlined, of the carcinogenic properties of many other extrinsic chemical and physical agents.

(6) Endogenous carcinogens

The carcinogenic hydrocarbons of the benzanthracene group are closely related chemically to some naturally occurring biological substances, including the bile acids, sex hormones and sterols. This relationship has naturally led to the suggestion that, under abnormal conditions, carcinogenic substances may be formed in the tissues themselves. Perhaps this happens in the obstructed duct contents in cases of cystic hyperplasia of the breast, in the gall-stones which are usually present in the cancerous gall-bladder, in the smegma retained under the prepuce of the cancerous penis, in the sebaceous contents of a cystic teratoma which has become malignant, in X-ray dermatitis of the skin or syphilitic inflammation of the tongue, or in those cirrhotic livers which are predisposed to carcinoma. Indeed smegma has been shown to be feebly carcinogenic.

Future research must include attempts to isolate pure carcinogenic substances from these and other precancerous conditions. Already a start has been made. For example, it has been found that suitable extracts of normal or diseased human livers have feeble carcinogenic properties, and that this applies also to cancerous livers or the cirrhotic livers of races unusually liable to hepatic cancer, e.g. the African Bantus. It is significant here that under suitable conditions desoxycholic acid and pure cholesterol have been found capable of inducing tumours.

(7) Co-carcinogenic and anti-carcinogenic agents

Some substances or physical agents, which alone are not carcinogens, when applied along with or after the application of carcinogenic agents may increase the carcinogenic effect ; these are called *co-carcinogens*. Other substances diminish the effect of carcinogenic agents ; these are *anti-carcinogens*. For example, in experimental carcinogenesis by hydrocarbons, croton oil is co-carcinogenic, while mustard gas is anti-carcinogenic. Again, the potency of chemical carcinogens varies considerably according to the solvent used. Since tars, mineral oils and other occupational carcinogens are usually complex and variable mixtures of substances, it is clear that their carcinogenic power will vary with their content of other ingredients which may reinforce or inhibit the action of their carcinogenic constituents.

An important co-carcinogenic agent in some cases is *trauma*. In 1924 Deelman first showed experimentally that wounds or other physical injuries in a tarred area might hasten tumour formation and determine its site. This was confirmed by later workers using pure carcinogens ; and the co-carcinogenic effect of local injury has been called the " Deelman phenomenon ". It probably accounts for some cases of human tumours apparently following injuries ; the injury may precipitate neoplasia in an area of tissue already prepared for it by the previous application of carcinogens. Trauma alone is probably never a cause of tumour formation.

(8) The time factor in carcinogenesis

In both human and experimental carcinogenesis, the latent period intervening between the first application of the carcinogenic agent and the appearance of the tumour is usually a long one.   Further, the agent may have ceased to act long before the tumour appears.   Thus, applications of a carcinogenic hydrocarbon may be made to the skin of mice for a few weeks and then discontinued ; and the resulting tumours may not appear until a further 6, 9 or 12 months have elapsed —that is, a considerable part of the life span of the animals.   So also in man ; the unprotected mule-spinner, X-ray technician or radium paint worker may have retired from his hazardous occupation 5, 10 or 15 years before his tumour appears.

Clearly, then, it is useless to look for the causes of human tumours in the habits and occupations of the victims just before the appearance of the tumours. The tumour of today is usually the result of carcinogenic and co-carcinogenic stimuli applied 10, 20 or 40 years ago ; sometimes these stimuli have been in operation continuously ever since, sometimes only intermittently, and sometimes there has been a long interval during which their application ceased.   Our medical histories of tumour patients are usually very defective in this respect ; detailed inquiry into the occupations and habits of the whole of a patient's previous life is seldom made.   Carefully planned inquiries of this kind and statistical evaluation of the results are necessary if we are to sift out from human environments those factors which are to blame for the many tumours the causes of which are still unrecognized.

OCCUPATION, SOCIAL CLASS AND TUMOURS

The association of particular kinds of tumours with certain occupations is sometimes too obvious to require statistical proof.   Many of these have already been mentioned, e.g. chimney-sweeps' cancer, mule-spinners' cancer, skin cancers in X-ray or radium workers, " aniline " cancers of the urinary tract, and osteosarcomas in radium paint workers.   Carcinoma of the lung in certain miners will be referred to on p. 488.

Apart from these more obvious examples, however, different occupations show statistically significant differences in the incidence of cancer generally and of particular kinds of tumours.   For example the general cancer death-rate is higher than the average for furriers, glass-blowers and tin and copper miners, lower than the average for teachers, clergymen and telephonists.   Or, as an example of occupational differences for a particular kind of tumour, the death-rate from cancers of the mouth and pharynx is much higher for barmen, furnacemen and dock labourers than for teachers, bankers and civil servants.   These differences must be due to differences in the environmental factors connected with the occupations ; and their discovery will depend on well-designed inquiries of the kind suggested in the preceding section.

Social factors other than occupation certainly play a part in the causation of tumours.   By grouping fatal cancer cases into several social classes according to occupations, it is found that some tumours are strikingly more prevalent in the lower classes than in the upper and vice versa.   Thus, cancers of the upper alimentary canal from the mouth to the pylorus are about twice as frequent in the lowest classes as in the highest ; and since this applies also to the wives of the men in different classes, it must be attributed more to factors connected with domestic and economic circumstances than to occupational factors. Uterine cancer also shows a distinct class gradient, being more frequent in the lower classes. Conversely, mammary cancer is more frequent in the upper than in the lower classes. These differences are almost certainly due to different reproductive and lactational habits at different social levels.

NATIONAL AND RACIAL DIFFERENCES IN THE INCIDENCE OF TUMOURS

Tumours of all kinds occur in all nations and races, but the relative frequencies of the different kinds vary. For example, in England and Wales, carcinoma of the breast accounts for about 20 per cent of all female deaths from cancer ; while in Italy it accounts for only 10 per cent, and in Japan for only 3 per cent. Carcinoma of the stomach causes about one-quarter of the deaths from cancer in England, more than one-half of them in Scandinavia, Holland and Switzerland, and about two-thirds of them in Czechoslovakia and in Japan. Carcinoma of the liver is a rare disease in Europeans, but is the commonest kind of cancer in Javanese and African Bantus. Malignant melanoma of the feet is much commoner in Indians, Cingalese and Sudanese than in white people. These national and racial differences in the relative frequencies of different kinds of tumours are clearly not due to genetic factors, but to differences of exposure to environmental carcinogenic stimuli connected with customs, habits, diet and occupation.

AGE INCIDENCE OF TUMOURS

(1) Cancer and old age

Cancer as a whole is predominantly a disease of the elderly. This is *not* because senile tissues are " predisposed " to cancer, as was once supposed, but because of the usually long induction period elapsing between exposure to carcinogenic agents and the development of tumours, as described above. About two-thirds of fatal cancers in Europeans occur at ages over 60 years, and about four-fifths of them at ages over 50 years.

(2) Age distribution and age incidence of tumours

If we graph the numbers of tumour cases (e.g. of carcinomas) occurring in various age periods, we obtain a distribution curve like that shown in Fig. 134. Such a graph of age distribution, however, takes no account of the age composition of the population from which the cases of carcinoma came, and therefore gives no indication of the relative liabilities of people of different ages to carcinoma. In order to do this we must compare the percentages of the number of cases of carcinoma in successive age periods with the percentages of the total population in those same age periods. When we do this, we find that the incidence of carcinoma rises continuously throughout life into old age. This is what was to have been expected ; the longer a person lives, the longer he is exposed to carcinogenic and co-carcinogenic agents in his environment, and the greater his chances of developing tumours. The decline of the curve in Fig. 134 after the seventh decade is simply due to the progressive falling off of the numbers of the population living in subsequent decades.

(3) Mean ages

Different kinds of tumours differ greatly in their age incidence. At one extreme are the embryonic tumours of infancy and early childhood, such as neuroblastomas, retinoblastomas and nephroblastomas, some of which actually develop during foetal life. At the other extreme are the latest appearing kinds of carcinoma, e.g. carcinoma of the prostate, the mean age of the victims of which at death is over 70 years. Even different kinds of carcinomas show widely different mean ages, e.g. carcinomas of the lung or cervix of the uterus occur at, and kill at, average ages about 6 years earlier than the average age for all carcinomas, while carcinoma of the prostate occurs at a mean age about 10 years later than the general average. The average age of the victims of carcinomas of all kinds in England is over 60. Some forms of sarcoma are commonest in youth ; e.g. nearly two-thirds of patients with osteosarcomas are between the ages of 10 and 30 years.

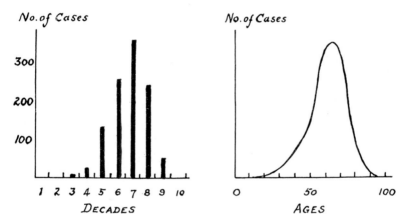

FIG. 134.—Age distribution of patients with carcinoma (all sites).

SEX INCIDENCE OF TUMOURS

Apart from the organs of reproduction and the breast, many other organs show pronounced sex differences of tumour incidence. Thus, carcinomas of the lip, tongue, other parts of the mouth, tonsil, pharynx, larynx and oesophagus are all much commoner in men than women, the general ratio being at least 10 to 1. Carcinoma of the stomach is twice as common in men as in women. The frequency of bronchial carcinoma in men is 4 or 5 times that in women ; that of carcinoma of the kidney about 3 times, and that of vesical and renal pelvic carcinoma also at least 3 times. The great preponderance of males in these classes of tumours is undoubtedly due to greater exposure to carcinogens connected with various occupations and male habits. On the other hand carcinoma of the thyroid occurs in women thrice as frequently as in men, and most patients with adrenal cortical tumours are females ; both of these differences may well be due to hormonal disturbances connected with sexual functions.

HEREDITY AND TUMOURS

By deliberate inbreeding, it is possible to obtain strains of mice with a very high or very low incidence of tumours of a particular organ. Such close inbreeding does not obtain in any human communities; and hereditary predisposition plays very little part in the causation of most kinds of human tumours. However, there are a few exceptions to this general rule. Familial retinoblastoma, multiple polyposis of the large intestine, and xeroderma pigmentosum, all of which will be considered in their proper places later, afford striking examples of inherited predisposition to particular kinds of tumours. In addition, there occur rare instances of many members of a family developing carcinoma of a particular organ. The most notable of these is the " Family G " first described by Warthin and later by Hauser and Weller. In this family, in the course of three generations, 41 persons out of 174 aged 25 or more produced 43 carcinomas; all of the 20 tumours in males were gastro-intestinal, while the 23 in females included 6 gastro-intestinal and 15 endometrial carcinomas; and the tumours appeared at relatively early ages. Napoleon Bonaparte and his father both died of proven cancer of the stomach, and 3 of his sisters and one brother probably also died of gastric cancer. Reports have appeared of families in which carcinoma of the breast has appeared in 3 or 4 successive generations, sometimes with bilateral growths, and sometimes at unusually early ages. However, these instances of familial cancer are exceptional; and most supposed " cancer families " exemplify only the laws of chance. By far the majority of human cancers are due to chemical or other carcinogenic stimuli which would be effective whatever the genetic inheritance of the exposed persons. Only occasionally is an inherited susceptibility clearly discernible.

TUMOURS IN ANIMALS

In mammals of most species, there occur many kinds of tumours, of varieties comparable with those of man. Birds, fish, amphibians and reptiles also exhibit a variety of tumours. The different species of mammals differ in the incidence of tumours in general and of particular kinds of tumours; e.g. guinea-pigs and rabbits develop tumours much less frequently than rats and mice, cats less frequently than dogs, and pigs and sheep less frequently than other domesticated animals.

Tumours are very common in middle-aged and old dogs. The commonest are fibro-adenoma and carcinoma of the mamma, carcinoma of the skin, carcinoma of the tonsil, tumours of the testis, melanoma, lymphosarcoma, and osteosarcoma. Carcinomas of the stomach or intestine are extremely rare; and this applies also to other species of mammals.

On the other hand, some tumours which are relatively rare in man are relatively common in certain other species. Thus, papillary carcinomas of the conjunctiva are among the commonest tumours of horses and cattle; testicular tumours of several kinds are very common in dogs; benign teratomas of the testis are common in horses; and embryonic renal tumours are relatively frequent in pigs, rabbits and rats.

The study of animal tumours is of interest in human pathology for several reasons. (a) Striking contrasts in the frequency of certain kinds of tumours in man and animals, e.g. the rarity of gastric and intestinal carcinomas in all mammals except man, may provide clues to the causative factors. (b) Histological study of certain animal tumours, e.g. those of the canine testis, has clarified the histogenesis of the corresponding human tumours. (c) Some animal tumours can be transplanted from animal to animal of the same species, and study of these readily transplantable tumours has led to valuable information regarding their growth and the factors modifying it.

HYPOTHESES AS TO THE NATURE OF NEOPLASIA

We have reviewed briefly all the known kinds of carcinogenic agents and the kinds of evidence by which we are searching for agents which have still to be detected. But these agents are only the exciting causes or evocators of the neoplastic change in cells. What is the nature of this all-important cellular response, which confers on the cells their peculiar powers of excessive unco-ordinated proliferation? And how do the many diverse carcinogenic agents bring it about? We do not know ; but the following hypotheses have been advanced.

(a) Cohnheim's cell-rest hypothesis

Cohnheim supposed that tumours arose from " cell-rests ", superfluous groups of embryonic cells which failed to mature and persisted in the adult. This idea, which is no longer tenable, is mentioned only for its historical interest. There is no evidence of the existence of embryonic " cell-rests " ; and there is clear evidence that, except for the truly embryonic tumours of foetal life and childhood, tumours arise from fully differentiated adult cells.

(b) The chronic irritation hypothesis

Virchow and many others adopted the view that tumours resulted from the persistant stimulation of the tissues by irritants. This hypothesis properly focused attention on the external factors responsible for tumours ; but it erred in emphasizing non-specific " irritation ", which was imagined to " provoke " or " goad " the tissues into anarchic growth. Study of pure chemical carcinogens has dispelled this misconception ; these substances are not " irritating " in the ordinary sense. In any case the chronic irritation hypothesis left quite unexplained the nature of the cellular " anarchy ".

(c) Parasitic hypotheses

Since the latter part of the last century, bacteria, protozoa, fungi and viruses, have all in turn been blamed as the indwelling causes of tumour growth ; but only the last of these remains for serious consideration. We have already noted the virus-induced tumours of laboratory animals (p. 42), and need only repeat here that there is no evidence of the presence of filterable agents in any human tumours or in the great majority of tumours of other animals.

(d) Nuclear hypotheses

Abnormalities of mitosis in rapidly growing tumours (p. 436) have led many workers to suppose that neoplastic change must involve some fundamental alteration in the composition or activities of the nucleus. Thus, a prevalent view has been that cancerous change consists in the sudden production of a new race of cells by gene mutation. This purely speculative idea is unacceptable : both the structural study of tumours and experimental carcinogenesis show that the neoplastic change does *not* take place suddenly, but in a gradual and cumulative manner, precancerous hyperplasia often passing insensibly into neoplasia, and benign non-invasive growth into malignant invasive growth, without any sudden changes in cell structure or proliferation.

(e) Cytoplasmic hypotheses

Many modern workers are turning to the view that the neoplastic habit of growth may depend on irreversible metabolic and cytoplasmic alterations occasioned by the application of carcinogenic agents, and differing in degree from tumour to tumour. According to this view, transmission of neoplastic qualities from tumour cells to their descendants indefinitely is attributable to permanent alterations, not in the nuclear genes, but in the cytoplasm. This idea is consistent with recent work on plasmagenes and cytoplasmic inheritance, and with the concept that the normal differentiation of cells into their various histological types is determined by stable particulate elements in their cytoplasm, elements which multiply and are transmitted to the daughter cells during cell division. Perhaps, then, neoplasia consists in a permanent and transmissible change in these cytoplasmic particles, a change which is evoked by specific carcinogenic stimuli, and which can occur gradually and to different degrees in different tumours. These ideas are still only speculative, but they are in keeping with modern biological concepts, and are preferable to the earlier hypotheses of cell-rests, cell anarchy evoked by irritation, intracellular parasites, and gene mutations.

SUPPLEMENTARY READING

References Chapter 28 (p. 415), and the following :

Brit. Med. Bull. (1947 and 1958). **4** and **14** give many valuable review articles on carcinogenesis.

Cotchin, E. (1956). *Neoplasms of the Domesticated Mammals.* C'wealth Agric. Bureaux, Farnham Royal. (A valuable review.)

Feldman, W. H. (1932). *Neoplasms of Domesticated Animals.* Philadelphia; W. B. Saunders. (A comprehensive monograph with many excellent figures.)

Hill, A. B. (1939). *Principles of Medical Statistics.* London; *Lancet.* (Students should read especially Chapters 14 to 16 on common statistical fallacies.)

Hueper, W. C. (1942). *Occupational Tumors and Allied Diseases.* Springfield, U.S.A.; Charles C. Thomas. (A comprehensive reference work.)

  – (1954). 'Recent developments in environmental cancer ''. *Arch. Path.*, **58**, 360, 475 and 645.

CHAPTER 30

THE MODE OF ORIGIN, STRUCTURE AND SPREAD OF TUMOURS

IN THIS chapter we survey certain general aspects of the structure, growth and spread of tumours. In doing this, we will have to anticipate the accounts of particular kinds of tumours given later, and the present chapter should be re-read in conjunction with these. Here we will consider the following :

> The mode of origin of tumours ;
> The shapes of tumours ;
> The microscopical structure of tumours ;
> The stroma of tumours ;
> The chemistry and metabolism of tumours ;
> The routes of invasion of malignant tumours ;
> The metastasis of tumours.

THE MODE OF ORIGIN OF TUMOURS

A widely held view is that each tumour has a simple unicentric origin, arising at a single point in time, from a single small focus of cells, and enlarging only by multiplication of these cells and their descendants. This view is false ; tumours arise from small or large *fields* of tissue and enlarge not only by cellular proliferation but also by progressive neoplastic conversion of tissue within those fields. This is clear from microscopic study of early tumours of many kinds, especially those of the epidermis and mucous membranes. These often show plainly gentle transitions from normal to neoplastic epithelium without any clear-cut demarcation.

Solitary localized squamous-cell carcinomas of the skin often have a field of origin demonstrably 1 or 2 centimetres in diameter. In cases of multiple growths and some extensive superficial carcinomas the potentially cancerous field is much greater than this. Thus fair-skinned people who are much exposed to sunlight from outdoor occupations often develop chronic sunburn, widespread keratoses and multiple carcinomas of the face, neck and other exposed parts. The whole epidermis of these parts is ready to become cancerous, and the multiple tumours which actually develop merely denote the foci of maximum cancer potential in these areas. An extreme instance is seen in the rare disease xeroderma pigmentosum, in which there is an inherited over-susceptibility of the skin to light, and at an early age the patient develops many carcinomas of exposed areas of skin. Carcinomas supervening on old scars, chronic ulcers or sinuses are never of simple unifocal origin, but show progressive cancerous change over more or less extensive, sometimes multiple, areas of epidermis clothing the scar or surrounding the ulcer (Figs. 135, 136). So also, basal-cell carcinomas of the epidermis are often demonstrably of multicentric or diffuse origin (Fig. 137) ; and subepidermal basal-cell growths often show clear signs of origin from several groups of sweat glands or pilo-sebaceous follicles simultaneously (Fig. 138).

429

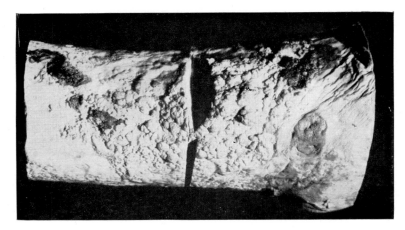

FIG. 135.—Cancerous varicose ulcer.

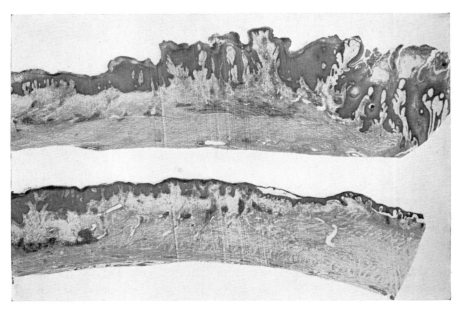

FIG. 136.—Representative sections from the tumour of Fig. 135, showing the widespread superficial character of the growth. (× 6.)

Similar structural evidence of progressive origin in a field of prepared tissue, or of multicentric origin, is seen also in many early carcinomas of the lip, tongue, stomach, intestine, urinary tract, and especially of the breast. Cheatle said of the breast, " the primary cancer process transforming epithelial into malignant cells may commonly operate on extensive duct surfaces " ; Nicholson said, " tumour formation is here multi-centric, or rather omni-centric " ; and Muir

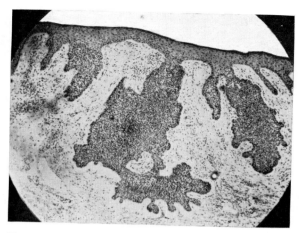

FIG. 137.—Multifocal development of basal-cell carcinoma of epidermis. (× 45.)

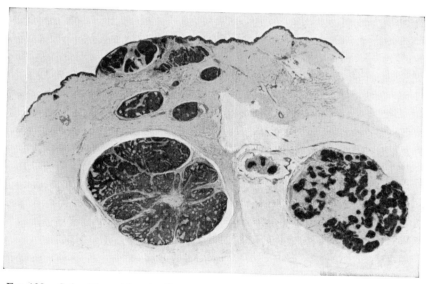

FIG. 138.—Subepidermal basal-cell carcinoma showing evidence of origin from at least two groups of sweat glands and their ducts. (× 8.)

said, " Malignancy is often not only of multicentric origin but can be seen to occur gradually and to affect groups of cells in a diffuse fashion " ; the neoplastic change " is regional rather than focal ". These conclusions will be endorsed by anyone who takes the trouble to examine a few cancerous breasts carefully.

Benign fibro-adenomas of the breast also often show clear signs of multicentric origin and additive as well as proliferative growth. The bed of breast tissue around a fibro-adenoma frequently contains multiple small satellite fibro-adeno-matous foci derived from hyperplastic breast lobules. As these enlarge they become added to the main mass, which therefore often has a lobulated form.

Tumours in paired organs sometimes afford striking instances of multicentric or widespread origin. Diffuse bilateral carcinoma of the human breasts sometimes occurs ; and dogs may show simultaneous similar carcinoma of all 4 or 5 pairs of mammae. Cystic tumours of the ovaries are frequently bilateral, and the papillary growths which arise in them may often be seen springing simultaneously from the linings of many separate cystic cavities in both organs.

These examples of widespread and multiple origin of tumours must suffice ; other instances will appear when the various classes of tumours are described. But before leaving the subject here, it may be remarked that, from our knowledge of experimental and occupational carcinogenesis, this mode of origin from a field of tissue is what might have been expected. Carcinogenic agents are applied, not to single cells or minute groups of cells, but to many cells over considerably areas. All of the cells in the area are acted on similarly, though not equally. The eventual neoplastic change will usually take place, not simultaneously throughout the prepared area of tissue, but in a particular sequence dependent on the gradients of the original carcinogenic stimuli and also the sites of applica-tion of subsequent co-carcinogenic stimuli. Tumour formation in the area may start at only one focus which may then extend peripherally, or it may start at several or many discrete foci which may long remain separate or may soon coalesce, or it may start almost simultaneously or in rapid succession over the entire area of tissue. Examples of all of these possible sequences are seen in both experimentally produced and human tumours.

THE SHAPES OF TUMOURS

The shape of a tumour depends on (a) its site of origin, especially whether from a surface or in the substance of a solid tissue ; (b) the extent of its field of origin ; (c) its mode of growth, whether purely expansive, both expansive and infiltrative, or mainly infiltrative ; and (d) moulding of the tumour by neighbouring structures. Figure 139 depicts diagrammatically the main forms of tumours as determined by these several factors.

The simplest form, the spherical, is seen when a benign tumour grows purely expansively from a single small focus in a homogeneous tissue. Even if the tumour is malignant and its growth partly infiltrative, it often retains a nearly spherical form, as is seen in many metastatic tumours in liver, lung, brain and other almost homogeneous tissues. If the enveloping tissues around a well-circumscribed growth are not physically uniform but are under directional pressure or tension, the tumour, which would otherwise have been spherical, becomes ovoid or flattened, as is seen in subcutaneous or intermuscular lipomas or fibromas. Cir-cumscribed tumours which have arisen from multiple neighbouring foci which have then coalesced, have a lobulated form, e.g. some fibro-adenomas of the breast, as already mentioned.

If the growth of a malignant tumour in solid tissues is mainly infiltrative, it often assumes the form shown in Fig. 139D, with more or less extensive ill-defined outrunners into the surrounding tissues, a common appearance in carcinomas of the breast. It was this appearance of claw-like processes and infiltrative grip on the tissues which suggested the ancient names " cancer " (L.=a crab) and " carcinoma " (Gk., *karkinos*=a crab). In some tumours, or parts of tumours, infiltrative growth is so diffuse and widespread in the tissues that these

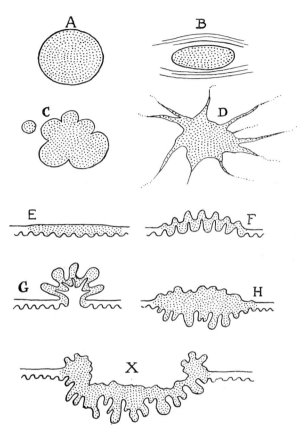

Fig. 139.—Diagram of shapes of tumours : A-D, in solid tissues ; E-X, on surfaces. A, encapsulated spherical tumour ; B, encapsulated tumour compressed to an ovoid shape by bounding tissues ; C, lobulated conglomerate tumour ; D, malignant tumour with infiltrating outrunners ; E, intra-epithelial tumour ; F, sessile papilloma ; G, pedunculated papilloma ; H, non-ulcerated carcinoma, and X ulcerated carcinoma.

show only general thickening without any clear limits or demarcated focus of obvious growth ; this is well exemplified in " leather-bottle " or " gizzard " cancer of the stomach (see Fig. 186).

The shapes of epithelial tumours of skin or mucous membranes are dependant on the extent of their fields of origin and the relative amounts of neoplastic conversion of tissue and proliferation of tumour cells which have taken place. Thus, neoplastic change in many or all of the cells of a surface epithelium *in situ*, with but little proliferative increase in the number of cells, produces a more or less extensive patch of altered, slightly thickened epithelium, not much elevated above the general surface level and not invading the underlying tissues ; as in Bowen's disease of the skin, a form of intra-epidermal carcinoma (Fig. 139E). Now suppose that, in a field of epithelium of similar extent, considerable proliferative increase in bulk accompanies the progressive neoplastic change. If the tumour is a benign papilloma and therefore non-invasive, it will assume the sessile form shown in Fig. 139F ; if malignant and invasive, the form shown in Fig. 139H. If a benign papilloma has arisen from a very small field of epithelium but has grown actively, it will be of the pedunculated form shown in Fig. 139G, with a narrow base corresponding in extent with the initial field of origin, and a projecting mass of more or less complexly branched structure. A malignant tumour, if not highly invasive, may at first have a similar gross form.

In many advanced malignant tumours of surfaces, proliferative and invasive growth predominates, and necrosis, ulceration and infection often supervene and destroy the original forms of the tumours (Fig. 139x).

The moulding of tumours by neighbouring tissues is often obvious. Visceral capsules often confine and mould tumours within the viscera ; periostea tend to confine and mould tumours in bones ; ligaments, tendons, fasciae and cartilages resist dislodgement or invasion by tumours, and so mould or groove their surfaces.

MICROSCOPICAL STRUCTURE OF TUMOURS

The structures of particular kinds of tumours are described in later chapters. Here, only certain general aspects of tumour structure need be noted.

(1) The degree of structural differentiation of tumours

As already noted in Chapter 28, most benign tumours show a high degree of differentiation, while many malignant tumours show imperfect differentiation or *anaplasia* of varying degrees. In general, there is a close parallelism between degree of anaplasia, rate of growth and degree of malignancy. Very anaplastic tumours grow and kill quickly, and many mitotic figures are found in the tumour cells. Yet, it must also be stressed that some malignant tumours have a highly differentiated, nearly normal structure, and their growth is influenced, in part at least, by the same factors which control the growth and organization of normal tissues.

(2) Metaplasia in tumour tissue

Tumours afford some of the most striking examples of metaplasia ; the tumour tissue undergoes differentiation into types not normally present in the organ. Carcinomas of the bronchi, endometrium, or gall-bladder often show small or large areas of squamous metaplasia, with stratification of the cells and the formation of spinous cells, squames and keratin. Rarely, similar squamous metaplasia occurs in glandular carcinomas of the stomach, intestine, breast or prostate (see Fig. 140c). The breast normally secretes no mucus, but in some mammary carcinomas the tumour cells produce abundant mucus, i.e. they undergo mucoid metaplasia. Some fibromas or other tumours of soft connective tissues undergo metaplastic ossification or chondrification.

(3) Individuality in tumours

A remarkable feature of the structure of tumours is that each tumour has its individual peculiarities, which tend to remain more or less stable throughout its life—in the primary growth, in metastases, and, in the case of transplantable animal tumours, in serial transplants. Thus, if a mammary carcinoma, soon after its inception, is poorly cellular, has much densely fibrous stroma, and grows slowly, it usually retains these " scirrhous " characteristics to the end ; or, if in its early stages it is a quickly growing, highly cellular, anaplastic tumour, it usually remains so, and spreads and kills quickly ; or, if it has any other special character—squamous metaplasia, mucoid metaplasia, a prominent glandular, papillary or cystic structure—it usually retains this throughout its course and in

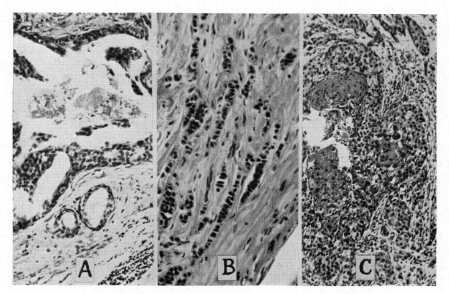

FIG. 140.—Variety of structure in different parts of one tumour—a mammary carcinoma. A, intra-ductal adenocarcinoma ; B, infiltrating scirrhous polyhedral-cell carcinoma ; C, cellular polyhedral-cell carcinoma with squamous metaplasia. (A, and B, × 120 ; C, × 80.)

whatever sites it may come to occupy. Sometimes, however, a tumour which starts off by growing slowly and differentiating well, later becomes more anaplastic and accelerates its growth. The reverse change—increased differentiation and slower growth with the lapse of time—is very rare. In general, metastatic tumours are rather less differentiated than their parent primary growth.

(4) Diversity of structure in a single tumour

Not only do the different tumours of one class show individual differences of structure, but a single tumour sometimes shows great diversity of structure in different parts (Fig. 140). This diversity does not conflict with the principle of tumour individuality just enunciated ; the range of structure in any given tumour is itself one of its individual peculiarities.

Since individual tumours often show a variegated structure, histological sub-division of any single species of tumour cannot be made with clear-cut

precision. To set out, for example, to label all bronchial carcinomas either
" adenocarcinoma ", " squamous-cell carcinoma " or " anaplastic carcinoma "
is futile. The entity is *bronchial carcinoma*, and all other prefixes or adjectives
are merely descriptive of the structural variants possible within the group, and
sometimes seen in one tumour. Certainly, in many tumours, one or other
variant is predominant or even exclusive, and we must learn what we can about
the prognostic significance of predominance of particular variants. But we
must avoid the error of regarding the variants as distinct species just because
they have been given distinct names.

(5) Secretory function in tumour cells

The cells of any tumour of glandular tissue, exocrine or endocrine, may produce
their appropriate secretion. The most familiar example of an external secretion
is the abundant mucus produced by many mucoid (so-called " colloid ") adenocarcinomas of the stomach, intestine or other organs. So also, liver-cell carcinomas may secrete bile, pancreatic carcinomas have been known to produce
digestive enzymes, and mammary fibro-adenomas may lactate.

Secretion by tumours, benign or malignant, of the endocrine glands is the
cause of most of the hyperhormonal states, e.g. acromegaly and giantism, hyperparathyroidism, adrenal virilism, hyperinsulinism, hyperadrenalinism, and
some cases of hyperthyroidism. Ovarian follicular tumours may produce excess
of oestrogens ; interstitial-cell tumours of the testis, excess of androgens ; and
chorion-epithelioma produces gonadotrophins and evokes the hormonal results
of pregnancy.

(6) Mitosis in tumour cells

In slowly growing, well-differentiated tumours, benign or malignant, mitotic
figures are few in number and usually normal in structure ; but rapidly growing
malignant tumours often contain numerous mitoses, and many of these are
abnormal. The abnormalities include multipolar mitoses, great variations in the
size and number of chromosomes, which may reach several hundred, delay and
irregularity in the movements of the chromosomes, and failure of cytoplasmic
division, leading to multinucleated giant tumour cells (Fig. 141). In very
anaplastic tumours the proportion of cells seen in mitosis at any given moment
(as in a micro-section) may reach 5 per cent or even 10 per cent.

Some workers have supposed abnormalities of mitosis to be the essential
change in neoplasia, indicative of genetic mutation in the cells (see p. 427). But
this is very doubtful ; for abnormal mitoses can be produced in growing or
regenerating tissues by the application of heat or chemical agents, so that their
presence in rapidly growing tumours, in which degenerative changes are common,
is not surprising.

(7) The stroma of tumours

All tumours have a non-neoplastic stroma composed of connective tissues
and blood-vessels derived from the surrounding tissues. In well-differentiated
growths, the stroma is often plentiful and of orderly, nearly normal pattern.
In cellular anaplastic growths, it is often very scanty and imperfectly organized.

(a) The blood-vessels in tumours

These comprise pre-existing vessels and newly formed vessels derived from them. The formation of new vessels is often abundant, especially at the growing margins of infiltrating tumours ; and the new-formed vessels are small and simple. In rapidly growing neoplasms they often consist of irregular channels lined by endothelium only or by naked tumour cells. The vascularity of tumours varies greatly ; e.g. densely fibrous (scirrhous) carcinomas are almost avascular, while some metastatic deposits of renal or thyroid carcinoma in bones are so vascular that they pulsate. Many malignant tumours, as they grow peripherally, undergo central necrosis, because of inadequate blood supply or " suicidal " occlusion of their blood-vessels.

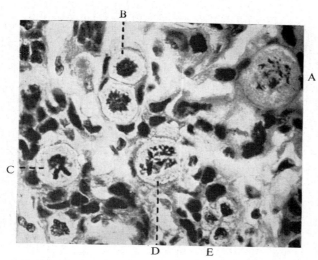

FIG. 141.—Numerous mitoses, including abnormal forms, in an anaplastic carcinoma of the tonsil. A, abnormally large cell with multipolar mitosis ; B, recently completed cell division ; C, multipolar mitosis with triradiate band of chromosomes ; D, another cell showing multipolar mitosis ; E, recently completed division into 3 daughter cells. (× 600.)

(b) The fibrous stroma of tumours

This also varies greatly in amount ; it is scanty in anaplastic growths, moderate in amount and orderly in arrangement in well-differentiated ones, and excessive and dense in " scirrhous " carcinomas (Gk., skiros=hard). In diffusely infiltrating tumours, there is often a very close intermixture of tumour cells and invaded connective tissues ; and the microscopical distinction between tumour and stroma cells is sometimes very difficult.

(c) The bony stroma of tumours

The reaction of bone to invasion by primary or secondary tumours varies greatly. In many cases the bone suffers rarefaction and absorption. In others it reacts like the fibrous stroma of a scirrhous cancer, and becomes more abundant and denser, a reaction called *osteoplasia*. This is very frequent in bones infiltrated by prostatic carcinoma, not uncommon in those infiltrated by mammary carcinoma, but relatively rare in other kinds of tumours. Another kind of proliferative reaction of bone to infiltrating tumours is the deposition of new bone by the expanded periosteum. This may occur either as "onion-skin" laminae parallel to the bone surface, or as fine or coarse spines projecting vertically from it (Fig. 142). Both of these are well seen in some metastatic neuroblastomas in bones

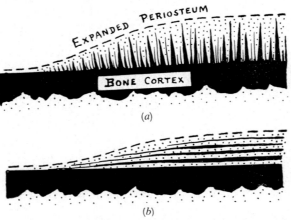

FIG. 142.—The two patterns assumed by new-formed sub-periosteal bone in tumours which expand periosteum. (*a*), vertical spines ; (*b*), "onion-skin" layers. (See Fig. 243.)

(p. 574), but similar new subperiosteal bone may form also in metastatic carcinomas. Quite often, tumour deposits in bone cause neither bone destruction nor osteoplasia, but simply infiltrate the marrow spaces and leave the bone trabeculae intact ; such bones appear radiographically normal.

(d) Leucocytes and phagocytes in tumours

Apart from inflammatory collections due to bacterial infection, leucocytes and phagocytes of various kinds often congregate in or around malignant tumours. There is no reason to suppose that they denote any specific defensive reaction to the tumours ; they are probably attracted by various products of degeneration or metabolites of the tumour cells. *Lymphocytes* and *plasma cells* are often plentiful at the invading margins of squamous-cell carcinomas of the skin, lip, tongue or other parts. *Neutrophil granulocytes* often collect around necrotic or mucoid foci in tumours. *Eosinophil granulocytes* are sometimes conspicuous in and around squamous-cell carcinomas of the uterine cervix, alimentary adeno-carcinomas, deposits of Hodgkin's disease, or other growths ; and in some cases there is also eosinophilia of the blood. *Macrophages* laden with lipoids or pigment are common in degenerating or haemorrhagic tumours ; and *foreign-body giant cells* also appear in areas of lipoid absorption or around fragments of degenerating keratin or bone.

(e) Metaplasia in the stroma of tumours

Sometimes, bony or cartilaginous metaplasia occurs in the fibrous stroma of a carcinoma, e.g. of the stomach, intestine (Fig. 143), uterus or prostate. It occasionally occurs also in the stroma of the relatively benign tumours of the

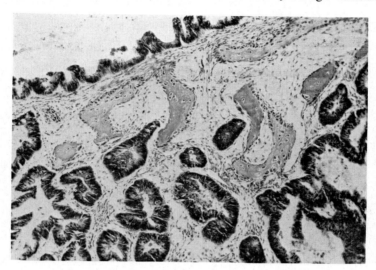

Fig. 143.—Bony metaplasia in the stroma of an adenocarcinoma of the rectum. (× 80.)

salivary glands. In mammary fibro-adenomas (especially in dogs), ossification or chondrification sometimes takes place in the fibromatous tissue which, however, is not strictly stromal but an essential neoplastic component of a mixed tumour (see p. 463).

(f) Sarcomatous change in the stroma of tumours

While occasional mixed carcino-sarcomas occur in the breast (p. 464), uterus (p. 517) and some other organs, it is very doubtful whether sarcomatous change ever occurs in the stroma of a carcinoma, at least in man. In most tumours which have been called " carcino-sarcoma ", the supposedly sarcomatous tissue is only diffusely arranged carcinoma. The great pleomorphism of anaplastic carcinomas and their capacity for simulating the appearances of other tumours are insufficiently appreciated.

(g) Nerves in tumours

There is no evidence that tumour cells are innervated. Nerves often persist deep in the substance of invading tumours ; and nerve fibres found in tumours are residues of these included nerves of the invaded tissues.

THE CHEMISTRY AND METABOLISM OF TUMOURS

(a) The chemical composition of tumours

No distinctive chemical constituents or metabolites have been found in tumour tissue. Tumours contain the same substances as the corresponding normal tissues, but in amounts varying with their degree of differentiation, their secretory activity, and the extent of degenerative changes in them.

(b) The carbohydrate metabolism of tumours

Under both anaerobic and aerobic conditions, rapidly growing tumours show marked glycolytic activity, i.e. the power of splitting sugar into lactic acid. This property is not peculiar to tumours, however, but is exhibited also by some normal tissues, e.g. testis, retina, intestinal mucosa, and bone-marrow. Further, many malignant tumours with moderate rates of growth do not show any significant peculiarities of carbohydrate metabolism.

(c) Constitutional effects of tumours

As already noted on p. 412, the cachexia accompanying some malignant tumours is readily explicable as the result of ulceration, haemorrhage, pain and interference with important bodily functions, and no specific cancerous poisons need be postulated. Changes in the blood or other body fluids of cancer sufferers are non-specific secondary effects ; and their study reveals nothing fundamental regarding the nature or growth of tumours and nothing diagnostic of the presence of " cancer ". Of the 50 or more chemical or serological " cancer tests " which have been advanced, many were patently absurd and none has proved of any value. Since spontaneous tumours are parts of their bearers, their tissues are unlikely to act as antigens, or to have immunological effects.

THE ROUTES OF INVASION BY MALIGNANT TUMOURS

Invading tumour cells take the following paths : (1) tissue spaces, (2) intracellular paths, (3) lymph vessels, (4) veins and capillaries, (5) arteries, (6) coelomic spaces, (7) cerebrospinal spaces, and (8) epithelial cavities.

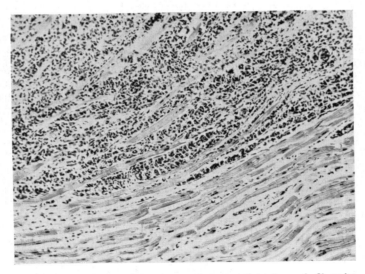

FIG. 144.—Infiltration of tissue spaces between skeletal muscle fibres by an anaplastic carcinoma. (× 100.)

(1) Infiltration of tissue spaces

Microscopically, tumour cells are often seen occupying all the available inter-cellular interstices in the tissues (Fig. 144). This is indeed the initial and funda-mental mode of invasion. It progresses most rapidly in soft tissues with many actual or potential interstices, e.g. muscle, the parenchyma of most viscera, central nervous tissue, or bone-marrow. Conversely, densely compact tissues, such as cartilage, tendons, ligaments and stout visceral capsules, act as more or less effective barriers to invading tumours.

(2) Intracellular infiltration

This occurs most strikingly in striated muscle fibres, which are sometimes seen to contain invading columns of tumour cells replacing the sarcoplasm and distending the sarcolemmal sheaths (Fig. 145).

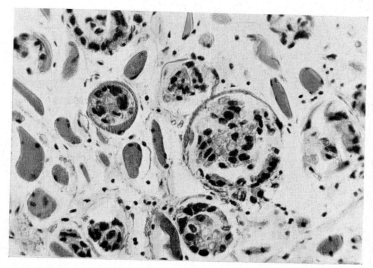

FIG. 145.—Invasion of skeletal muscle fibres by carcinoma of lung.
(× 280.)

(3) Infiltration of lymph vessels : lymphatic permeation

Lymphatics are frequently, and sometimes preferentially, invaded by carcinomas. Striking naked-eye examples of continuous lymphatic permeation are often seen in the serous membranes, especially the visceral pleura and the mesentery, where strands or networks of white growth occupy and distend the subserous lymphatic plexus (Fig. 146). Microscopic examination of the infil-trating margins of carcinomas often reveals the growth of tumour cells along perineural, perivascular or interstitial lymphatics (Fig. 147). In most cases lymphatic permeation extends only for short distances around the main tumour mass, but in others it is surprisingly extensive. Thus, from a small palpable growth in the breast, impalpably permeated lymphatics, perhaps indicated by scattered outcrop-nodules in the skin, may extend widely over the trunk. However, widespread permeation of this kind does not usually proceed in a simple centrifugal manner from the primary focus only : tumour occlusion of the regional lymph glands and main lymph vessels leads to retrograde embolic dissemination of tumour cells in the neighbouring lymphatic plexuses, thus establishing many fresh metastatic foci from which further permeation takes place.

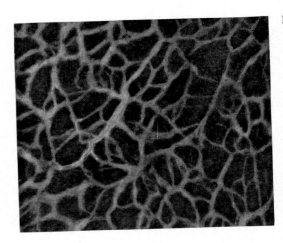

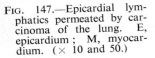

Fig. 146.—Surface view of visceral pleura, showing permeation of subpleural lymphatic plexus by carcinoma. (× 6.)

Fig. 147.—Epicardial lymphatics permeated by carcinoma of the lung. E, epicardium ; M, myocardium. (× 10 and 50.)

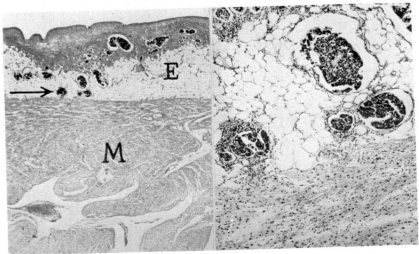

Lymphatic permeation is not confined to small vessels, but occurs also in large lymphatic trunks, such as the main tributaries of the cisterna chyli or the cisterna itself and the thoracic duct. The duct is invaded in about 6 per cent of all fatal cases of intra-abdominal carcinoma, the commonest primary growths being carcinoma of the stomach and carcinoma of the uterus. Invasion of the duct takes place from tumour deposits in the upper retroperitoneal lymph glands,

FIG. 148.—Diagram of thoracic duct and its main connections. B, broncho-mediastinal trunk ; C, coeliac lymph glands ; CC, cisterna chyli; I, intestinal trunk ; J, jugular trunk ; L, lumber lymph glands discharging into lumbar trunks ; M, mesenteric lymph glands ; RD, right lymphatic duct ; S, subclavian trunk ; SC, supraclavicular lymph glands ; T, thoracic lymph glands ; TD, thoracic duct ; X, collateral vessel accompanying duct.

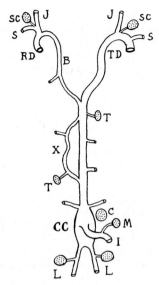

FIG. 149.—Carcinomatous infiltration of small vessels in diaphragm. Some of these are lymphatics ; many others are capillaries and small veins, some of which contain mingled tumour and thrombus. Peritoneum, muscle and pleura are all affected ; and the two serous membranes thickened (× 8.)

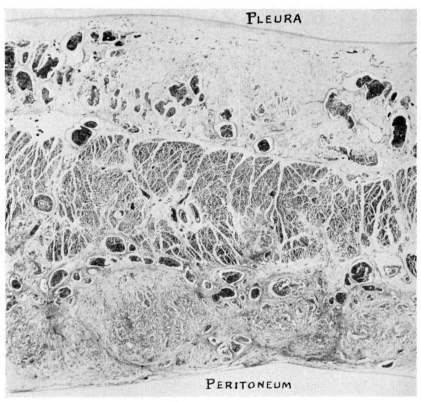

either directly or via the main lumbar or mesenteric tributaries of the cisterna (Fig. 148). The duct may be occupied and distended by tumour for part or the whole of its course, and the cylindrical or fusiform column of growth may be 1 or 2 centimetres in diameter. Permeation usually extends also into the tributaries of the occluded duct and to the neighbouring mediastinal and cervical lymph glands. From a completely occupied duct, the tumour may project into the great veins of the neck. Cancerous occlusion of the thoracic duct is the commonest cause (in non-filarial countries) of chylous ascites or chylothorax, which are due to obstructive ectasia and rupture of tributary lymphatics.

(4) Invasion of veins and capillaries

Infiltrating tumours frequently invade the walls of small or large veins and grow into their lumina (Figs. 149 and 197). Gross invasion of large veins often occurs from carcinomas of the kidney, lung, thyroid, stomach or liver, and from embryonic tumours of the kidney, teratomas of the testis, chondrosarcomas and other sarcomas. The invading tumours sometimes extend within the veins for long distances, e.g. renal carcinomas, testicular teratomas and chondrosarcomas have been observed to grow in continuity up the inferior vena cava into the right cardiac chambers and even through these into the pulmonary arteries. Tumour invasion of a vein often causes thrombosis within the vessel, which thus becomes filled with mingled tumour and clot.

Even more frequent than invasion of large veins is microscopic penetration of capillaries and small veins in invaded tissues—a venous and capillary permeation comparable to that in lymphatics. Sometimes cancerous permeation of small blood-vessels is so abundant as to give the lesions distinctive characters, as in " telangiectatic carcinoma of the breast ", in which affected areas of skin are red and erysipelas-like, due to extensive occlusion of small vessels by growth.

(5) Invasion of arteries

Unlike veins, arteries rarely suffer tumour invasion. Large and small arteries are often seen traversing massive tumours, yet their walls usually remain intact.

(6) Invasion of coelomic spaces

Visceral tumours spread in the serous cavities partly by direct extension and partly by metastatic implantation of detached fragments of growth. Some tumours show a special tendency to spread diffusely within coelomic spaces, e.g. some mucoid carcinomas of the alimentary system or ovary fill all parts of the abdominal cavity with jelly, and some primary and secondary growths in the lungs spread as diffuse sheets in the pleural cavity (see Fig. 183).

(7) Invasion of cerebrospinal spaces

The subarachnoid, ventricular, and especially the perivascular Virchow-Robin spaces are the main routes of spread of primary or secondary tumours in and around the central nervous system (Fig. 150).

(8) Invasion of epithelial cavities

Tumours sometimes extend for long distances within main ducts ; e.g. carcinoma of the kidney may invade the renal pelvis and grow down inside the ureter, even into the bladder ; or carcinoma of the uterus may grow back into the Fallopian tube. The invasion of minute epithelial spaces sometimes plays a part in the local extension of tumours. Thus infiltrating growths in the kidney

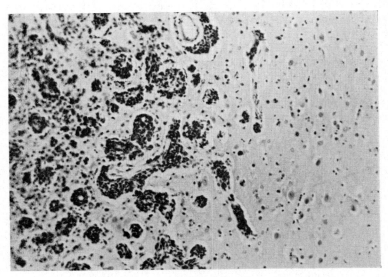

Fig. 150.—Metastatic melanoma of brain extending via perivascular Virchow-Robin spaces. (× 150.)

or testis may extend within the tubules ; and in the lungs tumours often spread from alveolus to alveolus via the septal pores.

THE METASTASIS OF TUMOURS

The preformed paths along which tumours spread in continuity are also the paths through which they metastasize by transport of detached tumour fragments. These paths are : (1) lymph vessels, (2) blood-vessels, (3) coelomic spaces, (4) cerebrospinal spaces, and (5) epithelial cavities.

(1) Metastasis via lymph vessels

(a) *The kinds of tumours which metastasize by lymphatics*

Carcinomas frequently metastasize to neighbouring lymph glands, e.g. from the breast to the axillary glands, from the tongue or pharynx to the upper deep cervical or submandibular lymph glands, or from the penis to the inguinal lymph glands. Lymphosarcomas and other malignant tumours of lymphoid tissue also frequently spread by metastasis from gland to gland. Excluding the tumours of lymphoid tissue itself, sarcomatous secondary growths in lymph glands are infrequent ; but occasionally osteosarcoma, fibrosarcoma, leiomyosarcoma or rhabdomyosarcoma produces lymph-nodal metastases. Malignant melanomas, neuroblastomas, nephroblastomas, and malignant teratomas often metastasize to lymph glands.

(b) Metastasis is embolic

That the metastasis of tumours to the regional lymph glands is initially by lymph-borne tumour emboli is shown by the usual absence of growth in the intervening tissues. Thus, in early cases of carcinoma of the breast with axillary metastases, no tumour cells are to be found in the lymphatics between the primary growth and the axilla, sometimes a distance of 6 inches or more. It is only later that the intervening lymphatics become permeated by tumour. Similarly, microscopical examination of the tissues between a small carcinoma of the tongue and its early metastases in the cervical lymph glands will fail to disclose any permeated lymph vessels connecting the two. A malignant melanoma of the toe may metastasize embolically to the inguinal lymph glands, a distance of three feet or more. Seminomas and other malignant tumours of the testis produce

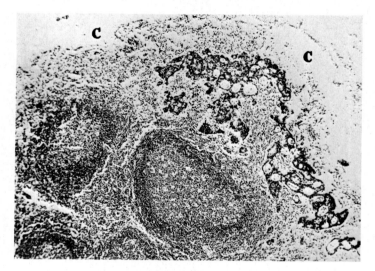

FIG. 151.—Early metastasis of adenocarcinoma of stomach in peripheral sinus of a lymph node. CC, capsule of node. (× 60.)

their first embolic metastases in the upper lumbar lymph glands, the destination of the testicular lymphatics. The afferent lymphatics of lymph glands open into their peripheral sinuses; and this is where early metastatic deposits are situated and should be searched for microscopically (Fig. 151).

(c) Further dissemination from lymph-nodal metastases

Lymph-nodal deposits of tumours are often the source of further dissemination by either the lymph-stream or blood-stream.

(i) *Lymphatic dissemination* takes place from gland to gland. This leads to occlusion of main afferent and efferent lymphatics and consequent diversion of the lymph flow to neighbouring unaffected glands; these now receive any lymph-borne tumour emboli dislodged from either the primary or secondary growths and new metastases develop in them. Retrograde diversions of the lymph flow may carry emboli to unexpected destinations, e.g. to the contralateral lymph glands in some cases of carcinoma of the tongue, tonsil, pharynx or breast.

(ii) *Haemic dissemination* from cancerous lymph glands is frequent and occurs in two ways, (a) via the lymphatic tributaries of veins, and (b) by direct invasion of veins. The most important instances of (a) are seen when the thoracic duct or its main tributaries are invaded ; tumour emboli are carried up the duct into the venous blood and so to the lungs. In most cases in which the thoracic duct is cancerous, the lungs contain many visible metastases or microscopic tumour emboli arrested in the pulmonary arterioles. As to (b), the direct invasion of veins from contiguous cancerous lymph glands, this is not unusual in the internal jugular vein from cancerous cervical glands, in the portal vein from the mesenteric or portal glands, in the superior vena cava or pulmonary veins from the hilar or mediastinal glands, and in the iliac veins or inferior vena cava from the pelvic or retroperitoneal glands.

(d) Lymph-nodal metastases mimicking primary tumours

A primary tumour which is still small and symptomless may nevertheless produce large secondary growths in lymph glands, and these may be mistaken clinically for primary tumours. Large deposits in the cervical lymph glands may be the first clinical signs of small unsuspected primary carcinomas of the tonsil, base of tongue, pharynx, oesophagus or lung. Most supposed " branchiogenic carcinomas " (i.e. carcinomas arising in branchial cysts) are really metastatic growths of this kind. Most, perhaps all, so-called " lateral aberrant thyroids " are really well-differentiated lymph-nodal metastases of small impalpable primary carcinomas in the thyroid. Metastatic growths in the axillary lymph glands may be the first or only signs of mammary, thyroid, bronchial or even intra-abdominal carcinomas. A small unsuspected melanoma of the lower limb or perineum may first announce itself by enlarged inguinal glands, or a small testicular tumour by an abdominal tumour due to metastases in the upper lumbar glands.

(2) Metastasis by the blood-stream

(a) The entry of tumours into the blood

Tumours gain access to the blood-stream in ways which have already been described, namely by invasion of veins (p. 444), or by invasion of the lymphatic tributaries of veins, especially the thoracic duct, cisterna chyli or their main tributaries (p. 442).

(b) The structure, size and number of blood-borne tumour emboli

Tumour emboli consist of small or large fragments of tumour or of mingled tumour and thrombus. Occasionally the emboli detached in invaded main veins are large enough to cause severe or fatal pulmonary embolism, or, if the pulmonary veins are invaded, severe arterial embolism of the limbs, brain or other parts. Much more frequently, emboli are small or microscopic in size (Fig. 152). Tumour emboli are often found arrested in pulmonary arterioles, usually in vessels between 50μ and 200μ in diameter, indicating this to be the commonest sizes of the emboli. The number of tumour particles circulating in the blood-stream must vary greatly from case to case and from time to time. Gross invasion of large vessels no doubt produces plentiful emboli in repeated showers ; and in cases with widespread secondary growths and multiple invasions of vessels in several different organs, superimposed metastatic cycles make the process of dissemination very complex and discharge many tumour particles into the circulation.

(c) *The three main destinations of blood-borne emboli*

Most tumour emboli, consisting of fragments of tumour or tumour and thrombus, are arrested in the first arterioles or capillaries into which they are

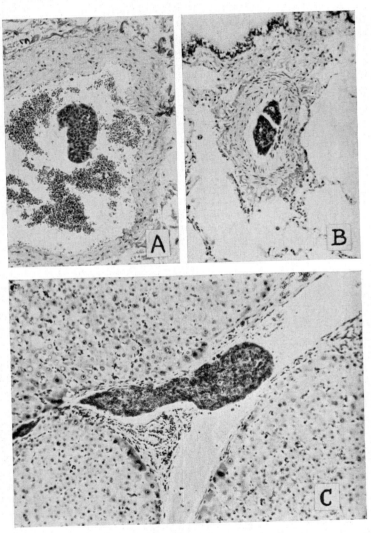

FIG. 152.—Tumour emboli in a case of widely disseminated carcinoma of breast.   A, free tumour embolus in a pulmonary artery ; B, an embolus arrested in a pulmonary arteriole ; C, embolus arrested at bifurcation of a portal branch-vein in liver. (× 100.)

carried. Most emboli liberated in systemic veins or in lymphatic tributaries of veins are arrested in the lungs ; those liberated in the portal venous system, in the liver ; and those liberated in the pulmonary veins or left heart chambers, in any tissue to which they may be carried by the systemic arterial blood.

(i) *Arrest of emboli in the lungs.*—As already mentioned, microscopic examination of the lungs in cases of disseminated tumour frequently reveals arrested tumour emboli in pulmonary arterioles. These are often present, not only in lungs containing obvious metastatic growths, but also in lungs devoid of visible metastases. In such cases the tumour cells may remain dormant in the lungs for long periods, failing to grow there into established secondary tumours, yet being the source of further dissemination by the arterial blood-stream to other organs.

(ii) *Arrest of portal emboli in the liver.*—In cases in which the portal vein or one of its main tributaries is invaded by a tumour, arrested tumour emboli can often be found in the small portal branch-veins in the liver.

(iii) *Arrest of emboli in systemic arterioles or capillaries.*—As might be expected, microscopic evidence of tumour embolism is to be found much less frequently in the systemic arterial system than in the lungs or liver. For, while all the venous blood traverses the lungs and all the portal blood traverses the liver, only a fraction of the arterial blood traverses any given piece of tissue. However, tumour emboli have been seen microscopically in the arterioles or capillaries of the heart, kidneys, adrenals, spleen, bone-marrow and thyroid ; and, as already mentioned, occasional examples of gross occlusion of main arteries by tumour emboli are seen.

(d) Unusual destinations of blood-borne emboli

To the foregoing rules as to the usual sites of arrest of tumour emboli, there are four possible or actual exceptions to consider :

(i) *Intracardial implantation.*—Rarely, fragments of growth carried in the venous blood become attached to the chordae or valve cusps of the right heart chambers and grow there ; or fragments carried from the pulmonary veins become implanted on the endocardium of the left side of the heart.

(ii) *Retrograde venous embolism.*—Retrograde venous embolism following occlusion of main veins with consequent reversal of flow, accounts for some peculiarly situated metastatic growths. Thus, cases have occurred in which carcinoma of the left kidney has invaded and occluded the left renal vein, and retrograde flow of blood down the left ovarian or spermatic vein has then carried tumour emboli into the pelvic, vaginal, vulval or pampiniform veins, and has established metastases in these situations. Similar retrograde venous embolism accounts for the occasional metastases of uterine chorion-epithelioma in the vulva, broad ligament or other parts of the pelvis.

(iii) *Paradoxical or crossed embolism.*—This means systemic arterial embolism by particles liberated in systemic veins but avoiding the lungs by traversing a patent foramen ovale. This must be admitted as a theoretical possibility, but its actual occurrence is very doubtful.

(iv) *Transpulmonary passage of single tumour cells.*—This has been suggested as the explanation of tumour dissemination in systemic tissues without visible metastases in the lungs. But, as already noted, seemingly unaffected lungs often contain microscopic tumour emboli which are abortive or quiescent ; and their presence renders it unnecessary to suppose that the metastases in other tissues arose from single tumour cells which passed uninterruptedly through the pulmonary capillaries—an unlikely event for tumour cells of most sizes.

(e) The frequency of blood-borne metastases in various sites

The three most frequent sites of blood-borne metastases are the liver, lungs and bones ; these contain metastatic deposits in about 40, 30 and 15 per cent respectively of fatal cases of malignant tumours of all kinds. The liver is the most frequent site of metastases—more frequent even than the lungs—because it receives tumour emboli in both its arterial and portal venous blood, and because carcinomas of the stomach and intestine are very common tumours, a high proportion of fatal cases of which disseminate to the liver via the portal blood-stream. In about 50 per cent of necropsies on cases of all malignant tumours in the portal area, hepatic metastases are present.

In cancer necropsies, blood-borne metastases are found in the adrenals in nearly 10 per cent of cases, in the kidneys in about 8 per cent, in the brain in about 6 per cent, in the heart in 5 per cent, and in other sites less frequently.

(f) The seed-soil concept of tumour metastasis

Various facts show that the different tissues of the body vary in their properties as " soils " for the embolic tumour " seeds " which are sown in them. The phenomenon of abortive tumour embolism in the lungs has already been referred to ; tumour cells arrested in the pulmonary arterioles may long remain dormant or may degenerate and disappear. In the liver, however, abortive tumour emboli are rarely or never seen. Liver tissue is clearly a favourable soil for the germination of most kinds of tumour seeds, while lung tissue is for some tumours an unfavourable soil. This is supported by study of the mitotic activity of metastatic growths in various sites ; mitosis is usually much more active in hepatic metastases than in the primary tumour or its metastases elsewhere, sometimes 5-fold or 10-fold. In contrast with the liver, skeletal muscles are remarkable for the rarity of metastatic tumours in them, in spite of their large aggregate blood-supply which must bring to them a correspondingly large share of tumour emboli distributed by the arterial blood. The contrasting properties of various tissues as soils for embolic tumour cells must depend largely on metabolic differences relevant to the requirements of these cells ; and doubtless, different kinds of tumours show different metastatic distributions because they have different chemical requirements. The precise factors concerned in the fate of transported tumour cells in various sites are still unknown ; perhaps they include oxygen supply, carbohydrate supply, and particular tissue enzymes or metabolites.

(g) The appearances of blood-borne metastatic tumours

Most blood-borne metastases appear as well-defined, rounded growths, which are usually multiple in the affected organ but sometimes single, and which vary greatly in size from microscopic foci to masses many inches in diameter. As they enlarge their shapes often become modified, sometimes by their contact with the confining capsule of the viscus or other less permeable structures, and sometimes by diffuse extension of the tumours themselves into the surrounding tissues with consequent loss of marginal definition. Carcinomas of kinds which prefer-entially extend by lymphatic permeation exhibit this character in their secondary sites also ; and affected organs may show widespread permeation or diffuse infiltration with or without visible evidence of the discrete metastatic foci whence this has proceeded. The colour and texture of metastatic growths depend on the nature of the primary tumours—whether papillary, cystic, mucoid, fibrous, bony, melanotic or haemorrhagic, as the case may be. The centres of enlarging secondary tumours often undergo yellow necrosis, cystic or haemorrhagic degen-eration, or fibrosis. These secondary changes account for the frequent umbilica-tion of metastatic tumours where they present on the surfaces of viscera.

(h) Blood-borne metastases in particular sites

Certain other special points about secondary tumours in particular organs are worthy of note.

(i) *In the liver.*—Metastatic tumours here are often very numerous and bulky. Liver weights of 10 pounds are not unusual, and weights twice and even thrice this have been recorded. Invasion of both the portal and hepatic veins by the tumours is common. Invasion of portal branch veins leads to further metastatic dissemination in the liver itself, the distribution of which depends on the size and position of the invaded vessels (Plate IX). Invasion of hepatic veins discharges emboli to the lungs.

(ii) *In the lungs.*—Multiple metastases are usually bilateral, well defined, spherical and of varying sizes. Occasionally, the lungs are studded uniformly by great numbers of small nodules, macroscopically mimicking miliary tuber-culosis. Ill-defined bronchopneumonia-like consolidation is a rare form of metastatic growth. Subpleural metastases often assume plaque-like or stellate shapes. Intrapleural dissemination of the tumours and serous or blood-stained pleural effusion are common. In the lung, small or large pulmonary veins are frequently invaded ; and occasionally a main bronchus is invaded, and the appear-ances of a primary bronchial carcinoma are simulated.

(iii) *In bones.*—Metastatic tumours in bones are almost always situated in red bone-marrow ; and their common sites are therefore the common sites of red bone-marrow, which in adults are the vertebrae, ribs, sternum, pelvic bones, diploe of the skull, and the proximal ends of the humerus and femur. Other bones are relatively infrequent sites of metastases, except in cases with wide-spread secondary growths in the skeleton. The effects of metastatic tumours on bone structure—whether causing destruction, osteoplasia, periosteal new bone formation, or none of these—are described on p. 437. Pathological fractures are common. Extensive replacement of bone-marrow by secondary growths may produce severe anaemia, often resembling pernicious anaemia except that there is usually a leucocytosis. This leucocytosis may be so pronounced (e.g. with leucocyte counts of 50,000 or more per cubic millimetre) that leukaemia may be simulated. Other effects of tumour replacement of bone-marrow include thrombocytopenia with a resulting haemorrhagic state, heterotopic myeloid haemopoiesis in the spleen or elsewhere, and very rarely Bence-Jones proteosuria. Widespread bone destruction may also cause hypercalcaemia, metastatic calcifica-tion in lungs or other parts, and parathyroid hyperplasia (see Fig. 119, p. 361).

(iv) *In the brain* (Fig. 153).—Metastatic tumours in the brain are usually multiple, but sometimes solitary. They occur in the cerebrum, cerebellum and brain stem in that order of frequency ; and in the cerebrum they are usually situated in grey matter or at the junction of grey and white matter, where the vascularity is greatest. They are of great clinical importance, not only because of their serious effects, but because they may occur from symptomless primary growths and so may easily simulate primary disease of the brain. This happens most often from bronchial carcinoma, which is notorious both for the great

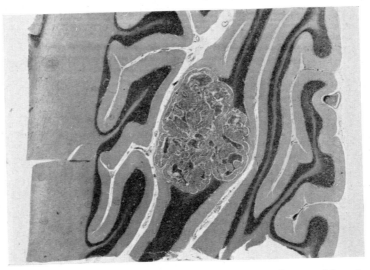

FIG. 153.—Small metastasis in cerebellum from carcinoma of breast.
(× 10.)

frequency with which it metastasizes to the brain and for the diagnostic difficulties occasioned by relatively symptomless primary growths.

(v) *In the adrenals.*—Metastatic tumours are almost always situated in the medulla, and are usually multiple and often bilateral. They are especially frequent from bronchial carcinoma. They rarely or never lead to total destruction of the adrenal cortex, and therefore rarely cause Addison's syndrome.

(vi) *In the kidneys.*—Metastatic tumours are almost always situated in the cortex, and are usually multiple and bilateral. They rarely attain clinically important sizes or produce symptoms.

(vii) *In the heart.*—Metastatic tumours in the myocardium occur in any part and are usually multiple. They may invade the endocardium of either the left or right chambers, whence they discharge prolific tumour emboli into the systemic arterial stream or to the lungs.

(viii) *In the spleen.*—The spleen is a less frequent site of metastasis than might be expected from its vascularity. The metastases are usually multiple, small and well defined ; but occasionally the spleen suffers diffuse metastatic infiltration with considerable enlargement.

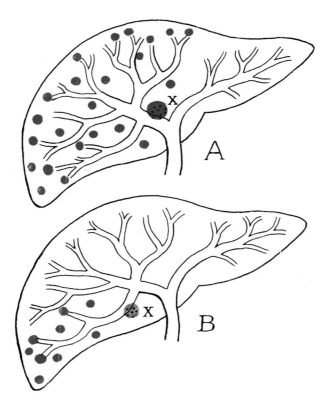

PLATE IX.—Diagram of intrahepatic metastasis. A, tumour X
has invaded a main branch of portal vein and produced
daughter tumours throughout lobe. B, tumour X has
invaded a smaller branch vein and produced daughter tumours
in a smaller area.

(*To face page 452*)

(ix) *Other sites.*—No tissue is exempt from metastatic tumours. These occur in the skin and subcutaneous tissues, thyroid, pituitary, oral or gastro-intestinal mucous membranes, eye, uterus, muscles, and any other tissues (Fig. 154). In a pregnant woman, they may even occur in the placenta ; and cases have been recorded in which maternal melanoma with placental deposits metastasized also to the foetus.

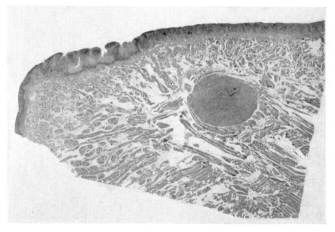

FIG. 154.—Metastasis in muscular substance of tongue from carcinoma of pharynx. (× 4.)

(*i*) Blood-borne metastases simulating primary tumours

It is not unusual for a small primary tumour to produce its first symptoms by its metastases in liver, lungs, bones, brain or some other site. We have already mentioned the clinical simulation of primary disease of the brain by the metastases of latent bronchial carcinoma ; and many similar diagnostic difficulties and errors occur. A large liver may be the first or only sign of a small gastric or intestinal carcinoma. A tumour or a pathological fracture of a bone may be the first sign of a carcinoma of the kidney, thyroid, lung, or even of the breast, or of a neuroblastoma in a child or adolescent. The first evidence of an otherwise symptomless gastric carcinoma may be a pernicious-anaemia-like state due to widespread metastases in bone-marrow. Pleural effusion, haemoptysis or other symptoms due to metastatic growths in the lung may be the first signs of gastric or other abdominal carcinomas, testicular tumours or chorion-epithelioma. The primary growths which most frequently mislead clinicians in this way are carcinomas of the lung, kidney, thyroid and stomach.

(3) Transcoelomic metastasis

Primary or secondary tumours of the thoracic or abdominal viscera frequently spread to the serous surfaces and disseminate in the serous cavities. Free tumour cells or cell-aggregates are often present in the effusions accompanying plueral or peritoneal carcinomatosis. In cases of glandular carcinoma, signet-ring cells, each with a large droplet of mucus, or even complete glandular acini may be recognizable in the fluid. Occasionally, free-floating carcinomatous vesicles are formed (Fig. 155). These free tumour particles are undoubtedly the source of the widespread and very numerous metastatic implants which develop on the serous surfaces in many cases ; and all stages in their implantation and growth can be traced microscopically.

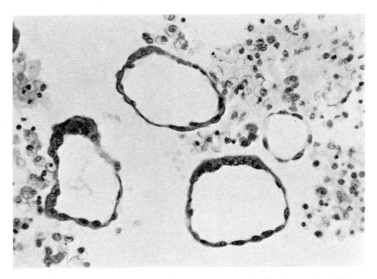

FIG. 155.—Carcinomatous vesicles in pleural effusion. (× 250.)

(a) *The distribution of transcoelomic metastases*

Doubtless owing to the effect of gravity in the usual erect or semi-recumbent postures, the earliest or most plentiful transperitoneal deposits are often found in the pelvis, especially in the recto-vesical or recto-uterine pouch. The ovaries are a frequent site of transperitoneal metastases ; and all stages can be traced in the implantation of tumours on the ovarian surface and their extension into the ovarian substance. Ovarian metastases often attain large sizes, sometimes of several pounds ; and, when the primary tumour in the stomach or elsewhere has produced few or no symptoms, the large ovarian growths have often been mistaken for primary ones. Other frequent sites of transperitoneal metastasis are along the attachment of the mesentery to the small intestine, the omentum, and the under-surface of the diaphragm ; but no part of the peritoneum is exempt, and even hernial sacs may be affected. In the pleural cavity, metastatic deposits may be found in all parts, but are especially abundant in the costo-phrenic recesses, on the surface of the diaphragm or along the paravertebral gutters. In the pericardial cavity, discrete nodular transplants are not common, and must be distinguished from outcrops from permeated epicardial lymphatics.

(b) *Results of secondary tumours in serous membranes*

The main results are (a) effusions, (b) adhesions and (c) the accumulation of mucus. *Effusions*, serous or haemorrhagic, are frequent, resulting partly from cancerous occlusion of lymphatics and veins and partly from irritative exudation. A persistent haemorrhagic effusion always strongly suggests malignant disease. *Adhesions* may become very extensive, binding the viscera densely together or to the parietes by mingled growth and reactive fibrosis. *Accumulation of mucus* often occurs in the peritoneal cavity, from mucus-secreting glandular carcinomas of the stomach, intestine or ovary. When these discharge great quantities of free mucus into the cavity, the condition has sometimes been called " pseudo-myxoma peritonei ", a condition which is also produced occasionally by a benign mucus-secreting cystic ovarian tumour which has ruptured or by a perforated mucocele of the appendix.

(c) Secondary tumours of serous membranes simulating primary ones

The first clinical signs of an otherwise symptomless visceral tumour may be produced by its secondary extensions in the adjoining serous cavity. " Pleurisy " with effusion may be the first sign of primary or secondary tumours of the lung. Ascites due to peritoneal dissemination may be the first sign of a gastric, ovarian or other intra-abdominal tumour. The mistaking of large secondary growths in the ovary for primary ones has already been mentioned. Bulky secondary deposits in the pouch of Douglas have occasionally led to a misdiagnosis of carcinoma of the rectum. Even after pathological biopsy or necropsy examination, tumours in the serous membranes have often been mistakenly called primary " endotheliomas " or " mesotheliomas ", when they were really only secondary growths from undetected primary tumours elsewhere. It is possible that primary coelomic tumours exist ; but it is certain that in most of the supposed instances which have been reported this diagnosis was erroneous.

(4) Metastasis in cerebrospinal cavities

Any malignant tumour, primary or secondary, in the brain or meninges, encroaching on any surface bathed by cerebrospinal fluid, may discharge tumour cells into the fluid and so cause implant metastases on near or remote surfaces of the meninges or ventricles. This occurs with various kinds of glioma, choroid plexus tumours, retinoblastoma which has invaded the cranial cavity via the optic nerve, pineal tumours, and metastatic tumours in the brain. The commonest sites of the transplants are around the cauda equina, on the dorsal surface of the spinal cord, and around the base of the brain. Tumour cells can often be identified in cerebrospinal fluid during life.

(5) Metastasis by implantation on epithelial surfaces

Medical records contain many supposed instances of metastasis by implantation of detached tumour cells on epithelial surfaces. But this interpretation must be rejected in most cases, in favour of metastasis by more likely and well-known routes or the development of multiple independent tumours.

(a) On skin and other squamous epithelia

Contact carcinomas of opposed epithelial surfaces are not uncommon on the skin, vulva, lips, tongue and cheek, larynx, or conjunctiva. Transplantation of tumour cells can be rejected for all of these ; the contact tumours are clearly due to multiple tumour formation in areas of similarly predisposed tissue.

(b) In the alimentary tract

The presence of mucus, digestive ferments and bacteria makes it highly improbable that tumour particles could ever engraft themselves in any part of the alimentary canal. Alleged instances of this, e.g. a carcinoma of the rectum following one of the proximal colon or a carcinoma of the colon following a carcinoma of the stomach, are all clearly due either to multiple tumour formation or to transcoelomic or vascular metastasis of the initial tumour.

(c) In the respiratory system

Very rare instances of intrabronchial implantation of tumours of the larynx or of adamantinoma have been recorded.

(d) In the female genital tract

There is no doubt that occasionally ovarian or other intra-abdominal carcinomas produce implants in the mucosa of the Fallopian tube, and perhaps also in the endometrium. But supposed instances of implant metastases of uterine tumours in the vagina or vulva are all unconvincing; these probably came by submucous extension or by local embolic metastasis in lymphatics or veins.

(e) In the urinary tract

Papillary growths in the renal pelvis or ureter are often accompanied by " secondary " growths in the lower part of the ureter or bladder. Are these due to surface implantation or to multicentric primary origin? It is impossible to be sure: multiple tumour formation is certainly frequent in the urinary tract, but it is possible that metastasis by implantation also occurs.

SUPPLEMENTARY READING

References Chapter 28 (p. 415), and the following:

Cheatle, G. L. (1920). " Benign and malignant changes in duct epithelium of the breast ", *Brit. J. Surg.*, **8**, 285.

Cowdry, E. V. (1940). " Properties of cancer cells ", *Arch. Path.*, **30**, 1245.

Foulds, L. (1940). " The histological analysis of tumours. A critical review ", *Amer. J. Cancer*, **39**, 1. (A valuable discussion of tumour structure, stroma, individuality, and carcino-sarcoma.)

Gray, J. H. (1939). " The relation of lymphatic vessels to the spread of cancer ", *Brit. J. Surg.*, **26**, 462.

Muir, R. (1941). " The evolution of carcinoma of the mamma ", *J. Path. Bact.*, **42**, 155.

Willis, R. A. (1952). *The Spread of Tumours in the Human Body.* London; Butterworths. (For details regarding spread and metastasis.)

CHAPTER 31

EPITHELIAL TUMOURS

IN THIS chapter we consider Group I of the classification adopted on p. 414, the tumours of epithelial tissues. This will be done regionally, describing first the benign and then the malignant growths of each organ or tissue. However, some introductory comments on the names of epithelial tumours are necessary.

THE NOMENCLATURE OF EPITHELIAL TUMOURS

(1) Papilloma

A *papilloma* is a benign non-invasive tumour of any surface epithelium. Papillomas are so called because their structure shows in exaggerated fashion the papillary patterns which characterize the normal surface epithelia ; i.e. they consist of projecting, elongated, sometimes branched, connective tissue cores clothed by the proliferating neoplastic epithelium (see Figs. 163, 164, 172, 200, 201, 203).

While it is justifiable and necessary to retain the name "papilloma" for well-differentiated non-invasive tumours, it is important to recognize that, in all situations in which papillary growths occur, no sharp separation of benign papillomas from papillary carcinomas is possible. In the skin, breast ducts, alimentary canal, ovary, genital and urinary tracts, tumours showing all degrees of transition between simple non-invasive papillomas and invasive carcinomas are encountered ; and a tumour which long remains a non-invasive papilloma may later become more aggressive, invasive, and therefore carcinomatous. It is better then to think of all these tumours as " papillary growths ", to recognize that many such growths have at least potentially malignant qualities, and to learn by their situations and structure which ones are likely to prove dangerous and which ones relatively innocuous.

(2) Adenoma

An *adenoma* is a benign tumour composed of epithelial glandular tissue. It usually consists of glands closely resembling those of the normal tissue, either exocrine or endocrine, and it often elaborates the secretion appropriate to the tissue (see Figs. 175, 190, 198, 211, 214). Unlike papillomas, adenomas of most tissues are much more easily distinguishable from the carcinomas of the same tissues ; i.e. adenoma and adenocarcinoma are in most organs fairly distinct kinds of growths, and we do not often see an adenoma become carcinomatous.

The term *fibro-adenoma* for the common adenomas of the breast deserves comment. It denotes the truly composite character of these growths, which contain both fibromatous connective tissue and adenomatous epithelial tissue. The connective tissue in these tumours is not a mere stroma to the epithelial elements, but has neoplastic properties of its own.

Cystadenoma is a convenient name for any benign glandular tumours with a prominently cystic structure, such as commonly occur in the ovary, and occasionally in the pancreas or breast.

(3) **Carcinoma**

Carcinoma embraces all malignant epithelial tumours. Since these vary greatly in structure according to the nature of the particular epithelia from which they spring and also according to the degree of differentiation which they attain, a number of qualifying adjectives are used to describe the main structural variants encountered. These are :

    (i) Squamous-cell carcinoma ;
    (ii) Basal-cell carcinoma ;
    (iii) Adenocarcinoma ;
    (iv) Carcinoma simplex, either (*a*) alveolar (i.e. clumped) or (*b*) diffuse.

    (i) *Squamous-cell carcinoma*, also called *epidermoid carcinoma, acanthoma* (i.e. spinous-cell carcinoma) and *epithelioma*, applies to any carcinoma in which there are recognizable squamous stratified characters, i.e. spinous-cells or keratinization (Figs. 156, 177, 206). Most tumours with these characters arise from

Fig. 156.—Typical squamous-cell carcinoma of skin. (× 40.)

epithelia which are normally stratified and squamous, namely those of the skin, oral cavity, pharynx, larynx, oesophagus, uterine cervix and vagina. But sometimes carcinomas of parts lined by other kinds of epithelia show partial or total squamous metaplasia ; this applies to some carcinomas arising from normally non-cornifying but stratified or pseudo-stratified epithelia, as in the nasal cavity, bronchi and urinary tract, and also to some adenocarcinomas, e.g. of the endometrium or gall-bladder.

    (ii) *Basal-cell carcinoma* is the name applied to certain slowly growing tumours of the epidermis or its appendages, which nevertheless fail to show keratinization or other epidermoid characters (see Figs. 137, 138, 173, 174). Some writers have extended the name to include somewhat similar growths of the oral, nasal and pharyngeal epithelia or the glands connected with them ; but the wisdom of this practice is doubtful.

(iii) *Adenocarcinoma* means any carcinoma in which glandular structure is recognizable (Figs. 157, 184, 188, 195, 199, 207, 213). This applies to most carcinomas of the stomach, intestine, pancreas, biliary tract, prostate, endo-metrium and thyroid, and to many of those of the breast, bronchi and kidney. Many adenocarcinomas are recognized at a glance by their obviously acinar or tubular structure, or by their secretion of mucus. Those in which mucus-secretion is abundant are called mucoid (or "colloid") adenocarcinomas. In

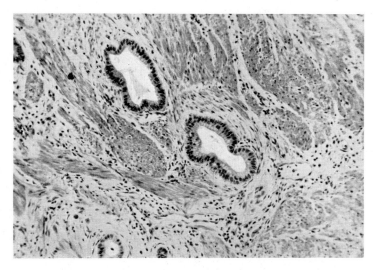

Fig. 157.—Well differentiated adenocarcinoma of stomach invading muscle coat. (× 120.)

some mucus-secreting tumours or parts of them, the cells are not arranged in acini but are scattered through the tissues, and each cell is rounded and distended by a droplet of mucus ; this structure is called " signet-ring-cell carcinoma ". Each " signet-ring cell " may be regarded as a detached, and therefore spherical, cancerous goblet cell (see Fig. 188). As mentioned above, some adenocarcinomas, e.g. of the endometrium or gall-bladder, exhibit partial squamous metaplasia ; these tumours are sometimes called " adeno-acanthomas ". Some adeno-carcinomas are conspicuously papillary or cystic.

(iv) *Carcinoma simplex* is applicable to any anaplastic carcinoma, or part of one, in which none of the foregoing characters is recognizable. The tumours consist merely of masses of cancerous epithelial tissue devoid of any differentiated features. Any epithelial organ may give origin to such tumours ; they are most common in the breast, lung, stomach and kidney. Sometimes carcinoma simplex consists of distinct clumps of polehedral epithelial cells, well defined in the meshes of their connective tissue stroma ; this structure is often called *polyhedral-* (or *spheroidal-*) *cell carcinoma*, or *alveolar carcinoma* (Fig. 158). If the cancerous

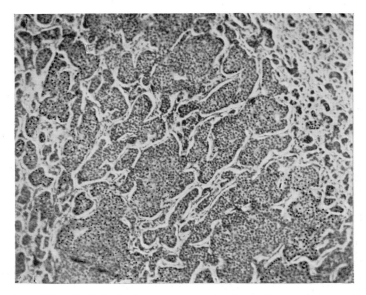

FIG. 158.—Typical alveolar carcinoma simplex of breast, composed of plentiful epithelial masses with a slight amount of stroma. (× 60.)

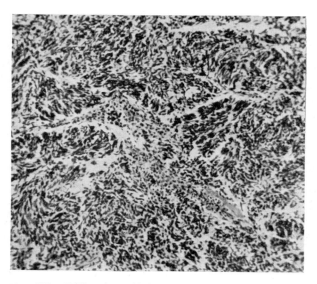

FIG. 159.—Diffuse bronchial carcinoma composed of small fusiform cells. (× 80.)

epithelium is plentiful and the stroma scanty, the tumour is soft and brain-like in consistency, whence the descriptive adjectives " encephaloid " and " medullary ". If the epithelium is scanty and the stroma abundant and fibrous, the tumour is hard or " scirrhous ". Sometimes carcinoma simplex is so anaplastic that its cells lose even their tendency to form distinct epithelial clumps and grow only as disorganized masses mingling confusedly with the invaded tissues ; this is *diffuse carcinoma simplex* (Fig. 159).

It is important to repeat that the foregoing names denote only varieties of structure and not distinct species of growths. Several of them may often be seen in a single tumour. Thus a carcinoma of the breast may show cystic papillary adenocarcinoma, acinar carcinoma, signet-ring-cell carcinoma, and alveolar carcinoma simplex in different places. In different parts of a bronchial carcinoma, squamous-cell structure, adenocarcinomatous structure, and both alveolar and diffuse carcinoma simplex may be found. The species are "mammary carcinoma" and " bronchial carcinoma " ; all the other names are merely descriptive of the structures found in particular members of the species (see also p. 435).

EPITHELIAL TUMOURS OF THE BREAST

Epithelial tumours of the breast comprise :
   (1) Fibro-adenoma ;
   (2) Duct papilloma and papillary cystadenoma ;
   (3) Carcinoma ;
   (4) Paget's disease of the nipple.

(1) Fibro-adenoma

Practically all adenomas of the breast are fibro-adenomas, the connective tissue component of which also has neoplastic qualities ; it is doubtful if a pure adenoma ever occurs. Fibro-adenomas are common tumours in young and middle-aged women ; they rarely develop before puberty or in old women, or in the male breast. They arise in areas of breast tissue which are the seat of lobular hyperplasia of adolescent type (" mazoplasia " or " adenosis "), and all transitions can be traced between hyperplastic breast lobules and young fibro-adenomas (Fig. 160).

(a) Macroscopic appearanc

Fibro-adenomas are well circumscribed, rounded or lobulated ; and on section they are usually uniform in texture, sometimes resembling soft fibrous tissue, and sometimes presenting on the cut surface a finely foliated pattern like a section of a crinkled cabbage. Old or large tumours are often more heterogeneous, with densely fibrous or cystic areas mingled with those of more usual appearance.

(b) Microscopic structure (Figs. 161, 162)

Two structural patterns occur in fibro-adenomas, pericanalicular and intracanalicular. But these are not two distinct kinds of tumour ; many fibro-adenomas show both kinds of structure. The *pericanalicular structure* shows simple ducts of nearly normal appearance set in a variable amount of loose-textured periductal connective tissue. Rarely, the fibromatous component is scanty and the tumour is almost a pure adenoma. Usually the fibromatous tissue exceeds the epithelial in bulk. The *intracanalicular structure* arises by proliferation of the fibromatous tissue in a coarsely papillary pattern, with consequent invagination and distortion of the epithelial spaces into interlocking dove-tailed and folded shapes, which give the cut surface its naked-eye appearance

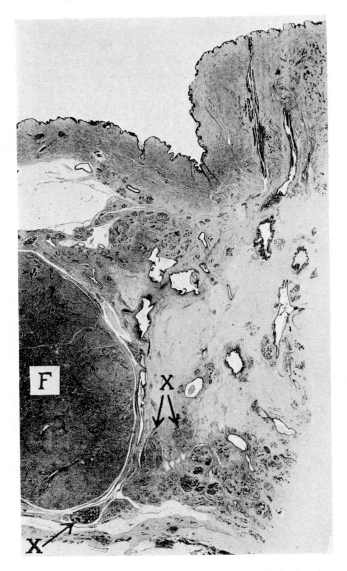

FIG. 160.—Part of large section of breast through nipple, showing a fibro-adenoma F, several satellite fibro-adenomatous foci XX, and lobular hyperplasia (adenosis) and dilatation of ducts elsewhere in breast. (× 4.)

of a crinkled cabbage. Young tumours show soft loose-textured fibromatous tissue, but in old ones this often becomes dense and hyaline in part. In some tumours the fibromatous tissue is distinctly mucinous and slimy, justifying the name "myxofibro-adenoma". Rarely in human tumours, but frequently in canine ones, cartilaginous or bony metaplasia takes place in the connective tissue component.

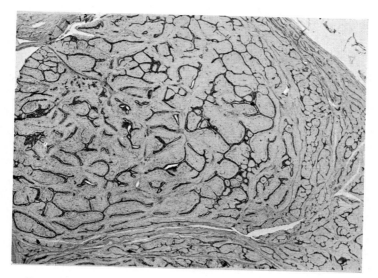

FIG. 161.—Fibro-adenoma of intracanalicular pattern. (× 10.)

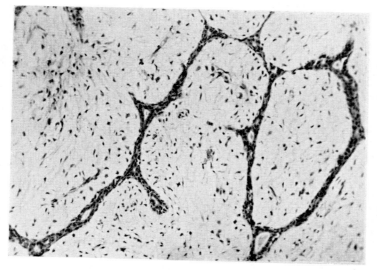

FIG. 162.—Enlarged view of Fig. 161. (× 120.)

(c) Giant fibro-adenoma

Occasionally, in a middle-aged or elderly woman, a previously small fibro-adenoma suddenly commences to grow rapidly and may attain a weight of several pounds in a few months. It often becomes cystic and degenerated in parts, and it sometimes suffers accidental ulceration and fungation through the distended overlying skin. Such tumours show luxuriant fibro-adenomatous structure, usually of the intracanalicular type, but with an unusually cellular sarcoma-like appearance of the fibromatous component. In spite of this, however, in most cases the growths remain non-invasive and non-metastasizing, and are cured by surgical removal.

(d) Malignant change in fibro-adenoma

While most huge rapidly growing fibro-adenomas are benign, occasional tumours of this kind are found to have undergone malignant change. This is usually in the fibromatous component, producing a fibrosarcoma or, rarely in the human breast, a chondrosarcoma or osteosarcoma. Carcinoma developing in a fibro-adenoma is exceedingly rare; so is composite carcino-sarcoma.

(2) Duct papilloma and papillary cystadenoma

Papillary tumours of the breast, like those of other regions, cannot be sharply separated into benign and malignant varieties. All gradations occur between solitary, slowly growing, highly organized papillomas of the main ducts and multiple, poorly organized, active papillary growths associated with, and showing transitions to, invasive carcinoma.

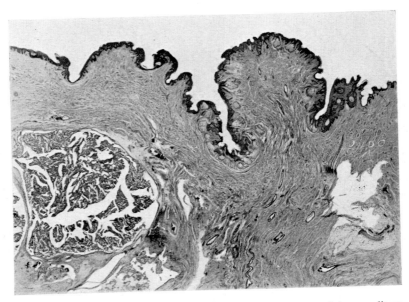

Fig. 163.—Part of large section of breast through nipple, containing a solitary well-circumscribed intracystic papilloma. (× 6.)

Most solitary benign papillomas occur in young or middle-aged women, and in the main ducts close to the nipple. They may be sessile or stalked, and the stalked ones may have one or several pedicles of attachment. Those with a single stalk may show a simple villous structure, but those with multiple stalks have reticular or lattice-like patterns (Figs. 163, 164). Simple distension of part of a duct by a single papilloma is less frequent than the formation of a " papillary

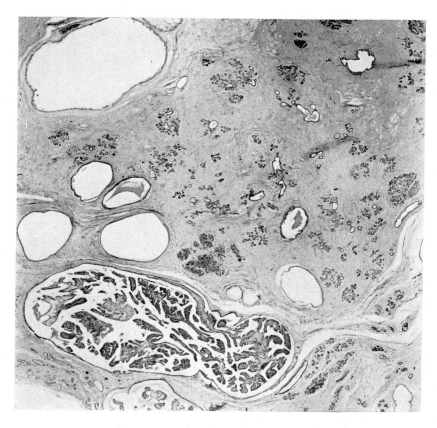

Fig. 164.—Part of large section of breast showing hyperplasia of ducts and lobules, many small cysts, and a papilloma in a dilated duct. (× 10.)

cystadenoma ", i.e. irregular clefts or cysts containing multiple papillary growths. Sessile, multiple and irregularly cystic growths are more likely than solitary pedunculated papillomas to become cancerous.

The chief symptom of papillomas of the main ducts is bleeding from the nipple ; and the particular duct involved can often be identified by gentle localized pressure exerted at different points in turn around the nipple, blood-stained fluid being expressed when pressure is applied over the affected duct.

(3) Carcinoma of the breast

Carcinoma of the breast is one of the commonest kinds of cancer in women —indeed the commonest in English women, in whom it exceeds both gastric and uterine carcinoma in frequency. It is infrequent before the age of 30 years, after which its incidence rises steeply. Only about 1 in 100 cases of mammary cancer is in a man. Single and sterile women are more liable to carcinoma of the breast than women who have borne and normally suckled children. Thwarted reproduction and thwarted lactation predispose to the disease ; and the high, and apparently increasing, prevalence of breast cancer in Western civilized communities is probably related to the unnatural reproductive life of a large proportion of the people.

(a) The relationship of carcinoma to cystic hyperplasia

The only well-recognized pre-cancerous lesion in the breast is cystic hyperplasia, often wrongly called " chronic cystic mastitis ". This is a non-inflammatory disease characterized by proliferation of duct epithelium, obstruction of the ducts with the formation of small or large cysts, and widespread fibrosis of the breast. Its precise causation is not clear, but there is little doubt that it is a hyperplastic process due to disturbed hormonal control, in which oestrogen excess is probably the main factor. Careful pathological examination of cancerous breasts discloses evidence of previous cystic hyperplasia (which has usually *not* been recognized clinically) in a considerable proportion of cases—somewhere between 20 and 50 per cent—often with clear transitions from one to the other. For good accounts of cystic hyperplasia and the evolution of carcinoma from it, the papers of Cheatle, Keynes, Charteris, and Muir should be consulted. Carcinoma supervening on hyperplasia is often multicentric in origin, arising from extensive reaches of one or more ducts or from the whole breast or even from both breasts. Should a woman with detectable hyperplasia of both breasts, however slight, develop carcinoma in one, the risk of carcinoma supervening in the other also is so great as to justify simple amputation of the second organ.

Some surgeons have denied the pre-cancerous potentialities of cystic hyperplasia ; but their denial rests on three false bases. (a) Under the fallacious term " chronic mastitis ", many surgeons continue to group all lumpy breasts irrespective of their pathology : hence they fail to distinguish the innocuous non-cystic lobular type of hyperplasia (" mazoplasia " or " adenosis ") which is common in adolescents and which often subsides, from the more dangerous cystic type of hyperplasia which usually develops after the age of 30, is irresolvable, and which may lead to gross cystic enlargement (Schimmelbusch's disease) or to carcinoma. (b) The second false reason for the denial of the pre-cancerous proclivities of cystic hyperplasia is that most women with cancerous breasts give no history of any clinically detected previous breast disease, and the clinician has not diagnosed the sequence "hyperplasia"→"carcinoma". But examination of excised breasts often shows widespread hyperplasia and sometimes multicentric carcinomas which were unsuspected clinically. For every case in which a *clinical* diagnosis of cystic hyperplasia antecedent to carcinoma is made, there are a score in which a *pathological* diagnosis of this sequence is plain from adequate study of amputated breasts. (c) The third reason for the idea that cystic hyperplasia is harmless is inadequate study of cancerous breasts. Microscopical examination of only

Fig. 165.—Part of large section of breast through nipple, showing widespread, finely cystic hyperplasia with supervening carcinoma in the part indicated by the arrows. (See Fig. 166.) (× 2½.)

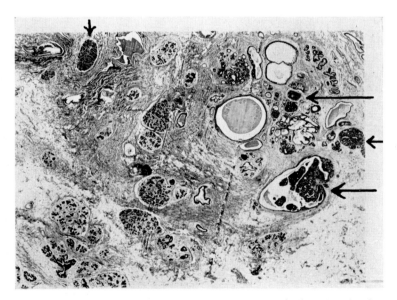

Fig. 166.—Enlarged view from the cancerous part of Fig. 165, showing cystic hyperplasia and supervening multifocal intra-ductal papillary and solid carcinoma, indicated by arrows. (× 10.)

one or two small sections, usually taken only from areas of obvious growth for diagnostic purposes, will often not suffice to reveal the relationship under discussion, and will thus lead to the false conclusion that it does not often exist. The more thoroughly the pathologist examines cancerous breasts, the more frequently he will discover cystic hyperplasia associated with carcinoma, with transitions from one to the other (Figs. 165, 166).

(b) The structure of mammary carcinoma

Different tumours show widely diverse kinds of structure, and this often applies also to different parts of a single tumour (see Fig. 140). The main structural variants can be grouped conveniently thus :

(i) *Intraductal and intra-acinar carcinoma* (Figs. 140, 167, 168, 171.)—Cheatle first stressed the frequency of pre-invasive carcinoma, still confined within duct or acinar boundaries. The epithelium of short or long reaches of the ducts shows carcinomatous change *in situ*, and the duct lumina are distended by the proliferating cancerous epithelium. This assumes various patterns : *papillary carcinoma*, not sharply separable from simple papilloma ; *cribriform* (or *laciform*) *carcinoma*, in which the epithelium is riddled by rounded or irregular spaces ; and *solid intraductal carcinoma* completely filling the distended ducts. All of these may be seen in one breast, and may coexist with invasive carcinoma. When terminal breast lobules become replaced and distended by growth, this is often called *intra-acinar* ; but in most cases it is the terminal ductules rather than the acini which are affected, and it is pointless to try to distinguish sharply between ductular and acinar growth.

(ii) *Extraductal infiltrative carcinoma* (Figs. 140, 158).—Invasive tumours may develop by escape of previously intraductal growths, or may be invasive from the beginning. Their structure is diverse, including *adenocarcinoma* of acinar, papillary or cystic types; *alveolar polyhedral-cell* (*spheroidal-cell*) *carcinoma simplex*, with either abundant dense stroma (" scirrhous " carcinoma) or scanty stroma (" medullary " or " encephaloid " carcinoma) ; and *diffuse anaplastic carcinoma* nearly or wholly devoid of epithelial arrangement. Several or all of these variants may appear in one tumour (see Fig. 140).

(iii) *Metaplastic types of carcinoma.*—These also are not separate types, but merely variants of the preceding types. They are : (*a*) *mucoid carcinoma*, including *signet-ring-cell carcinoma*, characterized by the secretion of much or little mucus by the epithelial cells, and a more or less gelatinous ("colloid") naked-eye appearance of the tumours ; and (*b*) *squamous-cell carcinoma*, a relatively rare type of structure (Fig. 140).

(c) The local spread of mammary carcinoma

Little need be added to what was said in Chapter 30. Mammary carcinoma readily invades and spreads along all the available paths—tissue spaces, lymphatics, blood-vessels and ducts. While the tumour is still confined to the breast itself, it produces no change in the overlying skin and remains freely mobile on the underlying muscle. Later, however, it becomes adherent to either or both of these ; and the skin may ulcerate. Centrally situated carcinomas frequently cause retraction of the nipple, due to stromal fibrosis and contraction accompanying the tumours. Permeation and local embolic spread in the deep dermal

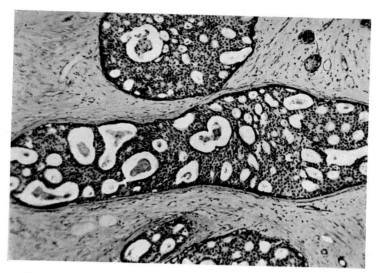

FIG. 167.—Intraductal carcinoma of cribriform pattern. (× 80.)

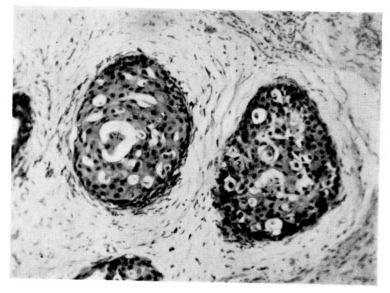

FIG. 168.—Solid intraductal carcinoma. (× 120.)

lymphatics are responsible for the lymphatic oedema of the overlying skin, to which the name " peau d'orange " is applied (p. 386), and also for the development of satellite tumour nodules in the surrounding skin and fascia in advanced stages of the disease. Another late result is *cancer en cuirasse* (" breast-plate cancer ") characterized by extensive diffuse infiltration, with great thickening and rigidity, of the skin and underlying tissues of much or all of the thoracic wall.

(d) Metastasis to regional lymph glands

The commonest site of metastases is the axillary group of lymph glands. Routine microscopic examination shows that these contain growth in from 50 to 70 per cent of surgically treated cases, and in about 80 per cent of necropsy cases. Of all the factors influencing post-operative prognosis, the condition of the axillary lymph glands is the most important : when the glands are unaffected, the 5-year survival rate is about 60 per cent ; but when the glands are affected, only about 25 per cent of the patients survive 5 years after operation. Careful dissection and microscopic examination are essential to determine the presence or absence of early axillary metastases (see Fig. 151) ; clinically impalpable glands may contain metastases, and enlarged glands may contain none. Occasionally, large axillary metastases attract attention before any abnormality is clinically apparent in the breast. It has been found that in some early operable cases, metastasis to the internal mammary lymph glands has already occurred, even before the axillary glands are diseased. Spread to the cervical lymph glands always denotes a late inoperable stage of the disease.

(e) Metastasis by the blood-stream

Remote metastases are present in about three-quarters of fatal cases. The lungs, liver and bones are their commonest sites. Sometimes they are already apparent at an early stage ; but more often they do not make their presence known until months or years—even as long as 15 or 20 years—have elapsed after otherwise successful surgical removal of the primary growth. Metastases in unusual situations, e.g. spleen, intestinal mucosa, pancreas or eye, are not rare.

(f) Hormonal factors influencing mammary carcinoma

The cells of some mammary carcinomas, like those of their parent tissue, are to some extent under hormonal control. This is shown by the retardation of growth effected in some cases by removal of the ovaries, adrenals or pituitary, or in other cases by the administration of oestrogens or androgens, and by the acceleration of growth which has often been observed during pregnancy and lactation. The exact way in which these factors influence the growth of the tumours still needs elucidation.

(4) Paget's disease of the nipple

In 1874 James Paget described 15 cases of the disease now named after him, This was a persistent spreading red dermatitis-like lesion of the nipple and areola which in every case was followed within one or two years by the development of a carcinoma of the usual kind in the substance of the breast, usually unconnected with the diseased nipple.

Microscopically, Paget's disease shows thickening of the epidermis and the presence in it of large pale rounded cells, scattered or in clumps, and often in mitosis (Figs. 169, 170). The main ducts of the nipple almost always show intraductal carcinoma *in situ*, usually in continuity with the diseased epidermis (Fig. 171). This intraductal carcinoma may or may not be in continuity with the carcinoma which usually develops deeper in the breast.

There are two opposing views as to the nature of Paget's disease. One, upheld especially by Muir and Inglis, is that the epidermal lesion is secondary to the disease in the main ducts and that the large " Paget cells " are carcinoma cells derived from the ducts and growing within the epidermis. The other view, upheld by Cheatle and Cutler and by the present writers, is that the disease is

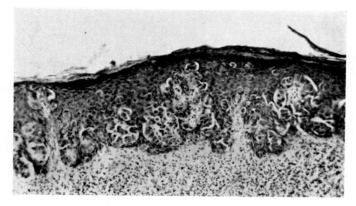

FIG. 169.—Paget's disease of nipple ; general view. (× 96.)

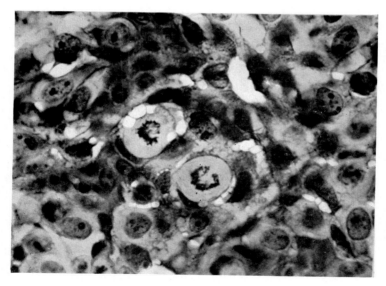

FIG. 170.—Mitotic figures in Paget cells in epidermis. (× 500.)

epidermal carcinoma *in situ* essentially similar to Bowen's disease of the skin (see p. 477), and that the intraductal carcinoma is concomitant. Perhaps the changes observed are the effects of carcinogenic agents applied simultaneously to a large field of tissue, including the epidermis, the main duct epithelium and the more peripheral mammary tissue. It is noteworthy that, when a mammary carcinoma invades the overlying skin from below, it does not produce Paget's disease.

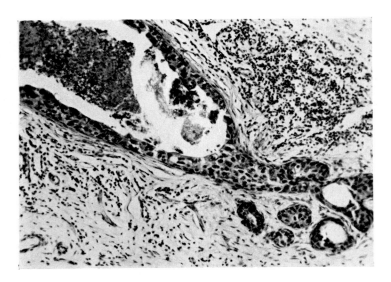

FIG. 171.—Intraductal carcinoma accompanying Paget's disease of nipple. (× 120.)

Paget's disease of the nipple, though itself not a destructive or metastasizing lesion, is a danger-sign calling for prompt radical removal of the breast because of the almost invariable development of intramammary carcinoma in association with it.

EPITHELIAL TUMOURS OF THE SKIN

(1) Causative factors and pre-cancerous lesions

Since the skin is an exposed tissue in which tumours can be watched from their inception, it is not surprising that it was the first tissue in which occupational tumours were recognized and that our knowledge of the causation of skin tumours is more advanced than that of tumours of other tissues. The recognized factors in causation are :

(a) Carcinogenic hydrocarbons

These are responsible for pitch, tar, creosote, mineral oil, paraffin and chimney-sweep's " warts " (papillomas) and cancers (see p. 416). These tumours are almost restricted to males, and are all squamous-celled. They occur on the forearm, hand, face, neck and scrotum, the last being the commonest site of oil cancers (notably mule-spinner's cancer) and chimney-sweep's cancer. Multiple tumours are frequent. It is probable that the kangri cancers of the abdominal wall or thigh in Kashmir natives are due, not solely to the chronic burns from the portable charcoal heater (or kangri), but to soot or tarry products.

(b) Sunlight

The frequency of skin tumours in fair-skinned farmers, gardeners, sailors and fishermen in sunny countries is undoubtedly due to the carcinogenic effects of actinic rays (see p. 420). The tumours occur on the face, neck, hands and forearms, are often multiple, and may be either squamous-celled or basal-celled. The multiple carcinomas of xeroderma pigmentosum are probably actinic.

(c) X-rays and radium

Cancers from these causes, always squamous-celled, may be due to occupational or therapeutic exposure. Occupational tumours, now rare, occur mainly on the hands.

(d) Arsenic

Chronic arsenical dermatitis (p. 338) may be due to occupation or to prolonged medicinal administration of arsenic. Supervening carcinomas are commonest on the limbs and face, are usually multiple, and of either squamous-cell or basal-cell type.

(e) Scars, ulcers, sinuses, etc.

Carcinoma sometimes supervenes on old scars from burns or inflammatory diseases, varicose ulcers, osteomyelitis sinuses, lupus vulgaris, psoriasis or other chronic skin lesions. It often develops in a widespread and progressive manner in the diseased area, and is almost always of squamous-cell type (see Figs. 135, 136).

(f) Excreted carcinogens

We must not overlook the possibility that carcinogenic agents excreted by the sweat or sebaceous glands may play a part in the causation of skin tumours, but nothing definite is yet known of this.

(2) Varieties of epithelial tumours of the skin

Following is a tabulation of the main varieties :
  (a) *Papilloma:* (i) Squamous or horny and unpigmented;
                  (ii) Pigmented.
  (b) *Squamous-cell carcinoma.*
  (c) *Basal-cell carcinoma:* (i) Superficial or rodent type ;
                          (ii) Subepidermal type ;
                          (iii) Mixed baso-squamous type ;
  (d) *Tumours of glandular structure,* arising from sweat, sebaceous or other glands.
  (e) *Intra-epidermal carcinoma.*
  (f) *Epithelial tumours of tibia or other long bones.*

(a) Papilloma

(i) *Squamous or horny papilloma.*—The simplest and most benign is the cutaneous horn, a solitary localized sessile overgrowth of the epidermis, surmounted by a short or tall conical mass of keratin. Early occupational carcinomas often show a pre-invasive " papillomatous " stage, e.g. tar or pitch " warts ". Infective papillary overgrowths of the epidermis, e.g. the common wart, molluscum contagiosum, and venereal condylomas, should not be classed with papillomas.

(ii) *Pigmented papilloma.*—Also badly called " seborrhoeic keratoses ", " acanthotic naevi " and " senile warts ", these are sessile or pedunculated, brown or grey, corrugated, slowly growing tumours, common in the skin of middle-aged and old people. They consist of thick folded layers of more or less pigmented epidermis composed of small cells of basal or baso-squamous type, enclosing rounded sharply defined horny foci (Fig. 172). They are unrelated to pigmented moles ; if they become malignant, a rare event, the tumour is a squamous-cell carcinoma.

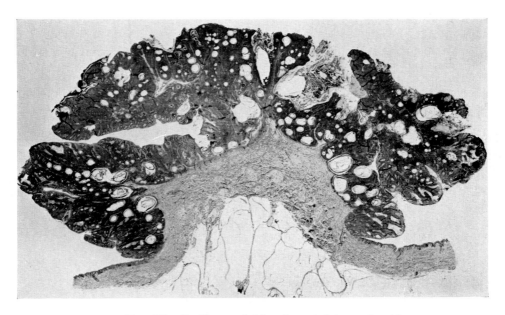

FIG. 172.—Papilloma of skin, pigmented type. ($\times$ 4.)

(b) Squamous-cell carcinoma

The familiar structure of well-differentiated cornifying squamous-cell or epidermoid carcinoma (" epithelioma ") is depicted in Figs. 136 and 156. Their downgrowths consist of epithelium with all the features of epidermis itself—a marginal layer of basal or germinal cells, a variable zone of large angular spinous cells, and a central core of squamous cells and keratin, this often appearing in cross sections of the downgrowths as rounded concentrically laminated " cell nests " or " pearls ". Rarely is melanin present in the epithelial cells. In anaplastic growths, which are uncommon, recognizable squamous characters may be lacking. Epidermoid carcinomas of the skin vary in their invasiveness ; most of them are of relatively low malignancy, growing and invading slowly and metastasizing late. Those of the face and neck are more malignant than those of the limbs. Metastasis when it occurs, is usually restricted to the regional lymph glands ; blood-borne metastases are rare.

(c) Basal-cell carcinoma

(i) *Of superficial origin.*—This common tumour arises from the epidermis itself, and often from the superficial parts of the hair follicles and dermal glands also, producing first a plaque-like thickening of the skin and later a slowly enlarging

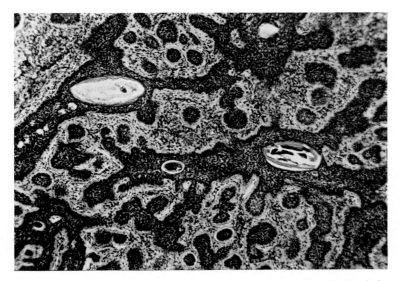

FIG. 173.—Basal-cell carcinoma, showing relationship to degenerating hair-shafts. (× 80.)

ulcer with a raised wire-like edge—a " rodent ulcer " or " Jacob's ulcer ". The growths are often multiple and sometimes numerous ; any part of the skin may be affected but the face and neck are the commonest sites. The structure varies (Figs. 137, 173, 174). Many tumours show multiple solid downgrowths from the basal layer of the epidermis, each consisting of small dark-staining polyhedral or elongate cells without squamous characters or keratin, the marginal cells often being cubical or columnar and forming a well-defined border to the clump. Minute rounded cysts often appear in the clumps, sometimes producing a honey-comb or pseudo-glandular pattern ; and reticular and folded patterns are also common. A few tumours are visibly pigmented, due to the presence of melanin in the tumour cells. The tumours are slowly invasive and destructive, and if untreated may eventually destroy most of the face, whence the name " rodent ". Metastases are extremely rare.

(ii) *Of sub-epidermal origin.*—Sub-epidermal basal-cell growths, arising from the pilo-sebaceous epithelium and sweat glands, are less common than those of more superficial origin. They may be solitary or multiple. When they are very numerous, they are distributed especially over the face, scalp and thorax, a condition which is sometimes familial. The tumours consist of well-defined masses of closely packed dark-staining cells generally resembling those of rodent carcinomas.

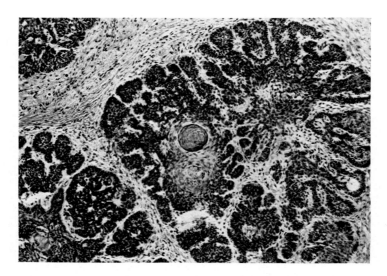

Fig. 174.—Baso-squamous carcinoma. (× 85.)

The shape and distribution of these masses often point clearly to their origin from multiple groups of sweat glands and ducts (Fig. 138). The larger masses show cysts, pseudo-glandular structures, and sometimes the reticular patterns more often seen in rodent carcinomas. To the more cystic varieties, Brooke's name " epithelioma adenoides cysticum " is often applied. The overlying epidermis is intact, though stretched over the tumours. These grow slowly and expansively, do not infiltrate surrounding tissues or ulcerate, and rarely produce metastases.

(iii) *Baso-squamous carcinoma.*—Most carcinomas of the skin are readily classified as of either squamous-cell or basal-cell type ; but occasional tumours of intermediate structure and behaviour occur. These show the general structure of basal-cell carcinoma but with distinct foci of spinous cells and keratinization (Fig. 174). Such tumours sometimes metastasize, both to the lymph glands and by the blood-stream.

(d) Skin tumours of glandular structure

The commonest of these are adenomas of sweat glands—hidradenomas—which form circumscribed dermal or subcutaneous growths, with a well-developed acinar structure often, however, with cystic, papillary or solid areas (Fig. 175).

Tumours showing the combined characters of hidradenoma and sub-epidermal basal-cell growth occur. Adenomas of the sebaceous, Meibomian, ceruminous, tarsal or lachrymal glands are seldom seen ; and adenocarcinomas of any of these glands or of the sweat glands are rare.

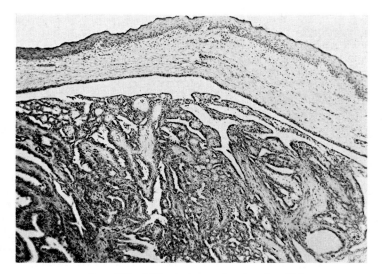

FIG. 175.—Hidradenoma of vulva. (× 60.)

(e) *Intra-epidermal carcinoma*

On p. 471, the view that Paget's disease of the nipple is a form of intra-epidermal carcinoma was referred to. Somewhat similar lesions occurring in other situations have been given various names—Bowen's disease, extra-mammary Paget's disease, and erythroplasia. However, these are not distinct diseases, but only variants of the entity intra-epidermal carcinoma, of which Bowen's disease is the most familiar form. This appears as single or multiple rough plaque-like thickenings of the epidermis, which slowly extend over considerable areas, in which after the lapse of several or many years invasive squamous-cell carcinoma develops. The thickened epidermis contains many atypical cells, including large and multi-nucleated cells, cells in mitosis, and pale rounded cells like those of Paget's disease of the nipple (Fig. 176). The lesion is essentially a slowly evolving pre-invasive form of epidermoid carcinoma.

(f) *Epithelial tumours of the tibia and other long bones.*—Rare cases have been reported in which a slowly growing epithelial tumour has developed in the shaft of the tibia, or very rarely the ulna or femur, without involvement of the overlying skin. The structure generally resembles that of a basal-cell carcinoma, but sometimes with a loose arrangement of the epithelial cells resembling that seen in the " adamantinomas " of the jaws. We do not know whether these rare growths

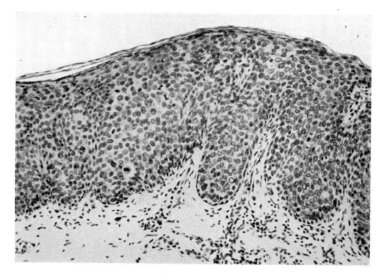

FIG. 176.—Bowen's disease—intra-epidermal carcinoma. (× 144.)

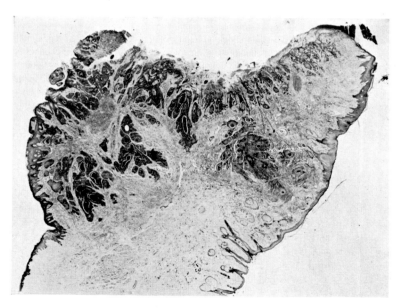

FIG. 177.—Vertical section of carcinoma of lip. × 6.)

take origin from developmentally misplaced epithelium or possibly from the epithelium of deep hair-follicles which has been driven into the periosteum by a blow.

EPIDERMOID CARCINOMAS OF THE LIPS, MOUTH, PHARYNX AND LARYNX

Simple papillomas of the lips, mouth, pharynx or larynx are rare ; most " papillomas " of these regions are only the pre-invasive stages of squamous-cell carcinomas. Carcinomas appear either as projecting, often papillary, growths, or as indurated ulcers, or as diffuse thickenings of the tissues, or as combinations of these.

(1) Epidermoid carcinomas of the lips

These frequent growths arise at the muco-cutaneous junction, and both the epidermis and mucosal epithelium contribute to their formation. They are 15 times commoner in men than in women, and are much more frequent on the lower than on the upper lip. There is good evidence that both pipe-smoking and exposure to sunlight are important causative factors. Most of the tumours are well differentiated squamous-cell carcinomas (Fig. 177) ; a few are anaplastic. Most of them grow slowly and metastasize late. Metastasis occurs chiefly to the submental and submandibular lymph glands, seldom by the blood-stream.

(2) Epidermoid carcinomas of the tongue

These arise on any part of the tongue. They occur in men 10 times as frequently as in women. Syphilitic glossitis is a causative factor of undoubted importance, preceding carcinoma in from 10 to 40 per cent of cases. Tobacco, alcohol, dental disease and hot food have also been blamed ; but, while they may indeed play a part, this is difficult to prove. Most of the tumours are cornifying squamous-cell carcinomas of varying degrees of differentiation ; a few are highly anaplastic.

Carcinoma of the tongue is a much more malignant tumour than carcinoma of the lip ; growth is often rapidly invasive, with early metastasis to the cervical lymph glands. Enlargement of these is sometimes the first sign of a small posterior lingual carcinoma. The cervical growths spread from gland to gland, and eventually form large fused masses enveloping the main vessels and frequently occluding or invading the internal jugular vein. Blood-borne metastases are present in about 40 per cent of fatal cases ; and in most cases with remote metastases, the jugular vein is invaded.

(3) Epidermoid carcinomas of other parts of the mouth

In Europeans, the cheeks, palate and floor of the mouth are relatively infrequent sites of carcinoma. But the cheeks are the commonest site of oral cancer in Indian, Cingalese and Filipino betel-chewers ; and the palate is a frequent site in certain Indians who smoke cigars with the lighted end in the mouth. The tumours are cornifying squamous-cell carcinomas, generally similar in behaviour to those of the tongue.

(4) Epidermoid carcinomas of the tonsils and pharynx

Again, these are several times more frequent in men than in women ; but post-cricoid carcinoma is remarkable in that it is commoner in women, in whom its development is often preceded (in some communities at least) by the Plummer-Vinson syndrome, i.e. anaemia, achlorhydria, and dysphagia due to atrophic changes in the oral and pharyngeal mucosa (see p. 397).

Most faucial and pharyngeal growths are epidermoid carcinomas of well-known varieties, though many of them are anaplastic and atypical (see Fig. 141). In some of the anaplastic tumours many lymphocytes are mixed with the poorly differentiated carcinomatous tissue, and such tumours are sometimes distinguished from the others under the caption " lympho-epithelioma ". But this is not a distinct species of tumour ; all gradations between well-differentiated squamous-cell carcinomas and " lympho-epitheliomas " are encountered.

In their production of metastases, pharyngeal and tonsillar carcinomas resemble those of the tongue. The upper deep cervical lymph glands are affected early, and glandular enlargement is the first symptom in about one-third of cases. Distant blood-borne metastases, usually consequent on invasion of the internal jugular vein, are present in about 40 per cent of fatal cases.

(5) Epidermoid carcinoma of the larynx

Almost all laryngeal carcinomas are of the ordinary cornifying squamous-cell type ; many of them in their early stages are papillary. Men are affected about 10 times as frequently as women, but the causative factors have yet to be identified. Most of the tumours start on the vocal cords or just below them. They usually grow slowly and metastasize late, metastasis often being deferred until the growth has extended through the laryngeal wall to the surrounding tissues.

TUMOURS OF THE EPITHELIUM OF THE DENTAL LAMINA

(1) Development of the dental lamina and tooth germs

The dental lamina is a sheet of epithelium which, during embryonic and early foetal life, grows from the oral mucosa into the substance of each jaw. From it two series of bud-like thickenings arise, the first rudiments of the deciduous and permanent teeth. Each epithelial tooth-germ or " enamel organ " enlarges, its attachment to the dental lamina becomes constricted, and its deep surface becomes indented by a condensed mass of mesenchyme, the dental papilla, which is the rudiment of the dentine and pulp of the tooth (Fig. 178). Only the deepest layer of the enamel organ, that in contact with the dental papilla, consists of ameloblasts and will form the enamel of the tooth : the rest of the enamel organ, and also the dental lamina, will undergo partial atrophy leaving, however, plentiful epithelial residues around the unerupted tooth. These para-dental residues persist into post-natal and even adult life, and it is from them that epithelial tumours and cysts within the jaws arise. These are of the following kinds :

(*a*) " *Adamantinoma* " (or " *ameloblastoma* "), a true neoplasm derived from the residues of the " enamel organ " ;

(b) *Intra-alveolar epidermoid carcinoma*, an ordinary squamous-cell carcinoma derived from the para-dental residues ;

(c) *Odontomes*, benign tumour-like malformations containing tooth tissues proper, dentine and enamel ; and

(d) *Epithelial cysts* of the jaws.

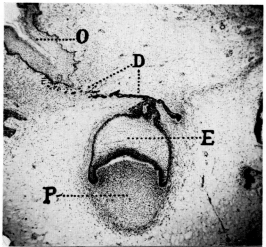

FIG. 178.—Vertical section through jaw of 15-weeks foetus showing development of a tooth-germ. O, oral mucosa ; D, dental lamina ; E, enamel organ ; P, dental papilla. (× 40.)

(2) Dental tumours and cysts

(a) " *Adamantinoma* " or " *ameloblastoma* "

These are the names applied to slowly growing, lowly malignant tumours of the jaws of rather characteristic structure resembling that of the developing enamel organ but never forming enamel. They appear chiefly in childhood and young adult life, and rarely after the age of 40. Microscopically (Fig. 179), they consist of well-defined epithelial masses of various sizes and shapes, the marginal cells of which are of columnar form with their nuclei away from their bases, and the central cells of which form an open meshwork with wide intercellular spaces closely similar to the stellate reticulum of the developing enamel organ. Accumulation of fluid within the epithelial clumps frequently produces small or large cysts which give a honeycomb or pseudo-glandular appearance. Other parts of the tumours may show little or no stellate reticulum and may closely resemble basal-cell growths of the skin ; while other areas may contain foci of squamous cells and keratin.

Though of slow growth, the tumours invade and destroy the surrounding bone; and, unless wide excision is carried out, post-operative recurrence is frequent. Metastasis to lymph glands or lungs occurs rarely.

The names " adamantinoma " and " ameloblastoma " are misnomers, since the tumours do not form enamel. This is not surprising, as they arise, not from the ameloblast layer of the tooth rudiment but only from the residual parts of the

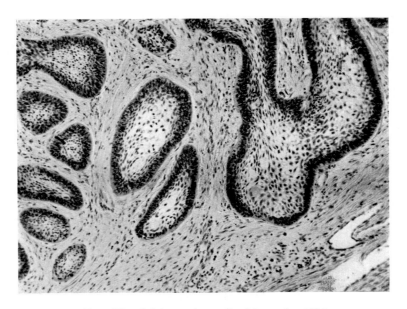

FIG. 179.—" Adamantinoma " of jaw. (× 100.)

" enamel organ ". These residues represent the stellate reticulum and the outer epithelial border of the tooth germ, parts which do not produce enamel.

(b) Intra-alveolar epidermoid carcinoma

Rarely, an ordinary squamous-cell carcinoma arises within the substance of the jaw, usually in an old person. Its origin must be from the para-dental residues, probably the persistent squamous-cell residues of the dental lamina.

(c) Odontomes

These are not true tumours, but rather masses of malformed or superfluous tooth tissues (Fig. 180). They contain dentine and enamel arranged in a disorderly manner ; but they have no powers of excessive continuous growth.

(d) Epithelial cysts of the jaws

These also are not tumours, but it is convenient to mention them here. Some of them arise from disturbances of eruption or other developmental malformations, and such cysts are often demonstrably connected with malformed or malerupted teeth or contain supernumerary teeth—dentigerous cysts. Others, simple dental cysts, are connected to the roots of dead or infected teeth, and are related to the common root-granulomas which are often present on the roots of such teeth and which contain plentiful epithelial tissue mingled with the granulation tissue. This epithelium is derived from the para-dental residues

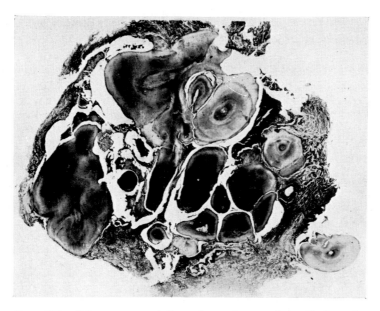

FIG. 180.—Odontome consisting of a cluster of small irregular masses mainly composed of dentine. (× 5.)

around the roots, and in a mildly infected root granuloma the proliferating epithelium can easily form a cyst with a squamous stratified lining. Such cysts, once formed, slowly enlarge and distend the jaw.

EPITHELIAL TUMOURS OF THE SALIVARY GLANDS

Epithelial tumours of the salivary glands are conveniently considered under the following heads :

(1) Simple adenomas ;

(2) Pleomorphic adenomas and adenocarcinomas (so-called " mixed " tumours) ;

(3) Anaplastic carcinomas ;

(4) Adeno-lymphomas.

It should be clearly understood, however, that (1), (2) and (3) are not sharply distinct, but merge into each other. The rare adeno-lymphomas, on the other hand, do form a distinctive group.

(1) Simple adenomas

A few salivary tumours are slowly growing, solid, encapsulated, and have a uniform glandular structure. They are therefore distinguished from the large group of " mixed " tumours under the name *adenoma* ; but this separation is arbitrary, for tumours of intermediate structure occur. Adenomas are, in fact, only the most slowly growing and best differentiated growths of the same kind as those now to be considered.

(2) Pleomorphic adenomas and adenocarcinomas (so-called " mixed tumours ")

Nine-tenths of all salivary tumours fall into this group. They are slowly growing, well circumscribed, but not strictly encapsulated, tumours of very variable appearance and consistency, occurring at all ages, most often in young adults. About 80 per cent of them arise in the parotid gland, about 10 per cent in the submandibular gland, only about 1 per cent in the sublingual gland, and the remaining 9 per cent in the small salivary glands of the palate, lips, cheeks and other sites.

(*a*) *Structure* (Figs. 181 and 182)

This is very diverse. It includes simple acinar or solid glandular structures resembling those of the adenomas, cribriform and cystic epithelial masses resembling those of cystic basal-cell growths of the skin, cornifying squamous-celled

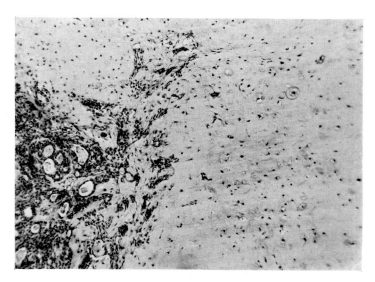

FIG. 181.—Glandular tissue fraying out into a mucinous matrix simulating cartilage in a salivary tumour. (× 72.)

foci, epithelial filaments and networks in a mucinous matrix (often mistakenly called " myxomatous "), lakes of mucus containing scattered isolated cells (often mistakenly identified as " cartilage "), and sometimes areas with frankly carcinomatous characters. Pleomorphism is thus a striking feature of the tumours;

several or all of the foregoing variants are often present in a single tumour, and obvious gradations from one kind of structure to another are frequent, affording clear evidence of the essential identity of the tumour parenchyma in all its forms.

The supposedly " myxomatous " and " cartilaginous " tissues call for further comment. The formation of these by sprouting and fraying-out of obviously epithelial structures into a mucinous matrix is frequently seen. The mucin is produced by the cells themselves, just as in mucoid adenocarcinomas of the stomach or intestine. Wide separation of isolated epithelial cells in lakes of mucin produces

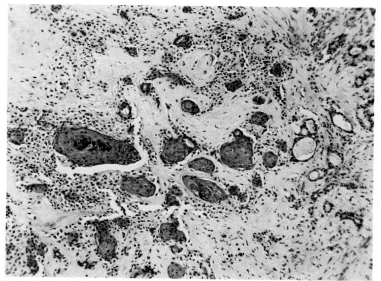

FIG. 182.—Cornified cell-nests along with glandular structures in a salivary tumour. (× 75.)

a spurious resemblance to soft cartilage. Many of the earlier pathologists believed that this was true cartilage, whence the erroneous name " mixed " tumour.

The connective tissue stroma of the tumours varies in amount and appearance. In some tumours it is well defined and fibrous ; in others it is mingled with the extravasated mucin and assumes the appearance of mucoid connective tissue. Occasionally, true cartilage or bone develops by metaplasia in the stroma.

(b) Behaviour

Although most of these tumours grow slowly and appear to be well circumscribed, they are not strictly benign ; for many of them recur after removal, some of them infiltrate the surrounding tissues, and a few of them metastasize to lymph glands or lungs. Recurrence is much more frequent from parotid growths than from those in other sites, mainly because surgical excision in the parotid is made as conservative as possible for the sake of the facial nerve. The histological structure of a salivary tumour is not a certain guide to its prognosis ;

tumours of any kind of structure may recur. Two variants which, however, are more dangerous than the average are " muco-epidermoid carcinoma " and cribriform or "adeno-cystic carcinoma" (sometimes badly called " cylindroma "). Occasionally, a relatively benign tumour, which has grown slowly for many years, suddenly commences to grow more actively and becomes invasive or metastasizes.

(3) Anaplastic carcinomas

Rarely, a salivary tumour is frankly malignant from its onset, growing rapidly, metastasizing widely, and consisting of cellular anaplastic tissue which may be difficult to recognize as epithelial. The better differentiated areas of such tumours sometimes show resemblances to the common pleomorphic tumours ; and conversely, the more malignant variants of structure in the pleomorphic group resemble those of the frank carcinomas. Clearly then, the latter do not form a distinct species, but are simply the most malignant of the whole class of salivary epithelial tumours.

(4) Adeno-lymphomas

Adeno-lymphoma, a relatively rare growth, alternatively called *papillary cystadenoma*, is an encapsulated benign one, which arises usually on the outskirts of the parotid gland and sometimes appears detached from it. It may be solid but is more often cystic ; and microscopically it shows a characteristic double-layered salivary-duct-like epithelium clothing intracystic papillary projections, and a stroma containing few or many lymphocytes which may be diffusely arranged or may form follicles. Of several views as to its origin, the most likely is that it arises from heterotopic salivary glandular tissue, which is known to occur sometimes in the pre-parotid lymph nodes.

EPITHELIAL TUMOURS OF THE NASAL AND PARA-NASAL CAVITIES

Epithelial tumours of the nasal cavity, maxillary antrum or other sinuses are infrequent. They include :

(1) *Papillary growths*, often called " papillomas ", but which frequently recur following removal or become invasive squamous-cell carcinomas later ;

(2) *Squamous-cell carcinomas*, sometimes well differentiated and cornifying, sometimes anaplastic ;

(3) *Adenocarcinomas*, arising from the nasal glands ; and

(4) " *Adamantinomas* " of the upper jaw, which have grown into the antrum.

EPITHELIAL TUMOURS OF THE BRONCHI AND LUNG

Bronchial carcinoma is by far the commonest tumour of the lung. It will be considered first, and the rarer *bronchial adenoma* later.

(1) Bronchial carcinoma

(a) Incidence

Bronchial carcinoma is a common disease, accounting for between 5 and 15 per cent of cases of carcinoma in most recent necropsy series. There has been much discussion as to whether the disease is increasing in frequency, but no definite conclusion is at present possible. Certainly many more cases are now being diagnosed as bronchial carcinoma than three or four decades ago; but this may well be due more to improved diagnosis than to an increase in the disease.   Even now, many cases of bronchial cancer are clinically misdiagnosed.

(b) Causation

All series of cases show a decided preponderance of males, with an average of about 80 per cent.   This strongly suggests extrinsic occupational or habitual factors as being important in causation.   Almost every known inhaled substance

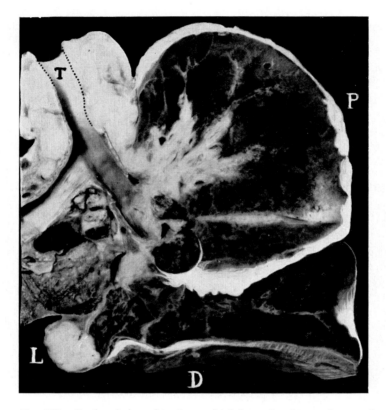

FIG. 183.—Sectional view of trachea and left lung, showing carcinoma of left main bronchus extending into upper lobe, which is also enveloped by pleural extension of growth P.   Trachea T is surrounded by confluent growth in lymph glands, hilar glands are enlarged and anthracotic and contain mottled patches of tumour, and an inferior mediastinal gland L is enlarged and replaced by growth.

and every possible infection of the lungs has, by one writer or another, been suggested as a causative factor ; but for most of them the evidence is inconclusive. There is no satisfactory evidence that tuberculosis, influenza, bronchitis or pneumonia predisposes to bronchial carcinoma.   Silicosis, asbestosis or siderosis is sometimes seen in association with carcinoma ; and mice exposed to dusts containing silica or iron oxide show an increased incidence of lung tumours : but it is improbable that pneumoconiosis plays an important part in the causation of lung carcinoma in general. Comparisons of the frequency of lung cancer in smokers and non-smokers show clearly that cigarette smoking is the main cause of modern lung cancer. It is possible also that the inhalation of carcinogenic hydro-carbons or other substances in soot, road dust, and urban smokes may be causative.

A remarkable occupational incidence of bronchial carcinoma occurs in the miners in certain mines in Saxony and Bohemia. The air in these mines contains not only iron, cobalt, nickel and silica dusts, but also arsenic and radioactive substances, especially radon. Each of these has been blamed as the main car-cinogenic factor ; but the weight of the evidence points to inhaled radon.

There is evidence from other occupations that inhaled arsenic, nickel and chromium compounds may cause bronchial carcinoma. Neither trauma nor heredity plays any signficant part in human lung cancer.

(c) *Site of origin* (Fig. 183)

Most, probably all, carcinomas of the lungs are bronchial carcinomas. At least three-quarters of them demonstrably arise in the large bronchi, either the main bronchi or the lobar bronchi in or near the lung hila. The more peripheral

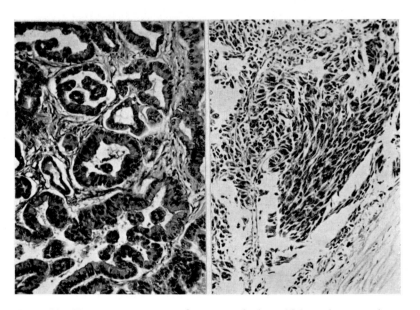

FIG. 184.—Two common types of structure in bronchial carcinoma—adeno-carcinoma and " oat-cell " carcinoma. (× 120.)

tumours do not differ in structure or behaviour from those of obviously bronchial origin ; there is no doubt that they too arise from bronchi, though an origin from a small peripheral bronchus is often not demonstrable. There are no tumours which we need suppose arise from the alveoli.

(d) Structure (Figs. 159, 184)

It is convenient and useful to distinguish three main types of structure—adenocarcinoma, squamous-cell carcinoma and anaplastic or undifferentiated carcinoma. While many bronchial carcinomas consist predominantly, sometimes exclusively, of one of these types of growth, it must be emphasized that there is only one entity, *bronchial carcinoma*, and that individual tumours may show various types of structure. Anaplastic structure is the most frequent, and is often of a rather characteristic " oat-celled " or spindle-celled kind, consisting of short fusiform cells arranged in compact clumps or diffusely. Some adenocarcinomas are mucoid. Some squamous-cell carcinomas undergo extensive friable keratinization, and may break down and discharge into the bronchi and so produce large cancerous cavities.

(e) Naked-eye appearance and local effects

In their size, rate of growth and spread, bronchial carcinomas vary greatly. Some attain huge sizes without producing remote metastases ; others remain small, and often symptomless, and yet produce large metastases. Some form well-defined bulky masses ; others produce localized thickening and replacement of the bronchial walls ; and a few produce diffuse ill-defined widespread bronchial and peri-bronchial infiltration with no clearly distinguishable focus of origin. Bronchial obstruction often develops and leads to collapse, bronchiectasis, pneumonia or abscess in the part of the lung peripheral to the growth.

(f) Local spread

The primary growths or lymph-nodal deposits contiguous with them frequently extend to involve the pleura, mediastinum, great vessels, pericardium or heart. Extensions in the pleura often cause persistent blood-stained effusion. Invasion of main pulmonary veins is frequent, and in some cases the intravenous growths extend into the left atrium. Invasion of the superior vena cava or innominate veins is not unusual. The pulmonary arteries are often compressed by the surrounding growth, but rarely invaded. Invasion of the pericardial cavity may take place around the great vessels and the base of the heart ; and in such cases the growth may spread widely in the epicardium, especially in lymphatics (see Fig. 147), even to the apex of the heart. The heart itself sometimes suffers gross invasion. Direct invasion of the oesophagus or vertebrae is rare. Apical tumours may invade the cervical sympathetic and the brachial plexus.

In not a few cases, one or other of the various complications and extensions just mentioned produces the first symptoms of disease, and may lead to such errors of diagnosis as " pleurisy ", " pneumonia ", " bronchiectasis ", " mediastinal tumour ", etc.

(g) Metastasis

Bronchial carcinoma metastasizes widely and frequently, and in many cases at an early stage while the primary growth is still small and symptomless. More diagnostic mistakes due to metastases are made in this disease than in any other.

(i) *Metastasis to lymph glands.*—The hilar or mediastinal glands contain deposits in nearly 90 per cent of necropsy cases. These often attain large sizes and may simulate primary mediastinal tumours. Extra-thoracic lymph glands also are frequently affected, and cervical or axillary masses are sometimes the first signs of disease.

(ii) *Metastasis in serous cavities.*—Extension to the pleura or pericardium, already referred to, may be followed by dissemination in these serous membranes. Most supposedly primary tumours of the pleura or pericardium are really secondary to undetected bronchial carcinomas.

(iii) *Metastasis by the blood-stream.*—Blood-borne metastases are present in three-quarters of necropsy cases. They are most frequent in the liver, adrenals, central nervous system and bones, in that order ; but no organ is exempt. Adrenal metastases are surprisingly frequent, being present in at least one-third of fatal cases. In about the same proportion of cases, the central nervous system, usually the brain, contains metastases. In many of these cases, primary cerebral disease is simulated and a relatively symptomless primary growth is overlooked. In all cases diagnosed as " cerebral tumour ", careful radiographic examination of the lungs should be a routine part of the clinical examination.

(2) Bronchial " adenoma "

This is the name applied to a group of slowly growing well circumscribed, solid tumours of the main bronchi, which develop usually in young people of

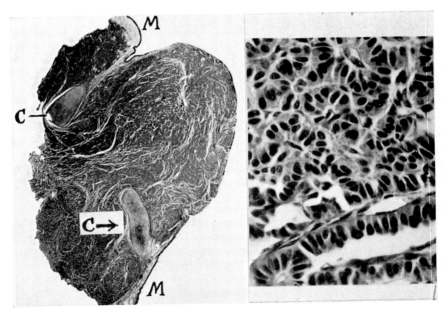

FIG. 185.—Bronchial " adenoma ". In the low-power view, MM is mucosa, CC cartilages surrounded by growth ; note projection of part of tumour into lumen. (× 6 and 400.)

either sex. In shape thay are " iceberg " tumours, with a part projecting into the bronchial lumen and usually covered by intact mucosa, and a larger part outside the bronchus in the surrounding hilar or lung tissue (Fig. 185). Some of them show well-differentiated acini or tubules, but most of them consist of solid epithelial columns or clumps with only occasional signs of lumina. There is little doubt that they arise from the mucous and mixed glands of the bronchi. They grow slowly and mainly expansively, but slight marginal infiltration is often present, and in rare cases metastasis has occurred. Their main effects are bronchial obstruction, collapse and infection of the lung.

CARCINOMA OF THE OESOPHAGUS

Adenomas and papillomas of the oesophagus are very rare ; *carcinoma* is the only common oesophageal tumour. It occurs in relatively old people, about 6 times as frequently in men as in women, and is most common in the middle third of the organ. Alcohol, tobacco, hot foods and other dietetic factors have been blamed as causative, but proof of the importance of any particular factor is difficult to obtain. Almost all the tumours are ordinary squamous-cell carcinomas, varying in their degree of differentiation. Adenocarcinomas are very rare. Bulky, stenosing or ulcerating growths all occur, or any combination of them ; and the main symptom is dysphagia. Invasion of surrounding parts, the lungs, bronchi, trachea, aorta, pleural cavity, heart or spine, is frequent. Septic pneumonia from invasion of the lungs or bronchi is a common cause of death. Fatal haemorrhage from perforation of the aorta is not unusual. Metastasis frequently occurs to neighbouring lymph glands ; and blood-borne metastases are present in about one-third of fatal cases, most often in liver, lungs or brain.

CARCINOMA OF THE STOMACH

Carcinoma is by far the commonest epithelial tumour of the stomach. Adenomas and papillomas are relatively rare, and will be discussed as precursors of carcinoma.

(1) Frequency and causation of gastric carcinoma

(a) Frequency

Cancer of the stomach is the commonest kind of malignant tumour in all European countries, U.S.A., other white populations derived from these, and also in the Japanese. In Great Britain and in Australia it accounts for about one-quarter of all cancer deaths ; in Scandinavia, Holland and Central European countries, for more than one-half of them, the highest proportion being Czecho-slovakia's 66 per cent. The greatest number of cases is seen in the sixth decade of life ; the disease is rare before the age of 30. Men are affected twice as frequently as women. People in the unskilled classes are more liable to the disease than those in the skilled and professional classes. It is surely very significant that, while gastric cancer is the commonest malignant tumour of man, it is one of the rarest in all other animals.

(b) Possible causative factors

The diets and dietetic habits of man are very diverse and complex ; so that statistical and experimental search for the possible causes of gastric diseases is like looking for a needle in a haystack. Let us briefly review the main factors known, or plausibly suggested, as of importance in the causation of gastric carcinoma.

(i) *Gastric ulcer.*—Occasional undoubted examples of carcinoma arising at the margins of a chronic ulcer are seen. But the proportion of cases of cancer showing acceptable evidence of previous ulcer is small—between 6 and 16 per cent in carefully studied series. The proportion of cases of chronic ulcer in which cancer later supervenes is much lower still—certainly well under 5 per cent.

(ii) *Chronic gastritis.*—The evidence that gastritis is a frequent pre-cancerous lesion is conflicting and inconclusive.

(iii) *Papilloma and adenoma.*—These infrequent tumours may be solitary or multiple, and there is good evidence that they often become carcinomatous.

(iv) *Pernicious anaemia and achlorhydria.*—There has been much recent discussion as to whether or not gastric carcinomas occur with undue frequency in people with pernicious anaemia or simple achlorhydria. The evidence is inconclusive. It must not be overlooked that otherwise symptomless gastric cancers sometimes produce severe pernicious-like anaemia (p. 495).

(v) *Ingested carcinogens.*—Man ingests, in addition to the ordinary food substances, a vast number of alcoholic drinks, drugs, dyes and antiseptics ; and, by roasting, grilling, frying and smoking his foods, he may also be introducing chemical carcinogens similar to those produced by Kennaway in artificial tars (p. 417). It is quite probable, but difficult to prove, that these factors are important in causing gastric cancer. The idea accords well with the rarity of alimentary cancers in animals, whose food is seldom cooked or smoked, and who do not consume drugs, dyes, flavouring and preserving agents, alcohol or tobacco juice.

(vi) *Food temperatures.*—We take our drinks and food at astonishingly high temperatures, even 65° or 75° C. Perhaps we pay for this by the great amount of chronic gastric disease—gastritis, ulcer and cancer—which we suffer.

(vii) *Heredity.*—We have already noted that occasional families show a special susceptibility to gastric carcinoma (p. 427).

(2) Sites of gastric carcinoma

The pyloric canal and antrum are the site of origin of nearly 50 per cent of gastric carcinomas ; the lesser curvature, of about 25 per cent ; while the remaining 25 per cent arise at the cardia or other parts, or involve the whole organ.

(3) Appearances of gastric carcinomas

As in other organs, carcinomas may appear (*a*) as bulky growths projecting into the stomach, (*b*) as ulcerated growths, (*c*) as diffuse infiltrations of the stomach wall, or (*d*) as combinations of any of these. Mucoid characters are frequent, and some tumours are conspicuously gelatinous. Diffusely infiltrating carcinoma sometimes affects most or the whole of the gastric walls, producing the " gizzard " or " leather-bottle " stomach, with great contraction and rigidity of the organ. In spite of the great thickening of the wall, often up to 2 centimetres or more, the various coats, especially the muscular layers, may remain clearly distinguishable on the cut surface (Figs. 186, 187).

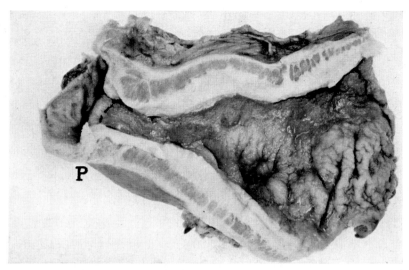

FIG. 186.—Diffuse carcinoma of pyloric half of stomach—" leather bottle " stomach. Note great thickening, especially of submucosa, and preservation of muscle coat in the growth. P, pylorus. (See Fig. 187.)

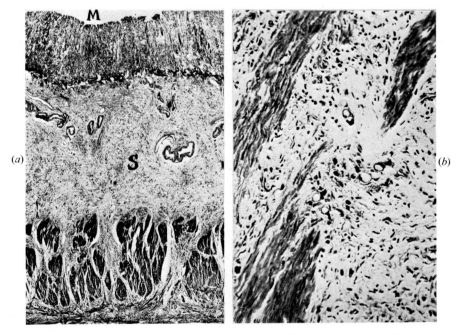

FIG. 187.—Micro-structure of stomach shown in Fig. 186. (a) (× 10) shows thickened mucosa M, submucosa S and muscle coat; (b) (× 180) shows much new-formed fibrous tissue between muscle bundles, and several signet-ring carcinoma cells near middle of field.

(4) Microscopic structure

Gastric carcinomas show the following varieties of structure (Figs. 157, 188) :—

    (1) *Adenocarcinoma*, showing distinct differentiation of acinar or tubular structure, sometimes with papillary characters also ;

    (2) *Mucoid adenocarcinoma*, a variant of the preceding ;

    (3) *Signet-ring-cell carcinoma*, a frequent type ;

    (4) *Spheroidal-cell carcinoma*, without obvious secretion; but (3) and (4) are often seen together, especially in diffusely infiltrating growths ;

    (5) *Metaplastic squamous-cell carcinoma* is rare ;

    (6) *Cellular anaplastic carcinoma* is frequent, forming soft necrotic or haemorrhagic growths devoid of glandular differentiation.

Again it must be emphasized that these varieties of structure are not distinct species, and that several or all of them may be seen in a single tumour.

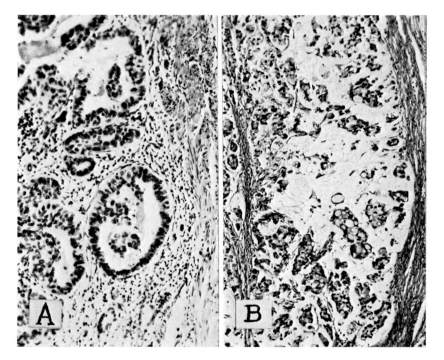

Fig. 188.—Common types of structure in gastric carcinoma. A, acinar adenocarcinoma (× 120) ; B, mucoid signet-ring-cell carcinoma (× 72).

(5) The direct spread of gastric carcinoma

Gastric carcinoma spreads via tissue interstices, lymphatics and veins in the wall of the stomach itself, and thence into the walls of the duodenum and oeso-phagus. Extension to the investing peritoneum and to the greater and lesser omenta is common, and is the usual source of trans-peritoneal dissemination. Perforation of the stomach into the peritoneal cavity through a carcinoma is much less common than through an ulcer. Direct cancerous invasion of the pancreas, bile duct, portal vein or its main tributaries, transverse colon, spleen or liver, is not unusual.

Lymphatic permeation is sometimes plainly visible, especially on the peritoneal surface of the stomach or the adjacent omenta. Microscopic study reveals that from some of the diffusely infiltrating growths very extensive, sometimes almost body-wide, lymphatic permeation has taken place. In such cases, however, permeation has usually not proceeded solely from the primary growth, but also from many embolic metastatic foci in lymph glands or lymphatic plexuses (see p. 442).

(6) Metastasis

(a) *Metastasis to regional lymph glands*

Even in operation specimens the perigastric lymph glands contain metastases in the majority of cases—between 50 and 70 per cent. In necropsy cases the percentage is 80 or 90. The presence or absence of lymph-nodal metastases is an important factor in prognosis.

(b) *Further lymphatic metastasis*

From the perigastric lymph glands the tumours spread to the coeliac, mesenteric, lumbar and mediastinal glands. Gastric carcinoma is the commonest tumour to invade the cisterna chyli and thoracic duct. Troisier's sign, i.e. enlargement of the supraclavicular lymph glands, usually on the left side, in cases of abdominal malignant disease, is most often due to carcinoma of the stomach, and in most cases is associated with invasion of the thoracic duct. Gastric cancer sometimes extends to the abdominal wall around the umbilicus, usually via the lymphatics of the round ligament of the liver.

(c) *Transperitoneal metastasis*

This is common, producing scattered discrete nodules on the peritoneal surfaces, plaque-like infiltrations, or massive accumulations of mucus, with or without ascites. Metastases in the *ovaries* are more frequent from gastric carcinoma than from any other tumour. Some ovarian metastases have the structure of acinar adenocarcinoma ; but many of them consist of diffusely infiltrating signet-ring-cell growth, and these are distinguished by the name " Krukenberg tumours ". The kind of gastric growth usually responsible for Krukenberg tumours is diffusely infiltrating spheroidal-cell or signet-ring-cell carcinoma. The primary growth is often small and symptomless, the large ovarian tumours being the first or main signs of disease.

(d) *Metastasis by the blood-stream*

The liver contains metastases in at least one-third of fatal cases. The lungs are involved in one-quarter of the cases, in some by discrete scattered metastatic growths, in others by widespread lymphatic permeation. Metastases develop in bones in 5 or 10 per cent of cases, in some of which they produce severe macrocytic anaemia or thrombocytopenia. Other tissues are less frequent sites of metastasis.

CARCINOMAS OF THE SMALL INTESTINE

Papillomas and adenomas of the small intestine are very rare. Carcinomas, also uncommon, are of two distinct kinds, (1) ordinary adenocarcinomas, and (2) argentaffin carcinomas.

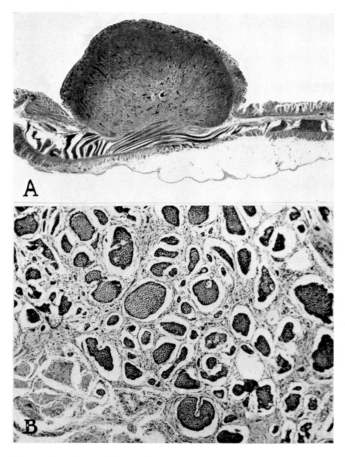

Fig. 189.—Argentaffin carcinoma : A, low-power view of section of whole tumour in small intestine ; note relationship to muscle coat. (× 4.) B, Characteristic micro-structure. (× 55.)

(1) Ordinary adenocarcinomas

These appear as solitary, annular stenosing, ulcerated or polypoid growths, and usually cause intestinal obstruction. The most frequent sites are the second part of the duodenum and the upper coils of the jejunum. Metastases occur in the neighbouring lymph glands or liver, but rarely elsewhere.

(2) Argentaffin carcinomas

These, which are nearly as frequent as the adenocarcinomas, are often called " carcinoid " tumours, a bad name since they are dangerous invasive and metastasizing growths. They should be called argentaffin or Kultschitzky-cell carcinomas since they arise from the specialized granular epithelial cells of the crypts of Lieberkühn, the yellow cells of Kultschitzky, which contain silver-stainable granules. Microscopically, most of the tumours have a distinctive and easily recognizable structure, consisting of well-defined solid clumps and strands of small closely packed polyhedral cells, with here and there small glandular lumina or distinct acini (Fig. 189).

Argentaffin carcinomas may be single, but are more often multiple and sometimes numerous. Both ileum and jejunum are affected. Young tumours form projecting nodules covered by intact mucosa ; older ones ulcerate and often spread to encircle the bowel and cause kinking and stenosis. Metastases occur frequently in the mesenteric lymph glands and liver, less often in other sites. Some of the tumours secrete 5-hydroxytryptamine (serotonin) and produce the " carcinoid syndrome "—attacks of flushing or cyanosis, diarrhoea, and eventually pulmonary valvular stenosis.

EPITHELIAL TUMOURS OF THE LARGE INTESTINE

(1) Papilloma and adenoma

Benign epithelial tumours with a villous surface are called *papillomas*, and those with a smooth or only corrugated surface *adenomas* ; but the two are not sharply

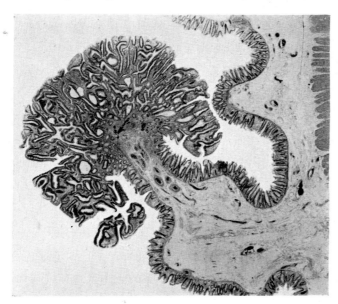

FIG. 190.—Sectional view of typical adenomatous polypus of large intestine. Compare with Fig. 196. (× 10.)

distinct, and they are both embraced by the term *polypi* (Fig. 190).   They consist of well-differentiated columnar goblet-celled epithelium like that of the colonic mucosa, clothing the villi of the papillomas, or forming glandular tubules in adenomas.   Young tumours are sessile, but larger ones develop strap-like pedicles clothed by normal mucosa.

Intestinal polypi are common tumours.   They may be solitary, few or numerous.   The rectum and pelvic colon are the commonest sites of solitary or few polypi, and are usually the most severely affected parts in cases with many polypi.   Patients with very numerous tumours—polyposis of the colon—often

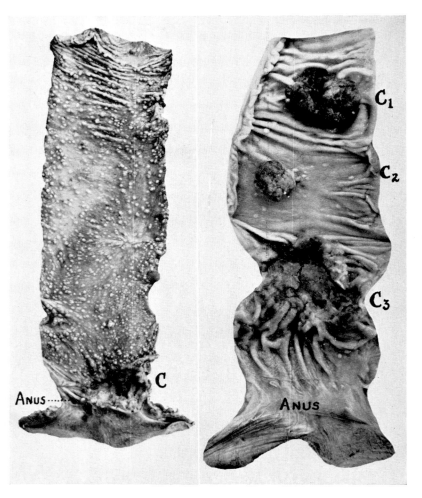

FIG. 191.—Polyposis of large intestine accompanied by ulcerated carcinoma C just above anus.

FIG. 192.—Three separate carcinomas of rectum, C_1 and C_2 polypoid, and C_3 ulcerated, associated with multiple small polypi.

give a family history of the disease.   This may arise in several successive generations, is transmitted by either sex, and affects males and females about equally.
The polypi are not congenital, but usually commence to appear during adolescence or early adult-hood.   What is inherited is, then, not polyposis as such,
but a special proneness of the colonic mucosa to neoplasia.   No family tendency
is apparent in cases with solitary or few polypi, and these may develop at any age.

All polypi of the large intestine in adults, whether solitary or multiple, are predisposed to carcinomatous change.   The predispostion is least in the solitary
polypus and greatest in familial polyposis (see below).

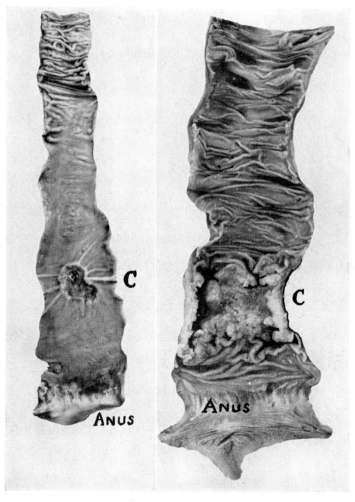

FIG. 193.—Early ulcerated   FIG. 194.—Advanced, ulcerated,
    carcinoma of rectum.   encircling carcinoma of rectum.

(2) Carcinoma

(a) Incidence

The large intestine is one of the most frequent sites of carcinoma—in many countries nearly as frequent a site as the stomach. The average age of patients is between 55 and 60 ; but carcinoma supervening on familial polyposis occurs in young people, in the second, third or fourth decades. Men suffer from carcinoma of the large intestine nearly twice as often as women. The disease increases in frequency from the proximal to the distal parts of the bowel, the

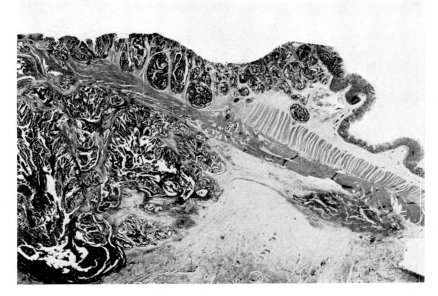

Fig. 195.—Sectional view of edge of deeply invading carcinoma of rectum, showing infiltration through muscle coat and massive extension in peri-rectal tissue. (× 3.)

rectum being the most frequent site. Multiple carcinomas are not uncommon, especially in cases of polyposis. Like gastric cancer, intestinal cancer is rare in all mammals except man.

(b) Causative factors

In most cases of intestinal carcinoma, no previous lesions or causative factors are recognizable. In a small proportion of cases, the disease supervenes on polypi or inflammatory lesions.

(i) *Polypi*.—The importance of polypi as pre-cancerous lesions is shown by three facts : (a) familial polyposis almost invariably terminates in carcinoma at a relatively early age ; (b) in cases of carcinoma other than those associated with familial polyposis, a few benign polypi are present in the neighbouring mucosa in more than one-half of the cases ; and (c) occasional examples of early carcinoma arising in a polypus are encountered. (See Figs. 191 and 192.)

(ii) *Inflammatory diseases.*—Carcinoma occasionally develops in cases of ulcerative colitis or diverticulitis, and in the former this occurs more often than is due to chance.

(iii) *Ingested carcinogens.*—It is possible that carcinogenic substances, ingested and carried through the alimentary canal, may be concentrated in the faeces. This possibility has yet to be investigated.

FIG. 196.—Sectional view of polypoid carcinoma of colon. Note traction and local hypertrophy of muscle coat. Compare with Fig. 190. (× 4.)

(c) *Structure and growth* (Figs. 191-196)

As in the stomach, so in the bowel, carcinoma may assume predominantly polypoid, ulcerative, stenosing or diffuse forms ; the last is the least common. Microscopically, most tumours are columnar-celled adenocarcinomas, of varying degrees of differentiation, and often mucus-secreting. Polypoid carcinomas often attain a large bulk within the bowel while still infiltrating its wall only slightly : but ulcerating and stenosing tumours soon extend through the muscular coat to the surrounding tissues. Here, perineural and other lymphatics, and veins, are often invaded.

(d) *Metastasis*

The regional lymph glands contain metastases in at least 50 per cent of surgically removed specimens ; and the presence or absence of lymph-nodal deposits is a major factor in prognosis. Peritoneal or ovarian metastases some-times occur early, and primary ovarian disease may be simulated. Blood-borne metastases are most frequent in the liver, which is affected in about one-half of fatal cases. The lungs and other parts are less frequently affected.

(3) Carcinoma of the appendix

Argentaffin carcinomas are the commonest tumours of the appendix. In most cases they are small solitary growths, found only incidentally during abdominal operations, and hence usually cured by appendicectomy. But in other cases they have infiltrated to the peritoneum, or have produced metastases in lymph glands or liver. Structurally they resemble their counterparts in the small intestine (see p. 497).

Adenocarcinoma of colonic type is rare in the appendix.

(4) Carcinoma of the anus and anal canal

Adenocarcinoma of the lower rectum may extend to the anal canal and perineum. Anal carcinoma proper, however, is squamous-celled and resembles epidermoid carcinoma of the skin. Metastases occur mainly in the inguinal lymph glands.

CARCINOMA OF THE LIVER

Circumscribed *liver-cell adenomas* (benign hepatomas) are occasionally seen; but they cannot be sharply distinguished from well-differentiated liver-cell carcin-omas. All transitions are seen between focal hyperplasia of liver tissue, adenoma and carcinoma.

(a) *Incidence and causation*

Primary carcinoma of the liver is rare in Europeans, accounting for only about 1 per cent of necropsy cases of malignant disease. It is much more frequent in some coloured races, e.g. the Bantu, Javanese, Chinese and Japanese, in some of whom, especially the Bantu, it is the commonest malignant tumour. Cirrhosis is present in a high proportion of cases of hepatic carcinoma, especially in native races with a high incidence of this disease, in whom cirrhosis is an almost invariable accompaniment of carcinoma. In these cases, it is clear that the cancerous change supervenes on focal regenerative hyperplasia. Carcinoma of the liver frequently supervenes in advanced cases of haemochromatosis (Fig. 197). Clonorchis infestation is a cause of bile-duct carcinoma in Hong Kong and China. Azodyes as selective hepatic carcinogens were discussed on p. 418.

(*b*) *Structure*

Two main types of structure are seen, (*a*) liver-cell carcinoma (malignant hepatoma), and (*b*) adenocarcinoma of bile-duct type (cholangioma). Many tumours are of one or the other type ; but a few show both types of structure. Liver-cell carcinomas show varying degrees of differentiation : some have cells which closely resemble liver cells and which secrete bile and stain the tumours yellow or green ; others show great anaplasia, cellular pleomorphism, multinucleated giant tumour cells, and degenerative changes and haemorrhage, even mimicking the structure of chorion-epithelioma. Cholangiomas resemble the adenocarcinomas of the extra-hepatic bile ducts (see below).

(*c*) *Spread and metastasis*

Portal branch veins are often invaded by growth, producing metastases in

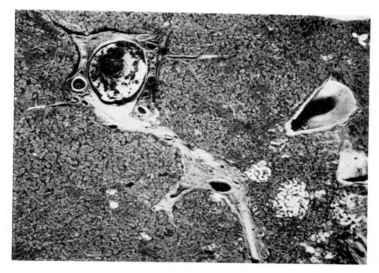

FIG. 197.—Gross tumour invasion of a large portal vein by hepatic carcinoma in a case of haemochromatosis. Multiple patches of white growth are seen in bottom right-hand corner. (Slightly enlarged.)

the liver itself (Fig. 197). Invasion of hepatic veins is also frequent and is a prelude to metastasis to the lungs. This has taken place in nearly one-third of fatal cases. Other blood-borne metastases, chiefly in bones, are less frequent.

CARCINOMA OF THE GALL-BLADDER

Simple papillomas of the gall-bladder are very rare. Carcinoma of the gall-bladder accounts for about 3 per cent of fatal cases of carcinoma in white people. Women are affected about thrice as frequently as men. Gall-stones are present in the majority of cases, and there is no doubt that the stones precede the tumours and play some part in their causation. The tumours are polypoid and infiltrating adenocarcinomas, often mucus-secreting, and sometimes (in about 10 per cent of cases) showing partial squamous metaplasia. Metastases occur frequently in neighbouring lymph glands, the liver, or peritoneum, and less frequently elsewhere.

CARCINOMA OF EXTRA-HEPATIC BILE DUCTS

Carcinoma of the main bile ducts is at least as common as carcinoma of the gall-bladder. Its usual site is in the common duct, in the lower part of which it is readily mistaken for carcinoma of the head of the pancreas. Unlike carcinoma of the gall-bladder, it is no more frequent in women than men, and gall-stones are present in only a minority of cases. The tumours are adenocarcinomas, sometimes with mucoid areas; squamous metaplasia rarely occurs. Metastasis is less frequent than with carcinoma of the gall-bladder, probably because early biliary obstruction proves fatal more quickly.

EPITHELIAL TUMOURS OF THE PANCREAS

Epithelial tumours of the pancreas comprise :—
    (1) Benign cystadenomas ;
    (2) Carcinomas of ducts and acini ;
    (3) Adenomas and carcinomas of islets of Langerhans.

(1) Benign cystadenomas

These are rare, well circumscribed tumours, composed of a honeycomb of spaces, lined by columnar or cuboidal epithelium resembling that of the pancreatic ducts, and containing mucinous secretion.

(2) Carcinomas of ducts and acini

Carcinoma of the pancreas occurs with about the same frequency as that of the biliary tract ; it affects men more often than women ; and it arises in any part of the organ. Most of the tumours are adenocarcinomas, and these are sometimes mucoid, papillary or cystic ; undifferentiated spheroidal-cell and diffuse anaplastic growths are uncommon. Carcinomas in the head of the pancreas often involve the bile duct and cause obstructive jaundice ; but those in the body or tail produce their first symptoms by their metastases or by attaining large sizes. Lymph-nodal and peritoneal metastases are frequent ; and blood-borne metastases are present in about two-thirds of fatal cases, mainly in the liver, but sometimes also in lungs, bones or brain.

(3) Adenomas and carcinomas of the islets of Langerhans

Islet adenomas are not very rare. They appear as small or large circumscribed growths, discovered incidentally at necropsy, or are clinically important by producing hyperinsulinism with attacks of hypoglycaemia.

Islet carcinomas are rarer than adenomas, from which, however, they differ only in being more active in growth and in producing metastases in the liver and elsewhere. They too produce hyperinsulinism.

ADENOMA AND CARCINOMA OF THE KIDNEY PARENCHYMA

(1) Cortical adenomas of the kidney

At necropsies we often discover in the renal cortex of middle-aged or old subjects well-defined, but usually not encapsulated, rounded or pyramidal, white or yellow, solid or finely cystic adenomas, varying in size from 1 millimetre to 2 centimetres or more in diameter. Though sometimes solitary, they are more often multiple and are often present in both kidneys. The kidneys almost always show some degree of chronic nephritic or ischaemic fibrosis.

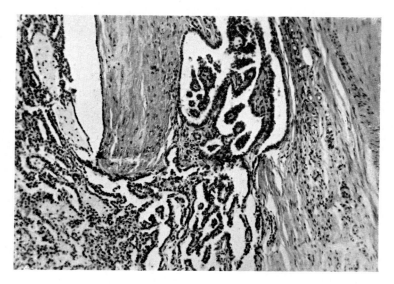

FIG. 198.—Adenoma of kidney, showing finely papillary, cystic structure and extension through capsule. (× 120.)

Microscopically (Fig. 198) adenomas consist of well differentiated but irregular tubular or cystic spaces, often with papillary ingrowths, lined by small solid cubical cells or by larger vacuolated cells laden with lipoids. Common secondary changes are granular calcification, collections of foam cells in the stroma, and intracystic haemorrhage and deposition of blood pigments. Most adenomas show no encapsulation, their marginal elements mingling with the neighbouring renal tubules. The larger and more actively growing adenomas cannot be distinguished sharply from carcinomas ; and there is no doubt that, in a proportion of cases, carcinoma arises from pre-existing adenoma.

(2) Carcinoma of the renal parenchyma

(a) Nomenclature

At the outset, we must mention, in order to discard, two unfortunate names which are frequently applied to the common renal carcinomas—" Grawitz tumour " and " hypernephroma ". Because many renal carcinomas contain lipoid-laden cells, resembling the clear cells of the adrenal cortex, Grawitz in 1884 advanced the suggestion that these tumours arise from heterotopic foci of adrenal cortex which are known to be present occasionally in the kidney. For some unaccountable reason, this quite speculative idea was widely accepted, and is still cited approvingly by some writers. Since, however, there is no real evidence in favour of it, but, on the contrary, overwhelming evidence against it, modern pathologists should cease using the inappropriate names " Grawitz tumour "

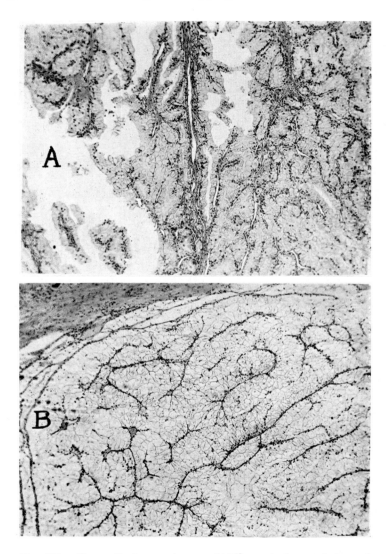

FIG. 199.—Clear-cell adenocarcinoma of kidney. A shows tubular and papillary structure ; B shows compressed structure without recognizable lumina. (× 60.)

and " hypernephroma ". The tumours so designated are *renal carcinomas*, arising from the epithelium of the convoluted tubules.

(b) Incidence and causation

Renal carcinoma accounts for 2 or 3 per cent of necropsy cases of carcinoma. It occurs in men twice or thrice as frequently as in women. Except for its relationship to adenoma in some cases, nothing is known of its causation.

(c) Naked-eye appearance

The tumours range in size from small adenoma-like growths to huge masses weighing many pounds. They usually appear well circumscribed, as if their growth were mainly expansive. Healthy parts of the tumours often have a characteristic yellow or orange colour, but in other growths or other parts of the same growth the tissue is white. Areas of fibrosis, necrosis, haemorrhage, cystic change, and sometimes calcification add to the variegated appearance of many of the tumours. Polypoid invasion of the renal pelvis, causing haematuria, or of the main renal veins is often seen ; and occasionally the tumour grows into the lumen of the inferior vena cava, even as far as the heart.

(d) Microscopic structure

This is very diverse, including the following variants :

(i) *Clear-cell adenocarcinoma* (Fig. 199). This, the most distinctive type of growth, is found in the healthy orange-yellow parts of the tumours. It may show distinct tubular or papillary structure, or it may be compressed together and devoid of recognizable lumina. The clear-cell appearance is an artefact due to the removal of lipoid substances from the tumour cells during the preparation of sections ; fat stains applied to frozen sections reveal the abundance of lipoids in the cells.

(ii) *Solid-cell papillary adenocarcinoma*, reminiscent of the common kind of structure in adenomas, may be found in parts of clear-cell tumours, and is the predominant or sole kind of structure in some growths. Occasional well differentiated, slowly growing, cystic tumours of this kind are appropriately called *papillary cystadenocarcinomas*.

(iii) *Anaplastic variants*, spindle-cell or pleomorphic-cell, and often diffuse and sarcoma-like, are frequent and have led to many erroneous diagnoses of " sarcoma " and " carcino-sarcoma ". Highly vascular and haemorrhagic areas may also produce confusing pictures. Several or all of these kinds of structure may be present in a single tumour.

(e) Metastasis

Metastases often occur both in lymph glands and by the blood-stream, the frequency of the latter being readily understood from the frequency with which carcinomas of the kidney invade the renal veins. The commonest sites of metastasis are the lungs, liver, bones and brain ; but no organ is exempt. Metastatic growths in unusual situations, e.g. skin, eye, oral or other mucous membranes, have often caused diagnostic difficulties ; and metastases in bones have often been mistaken for primary tumours. Left-sided renal tumours which have invaded the main renal vein sometimes liberate retrograde emboli down the ovarian or spermatic vein and so produce metastases in the pelvic organs or external genitalia.

EPITHELIAL TUMOURS OF THE URINARY PASSAGES

From the transitional stratified epithelium of the bladder, renal pelvis and ureter, papillomas and carcinomas arise which are similar in structure and behaviour in all these situations.   No sharp distinction between papillomas and carcinomas is possible ; all gradations from one to the other are seen, and their separation is arbitrary.   Haematuria is the commonest first symptom.

(a) Incidence and causation

Men are affected thrice as frequently as women.   The tumours arise in the bladder, renal pelvis and ureter, in that order of frequency.   In many cases the growths are, or become, multiple and sometimes very numerous.   The multiplicity is largely attributable to multifocal tumour formation, though metastasis by surface implantation probably also plays a part.

The only clear instance of an occupational incidence of urinary tract tumours is in aniline dye workers ; the chief carcinogenic agent responsible has been shown experimentally to be β-naphthylamine (p. 418).   It is quite possible, though still unproved, that some of the dyes used in sweets and cordials may be similarly carcinogenic.   It is believed that schistosomiasis of the bladder predisposes to carcinoma, but this is not quite certain and further investigation is needed.   The malformation, exstrophy of the bladder (ectopia vesicae) often shows mucoid glandular metaplasia of the exposed mucosa, and adenocarcinoma sometimes supervenes.   There is no evidence that chronic bacterial infections, —tuberculosis, pyelo-nephritis and cystitis—predispose to carcinoma.

(b) Naked-eye appearance

The tumours are single or multiple, sessile or pedunculated, papillary or solid, non-infiltrating or infiltrating, non-ulcerated or ulcerated, and there may be any combination or sequence of these characters.

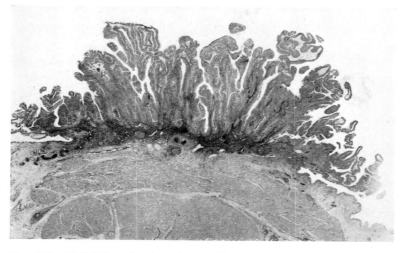

FIG. 200.—Well-differentiated, early, papillary carcinoma of bladder, as yet not invading muscle coat.   ($\times$ 6.)

(c) Microscopic structure (Figs. 200-202)

This may show :—

(1) *Highly organised villous papillary growth*, the structure of the typical " benign papilloma " ;

(2) *Irregular papillary growth*, showing more bulky, less typical epithelium —the structure of most papillary carcinomas ; invasion of the subjacent muscle coats may or may not have occurred ;

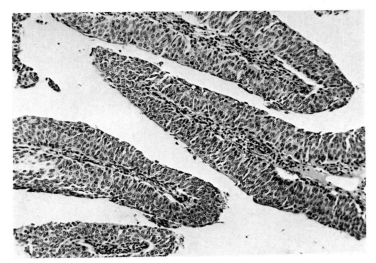

FIG. 201.—Structure of villous papilloma of bladder.   (× 120.)

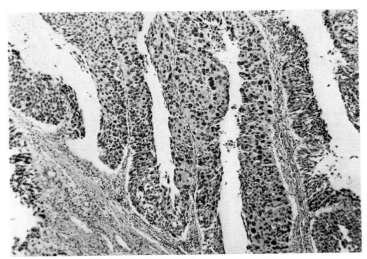

FIG. 202.—Structure of villous papillary carcinoma of bladder, showing cellular pleomorphism. (× 80.)

(3) *Cornifying squamous-cell carcinoma*, a frequent variant of (2) ;

(4) *Anaplastic carcinoma*, devoid of differentiated characters ;

(5) *Mucoid adenocarcinoma*, a rare form of growth, supervening on metaplastic *cystitis glandularis*, as in the exstrophied bladder, or from remnants of the urachus at the vertex of the bladder.

Again it must be stressed that (1) to (4) are not separate kinds of growth ; all gradations are seen.   But (5) is distinct from the other variants.

(d) Growth and prognosis

Some solitary papillomas are cured by local removal ; but in many cases multiple recurrent growths appear, and these often become increasingly active and finally invasive.

(e) Metastasis

Urinary tract carcinomas produce both lymph-nodal and blood-borne metastases, the latter occurring principally in the lungs, liver and bones.

EPITHELIAL AND RELATED TUMOURS OF THE OVARY

Ovarian tumours show great structural variety ; and a proper conception of their nature and origin depends on a proper conception of the histogenesis of the normal tissues of the ovary.   The old view that the follicular epithelium arises as downgrowths from the surface " germinal " epithelium, and is histogenetically distinct from the surrounding ovarian stroma, is now known to be mistaken. The cellular so-called " stroma " of the cortex of the ovary is *not* a mere supporting tissue, but is the essential ovarian parenchyma or plastic mother-tissue from which follicles are formed throughout adult reproductive life.   It produces both the epithelial granulosa-cells lining the follicles and the non-epithelial cells of the theca folliculi, which in their turn mature into luteal cells and finally the fibrous tissue of the corpus albicans.   If, then, tumours arise from this formative parenchyma of the ovary, we might expect their tissues to show evidence of differentiation into one or more of these several differentiated products of the ovarian follicles.   As we shall see, this expectation is fulfilled.

Excluding teratomas, which are considered on p. 585, ovarian tumours may be described, in order of frequency, under the following headings :

(1) Benign fibro-papillary growths and serous cystadenomas;

(2) Pseudomucinous cystadenomas ;

(3) Carcinomas corresponding to, or arising in, (1) and (2) ;

(4) Granulosa-cell, theca-cell and luteal-cell tumours ;

(5) Rarer special varieties of tumours, including :

(a) " Arrhenoblastoma " ;

(b) Brenner tumour ;

(c) " Dysgerminoma ";

(d) Others.

(1) Benign fibro-papillary growths and serous cystadenomas

These are the commonest ovarian tumours.   They appear as warty or cauliflower-like growths, springing from the surface of the ovary, or from the lining of small or large cysts in the organ, or from both ; and they are frequently bilateral. Their structure is essentially similar to that of the corrugations and crypts of the normal ovarian surface, of which the tumours appear to be but a neoplastic exaggeration. Branched connective tissue papillae are clothed by a single layer of cubical or columnar epithelial cells representing the "germinal" epithelium of the ovarian surface (Fig. 203). Papilliferous intra-ovarian serous

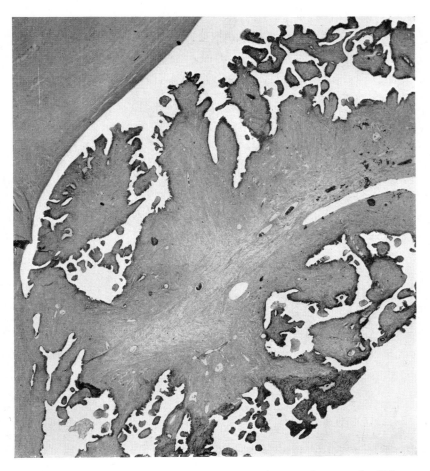

FIG. 203.—Section of papilliferous serous cystadenoma of ovary. (× 10.)

cysts, with watery contents, are clearly the same kind of growth, the cysts having arisen from the deeper parts of the surface crypts of the ovary. The epithelium secretes serous fluid, which accumulates within the cavities of cystic growths or may produce ascites from surface ones. Some tumours are gritty, because of the presence of scattered calcified granules.

The tumours grow slowly and non-invasively. But otherwise benign-looking growths sometimes disseminate widely in the abdominal cavity, producing a crop of similar non-invasive papillary growths on the peritoneal surfaces, a condition accompanied by marked ascites which recurs repeatedly after aspiration.

(2) Simple and papillary pseudomucinous cystadenomas

In their simple form these are multilocular honeycomb-like cystic growths occupied by mucinous fluid, and enclosed by thick fibrous outer cyst walls. They may attain huge sizes. The cysts are lined by columnar mucus-secreting epithelial cells, which in papilliferous tumours also clothe the numerous intra-cystic papillomas (Fig. 204). This epithelium is probably similar to that of the serous cystadenomas, but has acquired mucus-secreting powers by metaplasia. The mucin

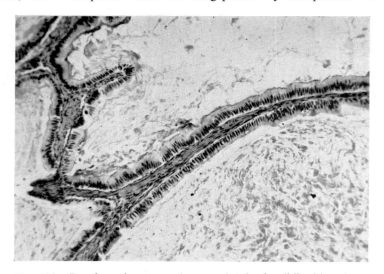

Fig. 204.—Pseudomucinous cystadenoma, showing loculi lined by a layer of mucous columnar cells. (× 110.)

differs in some minor respects from alimentary mucus ; hence the designation " pseudomucin ".

In their slow non-invasive growth and their usually good prognosis, pseudomucinous tumours resemble their serous counterparts, from which, however, they differ in not forming surface papillomas and therefore not causing secretory ascites. Rupture of the cyst walls by trauma or by the growth of prolific intra-cystic papillomas, however, may lead to spilling of quantities of mucinous secretion and mucus-secreting cells into the peritoneal cavity, a serious condition called " pseudomyxoma peritonei ".

(3) Carcinomas of serous or pseudomucinous alliance

No sharp distinction can be made between the more active papilliferous cystadenomas on the one hand and invasive papillary cystic carcinomas on the other. The carcinomas corresponding to, or arising in, cystadenomas constitute the majority of malignant tumours of the ovary. The varieties of structure seen in them include—well differentiated papillary adenocarcinoma, which in some cases shows granular calcification and is then sometimes called "psammo-carcinoma"; cystic mucoid adenocarcinoma, the malignant counterpart of pseudomucinous cystadenoma; solid spheroidal-cell or pleomorphic-cell carcinoma; and, rarely, squamous-cell structure due to metaplasia.

The tumours frequently become bilateral, grow to great sizes, and invade neighbouring viscera and the pelvic or abdominal walls. Peritoneal metastasis is frequent, leading to ascites, gelatinous accumulations and widespread adhesions. Metastases are present in the pelvic or abdominal lymph glands in the majority of fatal cases. Blood-borne metastases, mainly in lungs or liver, are infrequent.

(4) Granulosa-cell, theca-cell and luteal-cell tumours

These must be considered together, for they are akin and often combined. All the tumours of this group are essentially tumours of the formative ovarian

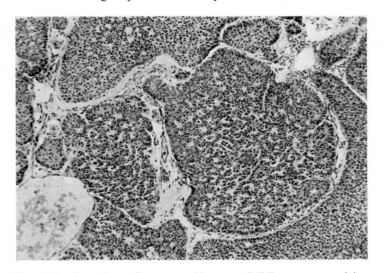

FIG. 205.—Granulosa-cell tumour with many Call-Exner spaces, giving rosetted appearance. (× 100.)

stroma, and they exhibit in varying degrees the several differentiated products of this tissue, namely granulosa cells, theca cells and luteal cells. Some tumours consist predominantly or exclusively of one or another of these elements; others show mixtures of two or three of them in varying proportions. The tissues of tumours of mixed structure often show clear evidence of progressive maturation changes comparable with those of normal follicles, namely luteinization of both granulosa and theca cells and the formation of bands of hyaline fibrosis resembling those of the corpus albicans. In a proportion of cases, the tumours secrete oestrogenic hormone and produce signs of hyperoestrinism.

(a) Naked-eye appearance

Pure granulosa-cell tumours are encapsulated growths composed of firm white or yellow solid tissue which, however, may contain cystic spaces. Most theca-cell tumours are wholly solid and consist of firm fasciculated tissue resembling fibrous tissue but with a faint or distinct yellow colour. Pure white dense "fibromas" of the ovary, which sometimes occur, are probably in most cases completely matured fibrous theca-cell tumours.

(b) Microscopic structure

The following types of structure are seen :—

(i) *Epithelial granulosa-cell structure.*—This may consist of compact masses of granulosa cells closely resembling those of normal follicles and containing Call-Exner spaces (Fig. 205), or of less typical epithelial trabeculae or networks.

(ii) *Theca-cell structure.*—This consists of interwoven bundles of spindle cells, sometimes plump and closely packed and resembling those of undifferentiated ovarian stroma, sometimes less cellular and more fibrous and resembling dense fibroma. The spindle cells often contain demonstrable lipoid droplets, which give the tumours their yellow colour. Cellular and fibrous areas may alternate.

(iii) *Luteal-cell structure.*—This shows large plump polyhedral lipoid-laden cells, which give the tissue a bright yellow colour.

(iv) *Mixed structure.*—This shows (i) and (ii) or (i), (ii) and (iii) intermingled in varying proportions.

(c) Endocrine effects

Functioning granulosa-cell and theca-cell tumours produce pronounced hyperplasia of the endometrium. If occurring before puberty, they induce precocious sexual development and menstruation ; if after the menopause, irregular and often severe uterine bleeding. Mammary enlargement and secretion have been recorded in a few cases.

(d) Malignancy

Granulosa-cell tumours are usually but not always benign ; post-operative recurrence or metastasis has occurred in a small proportion of cases. Theca-cell tumours are almost always quite benign and are cured by removal ; but it is possible that very rare spindle-celled sarcomas of the ovary may really be malignant theca-cell tumours.

(5) Sundry rare kinds of tumours

(a) " Arrhenoblastoma " (masculinizing tumour)

This is the name applied to rare epithelial tumours of the ovary accompanied by signs of masculinization, namely amenorrhoea, sterility, atrophy of uterus and breasts, hairiness, and enlargement of the clitoris. Microscopically, many of the tumours resemble granulosa-cell or mixed granulosa-cell and theca-cell tumours; and there is doubt that they belong to this group, but that for some unknown reason they have produced excess of androgens instead of oestrogens. This view is supported by the fact that androgens and oestrogens are closely related and easily interconvertible, and by the occurrence of cases of ovarian tumour (called " gynandroblastoma ") accompanied by signs of both hyperoestrinism and masculinization. Other masculinizing tumours are hilar-cell tumours, and are essentially the same as Leydig-cell tumours of the testis (p. 520).

(b) Brenner tumour or oöphoroma folliculare

These are the names applied to a usually benign tumour composed of clumps of large polyhedral epithelial cells set in a plentiful dense fibrous stroma, and usually not accompanied by hormonal disturbances. Brenner's original belief that his tumour arose from the ovarian follicles was probably correct; it is thus akin to granulosa-cell tumour.

(c) " Dysgerminoma "

This is the name applied to rare solid ovarian tumours, with a microscopic structure resembling that of seminoma of the testis, and devoid of endocrine effects. The idea that they may indeed be seminomas arising in testicular tissue in a bisexual gonad is supported by the fact that a relatively large number of them have been in patients with pseudo-hermaphroditism. However, their histogenesis is still unsettled ; and it may be that they are not really seminomas but only seminoma-like. They usually grow slowly, often to a large size, and they consist of uniform solid white tissue. In some cases surgical removal is curative ; but in others fatal recurrences and metastases develop.

(d) Others

Very rarely an *endometrial tumour*, either an adenocarcinoma or a mixed tumour, arises from endometriosis in the ovary. The so-called *mesonephromas* are probably only structural variants of cystadenomas or other kinds of tumours. A rare *distinctive tumour of young subjects* occurs, composed of epithelial tubules and open-textured meshworks; its histogenesis is still uncertain. It remains to add here that *secondary tumours* in the ovary sometimes produce oestrogenic or androgenic effects by evoking great theca-cell hyperplasia.

CARCINOMA OF THE FALLOPIAN TUBE

Tubal carcinoma, a rare tumour, is usually of well differentiated papillary type. In about one-third of the cases the tumours are bilateral. They distend the tubes, spread through their abdominal ostia, and frequently disseminate widely in the peritoneal cavity. Blood-borne metastases are rare. Tuberculous salpingitis is also present in a significant number of cases.

EPITHELIAL TUMOURS OF THE UTERUS

By far the most frequent epithelial tumours of the uterus are *carcinomas*. Some of the common endometrial and cervical polypi are probably benign tumours —adenomas of the endometrial or cervical glands ; but it is difficult or impossible to distinguish these from the localized forms of hyperplasia to which the uterine epithelia are liable. Here we will consider :

(1) Epidermoid carcinoma of the cervix ;
(2) Endometrial adenocarcinoma of the corpus ;
(3) Adenocarcinoma of the cervix ;
(4) Endometrial mixed tumours.

(1) Epidermoid carcinoma of the cervix

This is a very common tumour, and occurs three or four times as frequently as endometrial carcinoma. Women who have borne children are decidedly

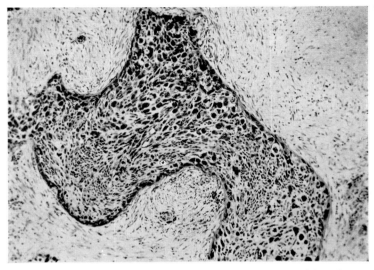

FIG. 206.—Great cellular pleomorphism in anaplastic epidermoid carcinoma of cervix. (× 80.)

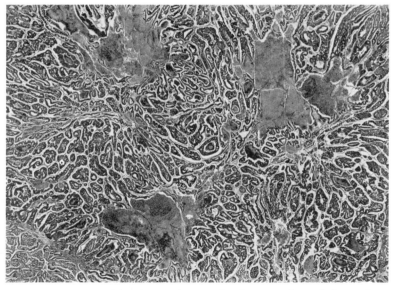

FIG. 207.—Endometrial adenocarcinoma showing complex tubular pattern. Note also areas of necrosis in tumour. (× 10.)

more liable to it than nulliparous women. The commonest age of onset of the disease is in the fifth decade. It usually arises from the squamous stratified epithelium of the vaginal surface of the cervix or external os ; and it appears as either a projecting papillary mass, an infiltrating and ulcerating growth, or a combination of these. Microscopically most of the tumours are easily recognizable squamous-cell carcinomas with varying degrees of keratinization, but some are anaplastic and diffuse (Fig. 206).

From the cervix, the tumour may extend to the body of the uterus, the vaginal walls and the parametrium ; and in advanced cases, the bladder or rectum may be invaded, and vesico-vaginal or recto-vaginal fistulae may develop. Obstruction of the ureters often leads to fatal ascending bacterial infection of the kidneys.

Metastases occur frequently in the parametrial and iliac lymph glands, whence the retroperitoneal groups of glands are later affected and even the thoracic duct may be invaded. Blood-borne metastases, mainly in liver or lungs, are less frequent.

(2) Endometrial adenocarcinoma of the corpus

In contrast to cervical carcinoma, endometrial carcinoma often occurs in single or nulliparous women, most often in the sixth decade of life. There is evidence that in some cases the disease supervenes on previous endometrial hyperplasia. Microscopically most of the tumours are well-differentiated adenocarcinomas, often with complex tubular or papillary patterns (Fig. 207) ; but a few are solid anaplastic growths. Squamous metaplasia is frequent in parts of the tumours, which are then sometimes called " adeno-acanthomas ".

In their early stages the tumours are confined to the endometrium itself. Later they fill and distend the organ and invade the myometrium. They may also extend into one or both Fallopian tubes, or through the cervical canal into the vagina. In general, endometrial carcinoma metastasizes late. This may take place to lymph glands, or by the blood-stream, or by trans-tubal carriage of tumour particles into the Fallopian tubes or onto the ovaries or pelvic peritoneum. Metastases in the vagina or vulva, sometimes attributed to surface implantation, are almost certainly due to direct extension or local embolic spread in the lymphatics or veins of the vaginal walls.

(3) Adenocarcinoma of the cervix

This relatively rare tumour arises from the cervical glands, and sometimes, like its parent tissue, it secretes mucus. Its behaviour generally resembles that of squamous-cell carcinoma of the cervix.

(4) Endometrial mixed tumours

There occur in middle-aged and elderly adults bulky tumours of the endometrium, containing both endometrial glandular tissue and a variety of non-epithelial tissues including undifferentiated mesenchyme-like tissue, mucoid tissue, smooth or striated muscle, and cartilage or bone. These tumours are thus composite carcino-sarcomas; but there occur also purely non-epithelial sarcomatous mixed tumours; the two varieties together form a single class of endometrial mixed tumours.

(5) Symptoms and diagnosis of uterine tumours

The chief symptom of all of the foregoing kinds of tumours is *haemorrhage.* Any unnatural uterine bleeding in a woman over 40 years of age, especially post-menopausal bleeding, calls for careful examination to determine its cause and to establish or exclude a diagnosis of tumour. To this end, diagnostic curettage of the uterine cavity, and microscopical examination of the curettings and of biopsy pieces from any suspicious lesions of the cervix, are often essential.

EPITHELIAL TUMOURS OF THE VAGINA AND VULVA

These include :—

(1) Squamous-cell carcinoma of the vagina ;

(2) Papillary adenocarcinoma of the vaginal vault, derived from remains of Gartner's duct ;

(3) Squamous-cell carcinoma of the vulva ;

(4) Basal-cell carcinoma of the vulva ;

(5) Adenoma and carcinoma of Bartholin's glands ;

(6) Tumours of the vulval sweat or apocrine glands, especially hidradenoma.

All of these, except (3), are rare. *Squamous-cell carcinoma* of the vulva occurs with about one-twentieth the frequency of uterine carcinoma. It occurs chiefly in elderly subjects, and is often preceded or accompanied by more or less extensive leucoplakia, a hyperplastic thickening and over-keratinization of the epidermis, which in this situation is a dangerous pre-cancerous lesion. Carcinoma super-vening on leucoplakia is often multiple or widespread in origin. Metastasis to the inguinal lymph glands often takes place early.

EPITHELIAL AND RELATED TUMOURS OF THE TESTIS AND EPIDIDYMIS

(1) Classification of testicular tumours

Of recent years, there has been great confusion regarding the histogenesis and classification of testicular tumours. This has been due to wide acceptance of an erroneous view that the commonest of these tumours—the seminomas—arise from teratomas, and to use of the unwarranted name " embryonal carcinoma ", under which various writers have confused seminomas with anaplastic cellular teratomas. Full discussion of the subject will be found in my " Pathology of Tumours " and need not be repeated here. Testicular tumours should be grouped and named thus, in order of frequency :—

(*a*) Seminomas ;

(*b*) Teratomas ;

(*c*) Tumours of Sertoli cells and excretory ducts ;

(*d*) Interstitial-cell tumours.

(*a*) *Seminomas*

These, the commonest tumours of the testis, arise from the spermatic cells of the seminiferous tubules ; they are *seminal carcinomas* or *spermatocytomas.* Contrary to a prevalent belief, they are unrelated histogenetically to teratomas, but they sometimes arise in testes already containing teratomas. They appear chiefly in the fourth and fifth decades of life, are decidedly commoner in incompletely descended than in normal organs, are occasionally multiple or bilateral, and occasionally familial. Injury has often been blamed as a causative factor but this is doubtful. Seminomas are common in dogs, in which multifocal origin from the seminal tubules is often well seen.

(i) *Naked-eye appearance.*—Most seminomas appear as well-defined, lobulated growths composed of uniform, rather soft, white tissue in which, however, yellow areas of necrosis are common.

(ii) *Microscopic structure.*—This is usually characteristic and easily recognized (Fig. 208). The tumours consist of well- or ill-defined clumps of large rounded or polyhedral cells, often closely resembling spermatocytes, set in a connective-tissue stroma which contains few or many lymphocytes and sometimes even lymphoid follicles. At the margins of young tumours, seminal tubules occupied by growth are often seen, and the appearances point clearly to progressive neoplastic change in the seminal epithelium.

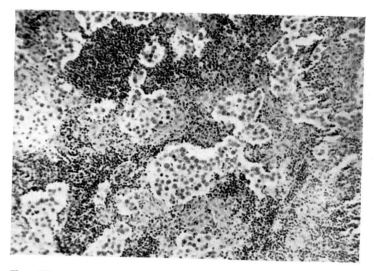

FIG. 208.—Seminoma with many lymphocytes in stroma. (× 120.)

(iii) *Extension and metastasis.*—Direct spread to the epididymis and cord is frequent. Discrete nodules sometimes appear along the cord above the main growth. Metastasis to the upper lumbar lymph glands *via* the lymphatics accompanying the spermatic artery occurs frequently and is the main factor in prognosis. Blood-borne metastases, principally in the lungs or liver, are less common.

(b) Teratomas

Teratomas of the testis will be discussed along with those of other organs in Chapter 35. Here they are mentioned only to complete the list of testicular tumours, and to insist that with rare exceptions they can and should be distinguished from seminomas. Some anaplastic teratomas contain much cellular undifferentiated epithelial tissue, which has often been confused with seminoma under the unfortunate name " embryonal carcinoma " ; but adequate examination will nearly always make the distinction clear. A seminoma sometimes arises in a testis in which a teratoma is also present ; the two growths are initially distinct, but may later coalesce. The average age of men with teratomas is about 30, i.e. 10 years earlier than the age of those with seminomas.

" *Chorion-epithelioma* " *of the testis.*—This name is applied to rare rapidly growing haemorrhagic tumours, part of which microscopically resemble true chorion-epithelioma. Most of these tumours are anaplastic teratomas, in which chorionic components predominate.

(c) *Tumours of Sertoli cells and excretory ducts*

The excretory ducts of the testis comprise the straight tubules, rete, efferent tubules, epididymis and ductus deferens. The Sertoli cells of the seminal tubules are also allied to the epithelium of the excretory ducts. Tumours of all of these non-seminiferous epithelia are rare. They are :

     (1) *Intra-testicular tumours :*
         (*a*) " Tubular adenomas " or Sertoli-cell tumours ;
         (*b*) Papillary adenocarcinomas of the rete and excretory tubules ;
     (2) *Extra-testicular tumours*, adenomas and carcinomas of the epididymis or of the appendix testis or other Müllerian vestiges.

(d) *Interstitial-cell tumours*

Tumours of the interstitial cells of Leydig are rare in man, though quite common in the dog. They are well circumscribed, solid, brown or orange growths, composed of masses of polyhedral cells with granular or vacuolated cytoplasm containing lipoid droplets or crystalloids. Most of them are benign, growing slowly and not metastasizing; but on rare occasions more active growth and metastasis have occurred. The tumours secrete androgens; and, when they develop in young children, they cause precocious puberty. Occasionally, gynaeco-mastia has occurred, showing that the tumours may secrete oestrogens as well as androgens.

(2) **Hormonal results of testicular tumours**

Besides the direct androgenic effects of Leydig-cell tumours just referred to, testicular tumours of other kinds also may produce hormonal disturbances. These include increased urinary excretion of gonadotrophins, sometimes sufficient to give a positive Aschheim-Zondek test as strong as that of pregnancy, and mammary hyperplasia and secretory activity (gynaecomastia). These results occur most often with rapidly growing anaplastic teratomas, especially the " chorion-epitheliomas ", indicating that these are truly chorionic in nature as well as in appearance.

CARCINOMA OF THE PROSTATE

Prostatic carcinoma is notable as occurring later in life than any other form of malignant disease, the average age at death of its victims being over 70, i.e. about a decade later than that of carcinomas in general. It is probable that simple

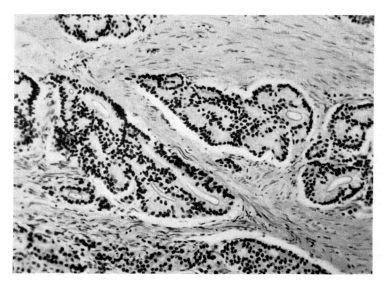

FIG. 209.—Adenocarcinoma of prostate. (× 150.)

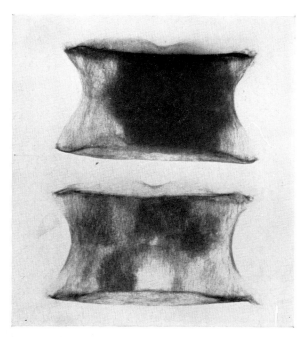

FIG. 210.—X-ray photograph of vertebrae containing osteoplastic metastases of prostatic carcinoma.
(Positive print)

hyperplastic enlargement of the prostate precedes carcinoma in a considerable proportion of the cases.

(a) Structure

Most tumours are adenocarcinomas with an acinar structure which is sometimes so well differentiated that it can be identified as prostatic by the experienced microscopist (Fig. 209). Other tumours are predominantly polyhedral-celled, either alveolar or diffuse. Squamous and mucoid metaplasia are both rare.

(b) Metabolic peculiarities

Well differentiated prostatic carcinoma shares with normal prostatic epithelium the power of secreting acid phosphatase, and cases with extensive metastases often show markedly increased amounts of this enzyme in the blood. Also like its parent tissue, the growth of prostatic carcinoma is encouraged by androgens and inhibited by oestrogens ; hence the temporary arrest of growth and prolongation of life obtained in some cases by castration and by treatment with oestrogens.

(c) Metastasis

Lymph-nodal metastases are present in a high proportion of fatal cases. Blood-borne metastases are also frequent, especially in the lungs, bones and liver. Metastases in bones are very common, and they are sometimes responsible for the first or main symptoms. Persistent " lumbago ", " sciatica " or other " rheumatic " pains in an elderly man should always excite suspicion of possible cancer of the prostate. The skeletal metastases are often osteoplastic, producing ivory-hard sclerosis of the affected bone with dense radiographic shadows, and sometimes prolific osteophytosis of the bone surfaces (Fig. 210). Contrary to what is often stated, the distribution of the metastases of prostatic cancer in the skeleton shows no special predilection for the pelvic and lumbar regions, but is similar to that of other tumours ; metastases often occur in the skull, upper parts of the spine, ribs, sternum and humerus, as well as in the bones of the lower parts of the trunk and the femur.

EPIDERMOID CARCINOMA OF THE PENIS AND SCROTUM

Epithelial tumours of the male external genitalia are almost all squamous-cell carcinomas of the epidermis ; carcinomas of the urethral mucosa are very rare, and so are basal-cell and glandular tumours of the epidermis in this region.

(1) Carcinoma of penis

This is a relatively infrequent tumour among Europeans, though much commoner among the Oriental races. Circumcision is undoubtedly the main factor responsible for racial and community differences in the incidence of penile carcinoma. This disease is almost unknown among the Jews who regularly practise circumcision during infancy ; and it is unusual in Mohammedans who circumcise during childhood or adolescence. In Europeans, it is much more frequent in uncircumcised than in circumcised persons, and phimosis is present in a high proportion of patients with this disease. The usual sites are on the glans or inner surface of the prepuce. In their early stages the tumours are usually papillary and superficial, less often ulcerating and infiltrative ; a rare mode of origin is in the form of " Paget's disease " or " erythroplasia ", a form of intra-epidermal carcinoma essentially similar to Bowen's disease of the skin (p. 477).   Advanced growths invade the urethra and the erectile tissue of the corpora cavernosa and spongiosa.   Metastases occur in the inguinal lymph glands in more than one-half of the cases.   Blood-borne metastases are rare.

(2) Carcinoma of scrotum

Chimney-sweeps' and mule-spinners' cancers are the classical examples of carcinoma of the scrotum, and a large proportion of cases of scrotal cancer are occupational.   The disease is very rare among professional and clerical workers and others with clean occupations.   The tumours usually start as warty growths, often multiple, which may long remain superficial, but eventually become invasive and ulcerated.   Metastasis to the inguinal lymph glands often occurs in the later stages.

EPITHELIAL TUMOURS OF THE THYROID GLAND

In the thyroid, as in many other organs, no sharp separation of hyperplasias, benign tumours and malignant tumours is possible.   Nodular areas of hyperplasia in goitrous thyroids become adenomatous, structural distinction between adenoma and carcinoma is sometimes impossible, and not a few " benign " adenomas eventually prove malignant by metastasizing.   However, it is convenient and necessary to assign thyroid overgrowths to one or another group, and in most cases the behaviour of the lesions corresponds with their histopathology.   The non-neoplastic hyperplasias of the thyroid, or goitres, are dealt with in Chapter 36 ; here we consider only the true tumours.   Most of these are adenomas or carcinomas ; mixed epithelial and mesenchymal tumours are very rare.

(1) Adenomas

These common tumours appear as rounded, well circumscribed, single or multiple, encapsulated masses of very variable appearance, some vesicular and thyroid-like, some solid and opaque, and many showing cysts, haemorrhage, fibrosis or calcification.   Most of them are in goitrous thyroids, and they are commoner in women than men.   Microscopically, they may consist of colloid-filled vesicles (Fig. 211), acini devoid of colloid (improperly called " foetal adenoma "), papillary structures with or without colloid, or all of these.   An unusual type of adenoma consists of large eosinophil cells lining small colloid-free acini or forming solid clumps and cords : this so-called " Hürthle-cell " type of growth will be considered under carcinoma.

Adenomas grow slowly, and sometimes come to a standstill and undergo retrogressive changes such as cyst formation and calcification (Fig. 212).   On the other hand, carcinoma may supervene in long-standing, apparently quiescent adenomas.   Papillary " adenomas " always possess low-grade potential or actual malignancy, often recurring repeatedly after local removal or eventually metastasizing.

(2) Carcinomas

(a) Frequency and causation

The frequency of thyroid carcinoma differs greatly in different communities according to the prevalence of goitre.   In goitrous regions, thyroid carcinoma

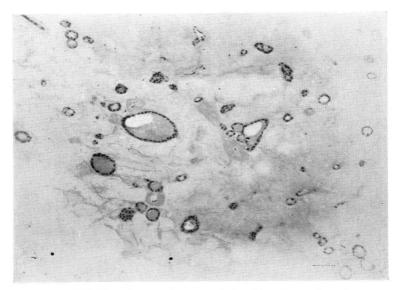

FIG. 211.—Colloid vesicular adenoma of thyroid with much oedematous
stroma. (× 90.)

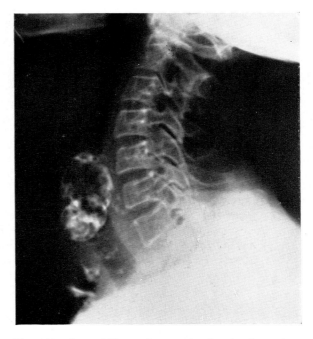

FIG. 212.—Lateral X-ray photograph of neck of a patient
with large calcified adenoma of thyroid.

may account for as much as 5 per cent of all carcinomas ; but in relatively goitre-free regions, for only one-tenth this number. Women are affected twice or thrice as frequently as men. Nodular goitre or adenoma precedes the development of carcinoma in a high proportion of cases ; and it is probable that most carcinomas actually spring from pre-existing adenomas. Thyroid carcinoma in young subjects is sometimes due to previous X-irradiation of the neck. Carcinoma of lingual thyroid tissue in the base of the tongue occasionally occurs.

(*b*) *Structure*

Individual tumours may show one or another of the following types of structure, several of which however are sometimes seen in one tumour :—

(1) *Colloid-containing vesicular adenocarcinoma*, more or less resembling normal or goitrous thyroid tissue (Fig. 213) ;

(2) *Non-colloid acinar adenocarcinoma*, of varying degrees of differentiation ;

(3) *Papillary adenocarcinoma*, often cystic ;

(4) *Carcinoma simplex*, either alveolar, or diffuse and pleomorphic-celled and easily mistaken for sarcoma ;

(5) *Squamous-cell structure*, a rare type ;

(6) *Large-eosinophil-cell structure.*

The eponym " Hürthle-cell " tumour, usually applied to this last type, is unfortunate, because it is doubtful if the tumours are related to the cells described by this histologist, and because in any case he was not the first to describe them. The tumours had better be called *large-eosinophil-cell tumours.* They do not arise from any special source but from ordinary thyroid epithelium ; they merely

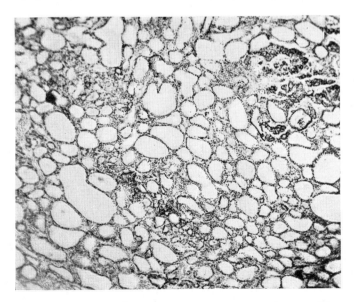

FIG. 213.—Colloid vesicular adenocarcinoma closely resembling normal thyroid tissue, from a pulmonary metastasis. (× 60.)

represent one of the many structural variants of which this epithelium is capable. They usually appear as well-defined, solid, homogeneous, opaque white or yellow growths ; and they consist microscopically of large eosinophilic finely granular or foamy polygonal cells, arranged in solid cords or around small acinar spaces. Many of the tumours are benign in that they grow slowly and do not recur after removal ; but in some cases they are malignant, recurring and metastasizing.

(c) Growth and degree of malignancy

Thyroid carcinomas differ greatly in their behaviour. The highly differentiated papillary growths grow slowly, metastasize mainly or solely to the neighbouring cervical lymph glands and spread from gland to gland over a period of years. Some of the vesicular adenocarcinomas arising in pre-existing adenomas produce prolific blood-borne metastases in lungs and bones. Most malignant of all are the anaplastic sarcoma-like growths which grow fast and to a huge size and metastasize quickly by the blood-stream.

(d) Metastasis

Metastasis to the cervical lymph glands is common from infiltrative and papillary types of carcinoma ; but is often absent in cases of " malignant adenoma " in which the tumour is still confined within the adenoma capsule, yet has metastasized by the blood-stream. Most, possibly all, supposed " lateral aberrant thyroids " are really lymph-nodal mestastases of papillary adenocarcinoma from small unsuspected primary tumours in the thyroid gland. In about three-quarters of fatal cases of thyroid carcinoma, blood-borne metastases are present in lungs, other viscera or bones. The source of these is often demonstrable as invaded small or large veins involved in the primary growth. The metastases in bones are sometimes mistaken clinically for primary tumours.

(3) Mixed tumours of the thyroid

Very rarely there occur mixed tumours containing both thyroid epithelial tissue and sarcomatous mesenchymal tissue, fibrous, osteogenic or cartilaginous. The metastases of these in the lungs usually contain the sarcomatous tissue only.

(4) Function in thyroid tumours

In only a small proportion of cases of adenoma or carcinoma of the thyroid, are symptoms of hyperthyroidism present.

EPITHELIAL TUMOURS OF THE THYMUS

Thymic tumours are rare. Most of them arise from the epithelial reticulum of the organ ; but, like their parent tissue, they show a rather diffuse structure and an intimate admixture of lymphocytes. Cornified concentrically laminated foci resembling Hassall's corpuscles are found in some of the tumours. Some thymic tumours are malignant carcinomas, producing both lymph-nodal and blood-borne metastases. But many others, including most of those associated with myasthenia gravis, are benign growths.

EPITHELIAL PITUITARY AND PARA-PITUITARY TUMOURS

Epithelial tumours of the pituitary gland and of the developmental residues associated with it can be grouped conveniently thus :

(1) *Tumours of the secreting tissue proper :*
   (*a*) Benign adenomas :
    (i) Chromophobe   (iii) Basophil
    (ii) Eosinophil   (iv) Mixed
   (*b*) Carcinomas.

(2) *Tumours of the para-pituitary residues :*
   (*a*) Simple cysts ;
   (*b*) Squamous-cell carcinomas and " adamantinomas ".

(1) Tumours of the secreting tissue proper

(*a*) *Pituitary adenomas*

These are benign growths, most of which, possibly all, arise from the anterior lobe. According to the staining properties of their cells, three main types of adenomas are distinguished—chromophobe (without distinctive staining properties), eosinophil and basophil ; but this distinction is somewhat arbitrary, since tumours of mixed type occur.

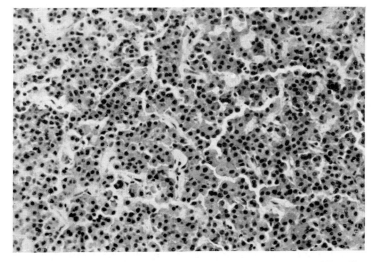

FIG. 214.—Pituitary adenoma composed entirely of eosinophil cells, from a case of acromegaly. (× 180.)

(i) *Chromophobe adenomas.*—These comprise about two-thirds of pituitary adenomas. They are of diverse micro-structure, often not much resembling pituitary tissue. They produce no symptoms attributable to secretory activity by the tumour cells ; their usual endocrine results being due to diminished pituitary functions or to compression of the hypothalamic region.

(ii) *Eosinophil adenomas.*—These resemble anterior-lobe tissue, but contain an excessive proportion of eosinophilic cells with characteristic granules (Fig. 214). They produce excessive growth hormone, with consequent acromegaly or acromegalic gigantism (p. 605).

(iii) *Basophil adenomas.*—These are rare, usually small growths, sometimes found in the endocrine disorder known as "Cushing's syndrome" (see p. 605).

(iv) *Growth of pituitary adenomas.*—These tumours grow slowly and non-invasively. Because of their intrasellar position, they usually cause early enlargement of the pituitary fossa and compression and hypofunction of the remainder of the gland. As they enlarge they project through the diaphragma sellae into the cranial cavity, where they compress the optic chiasma and tracts (causing progressive and often characteristic impairment of vision), indent the base of the brain, push into the third ventricle, and even extend through the foramina of Monro into the lateral ventricles.

(*b*) *Pituitary carcinomas*

Occasional pituitary glandular tumours are malignant, in that they invade and destroy the surrounding bones and meninges, or, rarely, produce blood-borne metastases.

(2) The para-pituitary residues and their tumours

The epithelial parts of the pituitary gland develop from an outgrowth of the embryonic oral cavity, Rathke's pouch ; and small residues of squamous-celled

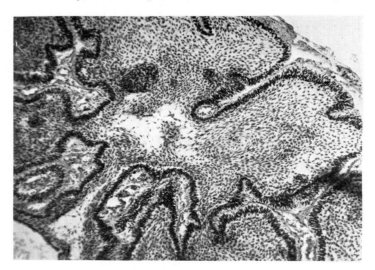

FIG. 215.—Suprasellar "adamantinoma". (× 85.)

or nondescript epithelium derived from the pouch are often present in the adult in the roof of the nasopharynx, in the body of the sphenoid, and especially in the capsule of the pituitary gland and around the infundibular stalk. In the last-named site, these residues sometimes give origin to cysts or tumours, which have often been lumped together under the unsatisfactory name " craniopharyngioma ". Most of these cysts and tumours are suprasellar, but some arise within the pituitary fossa. They themselves produce no endocrine secretion, but they often cause endocrine effects by compression of the pituitary gland or hypothalamic region.

(a) Cysts of the para-pituitary residues

These are lined by squamous stratified epithelium, in contradistinction to intra-pituitary cysts lined by columnar ciliated epithelium which arise from Rathke's cleft in the pars intermedia.

(b) Tumours of the para-pituitary residues

These are rare, slowly growing, often cystic, and sometimes calcified growths, which occur chiefly in children or young adults. They are dangerous tumours, invading the surrounding bones or brain and frequently recurring after removal, but they do not metastasize. Some of them have the ordinary structure of cornifying squamous-cell carcinomas, but most of them are " adamantinomas " resembling those of the jaws (Fig. 215). This resemblance is not surprising, seeing that in both situations the tumours spring from developmental residues of outgrowths from the embryonic oral cavity.

TUMOURS OF THE ADRENAL CORTEX

(1) Adenomas

Small cortical adenomas of the adrenals are often found incidentally at necropsy. They are well-circumscribed, rounded, yellow masses resembling the cortex of the gland, rarely more than 2 or 3 centimetres in diameter, single or multiple, and unaccompanied by any distinct signs of endocrine disturbances.

(2) Carcinomas

Sometimes also called " hypernephromas " (a name best discarded, because of its confusing misapplication to renal carcinomas), these are rare tumours which grow progressively to large sizes. They sometimes retain a well-differentiated adrenal-cortex-like structure and remain localized and without metastases, when they can also be regarded as giant adenomas ; but others are frankly malignant, exhibiting cellular pleomorphism, infiltration and metastases. They occur in both children and adults and are commoner in females than males.

Some adrenal cortical tumours are without endocrine results ; but in most cases the tumours secrete excessive androgens, oestrogens or other hormones, and produce clinically obvious effects on secondary sexual characters or metabolism (see p. 608).

TUMOURS OF THE PARATHYROID GLANDS

Parathyroid tumours are rare. They are usually single, occasionally multiple, appearing as well-circumscribed masses of uniform grey or brown tissue, in most cases only a few grammes in weight but occasionally as much as 200 grammes. Women are affected twice as frequently as men. Microscopically most of the tumours closely resemble normal parathyroid tissue, consisting of a mixture of clear cells and oxyphil cells arranged in clumps and cords. Most tumours are benign adenomas ; only occasionally has infiltration or metastasis occurred.

Most parathyroid tumours are accompanied by signs of hyperparathyroidism, namely decalcification and generalized fibro-cystic disease of bones, hypercalcaemia, excessive excretion of calcium and phosphates, and sometimes metastatic calcification (see p. 361 and p. 611). But occasionally a parathyroid tumour is found unaccompanied by obvious hormonal effects.

SUPPLEMENTARY READING

References Chapters 28 and 30 (p. 414 and p. 456) and the following :

Atlas of Tumor Pathology, Armed Forces Institute of Pathology, Washington, D.C., from 1949 onwards. (A series of *fascicles* with many excellent illustrations.)

Belisario, J. (1959). *Cancer of the Skin*. London; Butterworths. (Excellent illustrations.)

Bell, F. G. (1926). " Tumours of the testicle ", *Brit. J. Surg.*, **13**, 7 and 282.

Charteris, A. A. (1930). " On the changes in the mammary gland preceding carcinoma", *J. Path. Bact.*, **33**, 101.

Cheatle, G. L., and Cutler, M. (1931). *Tumours of the Breast*. London; Pitman. (A sound and beautifully illustrated classic.)

Collins, D. H. and Pugh, R. C. B. (Editors) (1964). *The Pathology of Testicular Tumours*. Edinburgh and London. (Suppl. to *Brit. J. Urol.*, **36**).

Cullen, T. S. (1900). *Cancer of the Uterus*. New York; D. Appleton. (A comprehensive work with an unsurpassed collection of illustrations.)

Doll. R., and Hill, A. B. (1964). " Mortality in relation to smoking: ten years' observations of British doctors ". *Brit. Med. J.*, **1**, 1399 and 1460.

Dukes, C. E. (1940). " Cancer of the rectum : an analysis of 1,000 cases ", *J. Path. Bact.*, **50**, 527.

Dunhill, T. P. (1931). " Carcinoma of the thyroid gland ", *Brit. J. Surg.*, **19**, 83.

Fraser, J. (1938). " Malignant disease of the large intestine ", *Brit. J. Surg.*, **25**, 647. (A good brief outline.)

Frissell, L. F., and Knox, L. C. (1937). " Primary carcinoma of the lung ", *Amer. J. Cancer*, **30**, 219. (A valuable, well-illustrated general account.)

Fry, R. M. (1928). " The structure and origin of the ' mixed ' tumours of the salivary glands ", *Brit. J. Surg.*, **15**, 291.

Hill, A. B., and Doll, R. (1956). " Lung cancer and tobacco ". *Brit. Med. J.*, **1**, 1160.

Illingworth, C. F. W. (1935). " Carcinoma of the gall-bladder ", *Brit. J. Surg.*, **23**, 4. (A good, brief, well-illustrated account.)

Joll, C. A. (1923). " Metastatic tumours of bone ", *Brit. J. Surg.*, **11**, 38.

Lattes, R. (1962). " Thymoma and other tumors of the thymus ", *Cancer* **15**, 1224. (Good review and illustrations).

Mackee, G. M., and Cipollaro, A. C. (1937). *Cutaneous Cancer and Precancer*. New York; *American Journal of Cancer*. (A valuable, well-illustrated monograph.)

Paget, J. (1874). " Disease of the mammary areola preceding cancer of the gland ", *St. Barts. Hosp. Rep.*, **10**, 87. (A classic which all students should read.)

Russell, W. O. *et al.* (1963). " Thyroid carcinoma ", *Cancer*, **16**, 1425. (A valuable study of whole-organ sections).

Willis, R. A. (1959). " Some uncommon and recently identified tumours." In *Modern Trends in Pathology* (Ed., D. H. Collins), Chap. 7. London; Butterworths.

CHAPTER 32

TUMOURS OF NON-HAEMOPOIETIC MESENCHYMAL TISSUES

INTRODUCTION

THIS CHAPTER is concerned with the tumours of Group II of the classification adopted on p. 414. For every specialized kind of connective, vascular and muscular tissue, there are corresponding tumours. We will discuss the tumours of each kind of tissue in turn, both its benign tumours and its sarcomas. In doing this we must not forget that our classification is somewhat arbitrary, and that the tissues of mesenchymal tumours, like their tissues of origin, readily undergo metaplastic conversion one into another, producing a variegated structure. Thus parts of fibromas may chondrify or ossify, osteosarcomas often contain cartilaginous or fibrous areas, myxomas cannot be sharply separated from fibromas, lipomas are often partly or even mainly fibromatous, synovial tumours are often predominantly fibrous and sometimes cartilaginous or bony, and predominantly vaso-formative (" angiomatous ") areas are found in mesenchymal tumours of other kinds.

Of course, it is necessary and useful to distinguish by appropriate names the various tumours showing, predominantly or wholly, differentiation of this or that kind. It is also true that the kind of differentiation exhibited by a tumour is in most cases wholly or mainly that of the tissue from which it sprang, e.g. that osteogenic tumours have usually sprung from bone, cartilaginous tumours from cartilage, fibroblastic tumours from connective tissue, and so on. The cell types of the several differentiated kinds of mesenchymal tissue naturally tend to be perpetuated in their tumours. But it is a mistake to assume that the named species, either of the normal tissues or of their neoplasms, are unrelated and immutable ; and it is foolish to participate in pointless discussions over the precise histogenesis of a tumour which displays several kinds of aberrant differentiation, or to call it by such names as " chondro-myxo-haemangio-endothelio-sarcoma ", as has actually been done.

Sarcoma

A sarcoma is a malignant tumour arising from any non-epithelial mesodermal tissue—fibrous, mucoid, fatty, osseous, cartilaginous, synovial, lymphoid, haemopoietic, vascular, muscular or meningeal. The simplest nomenclature specifies each form of sarcoma by an appropriate prefix—fibro-, myxo-, lipo-, osteo-, etc. It has not been customary to think of the leukaemias, Hodgkin's disease and plasma-cell tumours as sarcomas, or to speak of malignant meningeal or synovial tumours as " meningiosarcoma " or " synoviosarcoma ", but it would be quite justifiable to do so. The definition of sarcoma properly excludes chordoma, glioma and other malignant neural tumours, for these arise from non-mesenchymal tissues.

" Endothelioma " and " mesothelioma "

This is the place to mention these much abused terms. The name " endothelium " is usefully applied to the flat cells lining vascular and serous cavities, to which some would add the cells lining meningeal and synovial cavities. Now

531

tumours do arise from vascular, meningeal and synovial tissues ; but these several kinds of tumours already have appropriate names, and nothing is gained by forcing them all together under the further common label " endothelioma ", merely because they all represent tissues which happen to have flat-cell surfaces. Angiomas, synoviomas and meningiomas have no more in common than other varieties of mesenchymal tumours. As to " endotheliomas " or " mesotheliomas " of serous membranes, most tumours reported as such have really been secondary carcinomas from undetected primary tumours elsewhere (see p. 455). The names " endothelioma " and " mesothelioma " must therefore be used with great caution; they are often mis-applied.

FIBROMA AND FIBROSARCOMA

Mesenchymal tumours the cells of which recognizably differentiate as collagen-forming fibroblasts are fibromas or fibrosarcomas according to their behaviour. Between cellular anaplastic, highly malignant fibrosarcomas and slowly growing densely fibrous, benign fibromas structurally resembling normal fibrous tissue, there is a complete series of tumours of intermediate structure and behaviour. Structure is not always a sure guide to prognosis : some well-differentiated, densely collagenous tumours nevertheless prove to be sarcomatous, by infiltration or metastasis ; while some benign fibromas are moderately cellular and not very fibrous. However, most fibrosarcomas are readily detected microscopically as malignant tumours, by their cellular pleomorphism, numerous mitotic figures, or marginal infiltration.

Fibromas or fibrosarcomas may arise from fibrous connective tissues any-where in the body—in the skin, fascia, periosteum, visceral capsules or stroma, or nerve-sheaths. Nerve-sheaths are an important source of fibroblastic tumours ; a considerable number of these demonstrably arise from large nerves such as the sciatic or radial, and it is probable that many others arise from small nerves, though it may not be possible to show that they have done so.

Most benign fibromas appear as well-circumscribed firm masses, consisting of interwoven bundles of white fibrous tissue. Fibrosarcomas may show similar well differentiated fibrous structure, or may consist mainly of softer non-fasciculated tissue with areas of degeneration, haemorrhage or mucoid change.

A distinctive and rare form of fibroma is that known as " desmoid tumour " of the muscular aponeuroses of the abdominal wall. These growths appear clearly to owe their origin to injury ; for over 80 per cent of the patients have been parous women, and most of the remaining patients have had operations or injuries at the sites of the tumours. "Desmoids" grow to great sizes, sometimes many pounds, and they are not well circumscribed but infiltrate the adjacent muscles. In spite of these features, they are not really malignant ; for they show neither metastasis nor cellular anaplasia, and they are cured by adequate local removal.

MYXOMA AND MYXOSARCOMA

Just as for the histologist the presence of intercellular mucin is the only feature distinguishing mucoid from collagenous connective tissue, so for the pathologist this is the only feature distinguishing myxomas from fibromas. Myxomas and myxosarcomas are merely fibromas and fibrosarcomas in which mucin has developed in the intercellular matrix. Pure myxomas and myxosarcomas are rare. Fibroblastic tumours of nerve-sheaths, especially malignant ones, often show myxomatous areas. Most prominently mucoid tumours are only lowly malignant, prone to recur but seldom to metastasize.

LIPOMA AND LIPOSARCOMA

It is difficult to distinguish between true tumours and various non-neoplastic overgrowths of adipose tissue. The latter include local forms of adiposity associated with endocrine or metabolic disorders, and rare congenital adipose hamartomas or tumour-like malformations. However, some congenital lipomas are true tumours with powers of progressive growth.

(1) Benign lipomas and fibro-lipomas

Congenital lipomas are rare ; but acquired ones, usually in middle-aged people, are common. The most frequent sites are subcutaneous tissues, retro-peritoneal tissues and inguinal canal ; rarer sites are intermuscular, submucous (chiefly in the alimentary canal), and in solid viscera. Most lipomas are solitary ;

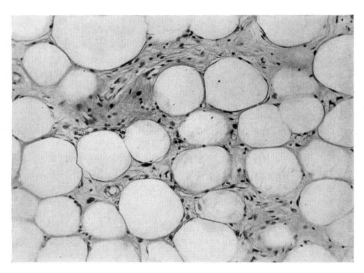

FIG. 216.—Fibro-lipoma. (× 120.)

but two or three tumours are not unusual, and occasionally they are numerous. When numerous, they are sometimes familial and sometimes associated with neurofibromatosis or other conditions of disturbed innervation.

Structure (Fig. 216).—Many lipomas closely resemble normal adipose tissue, consisting of lobules of fat cells with a variable amount of supporting connective tissue and blood-vessels. Some tumours, however, are fibro-lipomas, consisting of a mixture of fatty and fibroblastic tissue, the latter being, not merely an abundant stroma, but an integral part of the growth, and often showing cellular areas of rather atypical fibroblasts.

(2) Liposarcomas

Liposarcomas are rare.   In a few cases they have arisen in pre-existing benign lipomas, but this is not the general rule. Their structure ranges from well-differentiated tissue nearly resembling adipose tissue to highly cellular anaplastic growth, and myxomatous characters are frequent in them.   They often contain giant lipoblasts with multiple nuclei and multiple fat vacuoles.   Metastasis takes place chiefly by the blood-stream, especially to the lungs and liver.

TUMOURS OF CARTILAGE AND BONE

(1) Intermutability of cell-types in the skeleton and its tumours

Proper concepts of the nature and relationships of the various types of skeletal tumours depend on a sound knowledge of the changes which occur in the normal development and growth of the skeleton.   The formation of cartilage in young mesenchyme, endochondral ossification, direct osteogenesis in mesenchyme, periosteal ossification, and the absorption of bone or cartilage by osteoclasts—all of these normal processes (which are seen also in the repair of bone) have their counterparts in the growth of skeletal tumours.

In particular, we should appreciate the intermutability of the various types of cells concerned in skeletogenesis.   Skeletal fibroblasts, chondroblasts, osteoblasts and osteoclasts are not distinct immutable species of cells, but show transformations one into another during bone formation and bone absorption, in tumours as well as in normal growth and repair.   While some skeletal tumours show predominant differentiation of one or another cell type—fibroblastic, chondroblastic, osteoblastic or osteoclastic—others show variable admixtures of two or more of these.   Thus, some cartilaginous tumours ossify, many osteosarcomas contain cartilaginous or fibroblastic areas, and tumour osteoclasts are found in some osteogenic and cartilaginous growths.   Subdivision of skeletal tumours into several main groups is useful for purposes of description and prognosis ; but the names used denote only the predominant types of differentiation, and *not* specifically distinct tissues of origin.

(2) The classification adopted

For descriptive purposes, we can consider skeletal tumours under the following heads :

     (*a*) Multiple enchondroses ;
     (*b*) Multiple exostoses ;
     (*c*) Solitary exostosis, osteoma or osteochondroma ;
     (*d*) Solitary benign chondroma ;
     (*e*) Chondrosarcoma ;
     (*f*) Osteosarcoma ;
     (*g*) Osteoclastoma ;
     (*h*) Sundry other tumours of bones.

(3) Multiple enchondroses

Also called "dyschondroplasia", Ollier's disease and "multiple enchondromata", this rare disease is characterized by the presence from an early age of masses of cartilage within many or all of the cartilage bones, especially of the limbs. In most cases, with the lapse of time, the masses cease growing and either ossify or retrogress ; hence they are not regarded as true tumours. But in other cases they attain large sizes, especially in the bones of the hands and feet ; and chondrosarcoma occasionally supervenes.

(4) Multiple exostoses

Like enchondroses, these familiar growths, also called "multiple ossifying chondromata", are malformations rather than true neoplasms. They arise in early life, and grow as cartilage-capped bony excrescences during adolescence. When growth of the skeleton ceases and the epiphyses unite, the exostoses also cease to grow and their cartilage caps become completely ossified or reduced to a thin quiescent zone. The outgrowths are usually numerous and their distribution varies, but they are found only on chondral bones and are often largest on the juxta-epiphyseal parts of the long bones (Fig. 217). In the majority of cases there is a family history of the disease. About two-thirds of the patients are males. Chondrosarcoma sometimes springs from one of the exostoses, usually in early adult life.

(5) Solitary exostosis, osteoma or osteochondroma

Solitary cartilage-capped growths, generally similar in structure to multiple exostoses, are the commonest tumour of the skeleton. Like the multiple lesions, the solitary ones arise chiefly in young people, and affect chondral bones, especially their juxta-epiphyseal parts. Despite the resemblances, the solitary tumours are to be regarded as distinct from the multiple ones, especially as they often continue to grow after cessation of growth of the skeleton, to form bulky lobulated masses, thus deserving the name "osteoma" or "osteochondroma" rather than "exostosis". Chondrosarcoma sometimes supervenes.

(6) Solitary benign chondroma

Solitary benign chondromas arise most often in young adults in the metacarpals or phalanges of the fingers, less often in other bones or in the larynx. Benign chondroma is by far the commonest, indeed almost the only, central lesion to produce a single well-defined area of bone destruction in a phalanx or metacarpal (Fig. 218). With rare exceptions, well-differentiated cartilaginous tumours of the digits are benign ; but in

FIG. 217.—Multiple exostoses (from Virchow).

other bones no sharp line of distinction can be drawn between benign enchon-dromas and chondrosarcomas.

(7) Chondrosarcoma

No sharp line of distinction can be drawn between chondromas and chondro-sarcomas. Well-differentiated cartilaginous growths, whether enchondromas growing inside bones or ecchondromas growing from the bone surfaces, often grow slowly and attain large sizes without metastasizing ; but other similar tumours invade veins and produce metastases in lungs or elsewhere. Yet other tumours are frankly malignant in structure as well as in behaviour, showing cellular anaplasia and pleomorphism from their inception.

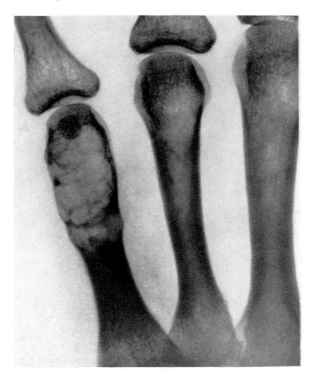

FIG. 218.—X-ray photograph of chondroma of fifth metacarpal.

The principal sites of chondrosarcomas are the long bones of the limbs, the pelvis and the ribs. In a small proportion of cases, chondrosarcoma arises from previously benign enchondrosis, and in a larger proportion of cases from pre-existing cartilaginous exostosis or osteochondroma.

(8) Osteosarcoma

Osteosarcomas are those in which the tumour parenchyma shows recognizable bony differentiation. In a bone tumour it is sometimes difficult to distinguish between reactive non-neoplastic bony tissue formed by expanded periosteum and bone formed by the tumour tissue itself. However, many sarcomas do produce bone, in metastases as well as in the primary growths.

(a) Incidence

Osteosarcoma is a relatively uncommon kind of tumour. It affects males more frequently than females. The greatest number of cases occurs in the second decade of life ; and about two-thirds of the patients are between the ages of 10 and 30. The commonest sites, in order of frequency, are lower end of femur, upper end of tibia or fibula, upper end of humerus, other parts of the long bones of the limbs and scapula. About one-half of the tumours arise from the bone ends close to the knee joint. In cases in which the epiphysis is still ununited, a sarcoma arises usually in the juxta-epiphyseal part of the shaft, and only rarely in the epiphysis itself. It is surely very significant that osteosarcoma is predominantly a disease of youth, when active growth of the bones is in progress ; and that its favourite sites are the sites of most active and prolonged skeletal growth.

(b) Causation

The facts just outlined strongly suggest that the causation of osteosarcoma is to be sought mainly in disturbances of bone growth and maturation. The nature of these disturbances is unknown. The only cases of bone sarcoma in which an extrinsic agent has been proved culpable are the rare occupational osteosarcomas of luminous watch-dial painters, in whom absorbed radioactive substances accumulate in the bones and set up a chronic radiation osteitis which ends in sarcoma formation, and rare sarcomas arising in bones previously exposed to X-rays (p. 420). Injury has often been blamed as a cause of bone sarcoma, but on very doubtful grounds. Paget's disease of bone (osteitis deformans) strongly predisposes to sarcoma of bone. Paget himself recorded 3 cases of sarcoma associated with the disease now named after him, and many subsequent instances have been reported. Most osteosarcomas in people over 50 years of age are in cases of Paget's disease.

(c) Naked-eye appearance (Fig. 219 (a) and (b))

The appearances of sarcomas of bone are very variable. Some start in the periosteum, some centrally ; but in most cases both periosteum and medulla are soon affected. Externally, in most cases, the tumour forms a smooth fusiform or slightly lobulated swelling confined within the expanded, and often thickened, periosteum. Internally it extends for variable distances along the medullary cavity, and effects variable degrees of rarefaction and replacement of the bony cortex. New bone formation by the expanded periosteum is sometimes prominent, usually in the form of a forest of sharp spinous outgrowths from the bone surfaces. Areas of irregular ossification in the tumour tissue itself add to the complexity of the structure and of the radiographic shadows.

(d) Microscopic structure (Fig. 220)

This is very variable, even in one tumour. It ranges from well-formed normal-looking woven bone to anaplastic spindle-cell or pleomorphic-cell growth devoid of osseous characters. Areas of tumour cartilage are not uncommon, and these may show clear signs of partial endochondral ossification. More or

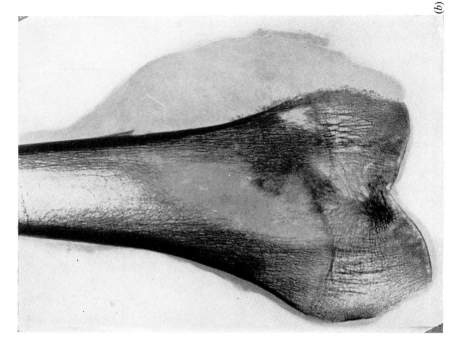

(b)

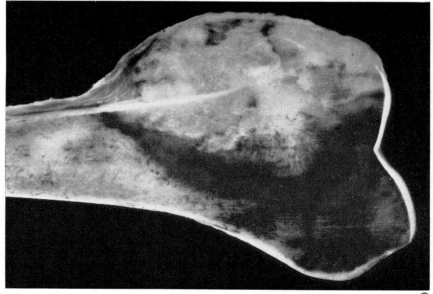

(a)

FIG. 219.—(a) Osteosarcoma of lower end of femur seen in vertical section (slightly reduced) ;   (b) X-ray photograph (positive print) of specimen shown in (a).   Note destruction of bone, periosteal formation of new bone, and projecting mass of tumour devoid of bone.

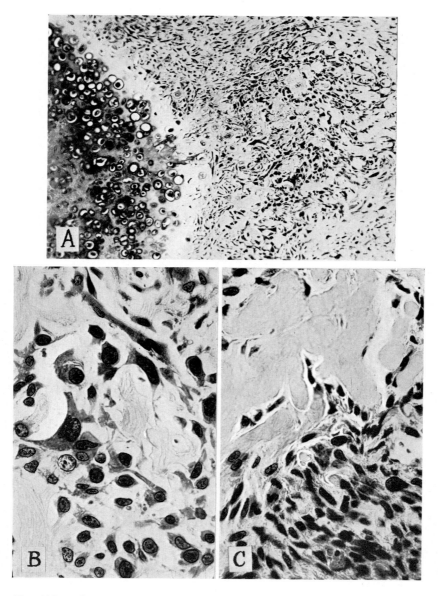

FIG. 220.—Micro-structure of osteosarcoma.

    A. Low-power view showing diverse structure, including well differentiated cartilage and cellular tissue with pre-osseous trabecular matrix. (× 70.)

    B. Details of the pleomorphic cells and trabeculae of A. (× 350.)

    C. From metastasis in lung, showing cellular growth with pre-osseous trabeculae. (× 350.)

less extensive areas of fibrosarcomatous or myxosarcomatous structure may occur in tumours, other parts of which are plainly osteogenic. Mingled with the tumour tissue, non-neoplastic residues of the affected bone, and also new-formed reactive bone derived from the periosteum, are often present.

The osteoblastic nature of the sarcoma cells is sometimes clearly displayed by their formation of osteoid tissue or well-formed bone in metastatic growths in the lungs or other sites ; and the neoplastic osteoblasts may produce much alkaline phosphatase, which is abundant in the tumour tissue and in excess in the patient's blood serum.

(e) Metastasis

The major risk of osteosarcoma is metastasis to the lungs, radiographic evidence of which should be sought in all cases prior to treatment and at intervals subsequently. In the majority of cases, pulmonary metastases appear within 2 years of the onset of the disease ; and they are present in nearly all fatal cases. Blood-borne metastases in other sites are relatively infrequent ; and lymph-nodal metastases are also uncommon.

(9) Osteoclastoma

This is the proper name for the tumour which was once wrongly called " myeloid sarcoma " and which some workers still prefer to call by the non-committal name " giant-cell tumour of bone." The completely bone-absorptive characters of the tumour, and the close resemblance of its giant cells to the large osteoclasts of non-neoplastic bone-absorptive lesions (e.g. in hyperparathyroidism), leave no room for doubt that the tumour consists of cells with the structural and functional attributes of osteoclasts.

(a) Incidence

Males and females are equally affected. Most tumours appear between the ages of 10 and 30 ; arising centrally in the ends of the long bones of the limbs, especially distal end of femur, proximal end of tibia or fibula, distal end of radius or ulna, and proximal end of humerus, i.e. the sites of most active growth. Other parts of the skeleton are less frequently affected. In cases in which the epiphysis is still ununited, the tumour arises, not in the epiphysis, but in the metaphysis. Thus, the distribution of osteoclastomas is generally similar to that of osteo-sarcomas.

(b) Naked-eye appearances (Fig. 221)

Young healthy tumours are uniformly red, brown or occasionally pale, and soft or friable ; but older tumours become variegated by fibrosis, cysts, haemorrhages, or yellow necrosis. The centrally situated, well circumscribed growth replaces the medulla and causes expansion and thinning of the cortex. At first the expansion is usually uniform in all directions, but later it may become irregular and eccentric. The radiographic appearances, of a well-defined " soap-bubble " or multilocular expanding radio-translucent lesion, are usually characteristic ; but they are not conclusively diagnostic, since similar appearances are sometimes produced by other central expanding non-osteogenic tumours. There may occur pathological fracture or palpable " egg-shell-crackling " of the expanded thin cortex.

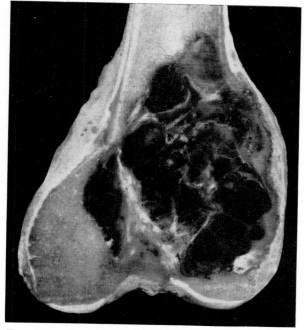

FIG. 221.—Cystic haemorrhagic osteoclastoma of lower
end of femur seen in vertical section. (Slightly reduced.)

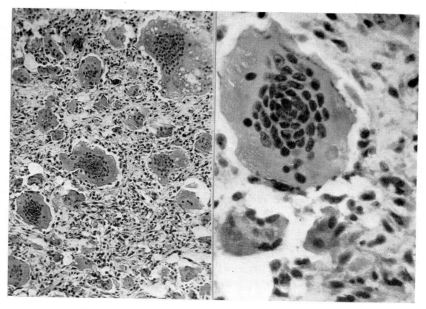

FIG. 222.—Micro-structure of osteoclastoma. (× 150 and 600.)

(c) Microscopic structure (Fig. 222)

Healthy parts of the growths show a mixture of multinucleated giant cells and mononucleated spindle cells. The giant cells are often very large, measuring 100μ or even 200μ in diameter, and containing scores or hundreds of nuclei usually centrally situated. They have processes which appear to anastomose with those of the smaller spindle-shaped cells around. There are often clear appearances of fusion of the spindle cells to form giant cells. Secondary changes include fibrosis, haemorrhage and pigmentation, and collections of lipoid-laden foam-cells.

(d) Behaviour

In the great majority of cases, the tumours pursue a benign course, growing slowly and non-invasively, producing no metastases, and being curable by adequate local removal. But a few tumours extend through the bone cortex and periosteum into the surrounding tissues or produce metastases in lymph glands, lungs or other parts. In their structure and rate of growth some of these malignant osteo-clastomas differ but little or not at all from benign ones ; but others show some degree of cellular anaplasia.

(e) Lesions confused with osteoclastoma

The following lesions are sometimes confused with osteoclastomas, from which, however, they should be sharply distinguished.

(i) *Osteoclastic areas in osteitis fibrosa.*—In the decalcified bones of hyper-parathyroidism (p. 611), and in localized " osteitis fibrosa " (p. 642), large masses of osteoclasts are sometimes found. These should not be called " osteo-clastomas ", as they are not tumours and they retrogress when the bone re-forms.

(ii) *Myeloid epulis.*—The local overgrowths of the gums, to which the name " epulis " is applied, are not true tumours, but masses of organizing chronic inflammatory granulation tissue. Some epulides are simply fibrous ; but others contain many osteoclast-like giant cells derived from the alveolar periosteum, and these are called " myeloid " epulides. They should not be called " osteo-clastomas ".

(iii) *Giant-cell synovial tumours* (see p. 543).—Some writers have likened these to the " giant-cell tumours " of bone (i.e. osteoclastomas)—a most unfortunate comparison, as the two are quite distinct.

(10) Sundry other tumours of bones

To complete the list of primary tumours in bone, we must add fibroma and fibrosarcoma, lipoma and liposarcoma, haemangioma, solitary plasmocytoma, myelomatosis, myeloid leukaemia, chordoma, and tibial " adamantinoma ", all of which are considered elsewhere. We must also add reticulum-cell sarcoma and " Ewing's tumour ", both of which require brief consideration here.

(i) *Reticulum-cell sarcoma of bone.*—This is the name applied to certain very rare, non-osteogenic medullary tumours which form large destructive growths composed of diffusely arranged, pleomorphic, rounded or polyhedral cells, and which are relatively slow to metastasize and have often been cured by amputation.

(ii) *" Ewing's tumour ".*—This is not a pathological entity ; it is a name applied (often very uncritically) to a syndrome, namely the syndrome of a non-osteogenic, round-celled, radio-sensitive tumour in a bone, usually a long bone, in a young person, and usually causing diffuse elevation of the periosteum. We now know that this syndrome can be caused by several different kinds of tumours, including metastatic neuroblastoma (p. 574), metastatic carcinoma, metastatic lymphosarcoma, and primary reticulum-cell sarcoma. In the majority of cases diagnosed as " Ewing's tumour ", the tumours have really been metastases from unsuspected primary growths elsewhere, in children and adolescents usually neuroblastomas.

TUMOURS OF SYNOVIAL TISSUES

There are two rather distinct groups of synovial tumours :

    (i) *Benign synoviomas*, moderately common, predominantly fibroblastic and giant-cell growths ;

    (ii) *Synoviosarcomas*, which are relatively rare.

(1) Benign synoviomas

These usually solitary tumours arise most frequently from tendon-sheaths, especially the flexor tendon-sheaths of the fingers. Less commonly they arise in joints, where they are often multiple and associated with villous hyperplastic chronic synovitis. They form well-circumscribed, firm or hard masses, most of which are only 2 or 3 centimetres in diameter when removed, and which may be either solid, white and fibrous, or may contain visible synovial clefts, or may show yellow or brown pigmentation.

Microscopically, they vary in structure. The commonest consist of fibrous tissue, or of giant-cell tissue with many large irregular multinucleated cells, or of mixtures of these with all transitions between fibroblasts and giant cells. Other common features include myxomatous areas with mucinous intercellular substance, clefts occupied by mucinous synovial fluid, and accumulations of lipoids and blood pigments which are taken up both by the tumour cells themselves and by wandering macrophages and which give some of the tumours their yellow or brown colour. Less commonly, the growths contain cartilage, bone or adipose tissue. These variations of structure are not surprising, in view of the nature of synovial membranes, which are modified connective tissue surfaces, the lining cells of which show transitions to neighbouring fibrous and cartilaginous tissues.

These tumours grow slowly and non-invasively, and are readily cured by simple excision.

(2) Synoviosarcomas

These are rare growths composed either of cystic mucinous synovial tissue or of solid pleomorphic-cell sarcoma devoid of synovial differentiation. Their main sites are around the knee joint, ankle joint, foot, wrist, forearm and elbow. They are very rare in the hand, the commonest site of the benign synoviomas. Most of those springing from joint capsules project externally and not into the joint cavity. The cystic mucinous variety of growth is often like an irregular spongework into the spaces of which there are papillary ingrowths clothed by plump cells, giving a pseudo-epithelial appearance. Metastasis takes place principally by the blood-stream to the lungs.

ANGIOMAS : TUMOURS AND TUMOUR-LIKE OVERGROWTHS OF VASCULAR TISSUE

Most so-called " angiomas " are not neoplasms but vascular malformations or " hamartomas " with an excess of vessels ; truly neoplastic angiomas are rare. Let us consider in turn :

> (i) *Angiomatoid malformations*
>> (*a*) Haemangiomas of particular tissues ;
>> (*b*) Multiple haemangiomas in two or more tissues ;
>> (*c*) Lymphangiomas.
> (ii) *Angiosarcoma.*
> (iii) *Glomangioma* and *Haemangiopericytoma.*

(1) Haemangiomas of particular tissues

(i) *Skin* (Fig. 223).—Any tissue may be the site of the vascular malformations which are called " haemangiomas ". These are most familiar in the skin, where

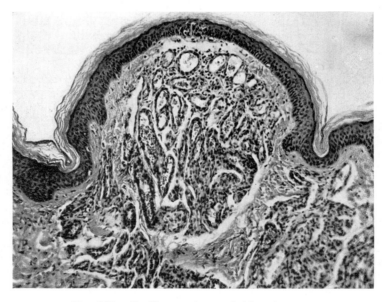

FIG. 223.—Capillary angioma of skin. (× 120.)

they are the commonest kinds of " birthmarks " or " congenital naevi ". (The name *naevus*, Latin for a birthmark, is a general term applied to any kind of localized blemish of the skin present at birth, but particularly to angiomas and pigmented moles.) Cutaneous vascular naevi may appear as flat red or purplish patches—" port-wine stains "—or as raised lobulated masses—" mulberry naevi ". According to whether they consist mainly of small thin-walled vessels, larger vascular spaces with septa between, or tortuous arterial vessels which may pulsate, they are called *capillary, cavernous* and *arterial* angiomas ; but these are not distinct types, and a single malformation may show all the varieties of structure. Angiomas are not sharply delimited ; the superfluous vessels mingle with the cutaneous glands and hair follicles, and also extend into the subcutaneous tissues, and even into the subjacent muscles or bones.

(ii) *Viscera.*—The commonest site of visceral angiomas is the *liver,* in which isolated small or large irregular areas of spongy cavernous angioma are often found. Much rarer is extensive diffuse angiomatosis of the entire liver, which sometimes proves fatal in infancy. Small cavernous angiomas of the *spleen* are not unusual ; they are prone to haemorrhages and cystic change, and are probably the origin of most splenic cysts. The *intestines, kidney, adrenal, placenta* and other viscera are rare sites of angiomas.

(iii) *Central nervous system.*—Haemangiomas are not unusual in the brain and cord or their coverings. The commonest are capillary angiomas, often cystic, in the cerebellum, which usually cause symptoms in young adult life and are curable by surgical removal. Localized cerebral angiomas are rarer ; and least common of all is extensive plexiform angiomatosis of the meninges with many large serpentine varicose vessels ramifying over the surface of the brain and cord.

(iv) *Muscles or bones.*—These are relatively rare sites of haemangiomas. Angiomas of bones sometimes produce characteristic X-ray pictures.

(2) Multiple haemangiomas in two or more tissues

Vascular hamartomas in a single tissue, e.g. skin, are often multiple. But two or more different tissues may also be affected simultaneously, e.g. skin and mucous membranes, liver and spleen, skin and central nervous system, retina and central nervous system, or widespread angiomas may occur in many tissues. Eponyms have been bestowed on some of the more notable syndromes, e.g. hereditary multiple angiomas of skin and mucous membranes (Osler) ; multiple angiomas of facial skin and meninges and brain (Sturge, Weber and Kalischer) ; angiomas of retina and nervous system, along with cystic and other anomalies in viscera (Lindau and von Hippel).

(3) Lymphangioma

This rare lesion occurs most frequently in the neck or axilla, where it forms a large spongy and cystic mass inextricably mingled with the muscles and other structures, and is called " cystic hygroma ". Similar lesions occur occasionally in the tongue, lip, mediastinum, mesentery, retroperitoneal tissues, or viscera.

(4) Angiosarcoma

All of the vascular lesions so far considered are benign malformations, which may be dangerous only because of their positions or because of accidental complications such as injury or infection. Occasionally, however, a genuinely malignant growth composed of recognizably vaso-formative tissue is seen ; in conformity with the rest of our terminology, this is best called *angiosarcoma* rather than " angioblastoma " or "haemangio-endothelioma ".   It is to be insisted, however, that we should not apply these names to highly vascular sarcomas of other kinds.   In order to identify a malignant tumour as truly angiomatous, there must be clear microscopical evidence of its essentially vaso-formative character; and this is often not easy to adduce.   True angiosarcomas are rare.

(5) Glomangioma and Haemangiopericytoma

Glomangiomas or glomus tumours are rare benign, usually small tumours which are clearly derived from the neuro-myo-arterial glomera of the skin and subcutaneous tissues.   They occur chiefly on the limbs, and are often associated with paroxysmal attacks of pain, which are cured by excision of the tumours. These consist of cavernous vascular spaces, in the walls of which there are sheets or masses of characteristic polyhedral epithelium-like glomus cells with or without smooth muscle fibres.   *Haemangiopericytoma* is a related, less well organized and more bulky tumour, arising in various sites from the pericytes of Zimmermann, and usually benign.

MENINGIOMA OR ARACHNOID FIBROBLASTOMA

Although almost all meningiomas are attached to the dura mater, their tissue of origin is not the dura, but the arachnoid villi and granulations (Pacchionian bodies) which project from the arachnoid into the dural venous sinuses.   In

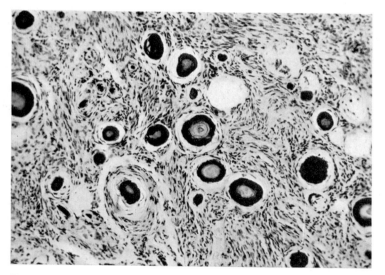

Fig. 224.—Meningioma with whorls and calcified spherules. (× 120.)

many of the tumours the cells and their arrangement closely resemble those of the normal granulations.   The fact that many meningiomas become fibrous and fibroma-like, and that a few of them become bony, is not surprising, seeing that the arachnoid is only a specialized kind of fibroblastic connective tissue.

(i) *Incidence.*—Meningiomas are among the commonest intracranial and intraspinal tumours. Most of them appear between the ages of 30 and 60. Their most frequent sites are near the main intracranial venous sinuses and in the spinal theca. Rarely, they are unattached to the dura, and may even lie in the lateral ventricles of the brain. They are usually single, but occasionally multiple.

(ii) *Naked-eye appearance.*—Meningioma is usually a well-defined, smoothly lobulated, firm mass attached to the dura by a broad base and indenting the brain. Less frequently it forms a thin flat dural plaque which may have a smooth or shaggy inner surface. Very rarely it forms an extensive sheet between the dura and the brain and cord—diffuse meningiomatosis. Most tumours are sharply delimited and non-invasive, but some show slow marginal invasion of the surrounding tissues, the dura, skull or brain. Those which invade the bone excite the formation of an externally projecting dense bony boss on the skull, which may attain large dimensions.

(iii) *Microscopic structure.*—This varies considerably from tumour to tumour. Most characteristic are whorls of plump fusiform cells, set in a vascular connective tissue stroma, the centres of the whorls forming hyaline spherules of collagen which frequently calcify into concentrically laminated grains or " psammoma bodies " (Fig. 224). Other tumours are predominantly fibro-collagenous, with or without whorls and sand-grains. Yet others are richly vascular and angioma-like. A few tumours show patchy or extensive ossification.

(iv) *Growth and behaviour.*—Most meningiomas are benign in that they grow slowly, are well circumscribed, and curable by local removal. But some are malignant, in that they grow more quickly, invade the dura, bone or brain, and occasionally produce metastases. Hyperostoses on the skull, due to invasion by underlying meningiomas, are permeated by the growths. Spinal meningiomas do not cause vertebral hyperostoses, because the spinal dura is unattached to the bone. Invasion of the venous sinuses of the dura sometimes occurs. Metastases have been seen in the lungs or liver on rare occasions. Tumours exhibiting these malignant features might properly be called *meningiosarcomas.*

LEIOMYOMA AND LEIOMYOSARCOMA

Tumours of smooth muscle, most of them benign, are common in the uterus, but relatively rare in all other sites. Let us consider those of the uterus first.

(1) Leiomyomas of the uterus

These common, well-circumscribed growths of the myometrium are sometimes solitary but more often multiple ; they may attain huge sizes. At first intramural, as they enlarge they may protrude into the peritoneal cavity or into the uterine cavity. Pedunculated intraperitoneal tumours may suffer torsion and strangulation ; submucous ones cause uterine haemorrhage, especially when they are in process of extrusion. The cut surfaces of healthy myomas are white and show characteristic fasciculation (Fig. 225) ; but large or old tumours may show hydropic degeneration with the formation of watery cysts, or red degeneration due to necrosis of the tissues and the seepage of blood into them, or more or less extensive calcification or even ossification.

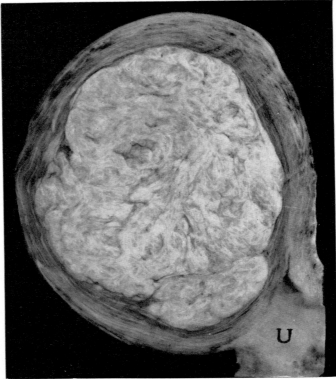

FIG. 225.—Sectional view of pedunculated intraperitoneal leiomyoma
attached to uterine horn (U). (Natural size.)

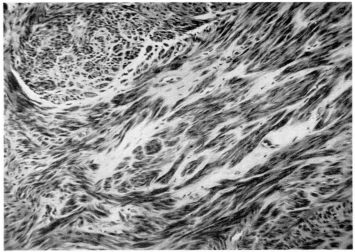

FIG. 226.—Micro-structure of leiomyoma. (× 110.)

Microscopically (Fig. 226), myomas consist of interwoven bundles of typical smooth muscle fibres, with intervening strands of connective tissue and blood-vessels. Most old tumours show fibrosis, which may become abundant, dense and hyaline, whence the commonly used names " fibro-myomas " or " fibroids ". Occasionally, some adipose tissue develops between the muscle bundles, and this may become so abundant that the tumour appears like a lipoma.

(2) Leiomyosarcoma of the uterus

Most sarcomas of the uterus are myosarcomas, and most of these arise in pre-existing myomas. A few of these sarcomas are so highly differentiated that they are difficult to distinguish microscopically from simple myomas, and so have been called by the contradictory name " malignant myomas ". But most sarcomas show anaplastic areas with cellular pleomorphism and excessive mitoses. Metastasis occurs chiefly by the blood-stream to the lungs, but lymph-nodal deposits also occur.

(3) Leiomyoma and leiomyosarcoma of other sites

These tumours occur less commonly in the oesophagus, stomach, intestines, skin, kidney, bladder, prostate and retroperitoneal tissues. The commonest benign non-epithelial tumour of the stomach is leiomyoma.

RHABDOMYOMA, RHABDOMYOSARCOMA AND RHABDOMYOBLASTIC MIXED TUMOURS

Tumours consisting of, or containing, striated muscle fibres are of several diverse kinds, all of which are rare : Let us discuss them under the following headings :

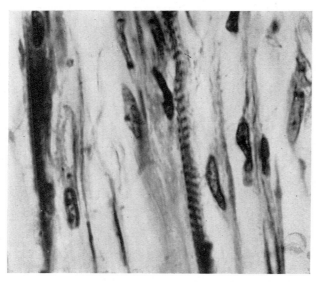

FIG. 227.—Striated fibres in rhabdomyosarcoma of middle-ear region in a child. (Iron-haematoxylin stain.) (× 800.)

(1) Rhabdomyosarcomas of skeletal muscles ;

(2) Rhabdomyosarcomas of the urogenital organs of children or young adults ;

(3) Rhabdomyosarcomas of other parts ;

(4) Mixed tumours containing rhabdomyoblastic cells ;

(5) " Rhabdomyoma " of the heart ;

(6) Granular-cell " myoblastoma ".

(1) Rhabdomyosarcoma of skeletal muscles

It is a striking fact that, in spite of the large mass of voluntary muscle in the body, benign rhabdomyomas are almost unknown and rhabdomyosarcomas are very rare.   Most of those which have occurred have been in children (Fig. 227), and the question at once arises whether the tumours spring from mature differentiated muscle fibres or from still immature myoblasts.   However, careful

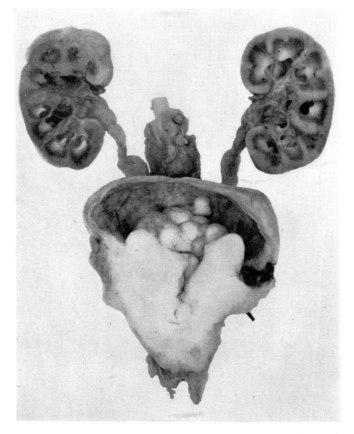

Fig. 228.—Polypoid embryonic sarcoma of base of bladder and prostatic region in a child 22 months old. Note obstructive hydro-ureter and hydronephrosis on both sides.

microscopic examination has shown that a few pleomorphic-cell sarcomas in adults are rhabdomyosarcomas, and a true benign rhabdomyoma of the tongue occurs as a great rarity.

(2) Rhabdomyosarcoma of the urogenital organs of children or young adults

In infants or children, and less frequently in adolescents or young adults, there occur rare polypoid sarcomas of the base of the bladder, vaginal walls, prostate or spermatic cord, in many of which cross-striated myoblastic cells can be demonstrated. These tumours often contain also much undifferentiated mesenchyme-like or myxomatous tissue, and some of them consist of this exclusively without recognizable myoblastic cells. There is little doubt that these tumours arise, not from differentiated muscular tissue, but from immature mesenchyme in the neighbourhood of the developing urogenital sinus; they are *embryonic sarcomas* in which rhabdomyoblastic differentiation frequently takes place (Fig. 228).

(3) Rhabdomyosarcoma of other parts

Very rare instances of rhabdomyosarcoma of the oesophagus, lung, breast or bile duct have been reported.

(4) Mixed tumours containing rhabdomyoblastic elements

Striated muscular tissue may occur as a component of teratomas (p. 589), embryonic tumours of the kidney or liver (p. 583), and mixed tumours of the uterus (p. 517). The presence of striated muscular tissue in the mixed tumours of the uterus, and also in the embryonic tumours of the kidney and liver, is of great interest because it shows clearly that rhabdomyoblasts can develop by aberrant differentiation in situations where normally no striated muscle is present.

(5) " Rhabdomyoma " of the heart

The rare so-called " congenital rhabdomyomas " of the myocardium are not true tumours but developmental anomalies. In many cases the patients are idiots with tuberose sclerosis of the brain, and often other developmental defects. Many of the subjects are still-born or die in early childhood, but survival into adult life has been recorded. The cardiac lesions, which are usually multiple or widespread, consist of areas of large malformed vacuolated and spider-shaped myocardial fibres.

(6) Granular-cell " myoblastoma "

This is the name which, unfortunately, has been applied to certain small benign nodular lesions of the tongue, consisting of large rounded cells with granular cytoplasm which appear to be derived from the lingual muscle fibres; and also to a group of tumour-like masses of cells of similar appearance which occur in the skin, breast, gums and elsewhere. The name " myoblastoma " is unfortunate for several reasons, namely—the nature of the cells is still uncertain; several quite different lesions have been included under this name; and it is also doubtful whether they are true tumours or non-neoplastic lesions with mesenchymal cells or macrophages laden with some as-yet unidentified metabolite.

SUPPLEMENTARY READING

References Chapter 28 (p. 415), and the following :

Adair, F. E., Pack, G. T., and Farrior, J. H. (1932). " Lipomas ", *Amer. J. Cancer*, **16,** 1104.

Atlas of Tumor Pathology. Ref. p. 530.

Alznauer, R. L. (1955). " Mixed mesenchymal sarcoma of the corpus uteri ". *Arch. Path.*, **60,** 329.

Catto, M. and Stevens, J. (1963). " Liposarcoma of bone ", *J. Path. Bact.*, **86,** 248.

Cushing, H. (1922). " The meningiomas (dural endotheliomas) ", *Brain*, **45,** 282.

Hanbury, W. J. (1952). " Rhabdomyomatous tumours of the urinary bladder and prostate ". *J. Path. Bact.*, **64,** 763.

Jacobson, S. A. (1940). " Critique on the inter-relationships of the osteogenic tumors ", *Amer. J. Cancer*, **40,** 375.

Jaffe, H. L. (1943). " Hereditary multiple exostosis ", *Arch. Path.*, **36,** 335.

King, E. S. J. (1931). " Concerning the pathology of tumours of tendon-sheaths ", *Brit. J. Surg.*, **18,** 594.

Lendrum, A. C., and Mackey, W. A. (1939). " Glomangioma ", *Brit. Med. J.*, **2,** 676.

Lichtenstein, L., and Jaffe, H. L. (1943). " Chondrosarcoma of bone ", *Amer. J. Path.*, **19,** 553.

Martland, H. S. (1931). " The occurrence of malignancy in radioactive persons ", *Amer. J. Cancer*, **15,** 2435. (An excellent account of occupational osteosarcomas.)

Pachter, M. R. and Lattes, R. (1963). " Mesenchymal tumors of the mediastinum ", *Cancer*, **16,** 74.

Rowbotham, G. F. (1939). " The hyperostoses in relation with the meningiomas ", *Brit. J. Surg.*, **26,** 593.

Russell, D. S. (1950). " Meningeal tumours: a review ", *J. Clin. Path.*, **3,** 191.

Stout, A. P. (1944). " Liposarcoma ", *Ann. Surg.*, **119,** 86.

Taylor, C. W. (1958). " Mesodermal mixed tumours of the female genital tract ". *J. Obst. Gyn.*, **65,** 177.

Willis, R. A. (1949). " The pathology of osteoclastoma or giant-cell tumour of bone ", *J. Bone Joint Surg.*, **31,** 236.

Wright, C. J. E. (1952). " Malignant synovioma ", *J. Path. Bact.*, **64,** 585.

CHAPTER 33

TUMOURS OF HAEMOPOIETIC TISSUES

THE HAEMOPOIETIC tissues are peculiar in that throughout life they continue to produce blood corpuscles and to discharge them into the circulation. Many of the neoplasms of these tissues do the same, leucopoietic tumours giving rise to leukaemia, and erythropoietic tumours to polycythaemia. However, not all tumours of the haemopoietic tissues produce circulating cells ; some of them form local masses of non-mobile cells which do not enter the blood-stream. We will consider in order : (*a*) tumours of lymphoid tissue, (*b*) myelomatosis and plasmocytoma, (*c*) myelogenous leukaemia and chloroma, and (*d*) primary polycythaemia.

THE TUMOURS OF LYMPHOID TISSUE

It is convenient to distinguish the following main types of lymphoid tumours :
  (1) Follicular lymphoma ;
  (2) Lymphosarcoma and lymphatic leukaemia ;
  (3) Hodgkin's disease ;
  (4) Reticulosarcoma.

These names, however, do not denote sharply distinct species of tumours, but only the main structural variants of tumours of the same fundamental kind, namely, " sarcomas of lymphoid tissue ". While some of these tumours fall readily into one or other of the four named subdivisions, others are of intermediate types and show all possible combinations and transitions. Lymphoid tumours occur at any period of life, more often in adults than in children ; they affect males twice or thrice as often as females. Nothing definite is known of their causes. They are among the commonest tumours of many kinds of animals.

(1) Follicular lymphoma

This is the name applied to rare, slowly growing tumours of lymph glands characterized by the differentiation of follicles throughout the growths. Occasional tumours of this kind are solitary and localized, and are cured by removal ; but more often the disease spreads to other glands or to the spleen, or it may terminate in active lymphosarcoma or leukaemia. It is the most slowly growing, best differentiated and least malignant form of lymphosarcoma.

(2) Lymphosarcoma and lymphocytic leukaemia

Lymphosarcoma means a malignant tumour of lymphoid tissue in which the tumour cells differentiate recognizably as lymphocytes. It includes both " lymphocytomas " with well-differentiated lymphocytes, and " lymphoblastomas " with immature cells. *Lymphocytic leukaemia* (usually called " lymphatic leukaemia ") means the presence of malignant lymphocytes or lymphoblasts in the circulating blood. It is not a different disease but merely a concomitant of some lymphoblastic tumours, a concomitant which was to have been expected from the natural

or more per cell, and few or many of them show mitosis. Eosinophil and neutrophil granulocytes and plasma cells mingle with the tissue in variable numbers. Older lesions show progressive fibrosis, which appears first in bands separating patches of more cellular Hodgkin's tissue, but which may later replace large areas of the tissue and may even convert an entire lymph gland into a hyaline fibrous mass. Groups of the large or multinucleated cells and few or many eosinophil leucocytes often persist here and there in the densely fibrous tissue.

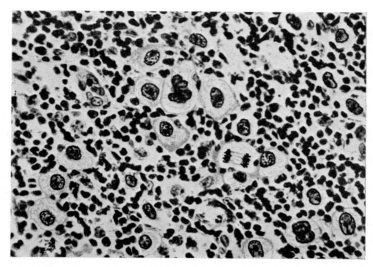

FIG. 231.—Micro-structure of Hodgkin's disease, showing the characteristic large cells, some multinucleated and one in mitosis. (× 375.)

The essential proliferating elements in Hodgkin's disease are the large multinucleated neoplastic reticulum cells, which are often called " Dorothy Reed " or "Sternberg-Reed" giant cells—unfortunate eponyms since the cells were described by several British workers before either Sternberg or Reed. Lymphocytes are regularly present not only in the lesions in lymphoid tissue, but also in secondary deposits in other tissues ; moreover, areas indistinguishable from active lymphosarcoma are sometimes present, and lymphocytic leukaemia occasionally supervenes. Hence it appears clear that tumour lymphocytes as well as reticular cells are produced in the lesions, and that we must look upon Hodgkin's disease as a stem-cell tumour with divergent differentiation in these two directions. The granulocytes in the lesions, however, are not neoplastic, but are merely reactionary cells such as occur in many other tumours. Eosinophilia occasionally occurs.

(iv) *Multifocal origin, spread, and metastasis.*—Microscopical study of early lesions shows that in some cases the disease arises multifocally or diffusely within a particular group of lymph glands. On the other hand, it is also clear that metastatic spread from gland to gland or by the blood-stream plays a major part in its extension. Even if one were prepared to suppose that all the lesions in lymphoid and other haemopoietic tissues are multiple autochthonous foci, this could scarcely apply to circumscribed deposits in the lungs, skin and other non-haemopoietic organs. In the liver, spleen and bone-marrow also, it is probable that most of the lesions are metastatic.

The distribution of secondary deposits varies greatly from case to case, After the lymph glands themselves, the commonest sites are spleen, liver, lungs, bones and skin, in that order of frequency ; but any tissue may be affected. The lesions in bones are often unsuspected clinically, but may cause radiographically obvious changes, local tumour, fracture, or vertebral collapse with paraplegia.

(v) *Gordon's test*.—Intracerebral inoculation of rabbits with fresh Hodgkin's tissue usually produces a characteristic encephalitis, whereas tissue from other lesions of lymph glands or from normal glands rarely does so. This result is not due to a virus, but to some chemical substance in the eosinophil granulocytes.

(4) Reticulum-cell sarcoma

This term, or " reticulosarcoma ", has been applied to two different varieties of sarcoma of lymphoid tissue, (*a*) undifferentiated sarcomas consisting of masses of closely aggregated immature cells with many mitoses, called also " syncytial reticulosarcoma " because cell boundaries are indistinct, and (*b*) rare spindle-cell or pleomorphic-cell sarcomas which tend to form reticulin and collagen fibres. Tumours of the variety (*a*) are really anaplastic undifferentiated lympho-sarcomas. Tumours of variety (*b*) are akin to Hodgkin's disease and to fibro-sarcomas of other tissues.

(5) Summary of the nature and relationships of lymphoid tumours

All primary tumours of lymphoid tissue are to be looked upon as related variants of one disease. The names used for the principal variants have descrip-tive and clinical value but do not denote distinct species. All kinds of transitions between the named varieties are seen. *Follicular lymphoma* is the least malignant, best differentiated variant of the group, with a structure approaching that of normal lymphoid tissue. *Lymphosarcomas* show predominant differentiation of neoplastic lymphocytes or their immediate precursors, lymphoblasts. *Lymphocytic leukaemia* is lymphosarcoma with a circulating metastasis. *Hodgkin's disease* is a sarcoma of lymphoid tissue in which both reticular and lymphoid elements are differentiated, the former often fibrifying. *Reticulosarcoma* is a name often applied to anaplastic syncytial growths devoid of differentiation, but better reserved for spindle-cell or pleomorphic-cell tumours showing reticular and fibrous differentiation.

PLASMOCYTOMAS AND MYELOMATOSIS

Tumours consisting of plasma cells are of the following three kinds :
(1) Plasma-cell myelomatosis ;
(2) Solitary plasmocytoma of bone ;
(3) Primary plasmocytoma of soft tissue.

(1) Myelomatosis

This relatively uncommon disease of adults is characterized by the appearance in many bones of multiple bone-absorptive tumours composed of plasma cells.

The skull, vertebrae, ribs, sternum, pelvis, femur and humerus are all commonly affected, the more peripheral bones of the limbs less frequently. The tumours usually appear as well-defined areas of soft grey tissue replacing the bone-marrow

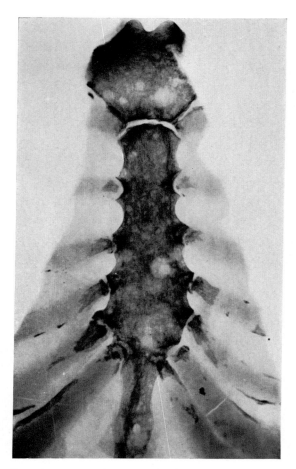

FIG. 232.—X-ray photograph (positive print) of plasma-cell myelomatosis of sternum.

and accompanied by absorption of the bone trabeculae ; and they produce well demarcated rounded radio-translucent areas in radiographs (Figs. 232, 233). Small tumours cause no tumefaction of the bones and are detected only by X-ray examination or by necropsy section of the bones. Larger tumours distend and destroy the cortex, produce palpable tumours, and may cause pathological fractures. Occasionally the marrow shows a diffuse infiltration without localized tumours.

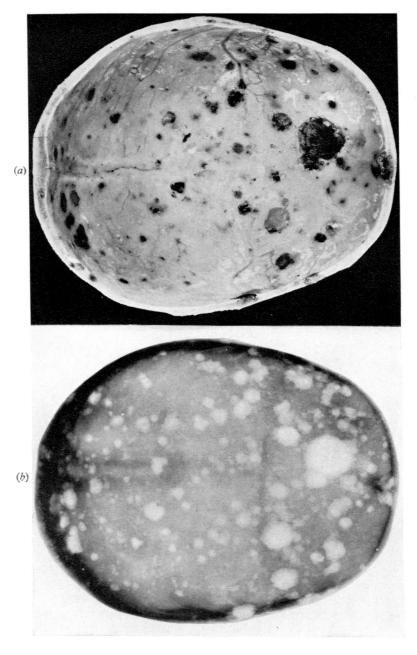

Fig. 233.—Plasma-cell myelomatosis of calvarium, same case as in Fig. 232:
(a), specimen; (b), X-ray photograph (positive print). (Half natural size.)

Microscopically, most tumours consist of typical plasma cells, some of which may be multinucleated (Fig. 234). In a few cases the cells are non-distinctive, and have been identified variously as myeloblasts, lymphoblasts or erythroblasts ; but probably in all cases they are only imperfectly differentiated plasma cells.

In many, but not all, cases Bence-Jones proteose appears in the urine at some time during the course of the disease. This is accompanied by other signs of disturbed protein metabolism, namely, an excess of particular gamma-globulins in the blood, the deposition of abnormal protein in various mesenchymal tissues constituting a variety of " amyloid disease", and the formation of amorphous or crystalline deposits in the renal tubules.

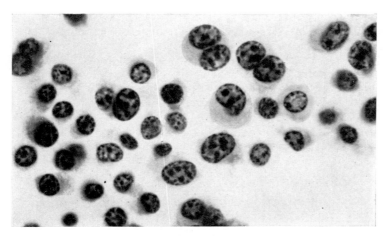

Fig. 234.—Smear preparation of myelomatous plasma cells, some of which are binucleated. ($\times$ 1000.)

In most cases of myelomatosis the tumours are restricted to the skeleton, but in a few cases localized metastases or diffuse infiltrations of plasma cells appear in the liver, spleen, lymph glands, kidneys or other organs. Occasionally, plasma cells appear in the circulating blood, constituting " plasma-cell leukaemia ".

(2) Solitary plasmocytoma of bone

Myelomatosis sometimes first declares itself by a single prominent tumour in bone, and this may precede the appearance of multiple growths by many months or even years. In addition, however, there occur also rare cases in which a large destructive solitary plasmocytoma of bone is not followed by tumours elsewhere in the skeleton but is cured by amputation or radiotherapy. A diagnosis of solitary plasmocytoma can be made only after careful radiographic examination of the rest of the skeleton has shown it to be normal and after sufficient time has elapsed to exclude the possibility that the one growth was a precocious manifestation of generalized myelomatosis. The principal sites of solitary plasmocytoma are femur, vertebrae, pelvis and humerus.

(3) Primary plasmocytomas of soft tissues

These rare growths are almost restricted to the mucous membranes and submucous tissues of the head and neck, the main sites being nasal cavity, nasopharynx, buccal cavity and larynx. Though generally benign and curable by local removal or radiation, these tumours sometimes metastasize to the cervical lymph glands.

MYELOGENOUS LEUKAEMIA

Myelogenous leukaemia is the over-population of the blood by malignant granulocytes following neoplastic change in the leucopoietic cells of bone-marrow. The truth of this concept is evident from the cytology of the disease, its lethality, and the analogy with lymphocytic tumours and leukaemia. As with lymphoid tumours, so with myeloid ones, the extent of the primary field of tissue undergoing neoplastic change is difficult to determine. We do not know whether the disease starts in a small area of bone-marrow somewhere in the skeleton and colonizes the rest of the marrow from circulating leukaemic cells, or whether the marrow of the entire skeleton suffers widespread neoplastic change from the beginning. All we know is that, by the time the patient dies, his bone-marrow shows universal leukaemic change with great excess of proliferating abnormal granulocytes.

(i) *Incidence.*—Myelogenous leukaemia is not a rare disease ; it appears in two forms, chronic and acute. The chronic form is the commoner, occurs chiefly in middle-aged adults, and often has a duration of several years. The acute form occurs chiefly in children and young adults, and often proves fatal within a few weeks or months. Both forms affect males about twice as frequently as females. Exposure to benzol and kindred volatile compounds, and to X-rays or radioactive substances, is believed to be causative in some cases.

(ii) *Changes in the blood* (Plate X (*b*) and Fig. 235).—The leucocyte count is excessive, ranging from just above normal to 1,000,000 per cubic millimetre or even more. The excessive cells are granulocytes and their precursors ; and these are not merely immature cells (myeloblasts and myelocytes), but often atypical cells defying classification and making it impossible to do precise differential counts. In acute leukaemia, or in acute terminal stages of chronic leukaemia, most of the cells are immature and abnormal and some of them show mitotic figures. With the progress of the leukaemia, the red corpuscles and haemoglobin show increasing secondary anaemia, which in acute cases is usually very severe. The blood-platelets also are diminished, and the patient may show haemorrhages in the skin, mucous membranes and other tissues.

(iii) *Changes in the tissues.*—We have already noted that *the bone-marrow* shows widespread infiltration by excess of the abnormal granulocytes and their precursors. Pale pink or creamy marrow is found not only in the usual sites of haemopoietic marrow but also in more distal bones. *The spleen* is usually much enlarged, sometimes weighing many pounds. The enlargement is due to the diffuse accumulation of leukaemic cells. Whether, in addition to colonization of the organ from cells in the blood, there is also myeloid metaplasia in the splenic tissues themselves, is doubtful ; but it seems clear that the former is the more important factor. *The liver* also is enlarged, pale, and diffusely infiltrated by leukaemic cells, which congregate and multiply especially in the portal capillaries (Fig. 236). *Lymph glands* show variable numbers of leukaemic leucocytes, but are usually not much enlarged. *The kidneys* frequently show some enlargement

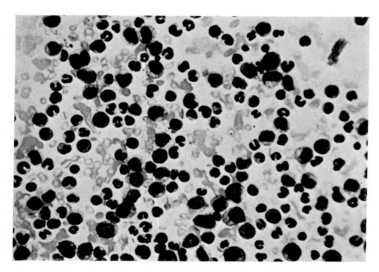

Fig. 235.—Blood film in myelogenous leukaemia, showing great excess of leucocytes (count 800,000 per c.mm.). (× 340.)

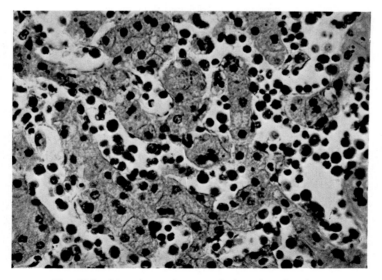

Fig. 236.—Liver in myelogenous leukaemia. (× 450.)

and pallor due to leukaemic infiltration. *Other parts*, including lungs, skin, testis and central nervous system, are less frequent sites of infiltration.

In most cases, secondary leukaemic deposits take the form of diffuse infiltrations in the tissue interstices, often with congregations of the cells within vessels. In some cases, however, in any of the sites mentioned, focal tumour-like masses of the cells also develop. These masses sometimes have a distinct greenish colour and are then called " chloromas ". Such tumours may be seen in any tissue, but are most common in periostea. They consist of immature myeloblast-like cells, and most cases in which they occur show many undifferentiated cells in the blood also.

PRIMARY POLYCYTHAEMIA

This disease, called also *polycythaemia vera, erythraemia* and *Vaquez-Osler disease*, occurs in middle-aged people of either sex, and is characterized by great increase in the number of circulating red cells, great erythroblastic activity of the red bone-marrow, and enlargement of the spleen. The red-cell count may reach 10 or 12 millions per cubic millimetre, and the haemoglobin percentage 150 or even 200. The blood-volume and viscosity are raised, and the left cardiac chambers become hypertrophied. All the organs are engorged with blood, and the patient's complexion is dark red or purplish. The enlargement of the spleen is due to the excessive destruction of red corpuscles proceeding in it. In a few cases myelogenous leukaemia is superadded, when the disease is called *erythroleukaemia*. These features strongly suggest that polycythaemia is a chronic neoplastic disease comparable with myelogenous leukaemia, but affecting the erythropoietic rather than the leucopoietic cells of the bone-marrow.

SUPPLEMENTARY READING

References Chapter 28 (p. 415), and the following :

Britton, C. J. C. (1969). Tenth edition. Whitby and Britton's *Disorders of the Blood*. London; Churchill. (On the leukaemias and polycythaemia.)

Churg, J., and Gordon, A. J. (1942). " Multiple myeloma with unusual visceral involvement ", *Arch. Path.*, **34**, 546.

Dolin, S., and Dewar, J. P. (1956). " Extramedullary plasmacytoma ". *Amer. J. Path.*, **32**, 83.

Ehrlich, J. C., and Gerber, I. E. (1935). " The histogenesis of lymphosarcomatosis ", *Amer. J. Cancer*, **24**, 1. (Careful study of 18 necropsy cases.)

Gall, E. A., and Mallory, T. B. (1942). " Malignant lymphoma. A clinico-pathologic survey of 618 cases ", *Amer. J. Path.*, **18**, 381.

Herbut, P. A., Miller, F. R., and Erf, A. L. (1945). " The relation of Hodgkin's disease, lymphosarcoma and reticulum cell sarcoma ", *Amer. J. Path.*, **21**, 233.

Lumb, G., and Prosser, T. M. (1948). " Plasma cell tumours ", *J. Bone Joint Surg.*, **30**, 124. (An excellent well-illustrated general account.)

Sikl, H. (1949). " Diffuse plasmocytosis with deposition of protein crystals in the kidneys." *J. Path. Bact.*, **61**, 149. (Also V. Neumann, in the same journal.)

Symmers, W. St. C. (1958). " Primary malignant diseases of the lymphoreticular system ". In *Cancer* (Ed. R. W. Raven), **2**, Chapter 24. London; Butterworths.

slow growth and well differentiated, they diffusely infiltrate the surrounding brain tissue, so that their limits are impossible to define and attempted removal is almost always incomplete. Some astrocytomas are so diffuse and ill-defined that they produce only a general enlargement of extensive areas of the brain— the whole brain-stem or a whole cerebral hemisphere—a condition appropriately named "diffuse gliomatosis". "Glioblastomas" grow rapidly and are fatal within a few months.

(2) Oligodendroglioma

Oligodendrogliomas are relatively uncommon tumours composed of small, compactly arranged neuroglial cells with lymphocyte-like nuclei, vacuolated cytoplasm (giving a "boxed" appearance), and short scanty processes. These growths often contain astrocytic cells as well. They occur chiefly in the cerebrum in adults, are relatively slowly growing and circumscribed, and hence are sometimes curable surgically. Some of them show calcification.

(3) Ependymoma

This is a rare, relatively benign, slowly growing glioma, the principal sites of which are the walls of the third or fourth ventricles and the spinal cord.

(4) Medulloblastoma

Medulloblastomas are not uncommon; indeed, in children, they are only slightly less frequent than the cerebellar astrocytomas. Most of them occur in infants or young children, more often males than females, and their usual situation is in the vermis of the cerebellum or roof of the fourth ventricle. They are much less frequent in young adults, and in the cerebrum. Microscopically, they are highly cellular, composed of small rounded or slightly elongated undifferentiated cells of uniform size, arranged usually diffusely but sometimes in clumps, rosettes or drifts. Some of them show partial astrocytic differentiation. The precise origin of the tumours is uncertain; but from their cytology and age incidence there is little doubt that they are truly embryonic tumours arising from immature neural tissues during foetal life or early childhood.

From their usual primary site in the fourth ventricle, many medulloblastomas spread directly or by metastasis into the basal and spinal subarachnoid spaces, often forming a diffuse layer of growth clothing the surface of the brain and cord.

Neuro-epithelioma or medullo-epithelioma

These are the names applied to very rare tumours usually of the region of the third or fourth ventricle in young children, composed of undifferentiated neuro-epithelial tissue resembling that of the embryonic medullary plate.

(5) Pinealoma

The tumours of the pineal gland are of two distinct kinds, both of which are rare—teratomas (p. 586) and pinealomas—the latter having a structure resembling that of the normal gland. Both kinds of tumours occur principally in children, much more often males than females; and both are often accompanied by precocious sexual and physical development. The reason for this is obscure; it is probably due to disturbances of neighbouring parts of the brain, especially the hypothalamus.

(6) Metastasis of gliomas

It is remarkable that, although many gliomas are highly malignant, invasive tumours, they rarely metastasize outside the central nervous system and meninges. This is largely attributable to their inability to invade blood-vessels ; vessels invaded by gliomatous tissue are not seen, either naked-eye or microscopically. The metastasis of gliomas is almost limited to the occasional formation of implant-metastases on the meningeal or ventricular surfaces, from free tumour particles in the cerebrospinal fluid.

PAPILLARY TUMOURS OF THE CHOROID PLEXUS

These rare growths occur in children more frequently than in adults, and in the fourth ventricle more frequently than in the lateral or third ventricles. They are vascular, friable, and sometimes gritty because of granular calcification ; and microscopically they consist of well-differentiated papillary growth closely resembling normal choroid plexus and containing calcified spherules. The tumours may grow to large sizes, filling and distending the ventricles ; and implant-metastases on the cranial or spinal meninges are not unusual.

NEURILEMMOMA AND OTHER TUMOURS OF NERVE-SHEATHS

There are good grounds for regarding the neurilemmal cells of Schwann as specifically neural cells, distinct from the endoneurial and perineurial connective tissue of the nerve-sheaths. One of these grounds is the existence of a specific Schwann-cell tumour or neurilemmoma which is distinct from the neurofibromas of nerves. However, it is best to consider both kinds of tumours here, for they may occur together.

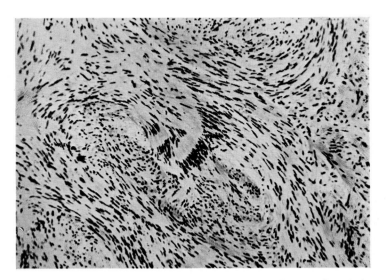

Fig. 238 —Neurilemmoma showing regimentation of cells. (× 120.)

(1) Neurilemmoma or Schwann-cell tumour

(i) *Incidence.*—This not uncommon tumour occurs at any age, in females rather more frequently than in males. Its commonest site is the eighth cranial nerve, where it is often called " acoustic-nerve tumour " or " cerebello-pontine angle tumour ". It rarely occurs on other intra-cranial nerves. Neurilemmomas of peripheral nerves occur chiefly in the limbs, head and neck, tongue and stomach. In most cases the tumour is solitary and unaccompanied by signs of generalized neurofibromatosis, but in a few cases the tumours are multiple or are accompanied by neurofibromatosis. Neurofibromatosis sometimes includes bilateral acoustic tumours ; and bilateral acoustic tumours are sometimes the main manifestation of familial neurofibromatosis.

(ii) *Structure.*—The tumour is usually rounded or ovoid, firm, well-circumscribed, and, if it arises from a large nerve, it shows the intact nerve bundles passing around it and not traversing it. Microscopically, it has the general appearance of a fibroma ; but the characteristic feature is the presence here and there or throughout the tumour of bundles or groups of spindle cells arranged in orderly parallel ranks, with their nuclei all at a similar level across each bundle, an arrangement aptly described as regimentation or palisading (Fig. 238). Intercellular fibres are present which, in young tumours, stain somewhat differently from those of ordinary fibrous tissue, but which in older tumours become more plentiful, dense and hyaline, and stain just like mesenchymal collagen.

(iii) *Behaviour.*—Neurilemmomas grow slowly and non-invasively, and are cured by local excision.

(2) Neurofibromatosis : von Recklinghausen's disease

This disease is a developmental disorder characterized by tumour-like overgrowth of the sheath-tissue of nerves, both its fibrous and neurilemmal components. The disease, which is often familial, usually announces itself in childhood or adolescence by the appearance of patches of pigmentation in the skin and the slow development of few or many, localized or widespread, tumour-like growths of cutaneous or deeper nerves. These slowly enlarge over a period of many years or throughout life ; not infrequently sarcoma supervenes in one of the affected nerves.

The distribution of the lesions varies greatly. In some cases they are numerous and widespread, affecting most or many nerves in all parts of the body. The most obvious are the skin lesions which form multiple soft nodular or pendulous projecting masses clothed by thin epidermis, a condition sometimes called " molluscum fibrosum " (Fig. 239). Other lesions include localized or diffuse plexiform swellings of deeper main peripheral nerves ; general enlargement of a whole limb or other part of the body in consequence of diffuse neurofibromatosis of all of its nerves, a condition called " elephantiasis neuromatosa " ; tumours of the cranial or spinal nerve roots, especially the acoustic nerves ; and tumours of autonomic and visceral nerves, especially of the stomach and intestine. Any or all of these manifestations of the disease may occur in one patient. Many other anomalies may be seen in association with neurofibromatosis, including meningeal or cerebral tumours, pigmented moles, lipomas, chromaffin tumours of the adrenals, and various skeletal or other malformations.

Structure of neurofibromatous lesions.—Unlike neurilemmomas, the excessive sheath tissue in neurofibromatous nerves is not well circumscribed but
involves the nerves diffusely for variable distances or mingles with the dermal
or other tissues.   Whorls and fasciculi in the growths often clearly correspond
to included nerve bundles (Fig. 240), and nerve-fibres in variable numbers traverse
the tissue.   The entire nerve suffers expansion and separation of all its constituent

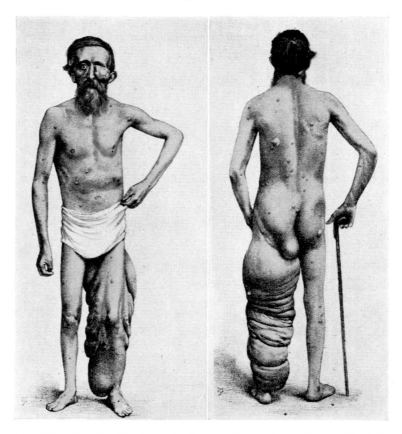

FIG. 239.—Von Recklinghausen's disease. (From Byrom Bramwell's *Atlas
of Clinical Medicine*, Edinburgh, 1892.)

bundles and fibres by a universal and diffuse increase in all its sheath tissues.
Excessive Schwann cells as well as the fibroblastic sheath tissues participate
in this.

True tumours in neurofibromatosis.—While the nerve-sheath growths in
this disease are primarily developmental anomalies or " hamartomas " and not
truly neoplastic, these lesions are strongly predisposed to the supervention of
true tumours—neurilemmal or fibroblastic.   In not a few cases of neurofibromatosis, localized, progressively growing neurilemmomas arise in the acoustic,

visceral or peripheral nerves ; or a localized progressively growing neurofibroma or neurofibrosarcoma arises in a particular nerve. The proneness of neuro-fibromatous nerves to sarcomatous change is well known. The structure of neurogenic sarcomas is very variable ; some are ordinary fibrosarcomas, many contain myxosarcomatous areas, some are anaplastic and pleomorphic-celled, and some contain compact masses of plump polyhedral cells which give parts of

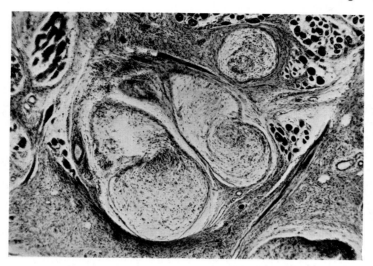

FIG. 240.—Section of tongue, showing large neurofibromatous whorls representing enlarged nerve bundles amidst the muscle fibres (darkly stained). (× 48.)

the tumours a resemblance to anaplastic carcinoma. Metastasis takes place chiefly by the blood-stream to the lungs and elsewhere, but lymph-nodal metastases also occur in some cases.

(3) Solitary neurofibroma and neurogenic sarcoma

Solitary neurofibromas and neurogenic sarcomas, structurally similar to those of von Recklinghausen's disease but unaccompanied by any other signs of it, are not uncommon, chiefly in young people. Whether they are solitary localized manifestations of neurofibromatosis or are unrelated to that disease, we do not know. Of course, before concluding that such a tumour is indeed solitary, a thorough search for other possible lesions of nerves should be made, for the manifestations of neurofibromatosis, especially in its early stages, are often few and scattered.

NEUROBLASTOMA AND GANGLIONEUROMA

(1) Nomenclature and incidence

Neuroblastoma means a tumour consisting of immature neuroblasts ; *ganglio-neuroma* means one consisting of well-differentiated nerve-cells and fibres. But the two types of growth are not distinct ; many tumours of transitional or mixed structure occur, and the two names denote merely the poorly differentiated and highly differentiated growths of the same species. These are truly embryonic tumours, arising from still immature nerve-cells at early ages, even in the foetus. Those in which the neuroblastic tissue continues to multiply entirely in the immature form are neuroblastomas ; those in which it matures to form well-differentiated nerve-cells as it grows are ganglioneuromas ; those in which both kinds of growth or partial maturation are seen may suitably be called " ganglio-neuroblastomas ".

(a) *Age incidence*

As we might expect, most of these tumours make their appearance in infancy or adolescence, and the neuroblastomas earlier than the ganglioneuromas. About three-quarters of neuroblastomas appear before the age of 4 years ; in not a

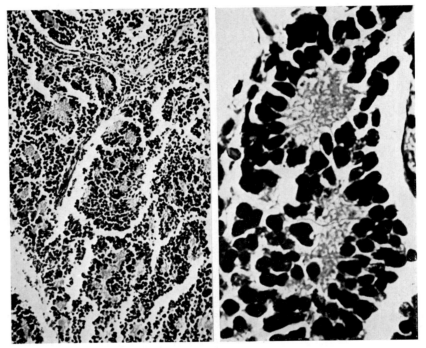

FIG. 241.—Neuroblastoma with plentiful rosettes, from a lymph-nodal metastasis of a primary growth in the lumbar sympathetic ganglia in a boy aged 5 years. (× 180 and 900.)

few cases, they are known to be present at birth ; adolescent and adult cases are rare. While many ganglioneuromas also appear during childhood or youth, in about one-third of cases their discovery is delayed until adult life, and in a few instances small quiescent growths are found incidentally in old people.

(b) *Sites of origin*

Contrary to what we might expect, nerve-cell tumours are extremely rare (if indeed they ever occur) in the central nervous system. Their most frequent sites are the adrenal glands and the main ganglia of the sympathetic system ; a few arise in the peripheral autonomic nerves in viscera or other tissues. Most tumours are solitary but a few are multiple.

(c) Frequency

While nerve-cell tumours are relatively uncommon, they compete closely with embryonic renal tumours as the most frequent malignant growths of infants and children. In the clinical diagnosis of any tumour of obscure nature in a young subject, neuroblastoma and ganglioneuroma must be considered.

(2) Structure

(a) Neuroblastoma (Fig. 241)

The most rapidly growing tumours consist of diffuse masses of undifferentiated rounded cells of uniform size with plentiful mitotic figures. The first recognizable evidence of differentiation is the grouping of the cells here and there in rosette-like clusters, the centre of each rosette being formed by a tangle of fine fibrils, which are young nerve fibres growing from the pyriform cells forming the rosette. These rosettes exactly resemble those of developing sympathetic ganglia. Still further differentiation produces enlarged angulated cells, recognizable as young nerve cells, and bundles of parallel nerve fibres.

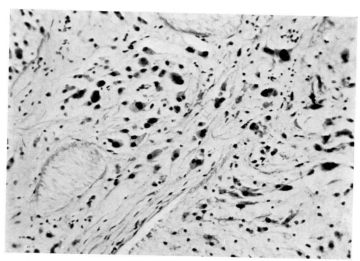

FIG. 242.—Ganglioneuroma of adrenal gland from a boy aged 2 years, showing well-differentiated nerve cells of various shapes and interlacing bundles of nerve fibres. (× 180.)

(b) Ganglioneuroma (Fig. 242)

The degree of cellular differentiation varies. Fully differentiated growths consist of groups of normal-looking, ovoid, pyriform or pyramidal nerve cells and plentiful bundles of non-medullated or medullated nerve fibres accompanied by Schwann cells. Imperfectly differentiated growths contain also many immature proliferating neuroblasts, with all transitions between these and the mature nerve cells.

(3) Behaviour

Completely differentiated ganglioneuromas are quite benign. All other tumours have varying degrees of malignancy, greatest in the undifferentiated neuroblastomas. These metastasize widely, both to lymph glands and by the blood-stream to the lungs, liver, bones and other organs. In many cases, metastases produce the first signs of otherwise small or symptomless primary tumours in the adrenal or sympathetic system. Some of these metastatic syndromes have been given eponyms, as follows :—

(i) *Pepper's syndrome* denotes great enlargement of the liver due to secondary neuroblastoma. It is seen chiefly in infants, and is sometimes present at birth.

(ii) *Hutchison's syndrome* denotes bulky metastases in the skull, again usually in infants or young children. Other bones also contain metastases, but these are less conspicuous than the calvarial ones.

(iii) *Ewing's syndrome*, referred to on p. 542, is readily produced by metastatic neuroblastoma, though this is not its only cause. Older children, adolescents, or even young adults may be affected.

FIG. 243.—Metastatic neuroblastoma in skull, expanding pericranium P and exciting formation of spines of new bone vertical to skull S. Bone is stained black with iron-haematoxylin. (× 5.)

Metastatic neuroblastoma in the calvarial or other flat bones often excites the formation of vertical spines of new bone by the expanded periosteum (Fig. 243) ; but in long bones the periosteal reaction usually produces " onion-skin " layers of new bone parallel to the bone surface (see Fig. 142). These appearances in radiographs of tumours in bones in young subjects strongly suggest neuroblastoma.

CHROMAFFINOMA

Tumours of the chromaffin tissue of the adrenal gland, also called " phaeochro-mocytomas ", are usually benign tumours, composed of polygonal or irregular cells resembling those of the adrenal medulla and giving the chromaffin reaction when immersed in fixatives containing potassium bichromate or other chromium salts. The functional activity of many of the tumours is proved by their causing persistent or paroxysmal hypertension and other symptoms of hyper-adrenalinism, and by the presence of large amounts of adrenaline in the tumour tissue.

TUMOURS OF CHEMORECEPTORS—CHEMODECTOMAS

Carotid-body tumours.—These are rare, smoothly lobulated, firm, white or grey growths, which are aptly described as " potato tumours ". They envelop the carotid arteries at the bifurcation, or the vessels lie in deep grooves in the tumour. The tissue consists of masses of large polyhedral epithelium-like cells separated by vessels and strands of connective tissue, thus mimicking the structure of the normal carotid body. Most carotid body tumours are benign, growing slowly, remaining encapsulated, and producing no metastases; but an occasional malignant tumour has been seen.

Other chemoreceptor tumours.—Tumours generally resembling those of the carotid body occasionally arise from the aortic bodies, glomus jugulare, ganglion nodosum, and possibly from small unnamed chemoreceptors in other parts of the body.

TUMOURS OF THE RETINA AND CILIARY BODY

Excluding the intra-ocular melanomas, tumours of the neural tissues proper of the interior of the eye are :

(1) *Retinoblastoma ;*

(2) *Tumours of the ciliary body or iris ;*

(3) *Glioma of the retina.*

(1) Retinoblastoma

(i) *Incidence.*—This rare tumour, formerly wrongly called " glioma " of the retina, is almost restricted to infancy and early childhood, and is sometimes known to be present at birth. While many cases appear to be isolated or sporadic, no relatives of the child being affected, there are many striking examples of a familial incidence and inherited liability to the disease, sometimes running through several generations. Multiple and bilateral tumours are frequent, especially in strongly inherited and familial cases. The bilateral tumours may appear simul-taneously, or one after the other. They are not due to spread or metastasis from one eye to the other, but are independent primary growths arising in similarly predisposed tissue.

(ii) *Structure.*—Many retinoblastomas consist largely of compact round-cell growth devoid of differentiated characters. Differentiation appears in the form of either rosettes or fibrillar areas or both. Usually these are present in parts only of an otherwise diffusely cellular tumour ; but some tumours show them throughout. Each rosette consists of a central spherical or ovoid cavity 10 to 60 microns in diameter, surrounded by radially arranged elongated cells, the central processes of which project through a distinct limiting membrane into the lumen of the rosette (Fig. 244). This structure is unmistakably similar to that of the rod-and-cone cell layer and external limiting membrane of the normal retina. The rosettes are, in fact, little retinal vesicles surrounded by rod-and-cone cells. In fibrillar areas the tumour cells become fusiform and each cell gives off a long wavy process at each end. These cells correspond to the bipolar nerve-cells of the plexiform or fibre layers of the normal retina.

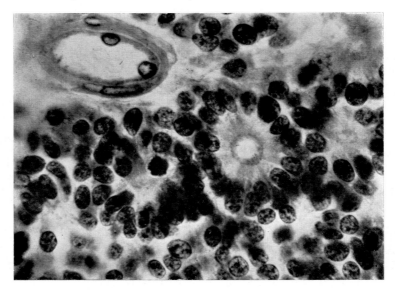

FIG. 244.—Retinoblastoma showing typical rosettes. (× 800.)

(iii) *Histogenesis.*—Retinoblastomas are essentially embryonic, arising from and consisting of immature retinal tissue. Those which continue to proliferate at the embryonic level fail to attain any structural differentiation ; but in those in which differentiation takes place, it produces rosettes of rod-and-cone cells or fibrillar areas of bipolar neural cells or both.

(iv) *Spread and metastasis.*—In the eye the tumour may form a nodule or plaque projecting into the vitreous from the retina ; but more frequently it dislodges the retina and grows in the subretinal space. Extra-ocular extension often takes place in the optic nerve or its sheath ; and by this route the growth may spread back to the chiasma and cranial cavity, where it sometimes spreads diffusely in the meninges to envelop the brain and spinal cord. Prognosis following removal of a retinoblastomatous eye depends mainly on whether or not the growth has reached the divided end of the optic nerve. From the globe the tumour may also spread directly to the orbital tissues via the perforating vascular canals of the sclera ; and from the orbit it may spread to the bones or may fungate on the face. Metastasis is infrequent, but blood-borne deposits occasionally appear in the viscera or bones. Spontaneous retrogression of a retinoblastoma is a rare but well-authenticated event.

(2) Tumours of the ciliary body or iris

These rare growths include :

(i) *Diktyomas*, embryonic ciliary tumours of childhood, structurally similar to retinoblastoma, but more benign, and sometimes producing thin-walled tumour vesicles floating freely in the aqueous humour ;

(ii) *Ciliary epithelial tumours*, composed of cubical or columnar, pigmented or unpigmented ciliary epithelium, arranged in papillary, tubular or reticular patterns.

(3) Glioma of the retina

True gliomas of the retina, comparable with the astrocytomas of the central nervous system, are very rare.   They consist of asteroid or fusiform cells with plentiful fibrillary processes ;   they grow slowly and non-invasively, and are cured by removal of the eye.

SUPPLEMENTARY READING

References Chapter 28 (p. 415), and the following :

Alpers, B. J., and Rowe, S. N. (1937).   " The astrocytomas ", *Amer. J. Cancer*, **30** 1.

Blacklock, J. W. S. (1934).   " Neurogenic tumours of the sympathetic system in children ", *J. Path. Bact.*, **39**, 27.   (Well-illustrated case reports and review.)

Cappell, D. F. (1929).   " Retroperitoneal ganglionic neuroma ", *J. Path. Bact.*, **32**, 43. (Beautiful photomicrographs.)

Russell, D. S., (1939).   " The pathology of intracranial tumours ", *Post-grad. Med. J.*, **15**, 150.   (An excellent outline.)

 — and Rubinstein, L. J. (1971).   *The Pathology of Tumours of the Nervous System.* London; Edw. Arnold.   (A comprehensive, authoritative and concise monograph.)

Snyder, C. H., and Vick, E. H. (1947). " Hypertension in children caused by pheochromo-cytoma ".   *Amer. J. Dis. Child.*, **73**, 581.

Stout, A. P. (1935).   " The peripheral manifestations of the specific nerve sheath tumor (neurilemoma) ", *Amer. J. Cancer*, **24**, 751.

Symington, T., and Goodall, A. L. (1953). " Studies in phaeochromocytomata ". *Glasgow Med. J.*, **34**, 75.

Wells, H. G. (1940).   " Occurrence and significance of congenital malignant neoplasms ", *Arch. Path.*, **30**, 535.   (Includes valuable accounts of neuroblastoma and retino-blastoma.)

Willis, R. A. (1962). *The Borderland of Embryology and Pathology.* London; Butterworths. (Chap. 11, on neuroblastoma and retinoblastoma.)

CHAPTER 35

SUNDRY SPECIAL CLASSES OF TUMOURS

MELANOMA

" MELANOMA " is the name applied to several varieties of malignant melanin-pigmented tumours. Just as there are several distinct kinds of normal melanin-producing cells—including the basal cells of the epidermis, the pigmented epithelium of the retina, and the dendritic pigment cells of the choroid and meninges—so there are several kinds of melanomas. A prevalent view that all melanomas must be histogenetically similar and derived from a single specific kind of cell, the " melanoblast ", is mistaken. " Melanoblasts " are of as many kinds as there are tissues which produce melanin. We will therefore consider melanomas according to their sites of origin, thus :

(1) Melanomas of the skin ;
(2) Melanomas of juxta-cutaneous mucous membranes ;
(3) Intra-ocular melanomas ;
(4) Meningeal melanosis and melanomas.

(1) Melanomas of the skin

(a) The Pigmented mole or naevus

The common pigmented moles or naevi, of which nearly all of us have a few, though often called " benign melanomas ", are not tumours but minor malformations. They are probably always congenital, and are discovered at birth or soon after. They vary in size from tiny brown spots to extensive areas covering a large part of the body ; they may be flat but are more often projecting and papillary, they may be darkly or only slightly pigmented, and either hairless or hairy. Their importance in tumour pathology is that they are the origin of most cutaneous melanomas in Europeans, and that they give the clue of the histogenesis of these tumours.

The typical features of a simple pigmented mole are seen in Fig. 245, which shows the papillary structure, the excessive pigmentation of the epidermis, and the presence of clusters of rounded " naevus cells " in the dermal papillae. These cells, which may be pigmented or unpigmented, are really detached epidermal cells ; in young moles from children, all stages of their splitting-off and separation from the epidermis are to be seen ; but in most moles in adults, this early process of separation has been completed and the naevus cells lie free in the dermis. In many moles the deeper parts of the dermis show bundles and whorls of spindle-cell tissue which resembles that of neurofibromas and probably is excessive nerve-sheath tissue.

(b) Histogenesis of cutaneous melanomas

When a melanoma starts in a pigmented mole, there is clear microscopic evidence that the tumour cells arise from both the naevus cells in the dermis and from the overlying epidermis. A widely held view that melanomas arise, not from the epidermis, but from nerve-sheath cells in the sensory end-organs of the skin, is not consistent with the microscopic structure of young moles and of early melanomas. There is no doubt that nerves and nerve-endings are implicated in the little malformations which constitute pigmented moles, and it is also probable that this disturbed innervation is in some way connected with the excessive pigmentation of the epidermis of the moles. But it still remains true that naevus cells and melanoma cells come from the epidermis. Many

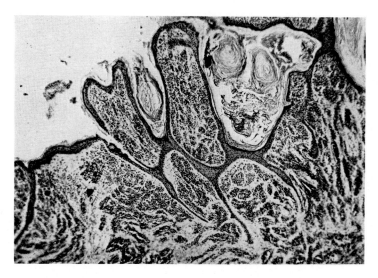

FIG. 245.—A simple pigmented mole, showing papillary structure and clusters of naevus cells in the dermal papillae. (× 60.)

writers have suggested that the heavily pigmented dendritic cells of the epidermis are the only cells to form pigment, and that they are the exclusive source of melanomas; but this is not yet certain.

(c) *Incidence and causation of melanomas of the skin*

Melanomas are fortunately not very common tumours. They occur at any age but are rare in childhood. In Europeans about three-quarters of them arise from pre-existing moles, and there is substantial evidence that injury plays a part in evoking malignant change in moles. Sub-ungual melanomas, however, arise without any signs of pre-existing moles. In Sudanese, Indians and other peoples who go bare-footed in tropical countries, the commonest sites of melanomas are the feet and legs, the tumours are usually not preceded by moles, and it seems likely that heat and injury of the skin are important causes. Melanomas have been produced experimentally in mice and dogs by prolonged painting with tar.

(d) *Structure of melanomas of the skin* (Fig. 246)

This is very diverse ; the cells may be rounded, fusiform or pleomorphic ; they may be arranged in epithelium-like clumps or diffusely ; they may be deeply

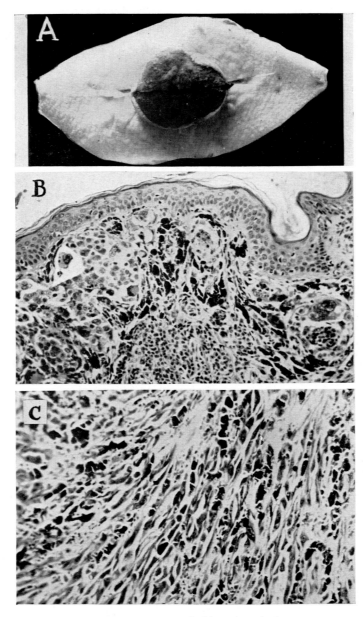

FIG. 246.—Malignant melanoma of skin. A, naked-eye appearance
(natural size). B, part near edge, showing epidermis underlain by
pleomorphic-cell growth with much melanin. C, deep part, showing
pigmented spindle-cell growth. (× 180.)

pigmented or unpigmented ; and a single tumour may show all of these structural variations. Great experience is needed in the microscopical diagnosis of early malignant change in pigmented moles ; suspicious features are increase in size and pleomorphism of the naevus cells, the presence of clumps of these atypical cells in the epidermis, mitoses, and unequal distribution of pigment in the suspicious areas. Amelanotic melanoma can cause great diagnostic difficulty ; but completely amelanotic tumours are rare.

(e) Spread and metastasis

No other tumour behaves so unpredictably as melanoma. A " mole " unsuspected of malignancy or so small as to be overlooked may produce wide-spread metastases. Or a " mole " may have been removed and forgotten years before metastases appear. Or, rarely, a frankly malignant growth is unexpectedly cured by excision, or has even been known to disappear spontaneously. Melanomas metastasize to the regional lymph glands, or by the blood-stream, or in both ways. Blood-borne metastases may develop in every organ, or in one or two organs only, or in unusual sites such as mucous membranes, endocardium or placenta. Maternal melanoma with metastases in the placenta has been known to metastasize also to the foetus. Cerebral or intestinal metastases from unsuspected primary growths have caused many diagnostic errors. Wide dissemination of secondary melanotic tumours may be accompanied by the appearance of large quantities of free melanin or melanogen in the blood and urine (melanaemia and melanuria).

(f) A note on " blue naevus "

This appears anywhere in the skin as a small well-circumscribed rounded, firm, blue or bluish-black papule, which is present at birth and usually remains unchanged throughout life. Microscopically it shows long fusiform or dendritic melanin-laden cells scattered in the deeper layers of the dermis. These are believed to be mesenchymal melanoblasts, similar to those of the " Mongolian spot ", a poorly defined brownish area sometimes congenitally present in the skin over the sacrum. " Blue naevus " and " Mongolian spot " are unrelated to ordinary pigmented moles, and do not give origin to malignant melanomas.

(2) Melanomas of juxta-cutaneous mucous membranes

Primary melanoma occurs occasionally in the nasal cavity, mouth, conjunctiva or episcleral tissues, rectum, vagina or cervix. This is not surprising, since these juxta-cutaneous mucous membranes are known to possess melanin-forming capacities, witness their pigmentation in Addison's disease. Hence, the histogenesis of melanomas in these situations is similar to that of the epidermal ones, i.e. from melanoblastic epithelium like that of the epidermis. Rectal melanoma is usually situated low down, involving the anal canal. In structure and behaviour the juxta-cutaneous melanomas resemble the cutaneous ones.

(3) Intra-ocular melanomas

After the skin, the next commonest site of melanomas is the uveal tract of the eye, usually the choroid, less frequently the ciliary body or iris. The histo-genesis of these tumours is uncertain : some workers have supposed them to be, like the cutaneous melanomas, essentially epithelial, derived directly or indirectly from the pigment epithelium of the retina ; but others believe them to arise from the dendritic pigment cells of the choroid, which they regard as mesenchymal melanoblasts, comparable with those in the leptomeninges, the structural homologue of the choroid.

In structure and behaviour, intra-ocular melanomas generally resemble the cutaneous ones ; but there are points of difference. The ocular melanomas are more often of diffuse spindle-cell structure, are very rarely amelanotic, and the liver is a particularly favourite site of their metastases. Secondary growths in this organ or elsewhere may appear many years—even 20 or 30 years—after removal of the eye with the primary growth. The hepatic metastases may attain great sizes, even 20 pounds or more ; and melanaemia and melanuria may be present.

(4) Meningeal melanosis and melanomas

Elongated branched pigment cells are normally present in the pia mater, especially of the brain stem and especially in dark-skinned races. These cells, which are usually regarded as mesenchymal melanoblasts, are the source of rare primary melanomas of the meninges. These tumours are heavily pigmented, focal or diffuse growths composed of fusiform or irregular cells. They may extend widely in the meninges, but they do not metastasize to other parts of the body.

CHORDOMA

Chordomas are rare tumours which arise from notochordal tissue in any part of the spinal axis. They are most frequent at the two ends of the axis, the sacro-coccygeal region and the base of the skull, where considerable residues of notochordal tissue are not unusual. In the base of the skull the notochordal residues

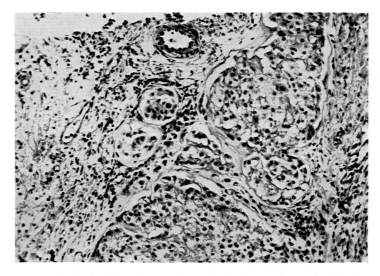

FIG. 247.—Typical chordoma composed of clumps of vacuolated cells.
(× 120.)

sometimes form visible gelatinous nodules called *ecchordoses*, projecting from the dorsum sellae into the cranial cavity.

The typical chordoma is a slowly growing, well-defined, lobulated mass of soft gelatinous tissue, often with areas of haemorrhage, cystic degeneration or calcification. It invades and destroys the neighbouring bone, and extends also into surrounding soft tissues. Sacro-coccygeal growths may attain weights of many pounds ; but the spheno-occipital tumours, which usually project into the cranial cavity, do not reach such large sizes.

Microscopically (Fig. 247), well-preserved chordoma tissue consists of masses of polyhedral or large irregular cells with plentiful cytoplasmic vacuoles, lying in a mucinous intercellular matrix. The structure recalls that of normal noto- chordal tissue.

While many chordomas grow slowly, their ultimate prognosis is uniformly bad. They always recur after attempted removal ; and in a small proportion of cases lymph-nodal or blood-borne metastases develop.

EMBRYONIC TUMOURS OF VISCERA

We have already discussed several kinds of truly embryonic tumours—medullo- blastoma and neuro-epithelioma of the brain, neuroblastoma and ganglioneuroma of the sympathetic system, retinoblastoma, and rhabdomyosarcomas of the pelvic organs and some other parts. Two kinds of embryonic tumours remain for consideration, *nephroblastoma* or embryonic renal tumour, and *hepatoblastoma* or embryonic hepatic tumour. Some general comments on embryonic tumours will be added.

(1) Embryonic tumours of the kidney : nephroblastomas

Many names have been applied to these well-known growths—" adeno- sarcoma ", " myosarcoma ", " mixed tumour ", " Wilms' tumour ", " embry- oma ", " embryonic nephroma ", etc. They arise from immature renal cortical tissue during foetal life or infancy.

(i) *Incidence.*—Nephroblastomas, though not very common, are one of the most frequent malignant tumours of infancy and early childhood. They have been seen in premature foetuses, in a considerable number of cases they are known to be present at birth, and the majority of them appear before the age of 3 years. Very rarely their appearance is delayed until adolescence or adult life. Males are affected about twice as often as females. The tumours are usually single but multiple and bilateral growths occur. Embryonic renal tumours occur also in pigs, sheep, rabbits, rats and fowls.

(ii) *Structure* (Figs. 248, 249).—The structure varies greatly ; in general it mimics that of embryonic renal tissue of varying degrees of differentiation, while certain heterotopic kinds of tissue may also appear. Undifferentiated cellular tissue is often plentiful, and forms the bulk or the whole of some tumours. Epithelial clumps, tubules and young glomeruli are frequently present ; and renal pelvic epithelium, sometimes showing cornification, may also be included in the tumours. Immature mesenchyme is often present ; and from it smooth and striated muscular tissue are frequently, and cartilage and bone less frequently,

differentiated.   The appearance of these heterotopic tissues in the tumours has
no special histogenetic significance ; it does *not* imply that fragments of myotome
and sclerotome have been included in the developing kidney, but is merely the
result of aberrant differentiation in the young nephric tissue itself.

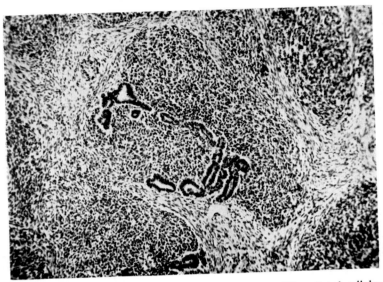

FIG. 248.—Nephroblastoma showing a mixture of undifferentiated cellular
tissue and differentiated tubules. (× 100.)

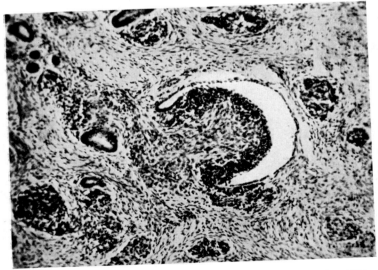

FIG. 249.—An immature glomerulus in a nephroblastoma. (× 120.)

(iii) *Growth and metastasis.*—The first symptom is either abdominal tumour or haematuria, the latter sometimes accompanied by the passage of fragments of growth. In a few cases the tumours are still confined to the kidney when first discovered and nephrectomy is curative. In most cases, however, they prove fatal either because they are inoperable or from metastasis. Invasion of large veins is sometimes evident in the excised kidney, and is an important factor in prognosis. Metastases are most frequent in the lungs, but occur also in liver, bones, lymph glands and peritoneum.

(2) Embryonic tumours of the liver : hepatoblastomas

Tumours comparable with the embryonic tumours of the kidney occur very rarely in the liver in the foetus or young infant. They consist of embryonic liver tissue, sometimes accompanied by undifferentiated mesenchyme, striated muscle, cartilage or bone.

(3) General comment on embryonic tumours

Most of the tumours of adults arise only after prolonged exposure of the tissues to extrinsic carcinogenic agents. This cannot apply to embryonic tumours ; they arise during actual organogenesis from tissues which are still immature. Their causes must be sought in disturbed embryonic chemistry ; and the factors responsible for the disturbances, though possibly environmental, will not be carcinogens in the usual sense, but factors concerned with the nutrition and maturation of embryonic tissues.

Since some of our tissues, e.g. parts of our skeletal and dental tissues, are still immature for many years after birth, it is probable that the same considerations may apply to tumours of these tissues which arise during youth. Perhaps the causes of osteosarcomas and osteoclastomas, which are chiefly tumours of young people, will be found more in disturbances of the chemistry of skeletal growth than in the direct application of extrinsic carcinogenic agents. Perhaps the " adamantinomas " of the paradental and parapituitary residues, most of which also appear during childhood or adolescence, are to be looked on as embryonic tumours, evoked in still immature tissue by factors which disturb the growth and maturation of that tissue. These are speculations ; but they are justified by the peculiar phenomenon of embryonic neoplasia, which clearly demands concepts of tumour causation very different from those arising out of studies of occupational and experimental carcinogenesis.

TERATOMAS

(1) Definition

A teratoma is a true tumour composed of multiple tissues of kinds foreign to the part in which it arises.

Teratomas are *true tumours*, that is, unlike simple malformations, they display some degree of progressive uncoordinated growth. They contain *multiple tissues;* and the more thoroughly they are examined, the greater the variety of tissues found in them. The notion that some of them contain only one kind of tissue or tissues of " only one germ layer " is almost always a confession of inadequate examination. Their tissues are *of kinds foreign to the part*. This distinguishes them from mixed tumours and embryonic tumours with heterotopic tissues. Thus, a nephroblastoma which contains skeletal muscle fibres and cartilage is not a teratoma, for these tissues arise by aberrant differentiation of the neoplastic embryonic nephric tissue. Nephroblastomas never contain completely exotic tissues, such as salivary, dental, respiratory or neural tissue, all of which are common in teratomas.

(2) Classification

In spite of a wide range of structure and behaviour, all teratomas constitute a single class, subdivision of which is arbitrary. In most cases distinction between benign and malignant teratomas can be made on structural grounds : those composed wholly of completely differentiated tissues are benign ; while those containing incompletely differentiated tissues, alone or along with well-differentiated ones, are malignant.

(3) The sites of teratomas

The main sites of teratomas are, in order of frequency, the ovary, testis, retroperitoneal region, anterior mediastinum, pre-sacral and coccygeal region, and base of skull. Rarer sites are the pineal gland, brain and neck. Teratomas are almost unknown in the face, posterior mediastinum, thoracic and abdominal walls or viscera (except the gonads), and the limbs.

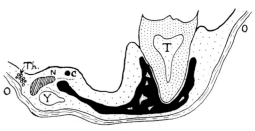

(4) Age incidence

Nearly all sacro-coccygeal, retroperitoneal, cervical, cranial and intra-cranial teratomas are known to have been present at birth or are discovered in early childhood ; and it is clear that they arise at early stages of embryonic development. Ovarian, testicular and mediastinal teratomas are discovered at rather later periods but still in early adult life, the mean ages for these three groups being about 33, 30 and 27 years respectively. Since many of these growths are large and clearly of long duration when first discovered, and since many examples of congenital tumours in these situations have been seen, there is little doubt that the gonadal and mediastinal teratomas also arise during early stages of development.

FIG. 250—Sketch of typical cystic teratoma (" dermoid cyst ") of ovary, with a skin-covered hairy knob and a tooth projecting into the cavity. Other components, identified microscopically, are shown in the diagram thus : bony socket of tooth T, black ; C, nodule of cartilage ; N, mass of neuroglial tissue ; Th., thyroid tissue ; Y, epithelial cyst ; nerves shown as dotted lines ; O, layer of ovary.

(5) The structure of teratomas

(i) *Naked-eye appearance.*— The two commonest sites of

teratomas, the ovary and testis, provide typical examples of both benign and malignant growths : most ovarian teratomas are of the well-differentiated, benign, cystic type, erroneously called " dermoid cysts " ; while most testicular teratomas are imperfectly differentiated, malignant, solid or polycystic tumours.

A typical example of the benign cystic teratomas of the ovary is depicted in Fig. 250. In most cases, the tumour contains a single main cyst into which a mass of solid tissue, or of solid tissue containing smaller cysts, projects from one part of the wall. This mass is often clothed by skin, which also lines the rest of the main cavity, and hairs or teeth or both frequently project from it. The contents of such skin-lined cysts is mainly sebaceous matter and detached hairs.

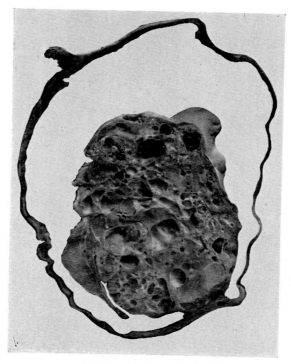

Fig. 251.—Malignant teratoma of ovary, consisting of a polycystic and solid mass projecting into the main cavity. This mass contained a great variety of immature tissues.

In other tumours, however, the main cyst is lined and the intracystic eminence is clothed, not by skin, but by mucus-secreting respiratory or alimentary epithelium or by nervous tissue and choroid plexus ; and the contents of such cysts accordingly consist of mucus or of clear cerebrospinal fluid. The intracystic eminence often contains easily recognizable masses of bone, cartilage or adipose tissue.

Malignant teratomas vary greatly in appearance. Some are completely solid, but often heterogeneous, perhaps with recognizable patches of cartilage or bone. Many contain multiple small cysts with intervening solid tissues. Others contain masses of solid or polycystic tissue projecting into the cavities of larger cysts, approaching the benign type of structure (Fig. 251).

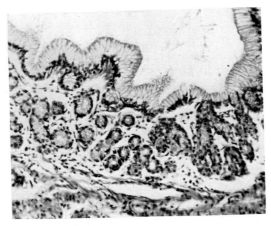

FIG. 252.—Gastric mucosa in a mediastinal teratoma.
(× 120.)

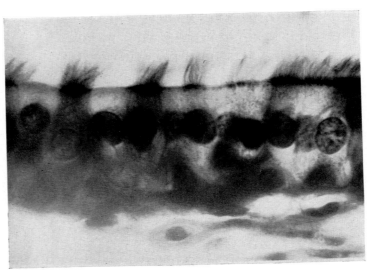

FIG. 253.—Ciliated respiratory epithelium in a teratoma (iron-haema-toxylin stain). (× 1000.)

(ii) *Component tissues* (Figs. 252-255).—In benign teratomas all of the tissues are of fully differentiated *adult* type. Malignant teratomas contain mixtures of differentiating *embryonic* tissues of all ages, with or without fully differentiated ones. Almost any kind of tissue may be present. The commonest tissues are skin and its appendages ; respiratory epithelium ; mucous and mixed glands ; teeth ; intestinal mucosa ; nervous tissue, including neuroglia, ependyma-lined cavities, choroid plexus, embryonic neuro-epithelial tubules, nerves and nerve ganglia ; cartilage, bone, vascular, connective and

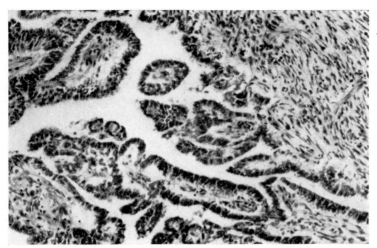

FIG. 254.—Papillary carcinoma-like growth and cellular connective tissue in a malignant teratoma of testis. (By courtesy of the late Prof. D. H. Collins.) (× 125.)

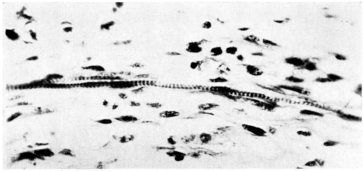

FIG. 255.—Slender striated muscle fibres in a malignant teratoma of testis (iron-haematoxylin stain). (× 500.)

adipose tissue ; smooth and striated muscle. Less common are gastric, pancreatic, pulmonary, thyroid, ocular, hepatic and renal tissue.

(iii) *Highly organized parts.*—In many teratomas the tissues are mixed together in an unorganized hotch-potch. But some tumours contain highly organized structures, such as imperfect digits, loops of intestine or bits of well-differentiated cerebellum or cerebrum. We should note also that those very familiar components, fully developed skin and teeth, are highly organized structures.

(6) Malignancy and metastasis of teratomas

Malignancy in teratomas is of two distinct kinds :

    (a) Total malignancy of embryonic teratomas ;

    (b) Malignancy of one component only of a previously benign teratoma.

(a) Total malignancy of embryonic teratomas

The total malignancy of most testicular and some other teratomas is clear from both their structure and the composition of their metastases. The structure of the ordinary malignant teratoma of the testis is not that of a benign cystic growth undergoing invasion by a single malignant component ; all of the component tissues are actively multiplying, immature and malignant, and all may appear in lymph-nodal or blood-borne metastases. Sometimes, however, a small, slowly growing, well-differentiated teratoma of the testis gives rise to bulky, rapidly growing, anaplastic, often chorion-epitheliomatous, metastases in the retroperitoneal lymph glands or lungs.

(b) Malignancy of one component of a previously benign teratoma

This is most often squamous-cell carcinoma arising in the skin of a " dermoid cyst " of the ovary ; but occasionally tumours of other kinds have been seen, e.g. adenocarcinoma of a glandular component, or argentaffin carcinoma of intestinal tissue in the teratoma.

(7) Nature and histogenesis of teratomas

Many and various hypotheses have been advanced regarding the nature of teratomas. The most popular of these, and the one still most widely held, is that a teratoma is a distorted foetus, either an included twin of its bearer or a parthenogenetic product of his or her own germ cells. There are many facts which show this prevalent view to be erroneous. From careful topographical study of teratomas, it is clear that they show no evidence of vertebrate organization —no signs of a spinal axis, no organs, no somatic distribution of parts. In spite of the occasional presence of highly organized structures, such as teeth or digits, teratomas are in fact completely non-foetiform.

If other facts were needed to show the falsity of the view that a teratoma is a foetus, the following could be mentioned—(a) the multiplicity of parts which are often present, e.g. hundreds of little patches of nervous tissue, hundreds of intestinal or respiratory cysts, hundreds of teeth ; (b) the mixture of adult and embryonic tissues of all ages, which is seen in many malignant teratomas ; (c) the fact that a teratoma is a neoplasm with powers of progressive growth ; and (d) the difficulty of explaining how an included twin or a parthenogenetic foetus gets into such situations as the pineal gland, brain, neck or anterior mediastinum.

We still do not understand the mode of origin of teratomas ; but recent advances in experimental embryology and the chemistry of early development show the crudity of our earlier speculations and provide a basis for a more scientific approach. The clue to teratoma formation must be sought in the study of organizers, the morphogenetic chemical substances of early development. It may be that teratomas arise from foci of plastic pluripotential tissue which, early in embryonic life, escaped from the influence of the primary organizer, so that they failed to participate in the normal architecture of the body and underwent instead chaotic differentiation along their own lines. The nature and cause of this escape are unknown, but increasing knowledge of embryonic chemistry will shed light on them. At the same time, there may come also an explanation of the neoplastic quality of teratomas, which no doubt is closely linked with the chemical disturbances responsible for their genesis. It is of great interest that, with few exceptions, teratomas arise in tissues which developmentally occupy immediately pre-axial, median or nearly median positions. This distribution strongly suggests the operation of disturbances emanating from the primitive embryonic axis, the notochord and other tissues derived from the invaginated primitive streak during gastrulation. It is these tissues which carry the sterol substances of the primary organizer, on which axiation and orderly somatic development initially depend. (See Chapter 39.)

CHORION-EPITHELIOMA

Chorion-epithelioma is unique in being a tumour of one individual transferred to another—a foetal tumour affecting the mother. It arises from the chorionic epithelium of the placental villi.

(a) Incidence

Fortunately, chorion-epithelioma is rare. It occurs in women of child-bearing age, usually between 25 and 45 years. It rarely develops from the normal placenta during a healthy pregnancy, but from retained placental tissue following delivery or more often following abortion. The interval elapsing between the gestation and the appearance of the tumour is usually only a few weeks or months, but occasionally it is several years.

(b) Site

The uterus is the primary site in fully 90 per cent of cases ; in most of the remainder the tumours arise in the Fallopian tubes following tubal gestations. Very rarely, chorion-epithelioma occurs in the vagina, vulva or other pelvic organs, or as disseminated growths in lungs, liver and other remote parts, without a demonstrable primary tumour in the uterus or tubes. The interpretation of these growths is often uncertain ; some of them may really be metastases from small primary uterine tumours which were later discharged or disappeared ; while others may arise from transported chorionic tissue which is known to be carried as emboli in the blood-stream even in normal pregnancy.

(c) Hydatidiform mole

" Mole " is an old-fashioned name applied to the uterine contents in any kind of early abortive gestation. The most important of such degenerated masses is hydatidiform mole, in which the placental villi become oedematous and swollen and converted into clusters of avascular grape-like bodies with thin stalks (Fig. 256). Early death of the foetus results and the mole is expelled ; but parts of the hydatidiform placental tissue, which tends to burrow into the uterine wall (" invasive mole "), are often left behind and these may become chorion-epitheliomatous. About 50 per cent of chorion-epitheliomas are preceded by recognized hydatidiform moles, about 25 per cent by ordinary abortions and about 25 per cent by apparently normal pregnancies ; but, since minor degrees of hydatidiform change easily escape recognition, it is probable that this change precedes chorion-epithelioma in the great majority of cases. It is sometimes difficult to distinguish between an invasive hydatidiform mole (" chorion-adenoma destruens ") and chorion-epithelioma. The causes of hydatidiform degeneration of the placenta are unknown ; there are good reasons for believing them to depend on defective development of the embryo itself rather than on any endocrine or nutritional abnormalities of the mother.

(d) *Structure of chorion-epithelioma* (Fig. 257)

To the naked eye, chorion-epithelioma appears as a friable, highly haemor-rhagic mass occupying the uterus and invading its wall to a varying depth. Micro-scopically, areas of well-preserved tumour tissue consist of irregular aggregates

FIG. 256.—Hydatidiform mole. (Natural size.)

of polyhedral cells and fused syncytial masses. These are sometimes distinctive enough to resemble the hyperplastic chorionic epithelium of retained or hydatid-iform placenta, but most tumours are anaplastic and show no real resemblance to placental tissue. Invasion of blood-vessels is a prominent feature, and large areas of the tumours consist of little more than blood-clot.

(e) *Metastasis*

The proclivity of chorion-epithelioma to invade veins explains the predomin-ance of blood-borne metastases. These occur especially in the lungs, where they may cause haemoptysis as the first symptom of the disease ; but they are frequent also in other organs, especially the spleen, intestines, brain, kidney and liver. Local blood-borne metastases, due to retrograde embolism in the vulvo-vaginal or other pelvic veins, often develop in the vulva, vaginal walls, bladder or broad ligaments ; and they are sometimes the first signs of disease.

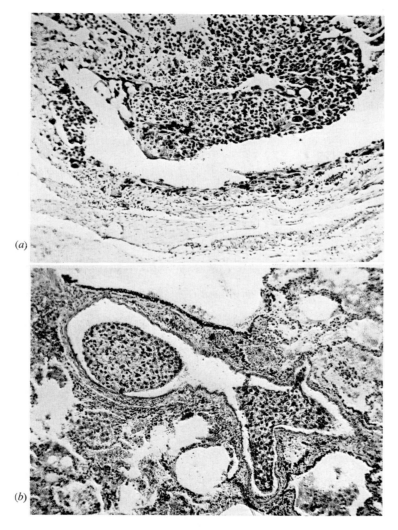

(a)

(b)

FIG. 257.—Chorion-epithelioma. (a) primary growth invading a uterine vein ; (b) tumour embolus arrested at bifurcation of a small pulmonary artery. (× 75.)

(f) Hormonal results

Both hydatidiform mole and chorion-epithelioma secrete gonadotropic hormone similar to that of the normal placenta, and so produce a strongly positive Aschheim-Zondek or Friedmann test. Rise or fall of the hormone titre in the urine in a case of mole or tumour under treatment provides a valuable indication of recurrence or non-recurrence of the disease. Extra-uterine chorion-epithelioma, like extra-uterine pregnancy, also produces decidual changes in the endometrium. Both hydatidiform mole and chorion-epithelioma are frequently accompanied by multiple luteal cysts of the ovaries.

SUPPLEMENTARY READING

References Chapter 28 (p. 415), and the following :

Allen, A. C. (1949). " A reorientation on the histogenesis and clinical significance of cutaneous nevi and melanomas ", *Cancer*, **2,** 28.

Cappell, D. F. (1928). " Chordoma of the vertebral column with three new cases ", *J. Path. Bact.*, **31,** 797.

Dawson, J. W. (1925). " The Melanomata, their morphology and histogenesis." Edinburgh. (Reprinted from *Edinb. Med. J.*, Oct., 1925.) (Numerous beautiful illustrations.)

Hou, P. C., and Pang, S. C. (1956). " Chorionepithelioma ". *J. Path. Bact.*, **72,** 95.

Kretschmer, H. L., and Hibbs, W. G. (1931). " Mixed tumors of the kidney in infancy and childhood ", *Surg. Gyn. Obst.*, **52,** 1. (A well-illustrated account.)

Nicholson, G. W. (1935). " Foetiform ovarian teratomata ", *Guy's Hosp. Rep.*, **85,** 8 and 379.

Nicholson, G. W. (1937). " Sacral teratoma with digits ", *Guy's Hosp. Rep.*, **87,** 46.

Wells, H. G. (1940). " Occurrence and significance of congenital malignant neoplasms ", *Arch. Path.*, **30,** 535.

Willis, R. A. (1935 and 1937). " The structure of teratomata ", *J. Path. Bact.*, **40,** 1; and **45,** 49.

  — (1962). *The Borderland of Embryology and Pathology.* London; Butterworths (Chap. 11, on embryonic tumours and teratomas).

CHAPTER 36

ENDOCRINE DISTURBANCES

THE NOW enormous subject of *endocrinology* cannot be compressed, even in summary form, into a text-book of general pathology. All that is attempted in this chapter is a brief introduction to the main diseases of the ductless glands and their well-established pathological effects. For the chemistry and functions of the hormones and for the rarer and less clearly understood effects of endocrine disturbances, Cameron's and Bloodworth's books are recommended.

SOME GENERAL PRINCIPLES

(1) Hypofunction and hyperfunction

The physiologist studies the effects of endocrine hypofunction by experimental removal of particular ductless glands, and the effects of hyperfunction by deliberate overdosage with pure hormones or glandular extracts. In the various endocrine diseases, nature presents us with a series of ready-made experiments in which hypofunction of particular glands has resulted from their injury or destruction by disease, or hyperfunction from their neoplastic or hyperplastic overgrowth. Most of the pathological endocrine syndromes are attributable to deficiency or excess of particular hormones, and are clearly comparable with those produced experimentally. The main endocrine diseases can therefore be grouped according to the ductless gland primarily affected and according to whether its secretion is deficient or excessive (see accompanying table).

(2) Causes of endocrine hypofunction and hyperfunction

(i) *Hypofunction.*—Any kind of injury or disease which destroys or impairs the activity of a ductless gland will cause deficiency of its hormone. Total or partial deficiency of the hormone may result from complete or partial surgical removal of the gland or from its complete or partial destruction by infarction, atrophy, inflammatory disease or a tumour. Consequently, similar endocrine effects may follow a variety of different lesions. Thus, myxoedema may result from total thyroidectomy, or from destruction of the thyroid by inflammatory disease, or, most frequently, from fibrosis and atrophy of unknown causes ; and hypopituitary syndromes may result from cysts or tumours which compress and destroy the gland, from infarction following thrombosis of its vessels, or from inflammatory disease.

(ii) *Hyperfunction.*—The lesions of the ductless glands responsible for hyper-secretion are in most cases either tumours or hyperplastic overgrowths, the relative frequency of these two varying from gland to gland. Thus, most cases of hyper-thyroidism are due to non-neoplastic hyperplasias of the thyroid (" goitres ") and only a few cases to tumours of the gland ; but most cases of acromegaly, hyperadrenalinism, adrenal virilism, hyperparathyroidism and hyperinsulinism are due to hormone-secreting tumours. Of course, to say that endocrine hyperfunction is due to a tumour or a hyperplastic enlargement of a particular

595

gland does not explain its cause ; in most cases, we do not know the cause of the tumour or hyperplasia.

TABULAR OUTLINE OF MAIN ENDOCRINE DISEASES

Organ	Hormones	Hypofunctional Disorders	Hyperfunctional Disorders
Thyroid	Thyroxine	(1) Cretinism (2) Myxoedema	Hyperthyroidism (" Toxic " goitre)
Pituitary (anterior)	Growth hormone, gonadotrophic hormones, and others	(1) Pituitary dwarfism and infantilism (2) Fröhlich's syndrome (3) Progeria (4) Simmonds' disease (pituitary cachexia)	(1) Gigantism (2) Acromegaly (3) Cushing's pituitary basophilism
Pituitary (posterior)	Oxytocin and vasopressin	Diabetes insipidus	
Adrenal (medulla)	Adrenaline		Hyperadrenalinism (from tumours)
Adrenal (cortex)	Various hormones, (a) androgenic, (b) oestrogenic, (c) metabolic (especially mineralo-corticoids—aldosterone)	Addison's disease	(1) Adrenal virilism (2) Adrenal feminism (3) Cushing's syndrome (4) Aldosteronism—Conn's syndrome
Ovary	Oestrogens (follicular) and progesterone (luteal)	(1) Primary dysgenesis (2) Castration effects (premature menopause) (3) Incomplete or temporary hypo-oestrogenism (4) Progesterone deficiency	(1) Hyper-oestrogenism with granulosa or theca-cell tumours (2) Androgenism with arrhenoblastoma (3) Hyper-oestrogenism without tumours
Testis (interstitial cells)	Androgens	(1) Primary dysgenesis (2) Castration effects (eunuchoidism)	Hyper-androgenism with interstitial-cell tumours
Placenta	Chorionic gonadotrophin (gives Aschheim-Zondek test), and oestrogenic hormone (emmenin)		Pregnancy effects of hydatidiform mole and chorion-epithelioma
Parathyroid	Parathormone	Tetany	Generalized fibro-cystic disease of bone
Islets of Langerhans	Insulin	Diabetes mellitus	Hyper-insulinism from tumours

(3) Age and effects of endocrine disturbances

The thyroid, pituitary, adrenal and gonadal hormones play important parts in growth and sexual development. Clearly, then, their deficiency or excess during foetal or early post-natal life will produce especially pronounced effects —effects which are distinct from, or additional to, those produced in adults by deficiency or excess of the same hormones. Indeed, in our nomenclature we recognize the distinction between the infantile and adult varieties of several of the endocrine diseases. Thus cretinism and myxoedema are but the infantile and adult varieties of the same disorder, hypothyroidism ; destruction of the anterior pituitary lobe during childhood causes pituitary dwarfism, during later life Simmonds' disease or pituitary cachexia ; hyperpituitarism in childhood or adolescence produces gigantism, in the fully grown adult acromegaly. The effects of a virilizing adrenal cortical tumour vary greatly with the age of the patient ; e.g. in an adult woman it may produce only slight reversal of secondary sexual characters, in a little girl it may produce precocious puberty (even at 3 or 4 years of age) and striking masculine changes, while in a foetus it may greatly alter the development of the genital system and produce pseudo-hermaphrodite characters.

(4) " Polyglandular " disturbances

Primary disease with secretory disturbance of one ductless gland often has reflex effects on the structure and function of other ductless glands.   For example, hypothyroidism is usually accompanied by some pituitary hyperplasia, and acromegaly or gigantism is often accompanied by enlargement of the thyroid, hyperplasia of the adrenal cortex, and atrophy of the gonads. These secondary changes in other ductless glands may produce their own effects ; e.g. it is probable that the excessive growth of facial hair in some female acromegalics is not due directly to the pituitary growth hormone, but to secondary adrenal cortical hyperplasia.   In this sense many endocrine diseases may be " polyglandular ". This term, however, is best avoided, for it is apt to be misinterpreted as meaning that two or more ductless glands are the seat of primary disease simultaneously. Simultaneous primary disease of multiple glands is extremely rare ; most cases of so-called " polyglandular " disturbance exemplify only the functional interrelationships of the various glands.   We return, then, to the principle with which we began, namely that each of the main endocrine disorders is a manifestation of hypofunction or hyperfunction of a particular gland.   We will deal with them in the order shown in the foregoing table.

DISEASES OF THE THYROID

(1) Hypothyroidism

(i) *Cretinism.*—Cretinism is a condition of hypothyroidism from birth or early childhood.   It occurs most frequently in communities where endemic goitre is prevalent and in the progeny of goitrous mothers ;   under these circumstances it is called *endemic cretinism*, and the cretinous child's thyroid is often goitrous also.   The essential cause is lack of iodine in both mother and child.   Occasionally, however, in any community, *sporadic cretinism* occurs in the offspring of healthy parents ;   this is the result of developmental deficiency, even absence, of the thyroid gland.   The typical cretin is dwarfed because of retarded skeletal growth, has a broad, dull face, sluggish mentality, dry rough skin, and a pot belly (Fig. 258).   Thyroid treatment instituted early produces striking improvement, especially in sporadic cases ; but if treatment has been long delayed, there is permanent physical and mental damage which no treatment can remedy.

(ii) *Myxoedema.*—Myxoedema is a condition of pronounced hypothyroidism in adults (Fig. 259).   It may follow total thyroidectomy or inflammatory disease of the thyroid, but in most cases it results from progressive atrophy and fibrosis

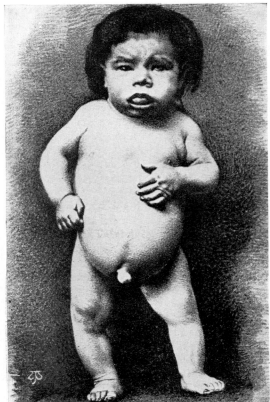

FIG. 258.—Cretinism. (From Byrom Bramwell's *Atlas of Clinical Medicine*, Edinburgh, 1892.)

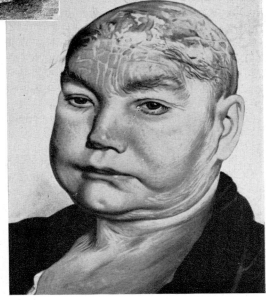

FIG. 259.—Myxoedema. (From Byrom Bramwell's *Atlas of Clinical Medicine*, Edinburgh, 1892.)

of unknown cause, possibly auto-allergic. It is much commoner in women than men. Its chief effects are (1) a puffy thickening of the skin and subcutaneous tissue, due to an accumulation of mucoid intercellular substance, whence the name " myxoedema ", though, unlike a true oedema, the swelling does not pit on pressure; (2) rough dry skin and loss of hair; (3) mental and physical sluggishness; (4) constipation; (5) anaemia; and (6) low basal metabolic rate. The administration of adequate doses of thyroid restores the patient's health and maintains it indefinitely.

(2) Hyperthyroidism

It is usual to distinguish two forms of hyperthyroidism, (1) *diffuse goitre with hyperthyroidism*, also called Graves' disease, Basedow's disease, exophthalmic goitre, diffuse toxic goitre, and primary thyrotoxicosis ; and (2) *nodular goitre with hyperthyroidism*, also called toxic adenoma, nodular toxic goitre, and secondary thyrotoxicosis. The first variety is symptomatically the more severe, often produces exophthalmos, and often develops acutely in a person in whom no goitre was previously apparent. The second variety seldom produces exophthalmos, and develops slowly in a person already known to have a nodular goitre. However, this distinction, though convenient, is not clear-cut ; cases of intermediate type occur.

Although we often speak of " toxic goitre " and " thyrotoxicosis ", these terms are metaphorical, meaning only " goitre with harmful endocrine effects ". These effects, enumerated below, are largely, perhaps wholly, due to hyperthyroidism ; they can all be produced experimentally in animals by overdosage with thyroxine.

(i) *Exophthalmic goitre* (*Graves' disease*) (Fig. 260).—This disease, which occurs chiefly in young and middle-aged adults, more often women than men, is usually of rapid but sometimes gradual, onset. The symptoms and signs include —(1) diffuse enlargement of the thyroid, usually of only slight or moderate degree and sometimes barely perceptible ; (2) exophthalmos, slight or pronounced ; (3) tachycardia, which in long-standing cases may be followed by auricular fibrillation and cardiac failure ; (4) tremor, nervousness and emotional instability ; (5) loss of weight ; and (6) increased basal metabolic rate, even plus 100 per cent.

In most cases the disease begins without any obvious cause, but in

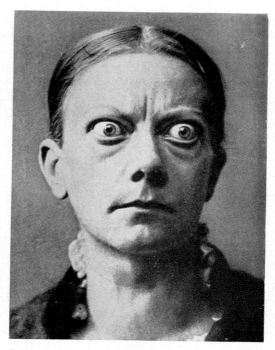

FIG. 260.—Exophthalmic goitre. (From Byrom Bramwell's *Atlas of Clinical Medicine*, Edinburgh, 1892.)

some it comes on after an illness, a period of physical or mental strain or a severe emotional disturbance. While there is no doubt that the main change in the disease is over-activity of the thyroid, the cause of this is unknown. Suggestions have been made that the primary fault is in the sympathetic nervous system or in the thyrotrophic hormone of the pituitary gland, or that auto-allergy plays a part—because the patients often have thyroid auto-antibodies.

To the naked eye the thyroid shows general enlargement, and, on section, patchy or uniform pallor and opacity with loss of the normal vesicular appearance. Microscopically, the main changes are loss of colloid and pronounced epithelial hyperplasia ; the epithelial cells are large and columnar, show many mitoses, and form irregular folds and papillary ingrowths into the vesicles (Fig. 261).

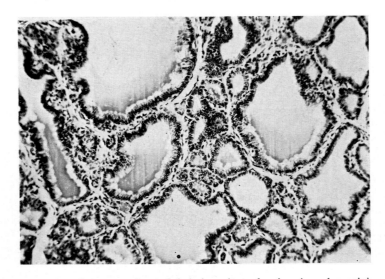

FIG. 261.—Structure of exophthalmic goitre, showing irregular acini, hyperplastic epithelium with papillary ingrowths, and deficiency of colloid. (× 110.)

In severe cases the appearance is very disorderly and closely resembles that of a malignant tumour. The stroma is infiltrated by few or many lymphocytes, which sometimes form distinct follicles. The reduced amount of colloid in the gland is reflected in chemical analyses, which show a marked diminution in its iodine content.

Changes occur also in other organs. These include hyperplasia of the thymus, degenerative changes and fibrosis of the liver, and, in long-standing cases, hypertrophy and dilatation of the heart. Pronounced osteoporosis, i.e. osteoclastic rarefaction of the bones, occurs in some cases.

(ii) *Nodular toxic goitre* (secondary Graves' disease).—A nodular simple goitre has usually been present for years before the appearance of signs of hyperthyroidism. These are generally similar to those in primary hyperthyroidism ; but slow in onset and milder, and exophthalmos is usually absent. In most cases the main effects are cardiac—tachycardia, auricular fibrillation and cardiac failure. The thyroid usually resembles that of simple nodular goitre, sometimes with recognizable patches of hyperplasia resembling that of primary Graves' disease.

(3) Simple (i.e. non-toxic) goitres

Simple goitres, also called " colloid " and " parenchymatous " goitres, are those which are unaccompanied by hyperthyroidism. Some are *diffuse* and some *nodular*; but these are not distinct types, and smooth goitres usually become nodular with the lapse of time. The structure of some simple goitres is closely like that of normal thyroid, but in most cases the vesicles vary greatly in size and many large vesicles distended with colloid are present (Figs. 262, 263). The nodularity of lumpy goitres is due to irregular distribution of colloid, areas of focal hyperplasia or adenomatoid change, the development of true adenomas in some cases, and secondary degenerative changes—cysts, haemorrhages, fibrosis and calcification.

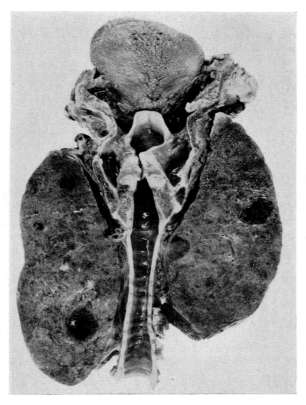

Fig. 262.—Simple colloid goitre, viewed from behind after vertical section. Note its extent in relationship to the pharynx, larynx and trachea. (Three-fifths natural size.)

Simple goitres are prevalent (i.e. endemic) in certain well-known localities, e.g. in Switzerland, New Zealand, the Great Lakes district in North America, and Derbyshire in England (" Derbyshire neck "). These goitrous districts show a low iodine content in the soil and water ; and endemic goitre can be prevented by the regular consumption of small quantities of iodides, e.g. mixed with the table salt. Iodine deficiency is undoubtedly the main factor causing simple

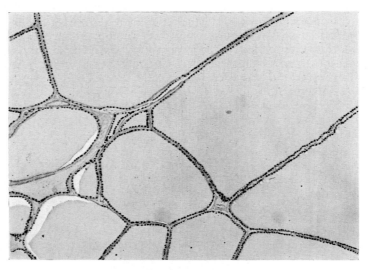

Fig. 263.—Structure of simple colloid goitre ; compare with Fig. 261. (× 60.)

goitre ; but other factors may also contribute, e.g. excess of calcium, excess of fat, vitamin A deficiency, chronic alimentary infections which interfere with iodine absorption, and puberty and pregnancy which increase the body's need for iodine. Goitres are much commoner in women than men. We have already noted that signs of hyperthyroidism may appear in a patient with a previously simple nodular goitre.

(4) The classification of goitres

In its restricted sense, " goitre " refers to the non-neoplastic non-inflammatory enlargements of the thyroid which we have just discussed. In its wider sense it embraces *all* enlargements of the gland. These may be grouped conveniently thus :

(1) *Simple goitre* (" colloid " or " parenchymatous " goitre) : (*a*) diffuse or (*b*) nodular ;

(2) *Toxic goitre* (i.e. with hyperthyroidism) : (*a*) diffuse—primary Graves' disease ; (*b*) nodular—secondary Graves' disease ;

(3) *Neoplastic goitre:* (*a*) adenoma ; (*b*) carcinoma ; (*c*) other tumours (see p. 523) ;

(4) *Inflammatory goitre:* (*a*) Hashimoto's disease and Riedel's disease ; (*b*) other inflammations.

Brief comment on this last group is called for.

(i) *Hashimoto's disease* (lymphadenoid goitre) is a rare condition characterized by uniform firm enlargement of the thyroid due to heavy infiltration by lymphocytes, partly diffuse and partly follicular, accompanied by some degree of fibrosis and parenchymatous atrophy (Fig. 265).

(ii) *Riedel's thyroiditis* (ligneous goitre), also rare, is characterized by lymphocytic infiltration, parenchymatous atrophy and dense fibrosis, slowly converting the

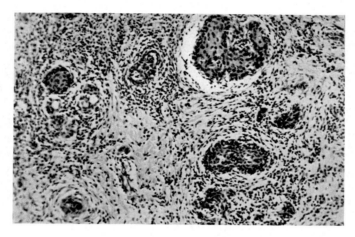

FIG. 264.—Squamous metaplasia of epithelial residues in Riedel's disease. (× 120.)

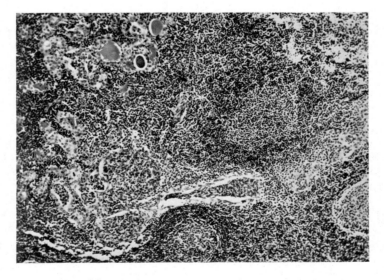

FIG. 265.—Hashimoto's lymphadenoid goitre. (× 65.)

thyroid into a hard mass which is adherent to surrounding structures and which is readily mistaken for a malignant tumour. The atrophic vesicles may show squamous metaplasia (Fig. 264), or may excite the formation of foreign-body giant cells.

Hashimoto's and Riedel's goitres are regarded by some workers as distinct diseases, and by others as different variants or stages of the same disease. Their causes are unknown; but serological studies have shown that in Hashimoto's disease the patient develops auto-immune antibodies against both thyroglobulin and thyroid cells (see p. 289). The parenchymatous atrophy associated with chronic thyroiditis is often accompanied by signs of hypothyroidism.

(iii) *Other inflammations* include tuberculosis, syphilis, and de Quervain's disease or virus thyroiditis, due to mumps virus in some cases.

DISEASES OF THE PITUITARY GLAND

The endocrine effects of disease of the pituitary gland are diverse and variable, because (*a*) it contains two distinct parts with distinct functions, the anterior and posterior lobes (indeed 3 parts, if we count the *pars intermedia*), (*b*) it produces at least 8 distinct hormones, and (*c*) because of its close relationship to the hypothalamus, space-occupying pituitary lesions (cysts and tumours) are apt to disturb hypothalamic functions also. However, most of the symptoms in the several syndromes produced by pituitary lesions *are* the effects of hyposecretion or hypersecretion of pituitary hormones.

(1) Anterior pituitary hypofunction

Causes.—Deficient secretion of the anterior pituitary hormones (namely the growth hormone, gonadotrophic, lactogenic, thyrotrophic, adrenotrophic and diabetogenic hormones) may result from (1) failure of the gland to develop properly, (2) compression or replacement by chromophobe adenoma, pituitary cysts, or para-pituitary cysts or tumours (p. 527), (3) inflammatory diseases of the gland, e.g., tuberculosis or syphilis, (4) ischaemic infarction due to vascular thrombosis (p. 385) or (5) compression of the gland by a bulging floor of the third ventricle in cases of chronic hydrocephalus (see p. 634). The effects depend on the age of the patient, the extent of the destruction, and the presence or absence of hypothalamic disturbance.

(a) Hypopituitarism in childhood

The effects of hypopituitarism are most marked when this occurs in infancy or childhood, for then deficiency of the growth and gonadotrophic hormones is apt to cause striking arrest of development of the skeleton and genital organs —pituitary dwarfism and infantilism. The following syndromes are recognized, though they are not always seen in complete and pure form.

(i) *Lorain or Lorain-Levi syndrome.*—The child shows dwarfism and sexual infantilism without adiposity or mental defect. Pituitary maldevelopment, inflammation, a tumour or cyst may be responsible.

(ii) *Fröhlich's syndrome (dystrophia adiposo-genitalis)* comprises stunted growth, genital hypoplasia, obesity, mental backwardness, and sometimes excessive sleepiness. The last three symptoms are probably the results of interference with the hypothalamic centres and third ventricle by the tumour, cyst or hydro-cephalus which is usually present. This is the commonest form of hypopituitarism.

(iii) *Progeria* is a very rare combination of dwarfism, sexual infantilism and premature senility. It is usually attributed to hypopituitarism but this is not certain.

(b) *Hypopituitarism in adults* (*Simmonds' disease* or *pituitary cachexia*)

Destruction of the pituitary gland in an adult may result from a tumour or its operative removal, from tuberculosis or other inflammatory disease, or from ischaemic post-partum necrosis (p. 385). The chief effects are emaciation, genital atrophy, weakness, subnormal basal metabolic rate, mental apathy and drowsiness.

(2) Anterior pituitary hyperfunction : hyperpituitarism

Three syndromes are attributable to hypersecretion of anterior pituitary hormones :

(i) *Gigantism*, due to excessive secretion of growth hormone by the eosinophil cells during childhood or adolescence ;

(ii) *Acromegaly*, due to the same excess in adults ;

(iii) *Cushing's pituitary basophilism*, due to excessive secretion of hormones by the basophil cells, mainly adrenocorticotrophic hormone (ACTH).

(i) *Gigantism.*—Growth to heights over 6 feet 6 inches is in most cases due to hypersecretion of the pituitary growth hormone, sometimes from hyperplasia, sometimes from an adenoma, of the eosinophil cells. The cause of the hyper-plasia is unknown. In some cases, the excessive growth commences in early childhood ; in others it is delayed until adolescence. The condition is almost restricted to males (Fig. 266). Heights of over 8 feet have been recorded. Acromegalic features are usually absent, but sometimes appear during the adolescent period.

(ii) *Acromegaly* (Gk.=" big extremities ").—This is due to hyperpituitarism in adult life, usually from an eosinophil-cell adenoma (p. 527). The hands, feet and facial parts become enlarged and coarse, mainly by overgrowth of the soft tissues. The calvarium and the mandible grow thicker and heavier. There is also *splanchnomegaly*, i.e. enlargement of the heart, liver and other viscera. Enlargement of the thyroid and hyperplasia of the adrenal cortex are often present, suggesting that there has been hypersecretion of the thyrotrophic and adreno-trophic hormones. As the disease progresses, loss of the sexual functions and atrophy of the genitalia usually ensue. Glycosuria is frequent.

(iii) *Pituitary basophilism.*—In some cases of Cushing's syndrome, an adenoma or focus of hyperplasia of basophil cells is present in the anterior pituitary lobe. Since the same syndrome occurs from adrenal cortical hyperfunction without a pituitary lesion (see p. 608), it is probable that when a pituitary basophil adenoma produces the syndrome it does so through stimulation of the adrenal cortex by an excess of adrenotrophic hormone. We can speak of " Cushing's syndrome " whether the lesion primarily responsible for it is in the pituitary or the adrenal ; the term " pituitary basophilism " should be reserved for the former. The chief features of the syndrome, due to hyper-secretion of adrenal glucocorticoids (cortisone, etc.), are adiposity of the trunk but not of the limbs, weakness, loss of sexual functions, in women hypertrichosis (excessive hair) and other signs of masculinization, hypertension, polycythaemia, and osteoporosis of the skeleton often with kyphosis.

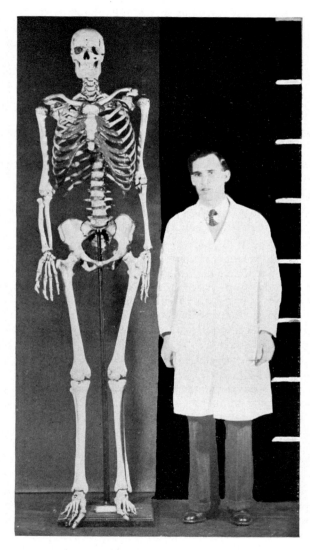

FIG. 266.—Skeleton of the Irish giant O'Bryne, prepared by John Hunter and preserved in the Royal College of Surgeons, London. The scale to the right is in feet ; the skeleton is 7 feet 6½ inches high.

(3) Diabetes Insipidus

Diabetes insipidus means pathological polyuria without glycosuria. It results from selective destruction of the posterior pituitary lobe, e.g. by metastatic tumours, or from hypothalamic injury in the close neighbourhood of the gland. It is relieved by injections of posterior lobe extracts, and there is little doubt that it is due to deficiency of vasopressin, which has an anti-diuretic action.

DISEASES OF THE ADRENAL MEDULLA

The adrenal medulla is not essential to life and can be removed without causing obvious ill-effects. Accordingly, there are no diseases in which the symptoms are clearly attributable to adrenaline deficiency. The only adrenal medullary disease with endocrine effects is hypersecreting *chromaffin-cell tumour* (p. 575), which produces paroxysmal attacks of hypertension, accompanied by tachycardia, pallor, sweating, nervousness, and glycosuria—effects which are all accounted for by the excessive secretion of adrenaline and noradrenaline.

DISEASES OF THE ADRENAL CORTEX

(1) Adrenal cortical hypofunction

In animals experimental removal of both adrenals causes rapidly fatal general weakness, with hypoglycaemia, increased excretion of sodium and chlorides, decrease of these substances in the blood, raised plasma potassium, dehydration, fall of blood-pressure and diminished basal metabolic rate. These effects are due to loss of the cortical hormones ; they can be relieved, and the animals kept in good health indefinitely, by regular injections of suitable cortical extracts.

(a) Acute cortical hypofunction in man

This occurs very rarely from bilateral haemorrhagic necrosis of the adrenals, a condition which is seen principally in infants with meningococcal septicaemia, diphtheria or other severe toxaemic infections.

(b) Chronic cortical hypofunction : Addison's disease

This not very uncommon disease is the chronic counterpart of the acute disease produced in animals by double adrenalectomy. It results from slow progressive destruction of both adrenals from any cause. The lesions responsible are, in order of frequency, bilateral fibro-caseous tuberculosis (p. 150), bilateral atrophy or developmental hypoplasia of unknown causes, and rarely other lesions including bilateral metastatic tumours. The disease is rare in children, and occurs principally in adults between 30 and 50 years old.

The main symptoms of Addison's disease are general weakness, subnormal blood-pressure, brownish pigmentation of the skin and oral mucosa due to excessive melanin, and vomiting, diarrhoea and loss of weight. As in the experimental animal, there is also sodium chloride depletion, raised plasma potassium, and hypoglycaemia. The disease is slowly progressive with, however, severe " crises " in which the symptoms are all aggravated. If untreated it is usually fatal within two years. Properly controlled treatment with sodium chloride and suitable preparations of cortical hormones (e.g., desoxycorticosterone), however, has greatly improved the prognosis.

(2) Adrenal cortical hyperfunction

The main endocrine effects of tumours and hyperplasias of the adrenal cortex (see p. 529) are attributable to hypersecretion of androgen, oestrogen, glucocorticoids, or aldosterone (mineralocorticoid). The main syndromes produced, though not always clear-cut, are the following:

(i) *Adrenal virilism or hyper-androgenism.*—This occurs usually in females, in whom it produces varying degrees of masculinization. In foetal life, pseudohermaphrodite characters may be produced. In a girl before puberty, the result is precocious puberty with male characters, frequently with prominent enlargement of the clitoris and labia. In adult females, there occur voice changes, hairiness of male distribution, adiposity, masculine contours, amenorrhoea, and atrophy of the breasts and internal genitalia. In male children, there is premature sexual, skeletal and muscular development (" infant Hercules " type). In male adults, as might be expected, endocrine results are often not apparent.

(ii) *Adrenal feminism or hyper-oestrogenism.*—This is rare. The most striking changes are in adult men, who show atrophy of the genitalia and mammary enlargement and secretion (gynaecomastia). In little girls premature feminine puberty is induced.

(iii) *Cushing's syndrome.*—Adrenal cortical tumours or hyperplasia are the commonest causes of Cushing's syndrome (p. 605).

(iv) *Aldosteronism (Conn's syndrome).*—This is characterized by hypertension, polyuria and muscular weakness, with disturbed plasma electrolytes and potassium depletion. Nearly all cases have an adrenal cortical tumour.

Hormone assays of the urine are essential in the investigation of these cases. It remains only to add that adrenal cortical hyperfunction does not always denote that a tumour is present; it is sometimes due to non-neoplastic hyperplasia, or even to hyperfunction without enlargement of the glands.

OVARIAN HORMONAL DISTURBANCES

(1) Primary anomalies of development of the ovaries

Failure of development of the gonads, of course, results in corresponding lack of gonadal sex hormones. *Turner's syndrome*, a distinctive form of dwarfism in girls with primary amenorrhoea and gonads represented by only functionless rudiments in the position of the ovaries, is due to sex-chromosomal anomalies— sometimes XO, sometimes mosaic mixtures such as XO/XY, XO/XX, etc. *Agenesis of the ovaries*, without stigmata of Turner's syndrome, also occurs and results in primary amenorrhoea and persistent infantile genitalia.

(2) Hypo-oestrogenism

(i) *Castration and premature menopause.*—Complete loss of ovarian oestrogens occurs physiologically at the menopause. It also results from the pre-menopausal removal of both ovaries or suppression of their function by X-rays or radium— a condition of premature artificial menopause. This is often accompanied by exaggerated menopausal symptoms—nervousness, hot flushes and irritability— which are probably related to an increased output of pituitary gonadotrophins, and which are relieved by giving oestrogens.

(ii) *Incomplete or temporary hypo-oestrogenism.*—Oestrogen deficiency has been blamed for many menstrual and other disorders of women, often on very slender evidence. The only indubitable effect of such deficiency is cessation of menstruation, *amenorrhoea.* Besides the physiological amenorrhoea of pregnancy or the menopause, this may also occur pathologically from many different endocrine or non-endocrine causes. Endocrine disturbances which are frequently accompanied by amenorrhoea include Fröhlich's syndrome, Simmonds' disease, acromegaly, Cushing's syndrome, myxoedema and hyperthyroidism. Non-endocrine causes include malnutrition from tuberculosis, diabetes or other disease, anaemia, drug addiction and emotional disturbances. In all cases the factor immediately responsible for amenorrhoea is deficient oestrogenic stimulation of the endometrium ; in many of the diseases just enumerated this is due in its turn to deficient stimulation of the ovaries by pituitary gonadotrophins.

(3) Progesterone deficiency

Some cases of habitual or threatened abortion may be due to deficiency of luteal hormone.

(4) Ovarian hyper-oestrogenism

On p. 514 we have already considered the hypersecretion of oestrogenic hormone by *granulosa-cell and theca-cell tumours* of the ovaries. There also, it was noted that some *arrhenoblastomas* are probably the same kind of tumours but with perverted formation of androgen in place of oestrogen, while others are hilar-cell tumours.

Metropathia haemorrhagica (" fibrosis uteri ") is the name applied to cases of excessive menstrual bleeding accompanied by general overgrowth of the endometrium and fibrosis and thickening of the myometrium. The ovaries usually contain multiple follicular retention cysts, and the uterine changes are almost certainly due to hyper-oestrogenic stimulation.

Hyperplasia of the mamma is easily induced experimentally in mice by oestrogen administration, and prolonged treatment may result in carcinoma (p. 419). Both cystic hyperplasia and non-cystic lobular hyperplasia (adenosis) of the human breast (p. 466) are undoubtedly due to hormonal disturbances ; excess of oestrogens is probably the main factor, though the ovaries do not usually show any distinctive lesions.

(It is important to appreciate that the ovaries are not the only source of oestrogens ; these are produced also by the adrenal cortex and the placenta, and in males probably by the interstitial cells of the testis also. Male urine contains oestrogens, derived from the adrenals or testis or both. Clearly then, in considering oestrogenic disturbances, we must not focus our attention exclusively on the ovaries.)

TESTICULAR HORMONAL DISTURBANCES

(1) Primary anomalies of development of the testes

Various forms of *male pseudohermaphoditism* occur. In the most common form, although the patient has a sex-chromosome constitution of XY, he has poorly developed cryptorchid testes and genitalia structurally approaching those of a female; he may be mistaken for a girl until non-occurrence of menstruation draws attention to his condition. Abnormal sex-chromosomal constitution also occurs in males; e.g. patients with *Klinefelter's syndrome*, small testes with aspermatogenesis often accompanied by gynaecomastia and mental deficiency, have one or more superfluous X chromosomes—XXY, XXXY, or a mosaic mixture such as XXY/XX.

(2) Hypo-androgenism

The testicular androgens are secreted by the interstitial cells of Leydig. Total loss of these occurs from castration or from destruction of both testes by injury, atrophy, inflammation or tumours,—eunuchoidism. If this occurs in childhood, puberty fails to appear and infantile characters persist. If it occurs after puberty, it may cause mental depression and some loss of sexual function and male characters, but in many cases the changes are slight.

(3) Hyper-androgenism

Hyper-androgenism from tumours of the interstitial cells of the testis was referred to on p. 520.

PLACENTAL HORMONAL DISTURBANCES

The only clearly recognized hormonal disturbance due to placental disease is the persistent secretion of chorionic gonadotrophins in cases of hydatidiform mole and chorion-epithelioma, the effects of which are a positive Aschheim-Zondek test, decidual reaction in the endometrium, and the formation of luteal cysts in the ovaries (see p. 593).

DISEASES OF THE PARATHYROID

The function of parathormone is to maintain a normal ratio of calcium and phosphate in the blood plasma. This it does in two ways, (*a*) by influencing the renal excretion of phosphates, and (*b*) by influencing osteoclastic activity in the bones. Excess of the hormone increases the excretion of phosphates, thereby causing hypophosphataemia, which in its turn induces a compensatory liberation of phosphate of calcium from the skeleton, with consequent hypercalcaemia ; and at the same time the excess of the hormone stimulates osteoclastic activity and so facilitates the mobilization of calcium phosphate from the bones. A prevalent view that the whole action of parathormone is on the kidneys is not true ; there is conclusive evidence of its direct action on osteoclasts.

(1) Hypoparathyroidism

This may result from surgical removal of all of the parathyroid glands, an accident which sometimes occurred in the early days of thyroidectomy but which is now rare. Temporary deficiency of parathyroid secretion may also follow removal of a hypersecreting parathyroid adenoma, the remaining glands being " asleep " and not awaking promptly to the functional stimulus following removal of the one gland which formerly relieved them of their work. Hypoparathyroidism produces hyperphosphataemia and hypocalcaemia, which, when it becomes sufficiently marked, manifests itself clinically as tetany. As already described on p. 360, parathyroid deficiency is not the only cause of tetany ; this may result from any condition which brings about hypocalcaemia.

(2) Hyperparathyroidism

In most cases this is due to a tumour of a parathyroid gland, usually an adenoma, rarely a carcinoma (p. 529) ; in a few cases it results from a non-neoplastic hyperplasia of all the glands. Its main effects constitute *von Recklinghausen's generalized fibro-cystic disease of bone.* This is characterized by severe decalcification of the skeleton, with great osteoclastic activity and fibrous replacement of bony tissue, deformities and fractures ; accompanied by hyper-calcaemia, hypophosphataemia, the formation of renal calculi, and metastatic calcification of other tissues (p. 361). Removal of the responsible tumour is followed by a return of the blood chemistry to normal and recalcification of the bones, but it does not cure established deformities.

DISEASES OF THE ISLETS OF LANGERHANS

(1) Hypo-insulinism : diabetes mellitus

Diabetes mellitus (="sweet polyuria") is a condition of hyperglycaemia and glycosuria due to deficient secretion of insulin by the islets of Langerhans. A typical diabetic syndrome sometimes occurs in cases of hyperthyroidism, hyper-pituitarism or cortical hyperadrenalism, because these hormones all oppose the action of insulin, and their excess therefore produces a *relative* insulin deficiency. However, in true diabetes mellitus there is an *absolute deficiency* of insulin due to disease or suppressed function of the islets. The disease is thus strictly comparable with experimental diabetes produced in animals by pancreatectomy, or by selective chemical injury of the islets by means of alloxan. It is of interest that permanent diabetes can also be produced in dogs by prolonged daily injections of pituitary extracts ; the constant excess of diabetogenic hormone exhausts the islets and causes their permanent degeneration.

The cause of diabetes

In most cases of diabetes, the cause of the injury to the islets is unknown. Only occasionally is the pancreas the site of obvious disease, such as chronic pancreatitis, extensive carcinoma or haemochromatosis ; in most diabetics the organ appears normal to the naked eye. Microscopically also, in a considerable proportion of cases the islets show no changes or only slight ones of doubtful significance ; in other cases they show fibrosis, hyalinization, or loss of the granular β-cells.

Diabetes is a common disease ; it is estimated that there are 200,000 diabetics in the United Kingdom and 700,000 in the U.S.A. The disease manifests itself at any age and in either sex. There is a distinct inherited predisposition, about one-third of diabetics having close relatives who are also diabetic. The incidence of the disease is particularly high among Jews.

Complications

The complications of untreated diabetes include ketosis resulting from im-paired fat metabolism (p. 342), hypercholesterolaemia and xanthomatosis (p. 348), arterial atheroma with consequent thrombosis and gangrene of the limbs (p. 374), diabetic retinitis which also is secondary to changes in the retinal arteries, in-creased susceptibility to infections, e.g. tuberculosis, and certain renal compli-cations, e.g. the glomerular sclerosis of Kimmelstiel and Wilson, and medullary necrosis especially of the papillae. Since the introduction of proper insulin and dietetic treatment, these complications are all much less frequent than formerly.

(2) Hyperinsulinism

The most familiar example of hyperinsulinism is insulin overdosage in diabetics under treatment. The counterpart of this occasionally occurs spontaneously as the result of an islet-cell tumour (p. 504). The chief effects of hyperinsulinism are due to hypoglycaemia (p. 344).

OTHER DISORDERS RELATED TO HORMONAL DISTURBANCES

(1) Diseases of the thymus

Although there is no proof that the thymus produces hormones, the following facts suggest that it may have endocrine functions :

(i) *Abnormal persistence of the thymus* is seen in cases of sexual infantilism from castration or pituitary disease.

(ii) *Hyperplastic enlargement of the thymus* occurs in hyperthyroidism.

(iii) *Myasthenia gravis* is usually accompanied by *thymic tumour or hyperplasia* (p. 526). Myasthenia is a dangerous disease of young people, characterized by excessive fatigue of the voluntary muscles and believed to be due to interference with the production or action of acetyl-choline at the myo-neural end-plates. Auto-allergy has been suspected in its pathogenesis. Thymectomy or X-ray irradiation of the thymic region effects good results in some cases.

(iv) *Rare cases of thymic tumour* associated with diabetes, adiposity and adrenal cortical hyperplasia have been described.

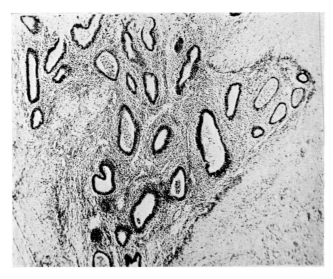

FIG. 267.—A focus of typical endometrial tissue in the wall of the small intestine ; kinking and intestinal obstruction had resulted. (× 50.)

(2) Diseases of the pineal gland

While it has recently been shown that the pineal gland contains a pigment-regulating substance, melatonin, and possibly other hormones, these have not been proved to be concerned in any diseases. Pineal tumours in childhood, either teratomas (p. 586) or pinealomas (p. 567), both of which are much more frequent in boys than girls, are frequently accompanied by precocious puberty. But this is attributed to effects on adjacent structures, especially the hypothalamus.

(3) Endometrial hyperplasia and endometriosis

The endometrium is under hormonal, especially oestrogenic, control ; oestrogen deficiency leads to its atrophy, and oestrogen excess to its proliferation. We know that oestrogen-secreting ovarian tumours produce great endometrial hyperplasia ; and we have also noted another kind of endometrial overgrowth, that of metropathia haemorrhagica (p. 609), which is attributed to hypersecretion of ovarian oestrogens. It is therefore pertinent to enquire whether other kinds of abnormal endometrial proliferation—those called *endometriosis*—may also be due to endocrine disturbances.

Endometriosis means the presence of endometrial tissue in situations where it does not occur normally. Three varieties are recognized :—

(i) *Endometriosis interna or adeno-myosis of the uterus itself.*—In this fairly common condition, strands of endometrium grow out from the uterine mucosa into the myometrium, the muscular tissue of which undergoes hypertrophy, so that the uterine wall is thickened. The change may involve the whole uterus uniformly ; or only a part of the wall may be affected, forming a localized thickening which, however, is never sharply circumscribed.

(ii) *Surgically transplanted uterine endometrium.*—Following an operation on the uterus, an area of endometrial tissue may develop in the operation scar in the abdominal wall. This is clearly due to accidental transplantation.

(iii) *Endometriosis externa.*—This is the spontaneous development of endometrial tissue outside the uterus (Fig. 267). The sites of such ectopic endometrium are the pouch of Douglas and other parts of the pelvic peritoneum, the recto-vaginal septum, the broad and round ligaments, tube and ovary, umbilicus, and the wall of the sigmoid colon, appendix or ileum. There has been much debate as to the origin of foci of endometriosis in these situations. One view is that they are implants of viable mucosa discharged retrogradely through the Fallopian tubes into the peritoneal cavity or carried as emboli in veins or lymphatics during menstruation. The other view is that they are genuinely heterotopic foci of endometrium derived from the peritoneal tissues by metaplasia.

The structure of aberrant endometrium is similar to that of the normal tissue, and it also undergoes proliferation and haemorrhage with each menstrual cycle. Since there is no exit for the blood and secretions produced, these accumulate in small cysts, and the lesion enlarges and becomes painful at each menstrual period. The surrounding tissues suffer progressive fibrosis, which in the wall of the intestine often causes kinking and obstruction. Though sometimes called " endometrioma ", the lesions are not neoplastic, but consist of ordinary endometrial glands and stroma which react to the same endocrine stimuli as the endometrium lining the uterus and which are abnormal only because they are misplaced. At the menopause, the lesions become quiescent and retrogress.

(4) Benign enlargement of the prostate

The ordinary benign prostatic enlargement of elderly men, though often called

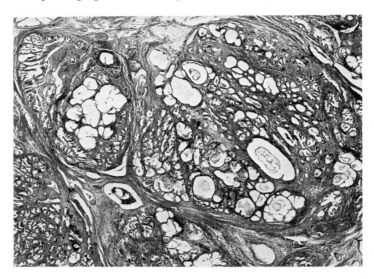

FIG. 268.—Sectional view of hyperplastic prostate showing lobular and cystic structure. (× 9.)

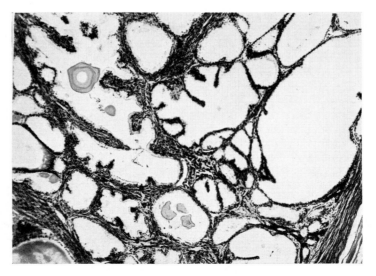

FIG. 269.—Enlarged view of Fig. 268, showing the irregular cystic tubules. (× 60.)

" adenomatous ", is not a tumour but a hyperplasia of the gland. This is believed to be due to some endocrine disturbance, probably a disturbance of androgen-oestrogen balance connected with involution of the sexual organs and functions. That the prostate is under endocrine control is shown by its periodic enlargment in certain animals during seasons of sexual activity or rut, by its atrophy following castration, and by the fact that administration of oestrogens to male animals induces a remarkable degree of squamous metaplasia in the organ. While, from these and other facts, it is almost certain that benign prostatic hyperplasia is a manifestation of endocrine imbalance, the precise factors involved have not yet been elucidated. Hence it is not surprising that the results of endocrine treatment of enlarged prostate have been disappointing and that its surgical removal is still the treatment of choice.

Microscopically (Figs. 268, 269), the enlarged prostate shows proliferation of both the glandular and intervening smooth-muscular tissue. In some glands an almost normal appearance is retained ; in others the hyperplastic glands show papillary irregularity and minor cystic change, or focal overgrowth of the muscular tissue may form well-defined leiomyomas 1 or 2 centimetres in diameter or larger. To the naked eye the gland may be smoothly and uniformly enlarged; but more often it shows a distinctly nodular structure both externally and on the cut surface. Not infrequently a large lobule of the gland—a so-called " middle lobe "—projects up into the bladder as a hemispherical eminence just behind, and over-hanging, the urethral orifice. The chief consequences of prostatic enlargement are elongation and compression of the prostatic urethra, which appears as a narrow median cleft between the two lateral halves of the gland, and urinary obstruction and its effects (p. 626).

(5) Gynaecomastia

Gynaecomastia, mammary hyperplasia in males, is probably always due to an imbalance of sex hormones, especially with excess of oestrogens, perhaps coupled in some cases with excess of pituitary mammotrophin. It occurs in a variety of different circumstances, some of which have already been mentioned.

(i) *Mammary hyperplasia in the new-born* (" mastitis neonatorum ") is a transient enlargement of the breasts in new-born infants, either male or female, probably due to the action of maternal oestrogens transferred through the placenta to the child.

(ii) *Mammary hyperplasia at puberty* (" puberal mastitis ") is a similar tem-porary enlargement and tenderness, doubtless connected with the endocrine adjustments of this period.

(iii) *Oestrogen-makers' gynaecomastia* sometimes occurs in men engaged in the manufacture of oestrogenic drugs.

(iv) *Gynaecomastia with diseases of the testis* has occurred not only from teratoma or " chorion-epithelioma " and from interstitial-cell tumour (p. 520), but also from castration and from atrophy of the testes following injury. Probably in all of these cases, the endocrine disturbance involved is an absolute or relative excess of oestrogens, either because oestrogens are formed by the tumours or because androgens are lost in consequence of the testicular disease.

(v) *Gynaecomastia with adrenal tumours*, mentioned on p. 608, is due to secretion of oestrogens by the tumours.

(vi) *Gynaecomastia with other tumours* has been reported occasionally with tumours of the pineal gland, mid-brain or hypothalamus, with acromegaly, and with mediastinal teratomas.

(vii) *Gynaecomastia with cirrhosis of the liver* has been recorded. It may be related to the fact that the liver is concerned with the inactivation of both androgens and oestrogens.

(6) Hypertension due to renin

The discharge of renin from the kidney into the blood for functional purposes (p. 309) conforms to our concepts of internal secretion. Hence, renin can be regarded as a hormone, the kidney as an endocrine gland, and hypertension due to renal ischaemia as an endocrine disorder.

SUPPLEMENTARY READING

Anderson, J. R. *et al.* (1964). " Autoimmunity and thyrotoxicosis." *Brit. Med. J.*, **2**, 1630.

Bloodworth, J. M. B. (Ed.) (1968). *Endocrine Pathology*. Baltimore; Williams & Wilkins: Edinburgh and London; E. & S. Livingstone. (A comprehensive up-to-date account.)

Cameron, A. T. (1947). *Recent Advances in Endocrinology*. London; Churchill. (A comprehensive monograph.)

Doniach, D., Hudson, R.V., and Roitt, L. M.(1960). " Human auto-immune thyroiditis ". *Brit. Med. J.*, **1**, 365.

Doniach, I. (1960). " Diseases of endocrine organs." In *Recent Advances in Pathology* (Ed., C. V. Harrison), Chapter 7. London; Churchill.

Dunn, J. S., and co-workers (1943). " Necrosis of the islets of Langerhans produced experimentally ", *J. Path. Bact.*, **55**, 245. (The discovery of alloxan diabetes.)

Gardiner-Hill, H. (1937). " Abnormalities of growth and development ", *Brit. Med. J.*, **1**, 1241 and 1302. (Includes a good outline of endocrine disturbances.)

Hartog, M. (1966). "Hormones and the endocrine system". *Brit. Med. J.* **1**, 225.

Knaggs, R. L. (1929). "Cretinism", *Brit J. Surg.*, **16**, 370. (Good account with interesting figures.

– (1935). " Acromegaly ", *Brit. J. Surg.*, **23**, 69. (Well illustrated account.)

Rolleston, H. (1937). " The history of endocrinology ", *Brit. Med. J.*, **1**, 1033. (This was the concluding article of a useful series of brief outlines of various aspects of endocrinology contributed by different authorities.)

Smith, J. F., Bolton, J. R., and Turnbull, A. L. (1955). " The renal complications of diabetes mellitus ". *J. Path. Bact.*, **70**, 475.

Symington, T. (1959). " The human adrenal cortex in disease." In *Modern Trends in Pathology* (Ed., D. H. Collins), Chapter 13. London; E. & S. Livingstone.

– (1969). *Functional Pathology of the Human Adrenal Gland*. Edinburgh and London; E. & S. Livingstone. (A comprehensive monograph.)

Vaux, D. M. (1938). " Lymphadenoid goitre : a study of 38 cases ", *J. Path. Bact.*, **46**, 441. (A beautifully illustrated account.)

Wilkins, L. (1957). Second edition. *The Diagnosis and Treatment of Endocrine Disorders in Childhood and Adolescence*. Illinois; Charles C. Thomas. (Many excellent illustrations.)

CHAPTER 37

OBSTRUCTION AND DILATATION OF HOLLOW ORGANS

INTRODUCTION

IN THIS chapter we consider briefly the effects of mechanical obstruction and distension of hollow organs or parts of organs. First let us clarify certain general terms.

(1) Obstruction

In pathology this has its ordinary dictionary meaning of blockage or impediment to the passage of the contents of an organ. In general the causes of obstruction are of three kinds : (*a*) abnormal contents of the organ, e.g. foreign bodies, calculi or parasites ; (*b*) diseases of the wall of the organ leading to its narrowing or occlusion, e.g. developmental stenosis, fibrosis resulting from injury or inflammation, or tumours ; and (*c*) compression of the organ from without, e.g. by neighbouring tumours or other masses, by fibrous adhesions or by incarceration in a hernial sac.

(2) Dilatation

This term also has its usual meaning of expansion or distension. In an obstructed organ, it results from accumulation of the contents and rise of the pressure within the organ. It may also result from weakening of the wall of the organ by disease, the weakened wall then yielding to the normal intra-visceral pressure, as in an aneurysm due to syphilitic aortitis or bronchiectasis due to chronic bronchitis.

(3) Diverticulum

This is a localized dilatation or outpocketing of the wall of a hollow organ. Some diverticula are developmental in origin, e.g. Meckel's diverticulum ; but acquired diverticula are due to localized weakening of the visceral wall, sometimes combined with raised intra-visceral pressure from obstruction.

(4) Hypertrophy

Just as the blacksmith's biceps enlarges by hypertrophy in response to the extra work it is called upon to do, so the muscular walls of hollow viscera hypertrophy in response to dilatation and raised intra-visceral pressure consequent upon chronic obstruction. The left cardiac chambers hypertrophy as a result of raised arterial blood-pressure, and the right chambers as a result of raised pulmonary blood-pressure ; the wall of the stomach hypertrophies in response to pyloric obstruction, the wall of the intestine in response to a stenosing growth or a chronic volvulus, and the wall of the bladder in response to prostatic enlargement or urethral stricture. In all of these cases the increased bulk of the muscle is, we believe, a hypertrophy and not a hyperplasia, i.e. the muscle fibres increase in size but not in number. The immediate stimulus evoking hypertrophy is the stretching of the muscle fibres ; stretched muscle fibres contract more vigorously, do more work, and therefore receive a greater blood supply than less active fibres. That the muscle of a distended viscus works harder than normal is shown by the vigorous peristalsis of the stomach or intestine in cases of chronic pyloric or intestinal obstruction, the peristaltic waves sometimes being plainly seen through the abdominal wall.

(5) Cysts

Cysts are any closed cavities in which secretions or other fluids accumulate because they have no exit. We are not concerned here with parasitic cysts, cystic tumours, or cysts due to developmental anomalies, all of which are discussed in other chapters ; but with *retention cysts* due to accumulation of secretions in obstructed glandular lumina or other epithelial cavities.

OBSTRUCTION OF THE ALIMENTARY CANAL

(1) Obstruction of the oesophagus

By far the commonest cause of oesophageal obstruction is carcinoma (p. 491). Infrequent causes are fibrous stricture following corrosive chemical injury (p. 332), chronic peptic ulceration (p. 95), the pressure of an aortic aneurysm (p. 382), and the condition of cardiospasm now to be described.

Cardiospasm or achalasia of the cardiac sphincter is a condition of progressive dilatation and hypertrophy of the oesophagus due to a peculiar type of obstruction at its lower end. This is not the result of any obvious organic lesion, but is attributed to achalasia or failure of the normal relaxation of the cardiac sphincter following deglutition, a failure possibly related to some disturbance of innervation. Food accumulates in the oesophagus, which becomes dilated and elongated. Its wall shows hypertrophy of the muscular coat, which may attain a thickness of 5 millimetres or more ; and the mucous membrane becomes thickened by chronic inflammation, and sometimes ulcerated. The patient wastes from starvation.

(2) Obstruction of the stomach

Except hour-glass stenosis due to chronic gastric ulcer, obstruction to the passage of food through the stomach and into the intestine is nearly always at the pylorus or in the first part of the duodenum. Its causes are pyloric or duodenal ulcer (p. 95), pyloric carcinoma (p. 492), rarely other chronic inflammations or tumours, and congenital pyloric stenosis now to be described. Pyloric obstruction from any cause results in accumulation of food in the stomach and periodic vomiting of large quantities of stagnant gastric contents. The organ becomes dilated and its muscular wall hypertrophied, and its peristalsis may be so active as to be visible or palpable on clinical examination.

Congenital hypertrophic pyloric stenosis

This is a disease of the first few weeks of infancy, most frequent in first-born children, more often boys than girls. The circular muscle of the pyloric canal is greatly hypertrophied, forming a firm barrel-shaped mass which may be palpable clinically. The canal itself is constricted, eventually causing almost total obstruction. The rest of the stomach becomes dilated and shows some secondary muscular hypertrophy, and visible peristalsis is often present. The cause of the disease is not known, but it is usually regarded as due to some neuro-muscular incoordination or achalasia at the pylorus.

(3) Obstruction of the intestine

Intestinal obstruction may result from many different causes, which can be grouped conveniently thus :

(*a*) *Developmental malformations*, e.g. imperforate anus, or stenosis or occlusion (atresia) of any other part of the intestine. Signs of obstruction of course appear soon after birth.

(*b*) *Impaction of foreign bodies*, e.g. gall stones (p. 327).

(*c*) *Diseases of the bowel wall itself:* (i) inflammatory strictures, e.g. from tuberculosis (p. 142), Crohn's disease (p. 311), or chronic dysentery (p. 179); (ii) tumours, especially stenosing carcinoma, which is the commonest cause of chronic intestinal obstruction (p. 502) ; (iii) endometriosis (p. 613).

(*d*) *Infarction of the intestine*, due to occlusion of the mesenteric vessels (p. 374).

(*e*) *Mechanical lesions of the bowel*, often accompanied by vascular strangulation (p. 370) : (i) strangulation in hernial sacs, (ii) intussusception, (iii) volvulus, (iv) strangulation by peritoneal bands or adhesions.

(*f*) *Congenital megacolon* (Hirschsprung's disease).

(*g*) *Paralytic ileus.*

Some of these lesions are commoner in the large intestine than in the small and vice versa ; some produce acute obstruction, and others chronic obstruction. Most of them have been described in other chapters ; but (*e*), (*f*) and (*g*) remain for notice here.

(i) *Hernia*

A hernia is any abnormal pouch-like protrusion of the peritoneum, containing intestine, omentum or other viscera. The two commonest kinds of hernia are the *inguinal*, due to developmental patency of the processus vaginalis of the peritoneum following descent of the testis, and the *femoral*, an acquired protrusion of the peritoneum through the weak spot afforded by the femoral canal, especially in women. Less common hernias are umbilical, ventral (usually through weak operation scars), sciatic, perineal, diaphragmatic, and intra-abdominal hernias (e.g. into the lesser sac, an abnormally large duodeno-jejunal fossa, or other exaggerated peritoneal pouches).

Intestinal obstruction in a hernia is due to strangulation of a loop of bowel by its constriction at the neck of the hernial sac. Once vascular constriction is started, a vicious circle is established ; the initial swelling due to venous obstruction increases the degree of constriction, and this in turn increases the vascular obstruction and swelling of the sac contents, which unless speedily released will suffer infarction (p. 371).

(ii) *Intussusception* (Fig. 270)

This is the telescoping of a length of intestine into the adjoining part. It usually takes place from above downwards, and its most frequent site is the terminal ileum—ileal or ileo-colic intussusception. The outer or receiving part is the *intussuscipiens* ; the invaginated intestine, consisting of entering and returning parts, is the *intussusceptum*. The most advanced part of the intussusceptum, where the entering and returning layers are continuous, is the *apex*. The entering and returning layers have their serous coats opposed, except where they are separated by the portion of the mesentery which is drawn in between them. The apex of the intussusceptum is the same from first to last ; as it advances, more and more of the outer intussuscipiens is invaginated to form the returning layer of the elongating intussusceptum. More and more tension is put on the invaginated mesentery, in consequence of which the intussusception becomes curved with its concavity towards the root of the mesentery. At the same time, obstruction of the mesenteric vessels leads to increasing strangulation of the invaginated parts, which become swollen, engorged and eventually gangrenous.

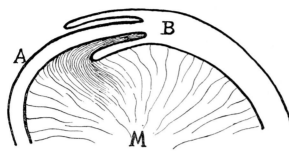

FIG. 270.—Diagram of intussusception : A, entrant part of intestine (intussusceptum) ; B, receiving part (intussuscipiens). Note how part of mesentery M which is invaginated with intussusceptum is crowded and taut.

Intussusception occurs most frequently in infants under a year old, more often boys than girls, in whom it is the commonest cause of intestinal obstruction. In infants intussusception is usually unassociated with any previous structural lesions of the bowel, and is evidently the result of irregular peristalsis probably due to faulty diet. Intussusception in older children and in adults is usually initiated by some local structural change, e g. a polypus, a polypoid carcinoma, a pedunculated lipoma, a metastatic growth, or an invaginated Meckel's diverticulum. The bowel grasps the swelling and propels it distally just as if it were a solid mass free in the lumen. It is possible that in infants also the initial cause may be a local lesion, namely a swelling of the Peyer's patches of the terminal ileum due to dietetic upset.

(iii) *Volvulus*

Volvulus is rotation of a loop of bowel through 180 degrees or more. Its most frequent site is the pelvic colon ; less frequently, a loop of the small intestine, or rarely the whole small intestine, is twisted. The twisted loop may quickly become strangulated and acute intestinal obstruction, gangrene and peritonitis may ensue. But in some cases of sigmoid volvulus the obstruction is incomplete and the condition is chronic ; the twisted pelvic colon slowly becomes greatly distended and its wall greatly hypertrophied, before complete obstruction or strangulation supervenes.

(iv) *Intestinal strangulation by peritoneal bands or adhesions*

A loop or several loops of bowel may be caught and strangulated beneath a band of adhesions left from previous appendicitis or other inflammatory disease, or by a fibrous cord associated with a Meckel's diverticulum or other developmental anomaly.

(v) *Congenital megacolon (Hirschsprung's disease)*

This rare disease is characterized by great dilatation and muscular hypertrophy of the colon, which from infancy or soon afterwards is unable to evacuate its contents normally, although there is no organic obstruction of its lumen. In some cases the dilatation extends distally to the pelvi-rectal junction, but the rectum is normal ; in others the rectum also shares in the dilatation. The child suffers from increasing constipation and abdominal distension and may die during childhood from anaemia and emaciation ; or, in less severe cases, he may survive until adult life. The obstruction is due to failure of development of Auerbach's plexus, which is found to be deficient at the level of the obstruction.

(vi) *Paralytic ileus*

This is the name applied to general atonic dilatation of the intestine with loss of peristalsis which sometimes follows abdominal operations and which also accompanies general peritonitis.

ACQUIRED DIVERTICULA OF THE ALIMENTARY CANAL

(1) Acquired diverticulum of the pharynx

The least uncommon kind of pharyngeal diverticulum arises in middle-aged people, more often men than women, as an acquired protrusion of the mucous membrane through the inferior constrictor muscle in the posterior mid-line. This is the weakest spot of the pharyngeal wall ; and the pouch probably develops because of some habit of neuro-muscular incoordination in deglutition whereby partial obstruction and raised intra-pharyngeal pressure occur. Once formed, the pouch enlarges and extends downwards behind the oesophagus, even as far as the mediastinum. Stagnant food accumulates in it and distends it, and increasing dysphagia results. The mucosa of the pouch becomes inflamed, and may ulcerate and perforate, leading to cervical or mediastinal abscess or cellulitis.

(2) Acquired diverticula of the small intestine

In middle-aged or elderly people it is not uncommon to find multiple rounded thin-walled diverticula, consisting of mucous and submucous coats only, bulging through gaps in the muscle coat along the line of attachment of the mesentery (Fig. 271). They are attributable to the thrust of the intra-intestinal pressure at

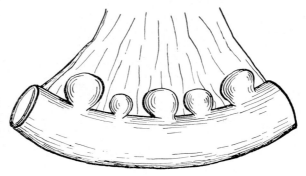

FIG. 271.—Sketch of diverticulosis of small intestine.

weak spots in the muscle, probably where the entrant vessels exert traction on the bowel wall. Perhaps this traction effect is greater when the vessels are atheromatous and less extensible than normal. Except for the peculiar X-ray shadows they may cause, the diverticula are of little clinical importance, for they rarely become inflamed or perforated.

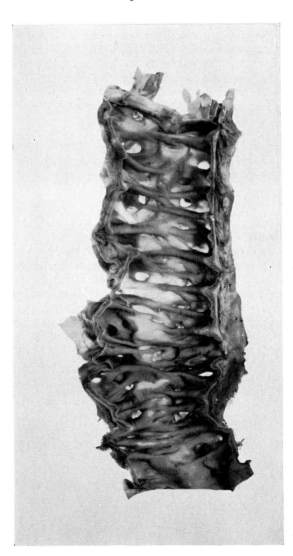

FIG. 272.—Diverticulosis of large intestine. Interior view of bowel showing trabeculation and multiple thin-walled diverticula seen partly by transmitted light. (Two-thirds natural size.)

(3) Acquired diverticula of the large intestine

Diverticula of the colon (Fig. 272) are present in at least 5 per cent of people over the age of 40. They are most frequent in the pelvic colon, and least frequent in the caecum and ascending colon. They form multiple, usually small pouches of mucosa protruding through the muscle coat into the serous coat or appendices epiploicae on any aspect of the bowel, and often containing pellets of inspissated faeces. They are symptomless unless they become infected and inflamed. Acute diverticulitis quickly leads to perforation, peri-colonic abscess or peritonitis. Chronic diverticulitis causes great thickening, fibrosis and constriction of the affected segment of colon, which may simulate carcinoma both clinically and in naked-eye appearance.

OBSTRUCTION OF THE BILIARY TRACT

The principal causes of obstruction of the bile ducts are gallstones (p. 326), carcinoma of the gall bladder or ducts themselves (p. 504), and tumours of neighbouring structures which compress or invade the ducts, especially carcinoma of the head of the pancreas (p. 504). The principal effects are dilatation

of the ducts and gall bladder above
the level of the obstruction (Fig.
273), obstructive jaundice (p. 358),
in long-standing cases fibrosis of
the liver (biliary cirrhosis), and
sometimes serious ascending bacterial infection (cholangitis) of the
dilated ducts and the development
of multiple abscesses in the liver
(p. 99). The liver of chronic
obstructive jaundice presents a
characteristic appearance ; it is
enlarged, finely granular on the
surface, stained a bright or dark
green colour, rigid and tough in
consistency, and traversed by
prominently dilated bile ducts
throughout ; and purulent cholangitis may also be present.

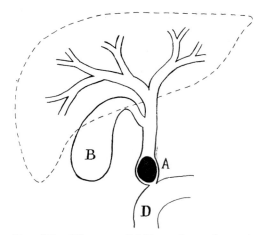

Fig. 273.—Diagram of biliary obstruction : A,
stone impacted in ampulla of Vater, causing
great dilatation of common bile duct, cystic and
hepatic ducts, and gall bladder B. D, duodenum.

OBSTRUCTION OF THE PANCREATIC DUCTS

Chronic obstruction of the pancreatic ducts may result from carcinoma, pancreatic calculi or gall-stones blocking the exits of the main ducts at the duodenal
papilla. The effects include dilatation of the ducts, varying degrees of atrophy of
the secreting acini, and chronic inflammation and fibrosis constituting *chronic
pancreatitis.* This is accompanied by impaired pancreatic digestion in the intestine,
and an excess of unsplit fat in the faeces. The islets of Langerhans usually
survive, so that diabetes does not occur.

Acute haemorrhagic necrosis of the pancreas may be mentioned here because
it is sometimes, though certainly not always, related to duct obstruction. It is a
disease of sudden onset with symptoms of a severe abdominal catastrophe—pain,
vomiting and shock—resembling those of acute perforation of a gastric or duodenal
ulcer. The pancreas is swollen, haemorrhagic, and shows more or less extensive
areas of necrosis, and the surrounding retroperitoneal tissues and omenta show
patches of yellow fat necrosis due to the escape of trypsin and lipase from the
pancreas. Diastase also escapes, and is absorbed and excreted in large amounts
in the urine, where its detection by its action on starch is a useful diagnostic test.
Although this disease is often called " acute pancreatitis ", it is not primarily
inflammatory ; it is believed to be due to sudden activation of the digestive
enzymes in the pancreas, with consequent rapid self-digestion. In cases of
pancreatic necrosis, gall-stones are frequently found in the biliary tract ; and in a
few cases, obstruction of the common exit of the pancreatic and bile ducts by a
stone is present, strongly suggesting that the factor causing activation of the
ferments in the pancreas is regurgitation of bile into the pancreatic ducts. In
other cases the initiating factor may be mild bacterial infection, or a focus of
infarction of the pancreas due to thrombosis of diseased vessels.

OBSTRUCTION AND DILATATION OF THE RESPIRATORY PASSAGES

Dilatation of the respiratory tract due to obstruction, raised internal pressure and chronic cough may occur in the larynx, bronchi or pulmonary alveoli, where it constitutes *laryngocele*, *bronchiectasis* and *vesicular emphysema* respectively.

(1) Laryngocele (cervical aerocele)

This is a rare air-containing diverticulum which results from bulging of the mucous membrane through the thyro-hyoid membrane into the neck. It occurs in glass-blowers, players of wind instruments and persons with chronic cough.

(2) Bronchiectasis

(a) Causes

Bronchial dilatation results from many different causes, most of which have already been considered, namely bronchopneumonia (p. 67), tuberculosis (p. 140), foreign bodies (p. 314), silicosis (p. 316), bronchial carcinoma (p. 489) and adenoma (p. 490). These all produce one or more of the following mechanical factors which tend to cause dilatation of the bronchi : (*a*) obstruction of the bronchial lumen, (*b*) raised intra-bronchial pressure, by accumulation of secretion or exudate, or by chronic cough, (*c*) inflammatory fibrosis and weakening of the bronchial walls, and (*d*) chronic pneumonia or fibrosis in the surrounding lung tissue, with consequent inequalities in the pressure or traction exerted on the bronchial walls from without.

While in some cases bronchiectasis is a congenital malformation or arises in bronchi which are congenitally weak, in the majority of cases it is acquired as a result of one or another of the diseases just named. Many cases date back to childhood and owe their origin to bronchopneumonia following whooping cough or measles (p. 188), though symptoms of well established bronchiectasis may not develop until some years later. In adults bronchiectasis may commence in influenzal or other kinds of bronchopneumonia ; or it may appear to be idiopathic, i.e. its originating cause is not known. Once bronchiectasis has developed, a vicious circle is established ; smouldering infection and discharges from the dilated bronchi maintain a chronic cough and cause repeated broncho-pneumonic attacks, and these aggravate the bronchial dilatation.

(b) Structure

Bronchiectasis may be bilateral or unilateral, widespread throughout a lung or confined to one lobe, most often the lower lobe. The dilatation is usually fusiform or irregular along the entire length of the affected bronchi, less frequently it forms saccular out-pockets. The cavities are lined by congested inflamed mucosa, with ciliated respiratory epithelium which, however, sometimes shows squamous metaplasia. The lung parenchyma between the diseased bronchi shows varying degrees of collapse, fibrosis, and chronic obstructive broncho-pneumonia (Fig. 274) ; and the whole lobe eventually becomes a shrunken, almost airless, fibrous mass traversed by the irregularly dilated bronchi.

(c) Symptoms and complications

These are readily understood from the pathology of the disease. The patient has a constant cough and produces large quantities of purulent, often putrid, sputum. Repeated bronchopneumonic exacerbations are common ; and other complications include pulmonary abscess, empyema, metastatic cerebral abscess, amyloid disease, and pulmonary osteo-arthropathy (p. 644).

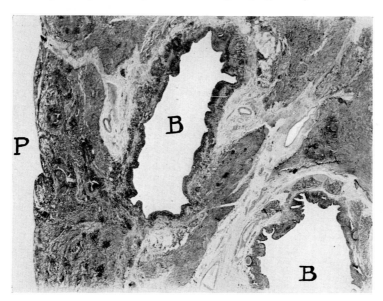

FIG. 274.—Advanced bronchiectasis. Section of lung, showing greatly dilated bronchi B,B in close proximity to pleural surface P, and collapse and fibrosis of surrounding lung tissue. (× 5.)

(3) Vesicular emphysema

Vesicular emphysema, as distinct from interstitial emphysema (see p. 320), means abnormal distension of the alveoli and alveolar ducts with air, producing a coarse honeycombed structure instead of the normal fine spongework of the lung. It results from violent or long-sustained rise of intra-alveolar and intra-bronchial pressure, especially if associated with irregularly distributed bronchiolar obstruction, as in asthma and chronic bronchitis. It also occurs in healthy areas of lung intervening between consolidated areas of bronchopneumonia or tuberculosis, when it is called " compensatory emphysema ".

General emphysema of the lungs has a striking effect on the shape of the thorax. As the blown-up lungs become more and more voluminous, the chest gradually becomes more and more rounded and barrel-shaped, like the chest in full inspiration. The diaphragm is depressed to a low level, the respiratory excursion and vital capacity are much reduced, and the patient quickly becomes breathless on exertion.

OBSTRUCTION AND DILATATION OF THE URINARY TRACT

(1) Causes of obstruction

Obstruction of the flow of urine may occur at any level of the urinary tract, from the renal pelvis to the urethra. Of the principal causes of obstruction, those described in other chapters are calculi (p. 328), tuberculosis (p. 148), tumours of the urinary system (p. 508), tumours compressing or invading the urinary tract from without, e.g. carcinoma of the uterus (p. 517), hyperplastic or cancerous enlargement of the prostate (p. 614 and p. 520), and gonorrhoeal or other strictures of urethra (p. 123). Another important cause is paralytic retention of the urine in the bladder owing to disturbance of the micturition reflexes by injury or disease of the spinal cord, e.g. from fracture-dislocations of the spine, spinal tuberculosis (p. 146), primary or secondary tumours of the spine or of the contents of the spinal canal, disseminated sclerosis and other primary degenerative diseases of the cord (p. 635). In a considerable proportion of cases of hydronephrosis and hydro-ureter in both children and adults, no organic obstruction can be found, and it is supposed that the cause is some disturbance of co-ordination of the musculature of the ureter or the uretero-vesical sphincter, analogous with the achalasias of the alimentary canal.

(2) Effects of obstruction at different sites

Obstruction in the renal pelvis or at the pelvi-ureteric junction, e.g. by stone, tuberculosis or tumour, causes distension of part or whole of the pelvis—hydronephrosis, or if the urine is infected, pyonephrosis. Obstruction in the ureter or at the ureteric orifice in the bladder causes hydro-ureter and hydronephrosis. Obstruction at the neck of the bladder or in the urethra, e.g. by stone, prostatic enlargement or stricture, causes first dilatation and hypertrophy of the bladder only, the uretero-vesical sphincters protecting the ureters and kidneys from back-pressure. Later, however, the increasing dilatation of the bladder renders these sphincters incompetent, and bilateral hydro-ureter and hydronephrosis then develop (see Fig. 228).

The degree of muscular hypertrophy occurring in the obstructed urinary tract is greatest in the bladder and least in the renal pelvis. In cases of long-standing vesical obstruction, e.g. from prostatic enlargement, the walls of the bladder often attain a thickness of 1 or 2 centimetres, and the muscle trebles or quadruples its bulk. On its interior the hypertrophied muscle bundles stand out prominently, like the trabeculae carneae of the heart. At the thinner parts of the walls between the muscular bands, diverticula sometimes develop by protrusion of the mucous membrane through the weak areas. A chronic hydro-ureter often shows considerable thickening of its wall from muscular hypertrophy ; but this is less noticeable in the walls of the hydronephrotic renal pelvis.

(3) Effects on the kidney

The effects of urinary obstruction on the kidney differ according to whether the obstruction is sudden and complete or gradual and partial. Following sudden complete obstruction, as from ligature of the ureter or its total blockage by an impacted stone, the kidney almost at once ceases excreting urine, because the rapid rise of intra-pelvic pressure (due to the little extra urine which is discharged by the organ) soon exceeds the maximum renal " excretion pressure ". Under these circumstances, the hydronephrosis consists of only a moderate distension of the pelvis and never attains a great size. Experiments on ligation of the ureter have shown that, if the obstruction is removed within two weeks, the kidney may recover its function ; but that if the obstruction is maintained for longer periods, the kidney suffers progressive and permanent atrophy, and the opposite kidney undergoes compensatory enlargement.

When obstruction to the outflow of urine from the kidney is incomplete or intermittent—and this applies to most cases of obstruction—the kidney continues to secrete and gradual distension of the pelvis may eventually produce a very large hydronephrosis, sometimes containing several pints. Not only is the renal tissue expanded into a thin crescentic zone around the distended pelvis, but the back-pressure in the renal tubules and glomeruli leads to their gradual atrophy and to fibrosis of the tissue. Eventually the secreting elements may disappear completely, leaving only a fibrous-walled sac.

(4) Complications of urinary obstruction

(a) Ascending bacterial infection

In cases of urethral or vesical obstruction, purulent *cystitis*, usually due to cocci or *E. coli*, frequently supervenes (p. 180). If the obstruction is severe or prolonged, the infection commonly ascends from the bladder into the dilated ureters and renal pelves, causing *pyelitis* and converting the hydronephrosis into

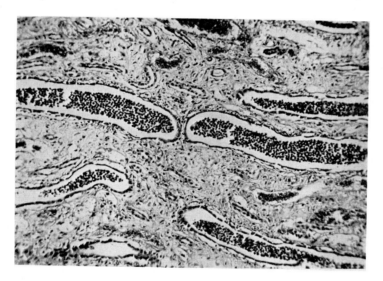

Fig. 275.—Ascending pyelonephritis ; collecting tubules dilated and filled by pus cells. (× 110.)

pyonephrosis ; it usually extends also into the renal tubules and causes *pyelonephritis* of varying degrees of severity (Fig. 275). Mild smouldering pyelonephritis with progressive destruction and fibrosis of the kidney may last for some months or years before eventually leading to renal failure or hypertension. When it is more severe it causes acute inflammatory swelling and radiating lines of suppuration or small abscesses throughout the kidney tissue, and rapidly fatal toxaemia and uraemia. This is a common cause of death from prostatic enlargement, urinary calculi, paralytic retention of urine, and carcinoma of the bladder, prostate or uterus. In some cases of pyelonephritis, the infection spreads also to the perirenal tissues, causing *perinephric abscess*.

(b) Calculus formation

Urinary obstruction from any cause, especially when complicated by chronic infection, is often accompanied by the secondary deposition of phosphatic calculi (p. 326).

(c) Renal failure (uraemia)

This is a frequent termination of obstructive diseases affecting both kidneys. It is similar to that occurring in chronic nephritis (p. 305). Renal function tests are of great value in the prognosis of obstructive disease and in the planning of its treatment.

OBSTRUCTION IN THE FEMALE GENITAL TRACT

(1) Obstruction of the Fallopian tubes

The only frequent cause of tubal obstruction is chronic salpingitis, usually gonococcal, leading to *hydrosalpinx* or *pyosalpinx* (see p. 123). It is an important cause of sterility.

(2) Obstruction of the uterus or vagina

This is rare. Its least uncommon cause is developmental malformation (imperforate hymen, absence of vagina or imperforate cervix) leading to retention of menstrual fluid which gradually distends the vagina or uterus—*haematocolpos* and *haematometra*. Much rarer is acquired stenosis of the vagina from injury or inflammation. Occasionally the cervical canal becomes stenosed or obliterated as a result of senile endometritis, carcinoma, or fibrosis following radium treatment of carcinoma, and the body of the uterus becomes distended with pus—*pyometra*.

OBSTRUCTION OF THE MALE GENITAL TRACT

Ligation of the vas deferens is followed by cessation of spermatogenesis and atrophy of the seminiferous tubules. The same result sometimes occurs from gonococcal or other inflammations of the epididymis ; and gonorrhoea is thus an important cause of male, as well as female, sterility. Syphilis, tuberculosis and tumours of the testis or epididymis may also cause suppression of spermatogenesis by obstruction of the efferent ducts. Some cysts of the epididymis (spermatoceles) contain milky fluid in which spermatozoa are present ; these are evidently due to local obstruction of the ducts of the epididymis or of the efferent ductules of the testis.

RETENTION CYSTS OF GLANDS

In the preceding paragraphs we have dealt with the effects of obstruction in the main visceral highways. We have now to add a note on obstruction of the small blind byways, the glands which open onto main epithelial surfaces.

(1) Skin

The commonest retention cysts of the skin are *sebaceous cysts* due to blockage of the orifices of pilo-sebaceous follicles. These may attain a large size by accumulation of sebum and desquamated squamous cells from the stratified epithelial lining which they develop. Simple cysts of sweat glands are rare.

(2) Breast

Retention cysts of the mammary ducts are common. In cystic hyperplasia, many small cysts are present (p. 466). If multiple large cysts develop throughout the breast, the condition is called Schimmelbusch's disease. Occasionally a solitary cyst or a few cysts may develop without much change in the remainder of the breast. Large cysts in the breasts are probably not specially liable to carcinoma ; the really dangerous precancerous lesion is the finely cystic type of hyperplasia which is often undetected clinically.

(3) Kidney

Retention cysts, solitary or few in number, are common in the kidneys of middle-aged and old people. They result from obstruction of renal tubules in consequence of chronic nephritis or ischaemic fibrosis ; and most granular contracted kidneys contain cysts. Their walls are lined by flattened atrophic epithelium, and the contents is clear yellow fluid. Occasionally a cyst becomes so large that it is clinically palpable.

(4) Liver

Small, solitary or few, thin-walled cysts, filled with watery fluid and lined by flat epithelium, are not uncommon in the liver. They are probably due to mild local inflammatory fibrosis with obstruction of small bile ducts. Retention cysts are rarely seen in cirrhotic livers.

(5) Ovary

Simple follicular and luteal cysts of the ovary may be regarded as retention cysts, since they are due to abnormal accumulation of fluid in follicles which should have discharged it. The causes of this retention are not clear ; they are probably related to disturbances of hormones concerned in the cyclical maturation of the follicles.

(i) *Follicular cysts.*—These cysts are usually small, unilocular, multiple and bilateral, with thin smooth walls lined by simple cubical epithelium, and clear watery contents.

(ii) *Luteal cysts.*—Luteal cysts are usually single, 3 or 4 centimetres in diameter, with a thick yellowish wall containing luteal cells and often lined in part by cubical or columnar epithelium, and containing clear or bloody fluid. Multiple bilateral luteal cysts, sometimes of large size, accompany hydatidiform mole and chorion-epithelioma (p. 594).

(iii) *Endometrial (chocolate) cysts.*—There has been much debate as to whether the fairly common " chocolate cysts " of the ovary (so called because they are full of dark brown thick fluid containing much altered blood pigment) are due to endometrial implants or to metaplastic changes in luteal cysts. The evidence is strongly in favour of the second view: luteal cells are commonly present in their walls, and no sharp distinction can be drawn between epithelialized luteal cysts and " chocolate cysts " with an endometrium-like lining. (See ovarian endometriosis, p. 613).

(6) Parovarian cysts

These are simple unilocular cysts lined by cubical epithelium ; they arise by abnormal accumulation of clear watery fluid in any of the closed Wolffian tubules which normally persist in the neighbourhood of the ovary, especially in the broad ligament. Attempts have been made to distinguish between cysts of different parts of the Wolffian remains ; but this is probably needless hair-splitting.

(7) Other retention cysts

Only passing mention need be made of the common *colloid cysts of the thyroid* in goitrous or adenomatous glands, the common *Nabothian cysts* of the mucous glands of the cervix uteri, and the relatively rare *cysts of Bartholin's glands*.

CARDIAC HYPERTROPHY AND DILATATION

(1) Causes

The causes of obstructive cardiac hypertrophy and dilatation have been described in other chapters. They are : (1) *systemic arterial hypertension*, which primarily imposes extra work on the left cardiac chambers (p. 369) ; (2) *pulmonary arterial hypertension*, due to obstructed circulation through the lungs from such diseases as silicosis, chronic tuberculosis, widespread bronchiectasis or emphysema ; this imposes extra work on the right cardiac chambers ; and (3) *chronic valvular disease*, either stenosis or incompetence, which imposes extra work primarily on the side of the heart containing the diseased valves, which is usually the left side (p. 81).

(2) Compensatory hypertrophy and failure of compensation

Like any other muscle the myocardium responds to extra work by hypertrophy ; and, as long as the extra load does not exceed the reserve capacity of the affected chambers for compensatory hypertrophy, an efficient circulation is maintained. But once the load exceeds this maximum limit, the hypertrophied chambers dilate, the output of blood by the heart (and therefore the venous return to it) falls, and signs of inefficient circulation, i.e. cardiac failure, appear.

As a specific example consider the effects of systemic arterial hypertension. The first result is progressive hypertrophy of the left ventricular muscle, which, if the valves and coronary arteries are healthy, may long maintain an adequate output in spite of the extra pressure opposing it. But, of course, the heart has a reduced reserve capacity, so that the patient's ability to undertake extra physical exertion without breathlessness and other signs of impaired circulation is increasingly limited. Eventually the compensatory hypertrophy of the left ventricle reaches its limit and the chamber begins to dilate. It can now no longer empty itself completely but still contains some residual blood at the end of systole ;

and by just this amount its capacity to receive blood from the left atrium is reduced. The atrium therefore in its turn undergoes first hypertrophy and then dilatation ; and, because it fails to take an adequate amount of blood from the pulmonary veins, pulmonary congestion ensues. The patient now lives under a constant risk of severe or fatal pulmonary oedema. The progressive inefficiency of the left cardiac chambers is aggravated by the fact that their dilatation has resulted in dilatation of the mitral orifice, so that the cusps eventually fail to meet completely and during ventricular systole some blood regurgitates back into the atrium, adding to its embarrassment and to the risk of pulmonary oedema. If the patient escapes this fate, the right cardiac chambers begin to show the effects of the obstructed pulmonary circulation. First the right ventricle, and then the atrium, undergo hypertrophy followed by dilatation ; and signs of progressive congestive cardiac failure appear (p. 385).

This sequence of embarrassment of one cardiac chamber extending step by step to preceding chambers is appropriately spoken of as " back-working ". The same principles apply whatever the site of the lesion which places an extra load of work on the heart. It is a good exercise for the student to work out for himself the nature of this extra load and its back-working effects in cases of aortic valvular incompetence, aortic stenosis, mitral incompetence, mitral stenosis, pulmonary fibrosis, and right-sided cardiac valvular defects (which, however, rarely occur alone).

A final point to notice is that in any kind of progressive heart failure, a vicious circle of impaired blood supply to the myocardium itself is established. The worse the myocardial failure, the worse the coronary circulation ; and the worse the coronary circulation, the worse the myocardial failure. This vicious circle is aggravated if the coronary arteries are also diseased, as they are in many cases of arterial hypertension.

OBSTRUCTION AND DILATATION OF THE VENTRICLES OF THE BRAIN : INTERNAL HYDROCEPHALUS

(1) Causes of internal hydrocephalus

The cerebrospinal fluid is secreted by the choroid plexuses, and makes its exit from the ventricular system through the foramina of Magendie and Luschka into the subarachnoid spaces, where it is absorbed into the venous circulation by the arachnoid villi. Internal hydrocephalus, i.e. distension of the ventricles by an accumulation of cerebrospinal fluid, is practically always due to mechanical obstruction at some point in the ventricular system. According to Professor Dorothy Russell's monograph, it seems probable that " the interposition of some obstruction in the cerebro-spinal pathway . . . is responsible for at least 99 per cent of all cases of internal hydrocephalus ". Obstructions are particularly apt to occur at the narrow parts of this pathway, namely the foramina of Monro, Magendie and Luschka, the aqueduct of Sylvius, and the subarachnoid spaces around the brain-stem at the level of the tentorial opening. The expectation that pathological changes in these situations will obstruct the free passage of fluid from the ventricles is fully borne out by experience, as Professor Russell's study shows. The supposition that impaired absorption of cerebrospinal fluid by the arachnoid villi may cause internal hydrocephalus has little to support it.

Which parts of the ventricular system are involved in hydrocephalus depends on the site of obstruction. Obstruction at a foramen of Monro causes hydrocephalus of the corresponding lateral ventricle only ; obstruction of the aqueduct causes dilatation of the third ventricle and both lateral ventricles ; and obstruction at the foramina of exit or in the subarachnoid spaces around the brain-stem causes distension of the whole ventricular system.

The causes of obstruction can be grouped conveniently as (a) malformations, (b) non-microbic inflammations, (c) microbic inflammations, and (d) tumours and other space-occupying masses.

(a) Malformations

Developmental errors are a frequent cause of internal hydrocephalus, and account for most cases of *congenital hydrocephalus*. In many cases the head is already much enlarged at birth and the infant is stillborn. In others the enlargement becomes noticeable after birth and the child may survive for some months or years, even into adult life, with a slowly enlarging skull which may attain a huge size. The commonest developmental errors are in the aqueduct, which may show stenosis, forking or a neuroglial septum. Other anomalies causing hydrocephalus include septa at the foramina of exit of the fourth ventricle, and the " Arnold-Chiari malformation ", which consists of a caudal tongue-like prolongation of the cerebellum adherent to a greatly elongated medulla oblongata, enclosing an elongated fourth ventricle and plugging the foramen magnum. In many cases this malformation is accompanied by spina bifida with or without meningocele or meningo-myelocele (p. 661).

(b) Non-microbic inflammations

Blood extravasated into the leptomeninges around the brain-stem, especially by birth injuries in infants, may excite a chronic meningitis which causes internal hydrocephalus by adhesive occlusion of the foramina of Magendie and Luschka. It is probable also that intra-ventricular haemorrhage in infants can cause a granular ependymitis which may lead to obstruction of the aqueduct and may be difficult to distinguish from a developmental anomaly. The evidence that meningeal haemorrhage in adults may lead to obstructive hydrocephalus is less convincing.

(c) Microbic inflammations

Ventricular dilatation—hydrocephalus or pyocephalus—accompanying chronic microbic meningitis or occurring as a sequel to healed meningitis results usually from sealing up of the foramina of the fourth ventricle or of the surrounding leptomeningeal spaces by inflammatory exudate or adhesions. A less frequent factor is obstruction of the aqueduct by inflammatory exudate or granulations. The infections responsible have been described in other chapters : they are meningococcal and other kinds of suppurative meningitis (Chapter 8), tuberculosis (p. 147), tertiary syphilis (rarely), and torulosis (p. 214). Of great importance is *post-meningitic hydrocephalus*, usually resulting from healed meningococcal meningitis. This is the commonest cause of chronic hydrocephalus in young children, though in many cases the original meningitis in infancy is mild and undiagnosed.

(d) *Tumours and other space-occupying masses*

Tumours of any kind, primary or secondary, encroaching on the ventricular system may cause hydrocephalus of its obstructed parts. Tumours of the brain-stem, cerebellum, or region of the third ventricle are particularly likely to do so, by obstructing the aqueduct or the third or fourth ventricle or by expanding the brain-stem and plugging up the tentorial opening. Hence, tumours in the posterior fossa very commonly cause hydrocephalus—more commonly than tumours of the cerebral hemispheres. Space-occupying lesions other than true tumours which may similarly cause hydrocephalus include abscesses, hydatid cysts, tuberculous or gummatous masses, intracranial dermoid cysts, and " colloid cysts " of the third ventricle.

(2) Changes in the hydrocephalic brain

Internal hydrocephalus in the adult brain, after the sutures of the skull have united, cannot attain a great size. Instead, even moderate distension of the ventricles quickly causes a dangerous rise of intracranial pressure. But in infants and young children, internal distension of the brain causes also distension of the still growing skull, which, by wide separation of its sutures, may attain a capacity of 2 gallons or more. In such cases the walls of the cerebral hemispheres are expanded into thin sheets, while the basal ganglia and brain-stem remain relatively normal and so suffice to maintain a vegetative life. Intracranial pressure is not greatly raised. Between these two extremes—the rigid adult skull within which great hydrocephalus is impossible, and the almost infinitely distensible infant skull which permits enormous expansion of the brain without a fatal rise of intracranial pressure—all degrees of hydrocephalus are seen, depending on the age of onset and the chronicity of the obstruction.

(3) Changes in the skull

(a) *In the adult*

The inexpansible adult skull shows only *internal pressure markings*. These are :

(i) *Convolutional impressions.*—Impressions made on the inner table of the vault by the pressure exerted by the cerebral gyri produce a characteristic " digital " or " beaten brass " appearance in X-ray photographs.

(ii) *Basal erosions and excavations.*—The cribriform plates are sunken in a bowl-like depression. The posterior clinoid processes are lost and the whole central sphenoid region is flattened out. Deep smooth-walled pot-holes develop in the bone of the middle fossae, where pedunculated protrusions of cerebral cortex are thrust through small apertures in the dura. The internal auditory meati are enlarged.

(b) *In the child*

The expanded cranial vault of the infant or child shows wide separation of the bones at the sutures, the intervals being occupied by fibrous tissue only ; closure of the bones is long delayed, and of course occurs only in long surviving cases in whom the distension ceases to progress. The wide sutures often develop many small Wormian bones. The bones of the face and skull-base are of normal size, so that they have a dwarfed appearance in comparison with the huge calvarium.

(4) Changes in other structures

(a) Eyes

In cases in which there is a dangerous rise of intracranial pressure from some progressive lesion, especially a tumour or other enlarging mass, the raised cerebro-spinal pressure is transmitted into the sheaths of the optic nerves, impeding the return of blood in the central vein of the retina and causing venous congestion and oedema of the optic disc and surrounding retina—*papilloedema* or " choked disc ".

(b) The pituitary gland

In some cases of chronic hydrocephalus, the distended floor of the third ventricle balloons out, indents the diaphragma sellae and compresses the pituitary gland. Various symptoms of hypopituitarism, especially Fröhlich's syndrome, may be brought about in this way.

SUPPLEMENTARY READING

Aird, I. (1936). " Intestinal obstruction ", *Edinb. Med. J.*, **43**, 375.

Edwards, H. C. (1934). " Diverticula of the small and large intestine ", *Lancet*, **1**, 169 and 221.

Negus, V. E. (1950). " Pharyngeal diverticula ". *Brit J. Surg.*, **38**, 129.

Raven, R. W. (1933). " Pouches of the pharynx and oesophagus ", *Brit. J. Surg.*, **21**, 235.

Russell, D. S. (1949). " Observations on the pathology of hydrocephalus." Special Report No. 265, Med. Research Council, London.

Turner, G. G. (1939). " Non-malignant stenosis of the oesophagus ", *Brit. J. Surg.*, **26**, 555.

Wakeley, C. P. G., and Willway, F. W. (1935). " Intestinal obstruction by gall-stones ", *Brit. J. Surg.*, **23**, 377.

Ward, R. O. (1938). " Fifty-three cases of vesical diverticula ", *Brit J. Surg.*, **25**, 790.

CHAPTER 38

SUNDRY DISEASES OF OBSCURE NATURE

ALTHOUGH the causal agents of many of the diseases discussed in earlier chapters (e.g. many endocrine diseases, blood diseases, " toxic " inflammations, and tumours) are still unknown or uncertain, we could allot these diseases to the main classes of pathological lesions to which they belong. Thus, we do not know the causal agent in many cases of nephritis, but we have no hesitation in placing them in the main class of " toxic " inflammations ; we do not know the ultimate cause of pernicious anaemia, but we understand enough of the way in which it is produced to discuss it as a dyshaemopoietic disturbance ; we do not know the causes of most tumours, but we recognize them as tumours and can place them in their proper class.

But, having dealt with as much of Pathology as possible by collecting related disorders together into main groups, we then have left a few diseases the very nature of which is still debatable and which therefore cannot be satisfactorily placed in any of the groups we have discussed. It happens that most of these diseases involve either the central nervous system or the skeleton. Perhaps this is not entirely fortuitous ; for both of these systems have peculiar structural and functional characters, and there is still much to learn of their growth and nutrition and the ways in which these may be disturbed.

MULTIPLE OR DISSEMINATED SCLEROSIS

(1) Clinical characters

This not uncommon disease affects both men and women, usually between the ages of 20 and 40. The symptoms vary greatly in different patients and in the one patient at different times. The commonest symptoms are tremor during deliberate movements (" intention tremor "), nystagmus, ataxia, staccato speech, loss of vision in circumscribed parts of the visual fields (scotomata), and increasing spastic paralysis. In the early stages the symptoms are often evanescent and variable, appearing in a succession of relapses and remissions ; but in later stages they progress steadily to the fatal ending. This sometimes occurs within a few months of the onset, but in most cases the disease lasts for many years.

(2) Pathological anatomy

(a) Naked-eye appearances

Multiple, firm, grey or yellowish, translucent areas of various sizes and shapes, are found scattered irregularly through the cord, brain stem, cerebrum, cerebellum and optic nerves, especially in the white matter (Fig. 276). These are usually well defined and discrete, and range in size from just visible dimensions to large irregular patches 1 or 2 centimetres in extent.

(b) Histology

The most recently developed patches, which are the yellowish ones, show degeneration of the myelin sheaths, and phagocytosis of the degenerative products by macrophages, accompanied by a few lymphocytes and plasma cells ; axons traversing the area are swollen and irregular but most of them are intact. The older grey patches show complete demyelination, and proliferated astrocytes which form a meshwork of glial fibres—an area of gliosis or glial scar ; the

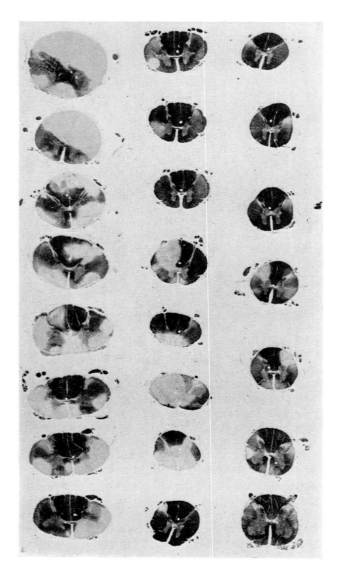

FIG. 276.—Disseminated sclerosis : a series of sections of brain-stem and cord stained by Weigert-Pal method, showing the scattered lesions of various sizes unstained. (Slightly enlarged.)

reactionary phagocytes and leucocytes have gone ; the axons may still be well preserved, or many of them may have degenerated and disappeared.

(3) Causation

In spite of a vast amount of work and speculation we are still ignorant of the cause of disseminated sclerosis. Bacterial toxins, virus infection, extrinsic poisons, dietetic deficiency, and allergic and auto-immune reactions have all been suspected; but nothing certain has been established. The change appears to be primarily degenerative rather than inflammatory; the slight inflammatory reaction which occurs in young patches is no more than is to be expected from the amount of demyelination present. Hence, special interest centres on toxic agents known to be capable of causing demyelination, e.g. arsphenamine, carbon monoxide, cyanides, etc. But whether such extrinsic poisons play any part in causing disseminated sclerosis, or whether a deficiency of copper or other " trace " metals may be to blame (as has been suggested), we still do not know.

NEUROMYELITIS OPTICA

This rare affection is characterized clinically by rapid loss of vision and symptoms of myelitis, and pathologically by very extensive demyelination and softening of the white matter of the cord and patches of demyelination in the optic nerves and brain. The disease may be related to, or possibly a special acute variety of, disseminated sclerosis.

SCHILDER'S DIFFUSE CEREBRAL SCLEROSIS

This is a rare acute or chronic disease of childhood characterized clinically by blindness, deafness, fits and mental failure. In acute cases, which may be fatal in a few weeks or even days, large areas of the cerebral white matter, especially in the occipital lobes, are soft and yellowish and show widespread demyelination and phagocytosis of the degenerating myelin by macrophages. In chronic cases, which may be slowly progressive during several years, the demyelinated areas show proliferated astrocytes and dense gliosis. In advanced cases almost the whole of the cerebral white matter may be replaced by firm grey semitranslucent glial tissue. The cause of the disease is unknown. Schilder called it " encephalitis periaxialis diffusa ", but it is doubtful if it is inflammatory in nature.

SYRINGOMYELIA

In syringomyelia, which usually appears in young or middle-aged adults, a cavity surrounded by a zone of glial tissue and containing clear colourless or yellow fluid develops in the spinal cord, usually in its cervical or upper thoracic parts, or occasionally in the brain-stem (*syringobulbia*). The cavity is not a mere dilatation of the central canal, though it may communicate with this. Its commonest situation is in the posterior grey horn, and it usually extends longitudinally through several segments of the cord ; occasionally two or more cavities are present at different levels. The change appears to be primarily a gliosis, with subsequent central degeneration and accumulation of fluid. It is slowly progressive and finally fatal. Its cause is unknown.

Symptoms vary with the site and size of the lesions. The chief results are paralysis and wasting of muscles innervated by the affected segments of cord, usually in the hand and the upper limb ; dissociated anaesthesia (pain and temperature sensation impaired, but touch retained), due to the usual situation of the lesion in the posterior grey horn or commissure ; spastic paresis of the lower limbs ; sometimes Charcot's arthropathy of the affected limb (p. 204).

AMYOTROPHIC LATERAL SCLEROSIS

This is a chronic progressive disease of middle-age, characterized by degenerative changes in both the upper and lower motor neurones. The pyramidal cells of the brain, the pyramidal tracts of the cord (especially the crossed fibres), the anterior horn cells and motor nerve roots all show atrophic and degenerative changes (Fig. 277). Varying combinations of spastic paralysis due to the upper neurone disease, and muscular atrophy due to the lower neurone lesions, are seen. In some cases the lower cranial motor nerves are the worst affected, wasting and tremor of the tongue and dysphagia from pharyngeal paralysis are prominent, and the syndrome is called *progressive bulbar paralysis*. The disease ends fatally in 2 or 3 years, usually from respiratory paralysis and bronchopneumonia. Its cause is unknown.

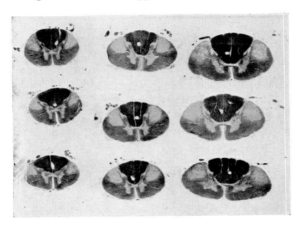

FIG. 277.—Amyotrophic lateral sclerosis : sections of cord stained by Weigert-Pal method, showing degenerated lateral pyramidal tracts unstained. (Slightly enlarged.)

OTHER DISEASES OF THE NERVOUS SYSTEM

There are many other rare degenerative diseases of the nervous system of obscure nature ; but they are outside the scope of a book on general pathology, and special works on neurology or neuropathology must be consulted. Many of these diseases are hereditary or familial ; and, although some of these do not manifest themselves until adult life, they are clearly attributable to inborn defects of the nervous tissues and are thus akin to malformations (see next chapter, p. 666).

PAGET'S DISEASE OF BONE : OSTEITIS DEFORMANS

(1) Clinical features

This not uncommon disease, first described by Paget in 1876, develops slowly in middle-aged or old people of either sex, and is characterized by thickening and change of texture of some or many of the bones and curvature of the weight-bearing bones of the lower limb and of the spine. In many advanced cases almost the entire skeleton is affected, but in earlier cases the changes may be confined to a few bones or even to a single bone or part of a bone. The first or worst affected bones are usually the long bones of the lower limb, the vault

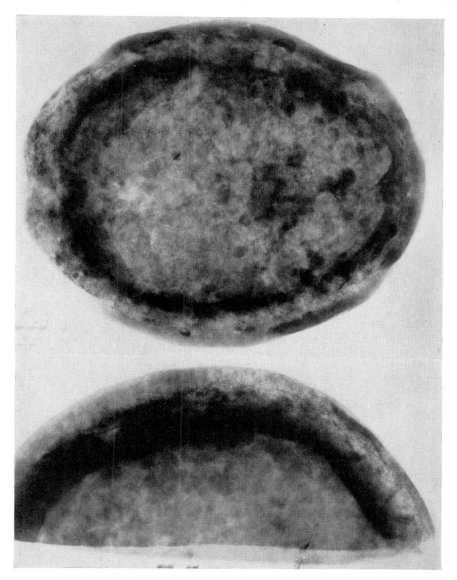

FIG. 278.—X-ray photographs (positive prints) of thickened skull vault in Paget's disease.

of the skull, spine, clavicle and humerus. Increase in size of the head may be the first symptom noticed; and aching pain in affected limbs is frequent. The disease progresses slowly, causing increasing cranial enlargement, bow-legs, dorsal spinal curvature, and reduction in height. Occasionally encroachment of the thickened bone on the spinal canal or posterior fossa of the skull causes compression of the spinal cord or cerebellum. Sarcoma of bone supervenes in about 5 to 10 per cent of advanced cases (p. 537).

(2) Pathological anatomy

(a) Naked-eye appearance of affected bones

These are thickened, rough-surfaced and deeply pitted or furrowed by the blood-vessels; e.g. in the skull, the meningeal vessels traverse deep grooves or completely submerged tunnels in the thickened inner table. In spite of thickening, the affected bones are, in earlier stages, softer and weaker than normal—hence the gradual bowing of weight-bearing bones. The weakness is due to striking

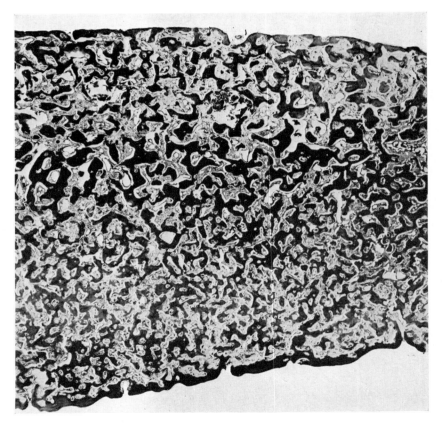

FIG. 279.—Sectional view of skull in Paget's disease, showing loss of demarcation of inner and outer tables and diploe, and uniform replacement by a spongework of bone with interspersed fibrous tissue. (× 10.)

changes in texture, the normal trabecular and Haversian architecture being lost and replaced by an abnormal spongy or pumice-like pattern, the meshes of which are filled by vascular connective tissue. In some bones, especially the skull, the distinction between cortex and medulla may be lost, the entire bone thickness presenting a hyperaemic, granular or spongy structure (Fig. 278). In later

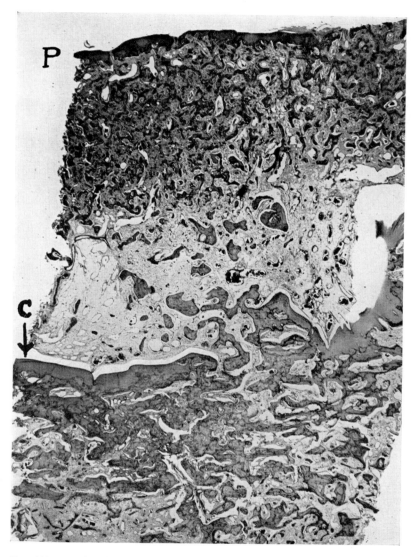

FIG. 280.—Sectional view of Paget's disease of ilium. The original cortical surface of the bone was at C ; the thick zone between this and the new surface P is new-formed subperiosteal bone mixed with fibrous tissue. The thickened ilium thus had double contours. (× 10.)

stages, however, the hyperaemia disappears, bony sclerosis ensues, and the affected bones become patchily dense and hard.

(b) Histology (Figs. 279, 280)

In early stages prominent features are absorption of existing bone by osteoclasts and the development of abundant intervening vascular connective tissue. Areas of aseptic necrosis of the bone trabeculae are sometimes evident. At the same time, new spongy woven bone is laid down in the connective tissue, and the new trabeculae enlarge by appositional growth or are in their turn absorbed. The picture is a peculiar mixture of constant bone dissolution and new bone formation, pointing to an inherent instability of the osseous tissue. In later stages, sclerotic parts of the affected bones attain a more stable structure but with an irregular distribution which is unrelated to directional stresses.

(3) Causation

The nature of Paget's disease is still obscure. It is certainly not inflammatory and the name " osteitis deformans " is therefore a misnomer. There is no evidence of disease of any of the ductless glands, nor of any nutritional deficiency. The irregular distribution of the lesions in many early cases, the age incidence, and the almost constant presence of pronounced arterial atherosclerosis in patients with Paget's disease, suggest that it may be a manifestation of ischaemic malnutrition of bone.

LOCALIZED FIBRO-CYSTIC DISEASE OF BONE

The name " osteitis fibrosa " is applied to two distinct conditions, (*a*) the generalized decalcification of bones which occurs from hyperparathyroidism (p. 611), and (*b*) localized fibro-cystic disease of unknown cause. To neither of these is the term " osteitis " appropriate, as the changes are not inflammatory ; a better name is *osteodystrophia fibrosa* or *fibrous dysplasia*.

Localized fibro-cystic disease of bone arises in most cases during late childhood or adolescence, usually in a single bone, rarely in two or more bones. Its commonest sites are the upper parts of the femur, humerus and tibia ; but any part of the skeleton may be affected. The lesion consists of a circumscribed area of osteoclastic absorption and fibrous replacement of the bone, which often contains a cyst or several cysts filled with watery fluid. The lesion enlarges slowly over a period of several years, and then may cease to extend and partial re-ossification may ensue. Pathological fracture is not uncommon, and following it firm bony union may occur.

LEONTIASIS OSSEA

This name is applied to rare cases of great overgrowth and thickening of the facial and cranial bones, giving the face a leonine appearance. The distribution of the change varies ; in some cases it is limited to the jaws, in others it affects all the bones of the face and skull and occasionally other bones as well, e.g. vertebrae or clavicles. The affected bone is sometimes spongy or pumice-like, but more often dense and heavy. The nature of the disease is unknown and possibly diverse ; in some cases it is probably a localized variety of Paget's disease, in others it may be akin to fibro-cystic disease, and in yet others it may be the result of chronic osteitis spreading from infected para-nasal sinuses or teeth-sockets.

OSTEOCHONDRITIS JUVENILIS

The term " osteochondritis " is applied to a peculiar disease of epiphyses during childhood or adolescence. The most familiar example is *osteochondritis of the upper femoral epiphysis* (Legg-Perthes-Calvé disease), which causes flattening, fragmentation, and mushroom-shaped deformation of the head of the femur. Somewhat similar lesions in other situations have given plenty of scope for eponymic nomenclature, e.g. *osteochondritis of the tibial tuberosity* (Osgood-Schlatter disease), *osteochondritis of the tarsal navicular* (Köhler's disease), *osteochondritis of the posterior epiphysis of the calcaneus* (Sever's disease), etc.

The cause of juvenile osteochondritis is uncertain ; it is regarded by most authorities as due to slight injuries with consequent interference with the blood supply of the affected epiphyses, but others believe that low-grade infection may also play a part. Strongly in favour of the traumatic hypothesis is the occurrence of some undoubtedly *post-traumatic rarefying diseases of bone* with characters very similar to those of juvenile osteochondritis. These include *Kienböck's disease* of the carpal lunate bone or less frequently of other carpal bones. *Kümmell's disease* of the vertebrae, and *Köhler-Freiberg disease* of the heads of the metatarsals.

MYOSITIS OSSIFICANS

Ossification in tendons and muscles occurs in two distinct forms : (*a*) localized traumatic myositis ossificans, due to injury, and (*b*) widespread progressive myositis ossificans, of unknown cause.

(1) Traumatic myositis ossificans

Following fractures, dislocations or bruises involving muscular or tendinous attachments, ossification in the reparative tissue sometimes spreads for considerable distances outside the bone confines into the muscles or fasciae. The histological appearances suggest that this may not be due solely to multiplication and migration of osteoblasts from the bone, but rather to metaplastic transformation of the neighbouring connective tissues under some diffusible chemical stimulus emanating from the bone. A haematoma in a muscle, unconnected with bone, sometimes undergoes ossification as it organizes. Intermittent strain in a muscle may lead to ossification in its attachments to bone, e.g. " rider's bone " growing from the linea aspera in the adductor region of the thigh.

(2) Progressive myositis ossificans

This rare disease commences in childhood or adolescence, usually in males. In most cases the muscles first affected are those of the back—longissimus, dorsi, latissimus, trapezius and rhomboids ; but later the disease slowly extends to many other muscles. In the early stages the muscles are stiff, swollen and tender, and show proliferative fibrosis. Subsequent ossification leads to progressive immobility ; and fatal bronchopneumonia finally supervenes. Although spoken of as " myositis ", the inflammatory nature of the disease is debatable. That there is some underlying developmental anomaly is shown by the presence of microdactyly in many cases.

HYPERTROPHIC PULMONARY OSTEO-ARTHROPATHY

The mildest degree of this condition, " clubbing " of the fingers and toes (Fig. 281), is often seen in cases of chronic phthisis, bronchiectasis, chronic empyema, or congenital malformations of the heart accompanied by cyanosis. It occurs

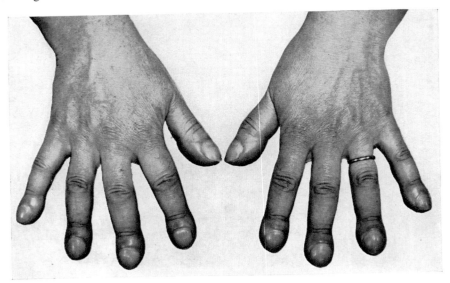

FIG. 281.—Clubbing of fingers.

FIG. 282.—Bones of leg from a case of elephantiasis, showing irregular thickening by formation of new subperiosteal bone. (From Virchow.)

infrequently in cases of bronchial carcinoma, subacute bacterial endocarditis, cirrhosis of the liver, ulcerative colitis, polyposis of the colon, amoebic dysentery, aortic or subclavian aneurysm, and some other diseases. " Clubbing " of the digits is symmetrical in nearly all cases ; but in cases of subclavian aneurysm, " clubbing " of the fingers is restricted to the side of the aneurysm. From the list of causes just given, it will be clear that the adjective " pulmonary " in the name of this condition is not always correct. However, about three-quarters of cases of " clubbing " are associated with chronic lung diseases.

" Clubbed " fingers show thickening of the connective tissues around the phalanges and joints, and eventually thickening of the bones themselves due to a gradual subperiosteal deposition of new bone. In advanced cases these changes extend proximally to the long bones of the limbs, which in rare instances become greatly enlarged by the subperiosteal formation of new bone.

The exact causation of the disease is still not clearly understood. Chronic congestive anoxaemia and bacterial toxaemia probably both play a part. The occurrence of unilateral "clubbing" accompanying subclavian aneurysm shows clearly the importance of local circulatory embarrassment ; and in many of the pulmonary and cardiac diseases in which " clubbing " occurs, chronic circulatory embarrassment is present. In elephantiasis due to filarial lymphatic obstruction also, the bones of the affected limb may show great subperiosteal overgrowth (Fig. 282). On the other hand it is difficult to see how " clubbing " in such diseases as ulcerative colitis or polyposis of the colon can be due to any circulatory disturbance in the limbs, and it seems necessary to postulate that " toxaemia " of some kind is to blame. But it must be admitted that we know nothing regarding the nature of this "toxaemia" nor of its mode of action on the skeleton.

SUPPLEMENTARY READING

Collins, D. H. (1949). *The Pathology of Articular and Spinal Diseases.* Edinburgh and London; E. & S. Livingstone. (Contains good accounts of Paget's disease, osteochondritis, etc.)
  – (1966). (Posthumous and prepared by O. G. Dodge.) *Pathology of Bone.* London; Butterworths. (Good account of Paget's disease).
Greenfield, J. G. *et al.* (1963). *Neuropathology.* London; Edw. Arnold.
Knaggs, R. L. (1926). *Inflammatory and Toxic Diseases of Bone.* Bristol; Wright. (Contains good accounts of osteitis fibrosa, Paget's disease, etc., with many excellent illustrations.)
Paget, J. (1876). " On a form of chronic inflammation of the bones (osteitis deformans) ", *Trans. Med. Chir. Soc.,* **60,** 37. (A classic which every student should read.)
Rodgers, R. E. (1941). " A case of unilateral clubbing of the fingers, with a summary of the literature ", *Brit. Med. J.,* **2,** 439.

CHAPTER 39

ANTENATAL PATHOLOGY

INTRODUCTION

WHAT HAPPENS to a human creature in its first 9 months of life, i.e. before it is born, matters a great deal to its subsequent health. At its very conception it may have the misfortune to receive from one or both of its parents abnormal (mutated) genes, which determine either that it will inevitably develop some definite malformation or disease or that it will be abnormally vulnerable to certain harmful environmental agencies—i.e. predisposed to the diseases caused by these agencies. Or, having received a healthy lot of genes from its parents, it may be subjected during its sojourn in its mother's womb to poisons, infections or nutritional deficiencies which injure it in various ways. Antenatal pathology is thus concerned with these two distinct subjects : (1) *genetic* or *hereditary abnormalities*, and (2) *abnormalities acquired in utero*.

This chapter attempts to outline the embryological principles necessary to a clear understanding of the nature and genesis of malformations and antenatal diseases. The writers believe this to be preferable to a detailed description of the numerous kinds of malformations or to a dissertation on genetical theory. For these, the works listed at the end of the chapter should be consulted. Let us discuss the general principles of antenatal pathology under the following heads :

Early embryonic development ;

Teratology, the pathology of the embryo ;

The influence of heredity in malformation and disease ;

Classification of malformations according to their modes of development ;

Extensions of the concept " malformation " ;

Extra-uterine pregnancy.

EARLY EMBRYONIC DEVELOPMENT

(1) Epigenesis

Nature shows us no process more amazing than the rapid transformation of a simple unicellular egg into a complex adult organism. Two different views of this transformation have been held—*preformation* and *epigenesis*. Preformation assumed that the egg itself had a complex structure in which all parts of the adult body were invisibly represented, and that embryonic development consisted only in an unfolding or rendering visible of this structure. Epigenesis, first clearly enunciated by William Harvey in 1651, holds on the contrary that the egg contains no prior representation of adult structure, but that development is a real genesis of complex structure, a progressive upbuilding of each new organism afresh.

The preformation hypothesis is now obsolete, and all biologists are unanimous that development is epigenetic. The fertilized egg (or zygote) is a protoplasmic system devoid of any spatial representation of future structure, but of such composition and potentialities, transmitted genetically to it from its parents, that it will respond in specific ways to the external stimuli which form its environment. If both the genetic equipment of the zygote and its environment are normal, normal development will ensue ; but abnormalities of either genes or environment will result in abnormal development. Development is an epigenetic series of reactions of a specific hereditary equipment to external stimuli, normal or abnormal.

(2) Early amphibian development

Since much of the relevant experimental work has been done on amphibian embryos, we begin with a brief reminder of the early development of the frog's egg. Although experimental proof of many points is still lacking, there is no doubt that similar general principles apply also to avian and mammalian eggs.

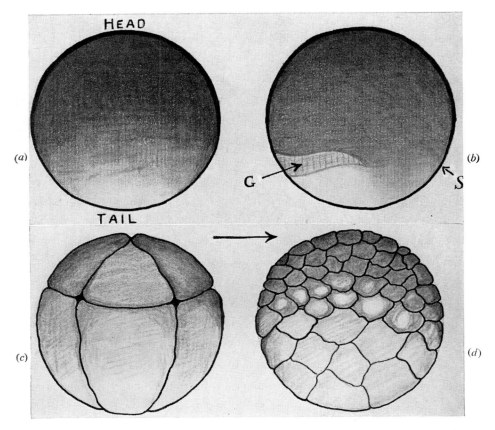

FIG. 283.—The frog's egg. (*a*), before fertilization. (*b*), after fertilization, S being meridian of sperm-entry (ventral) and G the grey crescent (dorsal). (*c*) and (*d*), segmentation forming blastula.

Even before fertilization the frog's egg has polarity (Fig. 283 (*a*)). One hemisphere is pigmented, the other pale ; the pigmented pole will become the future head, and the pale yolk-laden pole the hind end. Apart from this polarity, the cytoplasm of the egg is undifferentiated.

Fertilization does three important things ; it brings into the egg the paternal chromosomes, it activates the egg to begin its development, and it determines the plane of bilateral symmetry. The mid-ventral line of the embryo forms at the meridian where the sperm enters the egg ; while on the opposite or dorsal side, a little below the equator, there appears after fertilization a grey crescentic area which is of great importance in subsequent development (Fig. 283 (*b*)). After fertilization then, the zygote, though still an undivided cell, has all its main axes and its right and left sides determined.

Segmentation now ensues ; successive mitotic divisions convert the fertilized egg into a berry-like cluster of smaller cells or blastomeres, amidst which a small temporary cavity or blastocele appears. The cells of the pigmented hemisphere are smaller than the yolk-laden blastomeres of the lower hemisphere ; yolk appears to retard cell-division. At this stage the embryo is called a *blastula* (Fig. 283 (*c*) and (*d*)).

A highly important step in differentiation now takes place. The surface cells of the cranial hemisphere proliferate rapidly, spread downwards, and in the region of the grey crescent invaginate themselves into the interior. This invagination, called *gastrulation*, produces a new cavity, the archenteron or primitive gut,

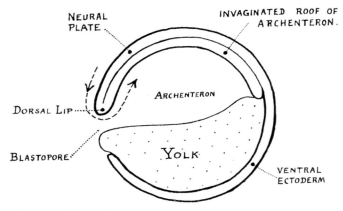

FIG. 284.—Median section of frog gastrula, showing invagination at dorsal lip of blastopore.

which opens to the exterior by an aperture, the blastopore (Fig. 284). The cranial or dorsal lip of the blastopore is the region of most active invagination, and, as we shall see, is a region of fundamental importance in early development. The diving in or invagination of surface cells at the dorsal lip of the blastopore is an autonomous process, for isolated pieces of the lip, when transplanted to the surface of another embryo, promptly dive inwards in the same way. The cells which invaginate in the mid-line at the dorsal lip give rise to important axial structures of the embryo, the primitive gut roof, the notochord and the paraxial mesoderm or future myotomes. Meanwhile, internally the endoderm and mesoderm are becoming differentiated ; so that gastrulation finally results in the delamination of the primary germ-layers, the ectoderm externally, the endoderm lining the archenteron, and the mesoderm between.

(3) Presumptive organ regions in the amphibian blastula

It is possible to map out on the surface of the blastula the regions which normally will become the various organs of the future embryo. By making small injuries or by staining small patches on the surface of the blastula with intra-vitam stains, and then following the subsequent movements and fates of the injured or stained regions, the amphibian blastula has been mapped completely in terms of organ rudiments. We may call these areas *presumptive organs.* Their distribution on the surface of the blastula is quite unlike their adult relationships, which are achieved by the extensive mass movements during gastrulation.

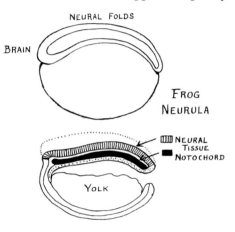

FIG. 285.—Frog embryo at neurula stage.

After gastrulation and delamination of germ-layers are complete, the arrangement of organ rudiments is essentially that of the adult. The plate of neural tissue which now lies along the dorsal side of the gastrula soon presents two parallel longitudinal folds which rise up and fuse dorsally to enclose the neural tube. When this has lost its connection with the surface, the entire surface of the embryo is clothed by epidermis, and the rudiments of all other structures lie internally in roughly their permanent positions. From the time the neural folds appear, the young embryo, which meanwhile is undergoing elongation, is called a *neurula* (Fig. 285).

(4) Early plasticity and later non-plasticity of tissues

When we speak of the " presumptive " organ rudiments in the blastula, we imply that, given normal development, the regions in question are destined to become particular organs. By transplantation experiments, however, it is found that, up to the middle stage of gastrulation, the prospective fates of most regions are *not* irrevocably fixed. A piece of presumptive neural tube, taken from its embryo and grafted into the side of another, will differentiate into epidermis in accord with its new surroundings, or into gills if it happens to have been grafted into the gill region of its host. Conversely a piece of presumptive epidermis grafted into the presumptive neural region of another embryo will differentiate into brain, spinal cord or eye according to its position. Even the germ-layers are interchangeable ; for, by suitably designed grafting, presumptive epidermis or presumptive neural tube can be made to differentiate into mesodermal tissues such as muscle or pronephros, and vice versa.   Up to the middle stage of gastrulation, then, the tissues are still plastic or undetermined as regards their final differentiation, for they will differentiate in accordance with whatever new surroundings they may be given, regardless of their origin and former surroundings. Thus, each part of an early embryo has potentialities for differentiation much wider than it ever displays in normal development.

During later gastrulation, however, the various regions begin to lose their plasticity.   Their prospective differentiation becomes irrevocably determined, and any given piece of tissue, wherever grafted, will now differentiate only into its predetermined structure.   A presumptive limb area from a late gastrula or neurula, grafted into an abnormal situation, will continue to form a limb ; presumptive epidermis will proceed to form skin wherever it may be transplanted ; and a presumptive eye rudiment grafted into the body cavity will become an eye there.   Some change, as yet invisible and presumably chemical in nature, has taken place in each region and has fixed its prospective fate.   This change, which is considerably prior to visible histological differentiation, is called *chemo-differentiation*.   It results in the embryo becoming a patchwork or mosaic of separately determined, though to some extent overlapping, regions ; and it is a progressive process, the mosaic coming to consist of more and smaller pieces as development proceeds.

(5) The primary organizer and axiation

One region of prime importance in the zygote and blastula is exempt from the rule that the different parts of the early embryo are indeterminate or plastic. This is the grey crescent of the zygote, which gives origin to the dorsal lip of the blastopore, and which from the beginning is destined irrevocably to invaginate beneath the surface and to form the essential axial structures of the embryo— the primitive gut roof, the notochord, and the paraxial mesoderm.

The grey crescent or dorsal lip is remarkable, not only because of its inflexible presumptive fate, but also because it has the power of inducing neighbouring plastic tissues to disregard their former presumptive fates and to participate in the formation of the main axial and paraxial structures of an organized embryo. Thus, as the classical experiments of Spemann and Mangold (1924) showed, if a small fragment of dorsal lip is grafted into the front or side of another embryo in the blastula or early gastrula stage, it at once invaginates itself and induces the neighbouring host tissues to take part in forming a secondary embryo in the body wall of the host.   The grafted fragment itself always forms the notochord and usually some of the paraxial mesoderm of the secondary embryo ; but the plastic host tissues around, under the inductive influence of the graft, produce neural tube, eyes, ears, mesodermal somites, and pronephric tubules.   In some experiments of this kind, the secondary embryo has become as perfectly formed as the primary.   Because of its remarkable power of inducing the formation of the main axial and paraxial parts of an embryo, the dorsal lip tissue has been given the name of " *primary organizer* ".   As might be suspected, the organizer is of fundamental importance in normal development, to which indeed it is essential.   An egg or blastula deprived of its grey crescent region, or a group of blastomeres isolated from a blastula before they have been affected by the primary organizer, grows and differentiates into a variety of tissues but these form only an unorganized medley.

One of the immediate effects of the primary organizer is to induce the formation of the neural plate and tube in the overlying ectoderm. All of the gastrular ectoderm underlain by invaginated organizer—notochord and paraxial mesoderm —will become neural. Even after the neural tube has formed, it continues to be influenced by the neighbouring notochord and paraxial mesoderm, removal of either of which from a young neurula leads to characteristic anomalies in the form of the neural tube. Further, the organizer's inductive effects on the over-lying ectoderm show distinct regional differences. The first tissue to be invaginated at the dorsal lip of the blastopore reaches furthest forward into the head region, and this tissue induces the formation of cephalic structures—brain, eyes, ears— from the overlying ectoderm. Later invaginated " trunk organizer " will induce only trunk structures and not brain or eyes. Thus regional differentiation is present in the organizer itself.

(6) Nature of the primary organizer

The action of the amphibian organizer is *not* species-specific ; it can induce the formation of a secondary embryo even when grafted into an embryo of a different species. Organizer material killed by heat, drying or alcohol still has inductive power ; so also has a piece of agar or gelatine which has been in contact with organizer tissue. Cell-free extracts of organizer material are potent, and the solubilities of the active ingredient suggest that it is a sterol. Clearly then the activity of the organizer is due to chemical substances which it elabo-rates.

(7) The organizer in birds and mammals

In birds and mammals the homo-logue of the amphibian blastopore is the *primitive streak* along with the *primitive node and pit*. These struc-tures (Fig. 286) give rise to the notochord and paraxial mesoderm ; and experiments prove that the avian primitive streak has inductive powers similar to those of the amphibian organizer.

Embryologists have long known that Hensen's node or primitive knot at the anterior end of the avian or mammalian primitive streak is an important centre of growth. From it the notochordal or head process grows forward beneath the ectoderm. The primitive pit, which appears

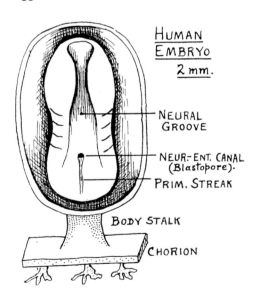

FIG. 286.—Dorsal view of human neurula 2 mm. long after removal of roof of amniotic cavity, showing neurenteric canal (blastopore) and primitive streak. (From Graf Spee.)

just behind the node and which becomes the neurenteric canal, is the homologue of the blastopore, and Hensen's node is its dorsal lip.

(8) Secondary induction ; dependent differentiation

The dorsal lip of the amphibian blastopore or the avian primitive streak is the *primary* organizer, because it initiates the whole complex process of somatic organization and because its inductive capacity is a primary one not dependent on any earlier organizer.   But, after primary organization has been determined during gastrulation, many tissues other than the derivatives of the primary organizer play the part of *secondary organizers* in that they exert inductive influences, presumably chemical in nature, on neighbouring plastic tissues, which therefore undergo *dependent differentiation*. The following are examples of dependent differentiation.

(a) Optic cup and lens

Spemann's first experiments in 1901 were concerned with this striking example of the dependence of the development of one structure on the presence of another. The lens of the vertebrate eye develops as a localized thickening of the embryonic ectoderm ; this sinks into the subjacent tissues and becomes detached from the overlying epidermis.   Spemann showed that formation of the frog's lens depends on the presence of the optic cup, the outgrowth from the brain which gives origin to the retina and optic nerve.   If the optic cup is removed before the lens has formed, no lens appears ; and an optic cup transplanted under the skin elsewhere induces the formation of a lens there.   Like the inductive power of the primary organizer the lens-inducing capacity of the optic cup is not species-specific ; a lens can be induced by the optic cup of a different species.

(b) Optic cup and conjunctiva

Just as the presence of an optic cup is necessary for the formation of a lens, so also it is necessary for the modification of the overlying epidermis into conjunctiva.   In the absence of optic structures beneath it, the presumptive conjunctival epidermis remains pigmented and opaque ; but the presence of the optic vesicle, or even of bits of vesicle, retina or lens, induces it to lose its pigment and become transparent.   Eye or lens grafted beneath the epidermis of other regions induces it also to modify into conjunctiva.

(c) Otic vesicle and capsule

The primary otic vesicle, which gives origin to the internal ear and labyrinth, develops by sequestration from the surface ectoderm.   If the vesicle is extirpated from an early embryo, the cartilaginous auditory capsule fails to develop ; while an ear vesicle transplanted to some other part of the embryo induces the formation of a capsule around itself in its new situation.

(d) Oral and nasal orifices

The orderly sinking in of the stomodaeal depression and the thinning and perforation of the stomodaeal membrane are dependent on contact of the ectoderm with the endoderm of the fore-gut.   The latter induces these changes when ectoderm of other regions is substituted for the normal stomodaeal ectoderm.

Formation of the choanal orifices depends on establishment of contact of the nasal-pit ectoderm with the underlying endoderm.

The foregoing are but a few examples of the inductive influences exerted by one tissue on another during development. There is no doubt that similar influences operate in many developing organs. It is probable that the formation of the tympanic membrane depends on the presence of the annular tympanic cartilage, that the differentiation of the pars nervosa of the pituitary depends on the presence of Rathke's pouch, that the formation of the dental papillae and of dentine depends on the presence of the epithelial tooth-germs, that the development of the renal tubules depends on the development of the ureteric outgrowths from the mesonephric duct, and that the formation of smooth muscle in the walls of hollow viscera depends on the growth of their epithelial linings. The post-gastrular embryo is a complex mosaic of interacting parts, in the progressive differentiation of which many secondary organizers are active. " There is, then, a whole hierarchy of organizers or morphogenetic hormones functioning in development " (Needham).

(9) Organ-fields

An important concept developed as the result of embryological experiments is that of organ-fields, i.e. parts of the embryo mosaic determined by chemo-differentiation to produce particular organs. Such fields are at first not sharply delimited from their neighbours ; adjacent fields overlap. As development proceeds, however, the overlap diminishes, each field becomes more and more distinctly demarcated, and at the same time progressive regional determination or chemo-differentiation takes place within the field itself. These points are well illustrated in the development of the limb-buds.

In the presumptive fore-limb area of the amphibian blastula or gastrula there is no definite spot essential to the formation of any particular part of the limb. Any part of the field has the power to form a whole limb, and the extent of the area capable of limb-formation is greater than that which actually forms the limb in normal development. Limb-forming potency is highest near the centre of the field and falls off progressively towards its periphery. Extirpation and grafting experiments show that a normal limb can be formed from a fraction of the limb-field. Deep cuts made in the early limb-buds result in several limbs growing out from the limb-area. Transplanted pieces of limb-fields often give rise to double or treble limbs. These facts show that, while in the early embryo the limb-field is determined as a whole to produce limb tissues, there is as yet no sharp delimitation of, or regional determination within, the field. It is only at later stages that the whole limb-field, which is one of the large mosaic pieces of the early embryo, becomes subdivided itself into a mosaic of minor fields each irrevocably destined to form a particular part of the future limb. When this stage has been reached, subdivision of the limb-rudiment will no longer result in the formation of two or more limbs. Each part will now give rise to a part only of a limb, a femur or a tibia or a foot as the case may be.

A further point of interest regarding the limb-fields is the phenomenon of dominance of an established limb-bud. A part of a limb-disc grafted onto the flank will develop into a limb there only if it is sufficiently removed from the developing limb area. If the site of grafting is close to this area, the transplanted limb-disc will be resorbed. Evidently the establishment of a dominant region in the limb-field inhibits the development of limbs from other parts of the field. The development of multiple limbs from grafted limb-discs must be due to the operation having disturbed the normal gradients of limb-forming potency within the field, permitting supernumerary centres of dominant activity to appear. These observations are of great interest as regards dichotomous growth and the development of supernumerary organs.

Since in the very early embryo a major field, like the limb-field, is ill-defined and overlaps surrounding regions, local extirpation or damage within it may not result in absence of the corresponding organ in the developed organism. Marginal overlapping parts of the field may suffice to produce the organ. This capacity of the early, still partly plastic, embryo to adjust itself to local damage or loss is called *regulation*. As mosaic subdivision proceeds, however, less and less regulation becomes possible ; adjustments of fields to compensate for local damage become more and more restricted, and local damage will now leave the organism with a permanent defect.

Experiments with the developing heart have shown similar results. In an amphibian neurula, extirpation of the presumptive heart-area may be followed by development of a normal heart from surrounding regions ; or part of the heart-field may be rotated through 180° and still produce a normal heart ; or a normal heart can be formed from a longitudinal half-rudiment ; or a split rudiment may produce two or three hearts ; or two rudiments grafted together can regulate to form a single heart. But after the neurular stage, regulation of this sort ceases to be possible, and local damage or dislocation of prospective heart tissue now produces permanent cardiac malformations.

Just as permanent deficiency of particular parts can be produced by experimental injury or extirpation of their rudiments in the mosaic embryo, so spontaneous malformations in which parts are absent or defective may in many cases be due to blasting of their rudiments by harmful agencies at critical periods in their early development.

TERATOLOGY, THE PATHOLOGY OF THE EMBRYO

(1) Teratology distinguished from foetal pathology

In 1904 Ballantyne stressed the distinction between the pathology of the foetus and the pathology of the embryo. In most respects the foetus, with its differentiated parts and tissues, resembles an adult in miniature, and the reactions of its tissues to injuries, poisons and infections are similar to those of the adult. But the embryo is structurally and functionally unlike an adult. " The great, almost the only, function of the embryo is to form tissues and organs, or, in one word, organogenesis." Hence, the pathology of the embryo is concerned with disturbed organogenesis ; it is *teratology*, the study of malformations. " Monstrosities and structural anomalies are the results of morbid agents acting upon the organism in the embryonic period of its existence, or, to write more exactly, upon such parts of the organism as are in the embryonic or formative stage " (Ballantyne).

This conception of the distinction between embryonic and foetal pathology has been confirmed experimentally. Many kinds of malformations can be produced by exposing embryos of suitable ages to abnormal environments. The abnormal conditions capable of causing malformations are very diverse, and include many agents which produce " diseases " in the developed organism, e.g. trauma, oxygen deficiency, abnormal temperatures, various chemical substances, and pathogenic organisms. Let us look at some of the experimental results, commencing with the grossest malformations of all, double monsters.

(2) Experimental production of double monsters

Double monsters have been produced in the following ways :

(i) *By interference with the zygote during segmentation.*—Double monsters have been produced in Amphioxus by shaking and disarranging the blastomeres, and in the frog by inverting the egg at the two-cell stage. It is to be noted here that if the blastomeres of Amphioxus or amphibian eggs at the two-cell stage are separated, each can develop into a complete organism, i.e. twins can be produced from one egg. Disarrangement of the early blastomeres doubtless causes double monsters by effecting incomplete isolation of the toti-potent cells.

(ii) *By partial constriction of the embryo during gastrulation.*—Spemann showed that if the newt blastula is constricted in the median plane during gastruation, the resulting embryo will show anterior duplication, i.e. it will have two

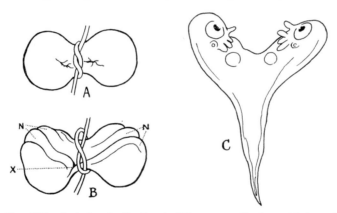

Fig. 287.—Anterior duplication in *Triton* as result of constriction of early gastrula in median plane. A, gastrula constricted by ligature so that each half contains part of blastopore. B, resulting neurula which shows bifurcation at point X, NN being neural folds of the duplicated anterior parts. C, resulting double-headed embryo. (After Spemann.)

heads joining with a single hind end at varying levels (Fig. 287).   The explanation is clear : the invaginating organizer, moving anteriorly from the dorsal lip of the blastopore, meets an obstacle in the constriction and therefore bifurcates, one-half of it going forward on each side of the constriction.   The Y-shaped invaginated organizer naturally induces a Y-shaped embryo, with duplicated head regions. For obvious reasons, posterior duplication cannot be produced by constricting the gastrula.   If the plane of constriction of a gastrula is not accurately median, but slightly oblique, then one head of the resulting double monster is normal and the other is incomplete and usually cyclopic. The greater the obliquity of the constriction, the greater the imperfection of the malformed head.

(iii) *By grafting two half-gastrulae together.*—Both anterior and posterior duplication can be obtained by grafting together gastrula halves—each with a complete blastopore—so that the directions of invagination of their organizers diverge or converge anteriorly. If they diverge, a Y-shaped embryo with anterior duplication is obtained ; if they converge, an embryo with posterior duplication.

(iv) *By oxygen deficiency or lowered temperature.*—By these means, Stockard caused retardation of development during early cleavage of trout and minnow embryos, and produced a great variety of major malformations, including double monsters. These results were traceable to irregular and delayed segmentation of blastomeres occasioned by the developmental arrest.

(v) *By chemical agents.*—Traces of chemical poisons or unnatural concentrations of various salts produce malformations similar to those of anoxia or chilling.

(vi) *By delayed fertilization.*—If fertilization of frog's eggs is delayed, various abnormalities result. Brief delay occasions an abnormal sex ratio in the resulting tadpoles, males predominating. Longer delay results in the production of double monsters and sometimes of teratoma-like growths. The double monsters are traceable to abnormalities of cleavage and to splitting of the blastopore lip at the onset of gastrulation.

All the foregoing methods of producing double monsters involve treatment of the embryos at very early (pregastrulation) stages. " When eggs are treated at later stages, as at the beginning of gastrulation, no double monsters will occur, their moment has passed " (Stockard). Duplication of the axis of the body presupposes duplication of the inducer of axiation, the primary organizer. Double monsters are malformations of primary axiation and can be produced therefore only by disturbances affecting the embryo prior to establishment of its axis. Once the axis is laid down, the embryo has become a mosaic of organ-fields, the future differentiation of each of which is determined ; and no amount of interference can produce more axis tissue.

(3) Experimental malformations in the mosaic embryo

After the stage of mosaic determination of a particular organ-field, extirpation or damage within that field will lead to permanent defects or deformities of the organ. Similar defects can be obtained by trauma, oxygen deficiency, cold or chemical poisons applied to the embryo at appropriate stages during or after gastrulation. By these means, the experimentalist can produce absence (agenesis) of limbs or other main parts, cyclopia, deformed mouths, nostrils, gills, brains or viscera. The kind of malformation produced depends, not on the kind of abnormal agent used, but on the time at which it is applied.

Thus, experimental work has verified Ballantyne's view that malformations " are the results of morbid agents . . . acting upon such parts of the organism as are in the embryonic or formative stage ". Stockard expressed the same principle when he spoke of " developmental arrests at *critical* stages ", that is at stages " when certain important developmental steps are in rapid progress. . . . The earlier the arrest the more numerous will be the type of defects found, and the later the arrest the more limited the variety of deformities, since there are fewer organs to be affected during their rapidly proliferating primary stages."

(4) Evidence of the action of environmental factors in mammalian teratology

Since malformations are so readily produced in genetically normal amphibian and fish embryos by environmental abnormalities, we must enquire whether the same is true of mammalian, including human, embryos. This enquiry is made difficult by the fact that, in the human being, we do not know of the existence of a malformation until 6 or 8 months after the embryonic period, the period during which teratogenic agents are most effective. By the time a malformed child is born, the agent responsible for the malformation may no longer exist, and the only clue to it may be the mother's hearsay evidence as to her health and nutrition during early pregnancy—evidence which is very unreliable.

From the mass of information which he collected, Ballantyne believed that malformations *did* occur with disproportionate frequency in the progeny of mothers with various infective and toxic diseases, including syphilis, tuberculosis, alcoholism, lead poisoning and other metallic poisonings. Although today we might question the statistical soundness of some of Ballantyne's conclusions, there is no doubt that his main thesis was correct, as shown by the following confirmatory facts.

Experimentally, malformations of many kinds can be produced in mouse, rat and other embryos by subjecting pregnant mothers, during early stages of pregnancy, to oxygen deficiency, various chemical agents (e.g. trypan blue), various vitamin deficiences, or X-rays.

German measles affecting a woman during the first 2 or 3 months of pregnancy is apt to cause certain characteristic defects in the child, including congenital opacities of the lenses (cataracts), patency of the ductus arteriosus, deaf mutism and microcephaly. Here then we have an example of a virus disease which selectively damages parts of the embryo and so causes, not the disease rubella as we see it in post-natal life, but certain malformations. The limb defects in the children of mothers who had taken *thalidomide* during pregnancy is a striking instance of chemically induced malformations.

There is substantial evidence that faulty implantation and placental development may lead to malformation of the embryo. Malformations are frequent in the embryos of ectopic gestations and in cases of hydatidiform mole or other disease of the placenta. Acardiac twins (p. 661) are a striking example of gross malformation resulting from abnormal circulation. As Streeter and Keith suggested, circulatory impairment may well be the cause of those developmental deficiencies of the limbs known as *intra-uterine amputations*, which were formerly wrongly attributed to " amniotic adhesions ". Keith extended the same idea to other kinds of malformations—facial clefts and scars of obscure origin, anencephaly, spina bifida and meningocele ; and advanced reasons for believing that "all such lesions are caused by a local necrosis probably due to a circulatory failure which may be placental in origin. The failure occurs along marginal areas where capillary formation is in progress ". Perhaps impaired circulation may underlie also such failures of coalescence as cleft palate, coloboma, hypospadias, and other local arrests of growth movements which Keith aptly likened to plastic surgical operations.

Malformed foetuses often show *multiple malformations* ; e.g. anencephaly, spina bifida and visceral anomalies often coexist, or malformations of the limbs or skeleton may be accompanied by cardiac or other visceral malformations. Such combinations strongly suggest the operation of damaging environmental agents rather than genetic defects.

Enough has been said to indicate the probable importance of external factors in the genesis of human malformations. But our knowledge is still very fragmentary, and a great deal of work remains to be done in this field. We must prosecute studies similar to Ballantyne's, but made in the light of our much greater knowledge of infection, nutrition and hormones, in an endeavour to discover the possible influence of all kinds of maternal and placental disease in producing teratological effects.

(5) Foetal and post-natal malformations

While it is certain that many major malformations take origin in the embryonic period, we must remember that the distinction between the embryonic and foetal periods is arbitrary. Many parts and tissues are still immature or " embryonic " in the foetal period, and indeed the developmental modelling and histological differentiation of some parts are not complete until long after birth. Thus, adult structure is not attained in the genital system until after puberty, in the skeleton until the epiphyses unite, and in the jaws until the permanent teeth erupt. The immaturity of such parts in childhood is, as it were, a projection of embryonic and foetal development into post-natal life ; and in such parts minor malformations theoretically may, and in fact do, sometimes arise from disturbances of late foetal or post-natal development. For example, destruction of an epiphysis will result in arrest of growth in length of the bone, and the stunted bone is as much a malformation as if it had resulted from damage inflicted on the primordium of the bone in the embryo. So also, dentigerous cysts and odontomes are malformations which may be brought about by disturbances of dental development during late foetal or post-natal life. The pseudo-hermaphrodite condition of the genitalia resulting from an adrenal cortical tumour in the foetus is another example of a malformation acquired relatively late in intra-uterine life.

(6) Foetal disease

This is the place to note that the foetus *in utero* may suffer from many of the infective and toxic diseases of post-natal life. The best known of these is syphilis ; but foetal smallpox, measles, pneumonia, endocarditis, streptococcal septicaemia, rheumatic fever, typhoid fever and tuberculosis also occur. Most of these infections reach the foetus through the placenta, but there is good evidence that foetal pneumonia may be caused by inspiration of infected or meconium-soiled amniotic fluid. The distribution and character of the lesions of foetal infections are influenced by the special conditions of intra-uterine life, or by the placental route of entry of the pathogens. Thus, the eruption of foetal smallpox is kept moist, is unaccompanied by crusts or pustules, and leaves minimal scars ; maternal erysipelas produces in the foetus, not erysipelas, but streptococcal septicaemia and endocarditis ; and the lesions of foetal tuberculosis are often most plentiful in the liver, not in the lungs.

(7) Importance of the placenta in antenatal pathology

As Ballantyne pointed out, the placenta has a four-fold importance in the pathology of the embryo and foetus.

(*a*) In the first place, the placenta may function as a barrier protecting the foetus from infections, toxaemias and other abnormalities of the maternal blood. Probably, however, the barrier function of the placenta has been over-estimated, and there are, as we have seen, abundant instances of the transfer of infections or poisons from mother to foetus.

(*b*) Secondly, the converse of (*a*), the placenta is the main—with the exception of the amniotic fluid, the only—pathway of noxious agents from mother to embryo and foetus. There is room for yet more experimental and clinical research on the permeability of the placenta to poisons, drugs and pathogenic organisms. (Maternal antibodies also traverse the placenta and enter the foetus.)

(*c*) Since the placenta is at once the alimentary, respiratory, and excretory organ of the foetus, the latter may be severely malformed or diseased without dying. Its intestines or bile ducts may be occluded, its kidneys entirely cystic, and its brain grossly defective ; but, as long as the placenta continues to function efficiently, this monster will survive and grow to full-term. The potential morbidity of malformations becomes actual only at birth, when for the first time the foetus is called on to perform all its own functions.

(*d*) The converse of (*c*), the foetus is so dependent on the integrity of the placenta that any serious placental disease at once imperils its life. The causes of foetal death, abortion and miscarriage are frequently to be found in the placenta. As already noted, further studies are needed to discover whether placental disease, not severe enough to destroy the embryo, may be an important cause of malformations.

THE INFLUENCE OF HEREDITY IN MALFORMATION AND DISEASE

So far, we have been discussing the production of malformations by environmental abnormalities external to the embryo. We must not forget, however, that the inherited qualities of the embryo itself are important in teratology. A great many familial and heritable structural and functional abnormalities are known which in variable proportions of cases depend on genetic faults. These include brachydactyly, polydactyly, syndactyly, congenital dislocation of the hip, club-foot, and many other anomalies of the limbs ; fragilitas ossium, achondroplasia, multiple exostoses and some other anomalies of the skeleton ; congenital cataract, night-blindness, colour-blindness, deformities of the iris, retinoblastoma, and other anomalies of the eyes ; albinism, piebaldness, congenital ichthyosis, xeroderma pigmentosum and some other skin anomalies ; cleft palate and hare-lip ; polyposis of the colon ; haemophilia, acholuric jaundice, diabetes, cystinuria, and some other metabolic disorders ; muscular dystrophies, Friedreich's ataxia, epilepsy, imbecility, neurofibromatosis, and a number of other nervous disorders. In some of these, such as brachydactyly, albinism and haemophilia, the genetic fault is of paramount importance ; the anomaly *will* appear, and will appear *only*, if the appropriate gene anomaly is present. In others, such as club-foot, congenital cataract, cleft palate and diabetes, the genetic predisposition is weaker and less constant.

This is not the place to discuss the principles of genetics or their application in detail to the many anomalies just referred to. These will be found in the books listed at the end of the chapter. Here it must suffice to point out that the two principles—(a) that malformations are often the result of the action of external pathogenic agents on the embryo, and (b) that genetic factors may determine, or predispose to, malformations—are not in conflict with, but are complementary to, one another. Roberts aptly speaks of " the unfolding of hereditary potentialities under the impact of environmental stimuli " ; and says that " it is upon the complex interactions of both that the result depends ". Even with such clear-cut genetically determined anomalies as brachydactyly, albinism or haemophilia, affected persons show these anomalies in different degrees, evidently because the faulty gene " is expressing itself against a background of thousands of other genes and of variable environmental influences ".

The relative influence of genetic and environmental faults is, for most of the common malformations, difficult to assess. Anomalies like brachydactyly and albinism, for which genes are regularly to blame, are far less common than those like cleft palate and spina bifida, for which hereditary predisposition can be demonstrated in only a small proportion of cases. We must recognize too that familial incidence of a disease is not necessarily proof that heredity is involved, because the members of families are exposed, from embryo life onwards, to similar environmental factors—nutritional, toxic and infective. " Inherited " syphilis in all the members of a family is not a genetic inheritance ; and the deformities of infantile rickets or cretinism, once regarded as striking instances of morbid inheritance, are now recognized as due to specific dietary deficiencies. The occasional familial incidence of mongolism, cardiac malformation, anencephaly and many other anomalies cannot be taken as proof that genetic factors are concerned ; such incidence may be due purely to maternal factors external to the embryo, related to the mother's age, nutrition, endocrine state, drug habits or infections.

In certain malformational syndromes, the body cells show abnormalities in the number of chromosomes. Thus in certain anomalies of sex development, e.g. Turner's syndrome and Klinefelter's syndrome (pp. 608 and 610), there are errors in the number of sex chromosomes (X and Y). In mongolism there is a supernumerary autosome (non-sex chromosome); and in many other rarer malformations autosomal anomalies have been present.

CLASSIFICATION OF MALFORMATIONS ACCORDING TO THEIR MODES OF DEVELOPMENT

Malformations of every conceivable kind, degree and combination occur, and no two of them are ever exactly alike. This is not the place to enumerate them all nor to describe them. But a classification according to the ways in which they arise, with some of the main examples of each type, will be helpful.

(1) Double monsters

These show complete or partial duplication of the axis. Double-headed monsters, with anterior duplication only, arise from a single embryo as a result of bifurcation of the invaginating organizer, as described on p. 655. All other double monsters result from the coalescence of twin embryos in various ways.

(2) Suppressed general development

This occurs principally in twin pregnancies, in which one twin thrives and the second is suppressed. The suppressed twin may be so attached by its umbilical cord to its healthy fellow that it is entirely dependant on the latter for its blood supply. Under these circumstances the defective twin has no heart and is called an *acardiacus* ; it may be also an *acephalus* (devoid of a head), or an *acormus* (a head without a body), or an *amorphus* (an externally formless skin-covered mass). Occasionally a single foetus, though possessing a circulation of its own, is so reduced that it appears externally amorphous ; it may be called a *pseudo-amorphus*.

(3) Agenesis of particular parts

A foetus may show total absence (*agenesis*) or great reduction of one or more of the limbs (*amelia*) or of parts of limbs. Absence of the central parts of the face results in fusion of the eyes in the midline (*cyclopia*). In addition the lower jaw may be absent (*agnathia*), and the ears are then closely approximated on the ventral aspect of the neck. Agenesis of the lower part of the anterior abdominal wall and of the ventral wall of the bladder results in *ectopia vesicae* (extroversion or exstrophy of the bladder).

(4) Failure of parts to coalesce or close

Early embryogenesis shows many instances of coalescence of growing parts, e.g. the dorsal closure of the neural canal, the fusion of the fronto-nasal, maxillary and mandibular processes to form the face and palate, the ventral closure of the abdominal wall at the umbilicus, the coalescence of the lower parts of the paired Müllerian ducts to form a single uterus and vagina, and the closure of the septa of the heart and the ductus arteriosus. Many malformations of these parts consist essentially in failures of coalescence.

Anencephaly is essentially non-closure of the cephalic part of the neural canal and therefore absence of a developed brain and of the vault of the skull. *Open spina bifida* is a similar failure of closure of some part of the spinal neural canal. In other cases the neural canal closes but its dorsal skeletal covering remains incomplete, and a mass of brain or spinal cord or a fluid-filled meningeal sac protrudes through the defect under the skin. In the head this constitutes either an *encephalocele* or *cranial meningocele* ; in the spine, it is a *myelocele* or *spinal meningocele*. If the defect in the spinal laminae is only small, there may be little or no obvious protrusion of the cord and its coverings, and the condition is called *spina bifida occulta*.

Failure of coalescence of the several parts of the face during the second month results in fissures in characteristic positions. The commonest of these are *hare lip* and *cleft palate* due to varying degrees of non-union of the nasal and maxillary processes.

If closure of the abdominal wall at the umbilicus is incomplete, the loop of ileum to which the vitello-intestinal duct is attached protrudes through the wall into an umbilical sac of peritoneum, constituting a *congenital umbilical hernia*. Rarely this sac is very large and contains most of the movable viscera of the abdomen.

The uterus and vagina develop by fusion of the lower parts of the bilateral Müllerian ducts. Non-fusion of the two ducts at various levels may result in *bicornuate uterus*, or *double uterus* opening into a single vagina, or *septate vagina*.

The septum between the right and left sides of the heart develops by fusion of several distinct partitions and cushions which grow inwards from the wall of the originally simple heart. Non-closure of these results in persistence of apertures of communication between the two sides. In the interatrial septum this takes the form of the common and unimportant *patent foramen ovale*. Less common and much more serious is *patent interventricular septum*, which is often accompanied by stenosis of the orifice of the pulmonary artery, displacement of the aorta to the right, and hypertrophy of the right ventricle—a combination known as " Fallot's tetralogy ", the commonest of the serious forms of congenital disease of the heart.

Patency of the ductus arteriosus, which normally closes soon after birth, commonly accompanies congenital pulmonary stenosis or aortic stenosis, or it may occur without any other malformation.

(5) Failure of parts to separate or to canalize

Syndactyly (webbed fingers and toes) results from non-separation of the digits, which in a seven-weeks embryo appear as thickenings in the webbed distal segment of the limb-bud but which normally begin to separate in the eighth week.

Imperforate anus results from non-canalization of the anal plug, a thickened part of the cloacal membrane which in the seven-weeks embryo separates the rectum from the anal depression on the surface. In some cases the imperforate anus is closed by only a thin membrane ; in others, the rectum ends blindly 2 or 3 centimetres above the anal dimple.

The vagina forms by canalization of the epithelial vaginal cords derived from the lower part of the Müllerian ducts. Failure of canalization results in various degrees of *vaginal atresia*, ranging from complete *absence of the vagina* to *imperforate hymen*.

(6) Abnormal coalescence of parts

We have already noted that arrested development of the central parts of the face may lead to approximation and fusion of the two eyes (*cyclopia*). So also, arrested development of the hind end of the body may lead to fusion of the two lower limb fields and therefore to " siren malformation " (*sympodia*). Minor instances of abnormal fusion of parts are *horseshoe kidney* resulting from coalescence of the lower poles of the right and left kidneys across the midline ; and *reno-adrenal fusion*, the kidney and adrenal being closely adherent and even enclosed within a single capsule.

(7) Supernumerary or accessory parts

Many parts of the body may show duplication. This applies both to large regions and to small parts of particular organs. Of the grosser forms of duplication, we have already mentioned *double-headed monsters* due to anterior bifurcation of the embryonic axis (p. 660), and *supernumerary limbs* which occur in all classes of vertebrates and are no doubt due to disturbances of the early limb-fields (p. 653). *Supernumerary digits* (polydactylism) are not unusual. Where there is a large series of similar structures, it is easy to understand how slight developmental disturbances might add to the number in the series, e.g. *cervical* or *lumbar ribs* and *extra vertebrae* in any part of the spinal axis.

Supernumerary breasts may occur at some part or other of the mammary line between the axilla and groin.

The ureteric bud, an outgrowth of the lower part of the Wolffian duct, frequently divides and produces either a *forked ureter* (with two renal pelves but a single vesical orifice) or completely *double* (or even *triple*) *ureters* with separate vesical orifices.

Supernumerary spleens, usually small rounded ones, are quite common in the neighbourhood of the main spleen. *Accessory adrenals,* usually small pieces of cortex only, occur beneath the renal capsule, in the retroperitoneal tissues, spermatic cord, broad ligament or wall of Fallopian tube.

(8) Aberrant (ectopic or heterotopic) organs and tissues

Many of the accessory spleens, pancreases and adrenals referred to above are necessarily aberrant. Occasionally a whole organ is developmentally misplaced. The commonest example is the *ectopic kidney,* which may be situated at an abnormally low level, even in the pelvis, or which on rare occasions may be displaced across the midline to fuse with its fellow and so produce a *unilateral composite kidney.*

Another important group of aberrant tissues are the *sequestration cysts* of the skin and mucous membranes. Skin-lined *dermoid cysts* due to developmental dislocation occur, (1) in the subcutaneous tissues, where they are sometimes difficult to distinguish from acquired cysts due to traumatic implantation and from sebaceous cysts ; (2) in the brain or spinal cord or meninges (especially in the mid-dorsal line in the region of the cerebellum), where they no doubt arise from inclusion of epidermis during the sequestration of the neural canal from the surface ectoderm ; (3) rarely in the abdominal viscera, where their mode of origin is uncertain and where teratomatous " dermoid cysts " might be mistaken for developmental inclusions.

Just as developmentally sequestrated epidermis forms cysts, so occasionally does sequestrated alimentary or other mucosal tissue. The most familiar examples are *enterogenous cysts,* lined by intestinal mucosa and usually possessing a muscle coat, which occur at the ileo-caecal junction and in the mesentery.

It remains to add here that, instead of closed sequestration cysts, open *skin-lined sinuses* and *mucosa-lined diverticula* also occur as developmental anomalies in the skin or viscera respectively.

(9) Abnormal persistence of structures which normally disappear

Persistent residues or vestiges of developmental structures are of two kinds, (1) vestiges which persist constantly or frequently, e.g. the para-dental, para-pituitary, parovarian and other Wolffian residues, and the urachal and notochordal residues, and (2) vestiges which persist only occasionally, e.g. the branchial, thyroglossal and vitello-intestinal residues. Those of the first group have all been considered in other chapters ; those of the second group remain for consideration here.

(a) Branchial cysts and sinuses

Although mammalian embryos never show branchial or gill clefts, depressions corresponding to these appear on the interior of the embryo pharynx and on the outer surface of the neck. Occasionally an epithelial *cervical sinus* opens on the side of the neck along the course of the sternomastoid muscle ; this arises by persistence of the cervical sinus of the embryo, a deep depression containing the third and fourth branchial regions. *Branchial cysts* occur as smooth ovoid cysts deep to the upper part of the sternomastoid muscle close to the carotid bifurcation. Some of them are lined by squamous stratified epithelium, others by pseudo-stratified ciliated epithelium, and their walls frequently contain much lymphoid tissue, so resembling the pharyngeal mucosa or tonsil. Some branchial cysts are certainly closed remains of the cervical sinus ; but others have an endodermal origin from remains of the pharyngeal recesses, especially the tonsillar recess.

(b) Thyroglossal residues and aberrant thyroid tissue

The thyroid arises from a median endodermal outgrowth of the embryonic pharynx, the original site of evagination of which is represented by the foramen caecum of the tongue. The lower part of this downgrowth proliferates to form the thyroid gland ; the thyroglossal duct or stalk usually disappears. Hollow residues of the duct or masses of thyroid tissue may, however, persist anywhere along its course. The least uncommon of these is a median *pyramidal lobe of the thyroid*. A mass of thyroid tissue is occasionally present also in the base of the tongue—an *accessory lingual thyroid* ; this may share in goitrous changes or may give origin to tumours (p. 525). *A persistent thyroglossal duct*, lined by pharynx-like epithelium associated with lymphoid tissue, lies in the midline and in close relationship to the body of the hyoid bone. Localized *thyroglossal cysts* also occur, in the tongue, hyoid or infrahyoid regions. Infection of a thyroglossal track or cyst often leads to the formation of a discharging sinus on the skin usually over or just below the larynx in the midline.

(c) Persistent vitello-intestinal structures

The vitello-intestinal duct connects the yolk-sac with the intestine of the embryo. It is obliterated during the formation and elongation of the umbilical cord. Rarely, it remains patent from the intestine into the cord, severance of which then leaves an *intestinal fistula*. More commonly, only its intra-abdominal part persists as a *Meckel's diverticulum* (regarding the eponym, see p. 676), a diverticulum of the lower 2 or 3 feet of the ileum, which is present in about 2 per cent of people. This is usually lined by intestinal mucosa, but in some cases it is lined by gastric mucosa, which may suffer peptic ulceration and haemorrhage (p. 95). Vitello-intestinal remnants occasionally persist at the umbilicus as exposed patches or closed cysts of intestinal or gastric mucosa—*umbilical heterotopias* (p. 96).

(10) Generalized anomalies of skeletogenesis

There are several generalized anomalies of the skeleton which are clearly due to developmental disturbances. These are:

(i) *Multiple enchondroses* (Ollier's disease) characterized by persistence of masses of cartilage within many or all of the cartilage bones of the body (see also p. 535).

(ii) *Multiple exostoses*, characterized by multiple cartilage-capped bony outgrowths from many or all of the chondral bones, an inherited genetic anomaly (p. 535).

(iii) *Achondroplasia* (chondrodystrophia foetalis) is another congenital anomaly which is often hereditary or familial. There is defective growth of the cartilage bones of the limbs, which remain abnormally short, while the trunk is well developed ("human Dachshund"). The membrane bones of the skull grow normally and the cranium is of normal size, but the chondral bones of the base of the skull are stunted and the bridge of the nose recedes. Achondroplasia is the commonest kind of dwarfism, and one with which we are all familiar.

(iv) *Fragilitas ossium* (osteogenesis imperfecta) is another congenital and frequently familial disease of unknown cause, characterized by abnormally thin, poorly calcified, brittle bones which fracture on the slightest strain. It may manifest itself in the foetus, which may suffer multiple fractures *in utero* and is often still-born. Or fractures may be deferred until birth or childhood. If the patients survive adolescence, the abnormal fragility of the bones may disappear, though deformities remain. Other features of the disease are "blue sclerae" due to the pigmented choroid showing through an abnormally thin poorly fibrous sclera, defective teeth, abnormal laxity of ligaments predisposing to sprains and dislocations, and deafness due to otosclerosis (condensation of bone around the ear). It seems likely that all of these changes are manifestations of an intrinsic defect in mesenchymal tissues.

(v) *Other inborn genetic anomalies* involving the skeleton, all of them rare, include *osteopetrosis* or " marble bones " (see p. 400); *cleidocranial dysostosis*, so named because its most obvious features are defects of the clavicles and skull, but in which many other bones also are abnormal; *arachnodactyly* (" spider-fingers ") or Marfan's syndrome, with anomalies not only of the skeleton but also of the heart and great vessels and eyes; and the *Laurence-Moon-Biedel syndrome*, comprising various skeletal anomalies along with obesity, genital infantilism, mental deficiency and other defects. There are many other even rarer inborn disorders of skeletogenesis.

(11) Pseudohermaphroditism : inter-sex

The determination of sex is primarily dependant on chromosomal constitution, female embryos developing from zygotes with an XX pair of sex chromosomes males from zygotes with an XY pair. But this primary determination is not absolute ; the subsequent development of the Müllerian and Wolffian systems is dependant on hormonal factors.

In embryos of the 7th week the rudiments of the male and female genital systems are of indifferent or neutral type, and the sexes are indistinguishable. The male embryo has Müllerian ducts, the female has Wolffian ducts, the external genitalia are alike, and the ovary and testis cannot be distinguished from one another. The subsequent differentiation of the sexes depends on the activity of diffusible substances akin to, possibly identical with, the sex hormones.

Thus, in cattle, in occasional cases of non-identical twins, one male and the other female, in which placental intercommunication develops between the two foetal circulations, androgenic hormone from the male embryo enters the circulation of the female embryo, in the genital tissues of which it induces the development of male characters and suppresses the development of the Müllerian system. Such twins, genetic females converted into imperfect males by hormonal influence, are known to breeders as " free-martins ". A similar condition occasionally results from the excessive production of androgens by an adrenal tumour in the human foetus (p. 608).

To conditions of this kind, in which the individual has gonads of one sex but other parts of the genitalia have developed so as to resemble more or less those of the opposite sex, we apply the name *pseudohermaphroditism*. In the examples just given the explanation of this abnormal development is known ; but in many other cases its cause is unknown. Recent studies show that in some cases there are abnormal numbers of sex chromosomes (p. 660). Pseudohermaphrodites may be of either sex. *Male pseudohermaphrodites*, those in whom testes are present but some or all of the rest of the genital organs resemble those of a female, are more common that *female pseudohermaphrodites* who possess ovaries but whose other genital organs more or less resemble those of a male.

True hermaphroditism, in which ovary and testis are both present in the same individual, is very rare in man.

EXTENSIONS OF THE CONCEPT " MALFORMATION "

The usual idea conveyed by the word " malformation " is of an obvious structural abnormality present at birth. However, as has already been noted in this and other chapters, it is necessary to extend the meaning of " malformation " to embrace :

(1) *delayed malformations*, which, though attributable to inborn defects of the tissues, may not make themselves evident until post-natal, even adult, life;

(2) *hamartomas*, minor malformations characterized by improper admixtures of tissues often with excess of a particular tissue ;

(3) *cellular and enzymic malformations*, inborn errors of structure of particular cells or errors of metabolism ;

(4) *certain tumours which are also malformations*.

It will be helpful to enumerate here the principal examples of these kinds of malformations.

(1) Delayed malformations

These are exemplified by many cases of multiple exostoses (p. 535), osteogenesis imperfecta (p. 665), neurofibromatosis (p. 569), and a variety of other rare diseases of the nervous system, including :

(i) *Friedreich's hereditary ataxia*, in which degeneration of the cerebellum and various tracts of the spinal cord usually first manifests itself during childhood ;

(ii) *Huntington's chorea*, in which a familial degeneration of the corpus striatum and other parts of the brain is deferred until adult life ;

(iii) *Tay-Sachs disease* (amaurotic family idiocy) in which the characteristic lipidosis of many groups of nerve cells may make itself clinically apparent in infancy or may be deferred until later childhood;

(iv) *Familial muscular atrophies and dystrophies of childhood,* which are of several eponymically named varieties, some of which show significant degenerative changes in the nervous system, while others appear to be primary disorders of the muscles themselves ;

(v) *Tuberose sclerosis* (epiloia) in which nodular areas of gliosis in the brain occasion progressive mental deficiency and epileptic fits during childhood, often accompanied by hyperplasia of the sebaceous glands of the skin of the face (" adenoma sebaceum "), and various visceral anomalies, especially cysts or hamartomas of the kidneys and so-called " rhabdomyoma " of the heart (p. 551).

(2) Hamartomas (Gk. = to go wrong).

These are exemplified by benign " angiomas " (p. 544), simple pigmented moles or naevi (p. 578), most of the lesions of neurofibromatosis (p. 569), enchondroses (p. 535), and some congenital lipomas (p. 533). In all of these, the tumour-like lesions contain developmentally superfluous tissue of one kind or another—superfluous vessels, nerve-sheath tissue, or adipose tissue, as the case may be. It is not surprising that various combinations of these hamartomas are encountered, e.g. neurofibromatosis and pigmented naevi, neurofibromatosis and lipomas, enchondroses and angiomas, etc.

(3) Cellular and enzymic malformations

We have already noted that familial acholuric jaundice (p. 402) and some other familial haemolytic anaemias (p. 403) are due to inborn anomalies of the erythrocytes. So also familial xeroderma pigmentosum (p. 473) is an inborn anomaly of the epidermis rendering it peculiarly photo-sensitive ; and familial polyposis of the colon denotes an inborn instability of the colonic epithelium. In some rare inborn metabolic disorders (e.g. glycogen disease, the lipidoses, cystinuria, alkaptonuria, and albinism, discussed in Chapter 25), it is not even possible to connect the anomaly with any particular kinds of cells ; and we suppose that it consists rather in a general defect or disturbance of a particular enzyme system—an enzymic " malformation ".

(4) Tumours which are also malformations

The several classes of embryonic tumours and the teratomas are simultaneously malformations and neoplasms, a combination of properties the implications of which we have already discussed (p. 585 and p. 590).

EXTRA-UTERINE OR ECTOPIC PREGNANCY

By these terms is meant implantation and development of the fertilized ovum in some situation other than the normal uterine one. *Tubal, tubo-ovarian* and *abdominal pregnancies* have been distinguished ; but the existence of abdominal pregnancy, except by rupture of a tubal one, is doubtful, and ovarian pregnancy is very rare. *Tubal pregnancy* is by far the commonest. The tube wall is ill-adapted to accommodate the rapidly enlarging gestation sac ; it quickly becomes distended and thin, and it is also invaded by placental tissue. Hence it usually ruptures in the first 2 or 3 months of the pregnancy, the rupture causing serious intraperitoneal haemorrhage (Fig. 288). Following rupture the foetus usually dies ; but occasionally it survives and the placenta acquires fresh vascular connections to surrounding parts (secondary abdominal pregnancy). Further rupture may ensue, or rarely the foetus reaches full-term in its intraperitoneal gestation sac. Unless it is removed then, it dies and remains as a foreign body, which may become infected and cause peritonitis or abscess, or may slowly undergo calcification and be converted into a *lithopaedion*.

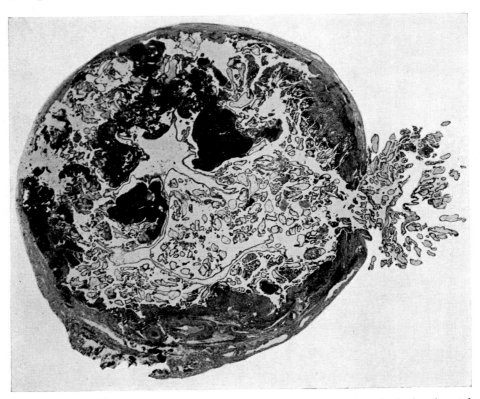

FIG. 288.—Cross section of ruptured tubal pregnancy, showing distension of tube by placental tissue and blood-clot and protrusion of placental tissue through ruptured part of wall. (× 6.)

SUPPLEMENTARY READING

Ballantyne, J. W. (1902 and 1904). *Manual of Antenatal Pathology and Hygiene*. Edinburgh; W. Green. (An encyclopaedic work, full of interest for the advanced student.)

Bremer, J. L. (1957). *Congenital Anomalies of the Viscera: Their Embryological Basis*. Cambridge (Mass.); Harvard University Press. Oxford; Oxford University Press.

CIBA (1960). *Foundation Symposium on Congenital Malformations*. (Eds. Wolstenholme, G. E. W., and O'Connor, C. M.). London; Churchill.

Harris, H. (1959). *Introduction to Human Biochemical Genetics*. Cambridge; Cambridge University Press.

Huxley, J. S., and de Beer, G. R. (1934). *The Elements of Experimental Embryology*. London; Hafner. (A fascinating book for the advanced student.)

Keith, A. (1940). " Concerning the origin and nature of certain malformations of the face, head and foot ", *Brit. J. Surg.*, **28**, 173.

— (1948). *Human Embryology and Morphology*. London; Edw. Arnold. (The most useful outline of embryology, with frequent reference to the mode of origin of malformations.)

Morison, J. E. (1963). Second edition. *Foetal and Neonatal Pathology*. London; Butterworths.

Needham, J. (1936). " New advances in the chemistry and biology of organized growth ", *Proc. Roy. Soc. Med.*, **29**, 1577. (A valuable brief outline of organizer phenomena and other aspects of embryology of importance in pathology.)

Penrose, L. S. (1963), second edition. *Outline of Human Genetics*. London; Heinemann.

Perlstein, M. A., and Le Count, E. R. (1927). " Pygopagus twins. The history and necropsy report of the Bohemian twins, Rosa-Josepha Blazek ", *Arch. Path.*, **3**, 171.

Potter, E. L. (1961), second edition. *Pathology of the Fetus and Infant*. Chicago; Year Book Publishing. London; Lloyd-Luke.

Roberts, J. A. F. (1940). *An Introduction to Medical Genetics*. London; Oxford University Press. (An excellent outline.)

Scheinfeld, A. (1939). *You and Heredity*. Philadelphia; Stokes. London; Chatto & Windus. (Entertaining and semi-popular, but thoroughly sound.)

Stockard, C. R. (1921). " Developmental rate and structural expression : an experimental study of twins, double monsters and single deformities, and the interaction among embryonic organs during their origin and development ", *Amer. J. Anat.*, **28**, 115. (A detailed experimental study of the production of malformations by environmental abnormalities.)

Willis, R. A. (1962), second edition. *The Borderland of Embryology and Pathology*. London; Butterworths. (A more detailed account of the subjects outlined in this chapter.)

APPENDIX A

NAMES

" Now many errors consist of this alone, that we do not apply names rightly to things." (Spinoza)

The learning, or teaching, of any branch of science consists largely in defining and understanding clearly the special names used in it. The student, or teacher, should deliberately develop the habit of precise definition of the terms he uses. When the meanings and implications of all the special terms in the subject are clear in his mind, then he knows the subject. To assist students in clarifying for themselves the meanings of the terms used in pathology, this appendix contains :

(1) A list of the most important Greek roots from which many biological and medical words are derived, with examples. An elementary knowledge of root meanings is of inestimable value, and takes little effort to acquire. *Medical Terms, Their Origin and Construction* by Ffrangcon Roberts (London, 1959) is an invaluable little book which every student should possess and absorb.

(2) A note on eponyms—words derived from the names of people who made important discoveries or descriptions in the subject.

(3) The names of bacteria.

(4) A note on words with a teleological flavour.

SOME GREEK ROOTS AND AFFIXES

(A few important Latin words with less obvious meanings are also included.)

a-, an-: Gk.=not, without.
    atrophy, anaemia, anoxia, aplasia, avascular, amorphous, asepsis
acro: Gk. *akros*=topmost, extremity
    acromegaly, acrocephaly
aden: Gk.=a gland
    adenitis, adenoma, adenosis
-aemia: Gk. *haima*=blood
    anaemia, hyperaemia, ischaemia, pyaemia, uraemia, hyperglycaemia
aisthesis: Gk.=sensation
    anaesthesia, hyperaesthesia, paraesthesia
-algia: from Gk. *algos*=pain
    neuralgia, myalgia, metatarsalgia, analgesia
ana-: Gk.=back, again
    anaplasia, anabolism, analysis, anatomy
andros, also *arrhenos:* Gk.=male
    androgen, gynandrous, arrhenoblastoma
angeion: Gk.=a vessel
    angioma, lymphangitis, cholangitis, sporangium
anthrax: Gk.=coal
    anthrax, anthracosis, anthracene
anti-: Gk.=against, opposite
    antibody, antigen, antitoxin, antidote, antisepsis
arthron: Gk.=a joint
    arthrology, arthritis, arthropod
-asis, see *-osis*

670

autos: Gk.=self
      autolysis, autopsy, autonomy, autogenous
blastos: Gk.=a sprout, hence something immature
      fibroblast, erythroblast, myeloblast, neuroblastoma, retinoblastoma, blastomyces, blastula, blastomere.
bolos: Gk.=a lump
      bolus, embolus, metabolism
carcino: Gr. *karkinos*=a crab or cancer
      carcinoma, carcinogenic
cele: Gk. *kele*=a swelling
      hydrocele, haematocele, mucocele, cystocele, meningocele, encephalocele
cephal: Gk. *kephale*=the head
      cephalic, encephalon, hydrocephalus, microcephaly, dibothriocephalus
chlor: Gk. *khloros*=green
      chlorosis, chloroma, chlorine, chlorophyll
chol: Gk. *khole*=bile
      cholic acid, cholangitis, cholelithiasis, cholaemia, cholera, melancholia
chondr: Gk. *khondros*=cartilage
      perichondrium, chondromucin, chondroma, ecchondrosis, chondrophyte
chrom: Gk. *khromos*=colour
      cytochrome, chromaffin, chromatolysis, chromoblasto myces, chromatophore, chromium
chym: Gk. *khumos*=juice
      chyme, parenchyma, mesenchyme, ecchymosis
coll: Gk. *kolla*=glue
      collagen, colloid, collodion
cyan: Gk. *kuanos*=blue
      cyanosis, *Ps.pyocyanea,* haemocyanin
cyst: Gk. *kustis*=a bladder
      cyst, cystitis, cholecystitis, endocyst, ectocyst
cyt: Gk. *kutos*=a cell or vessel
      cytology, leucocyte, erythrocyte, cytoplasm, syncytia, cytolysis
daktylos: Gk.=a digit
      dactylitis, syndactyly, polydactyly
derma: Gk.=skin
      dermis, epidermis, dermatology, dermatitis, dermatographia
dia-: Gk.=through, across
      diarrhoea, diabetes, diuresis, dialysis, diapedesis, diagnosis, diaphragm
dys-: Gk.=bad
      dyspepsia, dysentery, dysphagia, dyspnoea, dysuria, dysmenorrhoea
ectasis: Gk.=expansion
      bronchiectasis, telangiectasis, lymphangiectasis
ecto-: Gk.=outside
      ectoplasm, ectocyst, ectoderm, ectozoa
-ectomy (=*ec*+*tomy* q.v.), a cutting out or excision
em-, en-: Gk. and Lat.=in, into
      embolism, empyema, enchondroma, encyst, endemic, engraft
endo-: Gk.=within
      endocardium, endometrium, endarteritis, endogenous
enteron: Gk.=the bowel
      enteritis, enterolith, enterozoa, enterostomy, dysentery
epi-: Gk.=upon
      epicardium, epiphysis, epithelium
erythros: Gk.=red
      erythrocyte, erythropoiesis, erythema

ex-, ec-, e-: Lat.=out, from
        exude, exostosis, exfoliate, ecchondroma, effusion

filum: Lat.=a thread
        filiform, filamentous, filaria

fuse: Lat. *fundere, fusum*=to pour
        effuse, infuse, transfuse, perfuse

gaster: Gk.=stomach
        gastritis, gastrostomy, gastroscope, gastrulation, gasteropod

gen, genesis: from Lat. and Gk.=birth, origin, begetting
        gene, genetics, genital, histogenesis, carcinogenesis, fibrinogen, androgen,
        oestrogen

glia: Gk.=glue
        neuroglia, glioma

gnosis: Gk.=knowledge
        diagnosis, prognosis, stereognosis

graphos, gramma: Gk.=writing
        electrocardiography, electrocardiogram, pyelography, pyelogram, radiography,
        topography

gyne: Gk.=woman, female
        gynaecology, gynaecomastia, gynandromorph

haem: Gk. *haima*=blood
        haemorrhage, haemoglobin, haemoptysis, haematuria, haematocele,
        haemopoiesis

heteros: Gk.=other, different
        heterotopia, heterologous, heteromorphous

homos: Gk.=same
        homology, homoplastic, homozygous

hyalos: Gk.=glass
        hyaline, keratohyalin, hyaluronic acid

hydr, hygr: from Gk.=water, wet
        hydronephrosis, hydrocephalus, hydrophobia, hydrocele, hydatid, hygroma,
        hygroscopic

hyper-: Gk.=over, excessive
        hyperplasia, hypertrophy, hyperaemia, hyperostosis, hyperglycaemia

hypo-: Gk.=under, deficient
        hypoplasia, hypoglycaemia, hypotonus, hypostasis, hypodermic

idio-: Gk.=one's own
        idiopathic, idiosyncrasy

-itis: Gk. suffix=inflammation
        appendicitis, tonsillitis, hepatitis, etc.

lapsus: Lat.=a slip
        prolapse, relapse, collapse

leukos: Gk.=white
        leucocyte, leucoplakia, leukaemia, leucorrhoea

lipos: Gk.=fat
        lipoid, lipase, lipoma, lipaemia, lipuria

lithos: Gk.=a stone
        cholelithiasis, nephrolithiasis, enterolith, phlebolith, lithopaedion

lysis: Gk.=loosening or solution
        lysis, haemolysis, autolysis, paralysis, analysis

macro-: Gk. *makros*=large
        macrocyte, macrophage, macroglossia

malakos: Gk.=softening
        osteomalacia, myomalacia, malacoplakia

mastos: Gr.=breast
        mastitis, mastopathy, gynaecomastia, polymastia, mastoid

mega-, megalo-: from Gk. *megos*=great
        megacolon, megaloblast, megakaryocyte, acromegaly, splenomegaly

melan-: (from Gk. *melas, melanos*=black)
    melanin, melanoma, melancholia
meros: Gk.=a part
    merozoite, metamere, sarcomere, polymer
mesos: Gk.=middle, intermediate
    mesoderm, mesenchyme, mesentery, mesial
meta-: Gk.=with, after, beyond, often with the sense " change "
    metabolism, metastasis, metacarpal, metamorphosis, metaphysics
metros: Gk.=mother, womb
    endometrium, metritis, metrorrhagia
mikros: Gk.=small
    microscope, microcephaly, microphage
monos: Gk.=alone, single
    mononuclear, monoxide, monogamy
morphe: Gk.=form, shape
    polymorphonuclear, morphology, pleomorphism, amorphous
myc: from Gk. *mukes*=a mushroom, hence a fungus
    mycology, actinomyces, blastomyces, mycelium, mycetoma
myelos: Gk.=marrow
    myelitis, osteomyelitis, syringomyelia, myelin
my-, myo-: from Gk. *mus*=muscle
    myositis, myoma, myalgia, myasthenia, epimysium, myosin
myxa: Gk.=mucus
    myxoma, myxoedema, myxomycete
nekros: Gk.=a corpse
    necropsy, necrosis, necropolis
nephros: Gk.=a kidney
    nephritis, pyonephrosis, nephroblastoma
neuron: Gk.=a nerve
    neurone, neuritis, neurolemma, neuroblast, neuralgia, neuroglia
-oid, -ode: Lat. and Gk. suffix=like
    deltoid, mastoid, amyloid, fibroid, adenoids, cestode, nematode, plasmodium
oligos: Gk.=scanty
    oligaemia, oliguria, oligohydramnios
-oma, -ome: Gk. suffix, *-ma, -me,* expressing result of an action, used in pathology
    mainly for tumours
    adenoma, carcinoma, sarcoma, etc. ; atheroma, coma, coloboma, granuloma,
    odontome
ophthalmos: Gk.=eye
    ophthalmology, ophthalmoscope, exophthalmos, xerophthalmia
ops: Gk.=eye
    myopia, presbyopia, nyctalopia, hemianopia, cyclopia
orthos: Gk.=straight
    orthopnoea, orthopaedics, orthodox
-osis, -asis: Gk. suffix=a process or condition
    tuberculosis, pneumoconiosis, necrosis, scoliosis, exostosis, fibrosis, trichiniasis,
    amoebiasis
osteon: Gk.=bone
    osteology, osteitis, osteoma, osteoplasia, exostosis, osteoblast, osteoclast
otos: Gk.=ear
    otology, otoliths, otitis, otorrhoea
paed: from Gk. *pais, paidos*=a child
    paediatrics, lithopaedion, orthopaedics
para-: Gk.=beside
    parovarium, parametrium, paratyphoid fever, paralysis, parasite
pathos: Gk.=suffering
    pathology, pathogenic, encephalopathy, osteoarthropathy, sympathetic

ped, pod, pus: Gk. and Lat.=a foot
        pedicle, peduncle, pseudopodia, polypus, diapedesis, biped, pedigree

peri-: Gk.=around
        pericardium, periosteum, peri-arteritis, perisplenitis

phagos: Gk.=eating
        phagocyte, macrophage, bacteriophage, dysphagia, phagadaena, oesophagus

philos: Gk.=friend, lover
        eosinophil, basophil, philosophy

phlebos: Gk.=a vein
        phlebitis, phlebolith, phlebotomy

phobos: Gk.=fear
        phobia, hydrophobia, photophobia, claustrophobia, chromophobe

phoros: Gk.=a bearer
        chromatophore, spermatophore, phosphorus

phos, photos: Gk.=light
        photometry, photophobia, phosphorus

physa: Gk.=a puff of wind
        emphysema, physometra, *Physalia*

physis: Gk.=nature
        physics, physiology, physician

phyton: Gk.=a plant ; also *physis*=a growth
        osteophyte, chondrophyte, saprophyte, diaphysis, epiphysis

plast, plasia, plasm: from Gk. *plastos*=moulded
        plastic, protoplasm, plasmodium, aplasia, hyperplasia, metaplasia, anaplasia

pleo-: Gk.=more
        pleomorphism, pleocytosis

pneumon: Gk.=lung ; related to *pneuma*=air, and *pneo*=breathe
        pneumonia, pneumothorax, pneumatosis, dyspnoea, apnoea, hyperpnoea

poikilos: Gk.=various
        poikilothermic, poikilocytosis, poikiloderma

polios: Gk.=grey
        poliomyelitis, polio-encephalitis

poly-: Gk.=many
        polymorphism, polymorphonuclear, polypus, polyhedral

pro-: Lat. and Gk.=before
        prophase, prognosis, prothrombin, prolapse

protos: Gk.=first, primitive
        protoplasm, protozoa, protein

pseudo: Gk.=false
        pseudopodia, pseudohermaphrodite, pseudomyxoma

psyche: Gk.=soul, mind
        psychology, psychosis, psychopathology, psychical

ptosis: Gk.=a falling
        ptosis, gastroptosis, visceroptosis

pyon: Gk.=pus
        pyogenic, pyaemia, empyema, pyonephrosis, *Ps. pyocyanea*

rachis: from Gk. *rhakhis*=back, spine
        rachitis, rachischisis

rhinos: Gk.=nose
        rhinitis, rhinolith, rhinoscleroma, rhinorrhoea, rhinoceros

-rrhoea: from Gk. *rhoia*=a flow
        diarrhoea, otorrhoea, dysmenorrhoea, haemorrhoids

sarkos: Gk.=flesh
        sarcolemma, sarcoma, sarcophagus

schis, schiz: from Gk. *skhisma*=a split
        schistosome, schizomycete, schizont, schizogony

skleros: Gk.=hard
        sclera, sclerosis, scleroderma, rhinoscleroma

septikos: Gk.=putrid

    septic, asepsis, antisepsis, septicaemia, *Cl. septicum*

soma: Gk.=body

    somatic, somite, centrosome, schistosome

spora: Gk.=seed

    spore, sporocyst, sporozoite, sporotrichosis, sporozoa

stasis: Gk.=a standing, state

    stasis, haemostasis, metastasis

stercus: Lat.=faeces

    stercoral, stercolith, stercobilin

stoma: Gk.=mouth

    stomatitis, ankylostoma, gastrostomy (and all the other stoma-making operations)

syn-, sym-: Gk.=together

    synthesis, symptom, sympathetic, synapse, syncytia, sympus

syrinx : Gk.=a pipe

    syringe, syringomyelia, syringo-cystadenoma

tabes: Lat.=wasting

    tabes dorsalis, craniotabes

teratos: Gk.=a monster, deformity

    teratology, teratoma, teratogenic

thele: Gk.=a teat ; usually applied more generally in sense of " surface "

    epithelium, endothelium, polythelia

thesis: Gk.=thing laid down

    thesis, hypothesis, synthesis

thrix (see *trich*)

thrombos: Gk.=a lump, clot

    thrombus, thrombin, thrombocyte, thrombosis

tomia: Gk.=cutting, section

    microtome, myotome, anatomy, tracheotomy (and all other operations ending in *-tomy*=cutting into), appendicectomy (and all the other operations ending in *-ectomy*=a cutting out or excision)

topos: Gk.=a place

    ectopia, heterotopia, topography

trich: from Gk. *thrix, trikhos*=a hair

    trichophyton, trichina, streptothrix, trichobezoar

trope: Gk.=a turning

    tropism, chemotropic, lipotropic, gonadotropic

trophe: Gk.=food, nutrition

    trophic, atrophy, hypertrophy, dystrophy

ur, uro: from Gk. *ouron*=urine

    urology, uraemia, anuria, oliguria, haematuria, glycosuria

xanthos: Gk.=yellow

    xanthin, xanthoma, xanthochromia

xeros: Gk.=dry

    xerosis, xeroderma, xerophthalmia

zoon: Gk.=an animal

    protozoa, metazoa, spermatozoon, entozoa, ectozoa, haematozoa, merozoite

EPONYMS

The use of eponyms

Some eponyms are appropriate and useful ; they remind us (and we can too easily forget) that our subject has a human history sprinkled with the names of many famous discoverers. It is good that along with the cold impersonal Latin nomenclature of anatomy we should remember Hunter's canal, Peyer's patches, the Fallopian tube, the circle of Willis and the bundle of His. It is good that in the nomenclature of disease we should not forget Bright, Addison, Graves, Charcot or Paget. It is good that the important discoveries in bacteriology should be recalled by such names as the Klebs-Löffler bacillus, Leishman-Donovan bodies, *Cl. Welchii*, and *Rickettsia*. It is an entertaining exercise to compile lists of recognized eponyms in the various branches of medicine, and, with the aid of a good history of medicine such as Garrison's, to note the times and places of the relevant discoveries.

The misuse of eponyms

Eponymous nomenclature easily runs to absurd excess. It has become a fashion to attach to relatively trivial syndromes, symptoms, X-ray appearances or histological details, the names of the people supposed to have first recorded them. This indiscriminate use of proper names has three disadvantages. (1) It creates a flood of eponyms of no historical value or interest, which, once established, are difficult to eradicate. (2) Since many of these have been hastily and uncritically bestowed without adequate enquiry into the history of the subject, they are often found later to be inappropriate ; e.g. the "Dorothy Reed" cells of Hodgkin's disease were described by at least four persons before Dorothy Reed's account of them. (3) When, later, it is found that a hastily applied name has overlooked the claims of prior workers, cumbersome compound eponyms are invented, e.g. "Besnier-Boeck-Schaumann disease", "Hand-Christian-Schüller disease, and "Weber-Sturge-Kalischer syndrome". The creation of new eponyms should be avoided unless the discoveries which they recall are of outstanding importance and unquestionable originality. Let us take warning from the fact that even some famous eponyms overlook the claims of prior discoverers ; e.g. "Meckel's diverticulum" was described by John Hunter in 1763, about half a century before Meckel's description of it.

A capital letter for eponyms

If a man's name is worthy of the honour of an eponym, it is worthy of the dignity of its normal capital letter. A frequent practice of writing adjectival eponyms and eponyms denoting bacterial species with a small initial letter (e.g. "fallopian tube", "*Cl. welchii*", "*Leishmania donovani*") is, I hold, ugly and unnecessary. Moreover, those who mutilate men's names in this way are often inconsistent; e.g. in the 1957 edition of Anderson's well-known *Pathology* (ref. p. 685) Chapter 39 throughout spoke properly of "Müllerian duct", "Wolffian duct" and "Krukenberg tumour", but improperly of "fallopian tube", "graafian follicle" and "sertoli cells". To argue that for taxonomic consistency the species name of an organism *must* begin with a small letter, is absurd; *Cl. welchii* means "the clostridium of Welch", clearly demanding the capital "W". All eponyms, both adjectives and nouns, should be treated with respect as proper names and should retain their initial capitals. This practice, which is contrary to the prevailing one, has been followed throughout the present book. Place names also should retain their initial capital, to remind us that they *are* place names, e.g. *B. Aertrycke*, *Bacterium Tularense*, and *Trypanosoma Gambiense*.

THE NAMES OF BACTERIA

The following comments by one of our most eminent bacteriologists, Professor J. W. (now Sir James) Howie (*Lancet*, 1956, **2**, 951), will be comforting to medical students and should be heeded by their teachers:

Genera, Species, and Types

"Scientific students of bacteriology naturally try to arrange bacteria in a regular fashion into orders, tribes, families, genera, species, and types, to which, by rules which are continually debated, they attach what they regard as valid names. . . . Unfortunately, both the rules of the game and the bacteria themselves are not very stable. Moreover, bacteriologists tend to attach importance to the bacterial characters in which they themselves are interested: the fermentation chemist wishes to know which carbohydrates they will break down; the immunologist wishes to know the last details of their protein structure; the epidemiologist wishes to know their different habitats and how they get from one to another; and the doctor and his patients are concerned only in whether they are poisonous or harmless. For these reasons the names given to bacteria are liable to rather frequent changes as new work and changing perspectives constantly tend to adjust what seem to be their significant relationships among each other. Thus, even in the past twenty years, one organism has been named: *Bacillus typhosus, Eberthella typhi, Bacterium typhosum,* and now *Salmonella typhi* or *Salmonella typhosa* according to taste. In medical practice these changes are both confusing and unnecessary and I think that medical bacteriologists in communicating with medical practitioners should avoid using generic and specific names and content themselves with common names. In my own practice I have always reported this organism to doctors as "the typhoid bacillus". . . . Medical practitioners are interested only in the disease-producing potentialities of bacteria, and the stable common names that have served for so long are well known and understood. It is right that scientific bacteriologists should debate among themselves about the appropriate definitions of Mycobacterium, Actinomyces, Clostridium, Corynebacterium, and others, but their communications to medical men will be less open to misinterpretation if they refer to the tubercle bacillus, . . . the diphtheria bacillus, and so forth. . . . I commend these simple usages to medical bacteriologists; if they wish to flash their new genera under bright lights let them do it at each other."

WORDS WITH A TELEOLOGICAL FLAVOUR

Teleology (Gk. *telos*=an end) is the doctrine of " final causes " or the view that structures and processes are due to purpose or design. In pathology, as in all other biological subjects, we use many words, with teleological implications. Thus we speak of the " functions " of organs and cells, i.e. activities by which they play their proper part in the maintenance and economy of the body as a whole ; we think of repair, phagocytosis, inflammation and antibody formation as being " defensive " or " protective " reactions of the tissues ; we see the clotting of blood as an essential self-preservative measure ; and we regard hypertrophy, hyperplasia, leucocytosis, and the regeneration of erythrocytes following haemorrhage, as " useful " compensatory processes.

Some biologists strongly object to any suggestion of functional purpose or utility in tissue reactions and would abolish all terms with such suggestions. But this is a narrow viewpoint and one which it is really impossible to sustain. Even the most bigoted anti-teleologists can scarcely avoid speaking of the " functions " of organs and cells, and the very concept of function carries with it the idea of utility or appropriate activity. Moreover, when we use such terms as " functions ", " utility ", " defence ", " protection ", " compensation ", we *can* do so without becoming teleologists, metaphysicians or theists. When we say that the activity of liver cells in secreting bile and of red cells in carrying oxygen are " useful " functions, or that repair and blood-clotting are " self-preservative " processes, or that phagocytosis and inflammation are " defensive " reactions, we are *not* saying that we believe these to be the result of planning by an omniscient creator or that we think the tissues of the body are themselves endowed with intelligence. These words are equally consistent with the view that the structure and functions of cells and tissues and their reactions to injuries and other abnormalities of environment are all the result of evolution. Our blood clots, not because it was designed to clot, but because if our ancestors had not evolved the blood-clotting mechanism, they would have left no descendants. Our wounds heal, not because of a deliberate intention on the part of our tissues, but because animals which were devoid of this capacity would be automatically exterminated. We possess phagocytes, not because God deliberately introduced these helpful little animalcules into Adam on the sixth day of Creation, but because creatures which during their evolution failed to develop some means of dealing with extraneous microbes and foreign particles would not have survived in the struggle for existence.

Hence, we need not refrain from using the words " function ", " utility ", " protective ", etc., provided that we are clear in our minds as to their legitimate use in our subject. They need not, and should not, carry any particular philosophical or religious implications.

On the other hand, we *should* avoid crude teleological statements which imply that cells or even fluids have intelligence, e.g. that " phagocytes wage war on invading bacteria ", or that " fibrosis in the stroma of a scirrhus is an attempt to strangle the tumour cells ", or that in inflammation " the fibrin endeavours to limit the process by attempting to shut off the inflamed area ". Such statements are unfortunately common in current pathological writings ; indeed the third example just given is taken verbatim from the last edition of a well known text-book of pathology.

APPENDIX B

BRIEF NOTES ON SOME GREAT PATHOLOGISTS

Although HIPPOCRATES " the father of Medicine ", laid the firm foundations of clinical medicine in the 5th century B.C., and although GALEN made many discoveries in anatomy and physiology in the 2nd century A.D., the ancient Greeks and Romans knew almost no pathology. Hippocrates' speculative " doctrine of humors ", namely that all disease was due to disorders of the fluids of the body, was accepted in lieu of factual pathology for a period of nearly 2,000 years.

During the Renaissance period, BENIVIENI published his " Hidden Causes of Disease " (1507) which contained a few fragmentary records of necropsy findings ; LEONARDO DA VINCI (1452-1519) left in his amazing notebooks a few scattered observations on senile and atheromatous blood vessels and on cirrhosis of the liver : VESALIUS, author at the age of 28 of the first and greatest of all textbooks of anatomy (1543), noted senile changes in joints, the presence of omentum in scrotal hernia, and the distortion of viscera caused by corsets ; PARÉ (1510-1590), " the father of modern surgery ", described enlarged prostate as a cause of urinary obstruction, suggested the syphilitic nature of aneurysms, gave the first account of carbon monoxide poisoning, and wrote a fanciful treatise on malformations ; and other anatomists and surgeons recorded occasional examples of pathological lesions. In 1649 William HARVEY, the discoverer of the circulation of the blood, wrote of his intention of publishing a *Medical Anatomy*, " that I may relate from the many dissections I have made of the bodies of persons diseased, worn out by serious and strange affections, how and in what way the internal organs were changed in their situation, size, structure, figure, consistency, and other sensible qualities ". Unfortunately, Harvey did not carry out this intention ; and it was left to Morgagni to produce the first important book on pathology over a century later.

Giovanni MORGAGNI (1682-1771) of Padua was 79 years old when he published the results of his life-work in his " Seats and Causes of Diseases Investigated by Anatomy " (1761), the first text-book of pathological anatomy. It contains many necropsy records, correlated with the clinical histories, including cases of aneurysm, cardiac malformation, endocarditis, myocardial fibrosis, hepatic necrosis and cirrhosis, intussusception, pneumonia, pulmonary tuberculosis of all types, cerebral abscess, cerebral tumour, gastric carcinoma, uterine carcinoma, other tumours, and the internal thickening of the frontal bone (exostosis frontalis interna) in obese elderly women which is now called " Morgagni's syndrome ".

John HUNTER (1728-1793), a Scot who migrated to London, was the most versatile of all the great figures in the history of biology and medicine ; he was a great zoologist, embryologist, comparative anatomist, physiologist, pathologist, experimentalist and surgeon. His collection of anatomical and pathological specimens, the remains of which are preserved at the Royal College of Surgeons in London, is unique for the range of subject and the beauty of the preparations. He contributed to every branch of pathology, especially the general pathology of inflammation and repair, venereal disease, osteomyelitis, anomalies of growth and malformations. (For references to Hunter, see pp. 56, 70, 71, 74, 145, 175, 192, 196, 197, 198, 379, 606 and 676.)

Edward JENNER (1749-1823), a Gloucestershire doctor and a friend of John Hunter, was the discoverer of vaccination with cowpox as a protection against smallpox, and thus the founder of modern immunology. (See p. 256.)

René LAENNEC (1781-1826), the French inventor of the stethoscope, was also an expert pathologist, famous for his full clear descriptions of many intra-thoracic diseases, including pulmonary tuberculosis, pneumonia, gangrene, infarcts, emphysema, bronchiectasis, pleurisy and pneumothorax, and also of atrophic cirrhosis of the liver which is named after him (see p. 334). He died of the disease which he had studied so closely, phthisis.

Richard BRIGHT (1789-1858), physician to Guy's Hospital, London, is best remembered for his unsurpassed studies (1827) of the various types of toxic nephritis, which are now named after him. But he left also original and accurate descriptions of the pathological changes in many other diseases, especially diseases of the central nervous system and meninges, pancreatic diseases, and acute yellow atrophy of the liver.

Thomas ADDISON (1793-1860), also of Guy's Hospital, collaborated with Bright in a full and accurate study of appendicitis ; gave in 1849 (i.e. 20 years before Biermer) the first account of pernicious anaemia (hence Addison's anaemia—see p. 382) ; and in 1855 described the pathology of 11 cases of the syndrome of adrenal deficiency to which also his name is attached (see p. 607).

Carl ROKITANSKY (1804-1878), a Czech professor of pathology in Vienna, is said to have performed over 30,000 necropsies. From this huge material he made many fine original descriptions of disease and wrote his famous *Handbook of Pathological Anatomy*.

Rudolph VIRCHOW (1821-1902) of Berlin, an eminent anthropologist and politician, was also the greatest pathologist of the nineteenth century. Quickly grasping the importance for pathology of Schleiden and Schwann's doctrine of the cellular structure of organisms (1839), he applied it to the study of diseased tissues, a study which culminated in his epoch-making *Cellular Pathology* (1858) and his treatise on *Tumours* (1863-1867). He also established the doctrine of embolism and showed its importance in pyaemia and bacterial endocarditis ; he discovered leucocytosis and leukaemia ; and he enlarged our knowledge of the pathology of many other diseases. In histology, he discovered neuroglia (1846) and the perivascular sheaths of the cerebral blood vessels (Virchow-Robin spaces). He was also a great medical historian. (For references to Virchow, see pages 427, 535 and 644.)

Julius COHNHEIM (1839-1884), the most eminent of Virchow's pupils, made notable discoveries in histology and in the phenomena of inflammation. (See p. 42.)

Sir James PAGET (1814-1899), surgeon to St. Bartholomew's Hospital, London, and a close friend of Virchow, is notable for his studies of the disease of the nipple (1874) and the disease of bones (1877-1882), to which his name is applied (see pages 470 and 638), and for his excellent *Lectures on Tumours* (1851) and *Surgical Pathology* (1863).

Louis PASTEUR (1822-1895), the French founder of bacteriology, is memorable for his discoveries of the microbic nature of fermentation, diseases of silkworms, chicken cholera, anthrax, preventive vaccination of animals against the two last-named diseases, and successful vaccination of man and animals against hydrophobia (see p. 263).

Robert KOCH (1843-1910) was a German country doctor who became the greatest bacteriologist in the study of human disease. He proved the anthrax bacillus to be the cause of the disease (1876), discovered staphylococci (1878), the tubercle bacillus (1882), the Koch-Weeks bacillus of conjunctivitis (1883), and the cholera vibrio (1884) ; he prepared and studied the properties of tuberculin (see p. 152) ; and he established most of our modern methods of culturing and staining bacteria, and of sterilization by steam.

Paul EHRLICH (1854-1915), assistant to Koch, is equally famous as histologist, chemist, chemotherapist and immunologist. By special stains, he distinguished the three kinds of granulocytes, differentiated between lymphoid and myeloid tissue and between lymphoid and myeloid leukaemia, discovered mast cells, and devised supravital staining methods. Ehrlich's haematoxylin is still a favourite histological stain today. In the light of recent research, his famous theory of immunity is seen to have been very nearly true (see p. 280). He was the founder of chemotherapy, and the discoverer of the antisyphilitic remedies salvarsan and neo-salvarsan, called also " 606 " and " 914 " because these were their numbers in the series of substances which he put to therapeutic trial.

In these brief notes, only the greatest figures in the history of pathology—the founders of the subject and its main branches—have been mentioned. For further details of these and for information about many other discoverers in the special branches, the following books are recommended :

Garrison, F. H. (1929). *History of Medicine.* Philadelphia and London; W. B. Saunders.

Guthrie, D. (1945) (Rev. ed. 1958). *A History of Medicine.* London; Nelson.

Major, R. H. (1945) Third edition. *Classic Descriptions of Disease.* Springfield (Illinois) Charles C. Thomas.

Power, D'A. and others. (1936). *British Masters of Medicine.* London; Medical Press & Circular.

Sigerist, H. E. (1941). *Great Doctors.* London; Allen & Unwin.

APPENDIX C

OBSERVING, RECORDING AND READING

Pathology is largely an observational science calling for *thorough examination and accurate description* of diseased bodies, organs or microscopic preparations. From the beginning, the student of pathology should deliberately cultivate his powers of systematic observation and description, and should use every available surgical, post-mortem and museum specimen as an exercise of this art.

HOW TO EXAMINE AND DESCRIBE GROSS PATHOLOGICAL SPECIMENS

Whether for personal records, research purposes or exams., an orderly and systematic method of examining and describing specimens is a great asset. This should be carried out in the following three stages : (1) recognition of any normal anatomy present, (2) simple description of how the specimen differs from normal, (3) pathological interpretation of these differences.

(1) What normal anatomy is recognizable ?

In many museum specimens, the organ or part of the body affected can be recognized, and our description should begin with a statement to that effect. Thus, we should say, for example : "This is part of a left lung which has been cut vertically through the hilum, displaying the main bronchus and vessels ". Or : " The preparation consists of a slab of liver, cut horizontally through the porta and showing the main branches of the portal vein and the hepatic ducts in section ". Or : " This is an elliptical piece of skin and subcutaneous tissue 4 inches long by 2 inches wide, of unspecifiable locality ". Or : " No normal structures can be identified in the specimen, which consists of . . . ", then a description of the cyst, tumour, or whatever it is. (In searching for normal structures in a potted specimen, do not forget that the jar has a back and sides as well as a front.)

If you are making an original dissection of a corpse (a necropsy) or of a part, you of course know what you are examining. But again you must ascertain and describe clearly the relationships of any areas of disease to the normal structures. For example, if a retroperitoneal mass is present, you must define its exact position and its relationships to the kidneys, adrenals, pancreas, great vessels, sympathetic chain, muscles and vertebrae : or, if you are dissecting a limb containing a tumour, you must ascertain its situation with respect to the skin, muscles, nerves, vessels and bone, and which of these structures are involved.

(2) How does the specimen differ descriptively from normal ?

Having stated the anatomy of your specimen, it is a good plan to introduce your description of its pathology with : " It differs from normal in the following ways ". Then describe the lesion in simple non-pathological terms, such as any intelligent layman might use. For example, the specimen depicted in Fig. 85 of this book would be described thus : " The specimen consists of a slice of liver which differs from normal in that it contains a single, sharply circumscribed, rounded mass 6 centimetres in diameter, which shows a thin but distinct white capsule enclosing a quantity of degenerate-looking contents. This consists of folded greenish membranous material with a gelatinous appearance, and scattered areas of yellow opaque friable chalky-looking material. The liver tissue itself appears normal."

Notice that this description contains no pathological terms or diagnosis. It is a plain statement in non-technical language of the size, shape, texture and colour of the visible abnormality. In other cases the lesions present are more complicated or more numerous than in the foregoing example, and the description correspondingly longer and more detailed ; but the practice of making this description a simple non-technical one should still be adhered to.

(3) What is the pathological interpretation of the changes described ?

In some cases the pathological diagnosis of the lesion is immediately obvious from its description. This applies, for instance, to the example given above ; even without seeing this specimen, anyone with a reasonable knowledge of pathology would have no doubt from our description of it that it is a degenerated, probably bile-stained, hydatid cyst of the liver.

With other specimens, however, the nature of the lesion is not at once obvious, and indeed in many instances it is not possible to come to a certain diagnosis on the gross appearances alone. In such cases, a discussion of the *differential diagnosis* is necessary, i.e. we must consider all the reasonable possibilities and weigh the points for and against each of them. When confronted with a lesion of obscure nature, in order not to over-look any of the possibilities in the differential diagnosis, it is a good plan regularly to run over in one's mind the following list of the main classes of lesions (in the order in which they are considered in this book).

    (1) Traumatic lesions ;
    (2) Inflammations (acute or chronic) and parasitic lesions ;
    (3) Lesions due to extraneous foreign bodies or poisons ;
    (4) Circulatory disturbances ;
    (5) Nutritional and metabolic disorders ;
    (6) Haemopoietic disorders ;
    (7) Tumours (benign or malignant, primary or secondary) ;
    (8) Endocrine disorders ;
    (9) Mechanical lesions—(dilatation, hypertrophy, torsion, etc.) ;
    (10) Malformations.

Thus, suppose the specimen is a lung containing an ill-defined area of pale solid tissue of uncertain nature. We run over the above list and decide that 1, 5, 6, 8, 9 and 10 are certainly inapplicable. We are left with 2, 3, 4 and 7 as *possible* causes of the kind of pale solid area which our specimen shows, namely (*a*) an area of chronic inflammation (fibrous pneumonia, tuberculosis or some other chronic infection), or (*b*) an area of chronic inflammation due to a foreign body, or (*c*) an old organized infarct, or (*d*) an area of infiltrating malignant tumour. Then, from the detailed characters of the lesion—its size, shape, situation, relationship to bronchi, texture and colour— we argue the likelihood or unlikelihood of each of these possible diagnoses. In some cases this discussion will enable us to reduce the number of possibilities still further, and perhaps even to conclude that one of them is almost certainly the correct one. In other cases, we will come to the conclusion that the diagnosis is uncertain, that there are several possibilities, and that only microscopic examination can decide between them. Do not be distressed at leaving a diagnosis *sub judice* ; it is much better to do this than to run ahead of the available evidence.

HOW TO EXAMINE AND DESCRIBE MICROSCOPICAL SPECIMENS

The same rules that we applied in examining large specimens should also be applied in the examination of micro-sections, i.e. (1) identify any normal structures present, (2) give a non-technical account of the abnormalities observed, and (3) give a pathological interpretation—either a definite diagnosis or a discussion of the differential diagnosis of the several possible alternatives. Deliberately carry out (1) and (2) in three stages, in ascending order of magnification :

    (i) *First, always examine the section carefully with the naked-eye or with a hand-lens.* Note areas of obviously different structure, and make a rough outline sketch of the section to show these, for they must each receive microscopical attention presently. Sometimes the nature of the tissue is plain from the naked-eye examination alone, i.e. you can say at once that it is a section of skin, bone, lung, brain or eye. Examination with a good hand-lens is often of help; it may enable you to make or strongly suspect the diagnosis even before you use the microscope : see Figs. 36, 37, 49, 88, 115, 120, 122, 126, 147, 149, 153, 154, 160, 161, 163, 164, 172, 177, 187, 189, 190, 195, 196, 200, 203, 207, 243, 268, 274, 279, 280 and 288 of this book, all of which are hand-lens views.

(ii) *Next, make a thorough low-power microscopical examination of all parts of the section before using high powers.* Nine-tenths of all histo-pathological diagnosis can and should be done at magnifications of 100 or less. The distinct areas of the section, as you have sketched them naked-eye, should be explored one by one and a clear panoramic idea obtained of the relationships of the normal and diseased tissues.

(iii) *Finally, if necessary, use the high-power objectives.* Do this only for specific purposes—e.g. to identify polymorphs, lymphocytes, plasma cells or other kinds of small cells of which you are uncertain at low powers ; or to search for such cytological details as the cross-striae of muscle fibres, cilia, inclusion bodies, or mitotic figures ; or, with the oil-immersion lens, to detect bacteria, fungi or protozoa. Do not waste time in "wandering" haphazardly over a section with the high power ; but first pick out with the low power particular spots which you think will repay high-power study.

From your 3-stage examination you will have identified the normal tissues present and the areas of disease, and you will either have reached a definite diagnosis or be in a position to discuss the differential diagnosis. It may greatly aid your final description of a complex specimen to make another and enlarged sketch of it at this stage, marking in the identity of the tissues or the nature of the changes observed in the various areas which you had previously distinguished. Thus a sketch of the section shown in Fig. 17 of this book might look something like this :

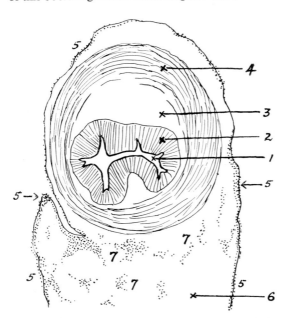

Fig. 289.—1. Lumen of appendix.
2. Mucous membrane.
3. Oedematous inflamed submucosa.
4. Muscle coat ; fibres separated by exudate.
5, 5. Inflammatory exudate (fibrin and cells) on serosa.
6. Adipose tissue of meso-appendix.
7, 7. Patches of haemorrhage and inflammation in the meso-appendix.

Such a sketch greatly simplifies a written description.

RECORDING YOUR OBSERVATIONS

The careful recording of personal observations is by far the best form of education.
To see a thing for yourself, and to see it clearly enough to describe it clearly, and to have a permanent personal record of what you have seen, is far better than to hear or read about the thing. The very act of description not only impresses the thing on your memory, but also calls for precise and detailed examination and so sharpens your powers of observation. If you are describing a specimen, you *see* far more in it than when you are merely looking at it.

Cultivate the art of description, then ; you will be astonished how quickly this will pay dividends in making you a better observer and in crystallizing your knowledge of the subject. Like all other arts, this one requires *cultivation* : the ability to describe natural objects clearly, simply and accurately can be acquired only by deliberate practice. But it is well worth the effort, for it is this more than anything else that distinguishes the scientist (and the artist) from the unscientific (and inartistic) mass of humanity.

(1) Personal observation books

Record your observations in one of the following ways :
(1) Keep a single large, strongly bound "day-book" in which you record consecutively dated records of all your personal observations—clinical, pathological and other ; and make a subject index at the back of the book. Do *not* burden your personal observation book with lecture notes or abstracts of other people's work, or (if you are a practising pathologist) your routine necropsy and biopsy reports, of which you can keep separate files.
(2) Keep several separate observation books, one for clinical medicine, one for pathology, one for any other special subject which interests you.
(3) Keep your observations on separate sheets which can then be filed in loose-leaf covers according to subjects.

The first method is the simplest, and has the advantage that all your observations are together and that your "day-book" will thus constitute an interesting permanent record of the development of your knowledge and interests. This is the method which one of the writers personally uses and can heartily recommend; he constantly refers back with profit to his "day-book" entries of 40 years ago, and wishes that he had begun to keep them earlier. These personal observation books are far more valuable than any text-book, for the information they contain is one's very own.

Let your observation books be a means also of cultivating an interest in *comparative pathology*. When your dogs, cats, fowls or canaries die, find out why by performing proper post-mortem examinations, if necessary make a microscopical study of the lesions found, and keep a record of your findings. If you know any veterinary surgeons, zoologists, trappers or fishermen, enlist their aid in collecting specimens from which, if your time and interest permit, you may enlarge your knowledge of animal pathology. If you have not time to examine such specimens at once, they can be put aside in 10% formalin solution until you are less pressed.

(2) Drawing

Drawings of three kinds can be useful in pathology, namely :
(1) *Diagrams* to aid our visualization of structures or processes, such as Figs. 1, 7, 10, 12, 25, etc. and the coloured plates in this book ;
(2) *Diagrammatic sketches of actual specimens* for record purposes, such as that on the preceding page and Fig. 250.
(3) *Drawings of actual specimens*, intended to portray them more or less realistically, such as Figs. 20, 22, 31, 32, 59, 123 and 271.

All students can and should make drawings of the first two kinds, but the extent to which they can make use of the third kind varies greatly with their natural aptitude for drawing. Those who can draw well should certainly use this ability to aid their studies ; but those who have no great aptitude for drawing should not waste their precious time (and should not be expected by their teachers to waste it) attempting to become artists.

(3) Photographs

The camera can be a great aid to the adequate recording of both gross and microscopical specimens; and those who are keen on photography may well employ it intelligently for this purpose. Good specimen photography is just as interesting a hobby as any other kind of still-life photography. But be judicious about this, and do not allow the means to become an end in itself.

(4) The Materials of research

Every accurate record of personal observations is a piece of research. Brilliant discoveries are not made at once by a sudden flash of inspiration, but are the outcome of the painstaking collection of accurately recorded data. The data which *you* collect may one day contribute to the making of a new discovery. No two cases of disease, and no two pathological specimens, were ever yet exactly alike; and, since there are still many obscurities in our pathological knowledge in all directions, the particular specimens which you examine *may* show features which will shed light on these obscurities. Hence, *meticulously thorough study of one unusual case or specimen is worth superficial study of a hundred others.*

Do not let your observation records become a burden or a duty. If they do, they are not worth keeping. Record only the things that *interest* you and which therefore give you pleasure to record; but, your interest having been excited by a thing, examine and record that thing as thoroughly and completely as possible. It is of such stuff that research material is made.

READING AND ABSTRACTING

(1) Journals

On every subject in pathology and medicine, a deluge of new papers flows unceasingly in hundreds of journals. It would be a physical impossibility for any one person to read and assimilate one-fiftieth part of these, even if he had nothing else to do. How, then, shall the busy under-graduate student, or the busy post-graduate pathologist or doctor, select the tiny fraction of current literature which it is possible for him to read? He can do so in one or both of two ways : (*a*) by regular perusal of certain selected current journals, or (*b*) by searching out selected papers, current or old, on subjects which specially interest him.

(a) Regular reading of selected current journals.

Deliberately select 2 or 3 or 6 important journals written in English (or in other languages which you know), to which you have easy access. In pathology and medicine, among the most important and easily accessible which you should try to include are *J. of Path. and Bacteriology, American J. of Path., Archives of Path., Brit. Med. J., Lancet* and *Brit. J. of Surgery.* Look through these regularly as they appear; and read and make a note of any articles which interest you or which you think will be useful for subsequent reference. You will soon find that only a small proportion of the articles need be read in detail; for many others, it suffices to glance through them and read the terminal summaries which most articles contain. The task of keeping up with 6 current journals is not really a very heavy one. Articles on important subjects in these 6 journals will keep you informed of what is going on in the various branches of pathology and medicine, and from them you will obtain also references to important papers on particular subjects in other journals.

(b) Searching for papers on particular subjects.

It often happens that a particular patient or specimen excites interest in some special disease, and that you wish to obtain more detailed information about it than is given by your textbooks. You may search for this in the following ways :

    (1) See if your textbooks give any references to other works on it; the reference lists for *supplementary reading* given at the ends of the chapters in this book

are intended to aid students in finding useful and easily accessible accounts of subjects on which they may want extra information.

(2) See if, in the abstracts of articles which you have already noted in current journals, there are any on the particular subject.

(3) Look up the index to each of the last 3 or 4 volumes of your favourite journals, or the *Quarterly Cumulative Index Medicus* (which all libraries have), for promising-looking references.

(4) Ask if your library has any recent works on that disease.

Once you have found a good recent account of the particular subject, it will give references to earlier papers which you may wish to see.

(2) Abstracts

When you read a journal article of value on a particular subject, do not let it pass from memory, but make an abstract of it on a sheet of paper headed by the writer's name, the title, and the date, volume and page of the journal. This abstract may be as long or as short as you please—perhaps only the title, perhaps the author's summary, perhaps your own detailed summary with excerpts—but, in any case, it will be a record of the paper which you can file away under subjects in a box file or loose-leaf book for future reference.

(3) Textbooks

The undergraduate student does not have time in which to familiarize himself with more than one good textbook of pathology. Yet it is well that he should dip into other books a little, if only to see that there is more than one way of presenting the subject; and both he and the postgraduate student may often wish to refer to several books on matters of detail on particular questions. The following textbooks, other than those already referred to in the text, are recommended to the student, because they are the ones to which I myself am most indebted for reliable information, stimulating viewpoints or helpful illustrations.

Anderson, W. A. D. and Scotti, T. M. (1968). Seventh revised eidition. *Synopsis of Pathology*. London; Kimpton.
— (Ed.) (1966). (2 vols.). Fifth revised edition. *Pathology*. London; Kimpton. (An encyclopaedic account.)
Cappell, D. F. (1964). Eighth edition. *Muir's Textbook of Pathology*. London; Edw. Arnold.
Collins, D. H. (Ed.) (1959). *Modern Trends in Pathology*. London; Butterworths.
Curran, R. C. (1966). *Colour Atlas of Histopathology*, London; Baillière, Tindall and Cussell.
Florey, H. (Ed.) (1954) 1958, 1970). *Lectures on General Pathology*. London; Lloyd-Luke.
Harrison, C. V. (Ed.) (1967). Eighth edition. *Recent Advances in Pathology*. London; Churchill.
MacCallum, W. G. (1950). Seventh revised edition. *A Textbook of Pathology*. London and Philadelphia; W. B. Saunders. (A fine presentation of pathology from the general viewpoint with unusually good illustrations.)
Mallory, F. B. (1923). *The Principles of Pathologic Histology*. London and Philadelphia; W. B. Saunders.
Wright, G. Payling (1950, 1954, 1958). *An Introduction to Pathology*. London; Longmans.
— and Symmers, W. St. C. (1967). (2 vols.). *Systemic Pathology*. London; Longmans. (An encyclopaedic work.)

APPENDIX D

BACTERIOLOGICAL METHODS
SPECIAL TESTS

Bile solubility test

This test is used to distinguish between the pneumococcus and *Strep. viridans*, and depends on the fact that lysis of pneumococci occurs in the presence of a bile salt. A few drops of a 10 per cent solution of bile salt are added to a broth culture of the organism, and the mixture (at pH 6·8–7·8) is incubated at 37°C. Rapid clearing occurs in the case of pneumococci but no change is produced with *Strep. viridans*. Positive and negative controls are included in the test.

Optochin test

This test is also used to distinguish between the pneumococcus and *Strep. viridans*, and depends on the fact that Optochin (ethyl hydrocuprein hydrochloride) is much more inhibitory to the growth of pneumococci than to that of streptococci. The test organism is streaked across a fresh-blood agar plate, together with positive and negative control organisms. A strip of sterile filter paper which has been soaked in a solution of Optochin in distilled water (1 in 4,000) and dried, is laid across the plate at right angles to the streaks, and the plate is incubated at 37°C overnight. The growth of streptococci is not inhibited; pneumococci show a zone of inhibition about 5 mm. wide on each side of the strip.

Coagulase test

This test depends on the fact that pathogenic strains of staphylococci produce an enzyme, coagulase, which coagulates blood plasma.

Slide method: A drop of normal saline is placed on each end of a microscope slide and a colony of the test organisms is emulsified in each drop. A loopful of undiluted human plasma is mixed with one of the drops and the slide gently rocked. Clumping of the organisms occurs rapidly if the strain is coagulase-positive; coagulase-negative strains show no change. The second drop of emulsified organisms without plasma is a control which ensures that spontaneous clumping of organisms does not occur.

Tube method: Five drops of a fresh broth-culture of the organism under test are added to 0·5 ml. of dilute plasma (1 in 10 in saline) in a small test tube, and this is incubated at 37°C along with a second tube which contains plasma only. Tubes are examined after 1 hour, and at intervals up to 24 hours. Coagulase-positive strains produce a clot, usually in 1–2 hours. The tube of plasma without the culture serves as a control and should show no coagulation. Known coagulase-positive and negative strains of staphylococci are included in the test as controls.

Deoxyribonuclease test

Most coagulase–positive staphylococci also produce the enzyme deoxyribonuclease, which is easily demonstrated in cultures growing on DNA agar containing methyl green, uncoloured zones being produced around the colonies as a result of liberation and diffusion away of the dye from the degraded DNA.

The Elek plate

The Elek plate is used to determine the toxigenicity or otherwise of strains of diphtheria bacilli, and may be used as a reliable substitute for animal inoculation experiments. It is essentially a precipitin test, and depends on the production of a precipitate when diphtheria toxin reacts with diphtheria antitoxin in an agar plate.

The clear agar medium is poured into a Petri dish, and before the agar sets a strip of filter paper which has been soaked in diphtheria antitoxin is laid across its surface. After setting and drying, the medium is streak-inoculated with the organism under test, at right angles to the filter paper strip, and is incubated for 48 hours. Pathogenic strains of diphtheria bacilli produce the toxin which diffuses away from the streak of culture. At the same time antitoxin diffuses away from the filter paper strip; and when toxin and antitoxin meet at the optimum concentration, precipitation occurs (Fig. 290). No precipitate is produced by non-toxigenic strains.

Bacteriophages and phage-typing

Bacteriophages are viruses which are parasitic and destructive for bacterial cells. Phages against most species of bacteria exist, but each kind of phage shows marked specificity for particular species. Phage activity may be demonstrated in fluid or on solid media; in broth cultures of bacteria the phages lyse the bacterial cells and thus cause clearing of the medium; on solid media the phages produce areas of clearing in the bacterial growth, which are called *plaques*. The specificity of different phages, and the ease with which their antibacterial activity can be demonstrated, has led to their use in the typing of bacteria, especially the staphylococci and Salmonellae. Of particular importance in these days of hospital cross-infection is the phage-typing of *Staph. aureus*.

Staphylococcal phage-typing.—The organism under examination is inoculated over the surface of a plate medium and the different phages are then spot-inoculated on the surface. The development of clear circular plaques in the confluent growth of the organism indicates where the phages are acting (Fig. 291). The several phages against *Staph. aureus*

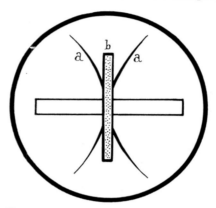

Fig. 290.—The Elek plate. Lines of precipitation (aa) are produced by a toxigenic strain of diphtheria bacillus growing in a streak culture (b) on the plate.

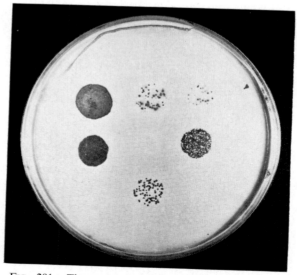

Fig. 291.—The agar plate is covered by a confluent culture of *Staphylococcus aureus*, in which plaques of clearing by different bacteriophages are seen.

are not all active against all strains of staphylococci; some strains are lysed by some phages but not by others. Phage-types of staphylococci are thus determined by the pattern of phages which attack them, and four *phage groups* (I, II, III and IV), each containing a number of phage-types are now recognized. Phage-typing of staphylococci is of epidemiological importance. Phage-types of group III are the strains most often encountered in hospital infections, and many of them are resistant to penicillin. Phage-type 80, which belongs to group I, appears to be the most virulent strain of *Staph. aureus*.

CULTURE OF BACTERIA

A great variety of media have been devised for the isolation and study of bacteria. Many organisms, such as staphylococci and coliform bacilli are not exacting in their growth requirements and grow well on the simplest of nutrient media. More fastidious organisms such as the gonococcus, the whooping-cough bacillus and the tubercle bacillus require the presence of a number of special growth factors. The optimum pH for the growth of most pathogenic bacteria is in the region of pH 7·5, so that most media are adjusted to this value. Some species will grow at a pH well above or below this optimum, for example, the cholera vibrio and lactobacilli, and use is made of this in the preparation of selective media for these organisms.

Fluid media

Many different fluid media are in general use. *Meat infusion broth* is a watery extract of meat to which is added 1 per cent of peptone and 0·5 per cent of NaCl. This simple medium is used for broth cultures of the less exacting organisms, and it also forms the basis of many other fluid and solid media. More fastidious organisms will grow in meat infusion broth if it is enriched with such substances as horse or human serum (*serum broth*), blood which has been denatured by heating (*heated-blood broth*), or small pieces of cooked meat (*Robertson's cooked-meat medium*). *Glucose broth* is meat infusion broth containing 1 per cent of glucose, and this medium is commonly used for blood culture. Fluid media may be rendered selective for certain organisms by incorporating substances which inhibit the growth of others. *Selenite broth*, for example, is used for the isolation of typhoid bacilli from faeces, since the sodium selenite inhibits the growth of coliform organisms. Similarly, Robertson's cooked-meat medium is rendered selective for the isolation of staphylococci from faeces by the addition to it of 10 per cent. of NaCl.

Fluid media may also be used as *indicator media*, specific changes being produced by organisms growing in them. For example, the bacilli of enteric fever, bacterial food poisoning, bacillary dysentery, *E. coli* and related organisms, which are all closely similar in appearance, staining properties and cultural characters, are distinguished by their different powers of fermenting various sugars and other carbohydrates with the formation of acid, or acid and gas. These fermentation tests are carried out in *peptone water* which contains the sugar and a pH indicator. Sterile milk containing an indicator (commonly litmus) may be used as an indicator medium for lactose fermentation by bacteria.

Solid media

Fluid media are rendered solid by adding to them 1–2 per cent of New Zealand agar, and are usually prepared as plate media in Petri dishes or as sloped media in test tubes. *Nutrient agar* is widely used for the surface culture of the less exacting organisms and consists simply of meat infusion broth solidified with agar. This medium will not support the growth of more demanding organisms, but it is used as the basis of a variety of enriched media. *Fresh-blood agar* is nutrient agar containing 10 per cent of whole horse-blood; it supports the growth of most pathogenic bacteria and also serves as an indicator medium in that some organisms cause lysis of the red cells. Haemolysis is of particular importance in distinguishing between the different types of streptococci (see p. 115). *Heated-blood agar* (*chocolate agar*) is nutrient agar containing 10 per cent of whole horse-blood which has been heated to 80°C. Heating causes lysis of the red cells and partial denaturation, so that the medium is of a smooth chocolate-brown colour. It is of particular value in the isolation of *Haemophilus influenzae*, and in the identification of the pneumococcus and some streptococci (see Chapter 11). Fresh-blood agar and heated-blood agar are the two most generally useful media in clinical bacteriology. Media used for the isolation of the whooping-cough bacillus are of special interest, since this organism is very sensitive to toxic substances in the medium. Care is taken, therefore, to select non-toxic ingredients for media (some brands of peptone, for example, are markedly inhibitory), and other substances are incorporated which " neutralize " inhibitory factors produced during growth of the organism. A high concentration of blood may be used for this purpose, as in Bordet-Gengou medium, or the blood may be replaced by finely divided charcoal, as in Turner's medium.

As with fluid media, solid media, usually in the form of plates, are extensively used as indicator and selective media. *Egg-yolk agar* is important for its use in the *Nagler plate* for the identification of clostridia (see p. 162); *desoxycholate-citrate agar* is used as a selective-indicator plate for the isolation of Gram-negative intestinal pathogens from the faeces. Some media are made selective by adjustment of the pH. Dieudonné's alkaline blood medium, which is used for the isolation of the cholera vibrio, inhibits the growth of coliform bacilli, and Kulp's medium, which is used for the isolation of lactobacilli, is acidic (pH 5) and inhibits the growth of many other bacterial species.

Antibiotics may also be used as selective agents. Penicillin is commonly incorporated in media for the isolation of the whooping-cough bacillus, and neomycin may be used in media for the isolation of anaerobes.

Not all solid media contain agar. Heat-coagulated egg media are used for the growth of the tubercle bacillus, *Löwenstein-Jensen* medium being an example; *Löffler's medium*, which is heat-coagulated ox serum containing one third its volume of 1 per cent glucose broth, is used for the isolation of diphtheria bacilli.

Determination of bacterial sensitivity

The sensitivity of an organism to an antibacterial agent may be determined either by a diffusion method or a dilution method. In diffusion techniques the antibiotic is applied to a localized area of a plate culture of the organism, from where it diffuses out into the medium. Absence of growth in the zone indicates sensitivity of the organism. The most commonly used method of this type employs inert tablets which have been impregnated with standard amounts of antibacterial agents. The whole surface of the plate is inoculated with the organism under test and the various antibiotic tablets are then placed on its surface. In dilution techniques a known amount of drug is incorporated in the medium, which may be either solid or fluid, and this is then inoculated with the test organism. This method is usually employed for sensitivity testing of the tubercle bacillus, the different drugs being incorporated in slopes of *Löwenstein-Jensen* or a similar medium.

ANAEROBIC CULTURE

Anaerobic bacteria will usually grow only in the absence of free oxygen, and a number of methods are available for producing suitable anaerobic conditions. The simplest method is to exclude oxygen from the culture. This may be done by *stab culture*, in which a deep tube of solid agar medium is inoculated by stabbing it with a straight wire charged with the organisms; anaerobic growth occurs in the depths of the medium. Oxygen may also be excluded by using *shake cultures*. Here, a deep tube of cool, molten agar is inoculated, and after gentle shaking, is allowed to set; anaerobic growth occurs some distance below the surface of the medium. Another commonly used method of growing anaerobes is to culture them in a deep fluid medium containing a reducing substance such as glucose or sodium thioglycollate. A medium of this type is *Robertson's cooked-meat medium* in which particles of cooked meat (horse or beef muscle) provide the necessary reducing substances. In order to obtain surface growth of anaerobes on agar plate media, it is necessary to produce an anaerobic atmosphere, and this is best done by using an anaerobic jar of the type devised by McIntosh and Fildes.

The McIntosh and Fildes anaerobic jar.—There are several types of this anaerobic jar, but all of them depend on the principle that platinum or palladium catalyse the combination of oxygen and hydrogen to form water. Anaerobic jars are cylindrical vessels made of metal or glass, flanged at the top to carry an air-tight lid which is held firmly in place by a clamp. The lid is provided with two taps, and carries on its undersurface a *catalyst capsule*. In the most modern type of anaerobic jar the catalyst capsule consists of a wire-mesh envelope containing pellets of alumina coated with finaly divided palladium, which are active at room temperature. In earlier types of the McIntosh and Fildes jar, the catalyst consists of palladinized asbestos which requires heating in order to activate it. Setting up the jar is simple. Cultures are placed inside the jar and the lid is clamped in place. Some of the air is evacuated with a vacuum pump and is replaced with hydrogen. Catalysis commences almost immediately and oxygen is rapidly removed from the atmosphere inside the jar. In order to ensure that anaerobic conditions are attained and maintained, the jar is provided with an external capsule containing a methylene-blue indicator; when exposed to the air the indicator is blue, but under anaerobic conditions the dye is reduced to a colourless form.

DESTRUCTION OF BACTERIA

Vegetative bacterial cells can be readily destroyed by a number of physical and chemical agencies, and in general, the destructive power of any effective agent depends on its concentration and the time for which it is allowed to act. Different species of organisms vary in their resistance, and bacterial spores are usually much more resistant to destruction than vegetative cells, higher concentrations and longer exposure times being required to kill them. An agent which kills bacteria is *bactericidal*; one which does not kill, but which prevents multiplication, is *bacteriostatic*. The preservation of meats by salting, and of fruits by dessication, are examples of bacteriostasis. Bactericidal agents in low concentrations may also be bacteriostatic. In order to achieve sterility, agents must be used which are not only bactericidal, but which are also *sporicidal*, and in hospital and laboratory practice, when possible, methods of sterilization are chosen which ensure complete destruction of all forms of bacterial life.

Physical agencies

Dry heat.—Dry heat in the form of a Bunsen burner flame is extensively used in the laboratory for sterilizing platinum loops and necks of test tubes and bottles. A red-hot searing iron is often used to " sterilize " the surface of an organ at autopsy prior to taking a specimen for bacteriological examination. Such drastic methods of sterilization are clearly inappropriate for most hospital and laboratory requirements, but *hot air ovens* are now widely used for the dry sterilization of syringes and glass-ware. Electrically heated ovens are more commonly used than gas ones and should be fitted with an automatic device such as a Cambridge Thermograph which records the temperature reached and the time of exposure. Dry heat at 160°C will destroy all forms of bacterial life in 1½ hours. Materials which cannot be dry-sterilized include fluids, all media, rubber, textiles and certain instruments such as cystoscopes.

Moist heat.—All vegetative bacteria are rapidly killed in water at 90–100°C, but below these temperatures different species show considerable variation in their heat-resistance. *Strep. faecalis*, for example, survives exposure to 60°C for 30 minutes, whereas the gonococcus and meningococcus are rapidly killed at about 50°C. Bacterial spores are much more resistant to heat than vegetative organisms, and though most spores will not survive boiling for longer than 15 minutes, the spores of some species may remain viable even after 5 hours' boiling. Wet steam under pressure effectively destroys all forms of bacterial life.

Boiling water was once used in hospitals for the sterilization of instruments, catheters, etc. This has now been abandoned in favour of other more effective methods.

Tyndallization entails exposure of material to steam at 100°C for 30 minutes on each of 3 successive days. All vegetative organisms are destroyed at the first exposure. If conditions are suitable, any surviving spores develop into vegetative organisms which are then destroyed by the second steaming. The third steaming ensures complete sterilization on the same principle. This method of intermittent sterilization has limited uses in the laboratory, but is of no value in hospital practice.

Wet steam under pressure is by far the best sterilizing agent, and is used for a great variety of articles, such as instruments, theatre linen, rubber gloves, etc. It is not suitable for heat-labile fluids and drugs, or for substances such as polythene and Perspex. For sterilization by steam under pressure an *autoclave* is required, of which the simplest is the home pressure-cooker. In pressure-cookers, and in some small autoclaves, steam is generated from water inside the container, and the temperature at which the water boils, and consequently the temperature of the steam, increases as the pressure inside the vessel rises. In large autoclaves saturated steam at the required pressure is admitted to the chamber from an external source. Exposure of articles to 15 lb. of saturated steam per sq. in. gives a temperature of 121°C and is sufficient to kill all bacterial spores in 20 minutes. In addition to the high temperature attained in this way, the latent heat of steam is given up to articles as the steam condenses on them.

Though this is not the place for a discussion on autoclave operation, attention is drawn to two important rules for efficient sterilization: (a) there must be no air present in the autoclave chamber, since an air-steam mixture is much less effective than pure saturated steam; and (b) the autoclave must be packed in such a way as to ensure penetration of the steam into every part of its contents. If the second rule is not fulfilled then the first rule is automatically broken. Articles must also be packed in such a way as to ensure that they do not become contaminated with bacteria after sterilization, when the autoclave is opened.

Pasteurization is used for the partial sterilization of fluids and is performed by heating the fluid to a given temperature for a given time. Its best known application is in the treatment of milk, but it is also used in the laboratory to eliminate contaminating vegetative organisms from mixtures with spore-forming bacteria.

Desiccation.—Slow drying is lethal to many vegetative organisms, two important exceptions being tubercle bacilli and staphylococci. The survival of these in dust poses problems of cross-infection which have been largely solved for tubercle bacilli, but which are still of concern in the case of staphylococci. Bacterial spores are resistant to desiccation.

Other physical agencies

Electromagnetic radiations, such as ultraviolet light and γ-radiation, are bactericidal in appropriate concentrations. Extensive use is made of γ-radiation for the sterilization of a variety of materials in bulk. *Ultrasonic vibrations* kill bacteria by disrupting them and are used chiefly in the study of the chemical composition of bacterial cells. *Filtration of fluids* which are adversely affected by heat is extensively used for their sterilization. Preparations for injection, such as toxoids, antitoxins, insulin, etc., are sterilized in this way. Various types of bacterial filter are available, candle-shaped filters constructed of diatomaceous earth or unglazed porcelain, filter discs made of asbestos, and membrane filters made of cellulose acetate. Candle filters are most commonly used for the sterilization of fluids in bulk; membrane filters are most useful for small volumes.

Chemical agencies

A great variety of chemicals have bactericidal and bacteriostatic activity and are called antiseptics or disinfectants. While all vegetative bacteria are readily killed by one or another of the different antiseptics in general use, bacterial spores usually survive chemical disinfection. Hence, only articles for which heat-sterilization is inappropriate are treated with antiseptics. In general, antiseptic solutions are only necessary for thermolabile materials, for use on tissues, and for domestic purposes. The choice of an antiseptic is determined by its ability to kill the organisms present, by its toxicity for the tissues, by its effect on articles exposed to it, by its compatability with pus, soap and other chemicals, and by its cost. The bactericidal activity of an antiseptic depends on its concentration and on the time for which it is allowed to act, so that reduction of the concentration or exposure time may result in bacteriostasis instead of sterilization. It will suffice here to mention briefly some of the compounds in common use.

Soaps.—Apart from the mechanical removal of bacteria by washing, soaps are bactericidal to some organisms such as *Strep. pyogenes* and the pneumococcus. Soap is ineffective against staphylococci, the tubercle bacillus and bacterial spores.

Detergents.—Cationic detergents, of which there are many with proprietary names, such as " Cetavlon ", are highly active bactericidal agents. Cetavlon, for example, kills staphylococci and *Ps. pyocyanea* in a dilution of 1 in 10,000 in 10 minutes. They are relatively ineffective against the tubercle bacillus and have no effect on bacterial spores. The antibacterial activity of detergents of these types is reduced in the presence of organic matter, and they are inactivated by soap.

Phenolic derivatives.—These include Lysol, carbolic acid, chlorxylenol and hexachlorophene. They have a wide range of activity and will kill most vegetative organisms, but have little effect on spores. Lysol and carbolic acid are caustic substances, which is a disadvantage. Chlorxylenol, of which " Dettol " is a well-known example, is non-caustic but its bactericidal properties are rather weaker than those of Lysol and carbolic acid. Hexachlorophene is commonly incorporated in toilet soap as a skin antiseptic.

Dyes.—Many dyes are strongly bacteriostatic, inhibiting the growth particularly of Gram-positive bacteria. Gentian violet, malachite green and the flavines are examples. They are sometimes used as skin antiseptics. The dyes are inactive against the tubercle bacillus and some anaerobic species, and are sometimes used as selective agents in media for the isolation of these organisms.

Halogens.—These are powerful antiseptics and act on a wide range of bacteria. Tincture of iodine and iodophors are particularly useful skin antiseptics. Chlorine is widely used for the sterilization of raw water in city water supplies, 0·2 parts per million being sufficient to destroy most vegetative bacteria. Bromine is too irritant to make it a useful antiseptic.

Metallic salts.—Sodium chloride in high concentrations is bacteriostatic, and advantage is taken of this in the preservation of meat. *Staph. aureus* is inhibited to a much less degree than most species, so that NaCl may be used as a selective agent in media for the isolation of staphylococci. Silver and mercury salts are strongly bactericidal, silver nitrate, mercuric chloride and Merthiolate being the compounds most commonly used.

Alcohols.—Ethyl alcohol is widely used as an antiseptic, usually in combination with some other agent such as iodine or chlorhexidine. It is most active at a concentration of 65 per cent.

Formaldehyde.—Though an efficient antiseptic, formaldehyde is too irritating to find much application. It is most frequently used in the gaseous form for the "decontamination" of large soiled objects such as mattresses.

Chlorhexidine (" *Hibitane* ").—This is a powerful antiseptic with a wide range of activity. It is non-toxic, but is inactivated by soap. It is widely used in aqueous and alcoholic solutions for pre-operative skin preparation.

INDEX

*Where several page numbers appear after a title,
those in heavier type are the more important ones*